Microbial Virulence Factors

Microbial Virulence Factors

Editor

Jorge H. Leitão

MDPI • Basel • Beijing • Wuhan • Barcelona • Belgrade • Manchester • Tokyo • Cluj • Tianjin

Editor
Jorge H. Leitão
Instituto de Biotecnologia e
Bioengenharia
Portugal

Editorial Office
MDPI
St. Alban-Anlage 66
4052 Basel, Switzerland

This is a reprint of articles from the Special Issue published online in the open access journal *International Journal of Molecular Sciences* (ISSN 1422-0067) (available at: https://www.mdpi.com/journal/ijms/special_issues/Virulence_Factors).

For citation purposes, cite each article independently as indicated on the article page online and as indicated below:

LastName, A.A.; LastName, B.B.; LastName, C.C. Article Title. *Journal Name* **Year**, *Article Number*, Page Range.

ISBN 978-3-03936-946-1 (Hbk)
ISBN 978-3-03936-947-8 (PDF)

© 2020 by the authors. Articles in this book are Open Access and distributed under the Creative Commons Attribution (CC BY) license, which allows users to download, copy and build upon published articles, as long as the author and publisher are properly credited, which ensures maximum dissemination and a wider impact of our publications.

The book as a whole is distributed by MDPI under the terms and conditions of the Creative Commons license CC BY-NC-ND.

Contents

About the Editor

Jorge H. Leitão, Ph.D., is associate professor at the scientific and pedagogical area of Biological Sciences of the Department of Bioengineering, Instituto Superior Técnico, Universidade de Lisboa, Portugal, and researcher at IBB- Institute for Bioengineering and Biosciences. After obtaining his Ph.D. in Biotechnology and Biosciences in 1996 by Instituto Superior Tecnico, his research work was focused on the biology and pathogeneis of bacteria of the Burkholderia cepacia complex (Bcc), in particular on the biosynthesis of exoplysaccharides. Bcc is a group of opportunistic pathogens causing life-threatening infections in patients suffering from Cystic Fibrosis, Chronic Granulomatous Disease, and immunocompromised patients. Current research interests are the post-transcription regulation of bacterial gene expression and the roles played by small non-coding regulatory RNAs and RNA chaperones on the biology and pathogenesis of bacteria of the Bcc. Other research interests include bacterial virulence, focusing on their exploitation as targets for the development of novel antimicrobial strategies; bacterial resistance to antimicrobials and the development of novel antimicrobials; and the molecular characterization of microbial populations of ecological, industrial or health interest.

International Journal of
Molecular Sciences

Editorial

Microbial Virulence Factors

Jorge H. Leitão

IBB–Institute for Bioengineering and Biosciences, Instituto Superior Técnico, Department of Bioengineering, Universidade de Lisboa. Av. Rovisco Pais, 1049-001 Lisboa, Portugal; jorgeleitao@tecnico.ulisboa.pt; Tel.: +351-21-841-7688

Received: 21 July 2020; Accepted: 25 July 2020; Published: 27 July 2020

Keywords: microbial virulence factors; bacterial pathogens; fungal pathogens; pathogenicity

Microbial virulence factors encompass a wide range of molecules produced by pathogenic microorganisms, enhancing their ability to evade their host defenses and cause disease. This broad definition comprises secreted products such as toxins, enzymes, exopolysaccharides, as well as cell surface structures such as capsules, lipopolysaccharides, glyco- and lipoproteins. Intracellular changes in metabolic regulatory networks, governed by protein sensors/regulators and non-coding regulatory RNAs are also known to contribute to virulence. Furthermore, some secreted microbial products have the ability to enter the host cell and manipulate their machinery, contributing to the success of the infection. The knowledge, at the molecular level, of the biology of microbial pathogens and their virulence factors is central in the development of novel therapeutic molecules and strategies to combat microbial infections. This is of particular importance in the present days with the worldwide emergence of microbes resistant to available antimicrobials, as well as of novel pathogens such as the SARS-CoV-2 responsible for the present pandemics. Advances in recent years in molecular biology, genomics and post-genomics technologies, and bioinformatics contributed to the molecular identification and functional analyses of a wide range of microbial virulence factors. The Special Issue of IJMS focused on virulence factors and their regulatory networks from microbes such as bacteria, viruses, fungi, and parasites, as well as on the description of innovative experimental techniques to characterize microbial virulence factors. A total of 18 papers was published in this Special Issue. The collection comprises state of the art papers on virulence factors and mechanisms from a wide range of bacterial and fungal pathogens for humans, animals, and plants, thus reflecting the impact of microorganisms in health and economic human activities and the importance of the topic.

Due to their impact on human health, bacterial pathogens that cause infections in humans have received a higher attention, with *Escherichia coli* as one of the most studied bacteria. Pokharel et al. investigated the roles played by the recently described serine Protease Autotransporters (SPATE) TagB, TagC, and Sha of *E. coli* on urinary infections using a 5637 bladder epithelial cell line [1]. Members of the SPATE family owe their proteolytic activity to the serine protease catalytic triad composed of an aspartic acid, a serine, and a histidine residue. Evidence is presented showing that the three SPATE proteins are internalized by bladder epithelial cells, leading to alterations of actin cytoskeleton distribution. Results presented indicate that Sha and TagC degrade mucin and gelatin, respectively [1]. The mutation analysis of the serine catalytic site showed that secretion of the three proteins is not affected, but impaired their entry into epithelial cells, affecting their cytotoxicity and proteolytic activity [1].

The presence of genes related to virulence factors including adhesins, siderophores, protectines or invasins, and involved in allantoin metabolism were investigated among 32 non-*E. coli Enterobacterales* isolates obtained from the feces of 20 healthy adults [2]. Similar studies analyzed virulent NECE strains from patients with an ongoing infection, and not commensal NECE from healthy subjects as in the present study [2]. Isolates were taxonomically characterized by 16S RNA sequencing and MALDI-TOF MS analysis, and profiled by pulsed-field gel electrophoresis. The genus *Klebsiella* was found as the most represented, followed by *Enterobacter* and *Citrobacter* [2]. The isolates were further characterized concerning the presence in their genomes of genes encoding selected virulence factors, as well as their phenotypes related to biofilm formation and resistance to a selection of antibiotics. Results point out that the isolates do not encompass particularly virulent strains and in most of the cases were susceptible to antibiotics [2].

Yang et al. investigated the role of the *Salmonella enterica* serovar Typhimurium (ST) *pdxB-usg-truA-dedA* operon on intracellular survival using deletion mutants constructed with the λ-Red recombination technology [3]. The *Salmonella* genus comprises several facultative intracellular pathogens capable of infecting both human and animal hosts. The ST deletion mutants was investigated in J774A.1 macrophage cells. The deletion mutants Δ*pdxB*, Δ*usg*, and Δ*truA* exhibited reduced replication abilities compared to ST and the deletion mutant Δ*dedA*. The *pdxB-usg-truA-dedA* operon is shown to contribute to ST virulence in mice, and to resistance to oxidative stress [3].

Aeromonas hydrophila is an aquatic Gram-negative bacterium, capable of causing serious and lethal infections to a wide range of hosts, including fish, birds, amphibians, reptiles, and mammals [4]. Dong et al. described the identification and functional characterization of the LahS global regulator of *A. hydrophila* [4]. LahS was identified after the screening of a Tn5-derived library of 947 *A. hydrophila* mutants for reduced hemolytic activity. The LysR family transcriptional regulator family member LahS was found to play a role in biofilm formation, motility, antibacterial activity, resistance to oxidative stress, and proteolytic activity, as well as essential for *A. hydrophila* virulence to zebrafish [4]. The comparative proteomics analysis performed by the authors confirmed the role of the protein as a global regulator in *A. hydrophila* [4].

Bacteria of the *Dickeya* genus comprise plant pathogens that affect crops such as potatoes. In order to succeed when infecting their hosts, *Dickeya* secrete several proteins with plant cell wall degrading activities, including pectinases, cellulases, and proteases [5]. To investigate the role played by the protease Lon on *D. solani* pathogenicity towards potato, Figaj et al. used a λ-Red-derived protocol to construct a *lon* deletion mutant [5]. Results presented indicate that the Lon protein plays a role in protecting the bacterium to high ionic and temperature stresses, affecting the activity of pectate lyases, the organism motility, and delaying the onset of infection symptoms in the potato host [5].

The plant pathogen *Candidatus Phytoplasma mali* is the causal agent of apple proliferation disease, that affects apple production in Northern Italy [6]. *Phytoplasma* are biotrophic, obligate plant and insect bacterial symbionts, with a biphasic life cycle comprising reproduction in phloem-feeding insects and in plants [6]. The paper of Mittelberger et al. focused on the effector protein PME2 (Protein in Malus Expressed 2), expressed by *P. mali* when infecting apples [6]. The in silico analysis of the PME2 protein sequence performed revealed that the protein has features of effector proteins of Gram-positive bacteria, with a predicted final localization at the cytoplasm or nucleus of the host [6]. Two main protein variants, PME2ST PME2AT, were found associated in infected apple trees from Italy and Germany. Using protein variants tagged with GFP, both variants were found to translocate to the nucleus of *Nicotiana* spp. protoplasts. A better understanding of the molecular mechanisms used by *P. mali* to manipulate its host will rely on genomics analysis, since no genetic manipulation is presently available for these organisms [6].

The necrotrophic fungal pathogen *Sclerotinia sclerotiorum* (Lib.) de Bary infects a wide range of plants causing devastating agricultural losses. The organism forms a typical structure named sclerotia when vegetative hyphae gather to form a hardened multicellular structure important in its development and pathogenesis, and that under favorable conditions germinate leading to vegetative hyphae or apothecia that will initiate novel disease cycles by producing ascospores [7]. Li et al. used a proteomics approach based on 2D gels followed by spot isolation and protein identification by MALDI-TOF to identify proteins differentially expressed between a wild-type strain and a deletion mutant on the gene *SsNsd1* encoding a type IVb GATA zinc finger transcription factor [7]. Although the gene encoding SsNsd1 was found as expressed at low levels during the hyphae stage, the mutant is unable to form the compound appressoria. The authors were able to identify a total of 40 proteins as differentially expressed, 17 with predicted functions and 23 as hypothetical proteins [7]. The authors emphasize the utility of the approach used to identify important proteins involved in the SsNsd1-mediated formation of appressorium.

In addition to other factors, the success of pathogens rely on cell-cell communication. Bacterial outer membrane vesicles (OMV) are recognized as an efficient means of bacteria-bacteria and bacteria-host communication, not only intra-species, but also interspecies [8]. Despite the lack of data on a possible role played by OMVs in bacterial-yeast communication, Roszkowiak et al. investigated the role played by *Moraxella catarrhalis* OMVs on the susceptibility of selected bacterial and fungal pathogens to the cationic peptide polymyxin B, and to the serum complement [9]. Using OMVs from *M. catarrhalis* strain 6, the authors found that these OMVs conferred protection against the cationic peptide polymyxin B to the non-typeable *Haemophilus influenzae*, *Pseudomonas aeruginosa*, and *Acinetobacter baumannii*. Furthermore, OMVs also protected serum-sensitive non-typeable *H. influenza* and promoted the growth of the serum-resistant *P. aeruginosa* and *A. baumannii* against the complement [9]. In addition, the results presented also show that OMVs facilitate the formation of hyphae by the pathogenic yeast *Candida albicans*, promoting its virulence [9]. As stated by the authors, this work might pave the way to uncover additional roles played by OMVs-dependent interactions in multispecies communities [9].

The RNA chaperone Hfq is a master regulator of gene expression in bacteria, mediating the interaction of small noncoding RNAs with their mRNA targets, including those related to virulence in Gram-negative bacteria [10]. Dienstbier et al. performed an integrated Omics comparative analysis of the Hfq regulon in the *Bordetella pertussis* human pathogen, responsible for respiratory tract infections, in particular of a whooping cough [11]. Based on the use of RNAseq, and gene ontology analysis, genes significantly upregulated in the *hfq* mutant fall into categories including "Translation", "Regulation of transcription", and "Transmembrane transport", while genes downregulated fall in the categories "Transmembrane transport", "Iron–sulfur cluster assembly", "Oxido-reduction process", "Pathogenesis", and "Protein secretion by the type III secretion system" [11]. Correlations of transcriptome, proteome, and secretome datasets are also presented [11]. Results presented corroborate the central role played by Hfq on the physiology and pathogenicity of *B. pertussis* [11].

In their brief report, Maisetta et al. performed the ex vivo evaluation of the bactericidal activity of combinations of the semi synthetic antimicrobial peptide lin-SB056-1 in combination with EDTA (Ethylenediaminetetraacetic acid) against endogenous *P. aeruginosa* present in the sputum from patients suffering from primary ciliary dyskinesia (PCD) [12]. The authors observed that the peptide and EDTA were almost inactive against PCD sputum endogenous *P. aeruginosa* when used alone, but exhibited a significant synergistic killing effect with a sputum sample-dependent efficacy [12]. EDTA, but not lin-SB056-1, was found to inhibit biofilm formation and the production of virulence factors including alginate, pyocyanin, and the metalloprotease LasA [12].

Various bacterial species have evolved various strategies to invade, survive, and multiply intracellularly in host cells. The paper of Denzer et al. presents an updated review of the mechanisms used by bacteria to invade the host cell, to manipulate their biochemical and gene expression machinery, and to multiply and escape from the host cell [13]. The authors present a thorough review of mechanisms used by intracellular pathogens, including the highjack of host immune defenses to enter into the host cell. Central attention is given to the various mechanism used to manipulate gene expression, including histone modification, control of host DNA methylation patterns, sabotage of host long non-coding RNAs, interfering with the host RNA transcription and translation, as well as with host protein stability [13]. The importance of the detailed molecular knowledge of pathogenesis mechanisms to the development of strategies to combat bacterial infections is highlighted [13].

The functions of grimelysin of *Serratia grimesii* and protealysin of *Serratia proteamaculans* that use actin as a substrate and promote bacterial invasion was reviewed by Khaitlina et al. [14]. The *Serratia* genus comprises facultative pathogens able to cause nosocomial infections or infections in immunocompromised patients, but nosocomial infections by *S. grimesii* or *S. proteamaculans* are low [14]. The paper focused on the discovery, properties and substrate specificity of the two proteases, their high specificity towards actin, and discussed their contribution to the invasiveness of *Serratia*, although further knowledge of the bacterium virulence factors and the cellular response mechanisms is required to fully understand the mechanism of *Serratia* invasion of the host cell [14].

The virulence factors that the bacteria use to cross the blood-brain barrier and cause meningitis is reviewed by Herold et al. [15]. Meningitis remains a worldwide problem often associated with fatalities and severe sequelae. After reviewing important traits of the central nervous system barriers to bacterial entrance, the authors review the various stages of the virulence processes of bacterial meningitis, including the processes of attachment and invasion, the routes used to enter the central nervous system, and the general mechanisms used to survive intracellularly [15]. The roles played by virulence factors produced by bacteria when crossing the central nervous system is also addressed, followed by the review of the specific traits of bacterial species more commonly associated with meningitis [15].

Coagulase-negative Staphylococci are a broad group of skin commensals that emerged as major nosocomial pathogens, with the species *S. epidermidis*, *S. haemolyticus*, *S. saprophyticus*, *S. capitis*, and *S. lugdunensis* as the most frequent pathogens [16]. In their paper, Argemi et al. reviewed the recent progress achieved in the pathogenomics of these species, based on published work supported by whole-genome data deposited in public databases [16]. As stated by the authors, the ever increasing amount of data available at the genomic, molecular, and clinical levels is expected to enhance the development of innovative approaches to characterize the pathogenicity of this bacterial group of pathogens [16].

Bacteria of the *Trueperella pyogenes* species are considered as belonging to the microbiota of animals skin and mucous membranes of the upper respiratory and urogenital tracts, but it is also an important opportunistic pathogen to animals, leading to important economic losses [17]. In their paper, Rzewuska et al. reviewed the taxonomy of the species, their pathogenicity to animals, and the various diseases associated, as well as their possible involvement in zoonotic infections, as well as the reservoirs and routes of transmission and infections [17]. The authors also present a thorough review of the main virulence factors used by the organism, including pyolysin, fimbriae, extracellular matrix-binding proteins, neuraminidases, and ability to form biofilms [17]. The availability of complete genome sequences and a better knowledge of *T. pyogenes* virulence factors, transmission routes, and epidemiology of infections is expected to lead to the development of effective vaccines, with particular hope deposited on DNA vaccines [17].

Candidiasis are on the rise worldwide, with *Candida albicans* and *Candida glabrata* as the more prevalent etiologic agents of these fungal diseases [18]. The paper by Galocha et al. thoroughly reviewed the distinct strategies used by the two *Candida* species to successfully cause human infections, starting by the adhesion and ability to form biofilms [18]. While *C. albicans* is dimorphic, growing as yeast or pseudohyphae, *C. glabrata* cannot undergo hyphal differentiation. As a consequence, *C. albicans* relies on the production of proteolytic enzymes and hyphal penetration to invade the host cell, while *C. glabrata* is thought to invade host cells by inducing endocytosis [18]. The authors extensively review the distinct mechanisms used by the two pathogenic to evade the host immune system, and succeed as pathogens. The detailed knowledge of the virulence mechanisms is critical to develop therapies that specifically target virulence traits of these two pathogenic yeasts [18].

Bacterial small non-coding regulatory RNAs (sRNAs) have emerged over the last decade as key regulators of post-transcriptional regulators of gene expression, being involved in a wide range of cellular processes, including bacterial virulence [19]. In their review, Pita et al. updated knowledge on sRNAs from two pathogens associated with respiratory infections and lung function decline of patients suffering from Cystic Fibrosis, *P. aeruginosa* and bacteria of the so-called *Burkholderia cepacia* complex (Bcc) [20]. As stated by the authors, the knowledge on *P. aeruginosa* sRNAs is far more extensive than from bacteria of the Bcc. After reviewing the main molecular characteristics of bacterial RNAs and their modes of action, including the role played by Hfq as a mediator of RNA-RNA interactions, the authors detail the description of the roles played by *P. aeruginosa* sRNAs known for their involvement in virulence traits of the bacterium. Despite the shorter information on Bcc sRNAs, the authors make a brief description of known sRNAs from Bcc [19]. The identification and functional characterization of additional sRNAs from these two pathogens will certainly enlighten our knowledge on their virulence traits.

The development of new tools to investigate microbial pathogenesis, at the molecular and cellular level, is of keen importance to comprehend how the microorganism can invade the host and cause infection. The paper from Hatlem et al. reviewed the basic molecular traits and applications of the SpyCatcher-SpyTag system, originally developed as a method for protein ligation [20]. The system consists of a modified domain of the SpyCatcher surface protein from *Streptococcus pyogenes* that recognizes the cognate SpyTag peptidic sequence composed of 13 amino acid residues [20]. Upon recognition, a covalent isopeptide bond is formed between a lysine side chain of the SpyCatcher and an aspartate of the SpyTag [20]. The authors describe in detail the fundamentals of the system and of related variants, emphasizing their uses in molecular studies of microbial virulence factors, surface proteins, membrane dynamics, as well as in the development of vaccines [20].

Microorganisms employ a wide array of virulence factors to successfully thrive and flourish with their hosts, leading this interaction to the development of infections that can often be fatal. The molecular knowledge of the virulence traits, associated with the recent availability of genomics data and bioinformatics tools for the more frequent human pathogens, is expected to lead in the near future of novel molecules and strategies to battle infectious diseases.

Funding: This research was funded by Fundação para a Ciência e a Tecnologia, through Project UIDB/04565/2020 from IBB—Institute for Bioengineering and Biosciences.

Acknowledgments: IBB—Institute for Bioengineering and Biosciences is acknowledged for funding.

Conflicts of Interest: The author declares no conflict of interest. The funders had no role in the design of the study; in the collection, analyses, or interpretation of data; in the writing of the manuscript, or in the decision to publish the results.

References

1. Pokharel, P.; Díaz, J.M.; Bessaiah, H.; Houle, S.; Guerrero-Barrera, A.L.; Dozois, C.M. The Serine Protease Autotransporters TagB, TagC, and Sha from Extraintestinal Pathogenic *Escherichia coli* Are Internalized by Human Bladder Epithelial Cells and Cause Actin Cytoskeletal Disruption. *Int. J. Mol. Sci.* **2020**, *21*, 3047. [CrossRef] [PubMed]

2. Amaretti, A.; Righini, L.; Candeliere, F.; Musmeci, E.; Bonvicini, F.; Gentilomi, G.A.; Rossi, M.; Raimondi, S. Antibiotic Resistance, Virulence Factors, Phenotyping, and Genotyping of Non-*Escherichia coli Enterobacterales* from the Gut Microbiota of Healthy Subjects. *Int. J. Mol. Sci.* **2020**, *21*, 1847. [CrossRef] [PubMed]

3. Yang, X.; Wang, J.; Feng, Z.; Zhang, X.; Wang, X.; Wu, Q. Relation of the *pdxB-usg-truA-dedA* Operon and the *truA* Gene to the Intracellular Survival of *Salmonella enterica* Serovar Typhimurium. *Int. J. Mol. Sci.* **2019**, *20*, 380. [CrossRef] [PubMed]

4. Dong, Y.; Wang, Y.; Liu, J.; Ma, S.; Awan, F.; Lu, C.; Liu, Y. Discovery of lahS as a Global Regulator of Environmental Adaptation and Virulence in *Aeromonas hydrophila*. *Int. J. Mol. Sci.* **2018**, *19*, 2709. [CrossRef] [PubMed]

5. Figaj, D.; Czaplewska, P.; Przepióra, T.; Ambroziak, P.; Potrykus, M.; Skorko-Glonek, J. Lon Protease Is Important for Growth under Stressful Conditions and Pathogenicity of the Phytopathogen, Bacterium *Dickeya solani*. *Int. J. Mol. Sci.* **2020**, *21*, 3687. [CrossRef] [PubMed]

6. Mittelberger, C.; Stellmach, H.; Hause, B.; Kerschbamer, C.; Schlink, K.; Letschka, T.; Janik, K. A Novel Effector Protein of Apple Proliferation Phytoplasma Disrupts Cell Integrity of *Nicotiana* spp. Protoplasts. *Int. J. Mol. Sci.* **2019**, *20*, 4613. [CrossRef] [PubMed]

7. Li, J.; Zhang, X.; Li, L.; Liu, J.; Zhang, Y.; Pan, H. Proteomics Analysis of SsNsd1-Mediated Compound Appressoria Formation in *Sclerotinia sclerotiorum*. *Int. J. Mol. Sci.* **2018**, *19*, 2946. [CrossRef] [PubMed]

8. Toyofuku, M.; Nomura, N.; Eberl, L. Types and origins of bacterial membrane vesicles. *Nat. Rev. Microbiol.* **2019**, *17*, 13–24. [CrossRef] [PubMed]

9. Roszkowiak, J.; Jajor, P.; Guła, G.; Gubernator, J.; Żak, A.; Drulis-Kawa, Z.; Augustyniak, D. Interspecies Outer Membrane Vesicles (OMVs) Modulate the Sensitivity of Pathogenic Bacteria and Pathogenic Yeasts to Cationic Peptides and Serum Complement. *Int. J. Mol. Sci.* **2019**, *20*, 5577. [CrossRef] [PubMed]

10. Feliciano, J.R.; Grilo, A.M.; Guerreiro, S.I.; Sousa, S.A.; Leitão, J.H. Hfq: A Multifaceted RNA Chaperone Involved in Virulence. *Future Microbiol.* **2016**, *11*, 137–151. [CrossRef] [PubMed]

11. Dienstbier, A.; Amman, F.; Štipl, D.; Petráčková, D.; Večerek, B. Comparative Integrated Omics Analysis of the Hfq Regulon in *Bordetella pertussis*. *Int. J. Mol. Sci.* **2019**, *20*, 3073. [CrossRef] [PubMed]

12. Maisetta, G.; Grassi, L.; Esin, S.; Kaya, E.; Morelli, A.; Puppi, D.; Piras, M.; Chiellini, F.; Pifferi, M.; Batoni, G. Targeting *Pseudomonas aeruginosa* in the Sputum of Primary Ciliary Dyskinesia Patients with a Combinatorial Strategy Having Antibacterial and Anti-Virulence Potential. *Int. J. Mol. Sci.* **2020**, *21*, 69. [CrossRef] [PubMed]

13. Denzer, L.; Schroten, H.; Schwerk, C. From Gene to Protein—How Bacterial Virulence Factors Manipulate Host Gene Expression during Infection. *Int. J. Mol. Sci.* **2020**, *21*, 3730. [CrossRef] [PubMed]

14. Khaitlina, S.; Bozhokina, E.; Tsaplina, O.; Efremova, T. Bacterial Actin-Specific Endoproteases Grimelysin and Protealysin as Virulence Factors Contributing to the Invasive Activities of *Serratia*. *Int. J. Mol. Sci.* **2020**, *21*, 4025. [CrossRef] [PubMed]

15. Herold, R.; Schroten, H.; Schwerk, C. Virulence Factors of Meningitis-Causing Bacteria: Enabling Brain Entry across the Blood–Brain Barrier. *Int. J. Mol. Sci.* **2019**, *20*, 5393. [CrossRef] [PubMed]

16. Argemi, X.; Hansmann, Y.; Prola, K.; Prévost, G. Coagulase-Negative Staphylococci Pathogenomics. *Int. J. Mol. Sci.* **2019**, *20*, 1215. [CrossRef] [PubMed]

17. Rzewuska, M.; Kwiecień, E.; Chrobak-Chmiel, D.; Kizerwetter-Świda, M.; Stefańska, I.; Gieryńska, M. Pathogenicity and Virulence of *Trueperella pyogenes*: A Review. *Int. J. Mol. Sci.* **2019**, *20*, 2737. [CrossRef] [PubMed]

18. Galocha, M.; Pais, P.; Cavalheiro, M.; Pereira, D.; Viana, R.; Teixeira, M.C. Divergent Approaches to Virulence in *C. albicans* and *C. glabrata*: Two Sides of the Same Coin. *Int. J. Mol. Sci.* **2019**, *20*, 2345. [CrossRef] [PubMed]

19. Pita, T.; Feliciano, J.R.; Leitão, J.H. Small Noncoding Regulatory RNAs from *Pseudomonas aeruginosa* and *Burkholderia cepacia* Complex. *Int. J. Mol. Sci.* **2018**, *19*, 3759. [CrossRef] [PubMed]

20. Hatlem, D.; Trunk, T.; Linke, D.; Leo, J.C. Catching a SPY: Using the SpyCatcher-SpyTag and Related Systems for Labeling and Localizing Bacterial Proteins. *Int. J. Mol. Sci.* **2019**, *20*, 2129. [CrossRef] [PubMed]

© 2020 by the author. Licensee MDPI, Basel, Switzerland. This article is an open access article distributed under the terms and conditions of the Creative Commons Attribution (CC BY) license (http://creativecommons.org/licenses/by/4.0/).

 International Journal of
Molecular Sciences

Article

The Serine Protease Autotransporters TagB, TagC, and Sha from Extraintestinal Pathogenic *Escherichia coli* Are Internalized by Human Bladder Epithelial Cells and Cause Actin Cytoskeletal Disruption

Pravil Pokharel [1,2], Juan Manuel Díaz [2,3], Hicham Bessaiah [1,2], Sébastien Houle [1,2], Alma Lilián Guerrero-Barrera [3] and Charles M. Dozois [1,2,4,*]

[1] Institut national de recherche scientifique (INRS)-Centre Armand-Frappier Santé Biotechnologie, Laval, QC H7V 1B7, Canada; Pravil.Pokharel@iaf.inrs.ca (P.P.); Hicham.Bessaiah@iaf.inrs.ca (H.B.); Sebastien.Houle@iaf.inrs.ca (S.H.)

[2] Department of Veterinary Medicine, Centre de recherche en infectiologie porcine et avicole (CRIPA), Faculty of Veterinary Medicine, Saint-Hyacinthe, QC J2S 2M2, Canada; jmdv93@hotmail.com

[3] Laboratorio de Biología Celular y Tisular, Departamento de Morfología, Universidad Autónoma de Aguascalientes (UAA), Aguascalientes 20131, Mexico; alguerre@correo.uaa.mx

[4] Department of Biology, Institut Pasteur International Network, Laval, QC H7V 1B7, Canada

*   Correspondence: charles.dozois@iaf.inrs.ca

Received: 1 March 2020; Accepted: 23 April 2020; Published: 26 April 2020

Abstract: TagB, TagC (*tandem autotransporter genes B* and *C*), and Sha (*Serine-protease hemagglutinin autotransporter*) are recently described members of the SPATE (serine protease autotransporters of *Enterobacteriaceae*) family. These SPATEs can cause cytopathic effects on bladder cells and contribute to urinary tract infection in a mouse model. Bladder epithelial cells form an important barrier in the urinary tract. Some SPATEs produced by pathogenic *E. coli* are known to breach the bladder epithelium. The capacity of these newly described SPATEs to alter bladder epithelial cells and the role of the serine protease active site were investigated. All three SPATE proteins were internalized by bladder epithelial cells and altered the distribution of actin cytoskeleton. Sha and TagC were also shown to degrade mucin and gelatin respectively. Inactivation of the serine catalytic site in each of these SPATEs did not affect secretion of the SPATEs from bacterial cells, but abrogated entry into epithelial cells, cytotoxicity, and proteolytic activity. Thus, our results show that the serine catalytic triad of these proteins is required for internalization in host cells, actin disruption, and degradation of host substrates such as mucin and gelatin.

Keywords: SPATEs; UTIs; cytotoxicity; serine proteases; 5637 bladder cells; mucin; gelatin; actin

---

1. Introduction

Urinary tract infections (UTIs) present a broad range of symptoms and include urosepsis, pyelonephritis (or upper UTI, with infection in the kidney), and cystitis (or lower UTI, with bacteria infecting the bladder) [1,2]. Uropathogenic *Escherichia coli* (UPEC) is the main cause of community-acquired UTIs (about 80–90%) [3], and the ability of UPEC to establish a UTI is due to the expression of a variety of virulence factors. These factors include type 1 and P fimbriae (pili), flagella, capsular polysaccharides, iron acquisition systems, and toxins including hemolysin, cytotoxic necrotizing factor (CNF), and serine protease autotransporters of *Enterobacteriaceae* (SPATEs) [4].

The bladder urothelium constitutes a physical barrier to ascending urinary tract infections [5]. UPEC can produce toxins that damage bladder tissue and can lead to release of host nutrients and counter host defenses and innate immunity. A pore-forming toxin HlyA, can lyse erythrocytes and nucleated host cells [6], induce apoptosis [7], promote exfoliation of bladder epithelial cells and cause extensive uroepithelial damage [8–11]. Another UPEC toxin, cytotoxic necrotizing factor 1 (CNF1), has been reported to mediate bacterial entry into host epithelial cells [12], induce apoptotic death of bladder epithelial cells [13], and potentially promote bladder cell exfoliation [13]. SPATEs such as Sat, Pic, and Vat were also shown to affect bladder or kidney epithelial cells [14–16].

An important step to understand the role of SPATEs in UPEC pathogenesis is to elucidate molecular mechanisms underlying their effect on the bladder epithelium and during urinary tract colonization. The proteolytic activity of SPATEs is mediated by a serine protease catalytic triad of aspartic acid (D), serine (S), and histidine (H), wherein serine is the nucleophile, and aspartic acid interacts with histidine [17]. Mutations within the catalytic triad have been shown to abolish proteolytic activity in a number of SPATEs [15,17–19].

Recently, members of our group identified three new SPATEs: TagB, TagC (*tandem autotransporter genes B and C*), and Sha (*Serine-protease hemagglutinin autotransporter*) in some strains of extra-intestinal pathogenic *E. coli* (ExPEC). In ExPEC strain QT598, *tagB* and *tagC* are tandemly encoded on a genomic island, and were present in 10% of UTI isolates and 4.7% of avian pathogenic *E. coli* (APEC) that we screened [20]. Further, Sha, which is encoded on a virulence plasmid in strain QT598 was present in 1% of UTI isolates and 20% of avian pathogenic *E. coli* [20]. The *tagBC* genes are also present in the genomes of sequenced UPEC strains such as multidrug-resistant CTX-M-15-producing ST131 isolate *E. coli* JJ1886 (Accession number CP006784), *E. coli* CI5 (Accession number CP011018), and multidrug-resistant uropathogenic *E. coli* strain NA114 (Accession number CP002797.2). When cloned into *E. coli* K-12, TagB, TagC, and Sha mediated autoaggregation, hemagglutination, and adherence to human HEK 293 renal and 5637 bladder cell lines, but did not contribute significantly to biofilm production [20]. Further, TagB and TagC exhibited cytopathic effects on the bladder epithelial cell line [20]. Following transurethral infection of CBA/J mice with a *tagBC* mutant or *sha* mutant, no significant difference in colonization was observed. However, the competitive fitness of a mutant derivative lacking all of the SPATEs present in QT598 was significantly lower in the kidney [20].

The purpose of this report was to more fully investigate the effects of the TagB, TagC, and Sha SPATEs on the 5637 bladder epithelial cell line focusing on the actin cytoskeleton. We also investigated potential entry of SPATE proteins within these bladder epithelial cells and whether they demonstrate mucinase or gelatinase activity.

2. Results

2.1. Processing and Secretion of TagB, TagC, and Sha Is Independent of the Serine Protease Motif

To evaluate the importance of the serine protease motif for processing and secretion of three novel SPATEs, we generated variant proteins of TagB, TagC, and Sha lacking the serine catalytic site. Plasmids expressing TagB, TagC, or Sha [20] were used as the templates for construction of site-directed mutant clones where the serine site was substituted for an alanine at residue S255, S252, and S258 respectively (Figures S1 and S2). Each of these three plasmids expressing mutant SPATEs, produced a high-molecular-weight protein (>100 kDa) in culture supernatants that corresponded to the expected size of the native protein, and also lacked breakdown products that are present in samples containing native SPATEs that exhibit some autoproteolytic activity (Figure 1A, asterisks). This demonstrated that the serine protease motif is not necessary for SPATE secretion and release from bacterial cells.

To further localize each of the SPATEs expressed from plasmids in *E. coli* BL21 on the bacterial cell surface, we used immunogold labeling and transmission electron microscopy. Polyclonal antibodies against the entire secreted Vat SPATE [20] were used as they were shown to strongly cross-react and recognize conserved epitopes of the other SPATEs. Thus, we used these polyclonal "SPATE antibodies" for the detection of other SPATEs in our experiments.

E. coli BL21 pBCsk+ expressing TagB, TagC, and Sha were immunogold-labeled (Figure 1B–D); demonstrating localization of these SPATEs on the bacterial surface, as well as release into culture supernatant. The inset depicts the heavily concentrated proteins on the bacterial surface that appears to cluster around each other and produce fiber-like aggregates on the cell surface (white chevron, Figure 1B–D). *E. coli* BL21 bacteria containing only the empty plasmid vector were not labeled after immunogold labeling with anti-SPATE antibodies and secondary anti-rabbit immunoglobulin conjugated to 10-nm gold particles (Figure 1E).

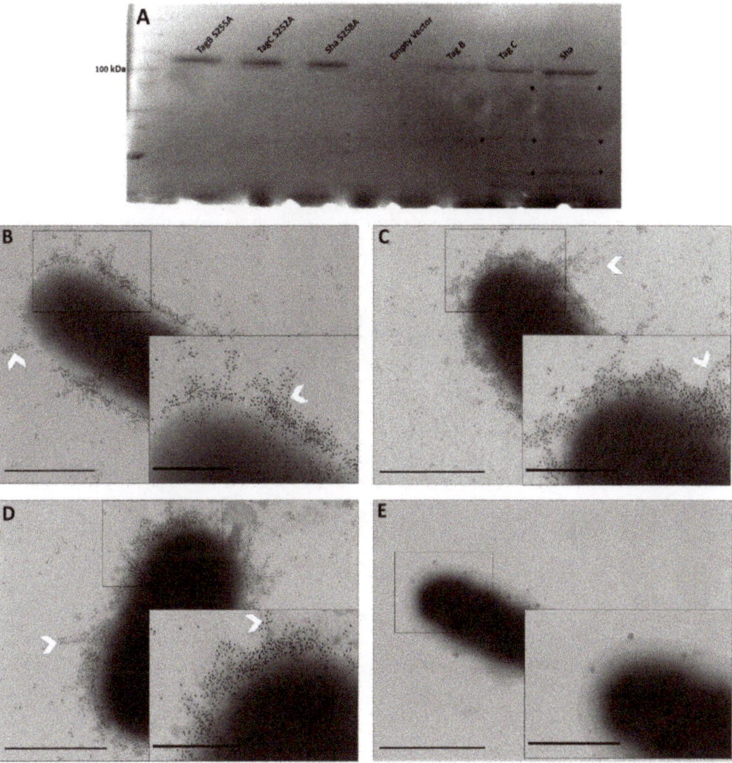

Figure 1. (**A**) Silver stained SDS-PAGE analysis of concentrated supernatants of *E. coli* BL21 expressing SPATE proteins. Filtered supernatants from clones expressing TagB, TagC, and Sha or the variant TagB S255A, TagC S252A, and Sha S258A proteins were concentrated through Amicon filters with a 50 kDa cutoff. Samples containing 5 µg of protein were migrated and stained with silver stain. (**B–E**) Immunogold Electron Microscopy (EM) of SPATEs (Serine protease autotransporters of *Enterobacteriaceae*) localized to the outer membrane and extracellular medium. Immunogold-TEM micrographs of SPATEs using SPATE-specific antiserum. Bacteria were cultured to 0.6 OD_{600nm} in Luria-Bertani medium. *E. coli* BL21 pBCsk+ expressing TagB (**B**), TagC (**C**), and Sha (**D**) labelled with immunogold particles. (**E**) *E. coli* BL21 pBcsk+ (vector only control) shows no immunogold staining. Insets represents boxed areas of higher magnification showing clustering of SPATE proteins. All images were acquired at ×17,000 magnification; scale bars represent 1 µm, and 0.5 µm (Insets).

2.2. The Serine Catalytic Motif of SPATEs Is Not Required for Autoaggregation or Hemagglutination Activity

To determine if the serine catalytic site of each of the three SPATEs is involved in autoaggregation or hemagglutination activity, we tested these phenotypes with our mutant SPATEs as described in [20]. Macroscopic analysis of autoaggregation of *E. coli* BL21 expressing TagB S255A, TagC S252A, or Sha S258A showed that bacterial cells settled at the bottom of the tube under static incubation similar to their respective native protein-expressing clones. The percentage of reduction in turbidity is given as a percentage of the initial OD_{600} value and was similar for both mutant and native proteins (Figure 2). The reduction in turbidity of the negative control *E. coli* BL21 pBCsk+ was significantly lower compared to clones expressing TagB, TagC, Sha, or AIDA-1 (positive control) (Figure 2). AIDA-1 (Adhesin Involved in Diffuse Adherence) of *E. coli* is a characterized self-associating autotransporter protein which mediates bacterial cell–cell interactions and autoaggregation [21].

These results show that inactivation of the serine catalytic site in each SPATE does not affect autoaggregation. It is therefore likely that other motifs or residues present in these proteins contribute to autoaggregation of bacterial cells.

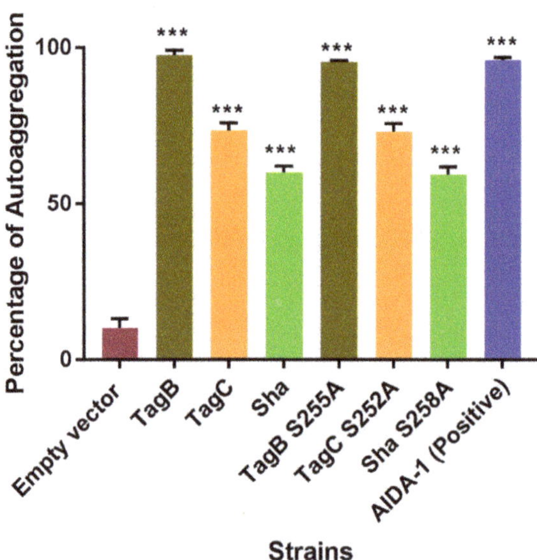

Figure 2. The autoaggregation phenotype is independent of the serine protease motif. Clones of *E. coli* BL21 expressing TagB, TagC, Sha, or their respective serine-site mutants were grown 18 h and adjusted to an OD_{600} of 1.5 and left to rest at 4 °C. Samples were taken at 1 cm from the top surface of the cultures after 3 h to determine the change in OD_{600}. Assays were performed in triplicate, and the rate of autoaggregation was determined by the mean decrease in OD_{600nm} after 3 h. *E. coli* BL21 pBCsk+ vector without insert (empty vector) was used as a negative control and the AIDA-1 autotransporter was the positive control for autoaggregation. Error bars represent standard errors of the means (*** $p < 0.001$ compared to empty vector using one-way ANOVA).

Likewise, Sha S258A showed similar hemagglutination of human blood as reported previously for the Sha native protein [20]. When cloned in the hemagglutination negative *E. coli* strain ORN172, there was no hemagglutination activity for either the native or mutant TagB or TagC proteins. In addition to autoaggregation and hemagglutination activity, the adherence capability of the serine catalytic site mutants of the SPATEs to 5637 human bladder epithelial cells was not affected (Figure S3). Hence, loss of the serine catalytic site did not affect autoaggregation, adherence, or hemagglutination phenotypes associated with each of the SPATEs compared to the native proteins.

2.3. Cytopathic Effect of TagB and TagC Requires the Serine Protease Motif

To assess the role of the serine protease motif for the cytopathic effect of SPATEs, extracts of supernatants of the different SPATEs (30 µg of protein per well) were incubated with human bladder epithelial cell line 5637 for 5h. Then the cells were fixed, stained with Giemsa stain, and observed by light microscopy. Cytopathic changes (dissolution in cytoplasm, enlargement of the nucleus

with vacuoles) observed under the microscope for TagB and TagC (Figure 3A) were absent from
cells treated with either TagB or TagC proteins lacking the serine protease active site. In addition,
no significant morphological changes were observed with cells treated with either Sha or the mutant,
Sha S258A, protein. No cytopathic effect was observed after treatment of cells with concentrated
filtered supernatant from *E. coli* BL21 pBCsk+ (empty vector) or media alone (Figure 3A). To examine
this cytopathic effect quantitatively, we measured lactate dehydogenase (LDH) release from epithelial
cells incubated with each of SPATEs or their respective catalytic site mutant proteins. There was release
of LDH after 5 h upon exposure of cells to TagB or TagC. However, the catalytic site mutant proteins
did not release LDH from cells (Figure 3B). Further, no LDH release was detected from cells treated
with either Sha or its catalytic site mutant variant (Figure 3B), indicating that cytotoxicity to human
bladder epithelial cells by TagB and TagC was dependent on the serine protease catalytic site.

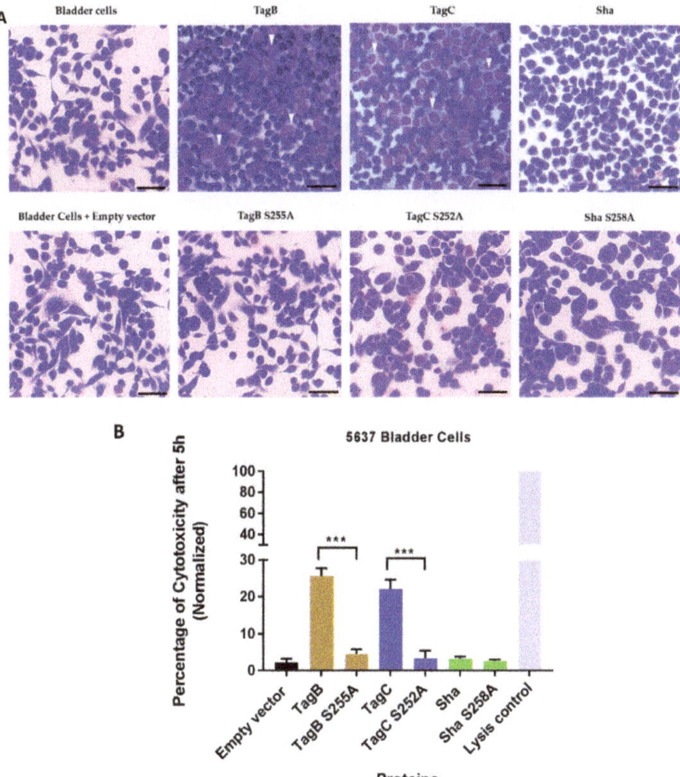

Figure 3. The serine catalytic site is necessary for the cytopathic effect of TagB and TagC. (**A**) Concentrated
supernatants containing 30 µg of protein per well derived from *E. coli* BL21 clones expressing TagB, TagC,
Sha, or their respective serine mutant variant proteins were incubated with monolayers of the 5637 human
bladder epithelial cell line for 5 h at 37 °C. Cytopathic effects (white triangle) were absent in cells treated with
the serine catalytic site mutant variants of TagB or TagC. The empty vector (pBCsk+) without insert was
used as a negative control. The scale bar represents 20 µm. (**B**) Cytotoxicity measured by LDH release from
5637 human bladder cells after incubation with supernatant filtrates of different clones (30 µg of protein per
well) at 37 °C for 5 h. Empty vector (pBCsk+) was used as a negative control and maximum LDH release
(positive control) was determined by treatment with lysis solution. Data are the means of three independent
experiments, and error bars represent the standard errors of the means. Significant differences between lysis
caused by native and mutant SPATEs were determined using Student's *t*-test with *** $p < 0.001$.

2.4. *Exposure to TagB, TagC, or Sha Alters Actin Distribution in Bladder Epithelial Cells*

Based on the cellular changes seen with bladder epithelial cells after exposure to TagB and TagC, we hypothesized that TagB and TagC could alter the distribution of cytoskeletal components such as actin, with actin being one of the most abundant intracellular proteins in the eukaryotic cell. So, to examine the effect on F-actin cytoskeleton organization, 5637 bladder cells were incubated with native and mutant TagB, TagC, or Sha (30 µg of protein per well) for 5 h at 37 °C, stained with fluorescently labeled phalloidin, and then observed under confocal microscopy. Cells treated with the supernatant extract from the empty vector containing clone were uniform, smooth-edged, and contained clearly visible actin stress fibers (yellow triangle) and strong actin staining around the cell (Figure 4A). By contrast, bladder cells treated with TagB showed reduced actin stress fibers and less actin staining (Figure 4A). Bladder cells treated with TagC, also had a pronounced effect on the cytoskeleton as demonstrated by the absence of actin stress fibers and reduced levels of actin staining. Sha treated cells showed a loss of actin stress fibers and the presence of punctate patterns of actin within the cytoplasm of the cells (yellow arrowheads, Figure 4A). By contrast, the TagB, TagC, and Sha mutants lacking the serine protease catalytic sites demonstrated no changes in the actin cytoskeleton and had actin stress fibers similar to negative control cells, indicating that the serine protease activity of these SPATEs mediates the changes in actin distribution within bladder cells. To quantify the level of phalloidin binding, we measured the staining intensity and distribution of fluorescence of phalloidin around each cell using ImageJ software [22]. Fluorescence intensity for cells was calculated using the channel for actin staining. In comparison with the negative control (empty vector), the density of F-actin staining was significantly lower in cells treated with TagB, TagC, or Sha. Cells treated with the serine catalytic site mutant proteins, demonstrated F-actin staining that was greater when compared to cells treated with the native SPATE proteins (Figure 4B). Overall, these results demonstrate that these SPATEs alter the cytoskeleton and reduce the distribution of actin in bladder epithelial cells.

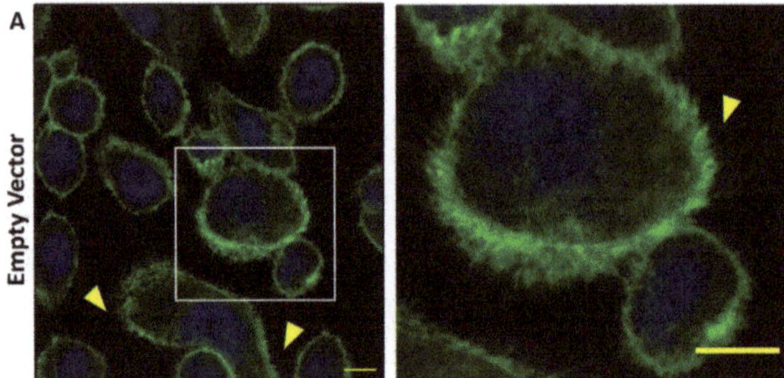

Figure 4. *Cont.*

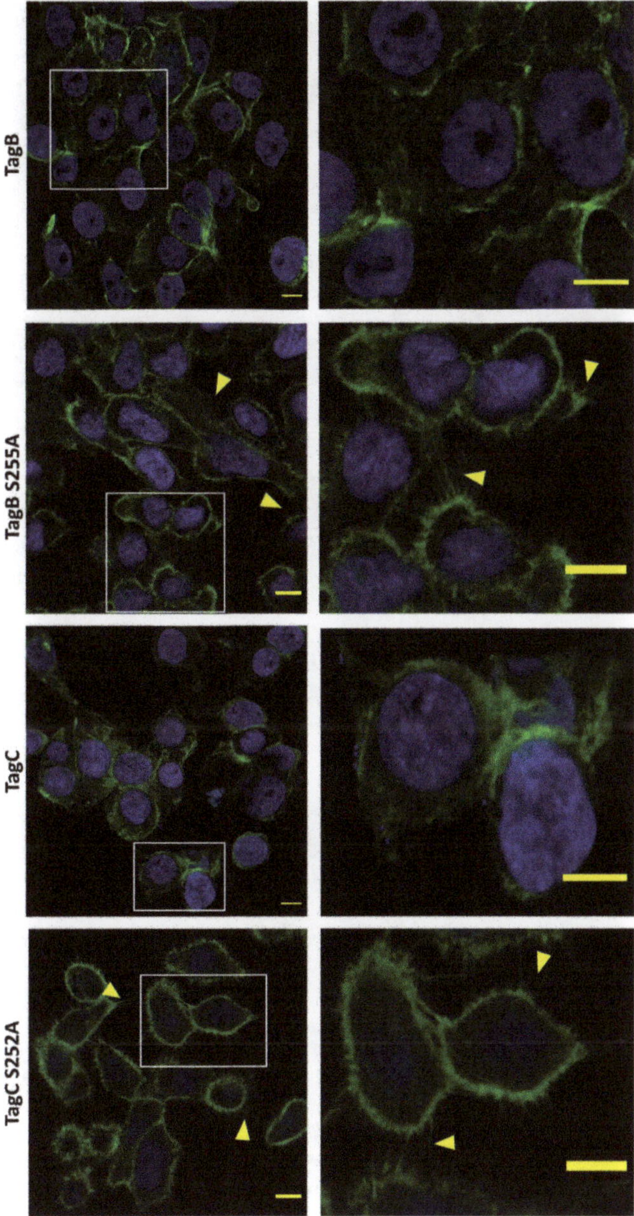

Figure 4. *Cont.*

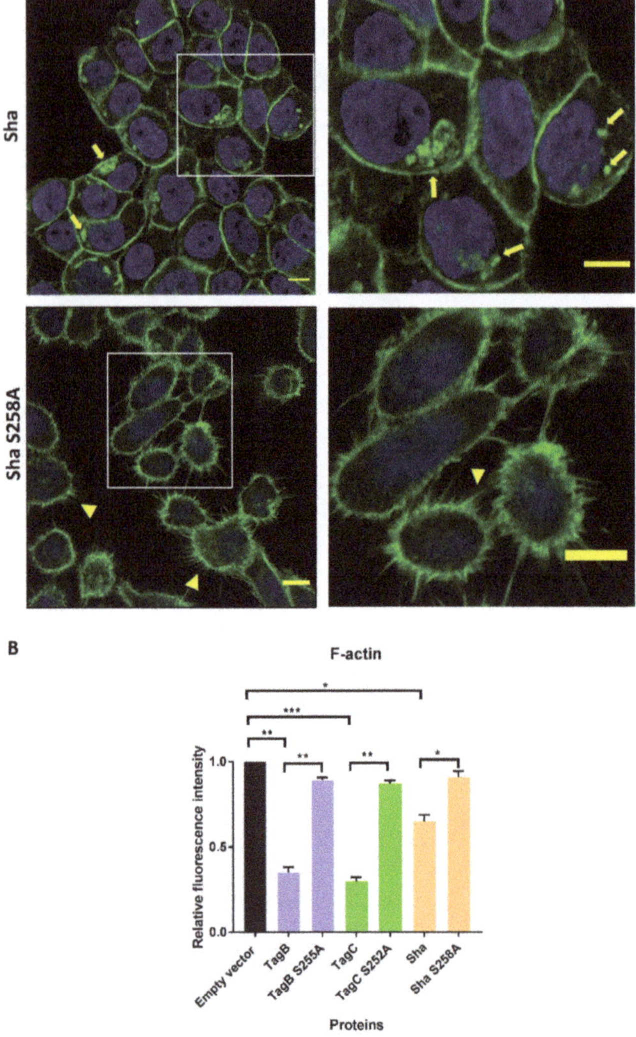

Figure 4. Effects of TagB, TagC, and Sha on the actin cytoskeleton of bladder epithelial cells is serine-protease-motif dependent. (**A**) Concentrated supernatant extracts (30 µg of protein per well) from *E. coli* BL21 clones expressing TagB, TagC, or Sha and their respective serine catalytic site mutants were incubated with monolayers of human bladder (5637) epithelial cells for 5 h at 37 °C. After incubation, cells were fixed and permeabilized. Actin was stained with fluorescently labeled phalloidin (green) and the nucleus was stained by DAPI (blue). Cells treated with the filtered supernatant of *E. coli* BL21 pBCsk+ without insert (empty vector) were used as a negative control. Slides were observed by confocal microscopy. Inset images from the left panels are magnified in the panels to the right. Bars represent 10 µm. (**B**) Quantitative analysis of fluorescence intensity of F-actin. Analysis of fluorescent intensity was done at the original magnification by measuring the mean gray value with ImageJ software [22] with an *n* value of at least 10 cells. Data values represent the mean ± SEM of at least three independent experiments. (* $p < 0.05$, ** $p < 0.01$, *** $p < 0.001$ one-way ANOVA with multiple comparisons).

2.5. SPATE Entry into Bladder Epithelial Cells Is Dependent on the Serine Protease Active Site

We previously showed that TagB and TagC demonstrated cytotoxicity as measured by lactate dehydrogenase (LDH) release from epithelial cells within 5 h [20]. This toxicity could be due to the interaction of the SPATEs with targets inside host cells. So, to gain insight into the potential internalization of these SPATEs, we employed immunofluorescence labeling of proteins followed by visualization using confocal or immunogold electron microscopy. Firstly, confocal Z-sections (optical slices) of 5637 bladder cells treated with SPATEs were examined to determine if SPATEs were translocated within cells. After 5 h of incubation, TagB, TagC, and Sha (red color) were found within cells as evidenced by cell sectioning analysis (Figure 5A). By contrast, the serine active-site mutant variants were unable to enter epithelial cells and were not detected (absence of red staining) (Figure 5A), suggesting that serine protease activity is needed for the entry of SPATEs within cells. Interestingly, TagB within cells also co-localized with actin (green color) in the outer border of the cell (Figure 5A). Further, cells incubated with serine mutant variants of SPATEs did not enter cells, and these cells also produced actin stress fibers (Figure 5).

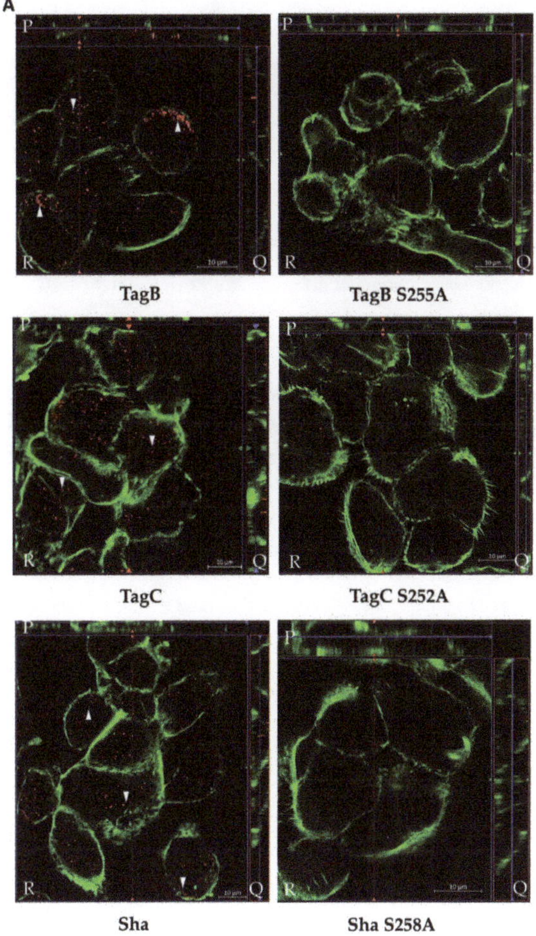

Figure 5. *Cont.*

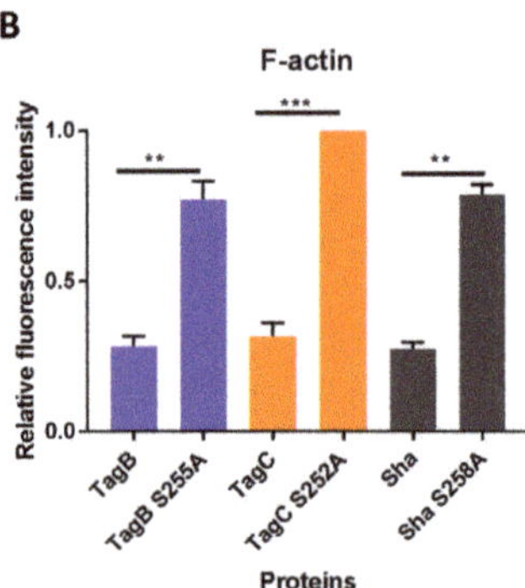

Figure 5. Intracellular localization of TagB, TagC, and Sha determined by confocal microscopy. (**A**) Z-stack imaging showing the localization of TagB, TagC, and Sha and their respective serine active site mutant variants during interaction with 5637 bladder epithelial cells after 5 h of incubation. SPATEs were detected by Alexa Fluor 594 (white arrowheads, red fluorescence) using anti-mouse secondary antibody and actin was stained by Alexa Fluor 488- phalloidin (green fluorescence). Images are displayed in a 3D section view with large Z-sections in the X-Y direction (R), Z-projection in the X–Z direction (P), and Z-projection in the Y–Z direction (Q). The green and red lines in R indicate the orthogonal planes of the X–Z and Y–Z projection. For each selected section, the signal was gathered from a span of 5 μm. Scale bar: 10 μm (**B**) Quantitative analysis of fluorescence intensity of F-actin in the cells treated with native or mutant SPATEs. Analysis of fluorescence intensity was done in green channel by measuring the mean gray value on ImageJ. Data represent the mean ± SEM of at least three independent experiments. Significant differences between fluorescence intensity of each native and mutant SPATE treated cell was determined using Student's t-test with ** $p < 0.01$, *** $p < 0.001$.

Analysis of thin-sections of SPATE-treated cells using immunogold staining and transmission electron microscopy (TEM) also confirmed the intracellular localization of all three SPATEs within cells. TagB and Sha were found in the cytoplasm, whereas TagC was present in the nucleus (Figure 6). However, in multiple independent experiments, we failed to detect the presence of serine mutant variants of TagB, TagC, or Sha within cells. The serine catalytic-site mutant proteins when visualized were almost exclusively observed on the extracellular surface of cells as seen in cells treated with TagB S255A (Figure 6D).

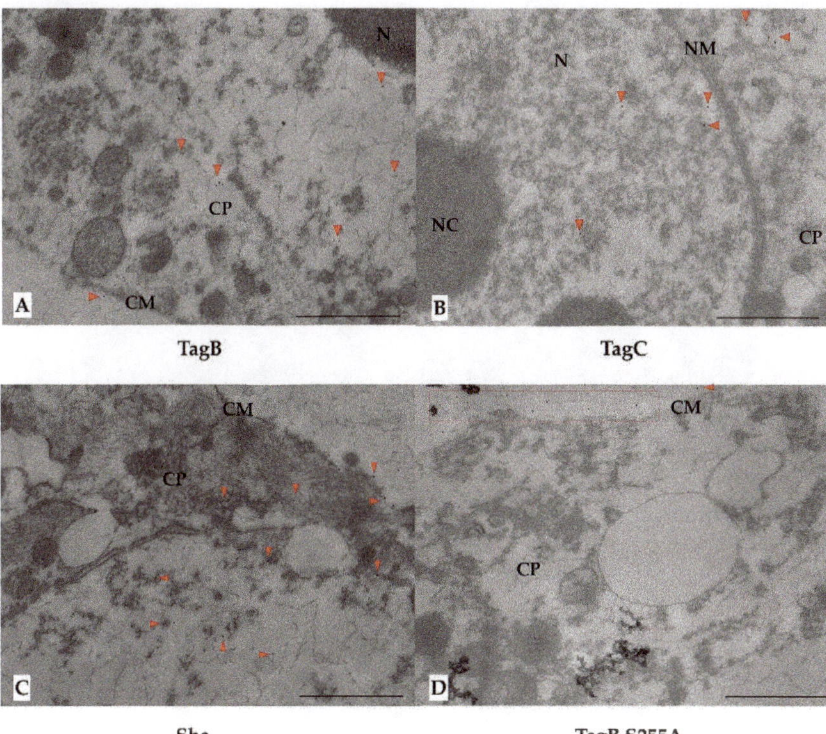

Figure 6. Transmission electron micrographs of 5637 bladder cells showing internalized SPATEs, immunolabelled with 10-nm-diameter gold particles after 5 h of incubation. Gold particles are highlighted with red triangles. (**A**) TagB is principally located in the cytoplasm (CP). (**B**) For Tag C, gold particles were associated with the nucleus (N) and cytoplasm (CP). (**C**) Sha was located mainly in the cytoplasm (CP). (**D**) For the serine mutant variants of TagB, TagC, and Sha, gold particles were only localized on the extracellular surface of cells (red box). Only the TagB S255A mutant protein localization is shown. Cell membrane (CM), Cytoplasm (CP), Nuclear Membrane (NM), Nucleus (N), Nucleolus (NC) Bars, 1 μm.

2.6. Sha Exhibits Serine Protease-Dependent Mucinase Activity

Epithelial cell damage caused by SPATEs was shown to require protease activity, and some other SPATEs were previously shown to demonstrate activity against host proteins such as mucin or gelatin [18,23]. Further, we also tested for mucinase activity, since two of the novel SPATEs identified in APEC QT598, Sha and TagB [20], belong to the class 2 SPATE family whose members have been shown to demonstrate mucinolytic activity. Clones of *E. coli* BL21 expressing each of the SPATEs were grown on agar plates containing 0.5% porcine gastric mucin for 24 h at 37 °C, followed by amido black-staining. Plates containing clones growing on discs expressing Sha revealed clear zones of mucin lysis (Figure 7A) and the lysis zone produced by Sha was intermediate when compared to clones expressing either Tsh (positive control) or Vat. Mucin containing plates had a clearing zone with a diameter of 3.9 ± 0.1 cm after exposure to Sha expressing bacteria, which was less than following exposure to Tsh expressing cells (4.2 ± 0.1 cm), but more than following exposure to Vat expressing cells (3.7 ± 0.2 cm). By contrast, TagB and TagC were mucinase-negative as evidenced by the absence of any clearing zones (Figure 7B). Further, the critical role of the serine catalytic site of Sha for mucinase activity was demonstrated with the clone expressing Sha S258A, which did not produce a zone of mucin lysis (Figure 7B). The clone containing only the empty vector (negative control) did not grow well in the presence of mucin and also demonstrated no clearing zone. When mucin was treated with culture filtrates of SPATE proteins (Figure 7C), it was not degraded by either TagB, TagC, or in the negative control (empty vector). Sha as well as Tsh and Vat degraded mucin, whereas the serine protease mutant of Sha, Sha S258A, did not. Hence, the serine catalytic site of Sha is required for mucinase activity.

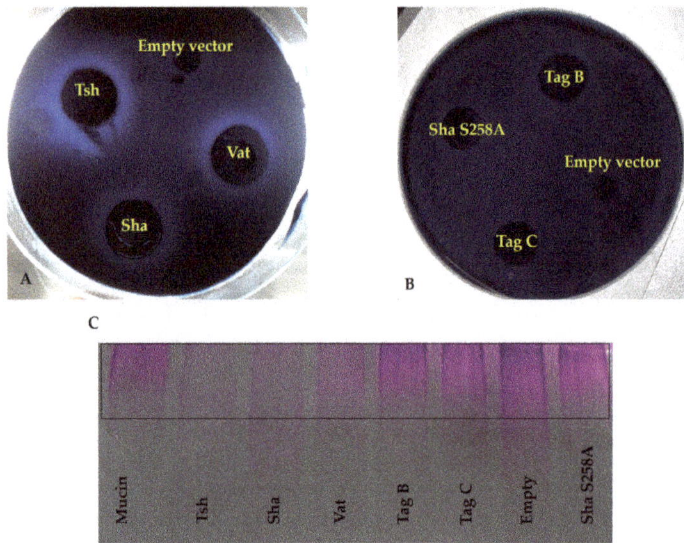

Figure 7. Sha demonstrates serine-protease dependent mucinase activity, but not TagB nor TagC. Mucinase activity was tested in a medium containing 1.5% agarose and 0.5% porcine gastric mucin. Filter discs inoculated with clones containing the empty vector, expressing Sha, Vat, Tsh, (**A**) TagB, TagC, or Sha S258A (**B**) were placed on the agar surface and incubated overnight at 37 °C. Mucin lysis zones were visualized by staining with 0.1% amido-black in 3.5 M acetic acid for 15 min, followed by destaining with 5% acetic acid and 0.5% glycerol for 6 h to overnight. (**C**) Zones of 0.5% porcine gastric mucin hydrolysis are visible in the stacking region of the SDS-PAGE gel (boxed area), concentrated supernatant extracts of SPATEs (5 μg of protein per well) were incubated at 37 °C for 48 h with mucin prior to migration. The gel was stained with a PAS glycoprotein staining kit.

2.7. TagC Exhibits Serine Protease-Dependent Gelatinase Activity

Some SPATEs were previously reported to degrade extracellular matrix proteins such as collagen and gelatin [23]. We previously demonstrated that TagB, TagC, and Sha could mediate increased adherence to chicken fibroblasts [20], which are cells that are associated with connective tissues and produce extracellular matrix proteins such as collagen. The hydrolyzed form of collagen—gelatin was used as a substrate to test for potential gelatinase activity from supernatant extracts containing SPATEs. Culture supernatant filtrate from *Pseudomonas aeruginosa* was used as a positive control, since it is known to demonstrate gelatinase activity [24]. Samples were incubated with 1% bovine gelatin for 48 h at 37 °C. Culture filtrates containing TagC as well as other SPATEs EspC, Tsh, and Vat demonstrated gelatinase activity (Figure 8A). By contrast, neither TagB nor Sha demonstrated gelatinase activity, since high-molecular-weight bands, indicating intact gelatin, remained after exposure to these SPATEs. Further, gelatinase activity from TagC was shown to be dependent on the serine protease motif, since the *E. coli* clone expressing a serine active site mutant protein, TagC S252A, did not generate a hydrolysis zone on medium containing 1% gelatin, whereas the TagC expressing clone did exhibit a hydrolysis zone (Figure 8B).

A

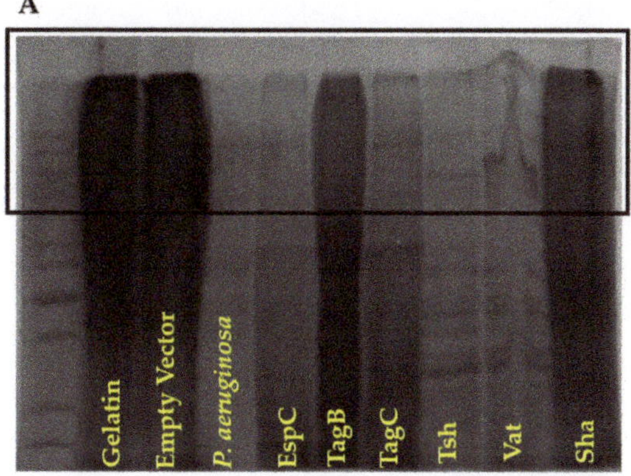

B

Figure 8. TagC demonstrates serine-protease dependent gelatinase activity, but not TagB nor Sha. (**A**) Zones of 1% bovine skin gelatin hydrolysis are visible in the stacking region of the SDS-PAGE gel (boxed area), concentrated supernatant extracts of SPATEs (5 μg of protein per well) were incubated at 37 °C for 48 h prior to migration. (**B**) Gelatinase activity of TagC was tested in a medium containing 1.5% agarose and 1% bovine skin gelatin. The disc inoculated with a clone expressing TagC or its serine catalytic site mutant variant, TagC S252A, were inoculated on the agar surface and were incubated for 48 h at 37 °C. Zones of gelatin lysis were visualized by staining with 0.1% amido-black in 3.5 M acetic acid for 15 min, followed by destaining with 5% acetic acid and 0.5% glycerol for 6 h.

3. Discussion

Colonization of the bladder is vital for UTI pathogenesis and UPEC deploys an array of virulence factors to infect and colonize the bladder, including secreted toxins [25]. Hemolysin A [8,9], UpxA (TosA) [26], cytotoxic necrotizing factor-1 (CNF-1) [27,28], and a variety of SPATEs (serine-protease autotransporters of *Enterobacteriaceae*) [29] are known toxins of host cells that are produced by some UPEC strains. The recent identification of new members of the SPATEs family present in some pathogenic *E. coli* and their cytotoxic activity on bladder cell lines [20], led us to further investigate mechanisms underlying the cytotoxic and proteolytic activity of the TagB, TagC, and Sha SPATEs on an established human urinary bladder cell line [30,31] and other properties of these virulence-associated proteins.

TagB, TagC, and Sha proteins demonstrated autoaggregating activity, and also promoted adherence of *E. coli* strain BL21 to the human HEK 293 renal and 5637 bladder human cell lines. Further, Sha also contributed to increased biofilm production [20]. SPATEs present on the bacterial surface are likely to contribute to the autoaggregation phenomenon. (Figure 1A). TagB and TagC also exhibited cytopathic effects on the bladder epithelial cell line. Further, we also previously determined that proteolytic activity of these SPATEs was strongly inhibited upon addition of serine protease inhibitor (PMSF), providing evidence for the importance of the serine protease motif in the activity of these SPATEs [20]. To further investigate the role of the serine protease activity, we generated catalytic site mutants of these three SPATEs. It is of note that the serine protease consensus motif (GDSGS) is conserved among different members of SPATEs [20,32–35]. Importantly, loss of the serine active site did not affect the processing or secretion of the SPATE proteins into the extracellular milieu (Figure 1A). Further, loss of the serine active site also eliminated any autoproteolytic activity (Figure 1A). Similarly, autoproteolytic activity has also been reported for other SPATEs including EspP, Sat, Pic [36], and for AspA autotransporter from *Neisseria meningitidis* [37]. Thus, from our results, it is clear that the processing of the passenger domain across the bacterial surface and autocatalytic activities of the TagB, TagC, and Sha is independent of the proteolytic serine site.

We investigated the role of the serine catalytic site of TagB, TagC, and Sha in autoaggregation or hemagglutination, as either SPATE protease activity on the bacterial or host cell surfaces could have possibly mediated these phenotypes. For instance, cleavage could have led to certain domains within the protein, leading to exposure of hydrophobic sites which could promote aggregation [38]. However, the serine protease site was not required for TagB, TagC, or Sha-mediated aggregation (Figure 2). These results indicate that other specific SPATE structural domains are likely to be responsible for aggregation. However, importantly, the autoaggregation phenotype is not a generalized phenotype of SPATEs, since in previous experiments, both the Tsh and Vat SPATEs did not demonstrate any autoaggregation phenotype [20]. Currently, the molecular mechanism of autoaggregation of TagB, TagC, and Sha is unknown. Unlike the three SPATEs described herein, loss of the active site serine of the Hap adhesin, a *Haemophilus influenzae* serine protease autotransporter, abrogated autoproteolytic processing leading to retention of this AT protein on the bacterial cell surface [39]. In fact, the increase in Hap present on the bacterial surface also increased aggregation, formation of microcolonies, and adherence of *H. influenzae* to host cells [40]. With regards to hemagglutination activity of the Sha protein, the serine active site was also dispensable. We found that Sha S258A hemagglutinated human blood with a similar titer to the native Sha protein. Similarly, a Tsh S259A variant protein was also able to bind to avian erythrocytes, turkey hemoglobin, collagen IV, fibronectin, and laminin [41]. Considering that the TagB S255A, TagC S252A, and Sha S258A variant SPATEs all retained the respective phenotypes present in the native SPATE proteins, this suggests that, despite lacking catalytic activity, that these variants are likely to have maintained a properly folded conformation.

In contrast to adherence or aggregation phenotypes, the presence of a serine protease motif was clearly required for cytotoxicity and entry of the SPATE proteins into bladder epithelial cells. In this study, the TagB S255A and TagC S252A mutant proteins were no longer cytopathic. Our results are similar to those described for other SPATEs [15,41,42] which have demonstrated a key role for the serine active site with regards to any native proteolytic or cytopathic activity of SPATEs on protein substrates or host cells.

Since TagC shares 60% identity/74% similarity with another SPATE, EspC, a non-LEE-encoded enterotoxin of enteropathogenic *E. coli* (EPEC) which causes cytotoxic effects and cleavage of cytoskeletal actin-associated protein [43]; we explored potential cellular targets in relation to the cytopathic effect observed in bladder epithelial cells. Following treatment with either TagB, TagC, or Sha, reorganization of the cytoskeleton and loss of actin stress fibers were seen in bladder epithelial cells (Figure 4). The effect of TagC was severe with faint staining remaining for actin compared to TagB interaction with cells. Exposure to Sha caused punctate localization of actin within the cytoplasm. Diminished actin staining and the formation of punctate actin accumulation suggests that each of these SPATEs are targeting the actin cytoskeleton or other cellular targets that lead to modifications in actin fiber formation or distribution within bladder cells. As expected, alterations in actin distribution were absent from bladder cells exposed to the serine catalytic site mutant variant proteins TagB S255A, TagC S252A, or Sha S258A, confirming the critical cytopathic role of serine protease activity.

Many pathogens exploit host actin for various stages of infection, including cellular invasion, intracellular replication, and dissemination by different mechanisms [44,45]. Specifically, during UTIs, UPEC utilizes the Rho family GTPase member Rac1 to mediate actin polymerization for *E. coli* bladder epithelial cell invasion [46]. It has been well documented that there is a relation between intracellular growth of UPEC in the bladder epithelium and the host F-actin cytoskeleton [47]. Based on the observation of actin rearrangement observed in bladder cells, it is also possible that the TagB, TagC, and Sha SPATEs might also contribute to UPEC invasion of the bladder epithelium, as these proteases may promote adhesion and loss of integrity of the protective epithelial barrier which could increase bacterial entry into epithelial cells as well as increase entry and systemic spread of the bacteria to other tissue sites during infection.

Before reaching the epithelial cell surface in the urinary tract, bacteria must cross the protective mucus layer that is coated with mucin [48]. Mucin serves as a primary antibacterial defense in the bladder and contributes to host innate defense by providing a barrier and by trapping bacteria [49]. Many pathogens can invade or reduce the viscosity of mucin by cleaving it [50–52]. Certain SPATEs, belonging to the Class 2 family, including Pic [18,53], PicC of *Citrobacter rodentium* [53], and Tsh, demonstrate mucinase activity [54]. We, therefore, tested whether any of the three novel SPATEs were mucinolytic, and only Sha was identified as a mucinase (Figure 7). The zone of mucinolytic activity of Sha was intermediate when compared to Vat and Tsh and, as has been shown for Pic [18], the serine catalytic site of Sha was required for mucinase activity. From this standpoint, it is interesting to note that in strain QT598, 3 of the 5 SPATEs (Tsh, Vat, and Sha) demonstrate mucinase activity [20] which might facilitate bacterial colonization by degrading mucus to overcome the mucous barrier at the interface of epithelial surfaces. TagC was also shown to degrade gelatin, which is the hydrolyzed form of collagen, although this activity was absent from Sha and TagB. Collagen is an abundant and ubiquitous extracellular matrix protein that forms an essential component of connective tissues [55]. From this standpoint, the TagC protease may contribute to tissue invasion and systemic spread of ExPEC by degradation of extracellular matrix proteins. As expected, the activity of TagC on gelatin was also dependent on the active serine catalytic site. Similarly, Pic [23] also demonstrated gelatinase activity that required an active serine catalytic site.

Previous reports have described different mechanisms of internalization of SPATEs and types of cytoskeletal damage in various epithelial cells in vitro. The Pet SPATE from enteroaggregative *E. coli* (EAEC) is internalized by a retrograde trafficking pathway [56] through the Pet host cell receptor, cytokeratin 8 [57]. Once internalized, Pet causes loss of actin stress fibers due to the breakdown of spectrin [58,59]. Internalization of EspC by EPEC requires the type 3 secretion system [60] and leads to cleavage of cytoskeletal proteins [43]. Sat is secreted by UPEC, enters the cell by an unknown mechanism, and localizes to the cytoskeletal fraction of fodrin/spectrin and integrin present within bladder and kidney epithelial cells [15]. In the present report, we have demonstrated that TagB, TagC, and Sha are also internalized in bladder epithelial cells by a mechanism that requires an active serine catalytic domain. We used confocal Z-sections to verify the intracellular localization of the SPATEs within human bladder epithelial cells. Of note, we observed the internalization of TagB, TagC, and Sha within bladder cells after 5 h and this was concomitant with diminished fluorescence staining of actin in the vicinity of the localized SPATEs. This observation was pronounced following exposure to TagB, and TagB was shown to be closely associated with actin. Furthermore, to confirm the internalization of SPATEs within bladder epithelial cells, we carried out immunogold TEM of cross-sections of cells to demonstrate SPATE proteins within epithelial cells. TEM demonstrated localization of TagB and Sha in the cytoplasm, whereas TagC targeted the nucleus. We speculate that, since TagC has previously been shown to promote nuclear enlargement [20], TagC may alter nuclear targets and elicit a significant increase in nuclear size. The entry of these SPATEs into host bladder cells does not require a type 3 secretion mechanism since it is absent from *E. coli* QT598 and *E. coli* BL21, and SPATE proteins from bacterial supernatants entered bladder epithelial cells directly. Future studies will elucidate the cytoplasmic or nucleo-cytoplasmic shuttling pathways that mediate the entry and trafficking of these three SPATEs. Importantly, the serine catalytic site was required for cell entry and cytotoxicity of all three SPATEs, since serine protease active site mutants were unable to enter cells or cause any cytopathic effects, further demonstrating a critical role for the serine catalytic site of these SPATEs.

Taken together, the TagB, TagC, and Sha SPATE proteins mediate multiple activities. These include adhesion, aggregation, cytopathic effects, mucinase and gelatinase activities that may collectively contribute to different stages of bacterial infection including initial colonization, invasion of host epithelia, and an increased potential for systemic infection.

4. Materials and Methods

4.1. Ethics Statement

This study was performed in accordance with the ethical standards of the University of Quebec, INRS. A protocol for obtaining biological samples from human blood donors was reviewed and approved by the ethics committee—*Comité d'éthique en recherche* (CER 19-507, approved November 19, 2019) of INRS.

4.2. Bacterial Strains, Plasmids, and Growth Conditions

E. coli clones expressing TagB, TagC, or Sha were described previously [20]. All DNA constructs were transformed into *E. coli* strain BL21 or the type 1 fimbriae *fim*-negative *E. coli* strain ORN172. Strains were grown at 37 °C on solid or liquid Luria-Bertani medium (Alpha Bioscience, Baltimore, MD, USA) supplemented with the appropriate antibiotics when required at concentrations of 100 µg/mL ampicillin, 30 µg/mL chloramphenicol, or 50 µg/mL of kanamycin. Strains, plasmids, and primers are listed in Table 1.

Table 1. Strains and plasmids used in this study.

Strains	Characteristic(s)	Reference
QT598	APEC O1: K1, serum resistant	[61]
ORN172	*thr-1 leuB thi-1Δ (argF-lac)U169 xyl-7 ara-13 mtl-2 gal-6 rpsL tonA2 supE44Δ (fimBEACDFGH)::Km pilG1*	[62]
BL21	*fhuA2 [lon] ompT gal [dcm] ΔhsdS*	New England Biolabs
QT1603	HB101 with AIDA-1 operon	[63]
QT4767	ORN172/pIJ553 (Expressing *sha*)	
QT5194	BL21/pIJ548 (Expressing *tagB*)	
QT5195	BL21/pIJ549 (Expressing *tagC*)	
QT5197	BL21/pIJ550 (Expressing *espC*)	
QT5198	ORN172/pIJ548 (Expressing *tagB*)	[20]
QT5199	ORN172/pIJ549 (Expressing *tagC*)	
QT5431	BL21/pIJ551 (Expressing *vat*)	
QT5432	BL21/pIJ552 (Expressing *tsh*)	
QT5433	BL21/pIJ553 (Expressing *sha*)	
QT5437	BL21 + pIJ554 (Expressing *tagB S255A*)	This study
QT5438	BL21 + pIJ555 (Expressing *tagC S252A*)	This study
QT5439	BL21 + pIJ556 (Expressing *sha S258A*)	This study
QT5598	ORN172 + pIJ554 (Expressing *tagB S255A*)	This study
QT5599	ORN172 + pIJ555 (Expressing *tagC S252A*)	This study
QT5600	ORN172 + pIJ556 (Expressing *sha S258A*)	This study
QT3046	*Pseudomonas aeruginosa* PA14	Eric Déziel, INRS
Plasmids		

Table 1. *Cont.*

pBCsk+	Cloning vector; Cm^r	Stratagene, La Jolla, CA
pIJ548	pBCsk+::*tagB*	
pIJ549	pBCsk+::*tagC*	[20]
pIJ551	pBCsk+::*vat*	
pIJ552	pBCsk+::*tsh*	
pIJ553	pBCsk+::*sha*	
pIJ554	pBCsk+::*tagB S255A*	This study
pIJ555	pBCsk+::*tagC S252A*	This study
pIJ556	pBCsk+::*sha S258A*	This study

4.3. Site-Directed Mutagenesis

Site-directed mutagenesis was performed using the Q5® Site-Directed Mutagenesis kit as specified by the manufacturer. pIJ548, pIJ544, and pIJ553 were used as a template for the construction of the serine catalytic site mutants TagB S255A (pIJ554), TagC S252A (pIJ555), and Sha S258A (pIJ556) at 25 to 50 ng per reaction with 10 pmol of each of the complementary primers. Primers used to generate the single point mutation substituting alanine for serine for TagB were 5′-TCCCGGTGACGCCGGCTCTCCT-3′ and 5′-GTACCGTAGGTTGAGAGTG-3′; TagC were 5′-AGGAGGAGACGCCGGTTCCGGA-3′ and 5′-GTCACTTCATTATAAAATCCACC-3′; and Sha were 5′-GGCTGGTGATGCCGGTTCTCCGC-3′ and 5′-TCACCATAGATCGGTAATAC-3′. Following mutagenesis, all constructs were verified by sequencing at the proteomics platform of the Institut de Recherche en Immunologie et en Cancérologie (IRIC) of the Université de Montréal (Montréal, QC, Canada).

4.4. Recombinant Protein and Antibody Preparation

Expression and purification of SPATE proteins from concentrated filtered culture supernatant fractions were obtained as described previously [20] and the extract was checked by silver staining before each assay. Antibodies against ~ 112 kDa Vat protein were used to generate a Vat-specific rabbit polyclonal antibody, according to a standard protocol [64] (Laboratorio de Biología Celular y Tisular, Departamento de Morfología, Universidad Autónoma de Aguascalientes (UAA), Aguascalientes, Mexico). Since SPATE proteins contain some highly conserved epitopes, anti-Vat antibodies were used to detect and label each of the SPATE proteins. The alignment of the passenger domain of Vat with TagB, TagC, and Sha share identities of 39%, 30%, 56%, respectively. Specific epitopes are not established but Vat-antibodies demonstrate multiple conserved residues (Figure S4) and strong immune cross-reactivity. Cross-reactivity of antibodies raised against other SPATEs have also been reported. Antibodies raised against Pet protein (45% identity with EspC and 60 gaps) cross-reacted with EspC [65]. Likewise in the supernatant of CFT073, anti-Pic (44% identity with Vat and 76 gaps) antibodies were used to detect Vat and PicU SPATEs [66]. Polyclonal antisera adsorption was done by incubating the filtered supernatant of *E. coli* BL21 pBCsk+ without insert with a 1:50 dilution of the Vat polyclonal antiserum for 1 h at room temperature under mild agitation followed by centrifugation at 2000× *g* for 5 min at 4 °C.

4.5. Autoaggregation and Hemagglutination Tests

Autoaggregation of bacterial cells was measured by a settling assay as performed previously [20]. The sedimentation of 10 mL of each culture of *E. coli* BL21 cells expressing native or serine active site mutant SPATEs were adjusted to an OD_{600nm} 1.5 from an overnight culture grown at 37 °C in liquid Luria-Bertani medium. Then, they were monitored for a reduction in turbidity from the top of the tube which was left at 4 °C for 3 h. The reduction of turbidity was plotted as a ratio against the initial turbidity.

For hemagglutination assays, human blood cells (RBCs) were washed and resuspended in PBS at a final concentration of 3% using a protocol adapted from [67]. The *E. coli fim*-negative K-12 strain ORN172 expressing either native or serine active site mutant SPATEs was grown overnight at 37 °C in Luria-Bertani medium, harvested and adjusted to an optical density (O.D.$_{600nm}$) of 60. Suspensions were serially diluted in 96-well round-bottom plates containing 20 µL of PBS mixed with 20 µL of 3% red blood cells and incubated for 30 min at 4 °C.

4.6. Epithelial Cell Culture

The 5637 bladder epithelial cell line was routinely cultured in RPMI 1640 medium (Thermo Fisher Scientific) supplemented with 10% heat-inactivated FBS at 37 °C in humidified 5% CO_2, and 2×10^5 cells/well were seeded into eight-well chamber slides (Thermo Fisher Scientific, Waltham, MA, USA) and allowed to grow to 75% confluence.

To determine cytopathic effects on bladder cells, a final concentration of 30 µg/mL of native SPATEs or the serine catalytic site mutants were added directly to monolayers and incubated for 5 h in RPMI 1640 medium at 37 °C with 5% CO_2. Cells were then washed twice with PBS (phosphate-buffered saline), fixed with 70% methanol, and stained with Giemsa stain. Cell morphology was analyzed at a magnification of ×20 with standard bright-field light microscopy. For the lactate dehydrogenase assay, supernatant from cells treated with native or mutant SPATEs were collected and the release of LDH in cell culture supernatants were quantified by using the CytoTox 96® Non-Radioactive Cytotoxicity Assay kit (Promega, Madison, WI, USA). Maximum LDH release (positive control) was determined by adding lysis solution (provided in the kit) to the non-infected cells.

For fluorescence actin-staining and immunostaining assays, cells were fixed with 3.0%–4.0% formaldehyde in PBS, washed, permeabilized by addition of 0.1% Triton X-100-PBS, stained with 0.05 µg of Alexa Fluor 488-phalloidin/mL (AAT Bioquest, Sunnyvale, CA, USA) at 37 °C for 1 h and counterstained with ProLong Gold/DAPI antifade reagent (Invitrogen, Carlsbad, CA, USA). After image acquisition using confocal microscope, the actin staining intensity was quantified by measuring mean gray value (mean pixel intensity) in ImageJ (https://imagej.nih.gov/ij/) [22]. The cells of interest as well as background with no fluorescence were selected manually to analyze the areas integrated intensities and mean gray value. The value was then corrected and total fluorescence (CTF) was calculated as CTF = Integrated Density − (Area of selected cell X Mean fluorescence of background readings). The averaged corrected mean gray value was used to generate relative quantitative comparison of fluorescence intensity.

SPATE protein localization in bladder cells was detected by immunofluorescence. Treated cells were fixed, permeabilized, and incubated with blocking solution (PBS with 5% BSA) for 1 h at 37 °C. Samples were then incubated with rabbit anti-SPATE polyclonal antibodies (UAA, Mexico) for 2 h at 37 °C. This was followed by incubation with secondary antibody Alexa Fluor 594-labeled goat anti-rabbit IgG antibody (Thermo Fisher Scientific, Waltham, MA, USA). Samples were mounted and imaged with the 60X objective of an LSM780 confocal microscope (Carl Zeiss microscopy Gmbh, Jena, Germany). Images were processed with ZEN 2012 software (Carl Zeiss microscopy Gmbh, Jena, Germany).

4.7. Electron Microscopy

Immunogold labeling of bacteria was carried out by culturing *E. coli* BL21 expressing different SPATEs in Luria-Bertani medium supplemented with 30 µg/mL chloramphenicol for 5 h. Bacterial suspensions (50 µL) were spotted on nickel-coated TEM grids. After 15 min, liquid was wicked away with bibulous paper and blocked with drops of PBS containing 1% ovalbumin for 15 min. A blocking solution was exchanged with a drop of SPATE antiserum diluted 1:100 in PBS. After 15 min, excess fluid was wicked away with bibulous paper and exchanged for PBS containing 1% ovalbumin drops for 5 min. The wash was repeated and then incubated in suitable goat anti-rabbit IgG (H+L), Alexa

Fluor 488–10 nm colloidal gold secondary antibodies (Thermo Fisher Scientific, Waltham, MA, USA) diluted 1:250 in incubation solution. After 15 min, grids were washed twice with PBS drops and rinsed twice with distilled water. Grids were dried with bibulous paper and imaged on a Philips CM-100 transmission electron microscope.

For immunogold labeling of epithelial cell thin sections, cells were fixed in 0.1% glutaraldehyde + 4% paraformaldehyde in cocodylate buffer at pH 7.2, and post-fixed in 1.3% osmium tetroxide in collidine buffer. After dehydration by successive passages through 25, 50, 75, and 95% solutions of acetone in water (for 15–30 min each) samples were immersed for 16–18 h in SPURR: acetone (1:1). Samples were then embedded in BEEM capsules using SPURR resin with the ELR-4221 kit (Polysciences Inc, Warrington, PA, USA) followed by placing the capsules at 60–65 °C for 20–30 h to polymerize the resin. After resin polymerization, samples were cut using an ultramicrotome (Ultratome) and were put onto Formvar and carbon covered-copper 200-mesh grids treated with sodium metaperiodate and were blocked with 1% BSA in PBS. Grids were then incubated with primary antibodies, washed, and incubated with goat anti-rabbit IgG (H+L), Alexa Fluor 488–10 nm colloidal gold secondary antibodies (Thermo Fisher Scientific, Waltham, MA, USA). After washing, samples were contrasted with uranyl acetate and lead citrate and subsequently visualized using a Philips EM 300 transmission electron microscope.

4.8. Cleavage of Protein Substrates

For mucinase activity, cultures of *E. coli* BL21 expressing SPATEs were incubated for 24 h at 37 °C on a medium containing 1.5% agarose and 0.5% porcine gastric mucin (Sigma-Aldrich, St. Louis, MI, USA). Plates were subsequently stained with 0.1% amido-black in 3.5 M acetic acid for 15 min, followed by destaining with 5% acetic acid and 0.5% glycerol for 6 h to overnight. Zones of mucin lysis were visualized as discolored halos around colonies. For the Periodic Acid Schiff (PAS) assay to detect mucin degradation [53], 5 µg of each SPATE protein were incubated with 5 µg of 0.5% porcine gastric mucin (Sigma-Aldrich) in 30 µL of MOPS buffer and incubated for 48 h at 37 °C. Treated samples were electrophoresed on an 8% SDS-PAGE gel and the gel staining was developed using a colorimetric Pierce™ Glycoprotein Staining kit (ThermoFisher Scientific, Waltham, MA, USA).

For gelatinase activity, 5 µg of each SPATE protein were incubated with 5 µg of bovine skin gelatin (Sigma-Aldrich, St. Louis, MI, USA) in 30 µL of MOPS buffer and incubated for 48 h at 37 °C. Samples were then boiled with Laemmli sample buffer, were electrophoresed on an 8% SDS-PAGE gel and then resolved by Coomassie blue staining. In addition, gelatinase activity was also tested by growing the clones on agar plates containing 1.5% agarose and 1% bovine skin gelatin for 48 h at 37 °C. Plates were subsequently stained with 0.1% amido-black in 3.5 M acetic acid for 15 min, followed by destaining with 5% acetic acid and 0.5% glycerol for 6 h to overnight. Zones of gelatin lysis consist of discolored halos around colonies.

4.9. Statistical Analysis

Experimental data were expressed as a mean ± standard error of the mean (SEM) in each group. The means of groups were combined and analyzed by a two-tailed Student *t*-test for pairwise comparisons and analysis of variance (ANOVA) to compare means of more than two populations. A *p* value of <0.05 was considered statistically significant. All data were analyzed with the Graph Pad Prism 7 software (GraphPad Software, San Diego, CA, USA).

5. Conclusions

In conclusion, TagB, TagC, and Sha are novel SPATEs that demonstrate different proteolytic activities on different substrates as well as distinct cytopathic effects on bladder epithelial cells. Additional molecular *in vitro* and *in vivo* studies are in progress in an effort to understand the link between protease activity of the TagB, TagC, and Sha SPATEs and how these proteases disrupt or alter the actin cytoskeleton during ExPEC infections. It will be of further interest to also investigate their potential interactions with other host cells or extracellular matrix proteins, and determine how these relatively large proteins (generally greater than 100 kDa) manage to enter host cells through serine protease activity and what specific trafficking pathways may be involved in their localization or association with specific cellular compartments.

Supplementary Materials: Supplementary materials can be found at http://www.mdpi.com/1422-0067/21/9/3047/s1.

Author Contributions: Conceptualization, Investigation, Data curation, Methodology, Validation, Writing-of the manuscript, P.P.; Methodology, Investigation, Software, Data curation, J.M.D., H.B., and S.H.; Writing—review and editing, J.M.D., H.B., S.H. A.L.G.-B. and C.M.D.; project administration, S.H., C.M.D.; Funding acquisition, Supervision, C.M.D. All authors have read and agreed to the published version of the manuscript.

Funding: Funding for this work was supported by NSERC Canada Discovery Grants 2014-06622 and 2019-06642 and scholarships from the CRIPA-FRQNT network Funds n°RS-170946 to P.P. and J.M.D.

Acknowledgments: We thank Arnaldo Nakamura for assistance in electron microscopy and Jessy Tremblay for assistance in confocal immunofluorescence microscopy imaging.

Conflicts of Interest: The authors declare no conflict of interest.

Abbreviations

AIDA-I	Adhesin involved in diffuse adherence
ANOVA	Analysis of variance
APEC	Avian pathogenic *E. coli*
CNF1	Cytotoxic necrotizing factor 1
DAPI	4′,6-Diamidino-2-phenylindole
EAEC	Enteroaggregative *E. coli*
EPEC	Enteropathogenic *E. coli*
ExPEC	Extra-intestinal pathogenic *E. coli*
LDH	Lactate dehydrogenase
LEE	Locus of enterocyte effacement
LPS	Lipopolysaccharides
OD	Optical density
PAS	Periodic acid–Schiff
PMSF	Phenylmethylsulfonyl fluoride
SDS-PAGE	Sodium dodecyl sulfate-polyacrylamide gel electrophoresis
SPATEs	Serine protease autotransporters of *Enterobacteriaceae*
TEM	Transmission electron microscopy
UPEC	Uropathogenic Escherichia coli
UTIs	Urinary tract infections

References

1. Foxman, B.; Brown, P. Epidemiology of Urinary Tract Infections: Transmission and Risk Factors, Incidence, and Costs. *Infect. Dis. Clin. North Am.* **2003**, *17*, 227–241. [CrossRef]
2. Foxman, B. Urinary Tract Infection Syndromes: Occurrence, Recurrence, Bacteriology, Risk Factors, and Disease Burden. *Infect. Dis. Clin. North Am.* **2014**, *28*, 1–13. [CrossRef] [PubMed]
3. Flores-Mireles, A.L.; Walker, J.N.; Caparon, M.; Hultgren, S.J. Urinary Tract Infections: Epidemiology, Mechanisms of Infection and Treatment Options. *Nature Rev. Microbiol.* **2015**, *13*, 269–284. [CrossRef] [PubMed]
4. O'Brien, V.P.; Hannan, T.J.; Nielsen, H.V.; Hultgren, S.J. Drug and Vaccine Development for the Treatment and Prevention of Urinary Tract Infections. *Microbiol. Spectr.* **2016**, *4*. [CrossRef]
5. Lewis, S.A. Everything You Wanted to Know About the Bladder Epithelium but Were Afraid to Ask. *Am. J. Physiol. Ren. Physiol.* **2000**, *278*, F867–F874. [CrossRef]
6. Keane, W.F.; Welch, R.; Gekker, G.; Peterson, P.K. Mechanism of *Escherichia coli* Alpha-Hemolysin-Induced Injury to Isolated Renal Tubular Cells. *Am. J. Pathol.* **1987**, *126*, 350.
7. Russo, T.A.; Davidson, B.A.; Genagon, S.A.; Warholic, N.M.; MacDonald, U.; Pawlicki, P.D.; Beanan, J.; Olson, M.R.; Holm, B.A.; Knight, P.R., 3rd. *E. coli* Virulence Factor Hemolysin Induces Neutrophil Apoptosis

and Necrosis/Lysis in Vitro and Necrosis/Lysis and Lung Injury in a Rat Pneumonia Model. *Am. J. Physiol. Lung Cell. Mol. Physiol.* **2005**, *289*, L207–L216. [CrossRef]

8. Smith, Y.C.; Rasmussen, S.B.; Grande, K.K.; Conran, R.M.; O'Brien, A.D. Hemolysin of Uropathogenic *Escherichia coli* Evokes Extensive Shedding of the Uroepithelium and Hemorrhage in Bladder Tissue within the First 24 Hours after Intraurethral Inoculation of Mice. *Infect. Immun.* **2008**, *76*, 2978–2990. [CrossRef]

9. Dhakal, B.K.; Mulvey, M.A. The Upec Pore-Forming Toxin A-Hemolysin Triggers Proteolysis of Host Proteins to Disrupt Cell Adhesion, Inflammatory, and Survival Pathways. *Cell Host Microbe* **2012**, *11*, 58–69. [CrossRef] [PubMed]

10. Wiles, T.J.; Mulvey, M.A. The Rtx Pore-Forming Toxin A-Hemolysin of Uropathogenic *Escherichia coli*: Progress and Perspectives. *Future Microbiol.* **2013**, *8*, 73–84. [CrossRef]

11. Ristow, L.C.; Welch, R.A. Hemolysin of Uropathogenic *Escherichia coli*: A Cloak or a Dagger? *Biochim. Biophys. Acta Biomembr.* **2016**, *1858*, 538–545. [CrossRef] [PubMed]

12. Visvikis, O.; Boyer, L.; Torrino, S.; Doye, A.; Lemonnier, M.; Lorès, P.; Rolando, M.; Flatau, G.; Mettouchi, A.; Bouvard, D. Escherichia coli Producing Cnf1 Toxin Hijacks Tollip to Trigger Rac1-Dependent Cell Invasion. *Traffic* **2011**, *12*, 579–590. [CrossRef] [PubMed]

13. Mills, M.; Meysick, K.C.; O'Brien, A.D. Cytotoxic Necrotizing Factor Type 1 of Uropathogenic *Escherichia coli* Kills Cultured Human Uroepithelial 5637 Cells by an Apoptotic Mechanism. *Infect. Immun.* **2000**, *68*, 5869–5880. [CrossRef] [PubMed]

14. Guyer, D.M.; Radulovic, S.; Jones, F.-E.; Mobley, H.L.T. Sat, the Secreted Autotransporter Toxin of Uropathogenic *Escherichia coli*, Is a Vacuolating Cytotoxin for Bladder and Kidney Epithelial Cells. *Infect. Immun.* **2002**, *70*, 4539–4546. [CrossRef]

15. Maroncle, N.M.; Sivick, K.E.; Brady, R.; Stokes, F.-E.; Mobley, H.L.T. Protease Activity, Secretion, Cell Entry, Cytotoxicity, and Cellular Targets of Secreted Autotransporter Toxin of Uropathogenic *Escherichia coli*. *Infect. Immun.* **2006**, *74*, 6124–6134. [CrossRef]

16. Heimer, S.R.; Rasko, D.A.; Lockatell, C.V.; Johnson, D.E.; Mobley, H.L.T. Autotransporter Genes Pic and Tsh Are Associated with *Escherichia coli* Strains That Cause Acute Pyelonephritis and Are Expressed During Urinary Tract Infection. *Infect. Immun.* **2004**, *72*, 593–597. [CrossRef]

17. Carter, P.; Wells, J.A. Dissecting the Catalytic Triad of a Serine Protease. *Nature* **1988**, *332*, 564. [CrossRef]

18. Gutierrez-Jimenez, J.; Arciniega, I.; Navarro-García, F. The Serine Protease Motif of Pic Mediates a Dose-Dependent Mucolytic Activity after Binding to Sugar Constituents of the Mucin Substrate. *Microb. Pathog.* **2008**, *45*, 115–123. [CrossRef]

19. Navarro-Garcia, F.; Gutierrez-Jimenez, J.; Garcia-Tovar, C.; Castro, L.A.; Salazar-Gonzalez, H.; Cordova, V. Pic, an Autotransporter Protein Secreted by Different Pathogens in the *Enterobacteriaceae* Family, Is a Potent Mucus Secretagogue. *Infect. Immun.* **2010**, *78*, 4101–4109. [CrossRef]

20. Habouria, H.; Pokharel, P.; Maris, S.; Garénaux, A.; Bessaiah, H.; Houle, S.; Veyrier, F.J.; Guyomard-Rabenirina, S.; Talarmin, A.; Dozois, C.M. Three New Serine-Protease Autotransporters of *Enterobacteriaceae* (SPATEs) from Extra-Intestinal Pathogenic *Escherichia coli* and Combined Role of Spates for Cytotoxicity and Colonization of the Mouse Kidney. *Virulence* **2019**, *10*, 568–587. [CrossRef]

21. Benz, I.; Alexander Schmidt, M. Aida-I, the Adhesin Involved in Diffuse Adherence of the Diarrhoeagenic *Escherichia coli* Strain 2787 (O126: H27), Is Synthesized Via a Precursor Molecule. *Mol. Microbiol.* **1992**, *6*, 1539–1546. [CrossRef] [PubMed]

22. Abràmoff, M.D.; Magalhães, P.J.; Ram, S.J. Image Processing with Imagej. *Biophotonics Int.* **2004**, *11*, 36–42.

23. Henderson, I.R.; Czeczulin, J.; Eslava, C.; Noriega, F.; Nataro, J.P. Characterization of Pic, a Secreted Protease Ofshigella Flexneri and Enteroaggregative *Escherichia coli*. *Infect. Immun.* **1999**, *67*, 5587–5596. [CrossRef] [PubMed]

24. Pickett, M.J.; Greenwood, J.R.; Harvey, S.M. Tests for Detecting Degradation of Gelatin: Comparison of Five Methods. *J. Clin. Microbiol.* **1991**, *29*, 2322–2325. [CrossRef] [PubMed]

25. Xicohtencatl-Cortes, J.; Saldaña, Z.; Deng, W.; Castañeda, E.; Freer, E.; Tarr, P.I.; Finlay, B.B.; Puente, J.L.; Girón, J.A. Bacterial Macroscopic Rope-Like Fibers with Cytopathic and Adhesive Properties. *J. Biol. Chem.* **2010**, *285*, 32336–32342. [CrossRef]

26. Vigil, P.D.; Wiles, T.J.; Engstrom, M.D.; Prasov, L.; Mulvey, M.A.; Mobley, H.L.T. The Repeat-in-Toxin Family Member Tosa Mediates Adherence of Uropathogenic *Escherichia coli* and Survival During Bacteremia. *Infect. Immun.* **2012**, *80*, 493–505. [CrossRef]

27. Rippere-Lampe, K.E.; O'Brien, A.D.; Conran, R.; Lockman, H.A. Mutation of the Gene Encoding Cytotoxic Necrotizing Factor Type 1 (Cnf 1) Attenuates the Virulence of Uropathogenic Escherichia coli. *Infect. Immun.* **2001**, *69*, 3954–3964. [CrossRef]

28. Davis, J.M.; Rasmussen, S.B.; O'Brien, A.D. Cytotoxic Necrotizing Factor Type 1 Production by Uropathogenic *Escherichia coli* Modulates Polymorphonuclear Leukocyte Function. *Infect. Immun.* **2005**, *73*, 5301–5310. [CrossRef]

29. Pokharel, P.; Habouria, H.; Bessaiah, H.; Dozois, C.M. Serine Protease Autotransporters of the Enterobacteriaceae (SPATEs): Out and About and Chopping It Up. *Microorganisms* **2019**, *7*, 594. [CrossRef]

30. Martinez, J.J.; Mulvey, M.A.; Schilling, J.D.; Pinkner, J.S.; Hultgren, S.J. Type 1 Pilus-Mediated Bacterial Invasion of Bladder Epithelial Cells. *EMBO J.* **2000**, *19*, 2803–2812. [CrossRef]

31. Mulvey, M.A.; Schilling, J.D.; Hultgren, S.J. Establishment of a Persistent *Escherichia coli* Reservoir During the Acute Phase of a Bladder Infection. *Infect. Immun.* **2001**, *69*, 4572–4579. [CrossRef] [PubMed]

32. Eslava, C.; Navarro-García, F.; Czeczulin, J.R.; Henderson, I.R.; Cravioto, A.; Nataro, J.P. Pet, an Autotransporter Enterotoxin from Enteroaggregative *Escherichia coli*. *Infect. Immune.* **1998**, *66*, 3155–3163. [CrossRef]

33. Brunder, W.; Schmidt, H.; Karch, H. Espp, a Novel Extracellular Serine Protease of Enterohaemorrhagic *Escherichia coli* O157: H7 Cleaves Human Coagulation Factor, V. *Mol. Microbiol.* **1997**, *24*, 767–778. [CrossRef]

34. Dozois, C.M.; Dho-Moulin, M.; Brée, A.; Fairbrother, J.M.; Desautels, C.; Curtiss, R. Relationship between the Tsh Autotransporter and Pathogenicity of Avian *Escherichia coli* and Localization and Analysis of the Tsh Genetic Region. *Infect. Immun.* **2000**, *68*, 4145–4154. [CrossRef] [PubMed]

35. Stein, M.; Kenny, B.; Stein, M.A.; Finlay, B.B. Characterization of Espc, a 110-Kilodalton Protein Secreted by Enteropathogenic *Escherichia coli* Which Is Homologous to Members of the Immunoglobulin a Protease-Like Family of Secreted Proteins. *J. Bacteriol.* **1996**, *178*, 6546–6554. [CrossRef]

36. Dutta, P.R.; Cappello, R.; Navarro-García, F.; Nataro, J.P. Functional Comparison of Serine Protease Autotransporters of Enterobacteriaceae. *Infect. Immun.* **2002**, *70*, 7105–7113. [CrossRef] [PubMed]

37. Turner, D.P.J.; Wooldridge, K.G.; Ala'Aldeen, D.A.A. Autotransported Serine Protease a of Neisseria Meningitidis: An Immunogenic, Surface-Exposed Outer Membrane, and Secreted Protein. *Infect. Immun.* **2002**, *70*, 4447–4461. [CrossRef]

38. Jonsson, P.; Wadström, T. Cell Surface Hydrophobicity Ofstaphylococcus Aureus Measured by the Salt Aggregation Test (Sat). *Curr. Microbiol.* **1984**, *10*, 203–209. [CrossRef]

39. Hendrixson, D.R.; De La Morena, M.L.; Stathopoulos, C.; St Geme, J.W., 3rd. Structural Determinants of Processing and Secretion of the Haemophilus Influenzae Hap Protein. *Mol. Microbiol.* **1997**, *26*, 505–518. [CrossRef]

40. Hendrixson, D.R.; St Geme, J.W., 3rd. The Haemophilus Influenzae Hap Serine Protease Promotes Adherence and Microcolony Formation, Potentiated by a Soluble Host Protein. *Mol. Cell* **1998**, *2*, 841–850. [CrossRef]

41. Kostakioti, M.; Stathopoulos, C. Functional Analysis of the Tsh Autotransporter from an Avian Pathogenic *Escherichia coli* Strain. *Infect. Immun.* **2004**, *72*, 5548–5554. [CrossRef]

42. Brockmeyer, J.; Spelten, S.; Kuczius, T.; Bielaszewska, M.; Karch, H. Structure and Function Relationship of the Autotransport and Proteolytic Activity of Espp from Shiga Toxin-Producing *Escherichia coli*. *PLoS ONE* **2009**, *4*, e6100. [CrossRef] [PubMed]

43. Navarro-Garcia, F.; Serapio-Palacios, A.; Vidal, J.E.; Salazar, M.I.; Tapia-Pastrana, G. Espc Promotes Epithelial Cell Detachment by Enteropathogenic *Escherichia coli* Via Sequential Cleavages of a Cytoskeletal Protein and Then Focal Adhesion Proteins. *Infect. Immun.* **2014**, *82*, 2255–2265. [CrossRef] [PubMed]

44. Gruenheid, S.; Finlay, B.B. Microbial Pathogenesis and Cytoskeletal Function. *Nature* **2003**, *422*, 775. [CrossRef]

45. Rottner, K.; Stradal, T.E.B.; Wehland, J. Bacteria-Host-Cell Interactions at the Plasma Membrane: Stories on Actin Cytoskeleton Subversion. *Dev. Cell* **2005**, *9*, 3–17. [CrossRef] [PubMed]

46. Martinez, J.J.; Hultgren, S.J. Requirement of Rho-Family Gtpases in the Invasion of Type 1-Piliated Uropathogenic *Escherichia coli*. *Cell. Microbiol.* **2002**, *4*, 19–28. [CrossRef] [PubMed]

47. Eto, D.S.; Sundsbak, J.L.; Mulvey, M.A. Actin-Gated Intracellular Growth and Resurgence of Uropathogenic *Escherichia coli*. *Cell. Microbiol.* **2006**, *8*, 704–717. [CrossRef]

48. Balish, M.J.; Jensen, J.; Uehling, D.T. Bladder Mucin: A Scanning Electron Microscopy Study in Experimental Cystitis. *J. Urol.* **1982**, *128*, 1060–1063. [CrossRef]

49. Parsons, C.L.; Greenspan, C.; Moore, S.W.; Grant, S. Mulholland. Role of Surface Mucin in Primary Antibacterial Defense of Bladder. *Urology* **1977**, *9*, 48–52. [CrossRef]

50. Lehker, M.W.; Sweeney, D. Trichomonad Invasion of the Mucous Layer Requires Adhesins, Mucinases, and Motility. *Sex. Transm. Infect.* **1999**, *75*, 231–238. [CrossRef]

51. Mantle, M.; Husar, S.D. Binding of Yersinia Enterocolitica to Purified, Native Small Intestinal Mucins from Rabbits and Humans Involves Interactions with the Mucin Carbohydrate Moiety. *Infect. Immun.* **1994**, *62*, 1219–1227. [CrossRef]

52. Finkelstein, R.A.; Boesman-Finkelstein, M.; Holt, P. Vibrio Cholerae Hemagglutinin/Lectin/Protease Hydrolyzes Fibronectin and Ovomucin: Fm Burnet Revisited. *Proc. Natl. Acad. Sci. USA* **1983**, *80*, 1092–1095. [CrossRef] [PubMed]

53. Bhullar, K.; Zarepour, M.; Yu, H.; Yang, H.; Croxen, M.; Stahl, M.; Finlay, B.B.; Turvey, S.E.; Vallance, B.A. The Serine Protease Autotransporter Pic Modulates Citrobacter Rodentium Pathogenesis and Its Innate Recognition by the Host. *Infect. Immun.* **2015**, *83*, 2636–2650. [CrossRef] [PubMed]

54. Kobayashi, R.K.T.; Gaziri, L.C.J.; Vidotto, M.C. Functional Activities of the Tsh Protein from Avian Pathogenic *Escherichia coli* (APEC) Strains. *J. Vet. Sci.* **2010**, *11*, 315–319. [CrossRef] [PubMed]

55. Ricard-Blum, S. The Collagen Family. *Cold Spring Harbor Perspect. Biol.* **2011**, *3*, a004978. [CrossRef]

56. Navarro-García, F.; Canizalez-Roman, A.; Burlingame, K.E.; Teter, K.; Vidal, J.E. Pet, a Non-Ab Toxin, Is Transported and Translocated into Epithelial Cells by a Retrograde Trafficking Pathway. *Infect. Immun.* **2007**, *75*, 2101–2109.

57. Nava-Acosta, R.; Navarro-Garcia, F. Cytokeratin 8 Is an Epithelial Cell Receptor for Pet, a Cytotoxic Serine Protease Autotransporter of Enterobacteriaceae. *MBio* **2013**, *4*, e00838-13. [CrossRef]

58. Navarro-García, F.; Sears, C.; Eslava, C.; Cravioto, A.; Nataro, J.P. Cytoskeletal Effects Induced by Pet, the Serine Protease Enterotoxin of Enteroaggregative *Escherichia coli*. *Infect. Immun.* **1999**, *67*, 2184–2192.

59. Dutta, P.R.; Sui, B.Q.; Nataro, J.P. Structure-Function Analysis of the Enteroaggregative *Escherichia coli* Plasmid-Encoded Toxin Autotransporter Using Scanning Linker Mutagenesis. *J. Biol. Chem.* **2003**, *278*, 39912–39920. [CrossRef]

60. Vidal, J.E.; Navarro-García, F. Espc Translocation into Epithelial Cells by Enteropathogenic *Escherichia coli* Requires a Concerted Participation of Type V and Iii Secretion Systems. *Cell. Microbiol.* **2008**, *10*, 1975–1986. [CrossRef]

61. Marc, D.; Dho-Moulin, M. Analysis of the Fim Cluster of an Avian O2 Strain of *Escherichia coli*: Serogroup-Specific Sites within Fima and Nucleotide Sequence of Fimi. *J. Med Microbiol.* **1996**, *44*, 444–452. [CrossRef]

62. Woodall, L.D.; Russell, P.W.; Harris, S.L.; Orndorff, P.E. Rapid, Synchronous, and Stable Induction of Type 1 Piliation in *Escherichia coli* by Using a Chromosomal Lacuv5 Promoter. *J. Bacteriol.* **1993**, *175*, 2770–2778. [CrossRef] [PubMed]

63. Charbonneau, M.-È.; Berthiaume, F.; Mourez, M. Proteolytic Processing Is Not Essential for Multiple Functions of the *Escherichia coli* Autotransporter Adhesin Involved in Diffuse Adherence (Aida-I). *J. Bacteriol.* **2006**, *188*, 8504–8512. [CrossRef] [PubMed]

64. Nakazawa, M.; Mukumoto, M.; Miyatake, K. Production and Purification of Polyclonal Antibodies. *Methods Mol. Biol. (Clifton, N.J.).* **2016**. [CrossRef]

65. Mellies, J.L.; Navarro-Garcia, F.; Okeke, I.; Frederickson, J.; Nataro, J.P.; Kaper, J.B. Espc Pathogenicity Island of Enteropathogenic *Escherichia coli* Encodes an Enterotoxin. *Infect. Immun.* **2001**, *69*, 315–324. [CrossRef] [PubMed]

66. Parham, N.J.; Srinivasan, U.; Desvaux, M.; Foxman, B.; Marrs, C.F.; Henderson, I.R. Picu, a Second Serine Protease Autotransporter of Uropathogenic Escherichia coli. *FEMS Microbiol. Lett.* **2004**, *230*, 73–83. [CrossRef]

67. Provence, D.L.; Curtiss, R., 3rd. Role of Crl in Avian Pathogenic *Escherichia coli*: A Knockout Mutation of Crl Does Not Affect Hemagglutination Activity, Fibronectin Binding, or Curli Production. *Infect. Immun.* **1992**, *60*, 4460–4467. [CrossRef]

© 2020 by the authors. Licensee MDPI, Basel, Switzerland. This article is an open access article distributed under the terms and conditions of the Creative Commons Attribution (CC BY) license (http://creativecommons.org/licenses/by/4.0/).

International Journal of
Molecular Sciences

Article

Antibiotic Resistance, Virulence Factors, Phenotyping, and Genotyping of Non-*Escherichia coli* Enterobacterales from the Gut Microbiota of Healthy Subjects

Alberto Amaretti [1,2,†], Lucia Righini [1,†], Francesco Candeliere [1], Eliana Musmeci [1], Francesca Bonvicini [3], Giovanna Angela Gentilomi [3,4], Maddalena Rossi [1,2] and Stefano Raimondi [1,*]

[1] Department of Life Sciences, University of Modena and Reggio Emilia, via Campi 103, 41125 Modena, Italy; alberto.amaretti@unimore.it (A.A.); lucia.righini@unimore.it (L.R.); francesco.candeliere@unimore.it (F.C.); eliana.musmeci@unimore.it (E.M.); maddalena.rossi@unimore.it (M.R.)

[2] Biogest-Siteia, University of Modena and Reggio Emilia, Modena, Viale Amendola 2, 42122 Reggio Emilia, Italy

[3] Department of Pharmacy and Biotechnology, Alma Mater Studiorum-University of Bologna, Via Massarenti 9, 40138 Bologna, Italy; francesca.bonvicini4@unibo.it (F.B.); giovanna.gentilomi@unibo.it (G.A.G.)

[4] Unit of Microbiology, Alma Mater Studiorum-University of Bologna, S. Orsola-Malpighi Hospital, Via Massarenti 9, 40138 Bologna, Italy

* Correspondence: stefano.raimondi@unimore.it; Tel.: +39-059-205-8595

† These authors contributed equally to the work.

Received: 2 March 2020; Accepted: 5 March 2020; Published: 7 March 2020

Abstract: Non-*Escherichia coli* Enterobacterales (NECE) can colonize the human gut and may present virulence determinants and phenotypes that represent severe heath concerns. Most information is available for virulent NECE strains, isolated from patients with an ongoing infection, while the commensal NECE population of healthy subjects is understudied. In this study, 32 NECE strains were isolated from the feces of 20 healthy adults. 16S rRNA gene sequencing and mass spectrometry attributed the isolates to *Klebsiella pneumoniae*, *Klebsiella oxytoca*, *Enterobacter cloacae*, *Enterobacter aerogenes*, *Enterobacter kobei*, *Citrobacter freundii*, *Citrobacter amalonaticus*, *Cronobacter* sp., and *Hafnia alvei*, *Morganella morganii*, and *Serratia liquefaciens*. Multiplex PCR revealed that *K. pneumoniae* harbored virulence genes for adhesins (*mrkD*, *ycfM*, and *kpn*) and enterobactin (*entB*) and, in one case, also for yersiniabactin (*ybtS*, *irp1*, *irp2*, and *fyuA*). Virulence genes were less numerous in the other NECE species. Biofilm formation was spread across all the species, while curli and cellulose were mainly produced by *Citrobacter* and *Enterobacter*. Among the most common antibiotics, amoxicillin-clavulanic acid was the sole against which resistance was observed, only *Klebsiella* strains being susceptible. The NECE inhabiting the intestine of healthy subjects have traits that may pose a health threat, taking into account the possibility of horizontal gene transfer.

Keywords: Enterobacterales; *Klebsiella*; *Enterobacter*; *Citrobacter*; virulence; antibiotic resistance; biofilm

---

1. Introduction

Important enteric pathogens belong to Enterobacterales, a bacterial order within the phylum Proteobacteria. This order encompasses permanent colonizers of the human gut that, in healthy conditions, constitute minor bacterial components of the microbiota. Opportunistic Enterobacterales can persist as gut commensals without inducing any infections, as long as the microbiota is balanced and the complex and dense bacterial community prevents their overgrowth. A bloom of Enterobacterales may occur as a result of disturbance of the microbiota, yielding pathogen-mediated infections and triggering inflammatory host responses.

Escherichia coli is the most studied among the Enterobacterales with regards to the traits that differentiate commensalism and pathogenicity. It normally colonizes the intestine but comprises both harmless commensals and different pathogenic variants that may instigate infections in the gut or in other tissues [1,2]. Virulent strains of *E. coli* isolated from infected patients attracted most research interest [3,4], but also fecal isolates from healthy subjects and environmental strains are the target of increasing attention, aiming to determine the pathogenic potential of a wider biodiversity reservoir [5,6].

Many non-*E. coli* Enterobacterales (hereinafter referred to as NECE) that can colonize the gut (e.g., *Klebsiella, Enterobacter, Citrobacter,* and *Serratia*) also present traits that can confer them virulence and pathogenicity or phenotypes that may result in severe heath concern, such as multidrug resistance [7–10]. The greatest efforts have been carried out to describe virulent strains, generally isolated from patients with an ongoing infection, while the pathogenic potential of NECE inhabiting the gut of healthy subjects has not been thoroughly investigated with genetic and phenotypic analysis, except for some genera [9]. The research herein presented aims to fill this gap, providing a genotypic and phenotypic description of the NECE population isolated from the feces of 20 healthy adults, and to complement a previous study that described the *E. coli* population of the same cohort of subjects [5]. A set of 32 NECE strains was isolated, taxonomically classified, subjected to PFGE genotyping, and described in terms of the genotypic determinants and phenotypic traits that may confer on them potential pathogenicity or invasiveness.

Information on the genes associated to virulence is detailed especially for *Klebsiella* spp., where a number of genes associated to harmful traits were identified, such as those encoding adhesins, siderophores (e.g., enterobactin, aerobactin, yersiniabactin), protectines, or invasins (responsible for mucoid phenotype and invasiveness), and involved in allantoin metabolism [7,11]. For other NECE genera, such as *Enterobacter, Cronobacter,* and *Citrobacter,* the knowledge of the genetic determinants associated to virulence and invasiveness is less comprehensive and, with few exceptions (e.g., *Citrobacter koseri*), mainly acquired from better characterized pathogens [12–15].

In addition to fimbrial and afimbrial adhesins, the production of surface cellulose structures and curli favors the adhesion of Enterobacterales and can exert a significant role in enteric biofilm-related infections [16,17]. Although not directly involved in pathogenic mechanisms, the acquisition of multiple antibiotic resistances favors the success of opportunistic Enterobacterales pathogens in invasion, survival, and spread, severely complicating the containment and treatment of infections [9,18]. Therefore, the occurrence of drug resistant bacteria within a commensal population and the possibility to exchange genetic material by horizontal gene transfer may represent a major health concern.

In the present study, multiplex PCR assays were utilized to screen the NECE, isolated from the feces of healthy subjects, for the presence of 17 main virulence genes associated to those of *Klebsiella* and *E. coli* [19–21]. From the point of view of the phenotype, the isolates were characterized for the ability to form biofilm and to yield curli and surface cellulose, were screened for the susceptibility to the most common antibiotics and for the ability to act as recipients in conjugation experiments, and biochemical tests were performed to compare the metabolic profile.

2. Results

2.1. Counting and Isolation of NECE

The selective differential medium HiCrome Coliform Agar (HCCA) was utilized to count and isolate *E. coli* [5] and NECE from fecal samples of 20 healthy adults. Total counts in HCCA ranged from 4.6×10^5 to 2.2×10^8 cfu/g (Figure 1). Blue colonies attributed to *E. coli* overcounted the pink ones attributed to NECE in all the samples except V11 (Figure 1; Supplementary Figure S1). NECE ranged between $< 10^4$ and 1.3×10^8 cfu/g and, except in V11, accounted for a minority of total Enterobacterales

(NECE + *E. coli*), the 75th percentile being the 5.7% (Figure 1). In some cases, NECE were not recovered, being outnumbered by *E. coli*. Spearman's rank correlation analysis excluded any significant correlation between NECE and *E. coli* counts.

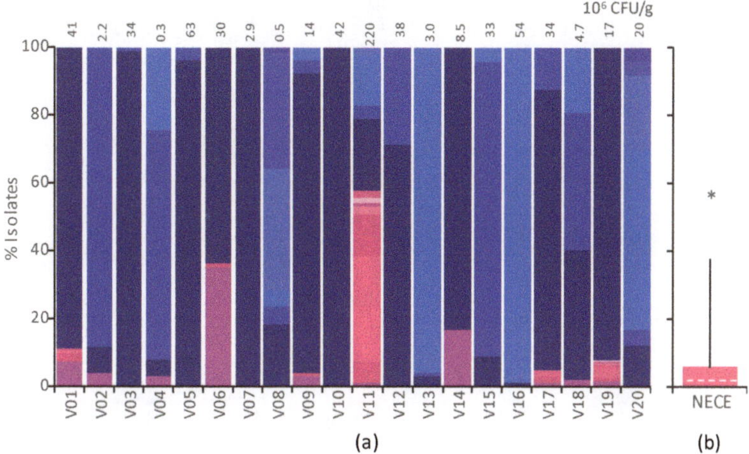

Figure 1. Counts of *Escherichia coli* and non-*Escherichia coli* Enterobacterales (NECE), enumerated onto HiCrome Coliform Agar (HCCA) plates. (**a**) Percentage of colonies attributed to *E. coli* (blue shades) and NECE (pink shades) in the feces of 20 subjects. For each subject, different shades indicate different biotypes according to enterobacterial repetitive intergenic consensus-PCR (ERIC-PCR) and random amplification of polymorphic DNA-PCR (RAPD-PCR) fingerprinting. The total count of Enterobacterales (*E. coli* + NECE) is reported in the top margin. (**b**) Distribution of the percentage of NECE colonies. The median (dashed line), the 25th and 75th percentiles (colored box), the 10th and 90th percentiles (whiskers), and outliers (*) are indicated.

2.2. Taxonomic Attribution and PFGE Genotyping

The isolates putatively attributed to NECE were clustered in 32 different biotypes utilizing ERIC-PCR (enterobacterial repetitive intergenic consensus-PCR) and RAPD-PCR (random amplification of polymorphic DNA-PCR) fingerprinting. A representative isolate of each biotype was assigned a taxonomic designation utilizing 16S rRNA gene sequencing and MALDI-TOF MS (Supplementary Table S1). The genus *Klebsiella* was the most represented (14/32), with the species *K. pneumoniae* (10 strains) and *Klebsiella oxytoca* (4) found in seven and three fecal samples, respectively. The genus *Enterobacter* was represented by eight strains belonging to *Enterobacter cloacae* (6), *Enterobacter aerogenes*, and *Enterobacter kobei* (1 strain each). Other strains belonged to *Citrobacter* (4 to *Citrobacter freundii* and 1 to *Citrobacter amalonaticus*), *Cronobacter* sp. (2), and to *Hafnia alvei*, *Morganella morganii*, and *Serratia liquefaciens* (1 strain each).

PFGE highlighted a wide diversity of the NECE isolates, which did not cluster according to the taxonomic attribution (Figure 2).

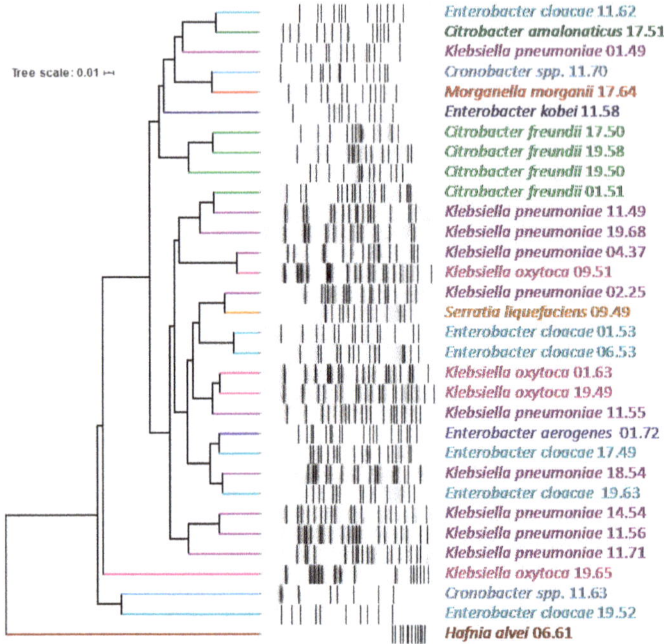

Figure 2. *Xba*I-PFGE pattern of NECE strains: unweighted pair group method with arithmetic means (UPGMA) dendrogram derived from Dice's coefficients, calculated based on the band profile. Strains are colored based on their MALDI-TOF MS taxonomic attribution.

2.3. Virulence Genotyping

PCR was used to investigate 17 virulence genes encoding adhesins (*fimH1*, *mrkD*, *kpn*, and *ycfM*), siderophores (enterobactin, *entB*; aerobactin, *iutA*; yersiniabactin, *irp-1*, *irp2*, *ybtS*, *fyuA*; catechols receptor, *iroN*; and other, *kfu*), protectines or invasins (*K2*, *magA*, *rmpA*, and *traT*), and involved in allantoin metabolism (*allS*).

Most strains of *K. pneumoniae* harbored the *mrkD*, *ycfM*, and *kpn* encoding adhesins and *entB* encoding enterobactin (Figure 3). Only *K. pneumoniae* 11.55 was positive to the main virulence genes involved in the synthesis of *Yersinia* siderophore, including *ybtS* (encoding for the synthase), *irp1* and *irp2* (for regulatory proteins), and *fyuA* (for the siderophore receptor). *irp2* was also detected in most of the other strains of *K. pneumoniae* although they lacked the counterpart *ybtS*. All the *K. pneumoniae* strains were negative to the genes *K2*, *magA*, and *rmpA* associated with hypermucoid phenotype and invasivity, except for *K. pneumoniae* 01.49 that was positive to *K2*. Similarly, other virulence genes, such as *allS*, *kfu*, and *iutA* occurred only once among the tested strains.

The strains of *K. oxytoca* harbored *entB* (three out of four isolates) but were negative to most of the other virulence genes. A sole strain harbored *ytbS*. Most of *Cronobacter* and *Enterobacter* isolates were characterized by the presence of the gene *irp2* but never harbored *ybtS* or other *Yersinia* siderophore genes. A few strains were positive to *mrkD* or *entB*.

The strains ascribed to *Citrobacter*, *H. alvei*, and *M. morganii* were negative to those virulence genes whose presence could not be excluded by primer-blast search. The strain of *S. liquefaciens* was positive to *ytbS*.

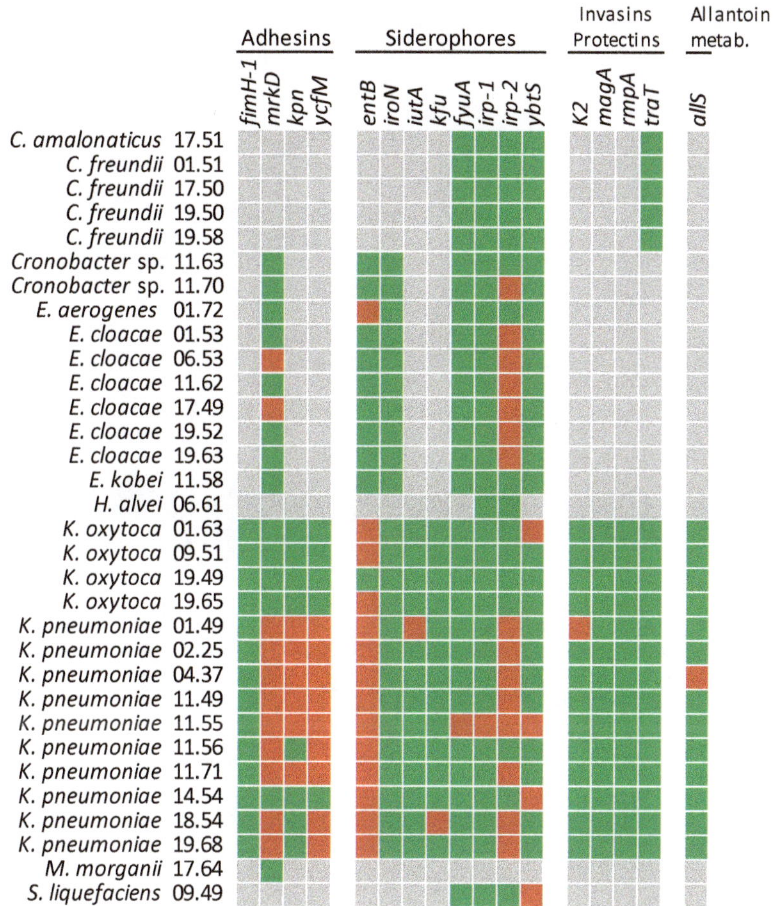

Figure 3. PCR assay of the NECE isolates for the presence of virulence genes. Colors: red, positive amplification; green, negative amplification; grey, PCR analysis not performed since the gene was putatively absent based on a primer-blast search.

2.4. Biofilm Formation and Production of Curli and Cellulose

NECE strains were tested for biofilm formation in minimal and rich media (M9 and LBWS, respectively; Supplementary Figure S2). The vast majority of the strains (26 out of 32), belonging to all the species except *E. aerogenes*, *H. alvei*, and *M. morganii*, formed biofilm in M9 (Figure 4; Supplementary Figure S2). Biofilm formation was less frequent in LBWS, being observed only in 10 strains of *K. oxytoca*, *K. pneumoniae*, and *S. liquefaciens*. The extent of biofilm production was always less abundant in the rich medium compared to M9 ($p < 0.05$).

Extracellular cellulose was detected in most of *Citrobacter* and *Enterobacter* strains, in five strains of *K. pneumoniae* and two out of four of *K. oxytoca*, in *H. alvei* and in *S. liquefaciens*. Curli were produced by nearly all *Citrobacter*, *Cronobacter*, and *Enterobacter* strains and by one isolate belonging to *K. oxytoca*. The isolates of *H. alvei* and *M. morganii* were also positive to curli. The strains belonging to *Citrobacter*, *E. cloacae*, *H. alvei*, and *K. oxytoca* 19.49 produced both cellulose and curli.

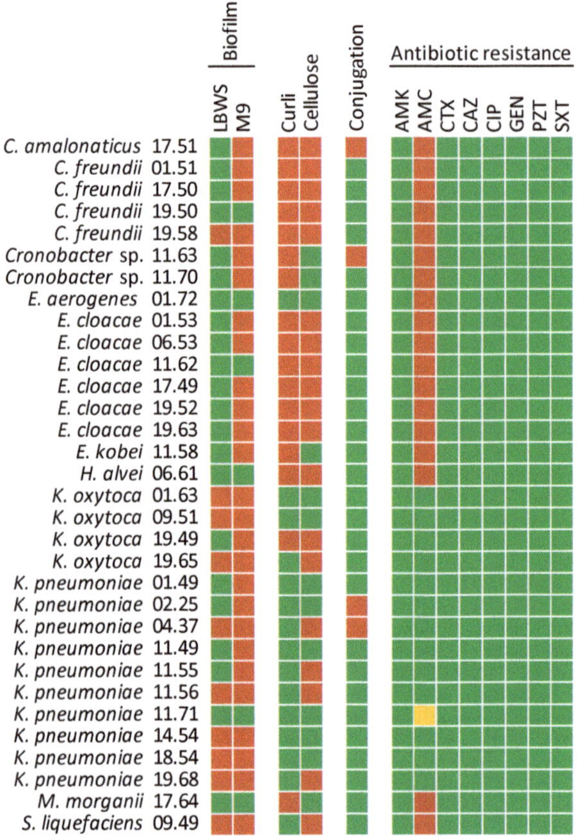

Figure 4. Phenotypic characterization of NECE isolates: biofilm formation in LBWS and M9 media, curli and cellulose production, conjugation, and antibiotic resistance. Colors: red, positive; green, negative. For antibiotics: red, resistant; green, susceptible, yellow, intermediate. Antibiotics: amikacin (AMK), amoxicillin–clavulanic acid (AMC), cefotaxime (CTX), ceftazidime (CAZ), ciprofloxacin (CIP), gentamicin (GEN), piperacillin-tazobactam (PZT), and trimethoprim-sulfamethoxazole (SXT).

2.5. Conjugation

The strains were challenged as conjugation recipients for receiving pOX38: Cm plasmid from *E. coli* N4i. Only two strains of *K. pneumoniae*, and single strains of *Cronobacter* and *Citrobacter amalonaticus* succeeded in plasmid acquisition (Figure 4).

2.6. Antibiotic Resistance

Phenotypic susceptibility to amikacin, amoxicillin–clavulanic acid, cefotaxime, ceftazidime, ciprofloxacin, gentamicin, piperacillin-tazobactam, and trimethoprim-sulfamethoxazole was assayed. Amoxicillin-clavulanic acid was the sole antibiotic against which few isolates presented some resistance, with all the strains of *Enterobacter*, *Citrobacter*, *Cronobacter*, *H. alvei*, *M. morganii*, and *S. liquefaciens* being resistant. All the 14 biotypes of *Klebsiella* spp. were sensitive to the whole set of tested antibiotics, with the exception of *K. pneumoniae* 11.71 that was partially resistant to amoxicillin–clavulanic acid, presenting a minimum inhibitory concentration (MIC) intermediate between resistance and susceptibility thresholds.

2.7. Biochemical Characterization

The fermentation of substrates and some distinctive enzymatic reactions and metabolic routes were assayed utilizing the API 20 E system (Figure 5). Generally, the NECE strains were positive to β-galactosidase. Most strains were capable of utilizing citrate, glucose, mannose, inositol, sorbitol, rhamnose, sucrose, melibiose, amygdalin, and arabinose. The main exceptions were *M. morganii* that could utilize only glucose, *H. alvei* that fermented a restricted number of sugars, and some *Citrobacter*, *Cronobacter*, and *Enterobacter* strains that exhibited specific substrate preferences. The majority of the isolates produced either lysine decarboxylase (*Klebsiella*) or ornithine decarboxylase (*Cronobacter* and *Enterobacter*). Urease was characteristic of *Klebsiella*, while arginine dihydrolase was found in most *Enterobacter*, *Citrobacter*, and *Cronobacter*. Acetoin was produced by *Cronobacter*, *Enterobacter*, *Hafnia*, and *Klebsiella*. Indole was produced by *K. oxytoca* and by few other strains, H_2S by two strains of *Citrobacter freundii*. Only *S. liquefaciens* was positive to gelatinase. All the strains except *M. morganii* and *S. liquefaciens* exhibited denitrifying activity, in most cases yielding nitrite. Nitrate reduction to N_2 was observed in *K. pneumoniae* and few other strains.

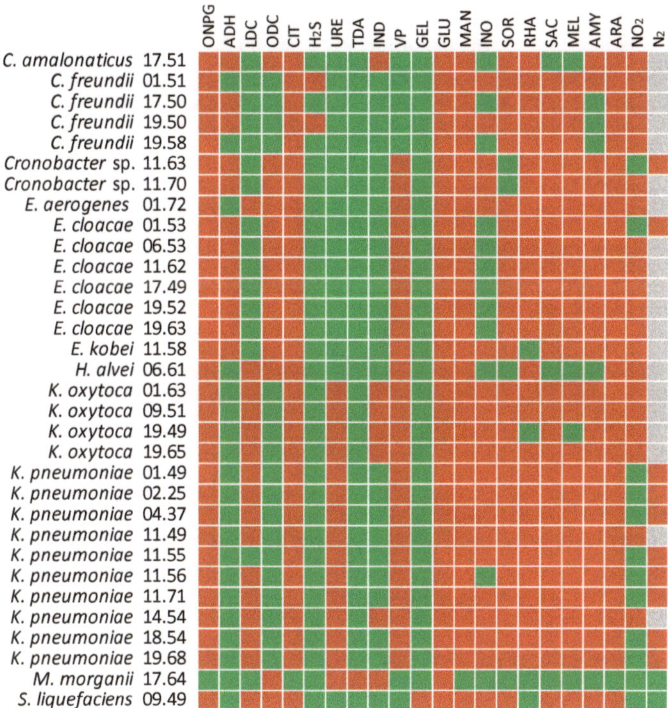

Figure 5. Biochemical reaction profiles of NECE isolates in the API 20 E assay: β-galactosidase (ONPG), arginine dihydrolase (ADH), lysine decarboxylase (LDC), ornithine decarboxylase (ODC), citrate utilization (CIT), production of hydrogen sulfide (H_2S), urease (URE), tryptophan deaminase (TDA), indole (Kovac's test, IND), acetoin (Voges-Proskauer test, VP), gelatinase (GEL), fermentation of glucose (GLU), mannitol (MAN), inositol (INO), sorbitol (SOR), rhamnose (RHA), sucrose (SAC), melibiose (MEL), amygdalin (AMY), arabinose (ARA), and reduction of nitrates to nitrites (N_2O) or nitrogen (N_2, tested only in case of negative N_2O). Colors: red, positive; green, negative.

3. Discussion

Thirty-two NECE strains were isolated from the feces of 20 healthy adults that did not present any dysbiosis, and thus as members of a relatively balanced gut microbiota. The load of Enterobacterales was in the order of millions or tens of millions per gram of feces, with a sole exception where they reached the magnitude of 10^8. NECE represented a small population of Enterobacterales, with *E. coli* being on average 20 times more abundant, and a minor component of the whole microbiota, being less than 0.1%. The genus *Klebsiella*, which is ubiquitous in nature, colonizing humans, animals, and plants, and frequently detected in waters, sewages, and soils, was the most represented, encompassing 14 of the 32 strains. *K. pneumoniae*, the most frequently isolated species (10 strains), was present in 7 out of 20 fecal samples, in accordance to the 35% of healthy adults from which it was isolated in a pioneering study [22].

K. pneumoniae encompasses opportunistic pathogens that can cause human infections in lungs, urinary tract, and bloodstreams, mostly to hospitalized and/or immunocompromised patients [23,24]. Virulence of *K. pneumoniae* is associated to the presence of capsule and pili, to the production of lipopolysaccharides and siderophores, to allantoin utilization, and to iron uptake systems, efflux pumps, and type VI secretion systems [7]. Surface molecules, such as capsular polysaccharides and lipopolysaccharides, are some of the major virulence factors that *Klebsiella* use to protect itself from the host innate immune apparatus. Furthermore, the capability to compete for iron has a pivotal role to the establishment of the infection. The genes involved in iron assimilation are generally clustered in pathogenicity islands, large chromosomal regions that were likely acquired by horizontal transfer. Virulence genetic determinants can be located both in the core or in the accessory genome [7], the former also including metabolic genes required for some species to cause disease.

Among the 10 *K. pneumoniae* isolates, all but one strain harbored *entB*, encoding a common catecholate siderophore located in the core genome, and *mrkD*, involved in the synthesis of type 3 pili that promote adherence to the surfaces [19]. The genes encoding other adhesins, such as *ycfM* and *kpn* were detected in nine and five strains, respectively. Only two *K. pneumoniae* strains were positive to the yersiniabactin gene *ybtS*, a common virulence factor associated with human infections [25,26], and one of the two also harbored *irp1*. These genes are involved in the synthesis of the siderophore yersiniabactin by virulent *Yersinia* strains, which harbor them within the high-pathogenicity island (HPI). HPI is widely distributed among members of the order Enterobacterales, including *E. coli*, *K. pneumoniae*, *Citrobacter* spp., *Salmonella enterica*, *Serratia liquefaciens*, and *Enterobacter* spp. [27,28]. In addition, *irp2*, another marker of HPI, was detected in the majority of the strains of *K. pneumonia* and *E. cloacae*, albeit they lacked the counterpart *ybtS*.

Although Enterobacterales are normally present as a low fraction of commensal bacteria in the healthy gut, their numbers can increase in the inflamed gut, and take advantage over other commensals [29]. Siderophores are major contributors of exploitative competition, since iron is an essential nutrient present in very low amounts in the gut and may play a role in virulence. Overall, *K. pneumoniae* 11.55 was the strain equipped with the broadest set of virulence genes involved in iron metabolism, including those encoding enterobactin (*entB*), yersiniabactin (*irp1*, *irp2*, and *ybtS*), and, the sole strain among the 32 isolates, also the *Yersinia* siderophore receptor (*fyuA*).

In agreement with their commensal behavior, none of the *K. pneumoniae* strains were positive for the two genes associated with invasive infections, i.e., the mucoviscosity-associated gene *magA* and the regulator of mucoid phenotype *rmpA* [30,31], that are associated with a hypervirulent and hypermucoid phenotype. In our isolates, the presence of other genetic determinants of virulence was sporadic or quite rare. In the gut of the healthy host, the Enterobacterales resides in the outer loose mucus layer, separated from the epithelial mono-cell barrier by an inner dense mucin coat [32]. During the infection, they penetrate the mucus layer, interact with the epithelial cells, and may breach the mucosal barrier. In case these enterobacteria strains reach other tissues, biofilm formation may act as a fitness factor concurring to pathogenesis [33]. Adhesion factors and extracellular matrix components are involved in formation of biofilms [34]. All but one strain produced biofilm in minimal medium M9, whereas this phenotype was less frequent when growth occurred in the rich medium LBWS. The higher extent of biofilm formation in M9 is consistent with the fact that this mineral medium is more challenging for these bacteria. Extracellular cellulose structures, determined in five strains, were consistent with the capability to form biofilm on both media. Curli were found in a sole strain that presented also cellulose structures (19.58 CA) and produced biofilm on both the substrates.

Based on genetic and functional features, some *K. pneumoniae* isolates present a higher potential to cause infections, albeit they are present at low charge in the microbiota of healthy hosts. The link between colonization and infection by *K. pneumoniae* in hospitalized patients has been demonstrated, with robust evidence that their own microbiota is the main source of the infective strains [35,36]. Thus, potentially more virulent *K. pneumoniae* strains may take advantage of critical conditions, becoming responsible for nosocomial infections [37].

The isolates of *K. oxytoca*, another prominent pathogen that may be involved in diseases and life-threatening infections [38,39], generally encoded the siderophore gene *entB*, but were negative to most of the other virulence genes. A sole strain harbored the gene encoding the yersiniabactin siderophore. Biofilm production was a general feature of the *K. oxytoca* strains, regardless of the presence of curli or cellulose structures.

Similar considerations are valid for the other NECE strains herein described, belonging to genera that may have clinical relevance, such as *Enterobacter*, *Citrobacter*, and *Cronobacter*. The isolates were generally capable of forming biofilm and producing curli and cellulose and were negative for most virulence genes. However, unlike *Klebsiella* that shares many virulence genes with *E. coli*, the genetic determinants of virulence of these genera have not been fully disclosed.

In this study emerged that NECE isolates from feces of healthy subjects are still quite susceptible to most of the antibiotics. This is important, since any treatment of opportunistic outbreaks of NECE (e.g., in case nosocomial infections) requires antibiotics, but resistance developments would seriously curb the therapeutic options [40]. All the isolates of *Klebsiella* were sensitive to the whole set of tested antibiotics. Amoxicillin-clavulanic acid was the sole antibiotic against which was detected some resistance: the strains ascribed to the genus *Enterobacter* and the isolates belonging to *Citrobacter*, *Cronobacter*, *Hafnia alvei*, *Morganella morganii*, and *Serratia liquefaciens* were all resistant to this combination of antibiotics. Interestingly, Enterobacterales presented the highest increase in terms of relative abundance in a short-term amoxicillin-clavulanic acid treatment in healthy adults [41]. Some genera belonging to this family, such as *Enterobacter* and *Citrobacter*, are recognized as intrinsically resistant [42], and may take advantage to this antibiotic treatment. In general, the profile of resistance was independent by the subject of the fecal sample, but two clusters encompassing sensitive or resistant strains were sharply differentiated by taxonomy.

PFGE genotyping was carried out to evaluate the genetic similarity among the bacterial isolates of this study and highlighted a wide dispersion of the strains, regardless their taxonomic attribution and phylogenetic relationships. The spreading of the strains regardless of species or genera may be attributed to the presence of plasmids, the horizontal acquisition of additional genes from diverse species of Enterobacterales, and to the exchange of mobile elements that rapidly integrate and promote DNA shuffling, in agreement with the capability of some strains to accept DNA by conjugation from *E. coli* as a donor. However, the biochemical profiling mostly clustered the strains in species (data not shown), confirming that energy production and conservation, and lipid, amino acid, and nucleotide metabolism are part of the conserved reactome, despite the genome plasticity.

The Enterobacterales encountered in this study are generally recognized as opportunistic pathogens, with some potential capability to cause disease, on the basis of predicted virulence factors. Except for *K. pneumoniae* hypervirulent strains, NECE virulence seems more associated to the host features than to the strain traits. According to the similar results obtained for the *E. coli* isolated from the same cohort of healthy subjects, the absence of antibiotic resistance for most of the tested antibiotics does not pose a serious challenge for infection control. This highlights the stratification of antibiotic resistance distribution among healthy and hospitalized/diseased subjects, with NECE associated risk increasing with both illness and antibiotic therapy.

4. Materials and Methods

4.1. Isolation and Enumeration of Enterobacterales

Fresh fecal samples were collected from 20 healthy adult subjects who gave written informed consent regarding their participation in the study in accordance with the protocol approved by the local research ethics committee (reference number: 974/2019/SPER-UNIMO-ENTEROPOP; Comitato etico dell'Area Vasta Emilia Nord, Italy). The subjects—10 males and 10 females aged 35 to 45, following a western omnivore diet, and who had not been treated with prebiotics and/or probiotics for 1 month and antibiotics for 3 months—were enrolled among the employees of the University of Modena and Reggio Emilia and their relatives and were not in relationship with the researchers.

Feces were homogenized (10% *w/v*) in isotonic Buffered Peptone Water (Sigma, Steinheim, Germany), then serial dilutions were spread onto plates of HiCrome Coliform Agar (HCCA, Sigma) and incubated at 37 °C. The medium differentiates *E. coli* (blue colonies) from NECE (salmon to red). For each subject, up to 48 colonies of putative NECE were picked and clustered into biotypes with ERIC-PCR [43] and RAPD-PCR [44] fingerprint presenting Pearson's similarity > 75%.

4.2. PFGE Genotyping

PFGE was performed according to PulseNet protocol (http://www.cdc.gov/pulsenet/PDF/ecoli-shigella-salmonella-pfge-protocol-508c.pdf). The genomic DNA was digested with 50 U of *Xba*I at 37 °C for 3 h. Fragments were resolved in a CHEF-DRIII apparatus (Bio-Rad, Hercules, CA, USA) using counter-clamped homogeneous electric field electrophoresis (24 h at 6.0 V/cm; initial switch time, 2.2 s; final switch time, 54.2 s). The run was digitally captured and analyzed with GelCompare II 6.0 software (Applied Maths NV, Sint-Martens-Latem, Belgium). Dice coefficient was computed to evaluate similarity between band profiles (position tolerance, 1%; optimization, 1%) and to derive an UPGMA dendrogram (unweighted pair group method with arithmetic means). Strains were ascribed to the same pulsotype if PFGE profile possessed >85% similarity.

4.3. Taxonomic Attribution

A strain for each biotype was taxonomically characterized by partial sequencing of the 16S rRNA gene sequencing, utilizing primers targeting the V1-V3 portion. Primer sequences and PCR conditions were set up according to Raimondi et al. [45]. The sequences, obtained from a service provider (Eurofins Genomics, Ebersberg, Germany), were compared to those in SILVA SSU database utilizing SINA Aligner v1.2.11 (https://www.arb-silva.de/aligner/).

In addition, the MALDI-TOF MS-based biotyping was carried using the MALDI Biotyper 3.1 system (Bruker Daltonics, Bremen, Germany). Sample preparation for MALDI-TOF MS was performed as previously described with minor modifications [46]. Briefly, colonies of fresh overnight culture were placed on a MALDI sample slide (Bruker Daltonics) and dried at room temperature. The sample was then overlaid with 1 µL of matrix solution (α-cyano-4-hydroxycinnamic acid in 50% acetonitrile and 2.5% trifluoroacetic acid) and dried at room temperature. A MALDI-TOF MS measurement was performed with a Bruker MicroFlex MALDI-TOF MS (Bruker Daltonics) using FlexControl software and a *Escherichia coli* DH5α protein extract (Bruker Daltonics) was placed on the calibration spot of the sample slide for external calibration. Spectra collected in the positive-ion mode within a mass range of 2000–20,000 Da were analyzed using a Bruker Biotyper (Bruker Daltonics) automation control and the Bruker Biotyper 3.1 software and library (a database with 5627 entries). High confidence species identification was accepted, if the log(score) was ≥2.00, low confidence species identification log(score) values (≥1.70 and <2.00) were accepted if the three best matches showed the same species name. Any results with log(score)<1.70 were considered as an unacceptable identification.

4.4. Profiling of Virulence Genes

All the isolates were screened by multiplex-PCR for the genes associated to 17 virulence factors: *allS, entB, fimH, fyuA, iroN, irp1, irp2, iutA, K2, kfu, kpn, magA, mrkD, rmpA, traT, ybtS*, and *ycfM*. Primer sequences and amplification conditions were set up according to El Fertas-Aissani et al. [19], Compain et al. [20], and Johnson et al. [21]. In order to assess the possibility to obtain the amplicon in the different NECEs species, a search with the primer-blast tool (https://www.ncbi.nlm.nih.gov/tools/primer-blast/) was performed for all the set of primers developed for *K. pneumoniae* and *E. coli*. The result of PCR amplification was reported only for the species in which the annealing and the possibility to yield an amplicon were predicted.

4.5. Biofilm and Phenotype Assays

Biofilm formation was quantified with crystal violet in the microtiter assay described in [47]. Two growth media were compared: Luria Bertani without salt (LBWS) and M9 (BD Difco, Sparks, MD, USA) containing 4 g/L glucose and 0.25 g/L yeast extract. Strains exhibiting a specific biofilm formation (i.e., the ratio between crystal violet absorbance at 570 nm and culture turbidity at 600 nm) > 1. The data herein reported are the means of three independent experiments, each carried out in triplicate.

The strains were screened for curli and cellulose production utilizing LBWS agar plates supplemented with the appropriate stain [48]. Red colonies in Congo red-supplemented plates were considered positive to curli. Colonies in calcofluor white-supplemented plates that emitted fluorescence due to UV exposure (315–400 nm) were considered positive to cellulose.

4.6. Solid Mating Conjugation Experiments

The strains were screened as recipients in conjugation experiments with the donor *E. coli* N4i pOX38:Cm (N4i: EcN immE7 Gmr; pOX38:Cm: Tra+ RepFIA+ Cmr) [5]. The donor and the recipient strains were cultured and put in contact onto LB plates under the conditions described in [5]. HCCA and HCCA with 20 µg/mL chloramphenicol were utilized to differentiate recipient, transconjugant, and donor colonies [5].

4.7. Antibiotic Susceptibility

The strains were tested for antimicrobial susceptibility with a Vitek2 semi-automated system (bioMerieux, Marcy-l'Étoile, France). Minimum inhibitory concentrations (MICs) were interpreted according to EUCAST (European Committee on Antimicrobial Susceptibility Testing—www.eucast.org) and susceptibility (S) or resistance (R) were defined based on the following thresholds (mg/L): amikacin, S < 8 and R > 16; amoxicillin/clavulanic acid, S < 8 and R > 8; cefotaxime, S < 8 and R > 2; ceftazidime, S < 8 and R > 8; ciprofloxacin, S < 0.5 and R > 1; gentamicin, S < 2 and R > 4; piperacillin + tazobactam, S < 8 and R > 16; trimethoprim/sulfamethoxazole, S < 40 and R > 80.

4.8. Biochemical Characterization

The strains were tested for distinctive enzymatic reactions and metabolic routes utilizing API 20 E test system (bioMerieux, France), according to the manufacturer's instructions.

Supplementary Materials: Supplementary materials can be found at http://www.mdpi.com/1422-0067/21/5/1847/s1.

Author Contributions: Conceptualization, G.A.G., F.B., M.R, and S.R.; data curation, S.R., F.C., and A.A.; investigation, A.A., F.C., L.R., E.M., S.R., and F.B.; methodology, S.R., L.R., F.B., M.R., and G.A.G.; project administration, M.R.; resources, M.R. and G.A.G.; supervision, M.R.; visualization, A.A.; writing—original draft, F.B., S.R., A.A. and M.R.; writing—review & editing, G.A.G. All authors have read and agreed to the published version of the manuscript.

Funding: Please add: This research received no external funding.

Conflicts of Interest: The authors declare no conflict of interest.

Abbreviations

NECE	non-*E. coli* Enterobacterales
HCCA	HiCrome Coliform Agar
MALDI-TOF MS	Matrix-assisted laser desorption ionization–time of flight mass spectrometry
ERIC	Enterobacterial repetitive intergenic consensus
RAPD	Random Amplification of Polymorphic DNA
UPGMA	Unweighted pair group method with arithmetic means

References

1. Leimbach, A.; Hacker, J.; Dobrindt, U. *E. coli* as an all-rounder: The thin line between commensalism and pathogenicity. *Curr. Top. Microbiol. Immunol.* **2013**, *358*, 3–32. [CrossRef] [PubMed]
2. Manges, A.R.; Geum, H.M.; Guo, A.; Edens, T.J.; Fibke, C.D.; Pitout, J.D.D. Global extraintestinal pathogenic *Escherichia coli* (ExPEC) lineages. *Clin. Microbiol. Rev.* **2019**, *32*. [CrossRef] [PubMed]
3. Clermont, O.; Christenson, J.K.; Denamur, E.; Gordon, D.M. The Clermont *Escherichia coli* phylo-typing method revisited: Improvement of specificity and detection of new phylo-groups. *Environ. Microbiol. Rep.* **2013**, *5*, 58–65. [CrossRef] [PubMed]
4. Croxen, M.A.; Law, R.J.; Scholz, R.; Keeney, K.M.; Wlodarska, M.; Finlay, B.B. Recent advances in understanding enteric pathogenic *Escherichia coli*. *Clin. Microbiol. Rev.* **2013**, *26*, 822–880. [CrossRef]
5. Raimondi, S.; Righini, L.; Candeliere, F.; Musmeci, E.; Bonvicini, F.; Gentilomi, G.; Starčič Erjavec, M.; Amaretti, A.; Rossi, M. Antibiotic resistance, virulence factors, phenotyping, and genotyping of *E. coli* isolated from the feces of healthy subjects. *Microorganisms* **2019**, *7*, 251. [CrossRef]
6. Sarowska, J.; Futoma-Koloch, B.; Jama-Kmiecik, A.; Frej-Madrzak, M.; Ksiazczyk, M.; Bugla-Ploskonska, G.; Choroszy-Krol, I. Virulence factors, prevalence and potential transmission of extraintestinal pathogenic *Escherichia coli* isolated from different sources: Recent reports. *Gut Pathog.* **2019**, *11*, 10. [CrossRef]
7. Martin, R.M.; Bachman, M.A. Colonization, Infection, and the accessory genome of *Klebsiella pneumoniae*. *Front. Cell. Infect. Microbiol.* **2018**, *8*, 4. [CrossRef]
8. Davin-Regli, A.; Pagès, J.M. *Enterobacter aerogenes* and *Enterobacter cloacae*; versatile bacterial pathogens confronting antibiotic treatment. *Front. Microbiol.* **2015**, *6*, 392. [CrossRef]
9. Liu, L.; Chen, D.; Liu, L.; Lan, R.; Hao, S.; Jin, W.; Sun, H.; Wang, Y.; Liang, Y.; Xu, J. Genetic diversity, multidrug resistance, and virulence of *Citrobacter freundii* from diarrheal patients and healthy individuals. *Front. Cell. Infect. Microbiol.* **2018**, *8*, 233. [CrossRef]
10. Gajdács, M.; Urbán, E. Resistance Trends and epidemiology of citrobacter-enterobacter-serratia in urinary tract infections of inpatients and outpatients (RECESUTI): A 10-year survey. *Medicina (Kaunas)* **2019**, *55*, 285. [CrossRef]
11. Lee, C.R.; Lee, J.H.; Park, K.S.; Jeon, J.H.; Kim, Y.B.; Cha, C.J.; Jeong, B.C.; Lee, S.H. Antimicrobial resistance of hypervirulent *Klebsiella pneumoniae*: Epidemiology, hypervirulence-associated determinants, and resistance mechanisms. *Front. Cell. Infect. Microbiol.* **2017**, *7*, 483. [CrossRef] [PubMed]
12. Azevedo, P.A.A.; Furlan, J.P.R.; Oliveira-Silva, M.; Nakamura-Silva, R.; Gomes, C.N.; Costa, K.R.C.; Stehling, E.G.; Pitondo-Silva, A. Detection of virulence and β-lactamase encoding genes in *Enterobacter aerogenes* and *Enterobacter cloacae* clinical isolates from Brazil. *Braz. J. Microbiol.* **2018**, *49* (Suppl. 1), 224–228. [CrossRef] [PubMed]
13. Mokracka, J.; Koczura, R.; Kaznowski, A. Yersiniabactin and other siderophores produced by clinical isolates of *Enterobacter* spp. and *Citrobacter* spp. *FEMS Immunol. Med. Microbiol.* **2004**, *40*, 51–55. [CrossRef]
14. Yuan, C.; Yin, Z.; Wang, J.; Qian, C.; Wei, Y.; Zhang, S.; Jiang, L.; Liu, B. Comparative genomic analysis of *Citrobacter* and key genes essential for the pathogenicity of *Citrobacter koseri*. *Front. Microbiol.* **2019**, *10*, 2774. [CrossRef]
15. Cruz, A.; Xicohtencatl-Cortes, J.; González-Pedrajo, B.; Bobadilla, M.; Eslava, C.; Rosas, I. Virulence traits in *Cronobacter* species isolated from different sources. *Can. J. Microbiol.* **2011**, *57*, 735–744. [CrossRef]
16. Zogaj, X.; Bokranz, W.; Nimtz, M.; Römling, U. Production of cellulose and curli fimbriae by members of the family Enterobacteriaceae isolated from the human gastrointestinal tract. *Infect. Immun.* **2003**, *71*, 4151–4158. [CrossRef]

17. Tursi, S.A.; Tükel, Ç. Curli-containing enteric biofilms inside and out: Matrix composition, immune recognition, and disease implications. *Microbiol. Mol. Biol. Rev.* **2018**, *82*. [CrossRef]

18. Partridge, S.R. Resistance mechanisms in Enterobacteriaceae. *Pathology* **2015**, *47*, 276–284. [CrossRef]

19. El Fertas-Aissani, R.; Messai, Y.; Alouache, S.; Bakour, R. Virulence profiles and antibiotic susceptibility patterns of *Klebsiella pneumoniae* strains isolated from different clinical specimens. *Pathol. Biol. (Paris)* **2013**, *61*, 209–216. [CrossRef]

20. Compain, F.; Babosan, A.; Brisse, S.; Genel, N.; Audo, J.; Ailloud, F.; Kassis-Chikhani, N.; Arlet, G.; Decré, D. Multiplex PCR for detection of seven virulence factors and K1/K2 capsular serotypes of *Klebsiella pneumoniae*. *J. Clin. Microbiol.* **2014**, *52*, 4377–4380. [CrossRef]

21. Johnson, J.R.; Porter, S.; Johnston, B.; Kuskowski, M.A.; Spurbeck, R.R.; Mobley, H.L.; Williamson, D.A. Host characteristics and bacterial traits predict experimental virulence for *Escherichia coli* bloodstream isolates from patients with urosepsis. *Open Forum Infect. Dis.* **2015**, *2*. [CrossRef] [PubMed]

22. Davis, T.J.; Matsen, J.M. Prevalence and characteristics of *Klebsiella* species: Relation to association with a hospital environment. *J. Infect. Dis.* **1974**, *130*, 402–405. [CrossRef] [PubMed]

23. Lau, H.Y.; Huffnagle, G.B.; Moore, T.A. Host and microbiota factors that control *Klebsiella pneumoniae* mucosal colonization in mice. *Microbes Infect.* **2008**, *10*, 1283–1290. [CrossRef] [PubMed]

24. Magill, S.S.; Edwards, J.R.; Bamberg, W.; Beldavs, Z.G.; Dumyati, G.; Kainer, M.A.; Lynfield, R.; Maloney, M.; McAllister-Hollod, L.; Nadle, J.; et al. Multistate point-prevalence survey of health care-associated infections. *N. Engl. J. Med.* **2014**, *370*, 1198–1208. [CrossRef]

25. Lan, Y.; Zhou, M.; Jian, Z.; Yan, Q.; Wang, S.; Liu, W. Prevalence of pks gene cluster and characteristics of *Klebsiella pneumoniae*-induced bloodstream infections. *J. Clin. Lab. Anal.* **2019**, *33*, e22838. [CrossRef]

26. Lawlor, M.S.; O'connor, C.; Miller, V.L. Yersiniabactin is a virulence factor for *Klebsiella pneumoniae* during pulmonary infection. *Infect. Immun.* **2007**, *75*, 1463–1472. [CrossRef]

27. Schubert, S.; Cuenca, S.; Fischer, D.; Heesemann, J. High-pathogenicity island of *Yersinia pestis* in Enterobacteriaceae isolated from blood cultures and urine samples: Prevalence and functional expression. *J. Infect. Dis.* **2000**, *182*, 1268–1271. [CrossRef]

28. Schubert, S.; Rakin, A.; Heesemann, J. The *Yersinia* high-pathogenicity island (HPI): Evolutionary and functional aspects. *Int. J. Med. Microbiol.* **2004**, *294*, 83–94. [CrossRef]

29. Raffatellu, M. Learning from bacterial competition in the host to develop antimicrobials. *Nat. Med.* **2018**, *24*, 1097–1103. [CrossRef]

30. Fang, C.T.; Chuang, Y.P.; Shun, C.T.; Chang, S.C.; Wang, J.T. A novel virulence gene in *Klebsiella pneumoniae* strains causing primary liver abscess and septic metastatic complications. *J. Exp. Med.* **2004**, *199*, 697–705. [CrossRef]

31. Yu, W.L.; Ko, W.C.; Cheng, K.C.; Lee, H.C.; Ke, D.S.; Lee, C.C.; Fung, C.P.; Chuang, Y.C. Association between *rmpA* and *magA* genes and clinical syndromes caused by *Klebsiella pneumoniae* in Taiwan. *Clin. Infect. Dis.* **2006**, *42*, 1351–1358. [CrossRef] [PubMed]

32. Johansson, M.E.; Sjovall, H.; Hansson, G.C. The gastrointestinal mucus system in health and disease. *Nat. Rev. Gastroenterol. Hepatol.* **2013**, *10*, 352–361. [CrossRef] [PubMed]

33. Flores-Mireles, A.L.; Walker, J.N.; Caparon, M.; Hultgren, S.J. Urinary tract infections: Epidemiology, mechanisms of infection and treatment options. *Nat. Rev. Microbiol.* **2015**, *13*, 269–284. [CrossRef] [PubMed]

34. Hobley, L.; Harkins, C.; MacPhee, C.E.; Stanley-Wall, N.R. Giving structure to the biofilm matrix: An overview of individual strategies and emerging common themes. *FEMS Microbiol. Rev.* **2015**, *39*, 649–669. [CrossRef] [PubMed]

35. Martin, R.M.; Cao, J.; Brisse, S.; Passet, V.; Wu, W.; Zhao, L.; Malani, P.N.; Rao, K.; Bachman, M.A. Molecular epidemiology of colonizing and infecting isolates of *Klebsiella pneumoniae*. *mSphere* **2016**, *1*. [CrossRef] [PubMed]

36. Gorrie, C.L.; Mirceta, M.; Wick, R.R.; Edwards, D.J.; Thomson, N.R.; Strugnell, R.A.; Pratt, N.F.; Garlick, J.S.; Watson, K.M.; Pilcher, D.V.; et al. Gastrointestinal carriage is a major reservoir of *Klebsiella pneumoniae* infection in intensive care patients. *Clin. Infect. Dis.* **2017**, *65*, 208–215. [CrossRef] [PubMed]

37. Gajdács, M.; Bátori, Z.; Ábrók, M.; Lázár, A.; Burián, K. Characterization of resistance in gram-negative urinary isolates using existing and novel indicators of clinical relevance: A 10-year data analysis. *Life (Basel)* **2020**, *10*, 16. [CrossRef]

38. Carrie, C.; Walewski, V.; Levy, C.; Alexandre, C.; Baleine, J.; Charreton, C.; Coche-Monier, B.; Caeymaex, L.; Lageix, F.; Lorrot, M.; et al. *Klebsiella pneumoniae* and *Klebsiella oxytoca* meningitis in infants. Epidemiological and clinical features. *Arch. Pediatr.* **2019**, *26*, 12–15. [CrossRef]

39. Pötgens, S.A.; Brossel, H.; Sboarina, M.; Catry, E.; Cani, P.D.; Neyrinck, A.M.; Delzenne, N.M.; Bindels, L.B. Klebsiella oxytoca expands in cancer cachexia and acts as a gut pathobiont contributing to intestinal dysfunction. *Sci. Rep.* **2018**, *8*, 12321. [CrossRef]

40. Gajdács, M.; Ábrók, M.; Lázár, A.; Burián, K. Comparative epidemiology and resistance trends of common urinary pathogens in a tertiary-care hospital: A 10-year surveillance study. *Medicina (Kaunas)* **2019**, *55*, 356. [CrossRef]

41. MacPherson, C.W.; Mathieu, O.; Tremblay, J.; Champagne, J.; Nantel, A.; Girard, S.A.; Tompkins, T.A. Gut bacterial microbiota and its resistome rapidly recover to basal state levels after short-term amoxicillin-clavulanic acid treatment in healthy adults. *Sci. Rep.* **2018**, *8*, 11192. [CrossRef] [PubMed]

42. Leclercq, R.; Cantón, R.; Brown, D.F.; Giske, C.G.; Heisig, P.; MacGowan, A.P.; Mouton, J.W.; Nordmann, P.; Rodloff, A.C.; Rossolini, G.M.; et al. EUCAST expert rules in antimicrobial susceptibility testing. *Clin. Microbiol. Infect.* **2013**, *19*, 141–160. [CrossRef] [PubMed]

43. Wilson, L.A.; Sharp, P.M. Enterobacterial repetitive intergenic consensus (ERIC) sequences in *Escherichia coli*: Evolution and implications for ERIC-PCR. *Mol. Biol. Evol.* **2006**, *23*, 1156–1168. [CrossRef] [PubMed]

44. Quartieri, A.; Simone, M.; Gozzoli, C.; Popovic, M.; D'Auria, G.; Amaretti, A.; Raimondi, S.; Rossi, M. Comparison of culture-dependent and independent approaches to characterize fecal bifidobacteria and lactobacilli. *Anaerobe* **2016**, *38*, 130–137. [CrossRef]

45. Raimondi, S.; Luciani, R.; Sirangelo, T.M.; Amaretti, A.; Leonardi, A.; Ulrici, A.; Foca, G.; D'Auria, G.; Moya, A.; Zuliani, V.; et al. Microbiota of sliced cooked ham packaged in modified atmosphere throughout the shelf life: Microbiota of sliced cooked ham in MAP. *Int. J. Food Microbiol.* **2019**, *16*, 200–208. [CrossRef]

46. Mencacci, A.; Monari, C.; Leli, C.; Merlini, L.; De Carolis, E.; Vella, A.; Cacioni, M.; Buzi, S.; Nardelli, E.; Bistoni, F.; et al. Typing of nosocomial outbreaks of Acinetobacter baumannii by use of matrix-assisted laser desorption ionization-time of flight mass spectrometry. *J. Clin. Microbiol.* **2013**, *51*, 603–606. [CrossRef]

47. Sicard, J.F.; Vogeleer, P.; Le Bihan, G.; Rodriguez Olivera, Y.; Beaudry, F.; Jacques, M.; Harel, J. N-Acetyl-glucosamine influences the biofilm formation of *Escherichia coli*. *Gut Pathog.* **2018**, *22*, 10–26. [CrossRef]

48. Vogeleer, P.; Tremblay, Y.D.N.; Jubelin, G.; Jacques, M.; Harel, J. Biofilm-forming abilities of shiga toxin-producing *Escherichia coli* isolates associated with human infections. *Appl. Environ. Microbiol.* **2015**, *28*, 1448–1458. [CrossRef]

© 2020 by the authors. Licensee MDPI, Basel, Switzerland. This article is an open access article distributed under the terms and conditions of the Creative Commons Attribution (CC BY) license (http://creativecommons.org/licenses/by/4.0/).

International Journal of
Molecular Sciences

Article

Relation of the *pdxB-usg-truA-dedA* Operon and the *truA* Gene to the Intracellular Survival of *Salmonella enterica* Serovar Typhimurium

Xiaowen Yang [1,2,3], Jiawei Wang [1], Ziyan Feng [1], Xiangjian Zhang [4], Xiangguo Wang [1,*] and Qingmin Wu [2,*]

[1] Animal Science and Technology College, Beijing University of Agriculture, Beijing 102206, China; yangxiaowen@icdc.cn (X.Y.); wangjiawei20190110@gmail.com (J.W.); zy.feng0215@gmail.com (Z.F.)
[2] Key Laboratory of Animal Epidemiology and Zoonosis of the Ministry of Agriculture, College of Veterinary Medicine, China Agricultural University, Beijing 100193, China
[3] State Key Laboratory for Infectious Disease Prevention and Control, Collaborative Innovation Center for Diagnosis and Treatment of Infectious Diseases, National Institute for Communicable Disease Control and Prevention, Chinese Center for Disease Control and Prevention, Beijing 102206, China
[4] Tianjin Xiqing District Animal Husbandry and Aquatic Industry Development Service Center, Tianjin 300380, China; zhangxiangjian2019@gmail.com
* Correspondence: xiangguo731@bua.edu.cn (X.W.); wuqm@cau.edu.cn (Q.W.); Tel./Fax: +86-10-8079-3721 (X.W.); +86-10-6273-3901 (Q.W.)

Received: 17 December 2018; Accepted: 15 January 2019; Published: 17 January 2019

Abstract: *Salmonella* is the genus of Gram-negative, facultative intracellular pathogens that have the ability to infect large numbers of animal or human hosts. The *S. enterica usg* gene is associated with intracellular survival based on ortholog screening and identification. In this study, the λ-Red recombination system was used to construct gene deletion strains and to investigate whether the identified operon was related to intracellular survival. The *pdxB-usg-truA-dedA* operon enhanced the intracellular survival of *S. enterica* by resisting the oxidative environment and the *usg* and *truA* gene expression was induced by H_2O_2. Moreover, the genes in this operon (except for *dedA*) contributed to virulence in mice. These findings indicate that the *pdxB-usg-truA-dedA* operon functions in resistance to oxidative environments during intracellular survival and is required for in vivo *S. enterica* virulence. This study provides insight toward a better understand of the characteristics of intracellular pathogens and explores the gene modules involved in their intracellular survival.

Keywords: *usg; truA*; *Salmonella enterica* serovar Typhimurium; oxidative stress; intracellular survival

---

1. Introduction

Salmonella is a genus of Gram-negative and facultative intracellular pathogens that consists of a large group of genetically similar organisms with the ability to infect many animal and human hosts [1]. In 2005, the *Salmonella* genus was divided into *S. bongori* and *S. enterica* species [2]. *Salmonella* spp. are the second most common causative-agents of gastrointestinal infections in humans, after *Campylobacter* spp. [3]. Some *Salmonella* serovars cause large outbreaks of gastroenteritis associated with contaminated meat and produced or processed food [4].

A previous study found that the *S. enterica usg* gene was associated with intracellular survival based on ortholog screening and identification [5]. Because the complemented STΔusg/p-usg and wild type strains showed significant variation in the infection assay, our lab suggested that the *usg* gene was located in an operon with other genes. Operon prediction for the *S. enterica* strain LT2 whole genome sequence [6] found that the *pdxB, usg, truA* and *dedA* genes were located in the same operon. The promoter of this operon was located upstream of the *pdxB* gene (http://www.softberry.com/berry.

phtml?topic=bprom&group=programs&subgroup=gfindb; accessed on 2 May 2018). Based on the genome annotation, the *pdxB* gene encodes 4-phosphoerythronate dehydrogenase, which is associated with de novo vitamin B6 biosynthesis [7]; the *usg* gene encodes putative aspartate-semialdehyde dehydrogenase, *truA* encodes RNA pseudouridine (38–40) synthase, and the *dedA* gene encodes a hypothetical protein.

In this study, the λ-Red recombination system was used to construct gene deletion strains and investigate whether the identified operon was related to intracellular survival. Furthermore, this study investigated the possible functions of the genes in the operon and assessed their roles in the virulence of *S. enterica*.

2. Results

2.1. The pdxB-usg-truA-dedA Operon Is Required for Intracellular Survival of S. enterica

The predicted operon showed that the *pdxB*, *usg*, *truA* and *dedA* genes were transcribed in the same direction, indicating that these genes may have co-transcription. The primers were used to amplify the cDNA of the upstream and downstream genes and the intergenic regions, and the DNA and RNA were used as controls. The results revealed that the genes in the operon were co-transcribed. The above results indicate that the operon contains four genes: *pdxB*, *usg*, *truA* and *dedA* (Figure 1). The expression of each gene, using RT-PCR, showed that the genes in the operon had no effect on other genes in the lysogeny broth (LB) medium.

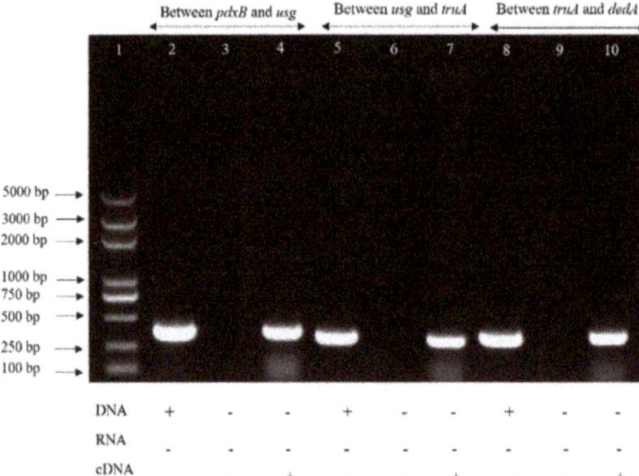

Figure 1. Analysis of co-transcription detection in the operon. Note: "+" means there was a fragment, "−" means there was no fragment. Lane 1 was a DNA marker, lanes 2, 5 and 8 were amplified using genome DNA; lanes 3, 6 and 9 were amplified using total RNA; lanes 4, 7 and 10 were amplified using cDNA.

The *pdxB*, *usg*, *truA* and *dedA* genes were located in the same operon and deletion strains for these genes were constructed using the λ-Red recombination system. The intracellular survival of the gene deletion strains was assessed using J774A.1 macrophage cells (Figure 2). At 12 and 24 h post-infection, the cells infected with the ST$\Delta pdxB$, STΔusg and ST$\Delta truA$ strains showed lower bacterial loads than the cells infected with the ST and ST$\Delta dedA$ strains ($p < 0.01$) (Figure 2A,B). The ST$\Delta pdxB$, STΔusg and ST$\Delta truA$ strains had reduced replication abilities inside the J774A.1 macrophages and, therefore, exhibited reduced virulence in vitro. However, the ST$\Delta dedA$ strain did not show reduced replication.

Subsequently, the complemented gene plasmids were electroporated into all of the gene deletion mutants. As shown in Figure 2C,D, the ST$\Delta pdxB$ strains electroporated with the recombinant plasmids p-*pdxB* and p-*usg* showed significant differences in bacterial loads ($p < 0.001$), whereas the ST$\Delta pdxB$ strains electroporated with the recombinant plasmid p-*truA* had almost the same bacterial loads in the macrophages. Similar results were found for the other complemented strains. These results indicated that the *truA* gene played a key role related to intracellular survival in this operon.

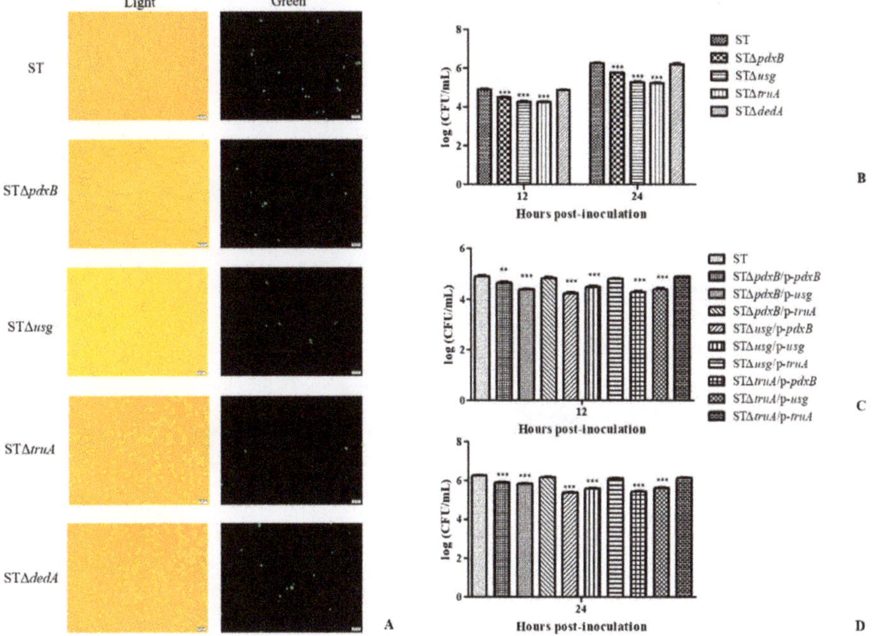

Figure 2. Intracellular survival of the operon gene deletion mutants and complemented strains in macrophages. (**A**) The infection process in J774A.1 macrophages by the strains was observed under a microscope at 12 h post-infection (40×). "Light" indicates that the cells were observed under natural light, and "Green" indicates that the cells were under a fluorescence microscope. (**B**) Bacterial loads of the gene deletion strains at 12 and 24 h post inoculation. (**C,D**) Bacterial loads of the complemented strains with different plasmids at 12 and 24 h post-inoculation. ** $p < 0.01$, *** $p < 0.001$.

In summary, these results showed that all genes in this operon, except for the *dedA* gene, confirmed that *S. enterica* has the ability to survive in macrophages and that the *truA* gene played a key role in this function.

2.2. The pdxB-usg-truA-dedA Operon Contributes to Virulence in Mice

Ten mice were intraperitoneally inoculated with a dose of 10^5 Colony-Forming Units (CFU) of the ST$\Delta pdxB$, STΔusg, ST$\Delta truA$, ST$\Delta dedA$ and wild type strains. The animals did not survive more than seven days after intraperitoneal inoculation with the wild type strain and not more than nine days post-inoculation with ST$\Delta dedA$. The mice in the group inoculated with ST$\Delta pdxB$ did not survive more than 15 days. Conversely, the survival rates of the groups inoculated with STΔusg and ST$\Delta truA$ were both 80% at 24 days post-inoculation (Figure 3A).

Five infected mice from each group were randomly selected at 6, 12, and 18 days post-inoculation. Their spleens were removed to assess the bacterial loads (Figure 3B). At six days post-inoculation, the bacterial loads were significantly lower for the gene deletion strains than for the wild type strain and a large reduction (above 2-log) in the spleen bacterial load was observed in the mice inoculated with STΔusg and ST$\Delta truA$ compared to the mice infected with the wild type strains. At 12 days post-inoculation, the spleen bacterial load increased in the mice inoculated with ST$\Delta pdxB$, which was similar to the load measured in the mice inoculated with the wild type strain at six days. The bacterial loads of the mice inoculated with ST∆*usg* and ST∆*truA* reduced slowly (Figure 3B). The results showed that *usg* and *truA* contributed to virulence in mice, which was consistent with the cell infection assay results.

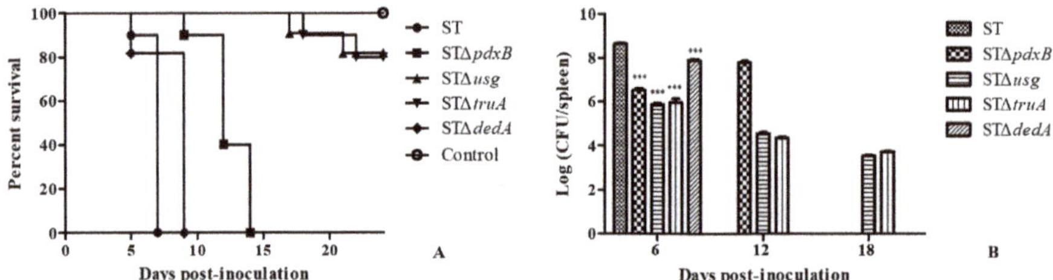

Figure 3. Survival of the gene deletion mutant and wild type strains in mice. (**A**) Survival curves of the gene deletion and wild type strains. (**B**) Bacterial loads of the spleens at six, 12, and 18 days post-inoculation with the gene deletion and wild type strains. *** $p < 0.001$.

2.3. The usg and truA Expression Levels Were Higher in the Oxidative Environment

The expression levels of genes in the operon were evaluated in the gene deletion and wild type strains in the routine growth medium and under oxidative conditions (Figure 4). As shown in Figure 3A, *pdxB*, *usg*, *truA* and *dedA* expression was induced in the wild type strain by H_2O_2 and was significantly higher in the samples treated with H_2O_2 than in the untreated controls (Figure 4A).

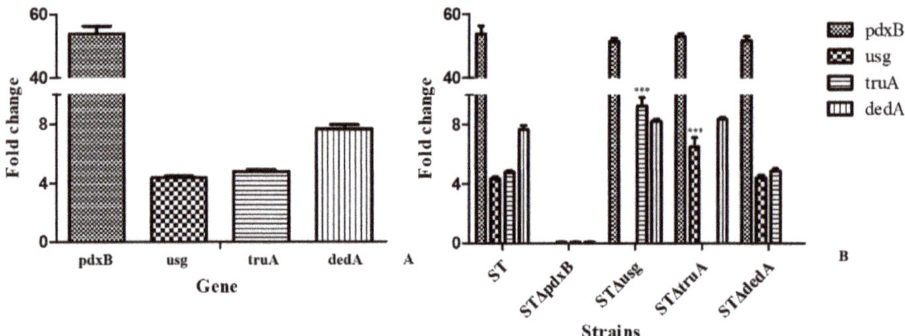

Figure 4. Gene expression levels in the bacterial strains. (**A**) Gene expression levels in the wild type strain under oxidative conditions. (**B**) Gene expression levels in wild type and gene deletion strains under oxidative conditions. *** $p < 0.001$.

Under oxidative conditions, the *usg*, *truA* and *dedA* genes were barely expressed when *pdxB* was deleted compared with the expression levels in the wild type strain. The *pdxB* gene expression levels were similar in the *usg*, *truA* and *dedA* gene deletion strains. When *dedA* was deleted, the expression levels of the other genes were not significantly different from the expression levels in the wild type strain. When *usg* was deleted, *truA* expression was induced by oxidative conditions and *truA* expression was significantly higher in these strains compared with the wild type strain. *Usg* expression also significantly increased when *truA* was deleted (Figure 4B).

2.4. The pdxB-usg-truA-dedA Operon Contributed to Resistance to Oxidative Conditions

The growth characteristics of the gene deletion and parent strains were determined in an LB medium. No significant variations were observed between the gene deletion and wild type strains (Figure 5A). These results suggested that the genes in the operon did not affect the in vitro growth of *Salmonella* spp. at normal temperatures. Under oxidative conditions, all of the strains grew slowly for the first 2 h. After 6 h, the growth of the wild type and STΔ*dedA* strains reached the plateau phase (Figure 3B). At this time, the STΔ*pdxB* strain was in the logarithmic phase, and the STΔ*usg* and STΔ*truA* strains had barely replicated. At 10 h post-infection, the STΔ*usg* and STΔ*truA* strains began to replicate, whereas the other strains were in the plateau phase (Figure 5B).

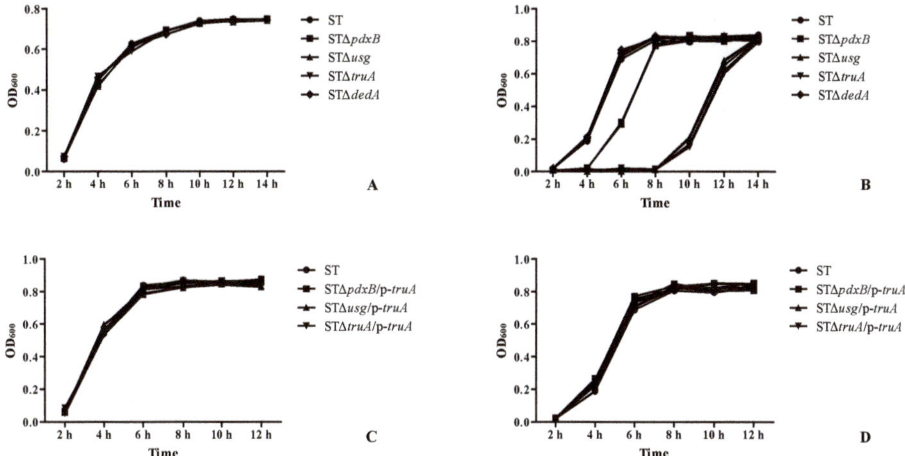

Figure 5. Growth characteristics of the gene deletion and complemented strains in the lysogeny broth (LB) medium and under oxidative conditions. (**A**) Growth characteristics of the gene deletion strains in the LB medium. (**B**) Growth characteristics of the gene deletion strains under oxidative conditions. (**C**) Growth characteristics of the strains complemented with the p-*truA* plasmid in the LB medium. (**D**) Growth characteristics of the strains complemented with p-*truA* plasmid under oxidative conditions.

The growth characteristics and oxidative resistance were also assessed for the STΔ*pdxB*/p-*truA*, STΔ*usg*/p-*truA* and STΔ*truA*/p-*truA* strains, which were complemented strains with the same recombinant p-*truA* plasmid (Figure 3C,D). These strains had characteristics similar to the wild type, even under oxidative conditions. All of the results suggested that the operon (except for the *dedA* gene) contributed to a resistance to oxidative conditions promoted the survival of *S. enterica* in macrophages.

3. Discussion

S. enterica is a common facultative intracellular pathogen. The main pathway used by macrophages to eliminate invading pathogens are endocytosis and digestion. *S. enterica* can exploit multiple aspects of host defenses to promote its replication in the host after adaptation to a variety of harsh environments, such as oxidative conditions [8]. A previous study found that the *usg* gene was related to intracellular survival and that the *pdxB*, *usg*, *truA* and *dedA* genes were located in the same operon. Another study found that *pdxB* was related to intracellular survival in *S. enterica*. This study found that a *pdxB* mutant strain was sensitive to oxidative conditions and reduced the bacterial load in macrophages. Another study found that *pdxB* in *E. coli* and *Pseudomonas aeruginosa* contained tightly bound NAD^+ and/or NADH [7,9,10] and that the nucleotide-binding domains of *pdxB* were homologous to the corresponding domains of D-3-phosphoglycerate dehydrogenase (PGDHs) from *E. coli* and *Mycobacterium tuberculosis* [7]. Because orthologs usually have conserved biological structures and functions [11,12], the results of this study suggest that *pdxB* in *S. enterica* has the same function. The *usg* and *truA* genes had similar results in the cell assay, similar expression levels under oxidative conditions, and inhibited TNF-α and IL-1β expression in macrophages. One study used random insertions of TnphoA-132 and found that *truA* was one target of glyoxal [13]. Another study showed that *truA* was associated with resistance to quinoxaline 1, 4-dioxides (QdNOs) in *E. coli*, which have been used in animals as antimicrobial agents and growth promoters for decades [14]. The results showed that *usg* and *truA* were related to intracellular survival. Although the *dedA* gene was in the same operon as the other genes, no effect on intracellular survival was observed for this gene.

There were many genes related to intracellular survival via oxidative resistance. The *sodA* gene deletion strain resulted in a slightly reduced growth rate, low SOD activity, increased susceptibility to reactive oxygen species and chicken serum, and no effect on the motility of the wild type strain [15]. One study found that three catalases (*KatE*, *KatG*, and *KatN*) and two alkyl hydroperoxide reductases (*AhpC* and *TsaA*) were related to oxidative resistance using silico genome analysis and gene deletion methods [16]. Large-scale profiling of *Salmonella* protein expression was performed under H_2O_2 treatment. The results showed that the abundance of 116 proteins were altered significantly among 1600 quantified proteins and that iron acquisition systems were induced to promote bacterial survival under oxidative stress [17]. Macrophages play important roles in the phagocytosis of pathogens and antigen presentation. Macrophages are immediately activated after phagocytosing pathogens, resulting in a variety of bactericidal mechanisms. These mechanisms include both oxidative and non-oxidative bactericidal mechanisms [8,18]. These poor survival environments lead *Salmonella* spp. to secrete effectors and generate a replicative compartment known as the Salmonella-containing vacuole (SCV). This study showed that genes in this operon (except for the *dedA* gene) conferred *S. enterica* with the ability to survive in macrophages and that the *truA* gene played a key role. The *usg* and *truA* expression levels were increased under oxidative treatment. All of these results suggested that this operon enhanced the intracellular survival of *S. enterica* by increasing resistance to oxidative environments and that *truA* played a key role in this function.

Operons are polycistronic clusters of genes transcribed from a promoter at the 5′ end of the cluster [19]. Several operons reportedly related to virulence have been identified in *S. enterica* [20–24]. Typically, genes in the same operon have similar functions and interact with each other. The amino acid sequences *pdxB*, *usg*, *truA* and *dedA* were uploaded to the STRING database [25] to predict protein–protein interactions. *Usg* and *truA* had the highest combined association score (0.935). Co-expression of *usg* and *truA* orthologs was also found in *Acinetobacter* sp. ADP1 and *Pseudomonas aeruginosa*. The combined association score of *dedA* was lower than the association scores for the other genes and no co-expression of *dedA* orthologs had been found to date. The results of our study are similar to our predictions.

4. Materials and Methods

4.1. Ethics Statement

All animal research was approved by the Beijing Association for Science and Technology. The approval ID is SYXK (Beijing) 2015–0028 (Validity period: 22 September 2015 to 22 September 2020), and the animal research complied with the Beijing Laboratory Animal Welfare and Ethics guidelines of the Beijing Administration Committee of Laboratory Animals.

4.2. Bacterial Strains and Media

All of the bacterial strains and plasmids used in this study are listed in Table 1. The *S. typhimurium* and *E. coli* strains, including the parental strain and the derived mutants, were routinely grown or incubated in an LB medium. Antibiotics were added at the following concentrations when required: ampicillin, 100 mg/L and chloramphenicol, 34 mg/L. All bacterial strains were frozen at −80 °C with 15–20% (*v/v*) glycerol.

Table 1. Strains and plasmids used in this study.

Strains or Plasmids	Description/Purpose	Source or Reference
Strains		
S. typhimurium ATCC14028 (ST)	Wild type (WT)	Guangdong Culture Collection Center
STΔ*pdxB*	Δ*pdxB* mutant of ST by the λ-Red recombination system	This study
STΔ*usg*	Δ*usg* mutant of ST by the λ-Red recombination system	Our lab
STΔ*truA*	Δ*truA* mutant of ST by the λ-Red recombination system	This study
STΔ*dedA*	Δ*dedA* mutant of ST by the λ-Red recombination system	This study
STΔ*pdxB*/p-*pdxB*	STΔ*pdxB* harboring the pBR322-*pdxB* plasmid, complement strains	This study
STΔ*pdxB*/p-*usg*	STΔ*pdxB* harboring the pBR322-*usg* plasmid, complement strains	This study
STΔ*pdxB*/p-*truA*	STΔ*pdxB* harboring the pBR322-*truA* plasmid, complement strains	This study
STΔ*usg*/p-*pdxB*	STΔ*usg* harboring the pBR322-*pdxB* plasmid, complement strains	This study
STΔ*usg*/p-*usg*	STΔ*usg* harboring the pBR322-*usg* plasmid, complement strains	Our lab
STΔ*usg*/p-*truA*	STΔ*usg* harboring the pBR322-*truA* plasmid, complement strains	This study
STΔ*truA*/p-*pdxB*	STΔ*truA* harboring the pBR322-*pdxB* plasmid, complement strains	This study
STΔ*truA*/p-*usg*	STΔ*truA* harboring the pBR322-*usg* plasmid, complement strains	This study
STΔ*truA*/p-*truA*	STΔ*truA* harboring the pBR322-*truA* plasmid, complement strains	This study
DH5α	For cloning	Takara
Plasmids		
pKD3, pKD46 and pCP20	λ-Red recombination system	Datsenko and Wanner, 2000 [26]
pBR322	For constructed complement strains	Virulent laboratory
p-*pdxB*	*pdxB* of ST product cloned into pBR322 for complementation assay	This study
p-*usg*	*usg* of ST product cloned into pBR322 for complementation assay	Our lab
p-*truA*	*truA* of ST product cloned into pBR322 for complementation assay	This study

4.3. Mice

BALB/c mice (aged 4 to 6 weeks) were purchased from the Weitong Lihua Laboratory Animal Services Center (Beijing, China), and bred in individually ventilated cage rack systems. All experiments involving animals followed the regulations of the Beijing Administration Office for Laboratory Animals.

4.4. Construction of Gene Deletion and Complemented Gene Deletion Mutant Strains

Deletion mutants and their complemented mutants were constructed for all genes in the operon. Gene deletion mutants were constructed using the λ-Red recombination system. After sequencing confirmation, the recombinant plasmids with the coding regions and their promoters were subsequently electroporated into every gene deletion mutant to complement the gene function. The complemented strains were selected from an LB medium containing ampicillin. The primers are shown in Table S1. The gene deletion and complemented gene deletion mutants were confirmed by PCR amplification and sequencing.

Total RNA was extracted from all strains using TRIzol (Invitrogen, Inc., Carlsbad, CA, USA) according to the manufacturer's instructions and treated with DNase (TaKaRa Bio, Inc., Dalian, China) before reverse transcription to remove DNA contamination. Total RNA was dissolved in diethypyrocarbonate (DEPC)-treated water, and the concentration and purity of the total RNA were estimated by reading the absorbance at 260 and 280 nm, respectively. cDNA was synthesized using the Prime Script™ RT Reagent Kit (TaKaRa Bio, Inc., Dalian, China), according to the manufacturer's instructions. The reverse transcription product was stored at −20 °C. PCR was performed with the primers shown in Table S1 to evaluate gene expression.

To determine if the genes in the operon were co-transcribed, the intergenic regions of genes were amplified by RT-PCR [27]. The RNA of the parental strain was extracted and cDNA was synthesized by reverse transcription using the method descibed before. The gene spacers were amplified using primers (Table S1) while the wild strain genomic DNA and RNA were used as controls. The amplification system and conditions were the same as before.

4.5. Cell Infection Assay

To investigate the intracellular survival of the strains, infection assays were performed using J774A.1 murine macrophages (Key Laboratory of Animal Epidemiology and Zoonosis of the Ministry of Agriculture, Beijing, China). The cells were cultured in 24-well plates and infected with each strain at a multiplicity of infection (MOI) of 10 CFU. Then, the infected plates were centrifuged at 1000 rpm for 5 min at room temperature and incubated at 37 °C in an atmosphere containing 5% (v/v) CO_2. After 20 min, the cells were washed three times with phosphate buffered solution (PBS) and incubated in a medium containing gentamycin (50 µg/mL) at 37 °C under 5% CO_2 until the end of the infection period. At 12 and 24 h post infection (p.i.), the cells were washed and lysed, and the numbers of bacteria exhibiting intracellular survival were determined through serial dilution and plating on an LB medium.

The pBR322 plasmid encoding the green fluorescent protein (GFP) was transferred into the parent and gene deletion strains using the electroporation method. Recombinant clones were selected from the LB medium containing ampicillin. Then, infection assays were performed as described above. After 12 h of incubation, the macrophages were washed three times with PBS. Infection of the J774A.1 macrophages by the strains was observed under a fluorescence microscope (Olympus, Tokyo, Japan).

4.6. Growth Characteristics and Oxidative Resistance Assay

The in vitro growth analysis of the deletion mutants and complemented strains was described previously. An oxidative resistance assay was performed as follows. One colony of each strain was inoculated into 3 mL of LB or LB with ampicillin medium and cultured overnight at 37 °C with shaking at 200 rpm. Subsequently the cultures were adjusted to the same concentration ($OD_{600} \approx 1.0$) and a 50 µL sample of each strain was inoculated into 5 mL of an LB or LB with ampicillin medium. Then, 30% H_2O_2 was add to the liquid medium (final concentration 4.4 mM) [28] to provide an oxidative environment. The cultures were incubated at 37 °C with shaking at 200 rpm, and the OD_{600} value was determined every 2 h using a BioTek microplate reader (Gene Company Limited, Hong Kong, China).

4.7. Gene Expression Levels in an Oxidative Environment

The gene expression levels under oxidative treatment were assessed by real-time PCR (RT-PCR). One colony of the gene deletion and parent strain was inoculated into 3 mL of LB medium and cultured overnight at 37 °C with shaking. The cultures were adjusted to the same concentration ($OD_{600} \approx 1.0$). A 50 µL sample of each strain was inoculated into 5 mL of LB medium and LB medium with H_2O_2. Total RNA was extracted from all strains, and cDNAs were obtained as described above. The cDNA samples were subjected to quantitative RT-PCR using the SYBR® Premix Ex Taq™ II Kit (TaKaRa Bio, Inc., Dalian, China). Each PCR reaction consisted of 2 µL of cDNA, 0.8 µL of each primer (10 µM), 10 µL of SYBR®Premix Ex Taq™ II, and 20 µL RNase-free water. The cycling conditions were a denaturation step, at 95 °C for 10 min, followed by 40 cycles of 95 °C for 10 s and 60 °C for 20 s. The specificity of the RT-PCR products was confirmed using a melting curve analysis. These reactions were repeated in triplicate for every sample as technical replicates. Gene mRNA quantification was performed using the $2^{-\Delta\Delta Ct}$ method to analyze the expression levels. The 16S rRNA expression level in *S. enterica* was used as a reference to normalize all values. The results presented in this study represent the averages from at least three separate experiments.

4.8. Virulence in BALB/c Mice

There were two experiments. The first experiment concerned the survival of the mice. Ten mice were intraperitoneally inoculated with a dose of 10^5 CFU of the gene deletion and wild type strains in 100 µL of phosphate-buffered saline (PBS) [29]; the control group included five mice intraperitoneally inoculated with 100 µL of PBS. The survival of the mice was observed over the next 24 days. The second experiment was the virulence of the gene deletion strains. Based on survival time, five mice were intraperitoneally inoculated with the same dose of the gene deletion and wild type strains. Five infected mice from each group were randomly selected at 6, 12, and 18 days post-inoculation. At each time point, the spleens were removed and homogenized individually in an aseptic manner in 1 mL of PBS and then serially diluted to isolate the bacteria. The results are presented as the mean number of CFU per spleen± the standard deviation (SD) in each group.

4.9. Statistical Analysis

The statistical analyses of the data, including the data from the growth curve analysis, cell infection study, oxidative resistance assay and virulence experiments, were performed using IBM SPSS Statistics version 23 (IBM, Armonk, New York, NY, USA, https://www.ibm.com/analytics/data-science/predictive-analytics/spss-statistical-software). A p value < 0.05 obtained through one-way analysis of variance (ANOVA) was considered significant. All graphics were drawn with GraphPad Prism 5 (GraphPad Software, La Jolla, CA, USA, https://www.graphpad.com/).

5. Conclusions

In conclusion, this study used the λ-Red recombination system to construct gene deletion strains and determine whether the identified operon was related to intracellular survival. Except for the *dedA* gene, all of the genes in this operon confirmed the ability of *S. enterica* to survive in macrophages, and the *truA* gene played a key role in resistance to oxidative conditions. Moreover, the genes in this operon (except for *dedA*) contributed to virulence in mice. These findings indicate that the *pdxB-usg-truA-dedA* operon functions in resistance to oxidative environments and contributes to intracellular survival; moreover, the operon is required for the virulence of *S. enterica* in vivo. In this study, clues were examined to gain a better understanding of the characteristics of intracellular pathogens and to explore the gene modules involved in the intracellular survival of intracellular pathogens.

Supplementary Materials: Supplementary materials can be found at http://www.mdpi.com/1422-0067/20/2/380/s1.

Author Contributions: X.Y. conceived of the experiments, interpreted the data, and supervised the research project; X.Y., Z.F., and J.W. performed the experiments and wrote the manuscript; X.Z., X.W., and Q.W. participated in the discussion and revised the manuscript; all authors approved the final draft.

Funding: This work was funded by the National Natural Science Foundation of China (No. 31372446 and No. 31802263), the Research of Key Technology for Prevention of Major Zoonosis in Dairy Cattle (No. 2015NZ0104), and the National Special Foundation for Science and Technology Basic Research (No. 2012FY111000).

Conflicts of Interest: The authors declare no conflict of interest.

References

1. Baker, S.; Dougan, G. The genome of Salmonella enterica serovar Typhi. *Clin. Infect. Dis.* **2007**, *45* (Suppl. 1), S29–S33. [CrossRef] [PubMed]

2. Tindall, B.J.; Grimont, P.A.; Garrity, G.M.; Euzeby, J.P. Nomenclature and taxonomy of the genus Salmonella. *Int. J. Syst.Evol. Microbiol.* **2005**, *55*, 521–524. [CrossRef] [PubMed]

3. Lamas, A.; Miranda, J.M.; Regal, P.; Vazquez, B.; Franco, C.M.; Cepeda, A. A comprehensive review of non-enterica subspecies of Salmonella enterica. *Microbiol. Res.* **2018**, *206*, 60–73. [CrossRef] [PubMed]

4. Nyachuba, D.G. Foodborne illness: Is it on the rise? *Nutr. Rev.* **2010**, *68*, 257–269. [CrossRef] [PubMed]

5. Yang, X.; Wang, J.; Bing, G.; Bie, P.; De, Y.; Lyu, Y.; Wu, Q. Ortholog-based screening and identification of genes related to intracellular survival. *Gene* **2018**, *651*, 134–142. [CrossRef]

6. Taboada, B.; Ciria, R.; Martinez-Guerrero, C.E.; Merino, E. ProOpDB: Prokaryotic Operon DataBase. *Nucleic Acids Res.* **2012**, *40*, D627–D631. [CrossRef]

7. Ha, J.Y.; Lee, J.H.; Kim, K.H.; Kim, D.J.; Lee, H.H.; Kim, H.K.; Yoon, H.J.; Suh, S.W. Crystal structure of D-erythronate-4-phosphate dehydrogenase complexed with NAD. *J. Mol. Biol.* **2007**, *366*, 1294–1304. [CrossRef]

8. Behnsen, J.; Perez-Lopez, A.; Nuccio, S.P.; Raffatellu, M. Exploiting host immunity: The Salmonella paradigm. *Trends Immunol.* **2015**, *36*, 112–120. [CrossRef]

9. Zhao, G.; Pease, A.J.; Bharani, N.; Winkler, M.E. Biochemical characterization of gapB-encoded erythrose 4-phosphate dehydrogenase of Escherichia coli K-12 and its possible role in pyridoxal 5′-phosphate biosynthesis. *J. Bacteriol.* **1995**, *177*, 2804–2812. [CrossRef]

10. Rudolph, J.; Kim, J.; Copley, S.D. Multiple turnovers of the nicotino-enzyme PdxB require alpha-keto acids as cosubstrates. *Biochemistry* **2010**, *49*, 9249–9255. [CrossRef]

11. Chen, F.; Mackey, A.J.; Vermunt, J.K.; Roos, D.S. Assessing performance of orthology detection strategies applied to eukaryotic genomes. *PLoS ONE* **2007**, *2*, e383. [CrossRef] [PubMed]

12. Fernandez-Breis, J.T.; Chiba, H.; Legaz-Garcia Mdel, C.; Uchiyama, I. The Orthology Ontology: Development and applications. *J. Biomed. Semant.* **2016**, *7*, 34. [CrossRef] [PubMed]

13. Lee, C.; Kim, J.; Kwon, M.; Lee, K.; Min, H.; Kim, S.H.; Kim, D.; Lee, N.; Kim, J.; Kim, D.; et al. Screening for Escherichia coli K-12 genes conferring glyoxal resistance or sensitivity by transposon insertions. *FEMS Microbiol. Lett.* **2016**, *363*. [CrossRef] [PubMed]

14. Guo, W.; Hao, H.; Dai, M.; Wang, Y.; Huang, L.; Peng, D.; Wang, X.; Wang, H.; Yao, M.; Sun, Y.; et al. Development of quinoxaline 1, 4-dioxides resistance in Escherichia coli and molecular change under resistance selection. *PLoS ONE* **2012**, *7*, e43322. [CrossRef] [PubMed]

15. Wang, Y.; Yi, L.; Zhang, J.; Sun, L.; Wen, W.; Zhang, C.; Wang, S. Functional analysis of superoxide dismutase of Salmonella typhimurium in serum resistance and biofilm formation. *J. Appl. Microbiol.* **2018**, *125*, 1526–1533. [CrossRef] [PubMed]

16. Hebrard, M.; Viala, J.P.; Meresse, S.; Barras, F.; Aussel, L. Redundant hydrogen peroxide scavengers contribute to Salmonella virulence and oxidative stress resistance. *J. Bacteriol.* **2009**, *191*, 4605–4614. [CrossRef] [PubMed]

17. Fu, J.; Qi, L.; Hu, M.; Liu, Y.; Yu, K.; Liu, Q.; Liu, X. Salmonella proteomics under oxidative stress reveals coordinated regulation of antioxidant defense with iron metabolism and bacterial virulence. *J. Proteom.* **2017**, *157*, 52–58. [CrossRef]

18. Buckner, M.M.; Finlay, B.B. Host-microbe interaction: Innate immunity cues virulence. *Nature* **2011**, *472*, 179–180. [CrossRef]

19. Blumenthal, T.; Davis, P.; Garrido-Lecca, A. Operon and non-operon gene clusters in the C. elegans genome. *WormBook* **2015**, 1–20. [CrossRef]

20. Eswarappa, S.M.; Panguluri, K.K.; Hensel, M.; Chakravortty, D. The yejABEF operon of Salmonella confers resistance to antimicrobial peptides and contributes to its virulence. *Microbiology* **2008**, *154*, 666–678. [CrossRef]

21. Zhang, Q.; Zhang, Y.; Zhang, X.; Zhan, L.; Zhao, X.; Xu, S.; Sheng, X.; Huang, X. The novel cis-encoded antisense RNA AsrC positively regulates the expression of rpoE-rseABC operon and thus enhances the motility of Salmonella enterica serovar typhi. *Front Microbiol.* **2015**, *6*, 990. [CrossRef] [PubMed]

22. Brandis, G.; Bergman, J.M.; Hughes, D. Autoregulation of the tufB operon in Salmonella. *Mol. Microbiol.* **2016**, *100*, 1004–1016. [CrossRef] [PubMed]

23. Qin, T.; Ken-Ichiro, I.; Ren, H.Y.; Zhou, H.J.; Yoshida, S. Legionella dumoffii Tex-KL Mutated in an Operon Homologous to traC-traD is Defective in Epithelial Cell Invasion. *Biomed. Environ. Sci.* **2016**, *29*, 424–434. [PubMed]

24. Godfrey, R.E.; Lee, D.J.; Busby, S.J.W.; Browning, D.F. Regulation of nrf operon expression in pathogenic enteric bacteria: Sequence divergence reveals new regulatory complexity. *Mol. Microbiol.* **2017**, *104*, 580–594. [CrossRef] [PubMed]

25. Szklarczyk, D.; Morris, J.H.; Cook, H.; Kuhn, M.; Wyder, S.; Simonovic, M.; Santos, A.; Doncheva, N.T.; Roth, A.; Bork, P.; et al. The STRING database in 2017: Quality-controlled protein-protein association networks, made broadly accessible. *Nucleic Acids Res.* **2017**, *45*, D362–D368. [CrossRef] [PubMed]

26. Datsenko, K.A.; Wanner, B.L. One-step inactivation of chromosomal genes in *Escherichia coli* K-12 using PCR products. *Proc. Natl. Acad. Sci. USA* **2000**, *97*, 6640–6645. [CrossRef] [PubMed]

27. Cui, G.; Wang, J.; Qi, X.; Su, J. Transcription Elongation Factor GreA Plays a Key Role in Cellular Invasion and Virulence of Francisella tularensis subsp. novicida. *Sci. Rep.* **2018**, *8*, 6895. [CrossRef] [PubMed]

28. Wang, M.; Qi, L.; Xiao, Y.; Wang, M.; Qin, C.; Zhang, H.; Sheng, Y.; Du, H. SufC may promote the survival of Salmonella enterica serovar Typhi in macrophages. *Microb. Pathog.* **2015**, *85*, 40–43. [CrossRef]

29. Gong, H.; Vu, G.P.; Bai, Y.; Chan, E.; Wu, R.; Yang, E.; Liu, F.; Lu, S. A Salmonella small non-coding RNA facilitates bacterial invasion and intracellular replication by modulating the expression of virulence factors. *PLoS Pathog.* **2011**, *7*, e1002120. [CrossRef]

© 2019 by the authors. Licensee MDPI, Basel, Switzerland. This article is an open access article distributed under the terms and conditions of the Creative Commons Attribution (CC BY) license (http://creativecommons.org/licenses/by/4.0/).

 International Journal of
Molecular Sciences

Article

Discovery of *lahS* as a Global Regulator of Environmental Adaptation and Virulence in *Aeromonas hydrophila*

Yuhao Dong, Yao Wang, Jin Liu, Shuiyan Ma, Furqan Awan, Chengping Lu and Yongjie Liu *

Joint International Research Laboratory of Animal Health and Food Safety, College of Veterinary Medicine, Nanjing Agricultural University, Nanjing 210095, China; 2015207019@njau.edu.cn (Y.D.); 2015107061@njau.edu.cn (Y.W.); 2017207019@njau.edu.cn (J.L.); 2017107059@njau.edu.cn (S.M.); 2015207039@njau.edu.cn (F.A.); lucp@njau.edu.cn (C.L.)
* Correspondence: liuyongjie@njau.edu.cn; Tel.: 86-25-84398606; Fax: 86-25-84398606

Received: 18 August 2018; Accepted: 4 September 2018; Published: 11 September 2018

Abstract: *Aeromonas hydrophila* is an important aquatic microorganism that can cause fish hemorrhagic septicemia. In this study, we identified a novel LysR family transcriptional regulator (LahS) in the *A. hydrophila* Chinese epidemic strain NJ-35 from a library of 947 mutant strains. The deletion of *lahS* caused bacteria to exhibit significantly decreased hemolytic activity, motility, biofilm formation, protease production, and anti-bacterial competition ability when compared to the wild-type strain. In addition, the determination of the fifty percent lethal dose (LD_{50}) in zebrafish demonstrated that the *lahS* deletion mutant ($\Delta lahS$) was highly attenuated in virulence, with an approximately 200-fold increase in LD_{50} observed as compared with that of the wild-type strain. However, the $\Delta lahS$ strain exhibited significantly increased antioxidant activity (six-fold). Label-free quantitative proteome analysis resulted in the identification of 34 differentially expressed proteins in the $\Delta lahS$ strain. The differentially expressed proteins were involved in flagellum assembly, metabolism, redox reactions, and cell density induction. The data indicated that LahS might act as a global regulator to directly or indirectly regulate various biological processes in *A. hydrophila* NJ-35, contributing to a greater understanding the pathogenic mechanisms of *A. hydrophila*.

Keywords: *Aeromonas hydrophila*; LysR-family; $\Delta lahS$; global regulator; virulence

1. Introduction

Aeromonas hydrophila is a gram-negative bacterium that is widely distributed in various aquatic environments, such as rivers, lakes, and swamps. In addition, *A. hydrophila* has a diverse host range, which includes fish, birds, amphibians, reptiles, and mammals [1]. Motile aeromonas septicemia (MAS) caused by this bacterium has caused serious damage to the aquaculture industry. In recent years, this bacterium has also been confirmed as an important pathogen of various human diseases, including diarrhea, sepsis, necrotizing fasciitis, meningitis, and hemolytic uremic syndrome [2,3]. *A. hydrophila* has a variety of factors that are associated with virulence, such as exotoxins, S-layers, extracellular enzymes and secretion systems [4]. Of these factors, hemolytic molecules may contribute to the reddening skin and systemic hemorrhagic septicemia; symptoms that are observed in diseased fish infected with *A. hydrophila*, and have been defined as one of the major virulence markers of this bacterium [5].

The expression of virulence factors in response to changes in environmental conditions is especially important for pathogenic bacteria, and this process is commonly governed by a complex network of regulatory elements [6]. Histone-like nucleoid structuring protein (H-NS) has been reported as a negative regulator of hemolytic activity by acting as a repressor of hemolysin gene expression in *Vibrio anguillarum* [7]. Sigma factor RpoE controls the development of hemolytic activity and virulence in *Vibrio harveyi* [8]. A transcriptional regulator of the MarR family, Eha, is required for the bacterial infection and the transcriptional regulation of important virulence factors in *Edwardsiella tarda*, including those that are associated with hemolytic activity [9]. Recently, the LysR family of transcriptional regulators (LTTRs), which is a well-characterized group of transcriptional regulators, has been shown to be global transcriptional regulators in a variety of bacteria [10,11]. LTTRs can regulate a diverse set of genes, such as those involved in metabolism, pilus synthesis, biofilm formation, antioxidant activity, acid resistance, toxin production, and drug efflux [12–17]. More often than not, LTTRs indirectly affect the expression of virulence factors by regulating virulence-related transcriptional regulators. For example, LrhA in enterohemorrhagic *Escherichia coli* positively regulates the expression of LEE genes by the regulation of their master regulators PchA and PchB [18]. The LysR family transcription factor LeuO in *Vibrio cholerae* regulates the transcription of the CarRS two-component regulatory system to control *almEFG* expression, which contributes to cationic antimicrobial peptide (CAMP) resistance by glycosylation of lipid A [19]. In *Haemophilus parasuis*, approximately 500 differentially expressed genes were identified, resulting from the deletion of *oxyR*, and these genes were involved in various biological processes [20]. Despite the importance of LTTRs, very little information is currently available regarding their regulation of virulence factors in *A. hydrophila*.

To identify the factors responsible for the hemolytic activity of *A. hydrophila*, in this study, we individually screened for reduced hemolytic ability in a Tn5-derived library of mutants that were previously generated by our group [21]. Interestingly, we identified a LysR-type transcriptional regulator (LahS) in *A. hydrophila* NJ-35. Our study showed that, in addition to hemolytic activity, LahS positively regulates a variety of virulence factors, including motility, biofilm formation, and protease activity, and it is involved in the anti-bacterial competition ability and virulence of this bacterium. This is the first study to evaluate the role of LahS in virulence and environmental adaptation in *A. hydrophila*, the results of which are of great importance in gaining a better understanding of the pathogenesis of this bacterium.

2. Results

2.1. Isolation of Transposon Mutants with Reduced Hemolytic Activity

Among 947 random EZ-Tn5 transposon mutants, a total of six mutants were identified as having reduced hemolytic activity (Figure 1). The EZ-Tn5 chromosomal insertion sites of the six mutants were analyzed by Tail-PCR, followed by BLASTn searches and sequence analyses. The data revealed that the six hemolysis-associated genes that were identified in this study included those encoding for PTS alpha-glucoside transporter subunit IIBC, a LysR family transcriptional regulator, an AraC family transcriptional regulator, and three hypothetical proteins (Table 1).

Table 1. Characteristics of the hemolysis-reduced mutants screened by transposon mutant library.

Mutant	Locus Tag	Insertion Site/Gene Length (bp) [a]	Function [b]
M80	U876_21575	2/480	hypothetical protein
M211	U876_23300	125/1587	PTS alpha-glucoside transporter subunit IIBC
M228	U876_22535	102/246	hypothetical protein
M307	U876_10550	440/909	LysR family transcriptional regulator
M402	U876_02250	330/516	hypothetical protein
M705	U876_01475	711/857	Arac family transcriptional regulator

[a] Insertion sites were identified using Tail-PCR and sequence analysis. [b] Putative functions were obtained from the NCBI BLAST online server [22].

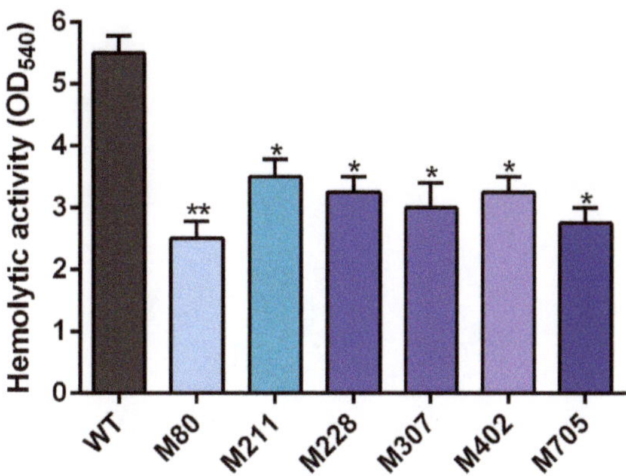

Figure 1. Hemolytic activity of the mutants compared to the wild-type strain. A total of six mutants were identified based on their hemolytic activities. * $p < 0.05$ or ** $p < 0.01$.

2.2. Effect of the LysR-type Transcriptional Regulator on A. hydrophila Hemolytic Activity

As shown in Figure 1, the M307 mutant exhibited decreased hemolytic activity. A sequence analysis of this strain revealed that the EZ-Tn5-disrupted in this strain encodes a protein belonging to LTTRs. This newly identified gene was designated as *lahS*. The open reading frame (ORF) of *lahS* is 909 bp, which encodes a hypothetical 34 kDa protein consisting of 302 amino acid (aa) residues. Secondary structure analysis showed that LahS protein has a conserved N-terminal DNA-binding helix–turn–helix (HTH) motif, which is located at amino acids 5–64 from N terminus (Figure 2A). In addition, there is a LysR substrate binding region that is located at amino acids 87–293, which has the ability to affect the binding capability of this protein. The two domains are connected via a long flexible linker helix that is involved in the oligomerization of the proteins (Figure 2B). The predicted three-dimensional (3D) structure of LahS showed that there are two α/β regulatory domains (RD1 and RD2) in the LysR substrate binding region, which are connected by two crossover regions (Figure 2B).

To verify whether the hemolytic phenotype of the M307 mutant was due to *lahS* inactivation, we constructed *lahS* mutant and complemented strains via homologous recombination. Similar to the M307 mutant, the *lahS* deletion mutant showed decreased hemolytic activity, and the hemolytic activity in the complemented strain CΔ*lahS* was restored to wild-type levels (Figure 3A). Through a BLAST analysis, *lahS* homologs were identified within the genomes of several *Aeromonas* species, including *Aeromonas salmonicida* strain A527 (89%), *Aeromonas veronii* strain TH0426 (83%), *Aeromonas caviae* strain 8LM (83%), and *Aeromonas schubertii* strain WL1483 (72%). Additionally, the transcription levels of both the upstream and the downstream genes, which encode an oxidoreductase and C4-dicarboxylate ABC transporter, respectively (Figure 3B), showed no significant differences between the Δ*lahS* and wild-type strains (Supplementary Figure S1). This finding indicated that the *lahS* mutation had no polar effect on the transcription of adjacent genes. Collectively, our findings indicate that LahS is a newly identified regulator of hemolytic activity in *A. hydrophila* NJ-35, and homologs of this regulator are present in a number of *Aeromonas* species.

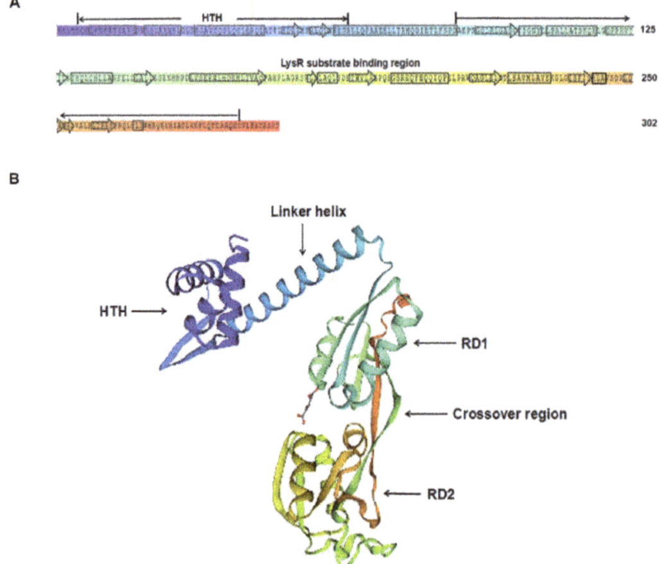

Figure 2. Structure of LahS protein. (**A**) Secondary structure of LysR family transcriptional regulator (LahS) protein. The N-terminal helix–turn–helix (HTH) domain and the LysR-substrate binding region were predicted using simple modular architecture research tool (SMART) software [23]. Rectangular box, α-helix; arrowheaded box, β-strand. (**B**) Predicted three-dimensional (3D) structure of LahS protein. The predicted maps were constructed with the SWISS-MODEL software [24]. Domains were shown in the corresponding position with the same colour coding, as in Figure 2A.

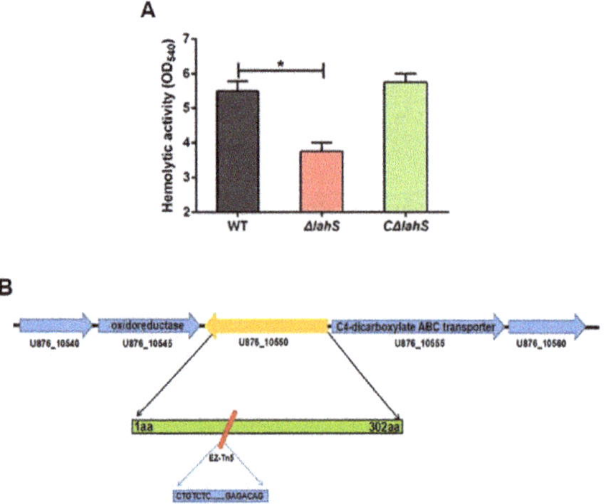

Figure 3. The *lahS* gene was involved in hemolytic activity of *A. hydrophila*. (**A**) Hemolytic activity of the wild-type and *lahS* mutant strains. *A. hydrophila* strains were grown overnight in Luria Bertani broth (LB) medium at 28 °C, after which 100 μL of supernatants were added to 1% sheep blood for 1 h at 37 °C. Hemolytic activity was expressed as the values measured at OD_{540}. * $p < 0.05$. (**B**) Schematic diagram of the EZ-Tn5 transposon insertion in the *lahS* gene. The gene cluster shows the location of the *lahS* gene in the NJ-35 genome. The red rectangle shows the transposon insertion site in *lahS*.

2.3. LahS Is Involved in Biofilm Formation of A. hydrophila NJ-35

CLSM analysis of the *A. hydrophila* NJ-35 wild-type, Δ*lahS,* and CΔ*lahS* strains showed that wild-type strain and CΔ*lahS* exhibited abundant biofilm formation, while the biofilm formation phenotype of the Δ*lahS* mutant is poor when compared to that of the wild-type strain (Figure 4A). This is consistent with our crystal violet staining results at 24 h. The ability of Δ*lahS* mutant to form biofilms was significantly decreased (by 38.8%) as compared to the wild-type strain ($p < 0.01$), whereas the biofilm formation was restored to the wild-type level in the CΔ*lahS* strain (Figure 4B). These data indicate that the transcriptional regulator LahS may directly or indirectly regulate biofilm formation of *A. hydrophila*.

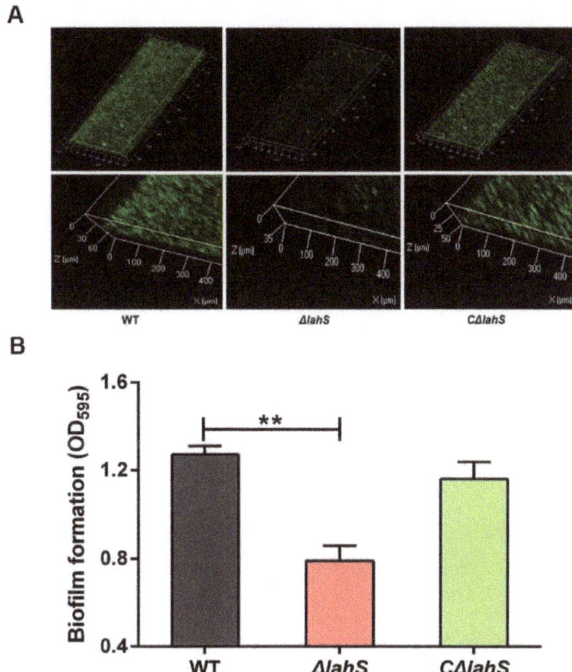

Figure 4. Biofilm formation of the wild-type and *lahS* mutant strains. (**A**) CLSM images of biofilms of the wild-type and *lahS* mutant strains. The viable cells exhibit green fluorescence. (**B**) Biofilm formation was measured by crystal violet staining using 96-well plates and was expressed as the values measured at OD_{595}. The data are presented as the means ± SEM from three independent experiments. ** $p < 0.01$.

2.4. LahS Influences Motility in A. hydrophila NJ-35

The swimming motility of *A. hydrophila* was measured by examining the distance cells migrated from the center to the periphery of the plate. As shown in Figure 5, migration diameters of 12.4 ± 0.6 and 8.4 ± 0.5 mm were measured for the wild-type and Δ*lahS* mutant strains, respectively, indicating that the swimming motility of the Δ*lahS* mutant was significantly decreased when compared with that of the wild-type strain. The motility phenotype was partially restored in the CΔ*lahS* strain. This finding indicates that LahS contributes to the motility of *A. hydrophila* NJ-35.

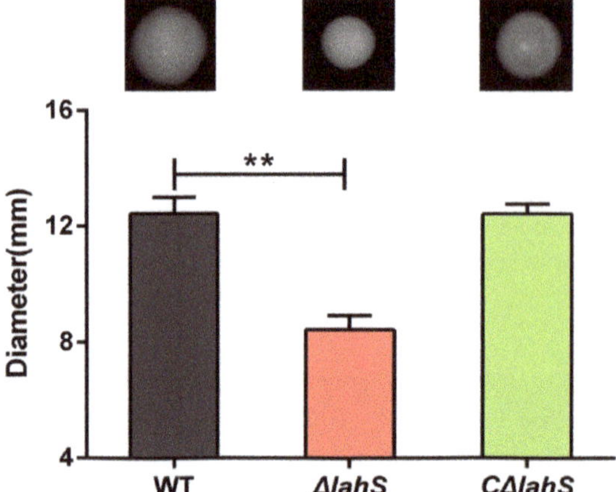

Figure 5. Motility of the wild-type and *lahS* mutant strains. Swimming ability was observed after culturing strains at 28 °C for 48 h on 0.3% Luria Bertani broth (LB) agar plates. The migration diameters were measured to assess the motility. The results are presented as the means ± SEM from three independent replicates. ** $p < 0.01$.

2.5. LahS Contributes to the Antibacterial Activity of A. hydrophila NJ-35

To determine whether the transcriptional regulator LahS influences the antibacterial ability of *A. hydrophila* NJ-35, we carried out growth competition experiments by co-culturing *A. hydrophila* strains with *E. coli* BL21. The *A. hydrophila* strains were ampicillin resistant, and the *E. coli* BL21 strain contained the plasmid pET-28a (+), which confers kanamycin resistance to allow for the selection of viable *E. coli* BL21 cells. As shown in Figure 6, co-culturing of *E. coli* BL21 with Δ*lahS* mutant showed about a six-fold rise in colony-forming unit (CFU) when compared to *E. coli* that was co-cultured with the wild-type strain ($p < 0.05$). The antibacterial capacity of the Δ*lahS* mutant was restored after complementing with *lahS* gene.

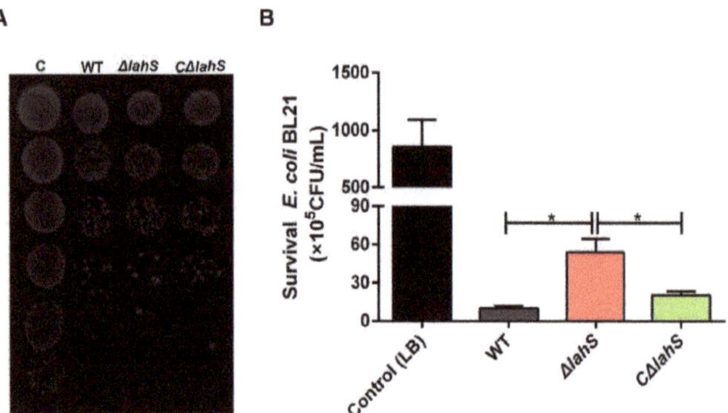

Figure 6. Competition ability of the wild-type and *lahS* mutant strains. (**A**) The image of competition ability is shown on Luria Bertani broth (LB) agar plate. (**B**) Quantification analysis of competition ability of the wild-type and *lahS* mutant strains. The respective bacterial strains were cultured together at a ratio of 1:1. The competition capability of *A. hydrophila* strains against *E. coli* BL21 was defined as the amount of observed *E. coli* survival after antagonism. Data are presented as the means ± SEM of three independent experiments. * $p < 0.05$.

2.6. LahS Plays a Role in the Resistance of A. hydrophila NJ-35 to Oxidative Stress

The antioxidant abilities of *A. hydrophila* strains were determined by treating each strain with H_2O_2. As shown in Figure 7, the $\Delta lahS$ mutant was hyposensitive to H_2O_2 as compared with the wild-type strain ($p < 0.001$). The viable cell number of the $\Delta lahS$ mutant was six-fold higher than that of the wild-type strain. This anti-oxidation ability was partially restored in the $C\Delta lahS$ strain. This result suggested that disruption of *lahS* enhanced the antioxidant capacity of *A. hydrophila* NJ-35.

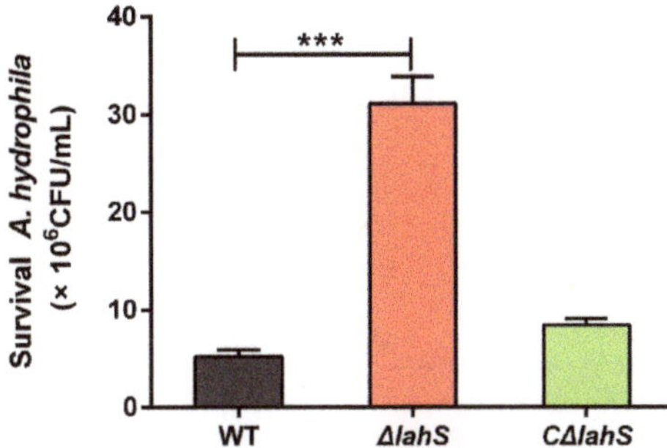

Figure 7. Oxidative stress resistance of the wild-type and *lahS* mutant strains. The effect of hydrogen peroxide on cell viability was examined to investigate the role of *lahS* in the resistance of *A. hydrophila* to oxidative stress. The H_2O_2 resistance levels were expressed as the colony-forming unit (CFU) of the viable *A. hydrophila* after treatment with H_2O_2. Data are presented as the means $\pm$ SEM of three independent experiments. *** $p < 0.001$.

2.7. Deletion of lahS Reduced the Protease Activity of A. hydrophila NJ-35

The culture supernatants of *A. hydrophila* strains were used to test for the presence of protease activity. The results in Figure 8 show that the $\Delta lahS$ mutant was significantly reduced for the production of protease (0.295 ± 0.013) as compared with that observed in the wild-type strain (0.409 ± 0.025) ($p < 0.05$). The protease activity was partially restored in the complemented strain $C\Delta lahS$. The result suggested that the *lahS* gene was involved in the protease activity of *A. hydrophila* NJ-35.

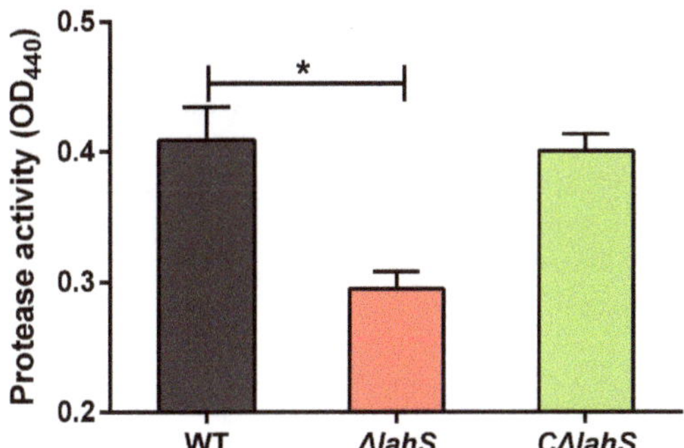

Figure 8. Protease activity of the wild-type and *lahS* mutant strains. The protease activity was detected using azocasein as a protease substrate and was measured at OD_{440}. Data are presented as the means $\pm$ SEM of three independent experiments. * $p < 0.05$.

2.8. LahS Is Essential for the Virulence of A. hydrophila in Zebrafish

To determine whether the *lahS* gene affected bacterial virulence, the LD_{50} values of the wild type and *lahS* mutant strains were compared while using zebrafish. The LD_{50} of *A. hydrophila* NJ-35 was

8.84×10^2 CFU, and all the fish were dead within three days. However, the *lahS* mutant had a LD_{50} of more than 10^5 CFU (Figure 9). Most of the dying fish showed typical clinical signs of hemorrhagic septicemia. Colonies of *A. hydrophila* were recovered from all dead fish, and no evident external lesions were observed in the surviving fish.

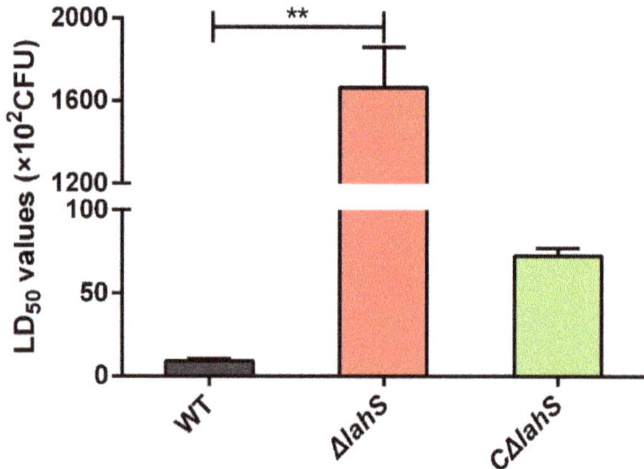

Figure 9. LD_{50} values of the wild-type and *lahS* mutant strains in zebrafish. The zebrafish were intraperitoneally (*i.p.*) injected with 10-fold serially diluted bacterial suspensions. The control group was *i.p.* injected with sterile phosphate buffered saline (PBS) only. The results are presented as the means ± SEM from three independent replicates. ** $p < 0.01$.

2.9. Comparative Proteomic Analysis

When considering that LahS functions as a transcriptional regulator, we investigated the differentially expressed profiles of the wild-type and the *lahS* mutant strains while using a label-free mass spectrometry method. The comparative proteomic analysis showed that a total of 2051 proteins matched to the Universal Protein Resource (UniProt). Thirty-four proteins were differentially expressed (change of > 1.5-fold) between the *lahS* mutant and wild type strain, including 10 upregulated and 24 downregulated proteins. A complete list of the names or locus tags of the 34 proteins is shown in Table 2.

To identify the relationship between the 34 different proteins, a hierarchical clustering method based on Pearson's correlation of variances was applied while using R studio. Figure 10A shows the hierarchical clustering of the 34 identified proteins, where an increasing color intensity indicates increasing protein expression levels. To further understand the functions of these identified proteins, we classified 34 differential proteins by Gene Ontology (GO) categories, including cellular component (CC), molecular function (MF), and biological process (BP) (Figure 10B). According to the analysis of CCs, the majority of the proteins were involved in cell projection and flagellum. According to the analysis of BPs, the primary functions were cellular detoxification and antioxygenation. According to the analysis of MFs, the assayed proteins regulated by LahS could be classified into six categories, as follows: heme binding, tetrapyrole binding, antioxidant activity, catalase activity, peroxidase activity, and oxidoreductase activity. In addition, to determine the accuracy of the mass spectrometry results, 14 of these 34 genes (six of the upregulated genes and eight of the downregulated genes) were randomly selected for further verification by qRT-PCR, the results of which showed that the quantitative PCR results were consistent with the proteomic data (Figure 10C).

Table 2. Differentially expressed proteins in *lahS* mutant compared to the wild type strain.

Locus Tag	Predicted Function	Fold Change
Up-regulated proteins		
U876_17600	Hypothetical protein	28.34991961
U876_01465	Catalase	11.31448355
U876_16540	Peroxidase	2.09906885
U876_04415	Cytochrome C biogenesis protein CcsA	2.043010753
U876_04895	Triose-phosphate isomerase	2.939215876
U876_07195	Fructose-6-phosphate aldolase	2.67060636
U876_05460	Penicillin-sensitive transpeptidase	6.074663579
U876_13305	Transcriptional regulator	2.236916529
U876_10585	Prolyl-tRNA synthetase	2.143335753
U876_23535	Chromosome partitioning protein ParA	2.1254663
Down-regulated proteins		
U876_07270	Flagellar hook protein FlgE	0.123192009
U876_16245	Flagellar hook-length control protein FliK	0.396639518
U876_07265	Flagellar hook capping protein FlgD	0.446402163
U876_14260	Flagellar hook-associated protein FliD	0.49944046
U876_14545	Phospho-2-dehydro-3-deoxyheptonate aldolase	0.021640145
U876_09970	Phospho-2-dehydro-3-deoxyheptonate aldolasee	0.067411226
U876_11610	LuxR family transcriptional regulator	0.279023788
U876_07335	Fis family transcriptional regulator	0.366104356
U876_14125	Glycine cleavage system protein T	0.014741915
U876_08310	Hypothetical protein	0.22058777
U876_14865	Cytochrome C	0.243506128
U876_17595	Hydroxylamine reductase	0.267712471
U876_13425	Thioredoxin reductase	0.280916369
U876_14870	Nitrate reductase	0.304735753
U876_22700	YeeE/YedE family protein	0.381751269
U876_14250	Hypothetical protein	0.401170621
U876_07930	Cytochrome c biogenesis protein CcmH	0.426902415
U876_15300	Hypothetical protein	0.43795805
U876_13000	Methionine gamma-lyase	0.448173659
U876_01200	Cytochrome C	0.45592439
U876_16185	RNA polymerase sigma 70	0.457438622
U876_16300	Peptidase C80	0.481165879
U876_01650	Single-stranded DNA-binding protein	0.487195726
U876_09765	Electron transporter HydN	0.494528201

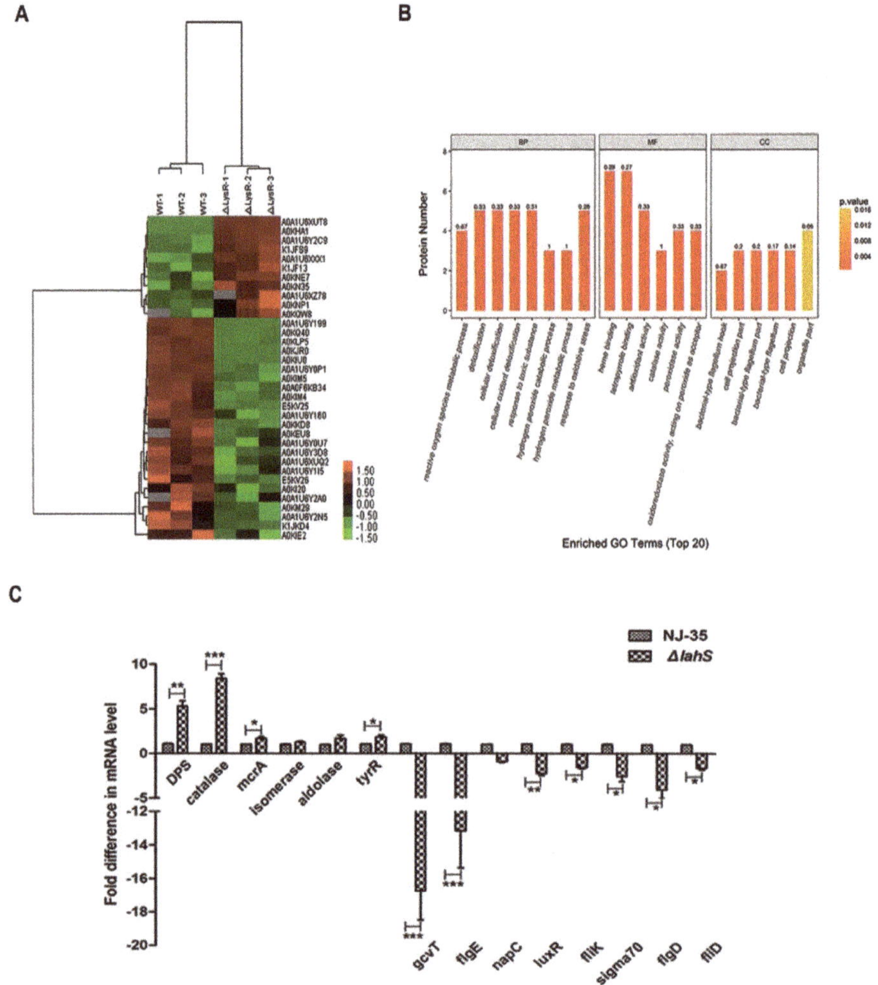

Figure 10. Label-free quantitative proteomics analysis of differentially expressed proteins between the wild-type and *lahS* mutant strains. (**A**) Heat map of the 34 identified proteins. Up- and down-regulated proteins are indicated in shades of green (increased) and red (decreased), respectively. (**B**) Gene Ontology (GO) classification of differentially expressed proteins. The differentially expressed proteins are grouped into three hierarchically structured terms: biological process, cellular component, and molecular function. (**C**) Relative mRNA expression levels of 14 genes coding the differentially expressed proteins in *lahS* mutant and wild-type strains. The results were expressed as n-fold increases with respect to the control. Data are presented as the means ± SEM from three independent experiments. * $p < 0.05$, ** $p < 0.01$ or *** $p < 0.001$.

3. Discussion

Hemolytic molecules are the major contributors to the hemorrhagic septicemia that is characteristic of *A. hydrophila* infections in fish, although this bacterium uses a variety of virulence factors [5]. However, prior to this study, little was known regarding the regulation of these hemolytic factors in this bacterium. In the present study, we identified a novel hemolysis-associated regulator (LahS) in *A. hydrophila* that belongs to the LTTRs. Interestingly, we observed that the LahS participated in different biological activities in this pathogen, including motility, biofilm formation, environmental adaptability, and virulence.

To investigate the global impact of the *lahS* deletion on protein expression in *A. hydrophila* NJ-35, we performed a label-free quantification analysis between the Δ*lahS* and wild-type strains. From this analysis, it was observed that LahS regulated the expression of a series of proteins that were involved in a wide range of biological processes, including oxidative stress, transcriptional regulation, DNA replication, and metabolism. LTTRs have been reported to regulate motility in many bacteria, but the molecular mechanisms are different in different strains. For example, LeuO in *V. cholerae* O1 modulates motility by cooperating with the nucleoid-associated protein H-NS to repress *vieSAB* transcription [25]. In *E. coli*, LrhA controls the transcription of flagellar, motility, and chemotaxis genes by regulating the expression of the master regulator FlhDC [26]. In this study, 34 proteins were observed to be differentially expressed in the *lahS* mutant as compared to the wild-type strain. Among the downregulated proteins, four proteins are involved in flagellin formation, including FlgE, FliK, FlgD, and FliD. In addition, the expression of ORF (U876_07335), which is 94% identical to FlrA from *A. hydrophila* AH-3 by a BLASTp analysis, was downregulated in the *lahS* mutant. FlrABC is known as the polar flagellum master regulator of a four-step hierarchical for the expression of flagellar genes in *A. hydrophila* [27]. Therefore, we speculate that LahS may control motility by positively regulating the expression of FlrABC.

Bacterial biofilm formation has been divided into three stages: the planktonic stage, the monolayer stage and the biofilm stage [28]. The transition between different stages is initiated by various environmental signals through the action of specific transcription factors. In *V. cholerae*, LeuO regulates the transition from the monolayer to the biofilm stage by the modulation of exopolysaccharide VPS gene transcription [29]. In addition to stage-specific regulatory functions, certain members of LysR family have roles in biofilm formation through a combination of direct and indirect regulation. For example, OxyR from *H. parasuis* regulates the DNA-binding transcriptional regulator FabR, which is directly involved in biofilm formation, as well as some membrane-related genes that are indirectly related to biofilm formation, such as an outer membrane assembly protein and an outer-membrane lipoprotein carrier protein [20]. In the present study, no proteins that are directly related to biofilm formation were identified among the differentially expressed proteins between the Δ*lahS* and wild-type strains, but RNA polymerase sigma 70 (σ^{70}), which has been reported to participate in the biofilm formation process [30], was observed to be downregulated. This led to us to speculate that deletion of *lahS* might downregulate the expression of the σ^{70} gene, which in turn, leads to a decrease in biofilm formation. In addition, a previous study confirmed that flagellar genes were actively transcribed during the planktonic stage, and motility and chemotaxis were important for the initiation of bacterial biofilm formation [31]. Therefore, there is another possibility that the reduction in motility may affect biofilm accumulation.

It is particularly noteworthy that we observed that the inactivation of *lahS* resulted in a significant downregulation of a LuxR family transcriptional regulator. LuxR regulators are key players in quorum sensing (QS), which coordinates the expression of a variety of genes, including those encoding virulence factors, motility, and biofilm formation [32]. Our previous study showed that the LuxR-type response regulator, AhyR, contributes to exoprotease and hemolysin production and the virulence of *A. hydrophila* [33]. In the present study, LahS was observed to be involved in motility, biofilm formation, interbacterial competition, hemolytic and proteolytic activities and virulence in zebrafish. All of these findings led us to speculate that the diverse role exhibited by LahS in *A. hydrophila* virulence might be achieved by upregulating the expression of a LuxR regulator. This speculation is further strengthened by a recent study that was conducted by Gao et al. [34], which indicated that a LysR-type transcriptional regulator (VqsA) is an important QS regulator in *Vibrio alginolyticus* and it plays essential roles in QS-regulated phenotypes, such as type VI secretion system 2 (T6SS2)-dependent interbacterial competition, biofilm formation, exotoxin production, and in vivo virulence. In this regard, it will be interesting to elucidate whether a correlation between LahS and QS exists in *A. hydrophila*.

Members of the LysR family have been widely reported to participate in regulating the antioxidant capacity and environmental adaptability, of which OxyR is the most thoroughly studied [35–37]. OxyR controls oxidative stress by acting as an activator [38] or a repressor [39] for defensive factors, such as catalase. In this study, the inactivation of *lahS* resulted in a substantial increase in the expression level of the catalases (U876_01465), which might partly explain the phenomenon that antioxidant capacity of the *lahS*-deleted strain was significantly higher than that of the wild-type strain.

In conclusion, our data suggest that the hemolysis-associated regulator LahS is positively involved in motility, biofilm formation, protease production, anti-bacterial competition ability, and virulence, but it negatively regulates the antioxidant capacity in *A. hydrophila*. The comparative proteomic analysis revealed that LahS directly or indirectly controls the expression of 34 proteins involved in flagellar assembly, cellular metabolism, oxidative stress, and environmental adaptability. It will be interesting to further examine the roles of LahS in virulence expression in *A. hydrophila* and to explore whether interfering with its function is an effective way to defend against this bacterial infection.

4. Materials and Methods

4.1. Strains, Plasmids and Growth Conditions

The bacterial strains and plasmids that were used in this study are listed in Table 3. *A. hydrophila* NJ-35, which belongs to the ST251 clonal group, was isolated from diseased cultured crucian carp in the Jiangsu province of China in 2010 [40]. The genome sequence of *A. hydrophila* NJ-35 has been published in GenBank (accession number CP006870). *A. hydrophila* and *E. coli* were cultured in Luria Bertani broth (LB) at 28 and 37 °C, respectively. When necessary, chloramphenicol (Cm) (Sigma Louis, MO, USA), kanamycin (Kan) (Sigma), or ampicillin (Amp) (Sigma) were added to the medium.

Table 3. Bacterial strains and plasmids used in this study.

Strain or Plasmid	Description [a]	Source or Reference
Strains		
NJ-35	Wild-type, isolated from diseased crucian carp, in China	Collected in our laboratory
SM10	*E. coli* strain, λpir$^+$, Kan^r	[41]
BL21	*E. coli* strain, F$^-$, ompT, hsdS (rB$^-$mB$^-$), gal, dcm (DE3)	Invitrogen
ΔlahS	*lahS* deletion mutant from NJ-35	This study
CΔlahS	ΔlahS complemented with pMMB-*lahS*	This study
Plasmid		
pET 28a (+)	Kan^r, F1 origin	Novagen
pYAK1	R6K-ori suicide vector, SacB$^+$, Cm^r	[42]
pYAK1-*lahS*	pYAK1 carrying the flanking sequence of *lahS*, Cm^r	This study
pMMB207	Low-copy-number vector, Cm^r	[43]
pMMB-*lahS*	Plasmid pMMB207 carrying the complete ORF of *lahS*	This study

[a] Characteristics of strains or plasmids. Kan^r, kanamycin resistant; Cm^r, chloramphenicol resistant.

4.2. Screening Transposon Insertion Mutants for Hemolytic Activity

A library containing 947 random EZ-Tn5 transposon mutants based on *A. hydrophila* NJ-35 was previously constructed in our laboratory [21]. All the mutants were assayed for hemolytic activity, as described previously [44], with some minor modifications. Briefly, *A. hydrophila* strains were grown overnight in LB medium at 28 °C, diluted to an optical density of 0.2 at 600 nm (OD$_{600}$), and pelleted by centrifugation at 10,000× *g* for 10 min. Next, the supernatants were filter-sterilized while using 0.22-μm (pore-size) membrane filters, and 100 μL of supernatant was double ratio diluted and dispensed into 96-well polystyrene plates. Subsequently, 100 μL of 1% sheep blood was added to each well and mixed completely. The plates were incubated for 1 h at 37 °C without agitation, followed by an overnight incubation at 4 °C. The plates were centrifuged at 1000× *g* for 10 min, after which 100-μL aliquots of supernatants were transferred into a new 96-well polystyrene plate. The OD$_{540}$ was monitored while using a spectrophotometer (BIO-RAD, Hercules, CA, USA).

4.3. Identification of Insertion Sites by Tail-PCR

The insertion sites of the EZ-Tn5 in the chromosome of *A. hydrophila* NJ-35 were identified by Tail-PCR. The primers that were used in this study are listed in Supplementary Table S1. Six specific primer pairs (SP1–SP6) were designed to amplify the DNA sequence from either end of the transposon. Degenerate primers AD1 to AD6 were paired with SP1 to SP6, respectively, to amplify the flanking sequences of the inserted EZ-Tn5. The detailed operation of the Tail-PCR was performed, as previously described [21]. After Tail-PCR, the specific products were gel purified and sequenced. The BLASTn program was used to compare the DNA sequence with the genome of the reference strain NJ-35 to confirm the sequences flanking the EZ-Tn5.

4.4. Construction of a lahS Mutant and Complemented Strains

The *lahS* deletion mutant strain was constructed, as previously described [42]. Briefly, two flanking regions of the target gene were amplified and ligated by PCR using two sets of primer pairs, *lahS-1/lahS-2* and *lahS-3/lahS-4* (Table 4). Next, the fusion fragment was generated with the primer pair *lahS-1/lahS-4* and then was inserted into pYAK1 and transformed into *E. coli* SM10 [41]. The recombination vector from the donor *E. coli* SM10 strain was conjugated into the recipient *A. hydrophila* NJ-35 strain. LB agar plates containing 100 µg/mL Amp and 34 µg/mL Cm were used to select for isolates in which the plasmid had integrated into the chromosome via recombination. The positive colonies were inoculated into LB broth supplemented with 20% sucrose to induce a second crossover event, and aliquots of this culture were subsequently spread onto LB agar plates containing 20% sucrose to generate the deletion of mutants. The *lahS* mutant was confirmed by sequencing the deleted region and flanking DNA in the mutated strain.

Table 4. Primers used for construction of *lahS* mutant and complemented strain.

Primer	Sequence (5′–3′)
lahS-1	CAGGTCGACTCTAGAGGATCCTGGAGGCAATGAAGGTGA
lahS-2	ATATCCAAACGCCAATGGGATGATCGG
lahS-3	TCCCATTGGCGTTTGGATATCCTGAACGATT
lahS-4	GAGCTCGGTACCCGGGGATCC CGCTCACGGTCAGGTAAC
lahS-C-F	GAGCTCGGTACCCGGGGATCCATGAAGCTGACCCTGCAAC
lahS-C-R	CAGGTCGACTCTAGAGGATCCTCAGGTGCGGCTGGC

The complemented Δ*lahS* strain was constructed using the shuttle plasmid pMMB207 [43]. The complete *lahS* gene and its putative promoter and terminator regions were amplified using the primer pair *lahS*-C-F/R and then ligated into the pMMB207 vector. The recombinant plasmid pMMB207-*lahS* was transformed into the Δ*lahS* mutant strain by bacterial conjugation to generate the complemented strain CΔ*lahS*, which was verified by PCR.

4.5. Biofilm Formation Assay

To compare the biofilm formation ability of strains NJ-35 and Δ*lahS*, confocal laser scanning microscopy (CLSM) was performed to analyze the three-dimensional architecture of biofilms as previously described [45]. Briefly, the stationary phase bacterial cultures were adjusted to an OD_{600} of 1.0 and then diluted 1:1000 in LB medium. Two milliliters of these dilutions were subsequently added into each well of 6-well plate containing pre-sterilized microscopic glass slides for biofilm growth. After incubating at 28 °C for 24 h to allow for biofilm development, the glass slides were carefully washed with phosphate buffered saline (PBS) to remove the planktonic cells. Next, 20 µL of fluorescein diacetate (FDA, Sigma, Louis, MO, USA) was added to each glass slide, which were then protected from light for 20 min. The slides were then observed by CLSM (Carl Zeiss LSM700, Oberkochen, Germany) to observe the biofilms using an argon laser. Seven random spots were measured for each of the three replicate glass slides. The image stacks were recorded under identical conditions (i.e., similar area and vertical resolution). Sterilized glass slides incubated with fresh LB medium were used as a control group.

Biofilm formation was also quantitatively measured by crystal violet staining, as previously described [46]. *A. hydrophila* strains were grown to an OD_{600} of 0.6–0.8 in LB broth at 28 °C and then diluted to an OD_{600} of 0.1. Next, 200-µL aliquots of suspensions (1:100 dilution in fresh LB) were dispensed into 96-well polystyrene plates, which were incubated at 28 °C for 24 h without shaking. The medium was decanted, and the wells were washed three times with sterile PBS. Next, the bacterial cells were fixed with 200 µL of methanol for 15 min, after which they were air dried at room temperature. After drying, 200 µL of a crystal violet solution (1% *w/v*) was added to each well and the cells were stained for 10 min. The wells were then rinsed with ddH_2O to remove the unbound crystal violet. The bound crystal violet was solubilized using 95% ethanol for 10 min, and the optical density was measured at OD_{595}. The assay was performed in three independent experiments.

4.6. Motility Assay

The swimming motility assay was performed using 0.3% agar plates, as previously described [47]. *A. hydrophila* strains were grown to the log phase in LB broth at 28 °C, and 1 µL of each culture (5×10^8 CFU/mL bacteria) was stabbed into motility assay agar plates. The plates were incubated at 28 °C for 48 h, after which motility was assessed by measuring the migration diameter of the bacterial cells. The assay was performed in three independent experiments.

4.7. Anti-Bacterial Competition Assay

The *E. coli* inhibition assay was performed as previously described with some modifications [48]. *A. hydrophila* and *E. coli* BL21 strains grown to an OD_{600} of 1.0 and then were concentrated 10 times. Cells were mixed at a ratio of 1:1, spotted onto 0.22-µm sterile filters on LB plates, and incubated at 28 °C for 3 h. *E. coli* BL21 cells that were mixed with equal volume of LB media was used as a control. Then, the spots were serially diluted in 1 mL LB medium, and the survival *E. coli* was quantified by serial dilution in LB and visualized on LB plates containing kanamycin. The assay was performed in three replicates.

4.8. Oxidative Stress Resistance Test

For oxidative stress tests, 4 mM H_2O_2 was freshly prepared before each experiment and was filter-sterilized using 0.22-µm (pore-size) membrane filters. Log-phase cultures were normalized to an OD_{600} of 0.5. Next, 400-µL aliquots were added to 100 µL H_2O_2 and the mixtures were incubated at 28 °C for 1 h. The oxidation was terminated with 2000 U of catalase for 10 min. The number of viable *A. hydrophila* was counted via serial dilution of the suspensions and plating on LB agar plates. The assay was performed in three independent experiments.

4.9. Protease Activity

A. hydrophila strains were grown overnight in LB medium at 28 °C and the OD_{600} values were normalized to 2.0 with fresh LB medium. Next, the cells were pelleted by centrifugation at $10,000 \times g$ for 10 min, and the supernatants were then filter-sterilized using 0.22-µm (pore-size) membrane filters. Subsequently, 250-µL aliquots of supernatants were added to 250 µL of 0.5% (*w/v*) azocasein in 50 mM Tris-HCl (pH 8.0) and then incubated at 37 °C for 2 h. After incubating, 500 µL of 10% (*w/v*) trichloroacetic acid (TCA) was added and the mixture was incubated on ice for 30 min. The cells were then pelleted by centrifugation at $10,000 \times g$ for 10 min, and 500 µL of the supernatants were added to 500 µL NaOH. The azodye was measured at OD_{440}.

4.10. Determination of LD_{50} in Zebrafish

The animal experiment was carried out in accordance with the animal welfare standards and guidelines of the Animal Welfare Council of China and was approved by the Ethical Committee for Animal Experiments of Nanjing Agricultural University, China [permit number: SYXK (SU).2017-0007]. Zebrafish were supplied by the Pearl River Fishery Research Institute, Chinese Academic of Fishery Science. The animal-challenge experiment was performed according to a previous study [49]. Log-phase bacteria were washed three times with sterile PBS and the suspensions were serially tenfold diluted from 5×10^2 to 5×10^7 CFU/mL. Ten zebrafish per group were intraperitoneally (*i.p.*) injected with 20 µL of the suspensions in PBS. An additional 10 zebrafish were injected with 20 µL of sterile PBS as the negative control. Mortality was recorded for seven days, and the LD_{50} values were calculated by the method of Reed and Muench [50].

4.11. Comparative Proteomic Analysis

A. hydrophila strains were grown overnight in LB broth at 28 °C, after which the bacterial cells were washed three times with PBS and then resuspended in lysis buffer. The bacterial suspensions were then ruptured by sonication at 4 °C and centrifuged ($12,000 \times g$, 30 min, 4 °C) to collect the supernatants. Next, 10% (*w/v*) TCA was added to the supernatants, which were incubated in ice-cold water for 30 min. The supernatants were centrifuged at $12,000 \times g$ for 10 min and then washed three times with ice-cold acetone. The proteins were harvested by centrifugation at $12,000 \times g$ for 10 min and analyzed using a label free mass spectrometry method. For quantitative proteomics analysis, 200 µg samples were digested with trypsin (1:50 *w/w*; Sigma Louis, MO, USA) at 37 °C for 24 h. Peptide mixtures were fractionated by nano-liquid chromatography and analyzed by mass spectrometry (MS). MS data were searched against uniprot_Aeromonas_hydrophila_27500_20170605.fasta (27,500 total entries, downloaded 5 June 2017). The identified proteins were analyzed by GO categories, KEGG enrichment, and clustering analyses.

4.12. Quantitative Reverse Transcription-PCR (qRT-PCR)

To validate the proteomic results, we used qRT-PCR to measure the transcription levels of randomly selected genes. The primer pairs used in this assay are shown in Supplementary Table S2. Total RNA was isolated using an E.Z.N.A. bacterial RNA isolation kit (Omega, Beijing, China). cDNA synthesis was performed while using a PrimeScript RT reagent kit (TaKaRa, Dalian, China). qRT-PCR was performed to quantify each target transcript using a QuantiTect SYBR green PCR kit (Qiagen, Valencia, CA, USA). The constitutively expressed *recA* gene was chosen as a reference gene for qRT-PCR, and the $2^{-\Delta\Delta C_t}$ method was used as previously described [51].

4.13. Statistical Analysis

Statistical analyses were performed while using GraphPad Prism 6. Student's *t* test was used for examining the differences between the wild-type and mutant strains. *p* values of <0.05 were considered significant.

Supplementary Materials: Supplementary materials can be found at http://www.mdpi.com/1422-0067/19/9/2709/s1. Figure S1: Fold difference in mRNA level; Table S1: Primers used for Tail-PCR; Table S2: Primers used for qRT-PCR.

Author Contributions: Y.L. and Y.D. conceived the study and drafted the paper. Y.D. and Y.W. performed most of the experiments described in the manuscript. J.L., S.M. and F.A. helped with the experiments. C.L. provided valuable suggestions of the manuscript. All authors read and approved the final manuscript.

Funding: This work was supported by the National Natural Science Foundation of China (31372454), the Three New Aquatic Projects in Jiangsu Province (D2017-3-1), the Independent Innovation Fund of Agricultural Science and Technology in Jiangsu Province (CX (17) 2027) and Priority Academic Program Development of Jiangsu Higher Education Institutions (PAPD).

Acknowledgments: We would like to thank the Shanghai Applied Protein Technology Co., Ltd., to provide technical support in our proteomics analysis.

Conflicts of Interest: The authors declare no conflict of interest.

References

1. Rasmussen-Ivey, C.R.; Hossain, M.J.; Odom, S.E.; Terhune, J.S.; Hemstreet, W.G.; Shoemaker, C.A.; Zhang, D.; Xu, D.H.; Griffin, M.J.; Liu, Y.J.; et al. Classification of a hypervirulent *Aeromonas hydrophila* pathotype responsible for epidemic outbreaks in warm-water fishes. *Front. Microbiol.* **2016**, *7*, 1615. [CrossRef] [PubMed]

2. Janda, J.M.; Abbott, S.L. The genus *Aeromonas*: Taxonomy, pathogenicity, and infection. *Clin. Microbiol. Rev.* **2010**, *23*, 35–73. [CrossRef] [PubMed]

3. Parker, J.L.; Shaw, J.G. *Aeromonas* spp. clinical microbiology and disease. *J. Infect.* **2011**, *62*, 109–118. [CrossRef] [PubMed]

4. Tomas, J.M. The main *Aeromonas* pathogenic factors. *ISRN Microbiol.* **2012**, *2012*, 256–261. [CrossRef] [PubMed]

5. Hamid, R.; Ahmad, A.; Usup, G. Pathogenicity of *Aeromonas hydrophila* isolated from the *Malaysian Sea* against coral (*Turbinaria* sp.) and sea bass (*Lates calcarifer*). *Environ. Sci. Pollut. Res. Int.* **2016**, *23*, 17269–17276. [CrossRef] [PubMed]

6. Balasubramanian, D.; Schneper, L.; Kumari, H.; Mathee, K. A dynamic and intricate regulatory network determines *Pseudomonas aeruginosa* virulence. *Nucleic Acids. Res.* **2013**, *41*, 1–20. [CrossRef] [PubMed]

7. Mou, X.; Spinard, E.J.; Driscoll, M.V.; Zhao, W.; Nelson, D.R. H-NS is a negative regulator of the two hemolysin/cytotoxin gene clusters in *Vibrio anguillarum*. *Infect. Immun.* **2013**, *81*, 3566–3576. [CrossRef] [PubMed]

8. Rattanama, P.; Thompson, J.R.; Kongkerd, N.; Srinitiwarawong, K.; Vuddhakul, V.; Mekalanos, J.J. Sigma E regulators control hemolytic activity and virulence in a shrimp pathogenic *Vibrio harveyi*. *PLoS ONE* **2012**, *7*, e32523. [CrossRef]

9. Gao, D.; Li, Y.; Xu, Z.; Sheng, A.; Zheng, E.; Shao, Z.; Liu, N.; Lu, C. The role of regulator Eha in *Edwardsiella tarda* pathogenesis and virulence gene transcription. *Microb. Pathog.* **2016**, *95*, 216–223. [CrossRef] [PubMed]

10. Heroven, A.K.; Dersch, P. RovM, a novel LysR-type regulator of the virulence activator gene *rovA*, controls cell invasion, virulence and motility of *Yersinia pseudotuberculosis*. *Mol. Microbiol.* **2006**, *62*, 1469–1483. [CrossRef] [PubMed]

11. Hernandez-Lucas, I.; Gallego-Hernandez, A.L.; Encarnacion, S.; Fernandez-Mora, M.; Martinez-Batallar, A.G.; Salgado, H.; Oropeza, R.; Calva, E. The LysR-type transcriptional regulator LeuO controls expression of several genes in *Salmonella enterica serovar* Typhi. *J. Bacteriol.* **2008**, *190*, 1658–1670. [CrossRef] [PubMed]

12. Stratmann, T.; Madhusudan, S.; Schnetz, K. Regulation of the *yjjQ-bglJ* operon, encoding LuxR-type transcription factors, and the divergent *yjjP* gene by H-NS and LeuO. *J. Bacteriol.* **2008**, *190*, 926–935. [CrossRef] [PubMed]

13. Blumer, C.; Kleefeld, A.; Lehnen, D.; Heintz, M.; Dobrindt, U.; Nagy, G.; Michaelis, K.; Emody, L.; Polen, T.; Rachel, R.; et al. Regulation of type 1 fimbriae synthesis and biofilm formation by the transcriptional regulator LrhA of *Escherichia coli*. *Microbiology* **2005**, *151*, 3287–3298. [CrossRef] [PubMed]

14. Wan, F.; Shi, M.; Gao, H. Loss of OxyR reduces efficacy of oxygen respiration in *Shewanella oneidensis*. *Sci. Rep.* **2017**, *7*, 42609. [CrossRef] [PubMed]

15. Kovacikova, G.; Lin, W.; Skorupski, K. The LysR-type virulence activator AphB regulates the expression of genes in *Vibrio cholerae* in response to low pH and anaerobiosis. *J. Bacteriol.* **2010**, *192*, 4181–4191. [CrossRef] [PubMed]

16. McCarthy, R.R.; Mooij, M.J.; Reen, F.J.; Lesouhaitier, O.; O'Gara, F. A new regulator of pathogenicity (*bvlR*) is required for full virulence and tight microcolony formation in *Pseudomonas aeruginosa*. *Microbiology* **2014**, *160*, 1488–1500. [CrossRef] [PubMed]

17. Shimada, T.; Yamamoto, K.; Ishihama, A. Involvement of the leucine response transcription factor LeuO in regulation of the genes for sulfa drug efflux. *J. Bacteriol.* **2009**, *191*, 4562–4571. [CrossRef] [PubMed]

18. Honda, N.; Iyoda, S.; Yamamoto, S.; Terajima, J.; Watanabe, H. LrhA positively controls the expression of the locus of enterocyte effacement genes in enterohemorrhagic *Escherichia coli* by differential regulation of their master regulators PchA and PchB. *Mol. Microbiol.* **2009**, *74*, 1393–1411. [CrossRef] [PubMed]

19. Bina, X.R.; Howard, M.F.; Ante, V.M.; Bina, J.E. *Vibrio cholerae* LeuO links the ToxR regulon to expression of lipid A remodeling genes. *Infect. Immun.* **2016**, *84*, 3161–3171. [CrossRef] [PubMed]

20. Wen, Y.P.; Wen, Y.P.; Wen, X.T.; Cao, S.J.; Huang, X.B.; Wu, R.; Zhao, Q.; Liu, M.F.; Huang, Y.; Yan, Q.G.; et al. OxyR of *Haemophilus parasuis* is a global transcriptional regulator important in oxidative stress resistance and growth. *Gene* **2018**, *643*, 107–116. [CrossRef] [PubMed]

21. Pang, M.D.; Xie, X.; Dong, Y.H.; Du, H.C.; Wang, N.N.; Lu, C.P.; Liu, Y.J. Identification of novel virulence-related genes in *Aeromonas hydrophila* by screening transposon mutants in a *Tetrahymena* infection model. *Vet. Microbiol.* **2017**, *199*, 36–46. [CrossRef] [PubMed]

22. Altschul, S.F.; Gish, W.; Miller, W.; Myers, E.W.; Lipman, D.J. BLAST: Basic Local Alignment Search Tool. Available online: http://blast.ncbi.nlm.nih.gov/Blast.cg (accessed on 11 May 2018).

23. Schultz, J.; Milpetz, F.; Bork, P.; Ponting, C.P. SMART. Available online: http://smart.embl-heidelberg.de/ (accessed on 28 July 2018).

24. Guex, N.; Peitsch, M.C. SWISS-MODEL. Available online: https://www.swissmodel.expasy.org/ (accessed on 7 September 2018).

25. Ayala, J.C.; Wang, H.; Benitez, J.A.; Silva, A.J. Molecular basis for the differential expression of the global regulator VieA in *Vibrio cholerae* biotypes directed by H-NS, LeuO and quorum sensing. *Mol. Microbiol.* **2018**, *107*, 330–343. [CrossRef] [PubMed]

26. Lehnen, D.; Blumer, C.; Polen, T.; Wackwitz, B.; Wendisch, V.F.; Unden, G. LrhA as a new transcriptional key regulator of flagella, motility and chemotaxis genes in *Escherichia coli*. *Mol. Microbiol.* **2002**, *45*, 521–532. [CrossRef] [PubMed]

27. Canals, R.; Ramirez, S.; Vilches, S.; Horsburgh, G.; Shaw, J.G.; Tomas, J.M.; Merino, S. Polar flagellum biogenesis in *Aeromonas hydrophila*. *J. Bacteriol.* **2006**, *188*, 542–555. [CrossRef] [PubMed]

28. O'Toole, G.; Kaplan, H.B.; Kolter, R. Biofilm formation as microbial development. *Annu. Rev. Microbiol.* **2000**, *54*, 49–79. [CrossRef] [PubMed]

29. Casper-Lindley, C.; Yildiz, F.H. VpsT is a transcriptional regulator required for expression of *vps* biosynthesis genes and the development of rugose colonial morphology in *Vibrio cholerae* O1 El Tor. *J. Bacteriol.* **2006**, *188*, 6044. [CrossRef]

30. Wang, K.; Liu, E.; Song, S.; Wang, X.; Zhu, Y.; Ye, J.; Zhang, H. Characterization of *Edwardsiella tarda rpoN*: Roles in σ^{70} family regulation, growth, stress adaption and virulence toward fish. *Arch. Microbiol.* **2012**, *194*, 493–504. [CrossRef] [PubMed]

31. Kirov, S.M.; Castrisios, M.; Shaw, J.G. *Aeromonas* flagella (polar and lateral) are enterocyte adhesins that contribute to biofilm formation on surfaces. *Infect. Immun.* **2004**, *72*, 1939–1945. [CrossRef] [PubMed]

32. Chen, J.; Xie, J.P. Role and regulation of bacterial LuxR-like regulators. *J. Cell. Biochem.* **2011**, *112*, 2694–2702. [CrossRef] [PubMed]

33. Bi, Z.X.; Liu, Y.J.; Lu, C.P. Contribution of AhyR to virulence of *Aeromonas hydrophila* J-1. *Res. Vet. Sci.* **2007**, *83*, 150–156. [CrossRef] [PubMed]

34. Gao, X.T.; Wang, X.T.; Mao, Q.Q.; Xu, R.J.; Zhou, X.H.; Ma, Y.; Liu, Q.; Zhang, Y.X.; Wang, Q.Y. VqsA, a novel LysR-type transcriptional regulator, coordinates quorum sensing (QS) and is controlled by QS to regulate virulence in the pathogen *Vibrio alginolyticus*. *Appl. Environ. Microbiol.* **2018**, *84*, e00444-18. [CrossRef] [PubMed]

35. Jo, I.; Chung, I.Y.; Bae, H.W.; Kim, J.S.; Song, S.; Cho, Y.H.; Ha, N.C. Structural details of the OxyR peroxide-sensing mechanism. *Proc. Natl. Acad. Sci. USA* **2015**, *112*, 6443–6448. [CrossRef] [PubMed]

36. Hennequin, C.; Forestier, C. *oxyR*, a LysR-type regulator involved in *Klebsiella pneumoniae* mucosal and abiotic colonization. *Infect. Immun.* **2009**, *77*, 5449–5457. [CrossRef] [PubMed]

37. Teramoto, H.; Inui, M.; Yukawa, H. OxyR acts as a transcriptional repressor of hydrogen peroxide-inducible antioxidant genes in *Corynebacterium glutamicum* R. *FEBS. J.* **2013**, *280*, 3298–3312. [CrossRef] [PubMed]

38. Chiang, S.M.; Schellhorn, H.E. Regulators of oxidative stress response genes in *Escherichia coli* and their functional conservation in bacteria. *Arch. Biochem. Biophys.* **2012**, *525*, 161–169. [CrossRef] [PubMed]

39. Jiang, Y.M.; Dong, Y.Y.; Luo, Q.X.; Li, N.; Wu, G.F.; Gao, H.C. Protection from oxidative stress relies mainly on derepression of OxyR-dependent KatB and Dps in *Shewanella oneidensis*. *J. Bacteriol.* **2014**, *196*, 445–458. [CrossRef] [PubMed]

40. Pang, M.D.; Jiang, J.W.; Xie, X.; Wu, Y.F.; Dong, Y.H.; Kwok, A.H.Y.; Zhang, W.; Yao, H.C.; Lu, C.P.; Leung, F.C.; et al. Novel insights into the pathogenicity of epidemic *Aeromonas hydrophila* ST251 clones from comparative genomics. *Sci. Rep.* **2015**, *5*, 9833. [CrossRef] [PubMed]

41. Melton-Witt, J.A.; McKay, S.L.; Portnoy, D.A. Development of a single-gene, signature-tag-based approach in combination with alanine mutagenesis to identify listeriolysin O residues critical for the *in vivo* survival of *Listeria monocytogenes*. *Infect. Immun.* **2012**, *80*, 2221–2230. [CrossRef] [PubMed]

42. Abolghait, S.K. Suicide plasmid-dependent IS1-element untargeted integration into *Aeromonas veronii* bv. *sobria* generates brown pigment-producing and spontaneous pelleting mutant. *Curr. Microbiol.* **2013**, *67*, 91–99. [PubMed]

43. Morales, V.M.; Backman, A.; Bagdasarian, M. A series of wide-host-range low-copy-number vectors that allow direct screening for recombinants. *Gene* **1991**, *97*, 39–47. [CrossRef]

44. Khajanchi, B.K.; Kozlova, E.V.; Sha, J.; Popov, V.L.; Chopra, A.K. The two-component QseBC signalling system regulates *in vitro* and *in vivo* virulence of *Aeromonas hydrophila*. *Microbiology* **2012**, *158*, 259–271. [CrossRef] [PubMed]

45. Du, H.C.; Pang, M.D.; Dong, Y.H.; Wu, Y.F.; Wang, N.N.; Liu, J.; Awan, F.; Lu, C.P.; Liu, Y.J. Identification and characterization of an *Aeromonas hydrophila* oligopeptidase gene pepF negatively related to biofilm formation. *Front. Microbiol.* **2016**, *7*, 1497. [CrossRef] [PubMed]

46. O'Toole, G.A.; Pratt, L.A.; Watnick, P.I.; Newman, D.K.; Weaver, V.B.; Kolter, R. Genetic approaches to study of biofilms. *Methods Enzymol.* **1999**, *310*, 91–109. [PubMed]

47. Grim, C.J.; Kozlova, E.V.; Sha, J.; Fitts, E.C.; van Lier, C.J.; Kirtley, M.L.; Joseph, S.J.; Read, T.D.; Burd, E.M.; Tall, B.D.; et al. Characterization of *Aeromonas hydrophila* wound pathotypes by comparative genomic and functional analyses of virulence genes. *mBio* **2013**, *4*, e00064-13. [CrossRef] [PubMed]

48. Liu, J.; Dong, Y.H.; Wang, N.N.; Li, S.G.; Yang, Y.Y.; Wang, Y.; Awan, F.; Lu, C.P.; Liu, Y.J. *Tetrahymena thermophila* predation enhances environmental adaptation of the carp pathogenic strain *Aeromonas hydrophila* NJ-35. *Front. Cell. Infect. Microbiol.* **2018**, *8*, 76. [CrossRef] [PubMed]

49. Pang, M.D.; Lin, X.Q.; Hu, M.; Li, J.; Lu, C.P.; Liu, Y.J. *Tetrahymena*: An alternative model host for evaluating virulence of *Aeromonas* strains. *PLoS ONE* **2012**, *7*, e48922. [CrossRef] [PubMed]

50. Stanic, M. A simplification of the estimation of the 50 percent endpoints according to the Reed and Muench method. *Pathol. Microbiol. (Basel)* **1963**, *26*, 298–302. [PubMed]

51. Livak, K.J.; Schmittgen, T.D. Analysis of relative gene expression data using real-time quantitative PCR and the $2^{-\Delta\Delta C_T}$ method. *Methods* **2001**, *25*, 402–408. [CrossRef] [PubMed]

© 2018 by the authors. Licensee MDPI, Basel, Switzerland. This article is an open access article distributed under the terms and conditions of the Creative Commons Attribution (CC BY) license (http://creativecommons.org/licenses/by/4.0/).

International Journal of
Molecular Sciences

Article

Lon Protease Is Important for Growth under Stressful Conditions and Pathogenicity of the Phytopathogen, Bacterium *Dickeya solani*

Donata Figaj [1,*], Paulina Czaplewska [2], Tomasz Przepióra [1], Patrycja Ambroziak [1], Marta Potrykus [3] and Joanna Skorko-Glonek [1,*]

[1] Department of General and Medical Biochemistry, Faculty of Biology, University of Gdańsk, Wita Stwosza 59, 80-308 Gdańsk, Poland; tomasz.przepiora@phdstud.ug.edu.pl (T.P.); patrycja.ambroziak@phdstud.ug.edu.pl (P.A.)

[2] Intercollegiate Faculty of Biotechnology, University of Gdańsk and Medical University of Gdańsk, Abrahama 58, 80-307 Gdańsk, Poland; paulina.czaplewska@ug.edu.pl

[3] Department of Environmental Toxicology, Faculty of Health Sciences with Institute of Maritime and Tropical Medicine, Medical University of Gdańsk, Dębowa 23A, 80-204 Gdańsk, Poland; marta.potrykus@gumed.edu.pl

* Correspondence: donata.figaj@ug.edu.pl (D.F.); joanna.skorko-glonek@ug.edu.pl (J.S.-G.)

Received: 10 April 2020; Accepted: 20 May 2020; Published: 23 May 2020

Abstract: The Lon protein is a protease implicated in the virulence of many pathogenic bacteria, including some plant pathogens. However, little is known about the role of Lon in bacteria from genus *Dickeya*. This group of bacteria includes important potato pathogens, with the most aggressive species, *D. solani*. To determine the importance of Lon for pathogenicity and response to stress conditions of bacteria, we constructed a *D. solani* Δ*lon* strain. The mutant bacteria showed increased sensitivity to certain stress conditions, in particular osmotic and high-temperature stresses. Furthermore, qPCR analysis showed an increased expression of the *lon* gene in *D. solani* under these conditions. The deletion of the *lon* gene resulted in decreased motility, lower activity of secreted pectinolytic enzymes and finally delayed onset of blackleg symptoms in the potato plants. In the Δ*lon* cells, the altered levels of several proteins, including virulence factors and proteins associated with virulence, were detected by means of Sequential Window Acquisition of All Theoretical Mass Spectra (SWATH-MS) analysis. These included components of the type III secretion system and proteins involved in bacterial motility. Our results indicate that Lon protease is important for *D. solani* to withstand stressful conditions and effectively invade the potato plant.

Keywords: protease Lon; *Dickeya solani*; plant pathogen; virulence factors; motility; type III secretion system; pathogenicity; resistance to stress; *lon* expression; pectinolytic enzymes

1. Introduction

Bacteria from the genus Dickeya, together with Pectobacterium, are classified as soft rot Pectobacteriaceae (SRP) [1]. They cause diseases of many economically important plants leading to significant financial losses all over the world [2,3]. *Dickeya solani* was first identified in 2005 and since then it has spread in Europe reducing yields of its primary host, potato [4,5]. *Dickeya solani* turned out to be a well-adapted and very successful pathogen, and due to its better adaptation in many regions, it has displaced another common potato pathogen, *Dickeya dianthicola*. Briefly, *D. solani* can infect the host plant with as little as 10 cells per tuber and has a broader temperature spectrum for infection compared to other SRP species [2]. SRP causes two types of plant diseases: blackleg and soft rot characterized by the blackening and wilting of a plant stem or tuber rot, respectively [5]. No effective methods to combat these pathogens have been developed so far [5]. The virulence factors that are primarily involved in the development of the disease symptoms caused by *D. solani* are plant cell wall degrading enzymes (PCWDE). This group of proteins encompasses pectinases, cellulases and proteases, whose joint action leads to maceration of plant tissues. Pectinases and cellulases are secreted via the type II secretion system (T2SS) and proteases are secreted via the type I secretion system (T1SS). However, an effective infection process requires also an array of other virulence determinants. These include the ability to actively move (motility) toward wounded tissue (chemotaxis), production of antioxidants, siderophores, secreted effectors and, finally, cellular factors that provide survival under unfavorable environmental conditions inside the host plant. *Dickeya* enters the plant via wounds or natural openings, and then it can penetrate the apoplast. There, the type III secretion system (T3SS), an essential virulence factor, is activated (reviewed in [6]). It consists of a pillus apparatus, so-called injectisome, connecting bacterial and plant cell cytoplasm [7] to directly inject effector proteins into the host. The effectors frequently manipulate host signaling pathways to disable the plant's defense systems to enable successful infection (reviewed in [8]).

For proper functioning of the cell, it is necessary to maintain its homeostasis, which is especially problematic under stressful conditions. Pathogenic bacteria are frequently exposed to unfavorable conditions, both in the host and during transmission between the hosts. In their life cycle, *Dickeya* may encounter several types of stress. These include changes in temperature and pH, exposure to oxidants, as well as osmotically active compounds [6]. All the stressors mentioned above can cause protein damage. To maintain cellular proteostasis, a dedicated set of proteins, termed protein quality control system (PQCS), is employed. It includes proteases and chaperones, that, among others, protect the cell against the accumulation of harmful protein aggregates and participate in regulatory proteolysis [9,10]. Lon and ClpP are two major cytoplasmic proteases, responsible for 70–80% of the ATP-dependent proteolysis in the cytosol [11]. Lon degrades aberrant proteins, which can arise in the cell not only as a consequence of stress but also under physiological conditions. Additionally, native proteins with the degradation tags (in general nonpolar amino acids exposed to the solvent) can undergo proteolysis by the Lon protease. Lon regulates many important cellular processes through the degradation of relevant substrates. For example, in *Escherichia coli*, cell division and synthesis of capsular polysaccharide are regulated by proteolysis of the cell division inhibitor SulA and transcription regulator RcsA, respectively [12,13]. Moreover, in various bacteria, functions of Lon can be implicated in such processes as motility, DNA replication, sporulation and, finally, pathogenicity [14].

Lon is crucial for virulence of many animal and plant pathogenic bacteria [8,14]. The *lon* mutant of *Brucella abortus* displayed decreased pathogenicity in BALB/c mice, however only at an early stage of infection [15]. The more pronounced effect was observed in the case of *Salmonella enterica* serovar Typhimurium *lon* mutants, which were highly attenuated in mice [16]. *Pseudomonas aeruginosa* Lon protease is necessary for effective bacterial infection in the mouse acute lung and amoeba model [17]. *Agrobacterium tumefaciens* depleted of a functional *lon* gene was unable to induce tumors in *Kalanchoe diagremontiana* [18]. The *lon* mutants of *Pseudomonas syringae* pv. *phaseolicola* and *P. syringae* pv. *tomato*, in turn, showed reduced disease symptoms in bean and tomato models, respectively [19].

To determine the role of the Lon protein in *D. solani*, we constructed a *D. solani* Δ*lon* strain using a modified lambda red recombination protocol. This allowed us to provide an insight into the functions played by Lon in *D. solani* pathogenicity and growth under stressful conditions.

2. Results

2.1. Construction of the D. solani IPO 2222 Δlon and the Complemented D. solani IPO 2222 Δlon/lon Strains

A *lon* deletion mutant of *D. solani* was constructed using the gene doctoring method, a modified protocol for lambda red recombination dedicated to the pathogenic bacterial strains. The gene encoding the Lon protease was substituted with the kanamycin resistance cassette, amplified from the pDOC-K plasmid. We confirmed the deletion of *lon* by three independent methods: (1) detection of the *lon* sequence on the *D. solani* chromosome by PCR using the primers complementary to the sequences flanking the *lon* gene; (2) detection of the *lon* transcript by real-time PCR; (3) immunodetection of the Lon protein using the anti-*E. coli* Lon rabbit antibodies. As can be seen in Figure 1, the *lon* gene and its product were not found in the strain *D. solani* IPO 2222 Δ*lon* but were present in the WT parental strain.

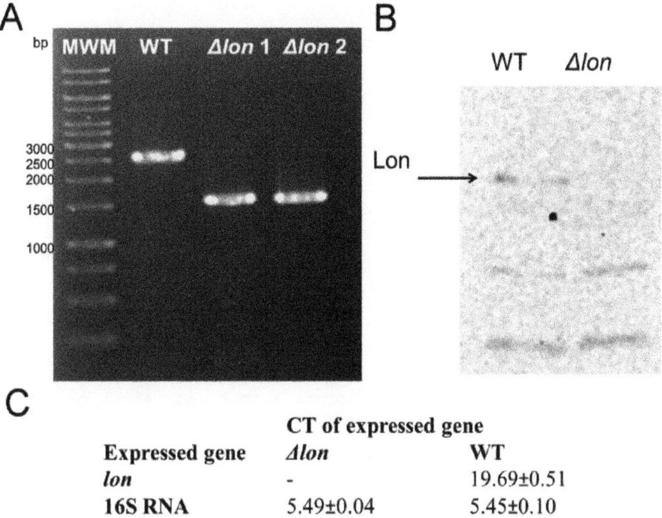

Expressed gene	CT of expressed gene	
	Δ*lon*	WT
lon	-	19.69±0.51
16S RNA	5.49±0.04	5.45±0.10

Figure 1. Confirmation of successful deletion of the *lon* gene in *D. solani* IPO 2222: (**A**) PCR analysis of genomic DNA isolated from the *D. solani* WT (wild type) and Δ*lon* mutant. The *lon* gene was replaced by a 1000 bp smaller kanamycin resistance gene. Δlon1 and Δlon2 denote two independent clones (**B**) immunodetection using the anti-Lon *E. coli* primary antibodies. (**C**) The qPCR analysis with the use of the *lon* gene-specific primers revealed that no cDNA amplification product was created within 50 cycles.

Single-copy complementation in the genome of the *D. solani* Δ*lon* strain was obtained using the *E. coli* MFD (Mu-free donor) *pir* conjugation strain. Immediately after the end of the *lon* gene, a marker gene encoding a pink Scarlett fluorescent protein was inserted. We confirmed the Δ*lon*/*lon* complementation at the gene and protein level (Figure S1).

A lack of the *lon* gene did not affect the growth of bacteria at the standard in vitro conditions (LB medium, 30 °C), as judged from the growth curves. The growth rates of the Δ*lon*, Δ*lon*/*lon* and WT *D. solani* cultures were comparable (Figure 2). Similar growth patterns of the Δ*lon* and WT strains have significantly facilitated the normalization of bacterial cultures in terms of age and cell density for subsequent stress sensitivity and pathogenicity tests.

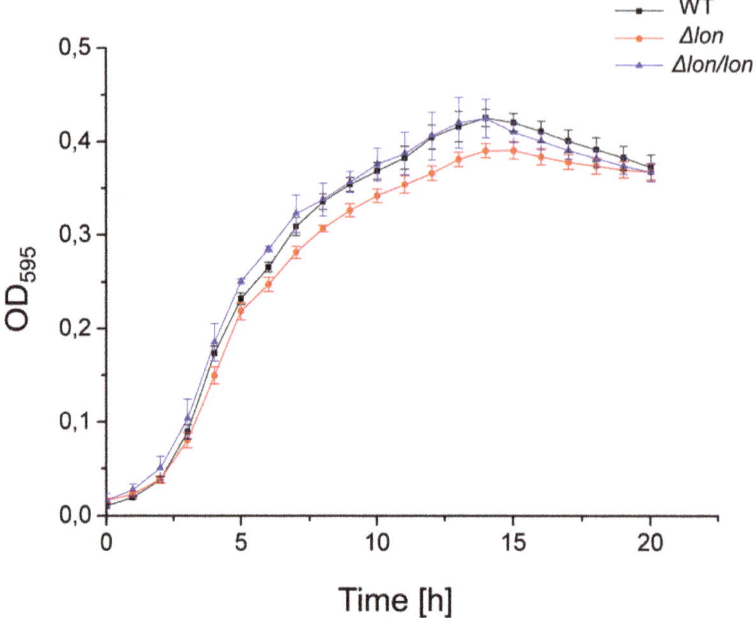

Figure 2. Growth of *D. solani* Δ*lon*. Curves were determined with the use of a plate reader at 30 °C. OD_{595} (optical density (595 nm)) values in LB medium were averaged for four replicates.

2.2. The Expression of the lon Gene is Upregulated under Certain Stressful Conditions

Under stress, a cell activates a variety of defense mechanisms that are manifested by increased expression of the key protective proteins [20–22]. To check the importance of Lon in the stress response, we performed the qPCR analysis to measure levels of transcription of the *lon* gene. We chose the common stressors, possible to affect *D. solani* during saprophytic and pathogenic life cycle: elevated temperature, nonionic and ionic osmotica, acidic pH and oxidants [23].

We found that transcription of *lon* was significantly upregulated in the exponentially growing bacteria under stressful conditions such as hyper osmosis, acidic pH and high temperature (Figure 3). The most pronounced effect was exerted by acidic pH and elevated temperature (over three $\log_2$ fold increase). A milder effect was caused by the presence of a nonionic osmoticum, sucrose (over two $\log_2$ fold increase). In contrast, the expression of the *lon* gene did not change significantly in cells in the stationary phase of growth. The exception was the upregulation (over 5.5 $\log_2$ fold increase) of the *lon* gene expression in cells treated with acidic pH. Interestingly, the changes in the *lon* transcript level upon treatment with ionic osmoticum, NaCl, were pronounced regardless of the growth phase, although they were not statistically significant (Figure 3). In contrast, oxidative stress did not affect the transcription of *lon* (Figure 3). Hence, the Lon protease is rather not a component of the oxidation response in *D. solani*.

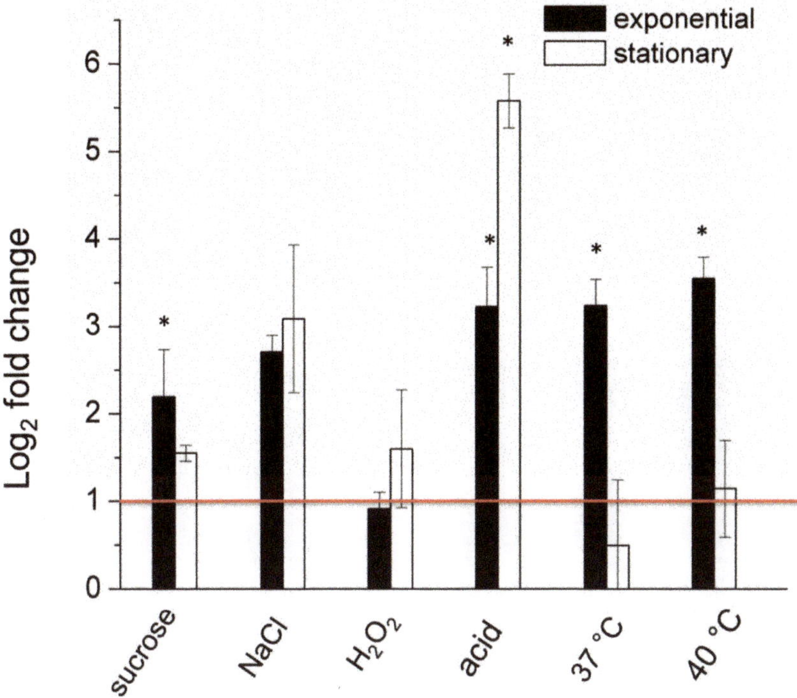

Figure 3. The relative log2 fold change of the expression levels of the *lon* gene in the *D. solani* cells under stressful conditions analyzed by qPCR. The data correspond to the means ± S.D. of three different samples, including three technical replicates. A red horizontal line indicates a relative two-fold increase in expression level. * indicates statistically significant (95% confidence interval) fold change in expression level according to the REST 2009 software.

2.3. Lon Protease Plays a Protective Role under Ionic and High-Temperature Stresses

Knowing that expression of *lon* is upregulated in cells in response to stress, we decided to check if the presence of the Lon protease in the cell is necessary for bacterial growth in the presence of selected stressors. Bacteria were exposed to the following stressful conditions: elevated temperature, ionic and nonionic osmotic shock, oxidative stress and low pH. We found that *D. solani* Δ*lon* was characterized by a decreased ability to form single colonies under three of five tested conditions. In particular, elevated temperature and presence of the nonionic osmoticum, sucrose, reduced viable cell counts by five and three orders of magnitude, respectively. Moreover, the Δ*lon* mutant colonies were very small under both tested conditions. The reintroduction of the *lon* gene into the *D. solani* Δ*lon* chromosome restored the wild-type phenotype of bacteria (Figure 4A). Hence, the strong phenotype of the mutant strain resulted from the lack of the Lon protease and not from putative additional suppressor mutations. The pronounced effect was also noticed under ionic osmotic stress: addition of NaCl resulted in a three-log reduction of cell counts of the Δ*lon* mutant with respect to the WT or complemented Δ*lon/lon* strain (Figure 4A). Acidic pH and oxidative stress affected all strains similarly (Figure 4A,B).

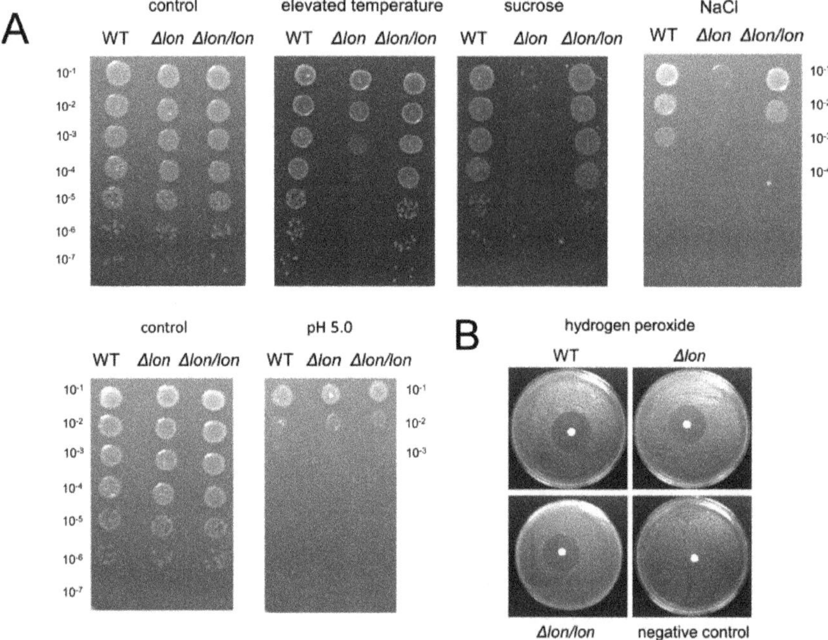

Figure 4. Growth of *D. solani* Δ*lon* under stressful conditions. (**A**) Overnight grown cultures were serially diluted and spotted on the LA (Luria Agar) agar plates, agar plates supplemented with 0.6 M sucrose or 0.3 M NaCl or on the LA medium adjusted to pH 5.0 when indicated. Bacteria grown on the LA agar plates at 30 °C refer to control. Disk diffusion assay with 1% hydrogen peroxide. As a negative control, sterile water was used (**B**). All plates were incubated at 30 °C except for the elevated temperature stress (39 °C).

2.4. Deletion of the lon Gene Delays the Onset of the Infection Symptoms

To test the importance of the Lon protease for pathogenicity of *D. solani*, we performed in vivo infection of the potato plants under greenhouse conditions. This kind of experiment shows the ability of bacteria to invade plants through the root system and produce blackleg symptoms. Although the deletion of the *lon* gene did not significantly reduce the occurrence of disease, an obvious delay in the development of the disease symptoms was observed. On the seventh day postinfection, only 30% of the potato plants treated with the Δ*lon* mutant bacteria showed blackleg symptoms, compared to 75% of symptomatic plants infected with WT *D. solani*. On the 17th day, the differences were much less pronounced, with 55% and 75% of symptomatic plants infected with Δ*lon* and WT *D. solani*, respectively (Figure 5).

To evaluate the effects of the Δ*lon* mutation on the ability of bacteria to macerate plant tissues, we used three models: potato tubers and leaves of chicory and Chinese cabbage. In no case did we observe differences in the degree of tissue maceration (Figure S2).

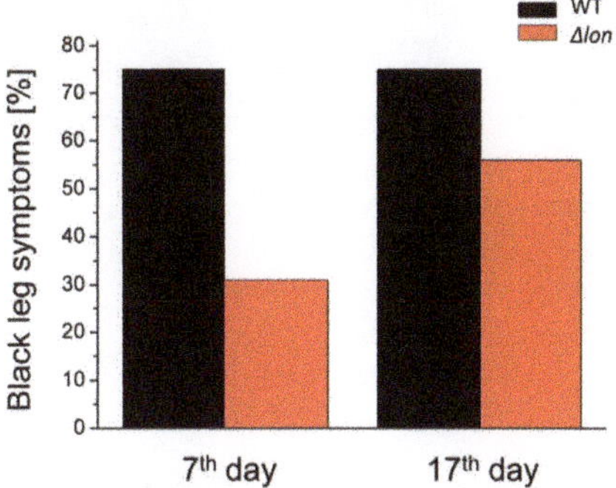

Figure 5. Pathogenicity of *D. solani* Δ*lon* in the whole potato plant model. Potato plants cv. Vineta were infected with WT and *lon* mutant strains. After 7 and 17 days of incubation at room temperature with a 16/8 photoperiod, the number of plants with blackleg symptoms was counted. Four plants watered with Ringer buffer represented a negative control. The number of infected plants with WT and Δ*lon* mutant was 16 for each strain.

2.5. The Deletion of lon Affects the Activity of Secreted Pectate Lyases

PCWDEs are the virulence factors that are directly responsible for the manifestation of disease symptoms. To check if altered pathogenicity of *D. solani* Δ*lon* results from changes in the level or activity of the enzymes that degrade the plant cell wall, we measured the activity of pectate lyases (major pectic enzymes), cellulases and proteases secreted from the mutant and WT *D. solani* cells. The enzymatic activity was assayed using PGA, modified cellulose CMC and casein, which are commonly used substrates for pectinases, cellulases and proteases, respectively. In the case of *D. solani* Δ*lon*, the secreted pectate lyase activity was 85% lower than that of the WT strain (Figure 6A). However, the activities of the remaining hydrolytic enzymes were not affected by the *lon* mutation (Figure 6B,C). The level of other secreted virulence factors, siderophores, also remained unchanged (Figure 6D).

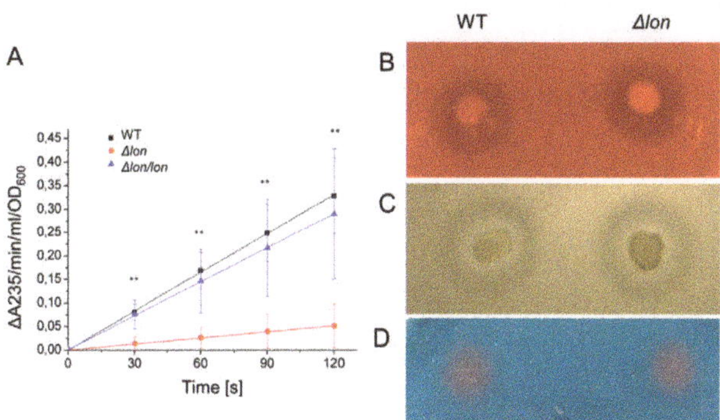

Figure 6. The activity of secreted virulence factors. (**A**) Pectinase activity was assayed as described in the Methods section with PGA as a substrate at 30 °C. ** $p < 0.01$ (*t*-test), $n = 5$. (**B**) Cellulases were assayed on M63Y with CMC. Seven microliters of bacterial cultures (10^8 CFU/mL) were spotted on the medium and incubated for 72 h. Plates were stained with 2% Congo red solution. (**C**) To monitor protease activity, 7 μL of bacterial cultures (10^8 CFU/mL) were spotted onto LA with skimmed milk and incubated for 48 h. (**D**) Siderophore activity was determined by spotting 10 μL of supernatant from overnight grown bacteria cultures onto chrome azurol S-agar plates. The picture was taken after 1 h of incubation at 30 °C. The experiments (**B–D**) were performed at least five times. The representative results are shown.

2.6. Lon Protease is Essential for Efficient Motility

Motility is one of the key factors for a successful invasion of the plant host. To determine if the observed delay of the blackleg symptoms development in the potato plants infected by *D. solani* Δ*lon* can be associated with altered bacterial motility, we examined two types of motility, swimming and swarming. While both types rely on the rotation of flagella, swimming is characteristic for an individual cell and is enhanced by chemotaxis. In contrast, swarming is common for a group of bacteria [24]. Indeed, the lack of Lon resulted in the altered motile phenotype of bacteria. The mutated strain showed considerable reduction in the swarm (Figure 7A) and 30% reduced swimming motility in the presence of galactose as a chemotactic agent (Figure 7B).

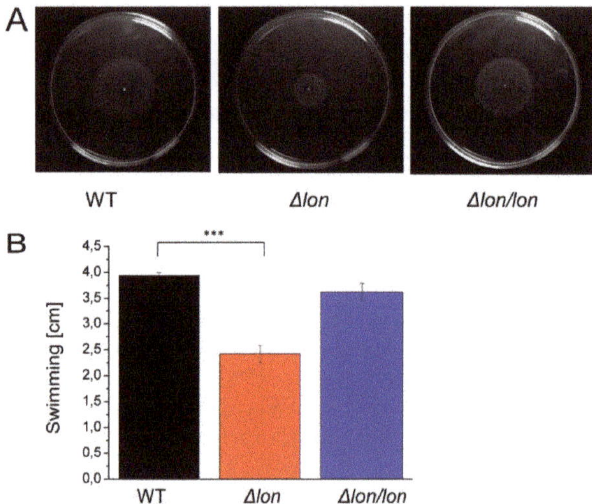

Figure 7. Swarming and swimming motility of *D. solani* Δ*lon*. (**A**) Representative pictures of bacteria swarming on 0.8% TSA after 12 h incubation at 30 °C. The experiments were performed at least five times. (**B**) Swimming was examined in a 0.3% MMA medium for 24 h at 30 °C. The diameter of the bacteria spreading area was measured. Presented data represent values for 5 biological replicates. *** $p < 0.001$ level (*t*-test).

2.7. Comparison of Proteomic Profiles of the D. solani Δlon and WT Cells under Physiological and Stressful Conditions

To gain more detailed insight into the properties of the Δ*lon* mutant cells, we compared the proteomes of the mutant and WT strains under physiological, as well as stress conditions, by the means of SWATH-MS (Sequential Window Acquisition of All Theoretical Mass Spectra) analysis. SWATH-MS is an advanced analysis method of proteomic data, recommended for quantification of identified peptides. It allows quantitative comparison of protein levels between different species or treatments due to the construction of a peptide spectral library [25].

It is well known that treatment with severe stressful agents can cause abnormal changes in the level of individual macromolecules [20] so we decided to subject the *lon* mutant to a rather mild stress—a short incubation at 40 °C.

The analysis identified a total of 635 proteins, for which at least two peptides per protein were quantified (Table S1). Deletion of the *lon* gene induced global changes in the *D. solani* proteome. We narrowed the number of differentially expressed proteins by applying the following cut-off criteria: $p < 0.05$, as well as fold changes below 0.5 or above 2.0. This resulted in 38 proteins with altered abundance in Δ*lon* compared to WT under physiological conditions and 60 proteins under stress conditions. Hence, the changes in the mutant proteome were more pronounced following the temperature shift than under control conditions, which may reflect the increased need for the Lon function during stress. In particular, the deletion of the *lon* gene resulted in upregulation of 17 or 41 proteins and downregulation of 13 or 19 proteins under physiological or stress conditions, respectively. Of these, 28 proteins shared a similar pattern of expression under both tested conditions (Figure 8A).

We grouped differentially expressed proteins into eight categories, depending on their physiological functions (Table 1). These include involvement in motility, iron metabolism, stress response, transport, general metabolism, transcription/translation, virulence and others. Percentage of proteins representing particular groups differs between control and induced conditions, however, the most abundant class encompasses proteins associated with general cell metabolism (Figure 8B). Consistent data were obtained for proteins involved in bacterial motility, namely all of them were repressed in Δ*lon* compared to WT. Among them, we identified flagellin, a structural component of bacterial flagella, and proteins responsible for chemotaxis (CheW, a positive regulator of CheA protein activity and CheA, signal transduction histidine kinase CheA). On the contrary, deletion of *lon* caused an increase in the cellular content of a group of proteins associated with virulence. They are all engaged in the T3SS and include hairpins, HrpN, a homolog of HrpW (Various polyols ABC transporter, permease component 2), as well as HrpA, Hrp pili protein. Deletion of *lon* differentially affected levels of proteins related to iron metabolism. We observed the upregulation of the proteins involved in the Fe-S cluster assembly (ISCU) and biosynthesis of achromobactin siderophore. However, the level of proteins involved in the synthesis of another siderophore, enterobactin, was decreased (for example isochorismate synthase enterobactin siderophore). The mutant strain was characterized by an increased content of several proteins engaged in transcription and translation. Among them, we could distinguish ribosomal proteins (50S ribosomal protein L27 and L7/L12), transcription factors (CytR, DksA) and RNA-binding protein Hfq. We obtained a similar trend regarding stress-related proteins, like ClpP protease and cold shock response proteins (CspE and CspG). A total of 77% of proteins associated with transport activity were downregulated, including histidine ABC transporter and efflux pump membrane transporter. The group named "others" comprises uncharacterized proteins or polypeptides which were not assigned to any other category. Among them, we identified putative membrane protein A0A2K8W3L1_9GAMM whose expression was increased more than 100-fold under both tested conditions. Protein blast indicated very close homology to periplasmic ComEA from many bacterial species, with the closest homology (100% coverage, 99% identity) to *Dickeya fangzhongdai*. ComEA is essential for DNA uptake in naturally competent bacteria, like *Bacillus subtilis* [26]. In Δ*lon*, we also observed an increased cellular level of S-ribosylhomocysteine lyase, which is associated with quorum sensing.

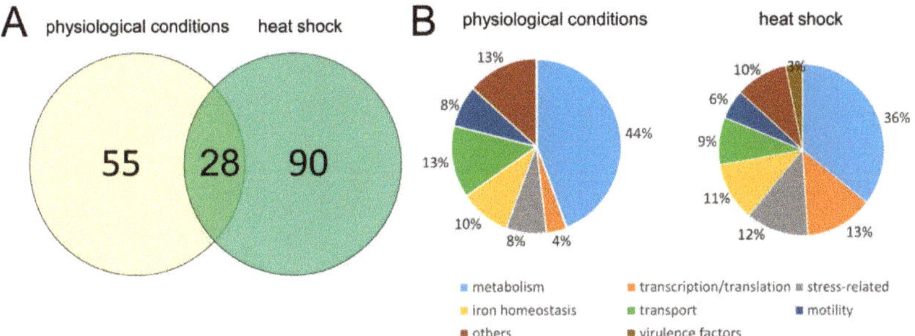

Figure 8. Number (**A**) and percentage (**B**) of proteins whose expression level was changed in the Δ*lon* mutant under physiological and heat shock conditions. Data are categorized into 8 groups depending on the protein function. Physiological and heat shock conditions refer to temperatures 30 °C and 40 °C, respectively.

Table 1. Summary of proteins whose level depends on Lon protease.

Protein		30 °C		40 °C	
Accession Number	**Name**	**x-Fold**	**Log$_2$x-Fold**	**x-Fold**	**Log$_2$x-Fold**
	Motility				
• Downregulated					
A0A2K8VVE7_9GAMM	Flagellin	0.35	−0.46		
A0A2K8VVK5_9GAMM	Protein phosphatase CheZ	0.41	−0.39		
A0A2K8W5V2_9GAMM	Methyl-accepting chemotaxis protein I (Serine chemoreceptor protein)	0.24	−0.63	0.21	−0.67
A0A2K8VVH9_9GAMM	Signal transduction histidine kinase CheA			0.34	−0.46
A0A2K8VXS1_9GAMM	Methyl-accepting chemotaxis protein I (Serine chemoreceptor protein)			0.43	−0.37
A0A2K8W5V2_9GAMM	Methyl-accepting chemotaxis protein I (Serine chemoreceptor protein)	0.24	−0.63		
A0A2K8VVG7_9GAMM	Flagellar motor switch protein FliG			0.46	−0.34
A0A2K8VVJ6_9GAMM	Positive regulator of CheA protein activity (CheW)			0.35	−0.45
	Iron metabolism				
• Downregulated					
A0A2K8VW36_9GAMM	Ferrichrome-iron receptor	0.50	−0.30	0.27	−0.56
A0A2K8VW52_9GAMM	2,3-dihydroxybenzoate-AMP ligase enterobactinsiderophore	0.30	-0.52	0.15	-0.82
A0A2K8W494_9GAMM	Nonspecific DNA-binding protein Dps/Iron-binding ferritin-like antioxidant protein/Ferroxidase			0.47	−0.32
A0A2K8VUB7_9GAMM	Ferrous iron transport protein B			0.45	−0.34
A0A2K8VW26_9GAMM	Isochorismatase enterobactin siderophore/Apo-aryl carrier domain of EntB			0.4	−0.34
A0A2K8VW34_9GAMM	Isochorismate synthase enterobactin siderophore	0.48	−0.32		
A0A2K8VW22_9GAMM	Enterobactin synthetase component F, serine activating enzyme			0.36	−0.44
• Upregulated					
A0A2K8W4W8_9GAMM	Achromobactin biosynthesis protein AcsASiderophoresynthetase superfamily, group B	7.58	0.88	3.044	0.48
A0A2K8VWX6_9GAMM	Iron-sulfur cluster insertion protein ErpA			4.90	0.69
A0A2K8W3W2_9GAMM	Ferric uptake regulation protein			9.87	0.99
A0A2K8VWQ7_9GAMM	Iron-sulfur cluster assembly scaffold protein IscU			3.67	0.56
A0A2K8W4W2_9GAMM	Achromobactin biosynthesis protein AcsD Siderophore synthetase superfamily, group A	2.77	0.44		

Table 1. *Cont.*

Protein		30 °C		40 °C	
Accession Number	Name	x-Fold	Log$_2$x-Fold	x-Fold	Log$_2$x-Fold
Stress-Related					
• Downregulated					
A0A2K8VUY7_9GAMM	Phage shock protein A	0.31	−0.51	0.35	−0.45
A0A2K8VZS1_9GAMM	Small heat shock protein IbpA	0.45	−0.35		
A0A2K8VZU2_9GAMM	Universal stress protein			0.49	−0.31
A0A2K8VUY1_9GAMM	Phage shock protein B OS=Dickeya solani			0.38	−0.41
• Upregulated					
A0A2K8VTF5_9GAMM	Protease II	4.44	0.65	2.94	0.47
A0A2K8VX21_9GAMM	Protein RecA			2.65	0.42
A0A2K8W3I9_9GAMM	ATP-dependent Clp protease proteolytic subunit			2.00	0.30
A0A2K8W3U6_9GAMM	Cold shock protein CspE			2.91	0.46
A0A2K8VZ71_9GAMM	Cold shock protein CspG			17.27	1.24
A0A2K8W1Q0	Osmotically inducible protein OsmY			2.50	0.40
A0A2K8VUA2_9GAMM	Protease HtpX			2.22	0.35
• Differentially expressed					
A0A2K8W260_9GAMM	Periplasmic protein related to spheroplast formation	0.40	−0.40	2.79	0.45
Transport					
• Downregulated					
A0A2K8W052_9GAMM	Phosphate-binding protein PstS	0.48	−0.32		
A0A2K8VVW1_9GAMM	Histidine ABC transporter, histidine-binding periplasmic protein HisJ	0.47	−0.33		
A0A2K8VU80_9GAMM	N-acetylneuraminic acid outer membrane channel protein NanC	0.27	−0.56		
A0A2K8VTK7_9GAMM	Oligopeptide ABC transporter, periplasmic oligopeptide-binding protein OppA			0.46	−0.34
A0A2K8VWQ4_9GAMM	Periplasmic substrate-binding transport protein			0.49	−0.31
A0A2K8W2K3_9GAMM	Inositol transport system sugar-binding protein	0.41	−0.39		
A0A2K8W3K7_9GAMM	Efflux pump membrane transporter	0.37	−0.43		
A0A2K8VZR0_9GAMM	Dipeptide-binding ABC transporter, periplasmic substrate-binding component			0.39	−0.41
A0A2K8W417_9GAMM	Cobalt/zinc/cadmium efflux RND transporter, membrane fusion protein, CzcB family			0.47	−0.337
A0A2K8VSS7_9GAMM	Methionine ABC transporter substrate-binding protein			0.44	−0.36
• Upregulated					
A0A2K8W0I4_9GAMM	Xylose ABC transporter, periplasmic xylose-binding protein XylF	3.06	0.49	2.85	0.45
A0A2K8VXC1_9GAMM	L-proline glycine betaine binding ABC transporter protein ProX	2.25	0.35	2.46	0.39
A0A2K8W3M0_9GAMM	Lead, cadmium, zinc and mercury transporting ATPase			3.14	0.50

Table 1. *Cont.*

Protein			30 °C		40 °C	
Accession Number	Name		x-Fold	Log$_2$x-Fold	x-Fold	Log$_2$x-Fold
	Metabolism					
• Downregulated						
A0A2K8W021_9GAMM	ATP synthase subunit delta		0.47	−0.33		
A0A2K8W4F2_9GAMM	Glutamate-1-semialdehydeaminotransferase		0.10	−1.01	0.16	−0.806
A0A2K8W580_9GAMM	Enoyl-acyl-carrier-protein reductase NADPH		0.38	−0.42		
A0A2K8W1R7_9GAMM	Alkyl hydroperoxide reductase protein C		0.40	−0.40		
A0A2K8VTW6_9GAMM	NAD(P) transhydrogenase subunit alpha		0.47	−0.33		
A0A2K8W040_9GAMM	ATP synthase epsilon chain		0.23	−0.64		
A0A2K8W444_9GAMM	6-phosphogluconolactonase		0.40	−0.40		
A0A2K8VVU5_9GAMM	NADH-quinone oxidoreductase				0.46	−0.33
A0A2K8VV27_9GAMM	Endo-1,4-beta-xylanase A		0.230	−0.53	0.13	−0.89
A0A2K8VXQ3_9GAMM	PTS system, cellobiose-specific IIB component				0.45	−0.35
A0A2K8W121_9GAMM	Biotin carboxyl carrier protein of acetyl-CoA carboxylase		0.41	−0.39		
A0A2K8W3J3_9GAMM	Cytochrome O ubiquinol oxidase subunit I				0.47	−0.32
A0A2K8VTY0_9GAMM	Superoxide dismutase [Cu-Zn]				0.38	−0.42
A0A2K8VU81_9GAMM	Sugar-binding protein		0.47	−0.33	0.37	−0.43
A0A2K8W4J6_9GAMM	Putative L-lactate dehydrogenase, Iron-sulfur cluster-binding subunit YkgF				0.36	−0.44
A0A2K8VWF7_9GAMM	Peptidyl-prolyl *cis-trans* isomerase				2.31	0.36
A0A2K8VT63_9GAMM	ATP phosphoribosyltransferase				0.45	−0.35
A0A2K8VVY8_9GAMM	Acetyl-coenzyme A carboxylase carboxyl transferase subunit beta				0.42	−0.38
A0A2K8VX73_9GAMM	Phosphoheptose isomerase				0.28	−0.55
A0A2K8W077_9GAMM	Bifunctional polymyxin resistance protein ArnA				0.20	−0.71
• Upregulated						
A0A2K8VTR1_9GAMM	3-hydroxypropionate dehydrogenase		4.17	0.62	4.24	0.63
A0A2K8VV57_9GAMM	Thioredoxin/glutathione peroxidase BtuE		3.29	0.52	5.75	0.76
A0A2K8VXQ4_9GAMM	Putative phosphatase/kinase		6.94	0.84	10.73	1.03
A0A2K8VTR5_9GAMM	SAM-dependent methyltransferase YafE (UbiE-like protein)		4.13	0.62	4.38	0.64
A0A2K8W193_9GAMM	Glyoxalase		2.42	0.38		
A0A2K8VYC9_9GAMM	Fructose-bisphosphate aldolase class II				2.80	0.45
A0A2K8W3S4_9GAMM	Thiol peroxidase, Bcp-type				2.57	0.41
A0A2K8VV22_9GAMM	Glutaredoxin				2.77	0.44
A0A2K8W3N7_9GAMM	Stomatin/prohibitin-family membrane protease subunit YbbK		2.66	0.43		
A0A2K8VXG5_9GAMM	Adenylate cyclase		2.37	0.38		
A0A2K8W3S1_9GAMM	Glycoprotein/polysaccharide metabolism		14.75	1.17	3.187	0.507
A0A2K8VVX6_9GAMM	Phosphatase YfbT				2.04	0.31
A0A2K8VYD6_9GAMM	Biosynthetic arginine decarboxylase				3.62	0.56

Table 1. *Cont.*

Protein		30 °C		40 °C	
Accession Number	**Name**	**x-Fold**	**Log$_2$x-Fold**	**x-Fold**	**Log$_2$x-Fold**
A0A2K8VSW1_9GAMM	Soluble aldose sugar dehydrogenase, PQQ-dependent PE = 4 SV = 1			2.38	0.38
A0A2K8VXS7_9GAMM	Aminotransferase	2.36	0.37		
A0A2K8VXE7_9GAMM	Sulfite reductase [NADPH] flavoprotein alpha-component	2.03	0.31		
A0A2K8VY35_9GAMM	3-isopropylmalate dehydratase large subunit			2.21	0.34
A0A2K8W254_9GAMM	Phosphopentomutase			2.60	0.41
A0A2K8VT52_9GAMM	Phosphoserine aminotransferase	2.47	0.39		
A0A2K8W0R3_9GAMM	ADP-L-glycero-D-manno-heptose-6-epimerase	4.45	0.65	5.20	0.72
A0A2K8VY36_9GAMM	3-isopropylmalate dehydratase small subunit			2.51	0.40
A0A2K8VWH1_9GAMM	Methylglyoxal synthase	2.27	0.36	2.85	0.45
• Differentially expressed					
A0A2K8W2T8_9GAMM	Phosphotransferase system, phosphocarrier protein HPr	0.49	−0.31	2.66	0.42
A0A2K8VYF1_9GAMM	Epimerase domain-containing protein	2.34	0.37	0.18	−0.75
A0A2K8W293_9GAMM	Thiol:disulfide interchange protein	0.46	−0.34	2.13	0.33
Virulence					
• Upregulated					
A0A2K8VU37_9GAMM	Various polyols ABC transporter, permease component 2			2.75	0.44
A0A2K8VUF2_9GAMM	Harpin hrpN (Harpin-Ech)			2.28	0.36
A0A2K8VUE5_9GAMM	Hrp pili protein hrpA (TTSS pilin hrpA)			10.64	1.03
Transcription/Translation					
• Downregulated					
A0A2K8VT33_9GAMM	Serine-tRNA ligase			0.47	−0.33
A0A2K8VW75_9GAMM	JmjC domain-containing protein			0.48	−0.32
• Upregulated					
A0A2K8VZG3_9GAMM	Transcriptional (Co)regulator CytR	3.47	0.54		
A0A2K8VZY6_9GAMM	DNA-directed RNA polymerase subunit omega			2.45	0.39
A0A2K8W0W8_9GAMM	50S ribosomal protein L7/L12			2.11	0.32
A0A2K8VUA3_9GAMM	Translation initiation factor 3			2.69	0.43
A0A2K8VX20_9GAMM	Ribosome hibernation protein YfiA			13.75	1.14
A0A2K8W224_9GAMM	50S ribosomal protein L27			2.03	0.31
A0A2K8W3J6_9GAMM	50S ribosomal protein L31 type B			3.42	0.53
A0A2K8VWY3_9GAMM	RNA polymerase-binding transcription factor DksA			2.07	0.32
A0A2K8VY97_9GAMM	ABC transporter, ATP-binding protein	2.45	0.39	2.04	0.31
A0A2K8VYE8_9GAMM	RNA-binding protein Hfq			2.70	0.43
A0A2K8VYF1_9GAMM	Epimerase domain-containing protein	2.34	0.37	0.18	−0.75

Table 1. *Cont.*

Protein		30 °C		40 °C	
Accession Number	Name	x-Fold	Log$_2$x-Fold	x-Fold	Log$_2$x-Fold
	Others				
• Downregulated					
A0A2K8W376_9GAMM	UPF0325 protein D083_3591	0.36	−0.45		
A0A2K8W2W2_9GAMM	Incl1 plasmid conjugative transfer protein TraF	0.30	−0.53	0.26	−0.58
A0A2K8W4R1_9GAMM	Uncharacterized protein	0.45	−0.34	0.33	−0.48
A0A2K8VTN2_9GAMM	Uncharacterized protein	0.41	−0.38		
A0A2K8VV11_9GAMM	Major outer membrane lipoprotein			0.50	−0.30
• Upregulated					
A0A2K8W3L1_9GAMM	Putative membrane protein OS=Dickeya solani	175.34	2.24	117.58	2.07
A0A2K8VWZ4_9GAMM	S-ribosylhomocysteinelyase			2.03	0.31
A0A2K8VZT1_9GAMM	Putative membrane protein			7.12	0.85
A0A2K8VTR0_9GAMM	Putative secreted protein			5.63	0.75
A0A2K8VT84_9GAMM	Uncharacterized protein	5.23	0.72	6.61	0.82
• Differentially expressed					
A0A2K8W4I6_9GAMM	Uncharacterized protein	4.19	0.62	0.33	−0.48

3. Discussion

For successful infection, a pathogen must have the capability to enter the host, overcome the host defense systems, acquire nutrients, multiply and disseminate. All these stages are associated with constant exposure to a variety of potentially harmful conditions, both in and outside the host. Hence, successful pathogens should have well-developed virulence mechanisms but also efficient stress response systems. Proteolytic enzymes were shown to play numerous crucial roles in bacterial virulence. They can directly act as virulence factors, but also can contribute to virulence by regulating the production of virulence factors and/or as components of protein quality control systems to provide cellular proteostasis. One of the latter cases is the Lon protease which is indispensable for stress tolerance and virulence of many bacterial species causing infectious diseases.

Our work showed that the Lon protease is necessary for the bacterium *D. solani* to resist exposure to stress, including ionic- and nonionic osmotic stress, as well as high temperature. This finding is in agreement with data obtained for other bacterial species. Lon has a well-documented role in bacterial viability under heat and salt stress [14]. In *E. coli*, expression of the *lon* gene depends on sigma32 (RpoH) transcription factor [27,28], activated under heat shock and osmotic stress [29,30]. High temperature stimulates expression of *lon* in *E. coli* [31] and *Francisella tularensis* LSV [32]. Ionic osmotic stress is responsible for the elevated level of *lon* expression in *B. subtilis* and *Dickeya dadantii* [33]. Consistently, in *D. solani*, *lon* expression was significantly elevated in exponentially growing bacteria following exposure to elevated temperature; a positive trend was also observed in case of salt stress. Heat shock is more harmful to bacteria in the logarithmic than the stationary growth phase [34], which may explain stronger stimulation of the *lon* gene in the logarithmically growing cells. Higher demand for Lon, suggested by the upregulation of *lon* under certain stress conditions, can explain reduced growth of the Δ*lon* bacteria under thermal and osmotic stress. Additionally, the elevated level of stress-related proteins in proteomes of bacteria treated with 40 °C, revealed by the SWATH-MS analysis, indicates the higher stress level in the mutant cells than the WT. Increased expression of RecA and protease HtpX suggests a higher frequency of DNA damage and probably impaired integration of membrane proteins, respectively, according to data published for *E. coli* [35,36]. Finally, the increase in the abundance of the second important cytosolic protease, ClpP, in the deletion strain, reveals the essential role of Lon protease in the quality control proteolysis in the cytoplasm. Most probably, ClpP takes over some of the Lon functions. However, it should be noted that the Δ*lon* strain showed a temperature-sensitive (TS) phenotype, so ClpP cannot substitute for Lon under heat shock conditions.

To our surprise, the Δ*lon* mutants were particularly vulnerable to treatment with sucrose. Nonionic osmotic agents, like sucrose, are considered less harmful for a cell than the ionic ones, (e.g., NaCl) [37]. We did not find data regarding the relationship between Lon and the response to osmotic stress caused by high sucrose in any bacterial species. Hence, this important function of Lon in resistance to nonionic osmotic stress needs to be elucidated. Although the expression of *lon* was strongly upregulated in the response of *D. solani* to low pH, we did not observe differences of growth between the mutant and WT strains. Possibly, the ClpP protease or other component of the protein quality control system takes over the duties of Lon under this type of stress. The involvement of the Lon protease in resistance to acidic stress is rather poorly investigated across different bacterial species. In *E. coli*, Lon is responsible for the degradation of the activator of acidic resistance, GadE, playing a role in the termination of the stress response [38]. Additionally, *S. enterica* serovar Typhimurium requires this protease to successfully cope with low pH [16]; however, the precise mechanism was not provided. Interestingly, the closely related species, *D. dadantii*, showed repression of *lon* expression in the low pH medium (although not statistically significant) [33], which is opposite to our findings. This may reflect interspecies differences but also certain differences in experimental design. As with *D. dadantii* [33], in *D. solani* the expression of *lon* was not affected by oxidative stress-induced with hydrogen peroxide. Hence, in Dickeya, the oxidation response most probably involves other components of PQCS.

Production of functional virulence factors is frequently dependent on specific proteolytic activity in the cell [39,40]. To verify the involvement of Lon in *D. solani* virulence, we checked the activity of the most abundant secreted virulence factors. We found that activity of the extracellular pectinases was reduced in the case of Δ*lon* mutant. Pectinases constitute a heterogeneous group of proteins. They differ in substrate specificity, abundance and role in virulence but all are secreted via T2SS [41]. At least 10 pectinases produced by *D. solani* have been identified so far [41]. The commonly used tests (including the one used in this work) measure a total pectinase activity and do not allow to distinguish between the individual pectinases. Analysis of the Δ*lon* and WT *D. solani* proteomes did not reveal differences in the cellular content of pectinases. However, we do not know if these proteins were efficiently transported outside the cell. As this is the first report of the function of Lon in the soft rot bacteria, there is no information regarding the relationship between Lon and PCWDE. No literature data is indicating the possibility of regulating the T2SS transport system by Lon. Moreover, the activity of extracellular cellulases, also T2SS dependent, remained unchanged in the *lon* mutant. Hence, the involvement of Lon in the regulation of T2SS is unlikely. Thus, further research is needed to clarify the Lon-dependent protease regulation of the secreted pectinase activity.

We did not observe any changes in the production of siderophores, although the lack of Lon protease lowered the abundance of several proteins engaged in iron metabolism. A higher level of proteins with function in iron-sulfur (Fe-S) protein biogenesis (IscU, ErpA) in the deletion strain may indicate them as potential substrates for the Lon protease. That is true in *Saccharomyces cerevisiae*, where a Lon homolog, Pim1, degrades Isu, a homolog of IscU [42]. The increase in the amount of the negative transcriptional regulator Fur is very interesting. Whether it is degraded by Lon is not known and no such Lon function was found in other bacteria. Almost a 10-fold increase of the Fur level in Δ*lon* may explain the decreased amount of certain Fur-dependent proteins involved in the synthesis of siderophores (e.g., enterobactin synthetase component F in *E. coli* [43]. On the other hand, the other enzymes from the siderophore biogenesis pathway were upregulated (like achromobactin biosynthesis protein AcsD), presumably compensating for the downregulated components to maintain iron homeostasis in the mutant cells.

A lack of Lon exerted a significant impact on the mobility of *D. solani*. We demonstrated that the cells deprived of Lon showed impaired swarming and swimming motility. This can be explained by the reduced levels of flagellin and positive regulators of chemotaxis in *D. solani* Δ*lon*, as revealed by the proteomic analysis. Depending on the bacterial species, the effects of the *lon* mutations on bacterial motility may be radically different. On one hand, the *lon* mutation can cause stimulation of motility. Good examples are *Proteus mirabilis* and *B. subtilis*, in which *lon* mutants showed better swarming [44,45]. In these bacteria, Lon degrades master activators of flagellin biogenesis- FlhD and SwrA, respectively. Hence, in the *lon* backgrounds, these activators became stabilized, leading to a hypermotile phenotype. In contrast, the *lon* mutant of *Erwinia amylovora* was characterized by a nonswarming phenotype [46]. In *E. amylovora*, a mutation in the *lon* gene resulted in the accumulation of RcsA/RcsB that negatively regulates transcription of *flhD*, the master regulator of flagellar biosynthesis. Finally, a lack of Lon may not affect bacterial motility at all, as shown for *S. entrica* serovar Typhimuirum [47]. The results obtained in this work suggest an indirect role of Lon in the motility of *D. solani*, analogous to that of *E. amylovora* Lon, as the *flhD* gene is present in *D. solani* genome and the *lon* mutant was characterized by a decreased flagellin content. Interestingly, *D. solani* Δ*lon* was characterized by the two-fold increased level of the CytR transcription factor, which in *Pectobacterium carotovorum* positively stimulates genes associated with motility: *fldH*, *fliA*, *fliC* and *motA* [48]. However, the increased content of the CytR protein in *D. solani* Δ*lon* obviously was not sufficient to compensate for the other Δ*lon* -dependent effects that lead to a reduced flagellin and chemotaxis protein levels, or CytR is not involved in regulation of motility in *D. solani*.

The phenotypes of *D. solani* Δ*lon* reported in this work suggested that the presence of Lon may be necessary for efficient infection of the potato plant. Indeed, we found that the process of development of blackleg symptoms in the plants infected with the mutant strain was markedly delayed in respect to infection with the WT *D. solani*. On the other hand, Lon was not essential for the maceration of plant tissues in vitro. Both types of infection tests differ fundamentally in terms of the availability of plant tissues for bacteria. In the whole plant model, the bacteria were placed in the soil, from where they must have got into the wounded tissue, in this case, roots. In this context, the motility and chemotaxis toward chemical signals (e.g., jasmonic acid) are crucial. Consequently, the nonmotile mutant strains are characterized by a lack or reduced virulence, as they may encounter severe problems with entering and/or spreading in the host [49]. In the slice or leaf model, bacteria were spotted directly into the wounded tissue, so chemotaxis and motility were less important. In the tissue model, the basis of infection's success lies in the production of PCWDE, iron homeostasis, and bacterial fitness under pH, oxidative and osmotic stresses (reviewed in [6]). As we did not observe any difference in the degree of maceration of the tuber or leaf tissues between the WT and mutant strains, we assumed that Lon was not essential for bacterial survival in the plant under experimental conditions. The secreted pectinase activity of the Δ*lon* mutant was reduced but was still enough for efficient plant maceration. Hence, we concluded that the observed delay in the potato plant infection process was most probably due to reduced motility of the Δ*lon* strain.

Lon is known to be engaged in the regulation of T3SS, however, the particular effects of *lon* mutations are species-dependent. In *P. aeruginosa* and *Yersinia*, the deletion of the *lon* gene results in the downregulation of T3SS [17,50]. However, the opposite effect was demonstrated in *E. amylovora* and *P. syringae* [46,51]. We found that in *D. solani*, Lon negatively affected the level of proteins associated with the type III secretion system: HrpN and a homolog of HrpW as well as HrpA (structural protein of T3SS pillus). These results are consistent with data obtained for *E. amylovora* and *P. syringae*. In these bacteria, Lon indirectly downregulates transcription of *hrpL* gene coding for HrpL, the RNA polymerase sigma factor, which is necessary for the initiation of transcription of T3SS genes. In *P. syringae* and *E. amylovora* this is mediated via degradation of the transcriptional activators of the *hrpL* gene, HrpR and HrpS, respectively [46,51]. In addition, Lon also indirectly regulates HrpS levels through RcsA proteolysis in the *E. amylovora* cells. RcsA is a component of RcsA/RcsB regulatory complex, which activates transcription of the *hrpS* gene [46]. Finally, Lon of *E. amylovora* degrades HrpA [46], which may also be true in the case of *D. solani*, as we observed an elevated level of this protein in the mutant strain. Interestingly, in *E. coli*, the Lon substrates, CspG and CspE proteins, positively regulate the *rcsA* expression [52]. In the *D. solani* Δ *lon* strain, CspG and CspE were upregulated. It cannot be ruled out that they can also be degraded by the Lon protease, which would additionally suppress the expression of T3SS.

In addition to the IscU, HrpA and CspG/E proteins discussed above, the proteome analysis revealed one more potential substrate for the *D. solani* Lon protease. We observed elevated levels of the RNA-binding protein Hfq, which is a known substrate for Lon in *P. aeruginosa* [53]. Fernandez and colleagues [53] suggested that the accumulation of Hfq can contribute to reduced motility of the *lon* strain. In the case of *D. solani*, these findings need verification.

The most pronounced effect of the *lon* deletion was observed in the case of the protein A0A2K8W3L1_9GAMM, a homolog of ComEA, whose level was more than 100-fold higher in Δ*lon*. ComEA is necessary for natural cell competence. However, *D. solani* was not reported to exhibit natural competence. Moreover, the calcium chloride transformation method of *D. solani* is highly inefficient and *D. solani* spp. genomes lack in general large plasmids [54]. Interestingly, in *Vibrio cholerae comEA* expression is activated by the transcriptional regulator CytR [55]. We also observed an increased level of CytR in the deletion strain, which may be the reason for the upregulation of the ComEA protein. However, the relationship between CytR and Lon protease requires further investigation.

Finally, SWATH-MS analysis revealed that the deletion of *lon* affected the balance among proteins involved in cellular metabolism and transport. This is not surprising as the housekeeping proteases, in general, regulate metabolic activities [56] and this is also true for Lon [14].

In light of the data presented in this work, the Lon protease is a protein that plays very important roles in *D. solani* physiology, both under physiological and stressful conditions. Lon was shown to be required for the full virulence of *D. solani* in the whole plant model. Lower pathogenicity of the Δ*lon* bacteria may result from impaired expression/activity of certain virulence factors, including motility and secreted pectinases, but also from decreased ability to withstand stressful conditions. To our knowledge, this is the first report that describes the function of the Lon protein in the bacterial species from the SRP group.

4. Materials and Methods

4.1. Materials

Restriction enzymes and dNTPs were purchased from EURx (EURx Sp. z o.o., Gdańsk, r, Poland); PrimeSTAR GXL polymerase for construction of deletion strain from Takara Bio Inc. (Shiga, Japan); T4 DNA ligase, T5 Exonuclease and Phusion High-Fidelity polymerase from New England Biolabs (USA) and reagents for media and buffers from Sigma-Aldrich (Saint-Louis, MI, USA), and Chempur (Piekary Śląskie, Poland). Oligonucleotides were synthesized by Eurofines Scientific (Luxembourg, Luxembourg) or Sigma-Aldrich (Saint Louis, MI, USA).

4.2. Bacterial Growth Conditions

Bacterial strains and plasmids used in this study are listed in Table 2. Bacteria were grown in the minimal medium M63Y (0.1 M KH_2PO_4, 15 mM $(NH_4)_2SO_4$, 9 µM $FeSO_4$, 1 mM $MgSO_4$, 1 mg/L vitamin B1 and 0.3% glycerol, pH = 7.0) [57], LB broth (1% tryptone, 0.5% yeast extract, 1% NaCl) or SOC (2% tryptone, 0.5% yeast extract, 10 mM NaCl, 20 mM glucose, 2.5 mM KCl, 10 mM $MgSO_4$) with shaking at 30 °C, unless indicated otherwise. For all analyses, overnight cultures were diluted 1:50 with M63Y or LB and cultured for the next 16 h until they reached the early stationary growth phase. Then, they were used in experiments or diluted again 1:50 with M63Y and grown for 4.5 h to reach a midexponential phase. The overnight *D. solani* Δ *lon* cultures were grown in the medium supplemented with kanamycin (0.1 mM); the cultures directly subjected to experiments were devoid of the antibiotic to provide comparable growth conditions of all bacterial strains.

Table 2. Bacterial strains and plasmids used in this study.

Strain	Genotype	Reference or Source
Escherichia coli DH5α	F– φ80lacZΔM15 Δ(lacZYA-argF)U169 recA1 endA1 hsdR17(rK–, mK+) phoA supE44 λ– thi-1 gyrA96 relA1	[58]
Escherichia coli DH5α *pir*	sup E44 ΔlacU169 (ΦlacZΔM15) recA1 endA1 hsdR17 thi-1 gyrA96, relA1 λpir phage lysogen	[59]
Escherichia coli MFD *pir*	MG1655 RP4-2-Tc::[ΔMu1::aac(3)IV-ΔaphA-Δnic35-ΔMu2::zeo] ΔdapA::(erm-pir) ΔrecA	[60]
Dickeya solani IPO 2222	WT	[61]
D. solani IPO 2222 Δ*lon*	Δlon	This work
D. solani IPO 2222 Δ*lon/lon*	Δlon/lon	This work
Plasmids	**Feature**	**Reference or Source**
pDOC-C	pEX100T, Sce1 -Sce1 *sacB* Amp^R	[62]
pDOC-K	pEX100T, Sce1-Kan^R -Sce1 *sacB* Amp^R	[62]
pACBSCE	I-Sce1 λ-Red Cm^R	[62]
pDFDOC-C-lon	pDOC-C Sce1-Kan^R-Sce1	This work
pRE112	pRE107 cm^R *sacB*	[63]
pmScarlet	pMB1 ori mScarlet Amp^R	[64]
pLonScar	pRE112 *lon mScarlet*	This work

Please define all abbreviations in table footer, if appropriate.

Growth curves were determined with the use of the EnSpire plate reader (PerkinElmer, Waltham, MA, USA) in a 24-well nontreated plate (#702011 Wuxi NEST Biotechnology Co., Ltd., Wuxi, China). Overnight grown cultures were diluted 1:50 with LB medium to a final volume of 1 mL. Bacteria were grown with orbital shaking (120 rpm) and OD_{595} measurements were taken every hour. Growth was monitored for 20 h at 30 °C. The final OD values were averaged for four biological replicates.

To induce stress, 4 µL aliquots of 10-fold serial dilutions of the stationary bacterial cultures in Ringer buffer (147 mM NaCl, 4 mM KCl, 2.24 mM CaCl$_2$ × 2H$_2$O) were spotted onto the LA agar plates (LB broth with 1.5% agar, control conditions and temperature stress), LA agar plates supplemented with 0.6 M sucrose (nonionic osmotic stress) or 0.3 M NaCl (ionic osmotic stress), or adjusted with malic acid to pH 5.0 and incubated for 20–48 h at 30 °C or 39 °C (heat shock). The Agar disk diffusion method was used to study the susceptibility of bacteria to oxidative stress. One-hundred microliters of overnight culture diluted 100 times was spread onto the LA medium. A sterile disk (6mm) of Whatman 1M paper was placed on an LA plate and then 8 µL of 1% hydrogen peroxide solution was spotted on it. The water-soaked disk served as a negative control. Plates were incubated at 30 °C for 24 h.

To analyze gene expression, the WT strain grown in M63Y to the midexponential or early stationary phase was subjected to the selected stress conditions for 15 min [33]. Briefly, NaCl and sucrose were added to the cultures to a final concentration of 0.3 M and 0.32 M, respectively. The shift of temperature was obtained by incubation of bacteria for 15 min in a water bath at 37 °C or 40 °C with shaking. As a control, bacteria grown in the absence of a stressor at 30 °C were used. To stabilize mRNA, a cold solution of 5% acid phenol (BioShop Canada Inc., Ontario, Canada), in 99.9% ethanol was added to the bacterial culture at the 1:9 ratio and bacteria were immediately put on ice.

4.3. Construction of the lon Deletion Strain

The deletion of the *lon* gene from the *D. solani* chromosome was performed according to gene doctoring protocol [62]. Briefly, primers with homology to the upstream/ downstream regions of the kanamycin resistance cassette from pDOC-K plasmid and 40 bp flanking region of the *lon* gene were designed (Table 3). Restriction sites for XhoI and KpnI were added at 5′ end of lonkan L and lonkan R primers, respectively. PCR reaction was performed with pDOC-K as a template. The PCR product and pDOC-C plasmid were digested with XhoI and KpnI restriction enzymes, then the PCR product was cloned into the backbone of pDOC-C, generating pDFDOC-C-lon. Plasmid pABSCE and pDFDOC-C-lon were electroporated into *D. solani* cells and the transformants were selected for resistance to chloramphenicol and ampicillin. The proper recombinants were selected based on a lack of ability to grow on a medium with 8% filtrated sucrose, as the pDFDOC-C-lon plasmid contains the *sacB* gene. To check this, 1 mL portions of LB medium with 0.5% glucose and appropriate antibiotics were inoculated with single colonies and incubated with shaking at 30 °C for 4 h. The culture was centrifuged (1167× *g*, 2 min), pellet resuspended in 1 mL LB medium supplemented with 0.1–2% arabinose and incubated at 30 °C with shaking until turbid. Bacteria were spotted on the LA plates supplemented with 8% sucrose (sterilized by filtration) and kanamycin and in parallel on LA with kanamycin. Next, the colonies that did not grow on the sucrose plates were tested for pDFDOC-C-lon and pABSCE plasmid loss by the selection of bacteria unable to grow on ampicillin and chloramphenicol. PCR with primers homologous to the flanking region of a *lon* gene (Table 3) was performed to confirm the deletion of the *lon* gene.

Table 3. List of primers and their characteristics.

Primer	Primer Sequence 5′-3′	Amplified DNA
lonkan L	CAGGGTACCTTCCCTTAACCTGGCGGAAACGAAACTAAGAGAGAGCTCTGACCGGTCAATTGGCTGGAG	kanamycin resistance gene with added sequences flanking the *D. solani lon* gene amplified from pDOC-K
lonkan R	GCACACTCGAGCCAGCCTTTTT TTCTCAGTGGTTTTTGCGATAGGTCACTAATATCCTCCTTAGTTCC	
lonsolani L	CGATTACCTATAGGCGAAACC	lon and kanamycin resistance gene amplified from *D. solani* and *D. solani* Δ *lon* gDNA, respectively
lonsolani R	CAGGCTCAACAGTGCTCTAAC	
1 L	AGTGAACTGCATGAATTCCCGTTGATCCAGATCTTGCGCGA	500 bp upstream from the start codon of *lon* gene amplified from *D. solani* gDNA
1 R	GTTCGGAACGCTCAGGGTTCATAGAGCTCTCTCTTAGTTTCGTTTCC	
2 lon L	ATGAACCCTGAGCGTTCCGAA	*lon* gene amplified from *D. solani* gDNA
2 lon R	CACGTTTCACTTTCCGGGTTCCTATTTTTTGGCTACCGACTTCAC	
3 scarlet L	GAGACCCGGAAAGTGAAAACGTG	*mScarlet* gene amplified from pmScarlet
3 scarlet R	TTACCGCCTTTGAGTGAGCTG	
4 L	CAGCTCACTCAAAGGCGGTAATGACCTATCGCAAAAACCAC	500 bp downstream from the stop codon of *lon* gene amplified from *D. solani* gDNA
4 R	ATGCGATATCGAGCTCTCCCAAAACCGTCCCACCTCAGATT	

4.4. Single-Copy Complementation

Complementation strain was obtained by reintroduction of the WT *lon* gene into its native locus on the chromosome in the *D. solani* Δ*lon* cells using conjugation strain *E. coli* MFD *pir*, according to [60]. To do this, a plasmid containing the WT *lon* gene with the *mScarlet* gene coding for a fluorescent pink protein as a marker was obtained by the Gibson assembly approach. Briefly, four insert fragments were amplified (Table 3) and mixed with the allelic exchange pRE112 vector cut with the SmaI restriction enzyme, and reaction mix [65]. pRE112 carries the *sacB* marker gene and chloramphenicol resistance gene. The total amount of DNA in the reaction was 150 ng.

The resulting reaction product was transformed into *E. Coli* DH5α *pir* and the subsequently isolated plasmid was named pLonScar. The pLonScar plasmid was introduced into *E. coli* MFD *pir* in the presence of 0.3 mM diaminopimelic acid (DAP, Sigma-Aldrich, Saint Louis, MI, USA) in the medium to allow the growth of bacteria [60]. The overnight cultures of *E. coli* MFD *pir* (pLonScar) and *D. solani* Δ*lon* were mixed in a 3:1 ratio (total volume 800 µL) and centrifuged (1677× g, 2 min). The pellet was suspended in 30 µL of LB and spotted on an acetate cellulose filter placed on the LA solid medium supplemented with chloramphenicol but without DAP to eliminate *E. Coli* MFD *pir*. After 24 h incubation at 30 °C, bacteria were recovered by shaking the filter in 1 mL M63Y. Bacteria were spotted onto the LA agar plates with chloramphenicol. Then, individual colonies were tested for loss of the pRE112 plasmid. Briefly, cultures in the middle logarithmic growth phase were serially diluted and spread on LA without NaCl but supplemented with 10% of 0.22 µL filtered sucrose. Only cultures unable to grow on medium with sucrose were subjected for further verification. Ultimately, pink colonies sensitive to chloramphenicol and kanamycin were considered as true recombinants.

4.5. Plasmid and Genomic DNA Purification

Genomic and plasmid DNA were isolated using Genomic Mini (A&A Biotechnology, Gdynia, Poland) and Plasmid Mini (A&A Biotechnology, Poland) kits, respectively, according to the manufacturer's protocols.

4.6. Preparation of Electrocompetent Cells and Electroporation

50 mL of SOC medium was inoculated with an overnight culture of *D. solani* at a 1:50 ratio. Bacteria were grown until OD_{595} of 0.45–0.5. The culture was centrifuged (7 min, 5063× g, 20 °C), pellet resuspended in 50 mL of deionized water, mixed thoroughly and forwarded to centrifugation (8 min, 5872× g, 20 °C). The cells were suspended in 25 mL of deionized water, mixed and centrifuged again (as above). The supernatant was precisely discarded, bacteria suspended in 1 mL of deionized water and split into 60 µL portions. Then, the cells were immediately used for electroporation [66]. Briefly, up to 50 ng of DNA was mixed with electrocompetent cells and transferred to 0.1 cm gap electroporation cuvettes (room temperature) for electroporation (1.25 kV). The bacterial suspension was diluted with 1 mL of SOC medium and incubated for up to 3 h for recovery. The 100 µL aliquots and the remaining bacteria (after 1 min 1677× g centrifugation and resuspension in 100 µL SOC) were plated onto the LA solid medium supplemented with an appropriate antibiotic and incubated for 24–36 h at 30 °C.

4.7. In Vivo Infection of the Potato Plants

The pot grown potato plants were obtained from sprouts. Briefly, the potato tubers of cultivar Vineta, obtained locally in Gdańsk, Poland, were stored in the dark until the development of sprouts (app. 3–4 months). Sprouts of a length of ca. 5 cm were carefully removed from tubers, planted into 7 cm square pots with potting soil (COMPO SANA ca 50% less weight) and placed on the windowsill for rooting and shoot development. After approximately two weeks, the rooted green plants were transferred to the humid growth chamber and grown under the white fluorescent light (48 × 5 W,

Mars Hydro Reflector 48 with 16:8 h light: dark photoperiod). The potato plants at least 10 cm high were subjected to pathogenicity tests. Four overnight cultures of WT and four of *D. solani* Δ*lon* grown in LB medium were diluted with Ringer buffer to OD_{595} ~0.125, corresponding to 10^8 CFU/mL. The roots of each potato plant were wounded with the scalpel about 2 cm from the stem. Plants were watered with 30 mL of bacterial suspensions and left for an hour. Then, the filtrate was discarded. Each bacterial culture was used to infect 4 plants. As a negative control, four plants treated with Ringer buffer were used. The experiment was carried out for 17 days and the percentage of plants with blackleg symptoms was estimated.

4.8. Pathogenicity on Potato Tubers, Chicory and Chinese Cabbage Leaves.

CFU/mL of each overnight bacterial culture was normalized to 10^8 with Ringer buffer and then 10-fold serially diluted. Potato tubers were sterilized with a 10% bleach solution for 20 min, then submitted to three washes with sterile deionized water for 20 min each. The tubers were cut into 1 cm thick potato slices; in each slice, a little hole was pierced with a sterile pipette tip. Chicory and Chinese cabbage leaves were washed with sterilized deionized water and incised with a sterile scalpel. Ten-microliter aliquots of bacterial culture of 10^7 CFU/mL were spotted onto the plant material. Ringer buffer was used as a negative control. The infection assays were performed in the humid boxes, at 30 °C for up to two days: one day for Chinese cabbage, two days for chicory leaves and for potato slices.

4.9. Determination of Motility

For swimming, a single bacterial colony (five replicates per strain) was inoculated into the semisolid agar plate with 0.3% MMA medium (40 mM K_2HPO_4, 22 mM KH_2PO_4, 0.41 mM $MgSO_4 \times 7H_2O$, 0.3% agar) supplemented with 1 mM galactose. The plates were incubated under aerobic conditions at 30 °C for 48 h. The diameter of the bacterial spreading area was measured.

To monitor swarming motility, a single bacterial colony (five replicates per strain) was inoculated into the plate with 0.5% TSA (tryptone soy broth) medium (Oxoid, Basingstoke, UK) supplemented with 0.5% agar). Plates were incubated under aerobic conditions at 30 °C for 12 h. Both tests were repeated two times.

4.10. Determination of Secreted PCWDE Activity

The measurement of pectinolytic activity was performed as described in [67]. Briefly, bacteria were cultured in the M63Y medium until an early stationary phase and centrifuged (13,148× *g*, 2 min). Then, 260 μL aliquots of supernatant were diluted with equal volumes of distilled water. Briefly, 500 μL samples of the diluted supernatant were mixed with 1.5 mL of PGA (polygalacturonic acid, Sigma-Aldrich, Saint Louis, MI, USA) buffer (100 mM Tris–HCl (pH 8.5), 0.35 mM $CaCl_2$ and 0.24% sodium polygalacturonate) warmed up to 30 °C. The reaction consisting in the formation of unsaturated products from polygalacturonate [68] was monitored spectrophotometrically by measurement of increase of absorbance at 232 nm for 2 min, every 30 s. Absorbance values obtained for PGA buffer were subtracted from values obtained for unsaturated product. The spectrophotometer was calibrated with distilled water. Pectynolytic activity was presented as $\Delta A235/min/mL/OD_{595}$. The experiment was repeated two times for each strain with at least three replicates.

The extracellular cellulase activity was assayed as described in [69]. Briefly, bacteria grown in the M63Y medium to the stationary phase were diluted with Ringer buffer to 10^8 CFU/mL. Seven-microliter aliquots of bacterial cultures were spotted on the agar plates with carboxymethyl cellulose CMC (M63Y medium supplemented with 1.5% agar and 1% CMC). The plates were incubated at 30 °C for 48 h and then subjected to staining with 2% Congo red solution for 20 min. The diameters of the arisen halo were measured. The experiment was repeated two times for each strain with five replicates.

To measure the extracellular protease activity, bacteria cultivated in the M63Y medium to the stationary phase were diluted with Ringer buffer to 10^8 CFU/mL. Seven-microliter aliquots of bacterial cultures were spotted on the milk agar plates (the LA medium supplemented with 5% skimmed milk). Plates were incubated at 30 °C for 48 h and the diameters of the arisen halo were measured. Each strain was tested two times with five replicates.

4.11. Siderophore Activity Assay

Ten-microliter aliquots of supernatants from the stationary cultures (grown in M63Y medium for 16 h) were spotted on the chrome azurol S-agar plates [70]. To prepare the chrome azurol S-agar medium the following solutions were prepared: (1) main medium, (2) 10% deferrated casamino acids (CAS), (3) 0.1 M $CaCl_2$, (4) filtered 1 mM $FeCl_3 \times 6H_2O$ in 10 mM HCl, (5) CTAB (cetrimonium bromide, Sigma-Aldrich, Saint Louis, MI, USA). To prepare the main medium solution the components, KH_2PO_4 (3 g), NaCl (0.5 g), NH_4Cl (1.0 g), $MgSO_4 \times 7H_2O$ (0.2 g), sucrose (4.0 g) and agar (15.0 g), were dissolved in 850 mL of 0.5 M Tris–HCl, pH 6.8 and sterilized. The deferrated casamino acids were prepared by the removal of ferrous ions with 3% 8-hydroxyquinoline in chloroform. The sterilized main solution was supplemented by 30 mL of 10% deferrated casamino acids, 10 mL of 0.1 M $CaCl_2$, 50 mL of 0.08 mM CAS, 10 mL of 1 mM filtered $FeCl_3 \times 6H_2O$ in 10 mM HCl and 40 mL of 2 mM CTAB (CAS, $FeCl_3$ and CTAB were mixed before adding to solution). Plates were incubated at 30 °C for 1 h and the intensity and diameter of the orange halo were compared. The experiment was performed two times for five replicates for each strain.

4.12. RNA Extraction

Bacterial RNA was extracted using the Total RNA Mini Plus RNA extraction kit (A&A Biotechnology, Gdynia, Poland) according to the manufacturer's instructions. The quantity and quality of the RNA samples were confirmed by measurement of absorbance at 260 nm and evaluation of A260/A280 (~2) and A260/A230 (>2) ratios, and by agarose gel electrophoresis. Samples of 5 µg of RNA were subjected to DNase treatment (A&A Biotechnology, Poland) by incubation of 20 µL reaction mixtures in the presence of DNase (1U/µL) at 37 °C for 25 min followed by incubation at 75 °C for 10 min. The samples served as a template for the reverse transcription reaction.

4.13. Reverse Transcription

cDNA was transcribed from 1.5 µg of RNA with the use of RevertAid First Strand cDNA Synthesis Kit (Thermo Fisher Scientific, Waltham, Massachusetts, USA), according to manufacturer's protocol. Obligatory step of denaturation of RNA with random hexamer primers mixture at 65 °C for 5 min was added.

4.14. Quantitative Real-Time PCR (qPCR)

qPCR analysis was performed as described in [23]. Briefly, diluted cDNA samples in a 1:2 ratio were used as qPCR templates. The qPCR reactions were carried out using the LightCycler 96 instrument (Roche Diagnostics, Rotkreuz, Switzerland). Primer3 software was used to design primers [71] (Table 4). Ten-fold dilution series of genomic DNA templates isolated from *D. solani* IPO 2222 were used to estimate the amplification efficiency of each pair of primers. qPCR reaction was carried out with FastStart Essential DNA Green Master (Roche Diagnostics, Rotkreuz, Switzerland). A 20 µL qPCR reaction mixture contained 0.5 µL of cDNA, 3–4.5 pmol of forward and reverse primers, 10 µL of PCR Mix. Thermal cycling parameters were as follows: preincubation at 95 °C for 5 min; 35–50 cycles of amplification and quantitation at 95 °C for 15 s, 62 °C for 20 s and 72 °C for 16 s. At the end of each cycle, melting curve analysis was performed (95 °C for 10 s, 65 °C for 60 s and 97 °C for 1 s). All qPCR reactions were performed for three biological replicates, with three technical repeats, negative no template control (NTC) and no-reverse transcriptase (NRT) controls. Cq (quantification cycle) values were averaged. The 16s rRNA gene was selected for normalization as it showed stability under all tested conditions. Pfaffl-ΔΔCT method with correction for PCR efficiency was used for the determination of the relative expression of the *lon* gene [72]. Statistical analysis was performed with the use of REST2009 software (v. 2009, Qiagen, Hilden, Germany) [73,74].

Table 4. Characteristics of primers used in gene expression analysis.

	FWD Primer Sequence (5′-3′)	REV Primer Sequence (5′-3′)	Amplicon Length [bp]	PCR Efficiency	R^2	Concentration [μM]
lon	TGGTCATTCCGTTGTT TGTTGGTC	CATCCGTTGAG GCTTCTTTCTGTG	111	1.97	1.0	0.3
16S rRNA	GCTCGTGTTGTGA AATGTTGGGTT	GCAGTCTCCCT TGAGTTCCCAC	94	1.96	1.0	0.225

4.15. Protein Electrophoresis and Immunodetection

SDS page electrophoresis and Western blotting were performed as described in [75,76]. Then, 7.5% polyacrylamide gels were used. Briefly, the Lon protein was detected with the anti-*Escherichia coli* Lon rabbit antibodies (#40219-T24, Sino Biological Inc., Beijing, China) at dilution 1:2000 followed by incubation with HRP conjugated secondary anti-rabbit antibodies (#31462 Thermo Fisher) diluted 1:50,000. Chemiluminescent signal was developed using a luminol/ p-coumaric acid (Carl Roth GmbH + Co. KG) mix (4 mL of 1.41 mM luminol, 400 μL of 6.7 mM p-coumaric acid in DMSO, 4 μL of 30% H_2O_2) and was recorded by Azure Biosystems c600 (Dublin, California, USA) imaging system.

4.16. Sample Preparation for Mass Spectrometry

The stationary growth phase cultures of *D. solani* cultivated in M63Y were subjected to high-temperature stress. Briefly, the cultures were transferred from 30 °C to 40 °C and incubated for 30 min with shaking. For control conditions, bacteria were cultivated at 30 °C. Five biological replicates of each strain were pooled and centrifuged (7000× *g*, 2 min). The pellets were lysed with the solution containing 4% SDS, 100 mM Tris/HCl pH 7.6, 0.1 M DTT (lysis solution) and incubated at 95 °C for 10 min. After cooling, cold acetone was added to the solution to precipitate the released proteins. The samples were incubated at -20 °C for about 2 h and then centrifuged for 20 min 20,000× *g*. The supernatant was decanted and the precipitate dried. The pellet was then dissolved in 8 M urea in 0.1 M Tris/HCl pH 8.5 [77].

4.17. Protein Digestion

First, the protein concentration was measured by measuring absorbance at 280 nm (MultiskanTM Thermo, Waltham, Massachusetts, USA) using the μDrop plate. Digestion was carried out according to the standard Filter Aided Sample Preparation (FASP) procedure [77]. Then, 100 μg of protein was used for each digestion and the procedure was carried out using microcons with 10 kDa mass cut-off membrane. Generated tryptic peptides were desalted with StageTips according to the protocol described by Rappsilber [78]. For each desalting step, 10 μg of the peptide was taken and desalted on StageTip containing three layers of 3 M Empore C18 exchange disks.

4.18. Liquid Chromatography and Mass Spectrometry

LC-MS/MS analysis was performed with the use of a Triple TOF 5600+ mass spectrometer (SCIEX Framingham, MA) coupled with the Ekspert MicroLC 200 Plus System (Eksigent, Redwood City, California, USA). All chromatographic separations were performed on the ChromXP C18CL column (3 μm, 120 Å, 150 × 0.3 mm). The chromatographic gradient for each IDA and SWATH runs was 3.5–20% B (solvent A 0% aqueous solution 0.1% formic acid, solvent B 100% acetonitrile 0.1% formic acid) in 60 min. The whole system was controlled by the SCIEX Analyst TF 1.7.1 software (version 1.7.1, Framingham, MA, USA).

4.19. SWATH Mass Spectrometry Experiments

All samples were acquired in triplicates. Experiments were performed in a looped product ion mode.

A set of 25 transmission windows (variable wide) was constructed and covered the precursor mass range of 400–1200 *m/z*. The collision energy for each window was calculated for +2 to +5 charged ions centered upon the window with a spread of two. The SWATH-MS1 survey scan was acquired in high sensitivity mode in the range of 400–1200 Da in the beginning of each cycle with the accumulation time of 50 ms, and SWATH-MS/MS spectra were collected from 100 to 1800 *m/z* followed by 40 ms accumulation time high sensitivity product ion scans, which resulted in the total cycle time of 1.11 s.

4.20. Data Analysis

Database search was performed with ProteinPilot 4.5 software (Sciex, v.4.5 AB, Framingham, MA, USA) using the Paragon algorithm against the UNIPROT *Dickeya solani* database with an automated false discovery rate, and standard parameters [79,80]. Next, a spectral library was created with the group file data processing in PeakView v. 2.2 (Sciex), with parameters as described in detail by Lewandowska [79]. Files from SWATH experiments for each sample were downloaded to PeakView (Sciex, v.2.2, Framingham, MA, USA) software and processed with the previously established library. Resulting data were exported to the .xml file and exported to Marker View software. All data were normalized using total area sums (TAS) approach, grouped as wild type and tested samples and *t*-test was performed. Samples were compared to each other, coefficient of variation (CV%) was calculated, and proteins with a *p*-value lower than 0.05 with fold change 2 were considered as differentially expressed in examined samples. The mass spectrometry proteomics data have been deposited to the ProteomeXchange Consortium via the PRIDE [81] partner repository with the dataset identifier PXD018297.

Supplementary Materials: The following are available online at http://www.mdpi.com/1422-0067/21/10/3687/s1, Figure S1: Verification of the constructed Δ*lon/lon* strain, Table S1: SWATH-MS raw data, Figure S2: Ability of *D. solani* Δ*lon* to cause maceration of plant tissues: (A) potato tubers (B) chicory leaves and (C) Chinese cabbage leaves.

Author Contributions: Conceptualization, J.S.-G.; Data curation, D.F.; Formal analysis, D.F., P.C. and J.S.-G.; Funding acquisition, J.S.-G.; Investigation, D.F., P.C., T.P. and P.A.; Methodology, D.F., P.C. and J.S.-G.; Supervision, J.S.-G.; Visualization, D.F.; Writing—original draft, D.F., P.C. and J.S.-G.; Writing—review and editing, D.F., T.P., P.A., M.P. and J.S.-G. All authors have read and agreed to the published version of the manuscript.

Funding: This research was funded by the National Science Centre, Poland, NCN OPUS-7 UMO-2014/13/B/NZ9/02021.

Acknowledgments: We thank Guy Condemine, Erwan Guegen, Nicole Hugouvieux-Cotte-Pattat and Vladimir Shevchik from Claude Bernard University Lyon 1, for the knowledge support in complementation strain construction. The strains *E. coli* DH5α *pir*, *E. coli* MFD *pir* and pRE112 vector, which we used in our research, were donated by Erwan Guegen. We also thank Daniel Stukenberg and Georg Fritz for providing the mScarlet plasmid.

Conflicts of Interest: The authors declare no conflicts of interest.

Abbreviations

CAS	casamino acids
CFU	colony forming units
CMC	carboxymethyl cellulose
C_q	quantification cycle
DAP	diaminopimelic acid
LA	Luria Agar
MFD	Mu-free donor
OD	optical density
PGA	polygalacturonic acid
PCWDE	plant cell wall degrading enzymes
PQCS	protein quality control system
SRP	soft rot Pectobacteriaceae
SWATH-MS	Sequential Window Acquisition of All Theoretical Mass Spectra
T1SS	type I secretion system
T2SS	type II secretion system
T3SS	type III secretion system
WT	wild-type

References

1. Adeolu, M.; Alnajar, S.; Naushad, S.; S Gupta, R. Genome-based phylogeny and taxonomy of the "*Enterobacteriales*": Proposal for *Enterobacterales* ord. nov. divided into the families *Enterobacteriaceae, Erwiniaceae* fam. nov., *Pectobacteriaceae* fam. nov., *Yersiniaceae* fam. nov., *Hafniaceae* fam. nov., *Morganellaceae* fam. nov., and *Budviciaceae* fam. nov. *Int. J. Syst. Evol. Microbiol.* **2016**, *66*, 5575–5599. [PubMed]

2. Toth, I.K.; Saddler, G.; Hélias, V.; Van Der Wolf, J.M.; Lojkowska, E.; Pirhonen, M.; Lahkim, L.T.; Elphinstone, J. Dickeya species: An emerging problem for potato production in Europe. *Plant Pathol.* **2011**, *60*, 385–399. [CrossRef]

3. Mansfield, J.; Genin, S.; Magori, S.; Citovsky, V.; Sriariyanun, M.; Ronald, P.; Dow, M.; Verdier, V.; Beer, S.V.; Machado, M.A.; et al. Top 10 plant pathogenic bacteria in molecular plant pathology. *Mol. Plant Pathol.* **2012**, *13*, 614–629. [CrossRef] [PubMed]

4. Sławiak, M.; Van Beckhoven, J.R.C.M.; Speksnijder, A.; Czajkowski, R.; Grabe, G.; Van Der Wolf, J.M. Biochemical and genetical analysis reveal a new clade of biovar 3 Dickeya spp. strains isolated from potato in Europe. *Eur. J. Plant Pathol.* **2009**, *125*, 245–261. [CrossRef]

5. Czajkowski, R.; Perombelon, M.C.M.; Van Veen, J.A.; Van Der Wolf, J.M. Control of blackleg and tuber soft rot of potato caused by Pectobacterium and Dickeya species: A review. *Plant Pathol.* **2011**, *60*, 999–1013. [CrossRef]

6. Reverchon, S.; Nasser, W. Dickeyaecology, environment sensing and regulation of virulence programme. *Environ. Microbiol. Rep.* **2013**, *5*, 622–636. [CrossRef]

7. Wagner, S.; Grin, I.; Malmsheimer, S.; Singh, N.; E Torres-Vargas, C.; Westerhausen, S. Bacterial type III secretion systems: A complex device for the delivery of bacterial effector proteins into eukaryotic host cells. *FEMS Microbiol. Lett.* **2018**, *365*, 365. [CrossRef]

8. Figaj, D.; Ambroziak, P.; Przepiora, T.; Skorko-Glonek, J. The Role of Proteases in the Virulence of Plant Pathogenic Bacteria. *Int. J. Mol. Sci.* **2019**, *20*, 672. [CrossRef]

9. Flanagan, J.M.; Bewley, M.C. Protein quality control in bacterial cells: Integrated networks of chaperones and ATP-dependent proteases. *Genet. Eng. (N.Y.)* **2002**, *24*, 17–47.

10. Ventura, S.; Villaverde, A. Protein quality in bacterial inclusion bodies. *Trends Biotechnol.* **2006**, *24*, 179–185. [CrossRef]

11. Maurizi, M.R. Proteases and protein degradation inEscherichia coli. *Cell. Mol. Life Sci.* **1992**, *48*, 178–201. [CrossRef] [PubMed]

12. Schoemaker, J.M.; Gayda, R.C.; Markovitz, A. Regulation of cell division in Escherichia coli: SOS induction and cellular location of the sulA protein, a key to lon-associated filamentation and death. *J. Bacteriol.* **1984**, *158*, 551–561. [CrossRef] [PubMed]

13. Torres-Cabassa, A.S.; Gottesman, S. Capsule synthesis in Escherichia coli K-12 is regulated by proteolysis. *J. Bacteriol.* **1987**, *169*, 981–989. [CrossRef] [PubMed]

14. Tsilibaris, V.; Maenhaut-Michel, G.; Van Melderen, L. Biological roles of the Lon ATP-dependent protease. *Res. Microbiol.* **2006**, *157*, 701–713. [CrossRef] [PubMed]

15. Robertson, G.T.; Kovach, M.E.; Allen, C.A.; Ficht, T.A.; Roop, R.M. The Brucella abortus Lon functions as a generalized stress response protease and is required for wild-type virulence in BALB/c mice. *Mol. Microbiol.* **2002**, *35*, 577–588. [CrossRef] [PubMed]

16. Takaya, A.; Suzuki, M.; Matsui, H.; Tomoyasu, T.; Sashinami, H.; Nakane, A.; Yamamoto, T. Lon, a Stress-Induced ATP-Dependent Protease, Is Critically Important for Systemic Salmonella enterica Serovar Typhimurium Infection of Mice. *Infect. Immun.* **2003**, *71*, 690–696. [CrossRef] [PubMed]

17. Breidenstein, E.B.M.; Janot, L.; Strehmel, J.; Fernández, L.; Taylor, P.K.; Kukavica-Ibrulj, I.; Gellatly, S.; Levesque, R.C.; Overhage, J.; Hancock, R.E. The Lon Protease Is Essential for Full Virulence in Pseudomonas aeruginosa. *PLoS ONE* **2012**, *7*, e49123. [CrossRef]

18. Su, S. Lon protease of the -proteobacterium Agrobacterium tumefaciens is required for normal growth, cellular morphology and full virulence. *Microbiology* **2006**, *152*, 1197–1207. [CrossRef]

19. Lan, L.; Deng, X.; Xiao, Y.; Zhou, J.-M.; Tang, X. Mutation of Lon Protease Differentially Affects the Expression ofPseudomonas syringaeType III Secretion System Genes in Rich and Minimal Media and Reduces Pathogenicity. *Mol. Plant Microbe Interact.* **2007**, *20*, 682–696. [CrossRef]

20. Guo, M.; Gross, C.A. Stress-induced remodeling of the bacterial proteome. *Curr. Boil.* **2014**, *24*, R424–R434. [CrossRef]

21. Ron, E.Z. Bacterial Stress Response. In *The Prokaryotes: Prokaryotic Physiology and Biochemistry*; Rosenberg, E., DeLong, E.F., Lory, S., Stackebrandt, E., Thompson, F., Eds.; Springer: Berlin/Heidelberg, Germany, 2013; pp. 589–603. ISBN 978-3-642-30141-4.

22. Gottesman, S. Trouble is coming: Signaling pathways that regulate general stress responses in bacteria. *J. Boil. Chem.* **2019**, *294*, 11685–11700. [CrossRef] [PubMed]

23. Przepiora, T.; Figaj, D.; Radzinska, M.; Apanowicz, M.; Sieradzka, M.; Ambroziak, P.; Hugouvieux-Cotte-Pattat, N.; Lojkowska, E.; Skorko-Glonek, J. Effects of stressful physico-chemical factors on the fitness of the plant pathogenic bacterium Dickeya solani. *Eur. J. Plant Pathol.* **2019**, *156*, 519–535. [CrossRef]

24. Kearns, D.B. A field guide to bacterial swarming motility. *Nat. Rev. Genet.* **2010**, *8*, 634–644. [CrossRef] [PubMed]

25. Ludwig, C.; Gillet, L.C.; Rosenberger, G.; Amon, S.; Collins, B.C.; Aebersold, R. Data-independent acquisition-based SWATH - MS for quantitative proteomics: A tutorial. *Mol. Syst. Boil.* **2018**, *14*, e8126. [CrossRef]

26. Inamine, G.S.; Dubnau, D. ComEA, a Bacillus subtilis integral membrane protein required for genetic transformation, is needed for both DNA binding and transport. *J. Bacteriol.* **1995**, *177*, 3045–3051. [CrossRef]

27. Goff, S.A.; Casson, L.P.; Goldberg, A.L. Heat shock regulatory gene htpR influences rates of protein degradation and expression of the lon gene in Escherichia coli. *Proc. Natl. Acad. Sci. USA* **1984**, *81*, 6647–6651. [CrossRef]

28. Grossman, A.D.; Erickson, J.W.; Gross, C.A. The htpR gene product of E. coli is a sigma factor for heat-shock promoters. *Cell* **1984**, *38*, 383–390. [CrossRef]

29. Grossman, A.D.; Straus, D.B.; A Walter, W.; A Gross, C. Sigma 32 synthesis can regulate the synthesis of heat shock proteins in Escherichia coli. *Genes Dev.* **1987**, *1*, 179–184. [CrossRef]

30. Bianchi, A.A.; Baneyx, F. Hyperosmotic shock induces the σ32 and σE stress regulons of Escherichia coli. *Mol. Microbiol.* **1999**, *34*, 1029–1038. [CrossRef]

31. Richmond, C.S. Genome-wide expression profiling in Escherichia coli K-12. *Nucleic Acids Res.* **1999**, *27*, 3821–3835. [CrossRef]

32. He, L.; Nair, M.K.M.; Chen, Y.; Liu, X.; Zhang, M.; Hazlett, K.R.O.; Deng, H.; Zhang, J.-R. The Protease Locus of Francisella tularensis LVS Is Required for Stress Tolerance and Infection in the Mammalian Host. *Infect. Immun.* **2016**, *84*, 1387–1402. [CrossRef]

33. Jiang, X.; Zghidi-Abouzid, O.; Oger-Desfeux, C.; Hommais, F.; Greliche, N.; Muskhelishvili, G.; Nasser, W.; Reverchon, S.; Xuejiao, J.; Ouafa, Z.; et al. Global transcriptional response of Dickeya dadantii to environmental stimuli relevant to the plant infection. *Environ. Microbiol.* **2016**, *18*, 3651–3672. [CrossRef] [PubMed]

34. Cebrián, G.; Condón, S.; Mañas, P. Physiology of the Inactivation of Vegetative Bacteria by Thermal Treatments: Mode of Action, Influence of Environmental Factors and Inactivation Kinetics. *Foods* **2017**, *6*, 107. [CrossRef] [PubMed]

35. Witkin, E. RecA protein in the SOS response: Milestones and mysteries. *Biochimie* **1991**, *73*, 133–141. [CrossRef]

36. Sakoh, M.; Ito, K.; Akiyama, Y. Proteolytic Activity of HtpX, a Membrane-bound and Stress-controlled Protease fromEscherichia coli. *J. Boil. Chem.* **2005**, *280*, 33305–33310. [CrossRef] [PubMed]

37. Shabala, L.; Bowman, J.P.; Brown, J.; Ross, T.; McMeekin, T.; Shabala, S. Ion transport and osmotic adjustment inEscherichia coliin response to ionic and non-ionic osmotica. *Environ. Microbiol.* **2009**, *11*, 137–148. [CrossRef] [PubMed]

38. Heuveling, J.; Possling, A.; Hengge, R. A role for Lon protease in the control of the acid resistance genes of Escherichia coli. *Mol. Microbiol.* **2008**, *69*, 534–547. [CrossRef]

39. Ingmer, H.; Brøndsted, L. Proteases in bacterial pathogenesis. *Res. Microbiol.* **2009**, *160*, 704–710. [CrossRef]

40. Frees, D.; Brøndsted, L.; Ingmer, H. Bacterial Proteases and Virulence. *Membr. Biog.* **2013**, *66*, 161–192.

41. Hugouvieux-Cotte-Pattat, N.; Condemine, G.; Shevchik, V. Bacterial pectate lyases, structural and functional diversity. *Environ. Microbiol. Rep.* **2014**, *6*, 427–440. [CrossRef]

42. Ciesielski, S.J.; Schilke, B.; Marszalek, J.; Craig, E. Protection of scaffold protein Isu from degradation by the Lon protease Pim1 as a component of Fe–S cluster biogenesis regulation. *Mol. Boil. Cell* **2016**, *27*, 1060–1068. [CrossRef] [PubMed]

43. Hunt, M.D.; Pettis, G.S.; A McIntosh, M. Promoter and operator determinants for fur-mediated iron regulation in the bidirectional fepA-fes control region of the Escherichia coli enterobactin gene system. *J. Bacteriol.* **1994**, *176*, 3944–3955. [CrossRef] [PubMed]

44. Clemmer, K.M.; Rather, P.N. The Lon protease regulates swarming motility and virulence gene expression in Proteus mirabilis. *J. Med. Microbiol.* **2008**, *57*, 931–937. [CrossRef] [PubMed]

45. Mukherjee, S.; Bree, A.C.; Liu, J.; Patrick, J.E.; Chien, P.; Kearns, D.B. Adaptor-mediated Lon proteolysis restricts Bacillus subtilis hyperflagellation. *Proc. Natl. Acad. Sci. USA* **2014**, *112*, 250–255. [CrossRef] [PubMed]

46. Lee, J.H.; Ancona, V.; Zhao, Y. Lon protease modulates virulence traits in Erwinia amylovora by direct monitoring of major regulators and indirectly through the Rcs and Gac-Csr regulatory systems. *Mol. Plant Pathol.* **2017**, *19*, 827–840. [CrossRef]

47. Takaya, A.; Tomoyasu, T.; Tokumitsu, A.; Morioka, M.; Yamamoto, T. The ATP-Dependent Lon Protease of Salmonella enterica Serovar Typhimurium Regulates Invasion and Expression of Genes Carried on Salmonella Pathogenicity Island 1. *J. Bacteriol.* **2002**, *184*, 224–232. [CrossRef]

48. Haque, M.M.; Oliver, M.H.; Nahar, K.; Alam, M.Z.; Hirata, H.; Tsuyumu, S. CytR Homolog of Pectobacterium carotovorum subsp. carotovorum Controls Air-Liquid Biofilm Formation by Regulating Multiple Genes Involved in Cellulose Production, c-di-GMP Signaling, Motility, and Type III Secretion System in Response to Nutritional and Environmental Signals. *Front. Microbiol.* **2017**, *8*.

49. Josenhans, C.; Suerbaum, S. The role of motility as a virulence factor in bacteria. *Int. J. Med. Microbiol.* **2002**, *291*, 605–614. [CrossRef]

50. Jackson, M.W.; Silva-Herzog, E.; Plano, G.V. The ATP-dependent ClpXP and Lon proteases regulate expression of the Yersinia pestis type III secretion system via regulated proteolysis of YmoA, a small histone-like protein. *Mol. Microbiol.* **2004**, *54*, 1364–1378. [CrossRef]

51. Bretz, J.; Losada, L.; Lisboa, K.; Hutcheson, S.W. Lon protease functions as a negative regulator of type III protein secretion in Pseudomonas syringae. *Mol. Microbiol.* **2002**, *45*, 397–409. [CrossRef]

52. Ventrice, C. Functional Characterization of the csp Homologs of *E. coli* K-12. Electronic Theses and Dissertations. Master's Thesis, Duquesne University, Pittsburgh, PA, USA, 2004.

53. Fernández, L.; Breidenstein, E.B.M.; Taylor, P.K.; Bains, M.; De La Fuente-Núñez, C.; Fang, Y.; Foster, L.J.; Hancock, R.E. Interconnection of post-transcriptional regulation: The RNA-binding protein Hfq is a novel target of the Lon protease in Pseudomonas aeruginosa. *Sci. Rep.* **2016**, *6*, 26811. [CrossRef] [PubMed]

54. Golanowska, M. Characterization of Dickeya solani strains and identification of bacterial and plant signals involved in induction of virulence. In *Microbiology and Parasitology*; INSA de Lyon: Lyon, France, 2015.

55. Antonova, E.S.; Bernardy, E.; Hammer, B.K. Natural competence in Vibrio cholerae is controlled by a nucleoside scavenging response that requires CytR?dependent anti?activation. *Mol. Microbiol.* **2012**, *86*, 1215–1231. [CrossRef] [PubMed]

56. Porankiewicz, J.; Wang, J.; Clarke, A.K. New insights into the ATP-dependent Clp protease: Escherichia coli and beyond. *Mol. Microbiol.* **1999**, *32*, 449–458. [CrossRef] [PubMed]

57. Miller, J.H. *A Short Course in Bacterial Genetics: Handbook*; Cold Spring Harbor Laboratory Press: Cold Spring Harbor, NY, USA, 1992; ISBN 978-0-87969-349-7.

58. Hanahan, D. Studies on transformation of Escherichia coli with plasmids. *J. Mol. Boil.* **1983**, *166*, 557–580. [CrossRef]

59. Platt, R.; Drescher, C.; Park, S.-K.; Phillips, G.J. Genetic System for Reversible Integration of DNA Constructs and lacZ Gene Fusions into the Escherichia coli Chromosome. *Plasmid* **2000**, *43*, 12–23. [CrossRef]

60. Ferrières, L.; Hémery, G.; Nham, T.; Guérout, A.-M.; Mazel, D.; Beloin, C.; Ghigo, J.-M. Silent Mischief: Bacteriophage Mu Insertions Contaminate Products of Escherichia coli Random Mutagenesis Performed Using Suicidal Transposon Delivery Plasmids Mobilized by Broad-Host-Range RP4 Conjugative Machinery. *J. Bacteriol.* **2010**, *192*, 6418–6427. [CrossRef]

61. Van Der Wolf, J.M.; Nijhuis, E.H.; Kowalewska, M.J.; Saddler, G.S.; Parkinson, N.; Elphinstone, J.G.; Pritchard, L.; Toth, I.K.; Lojkowska, E.; Potrykus, M.; et al. Dickeya solani sp. nov., a pectinolytic plant-pathogenic bacterium isolated from potato (Solanum tuberosum). *Int. J. Syst. Evol. Microbiol.* **2014**, *64*, 768–774. [CrossRef]

62. Lee, D.J.; Bingle, L.E.; Heurlier, K.; Pallen, M.J.; Penn, C.W.; Busby, S.; Hobman, J. Gene doctoring: A method for recombineering in laboratory and pathogenic Escherichia coli strains. *BMC Microbiol.* **2009**, *9*, 252. [CrossRef]

63. Edwards, R.A.; Keller, L.H.; Schifferli, D.M. Improved allelic exchange vectors and their use to analyze 987P fimbria gene expression. *Gene* **1998**, *207*, 149–157. [CrossRef]

64. Stukenberg, D.; (Zentrum für Synthetische Mikrobiologie, Marburg, Germany). Personal communication, 2019.

65. Isothermal Reaction (Gibson Assembly) Master Mix. *Cold Spring Harb. Protoc.* **2017**, *2017*, 090019.

66. Tu, Q.; Yin, J.; Fu, J.; Herrmann, J.; Li, Y.-Z.; Yin, Y.; Stewart, A.F.; Müller, R.; Zhang, Y. Room temperature electrocompetent bacterial cells improve DNA transformation and recombineering efficiency. *Sci. Rep.* **2016**, *6*, 24648. [CrossRef] [PubMed]

67. Collmer, A.; Ried, J.L.; Mount, M.S. Assay methods for pectic enzymes. In *Methods in Enzymology*; Elsevier BV: Amsterdam, The Netherlands, 1988; Volume 161, pp. 329–335.

68. Tardy, F.; Nasser, W.; Robert-Baudouy, J.; Hugouvieux-Cotte-Pattat, N. Comparative analysis of the five major Erwinia chrysanthemi pectate lyases: Enzyme characteristics and potential inhibitors. *J. Bacteriol.* **1997**, *179*, 2503–2511. [CrossRef] [PubMed]

69. Andro, T.; Chambost, J.P.; Kotoujansky, A.; Cattaneo, J.; Bertheau, Y.; Barras, F.; Van Gijsegem, F.; Coleno, A. Mutants of Erwinia chrysanthemi defective in secretion of pectinase and cellulase. *J. Bacteriol.* **1984**, *160*, 1199–1203. [CrossRef] [PubMed]

70. Schwyn, B.; Neilands, J. Universal chemical assay for the detection and determination of siderophores. *Anal. Biochem.* **1987**, *160*, 47–56. [CrossRef]

71. Primer3 Input (Version 0.4.0). Available online: http://bioinfo.ut.ee/primer3-0.4.0/ (accessed on 31 March 2020).

72. Pfaffl, M.W. A new mathematical model for relative quantification in real-time RT-PCR. *Nucleic Acids Res.* **2001**, *29*, 45. [CrossRef] [PubMed]

73. Pfaffl, M.W. Relative expression software tool (REST(C)) for group-wise comparison and statistical analysis of relative expression results in real-time PCR. *Nucleic Acids Res.* **2002**, *30*, 36. [CrossRef]

74. REST-BioInformatics in Real-Time PCR. Available online: http://rest.gene-quantification.info/ (accessed on 31 March 2020).

75. Laemmli, U.K. Cleavage of Structural Proteins during the Assembly of the Head of Bacteriophage T4. *Nature* **1970**, *227*, 680–685. [CrossRef]

76. Oberfelder, R. Detection of proteins on filters by enzymatic method. In *Methods in Nonradioactive Detection*; Howard, G.C., Ed.; Appleton & Lange: Norwalk, CT, USA, 1993; pp. 83–85. ISBN 978-0-8385-6946-7.

77. Wiśniewski, J.R.; Zougman, A.; Nagaraj, N.; Mann, M. Universal sample preparation method for proteome analysis. *Nat. Methods* **2009**, *6*, 359–362. [CrossRef]

78. Rappsilber, J.; Mann, M.; Ishihama, Y. Protocol for micro-purification, enrichment, pre-fractionation and storage of peptides for proteomics using StageTips. *Nat. Protoc.* **2007**, *2*, 1896–1906. [CrossRef]

79. Lewandowska, A.E.; Macur, K.; Czaplewska, P.; Liss, J.; Lukaszuk, K.; Ołdziej, S. Human follicular fluid proteomic and peptidomic composition quantitative studies by SWATH-MS methodology. Applicability of high pH RP-HPLC fractionation. *J. Proteom.* **2019**, *191*, 131–142. [CrossRef]

80. Gillet, L.C.; Navarro, P.; Tate, S.; Röst, H.; Selevsek, N.; Reiter, L.; Bonner, R.; Aebersold, R. Targeted data extraction of the MS/MS spectra generated by data-independent acquisition: A new concept for consistent and accurate proteome analysis. *Mol. Cell. Proteom.* **2012**, *11*. [CrossRef] [PubMed]

81. Perez-Riverol, Y.; Csordas, A.; Bai, J.; Bernal-Llinares, M.; Hewapathirana, S.; Kundu, D.J.; Inuganti, A.; Griss, J.; Mayer, G.; Eisenacher, M.; et al. The PRIDE database and related tools and resources in 2019: Improving support for quantification data. *Nucleic Acids Res.* **2018**, *47*, D442–D450. [CrossRef] [PubMed]

© 2020 by the authors. Licensee MDPI, Basel, Switzerland. This article is an open access article distributed under the terms and conditions of the Creative Commons Attribution (CC BY) license (http://creativecommons.org/licenses/by/4.0/).

International Journal of
Molecular Sciences

Article

A Novel Effector Protein of Apple Proliferation Phytoplasma Disrupts Cell Integrity of *Nicotiana* spp. Protoplasts

Cecilia Mittelberger [1], Hagen Stellmach [2], Bettina Hause [2], Christine Kerschbamer [1], Katja Schlink [1], Thomas Letschka [1] and Katrin Janik [1,*]

[1]   Applied Genomics and Molecular Biology, Laimburg Research Centre, 39040 Auer/Ora (BZ), Italy
[2]   Jasmonate Function & Mycorrhiza, Leibniz Institute of Plant Biochemistry, 06120 Halle, Germany
*   Correspondence: katrin.janik@laimburg.it

Received: 7 September 2019; Accepted: 14 September 2019; Published: 18 September 2019

Abstract: Effector proteins play an important role in the virulence of plant pathogens such as phytoplasma, which are the causative agents of hundreds of different plant diseases. The plant hosts comprise economically relevant crops such as apples (*Malus* × *domestica*), which can be infected by 'Candidatus Phytoplasma mali' (P. mali), a highly genetically dynamic plant pathogen. As the result of the genetic and functional analyses in this study, a new putative P. mali effector protein was revealed. The so-called "Protein in *Malus* Expressed 2" (PME2), which is expressed in apples during P. mali infection but not in the insect vector, shows regional genetic differences. In a heterologous expression assay using *Nicotiana benthamiana* and *Nicotiana occidentalis* mesophyll protoplasts, translocation of both PME2 variants in the cell nucleus was observed. Overexpression of the effector protein affected cell integrity in *Nicotiana* spp. protoplasts, indicating a potential role of this protein in pathogenic virulence. Interestingly, the two genetic variants of PME2 differ regarding their potential to manipulate cell integrity. However, the exact function of PME2 during disease manifestation and symptom development remains to be further elucidated. Aside from the first description of the function of a novel effector of P. mali, the results of this study underline the necessity for a more comprehensive description and understanding of the genetic diversity of P. mali as an indispensable basis for a functional understanding of apple proliferation disease.

Keywords: phytoplasma; effector protein; apple; pathogenicity; virulence; apple proliferation

---

1. Introduction

Phytoplasma are small, biotrophic bacteria that cause hundreds of different plant diseases and are involved in their infection cycle not only in plant hosts, but also in insect vectors. 'Candidatus Phytoplasma mali' (P. mali), the causal agent of apple proliferation (AP) disease, has caused significant economic losses in apple production in Northern Italy (one of Europe's main production areas) in the last decades [1]. Phytoplasma are obligate plant and insect symbionts that exhibit a biphasic life cycle comprising reproduction in certain phloem-feeding insects as well as in plants [2,3]. Within their plant host, phytoplasma colonize the phloem. By ingestion of phloem sap, insect vectors acquire the phytoplasma, with the colonization of those insects enabling the transmission of the pathogen between host plants [3,4]. Although several concepts of phytoplasma effector biology were able to be unraveled for the 'Candidatus Phytoplasma asteris' strain Aster Yellow Witches' Broom (AY-WB) in the model plant *Arabidopsis thaliana* [5–12], the understanding of effector-driven changes induced by P. mali remain limited. Genetic and functional homologues of AY-WB phytoplasma protein SAP11 could be identified in P. mali [10,13]. Recently a novel effector was described that exhibits E3 Ubiquitin ligase function and affects the plant's basal defense [14]. Furthermore, the immunodominant membrane protein Imp of

P. mali was shown not to be involved in symptom development but is considered to play a role during plant colonization [15]. A role of phytoplasmal HflB proteases and an AAA+ ATPase in AP virulence has been hypothesized but not yet clarified [16–18]. P. mali encodes genes for a Sec-dependent protein secretion system, whereas genes encoding components of other secretion systems, such as the type three secretion system, are mainly lacking [19,20]. Secreted phytoplasma proteins may directly interact with cellular host components and thus manipulate the cell's metabolism [3]. Potential effector proteins may thus be identified by the presence of a characteristic *N*-terminal secretion signal.

The aim of this study was to characterize the function of the phytoplasmal "Protein in *Malus* Expressed 2" (PME2) from P. mali that exhibits genetic features indicating that it acts as an effector protein in plants. To unravel PME2s potential role as an effector, this study analyzed (1) whether it is genetically conserved; (2) whether it is expressed during infection; (3) where it is translocated within the plant cell; and (4) if it induces morphological changes within the expressing plant cells.

To address these questions, we analyzed the expression of PME2 in P. mali-infected *Malus* × *domestica* leaf and root tissue, and in infected *Cacopsylla picta* (i.e., insect species transmitting P. mali). In infected *Malus* × *domestica* we found two distinct genetic variants of *pme2*. In addition, heterologous overexpression of PME2 in mesophyll protoplasts of *Nicotiana* spp. was used to gain insights into the subcellular localization of PME2 as well as its effects on plant cell integrity. These data were complemented by the expression of PME2 in yeast. With the data presented here, the first steps into unraveling the molecular mechanism of PME2 function were taken, but further experiments in the future will be indispensable.

2. Results

2.1. In Silico Analysis of PME2 Indicates Effector Potential

Bioinformatic analysis of conserved hypothetical proteins encoded in the P. mali genome [19] revealed that CAP18323.1, encoded by the gene *atp_00136*, contains interesting features that might confer effector function. Neural networks and hidden Markov prediction models (Transmembrane Helices Hidden Markov Model; TMHMM) were applied to analyze CAP18323.1 for the presence of a signal peptide and the presence of transmembrane regions (SignalP v. 3.0 [21], TMHMM [22]). Since phytoplasma phylogenetically belong to Gram-positive bacteria [3], a prediction algorithm trained on this bacterial group was applied. The N-terminal amino acid-stretch 1–31 contains a signal peptide that is supposed to confer Sec-dependent secretion of the protein (Figure 1). Further transmembrane regions were not predicted, indicating that CAP18323.1 is not inserted in a membrane. Upon translation, *N*-terminal signal peptides are cleaved [23]. At the C-terminal part of CAP18323.1 an importin α/β-dependent nuclear localization site (NLS) and a nuclear export signal (NES) were predicted [24,25]. The absence of transmembrane regions in the mature protein, the predicted localization in the plant cytoplasma or the nucleus (WoLF PSORT, [26]), and the small size of about 16 kDa (Analysis Tool on the ExPASy Server, [27]) indicate that CAP18323.1 may exhibit an effector function (Figure 1).

2.2. Atp_00136 (Pme2) *is Expressed in P. Mali-Infected Malus* × *Domestica but not in the Insect Vector C. Picta*

Subsequently, it was analyzed whether *atp_00136* was expressed in apple trees infected with P. mali. Leaf and root samples of P. mali-infected and non-infected *Malus* × *domestica* cv. "Golden Delicious" trees were taken in May and October. Expression of *atp_00136* was analyzed with *atp_00136*-specific primers and *Malus* × *domestica* cDNA derived from mRNA. Expression of *atp_00136* was confirmed in P. mali-infected leaf and root tissue by the detection of distinct amplicons at the expected size in the respective samples (Figure 2).

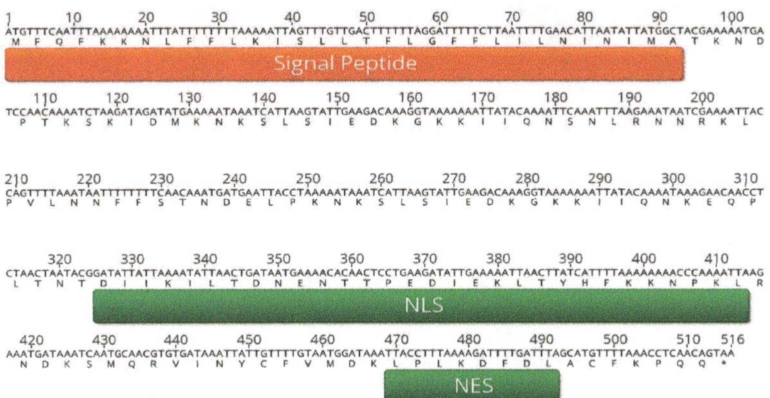

Figure 1. Results of the in silico analysis. Sequence analysis of *atp_00136* revealed the presence of an *N*-terminal signal peptide (indicated in red), as well as a nuclear localization signal (NLS), and a C-terminal nuclear export signal (NES), both indicated in green. Graphs were generated with Geneious Prime 2018 version 11.1.4.

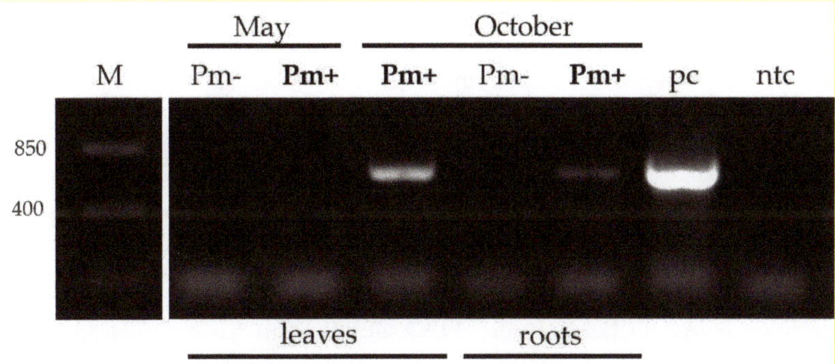

Figure 2. Expression of *pme*2 (CAP18323.1) in '*Candidatus* Phytoplasma mali' (P. mali)-infected *Malus* × *domestica*. Transcripts of *pme*2 were detected by PCR using cDNA from *Malus* × *domestica* infected with P. mali. A discrete band of the size indicative for the *pme*2 transcript was detected in P. mali-infected (**Pm+**) but not in non-infected (Pm–) leaves and roots harvested in October. DNA derived from an infected *Malus* × *domestica* served as a positive control (pc) and water as the non-template control (ntc).

Using quantitative PCR (qPCR) the expression levels of *atp_00136* and P. mali in the samples were quantified. The results show that *atp_00136* is only expressed in tissue colonized by P. mali (Table 1). Since identified expressed genes were named in a chronological manner, *atp_00136* was named "Protein in *Malus* Expressed 2" (*pme2*) based on the general recommendations for bacterial gene nomenclature [28].

To analyze if *pme2* was expressed in the transmitting insect vectors during infection, three P. mali-infected *C. picta* individuals were analyzed for the expression of the potential effector. In the RNA/cDNA of all infected individuals, P. mali-specific transcripts of the ribosomal protein *rpl22* were detected, but expression of *pme2* was not detectable.

Table 1. Detection of *atp_00136* in cDNA samples from infected and non-infected leaf tissue from May and October 2011. In May phytoplasma were only detectable in the roots but not in the leaves. *atp_00136* was only detectable in P. mali-infected and colonized tissue. Cq values are given as the mean value of three repeated qPCR runs.

Month	Status	Pool	cDNA Integrity (*tip41*)	Phytoplasma (*16S*)	*atp_00136*
May	non-infected	1	26.38	N/A	N/A
		2	26.38	N/A	N/A
Oct	non-infected	3	26.58	N/A	N/A
		2	26.59	N/A	N/A
		3	26.58	N/A	N/A
May	infected	1	26.56	N/A	N/A
		2	26.53	N/A	N/A
		3	26.61	N/A	N/A
Oct	infected	1	26.71	23.67	28.00
		2	26.44	23.18	27.66
		3	26.48	23.34	28.34

2.3. Genetic Variability of Pme2

Cloned amplicon sequencing revealed that the prevalent variant of *pme2* from infected trees in South Tyrol (North-East Italy) differs compared to the *pme2* sequence of the P. mali AT strain from Germany [19]. In a total of 20 samples from naturally infected apple trees in the regions Burggraviato and Val Venosta, a prevalent, conserved sequence of *pme2* was identified (*pme2*$_{ST}$; accession number MN224214). This conserved variant exhibits a single nucleotide polymorphism (SNP) in the sequence stretch before the NLS, and two SNPs within and one SNP after the NLS compared to the AT strain (Figure 3). All four SNPs in the *pme2*$_{ST}$ variant lead to nonsynonymous missense substitutions at the protein level as compared to the *pme2* sequence published previously [19] (*pme2*$_{AT}$). The NLS of *pme2*$_{ST}$ has a slightly higher prediction score than the NLS of *pme2*$_{AT}$. The most striking difference between *pme2*$_{AT}$ and *pme2*$_{ST}$ is a stretch of 120 bp in *pme2*$_{ST}$ which is absent in *pme2*$_{AT}$. This stretch is a partial duplication of a fragment also present in *pme2*$_{AT}$ (Figure 3). In three *Malus* × *domestica* samples, a very sporadic sequence of *pme2* could be detected that did not contain the *pme2*$_{ST}$ characteristic sequence duplication but showed strong sequence similarity to *pme2*$_{AT}$. The sporadic sequence contains six SNPs at positions 218 (A > T), 220 (A > G), 322 (A > G), 331 (A > C), 344 (C < T), and 427 (T > G) that lead to nonsynonymous missense mutations (accession number MN224215) compared to *pme2*$_{AT}$. However, in the trees in which these very sporadic *pme2* sequences were found, *pme2*$_{ST}$ could also be detected, indicating the presence of a mixed population of different P. mali strains.

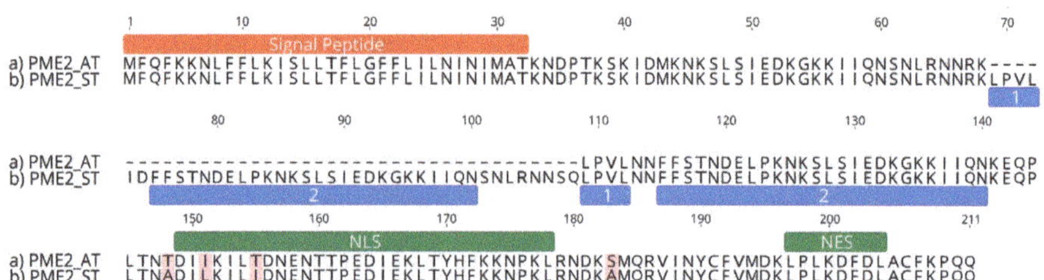

Figure 3. Sequence comparison of PME2$_{ST}$ and PME2$_{AT}$. The protein variants (**a**) PME2$_{AT}$ and (**b**) PME2$_{ST}$ contain the same *N*-terminal signal peptide sequence (red). PME2$_{ST}$ (**b**) contains a duplicated amino acid stretch (the replicative sequences 1 and 2; marked in blue) of a partial sequence also present in PME2$_{AT}$ (**a**). Both variants show slight differences in and directly before the nuclear localization signal sequences (NLS, green). The nuclear export signal sequence (NES, green) is identical in both protein variants. Amino acid differences of PME2$_{ST}$ to the PME2$_{AT}$ variant are shown in black, whereas similarities are shown in grey. Graphs were generated with Geneious Prime 2018 version 11.1.4.

2.4. *PME2$_{ST}$ and PME2$_{AT}$ Translocate to the Nucleus of* Nicotiana *spp. Protoplasts*

To identify the subcellular localization of the PME2 protein in the plant cell, mesophyll protoplasts of *Nicotiana occidentalis* and *N. benthamiana* were transformed, with expression vectors coding for PME2$_{AT}$ and PME2$_{ST}$ tagged with GFP or mCherry-fluorescent protein to allow subcellular tracking. The *N*-terminal signal part was not considered for these studies, since it is removed from the processed, mature CAP18323.1 protein. *N. occidentalis* and *N. benthamiana* can be infected with P. mali. Upon infection, both *Nicotiana* species show disease-specific symptoms and are thus appropriate model plants for P. mali effector studies [15,29]. Confocal microscopy analysis revealed that overexpressed PME2$_{AT}$ and PME2$_{ST}$ are translocated to the nucleus of *Nicotiana* spp. protoplasts. This translocation was independent of the used tag and *Nicotiana* species (Figure 4 and Figures S1–S3). The in vivo results therefore confirm the in silico prediction that PME2$_{ST}$ and PME2$_{AT}$ are translocated to the nucleus of potential host cells.

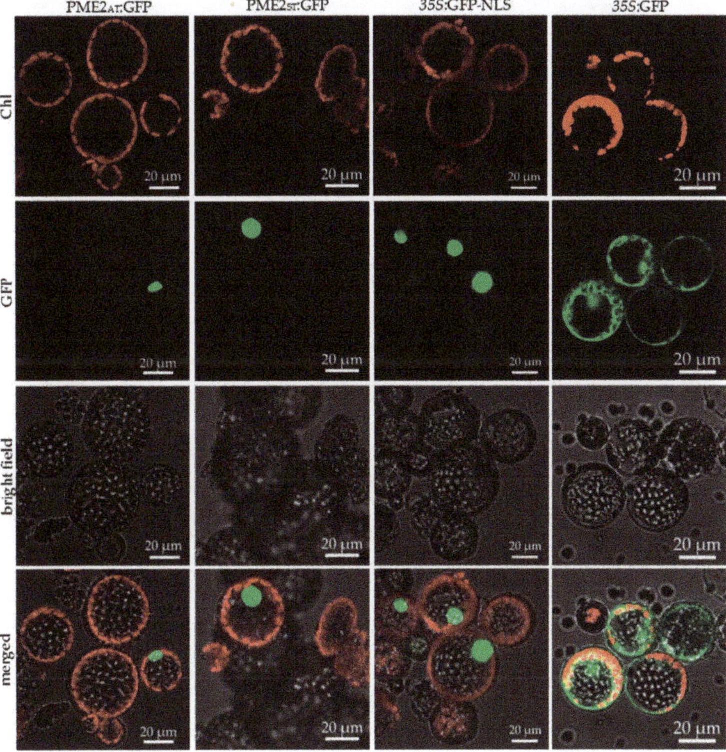

Figure 4. PME2$_{ST}$ and PME2$_{AT}$ are translocated to the nucleus of mesophyll protoplasts. Mesophyll protoplasts of *Nicotiana benthamiana* were transformed with the plasmid pGGZ001 encoding C-terminal GFP-tagged PME2$_{ST}$ (first column), PME2$_{AT}$ (second column), GFP *N*-terminally fused to a NLS sequence (third column), or GFP only as a control for nuclear localization (fourth column). Expression of the transgenes was under the control of a *35S* promoter. The upper panel shows autofluorescence of chloroplasts (Chl), the second panel the signal derived from the GFP, and the third panel the bright field image and the last panel an overlay of all images (merged). Microscopic analysis was performed with a Zeiss LSM 800. Corresponding images after expression of mCherry-tagged PME2 and of use of *Nicotiana occidentalis* mesophyll protoplasts are presented in Figures S1–S3. Bars represent 20 µm.

A leaf infiltration assay using *Agrobacterium* strain EHA105 transformed with PME2 encoding expression vectors did not result in detectable expression or phenotypic alterations of either PME2$_{AT}$ or PME2$_{ST}$ in both *Nicotiana* species. Nonetheless, positive controls expressing the fluorophore tag only and leaves infiltrated with the P. mali SAP11-like effector protein ATP_00189 [13] as control showed strong signals (Figure S5), indicating that PME2 expression might be somehow blocked or is immediately degraded by the plant.

2.5. PME2$_{ST}$ but not PME2$_{AT}$ Affect Cell Integrity of Nicotiana spp. Protoplasts

Protoplasts transformed with the PME2$_{ST}$ expression vector often showed shrinkage, and only about 50% of the *N. benthamiana* protoplasts were viable 20 h post-transformation compared to the transformation control expressing the fluorophore only or a GFP with NLS (Figure 5a). The shrunk cells lysed and only the remaining cell debris was microscopically detectable (Figure 4). The effect on protoplast integrity was observed in protoplasts expressing PME2$_{ST}$:GFP and PME2$_{ST}$:mCherry, and thus was independent of the fluorophore used as a tag for microscopic analyses. Similar results were obtained using *N. occidentalis* as the heterologous PME2$_{ST}$ expression system. The mCherry-tagged PME2$_{ST}$ induced a weak but significant reduction of viability in *N. occidentalis* protoplasts (Figure 5b). The GFP-tagged PME2$_{ST}$ showed the same tendency but to a stronger extent, i.e., it reduced cell viability by about 50%, which is similar to the effect seen in *N. benthamiana* protoplasts. Cell viability stain with fluorescein diacetate (FDA) showed similar results, i.e., that *N. benthamiana* protoplasts transformed with the PME2$_{ST}$-expressing vector showed a significantly reduced viability (Figure 6). Shrunk cells were positive for propidium iodide (PI) staining (Figure S4), indicating that these cells were dead.

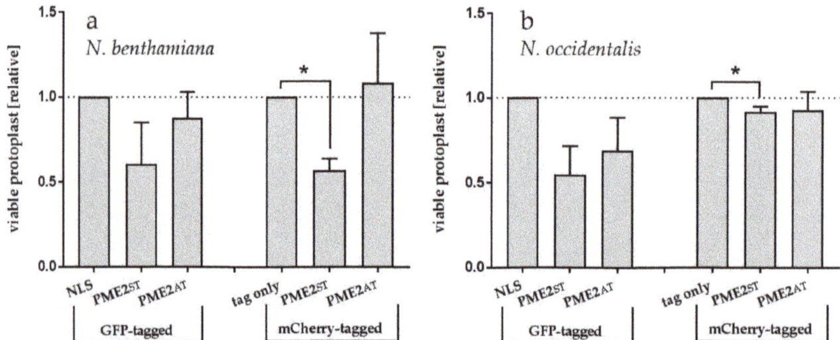

Figure 5. PME2$_{ST}$ overexpression reduces viability of (**a**) *N. benthamiana* and (**b**) *N. occidentalis* mesophyll protoplasts. For each assay, 20,000 mesophyll protoplasts were transformed with the plasmid pGGZ001 encoding PME2$_{ST}$, PME2$_{AT}$ (tagged with GFP or mCherry), the GFP-tagged control for nuclear localization (NLS), or the mCherry tag (tag only) and viable protoplasts were counted. Overexpression of the transgenes was under the control of a *35S* promoter. Data represent the mean viability +/– SE of 3–4 independent experiments. The respective control (NLS or tag only) was set at 1 to allow comparison between different experiments. Differences between the groups were determined applying a one way-ANOVA analysis. Significant differences ($p < 0.05$) between groups are indicated with an asterisk (*).

Interestingly, PME2$_{AT}$ did not have an effect on protoplast integrity in *N. benthamiana* nor in *N. occidentalis* protoplasts (Figure 5).

2.6. A Yeast Two-Hybrid Screen Was Unsuitable for the Elucidation of PME2$_{ST}$ Function

Upon expression of PME2$_{ST}$, the yeast reporter strain *Saccharomyces cerevisiae* NMY51 showed several macroscopic aberrations in colony growth (Figure 7a). However, at the microscopic level when visualizing the yeast cell wall with calcofluor white, no phenotypic differences between yeast cells expressing $PME2_{AT}$, $PME2_{ST}$, and empty bait vector pLexA-N could be detected (Figure 7b). Considering the effect of mere $PME2_{ST}$ expression on growth of the yeast reporter strain, the relevance of any identified interaction in a yeast two-hybrid screen remains highly questionable and the assay was therefore not performed.

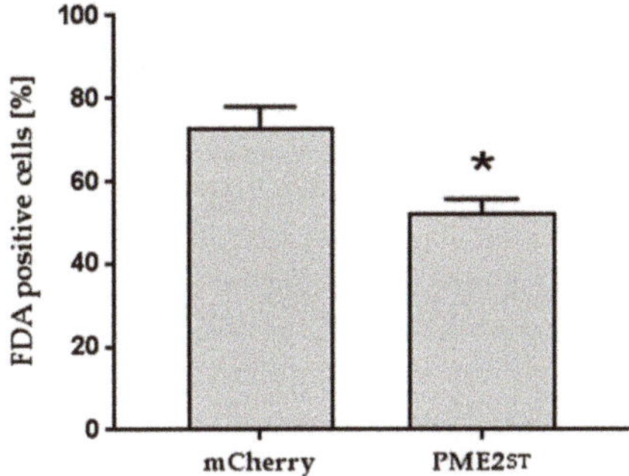

Figure 6. $PME2_{ST}$ overexpression reduces viability of *N. benthamiana* mesophyll protoplasts. Mesophyll protoplasts were transformed with the plasmid pGGZ001 encoding mCherry-tagged $PME2_{ST}$ (PME_{ST}) or the mCherry tag only (mCherry) and stained with fluorescein diacetate (FDA) to detect viable cells. Data represent the mean percentage of FDA-positive stained cells +/− SE ($n = 3$). The statistical difference between the two groups was determined by using a Student's *t*-test and is indicated with an asterisk (*$p < 0.05$).

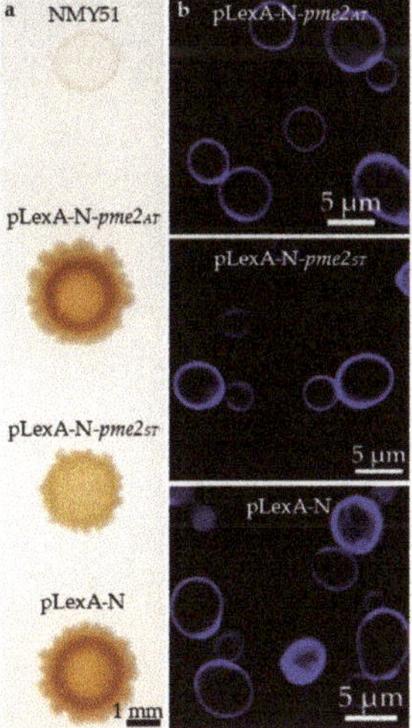

Figure 7. $PME2_{ST}$ overexpression in the yeast reporter strain NMY51 leads to macroscopic aberrations. *Saccharomyces cerevisiae* strain NMY51 was transformed with the yeast two-hybrid (Y2H) bait vector pLexA-N, which encodes tryptophan auxotrophy, expressing $PME2_{AT}$, $PME2_{ST}$, or the empty vector only, and drop-plated onto SD-trp plates. In comparison to the empty pLexA-N vector and the vector expressing PME_{AT}, colonies expressing PME_{ST} showed reduced growth and remained white (**a**). Yeast cells stained with calcofluor white did not show any phenotypic differences on single-cell level (**b**). Calcofluor white fluorescence was visualized on a confocal microscope (LSM 800, Carl Zeiss AG, Oberkochen, Germany).

3. Discussion

The results of this study show that PME2$_{ST}$ (a variant of CAP18323.1 previously annotated as "conserved hypothetical protein") affects plant cell integrity. Based on our findings and the definition that effectors are secreted pathogen proteins altering host-cell structure and function [30], we propose defining PME2 as a phytoplasmal effector. Interestingly, two different variants of PME2 were identified and both variants translocate to the nucleus of plant cells, but only the newly described regional variant PME2$_{ST}$ subsequently affects protoplast integrity. The small size of about (at a maximum) 21 kDa (PME2$_{ST}$: 21 kDa and PME2$_{AT}$: 16 kDa; both considering the mature protein without the signal peptide) indicates that PME2 can be translocated from the phloem and target adjacent tissues or be distributed systemically in the plant [3]. Subcellular localization using microscopy requires the use of fluorescent tags that are attached to the protein of interest. Tagging can affect subcellular localization of the protein; however, we used two different tags (GFP and mCherry) to analyze whether tagging influences the target localization. In cells expressing the tag only, a localization of the fluorescent signal in the cytoplasm could be observed. PME2 was localized only in the nucleus and since no signal was visible in other cell compartments, it can be assumed that the observed localization is effector-mediated (see also [31]). Only protoplasts transformed with PME2$_{ST}$ showed significant cell disruption as indirectly quantified by counting the remaining viable cells and FDA staining of the protoplasts. Shrunk cells were positive for the PI stain but did not show a GFP signal. The lack of the GFP-signal might be caused by a disruption of the nucleus, protein degradation, and/or leakage of the signal into the surrounding medium. Cells expressing PME2$_{ST}$ were intact, indicating that the effector either exhibits a dose-dependent or delayed effect on cell integrity.

Both variants of PME2 contain an *N*-terminal signal peptide, a nuclear localization signal (NLS), and a nuclear export signal (NES). It is a common feature of nuclear proteins to contain both NLS and NES and these signals coordinate the translocation of the protein between nucleus and cytoplasm [32]. Nuclear targeting of proteins containing a classical NLS is mediated by the importin α/β heterodimer through NLS-dependent binding to the importin α subunit and importin β–mediated attachment to the nuclear pore complex [33,34]. The SNPs in the NLS region of PME2$_{ST}$ lead to a (slightly) higher sequence-based NLS prediction; thus, the differences might show a stronger translocation to the nucleus. The NES signal (which indicates that shuttling of PME2 between nucleus and cytoplasm might occur) is the same in both variants. Even though nucleocytoplasmic distribution is predicted, PME2 was only detected in the nucleus. Many proteins containing NLS and NES appear to be localized in the nucleus because the rate of import to the nucleus is higher than the rate of export to the cytoplasm [35]. It remains thus unclear if PME2 is strictly limited to the nucleus or if a constant shuttling between nucleus and cytosol occurs.

Bacterial effectors that translocate to the nucleus, the so-called nuclear effectors, can affect master switches of the host immune machinery or alter host transcription to the benefit of the pathogen [31]. Effectors from different phytoplasma species target plant–host transcription factors or affect gene expression on the transcriptional level to alter the host metabolism to their own benefit [4,12,13,36,37]. However, none of these effectors have yet been reported to exhibit such detrimental effects during in planta expression. The effector protein BR1 of the phloem colonizing squash leaf curl geminivirus shuttles between the cytoplasm and the nucleus of protoplasts [38]. Upon binding to the second movement protein BL1, BR1 shuttles to the cytoplasma [39] and the concerted action between BR1 and BL1 mediates cell-to-cell movement of the virus within the phloem and to adjacent cells [35,38,40,41]. To unravel BR1 function it was necessary to identify its interaction partner, a general approach to investigate effector function. Yeast two-hybrid (Y2H) screens have been successfully applied to determine phytoplasmal effector targets on the molecular level [10,13]. These screens allow the screening of a protein of interest (effector) against a library containing hundreds of thousands of different potential interaction partners of a certain host species [42,43]. Successful interaction is monitored by a genetic reporter system that complements certain auxotrophies in the recombinant yeast reporter strain. However, a Y2H with PME2$_{ST}$ is not suitable since PME2$_{ST}$ expression strongly affected the Y2H yeast reporter strain. This effect on yeast cells further supports the finding that PME2$_{ST}$ exhibits a strong effect not only on plant, but also on yeast cells, even though the latter do not have relevance as phytoplasma host cells. Since PME2$_{ST}$ exhibits such a strong effect on the expressing host and non-host cells, alternative approaches must be applied to unravel its molecular function. *Nicotiana* spp. leaf infiltration assays with recombinant *Agrobacterium* strains expressing PME2 failed. It remains furthermore elusive as to whether PME2 exhibits effects on the host plant phenotype. Considering the PME2$_{ST}$ effects on protoplasts it can be assumed that a systemic overexpression would lead to overwhelming deleterious effects in transgenic plants that express this effector. The results show that PME2 is expressed in roots and leaves of infected *Malus* × *domestica*, but not in infected individuals of its insect vector *C. picta*, underlining the hypothesis that PME2 plays a role as an effector protein in plant cells. However, it needs further clarification if expression is fine-tuned in a tempo-spatial manner in the plant host.

A neatly coordinated and local expression during infection might have very local effects and might not lead to cell disruption as seen in heterologous overexpression experiments. It is hypothesized that phytoplasma are able to degrade plant cell walls or generate holes in plant cell membranes to expedite cell-to-cell effector translocation [4]. Infection with P. mali induces cytochemical modifications and injuries of the affected phloem cells [44,45]. It is speculated that plasma membrane integrity is affected by until-now unknown P. mali effector(s) and that plasma membrane disruption is involved in the observed phloem damage induced by virulent P. mali strains [45]. However, since molecular indications are lacking, interpretation of the mode of PME2 action remains speculative. Subsequent approaches to analyze PME2 function should comprise assays that do not depend on functional living cells.

Since PME2 is translocated to the nucleus it is possible that it directly targets the host DNA by mimicking DNA regulatory elements, such as transcription factors or repressors. Some plant pathogen effectors bind host DNA and thus modulate gene expression [46,47]. An example of these effectors are TAL effectors of the plant pathogen *Xanthomonas*. TAL effectors bind promoter elements and regulate plant host expression to the benefit of the pathogen [48–51]. The effector AvrBs3 of *Xanthomonas* translocates to the nucleus where it acts as a transcription factor and affects the size of mesophyll cells [52]. Bioinformatic prediction and sequence comparison did not indicate that PME2 has similarity with currently known transcription factors or other gene expression regulating factors in plants.

Both P. mali strains from which the two different PME2 variants were derived cause infection and typical disease symptoms in *Malus* × *domestica*. Thus, the effect of PME2$_{ST}$ on cell integrity seems to be dispensable for infection and symptom development but might affect strain virulence. However, a direct comparison between the two strains regarding their virulence is missing. It might also be possible that another effector of P. mali strain AT (unknown at the time of this research), mimics and thus complements the function that PME2$_{AT}$ is lacking.

Some P. mali strains strongly differ regarding their virulence potential in *Malus* × *domestica* and several studies addressed the genetic identification of virulence factors or certain genetic determinants that account for these differences [17,18,53–55]. Since phytoplasma cannot be genetically manipulated, determining the importance of an effector during infection often involves tortuous experimental paths. In this study we provide the first characterization of the P. mali effector PME2 and its effect on cells of potential plant hosts. We report an interesting difference between two variants of PME2 that occur in Italy and Germany, claiming that further full genomic sequence analysis is required to better understand how P. mali manipulates its host on the molecular level.

4. Materials and Methods

4.1. Verification of Pme2 *Expression in* Malus × Domestica *and* C. Picta

For the verification of *pme2* expression by PCR in infected apple root and leaf samples RNA was extracted from the plant tissue as described in [13]. Extracted RNA was subjected to DNase treatment using DNAfree Turbo reagent (Ambion, Austin, TX, USA) and cDNA synthesis was performed using the SuperScript™ VILO™ cDNA Synthesis Kit (Invitrogen, Waltham, MA, USA). The generated cDNA was diluted 1:200 in nuclease free water and cDNA integrity was checked in all samples by performing a control PCR targeting the house-keeping gene transcript putative tip41-like family (transcript identifier: Mdo.1349) using the primers 5′-ACATGCCGGAGATGGTGTTTGG-3′ (forward) and 5′-ACTTCCAGAGTACGGCGTTGTG-3′ (reverse). Contamination with genomic DNA was checked by performing a PCR with primers amplifying a fragment within the non-coding region trnL of chloroplast DNA using the primers B49317 and A49855 [56]. No DNA contamination was detected in any of the cDNA samples, and the amplification of the putative tip41-like transcript fragment was positive, thus confirming the integrity of the generated cDNA. PCR reactions to verify *pme2* expression were set up in a total reaction volume of 10 µL, using 2 µL of diluted cDNA (1:200) as template, 0.05 µL GoTaq® DNA Polymerase (Promega, Madison, WI, USA), 2 µL of 5X Green GoTaq® Reaction Buffer (Promega, Madison, WI, USA), 0.2 µL dNTP-mix (40 mM), 1 µL of forward primer ATP00136_forw_EcoRI (10 µM, 5′-CCCCCCGAATTCATGTTTCAATTTAAAAAAAATTTA-3′), and 1 µL of reverse primer ATP00136_rev_SalI (10 µM, 5′-CCCCCCGTCGACATTATTACTGTTGAGGTTTAA-3′). Cycling conditions were applied as follows: 95 °C for 5 min followed by 40 cycles of 95 °C for 1 min, 44.9 °C for 1 min, 72 °C for 1 min, and a final elongation step at 72 °C for 5 min. PCR products were visualized on 1% agarose gel. Additionally, *pme2* expression level was detected by qPCR based on SYBR-Green chemistry using the primer pair ATP00136_GW_fwd (5′-CACCATGACGAAAAATGATCCAACAAA-3′)/ATP00136_nostopp_rev (5′-CTGTTGAGGTTTAAAACAT-3′) in a total reaction volume of 20 µL using 4.0 µL of diluted cDNA (1:200) as a template together with 10.0 µL 2× SYBR FAST qPCR Kit Master Mix (Kapa Biosystems/αmann-La Roche, Basel, Switzerland), 1.0 µL of each primer (10 µM), and 4.0 µL of nuclease free water. qPCR conditions were as follows: an initial denaturation step at 95 °C for 20 s followed by 34 cycles of 95 °C for 3 s and 60 °C for 30 s and a melting curve ramp from 65 to 95 °C, at increments of 0.5 °C every 5 s (CFX384 Touch Real-Time PCR Detection System; BioRad, Hercules, CA, USA). Data analysis was performed using the CFX ManagerTM software (BioRad, Hercules, CA, USA).

To control whether *pme2* is expressed in infected individuals of the insect vector *C. picta*, RNA of six potentially infected and two uninfected F1 individuals was extracted with the ZR Tissue & Insect RNA MicroPrep™ kit (ZymoResearch, Irvine, CA, USA) according to the manufacturer's instructions. Extracted RNA was subjected to DNase treatment using DNAfree Turbo reagent (Ambion, Austin, TX, USA) and RNA integrity was controlled with an RNA ScreenTape on a TapeStation 2200 (both Agilent, Santa Clara, CA, USA). cDNA was synthesized with the iScript™ cDNA Synthesis Kit (BioRad, Hercules, CA, USA). Together with the cDNA synthesis a control was performed lacking the reverse transcriptase (-RT). Here, 2 µL of diluted cDNA (1:200) were used as template in a total qPCR reaction volume of 10 µL, together with 5 µL 2× SYBR FAST qPCR Kit Master Mix (Kapa Biosystems/Hoffmann-La Roche, Basel, Switzerland), 2 µL of nuclease free water, and 0.5 µL of forward and reverse primer (10 µM). The primer combination qPSY-WG-F and qPSY-WG-R, targeting the species-specific *wingless* gene [57], was used to determine cDNA integrity. P. mali infection was detected in three of the six individuals with primer pair rpAP15f-mod and rpAP15r3, targeting the ribosomal protein gene *rpl22* [58]. *Pme2* expression was checked with primer pair ATP00136_GW_fwd and ATP00136_nostopp_rev using the same qPCR conditions as described for the qPCR detection in *Malus × domestica* leaf samples.

4.2. Amplification, Subcloning, and Sequencing of atp_00136

DNA was purified from leaves from P. mali infected *Malus x domestica* cv Golden Delicious trees (10 trees from Burggraviato and 10 trees from Val Venosta) using the DNeasy Plant Mini kit (Qiagen, Hilden, Germany) following the manufacturer's instructions. DNA was diluted 1:10 in water and 2 µL template were used in a total PCR reaction volume of 50 µL as follows: *atp_00136* was amplified using 0.02 U/µL Phusion High-Fidelity DNA Polymerase (Thermo Fisher Scientific,

Waltham, MA, USA) using HF-buffer supplied by the manufacturer, 400 µM dNTPs, and 0.5 µM of each primer (forward: 5′-CCCCCCGAATTCATGTTTCAATTTAAAAAAAATTTA-3′; reverse: 5′-CCCCCCGTCGACATTATTACTGTTGAGGTTTAA-3′). DNA was denatured at 98 °C for 30 s followed by 30 cycles of denaturation for 10 s at 98 °C, amplification for 30 s at 49.3 °C, and elongation at 72 °C for 30 s. The PCR was finalized by a terminal elongation step at 72 °C for 5 min. The PCR product was purified using the Illustra GFX PCR DNA and Gel Band Purification Kit (GE Healthcare, Chicago, IL, USA) and 1 µg of purified PCR product was digested with 4 U EcoRI and SalI following the manufacturer's instructions (Thermo Fisher Scientific, Waltham, MA, USA), ligated into equally digested pUC19 using T4-Ligase (Thermo Fisher Scientific, Waltham, MA, USA) and transformed into MegaX DH10B™ T1R cells (Life Technologies, Carlsbad, CA, USA). At least five clones from each tree were sequenced with pUC19 specific primers (GATC Biotech, Constance, Germany) and analyzed to see different variants of the gene indicating a mixed infection.

4.3. Subcloning of Pme2 into GreenGate Expression Vectors

The genes *pme2*$_{ST}$ and *pme2*$_{AT}$ were subcloned into the GreenGate-entry module pGGC000 [59] using the primer pair ATP00136pP_CBsaI_fw (5′-AACAGGTCTCAGGCTCCATGACGAAAAATGATCCAACAAA-3′) and ATP00136pP_DBsaI_rv (5′-AACAGGTCTCACTGACTGTTGAGGTTTAAAACAT-3′). Using different components from the GreenGate-kit plant, transformation constructs coding for *pme2*$_{AT}$-*linker-GFP* or *pme2*$_{AT}$-*linker-mCherry* and *pme2*$_{ST}$-*linker-GFP* or *pme2*$_{ST}$-*linker-mCherry*, driven by the *35S* promoter and flanked at the 3′-end by the *RBCS* terminator, including kanamycin as the plant resistance marker, were designed. The following modules were assembled by GreenGate reaction in a total volume of 15 µL: 150 ng pGGA004 (*35S*), 150 ng pGGB003 (B-dummy), 150 ng pGGC000-*pme2*$_{AT}$ or pGGC00-*pme2*$_{ST}$, 150 ng pGGD001 (*linker-GFP*) or pGGD003 (*linker-mCherry*), 150 ng pGGE001 (*RBCS*), 150 ng pGGF007 (*pNOS:Kan*^{R}*:tNOS*), and 100 ng pGGZ001 (empty destination vector). Subsequently, 1.5 µL 10× CutSmart Buffer (New England Biolab, Ipswich, MA, USA), 1.5 µL ATP (10 mM), 1.0 µL T4 DNA Ligase (5 u/µL) (Thermo Fisher Scientific, Waltham, MA, USA), and 1.0 µL BsaI-HF®v2 (20,000 u/mL) (New England Biolab, Ipswich, MA, USA) were added to the module mixture, and 30 cycles of 2 min at 37 °C and 2 min at 16 °C each, followed by 50 °C for 5 min and 80 °C for 5 min were performed. Subsequently, 5 µL of the reaction mixture were used for heat-shock transformation of *ccdB*-sensitive One Shot® TOP10 chemically competent *Escherichia coli* (Invitrogen, Carlsbad, CA, USA). For the assembly of positive controls, the modules pGGC012 (*GFP-NLS*) or pGGC014 (*GFP*) or pGGC015 (*mCherry*) were used instead of the above mentioned pGGC000 modules. The correct assembly of the plant transformation constructs was confirmed by sequencing. Plasmid-DNA for protoplast transformation was obtained as described elsewhere [60], using the NucleoSnap® Plasmid Midi preparation kit (Macherey-Nagel, Düren, Germany) and PEG precipitation.

4.4. Protoplast Isolation and Transformation

Protoplasts of *N. benthamiana* and *N. occidentalis* were isolated from four- to five-week-old plants, cultivated under long photoperiod conditions (16 h/8 h, 24 °C/22 °C, 70% rH) and transformed as described in [60] using 10 µg plasmid-DNA per 20,000 protoplasts. After 18 h, at least 100 protoplasts of each transformation were checked for the occurrence of GFP or mCherry-fluorescence using a confocal laser scanning microscope (LSM800, Zeiss, Oberkochen, Germany) with an excitation wavelength of 488 nm for GFP and 561 nm for mCherry. The detection wavelength of GFP was set between 410 nm and 575 nm and of mCherry between 575 nm and 650 nm. Autofluorescence of chlorophyll was detected between 650 nm and 700 nm. After 20 h the number of intact protoplasts/mL was determined by counting in a Fuchs-Rosenthal chamber. Protoplast transformation and viability determination was repeated independently four times.

Only experiments in which at least 20% of the protoplasts in the control setup were viable after transformation were considered for further evaluation. Significant outliers were removed from the data set using the GraphPad QuickCalcs Outlier calculator online tool (https://www.graphpad.com/quickcalcs/Grubbs1.cfm; status of information 16th September 2019). Greisser Greenhouse correction on raw data and one-way-ANOVA with a Tukey Posttest were performed to analyze statistical differences between groups (GraphPad Prism 7.01., GraphPad Software, San Diego, CA, USA). To allow a better visual comparison, data were normalized to each respective control, which was set to 1.

Additionally, protoplast viability was visualized by propidium iodide (PI) and counted by fluorescein diacetate (FDA) staining in three independent repetitions. For the first, 20 µL of protoplasts transformed with GFP tagged expression vectors were mixed with 20 µL of PI solution (10 µg/mL PI in 0.65 M mannitol). FDA staining was done according to [61] using 20 µL of protoplasts transformed with either mCherry tagged $PME2_{AT}$ expression vectors or a vector expressing only mCherry and 20 µL of FDA solution (0.1 mg/mL FDA in 0.65 M mannitol). Fluorescence of PI, mCherry, and GFP was recorded using a LSM800 confocal laser scanning microscope (Zeiss, Oberkochen, Germany) with excitation and detection wavelengths for GFP and mCherry as described above and for PI excitation at 561 nm and detection between 560 nm and 640 nm.

4.5. Nicotiana spp. Leaf Infiltration

For subcellular localization of PME2, the two GreenGate expression vectors, as well as GFP and GFP-NLS expression vectors as positive controls, were subcloned by electroporation into *Agrobacterium tumefaciens* strain EHA105. As an additional control, we subcloned a GreenGate expression vector expressing the SAP11-like *P. mali* effector protein ATP_00189 [13] with an *N*-terminal fused GFP tag into *A. tumefaciens* strain EHA105. The transgenic *A. tumefaciens* clones were cultured for 2 days at 28 °C in liquid selective LB medium. Subsequently, 0.5 OD/mL were resuspended in infiltration medium (10 mM $MgCl_2$, 10 mM MES, 200 µM acetosyringone, pH 5.7) and infiltrated with a blunt syringe into leaves from four- to five-week-old *N. occidentalis* and *N. benthamiana*. Fluorescence was recorded after 48 h and 72 h using the confocal laser scanning microscope (LSM800, Zeiss, Oberkochen, Germany) with excitation for GFP at 488 nm and detection between 410 nm and 546 nm and excitation for mCherry at 561 nm and detection between 562 and 624 nm.

4.6. Expression in Yeast

For a potential Y2H, $pme2_{AT}$ and $pme2_{ST}$ were subcloned into bait-vector pLexA-N as described in [13,62] with primer pair ATP00136_forw_EcoRI/ATP00136_rev_SalI. The bait-plasmids pLexA-N-$pme2_{ST}$ and pLexA-N-$pme2_{AT}$ were transformed into *S. cerevisiae* strain NMY51. Growth aberrations of yeast colonies on selective SD-trp plates were observed and recorded by photographing.

For calcofluor white staining, yeast cells were grown overnight in SD-trp liquid media. Subsequently, 2 mL of the overnight culture were centrifuged, supernatant removed, and the cells resuspended in clear phosphate-buffered saline (PBS) buffer. Then, 10 µL of a 5 mM calcofluor white solution (Biotium, Fremont, California) were added to the cell suspension and incubated for 20 min at room temperature. The yeast cell wall was visualized by a confocal laser scanning microscope (LSM800, Zeiss, Oberkochen, Germany) with excitation at 405 nm and detection wavelength between 400 nm and 560 nm.

5. Conclusions

In this study we identified and characterized the novel *P. mali* effector protein PME2. This effector contains an NLS and an NES sequence and translocates to the nucleus of *N. benthamiana* mesophyll protoplasts. Two naturally occurring genetic variants of PME2, namely $PME2_{ST}$ and $PME2_{AT}$, differ regarding their ability to induce cellular modifications in yeast and plant cells. When overexpressed, the variant $PME2_{ST}$ affects yeast growth and reduces the viability of *Nicotiana* spp. mesophyll protoplasts. These findings indicate that PME2 might play a role for *P. mali* virulence in plants. Despite the similarities between both PME2 variants, this effect was not observed in yeast or protoplasts expressing PME2$_{AT}$. The results of our study show for the first time that a phytoplasmal effector causes detrimental effects when overexpressed in protoplasts.

Supplementary Materials: Supplementary materials can be found at http://www.mdpi.com/1422-0067/20/18/4613/s1.

Author Contributions: Conceptualization, K.J. and K.S.; methodology, K.J., C.M., K.S., H.S., and B.H.; validation, C.M. and K.J.; formal analysis, C.M., K.J., and C.K.; investigation, C.M. and K.J.; resources, K.J., T.L., and K.S.; data curation, C.M. and C.K.; writing—original draft preparation, C.M. and K.J.; writing—review and editing, C.M., H.S., B.H., C.K., K.S., T.L., and K.J.; visualization, C.M. and K.J.; supervision, K.J., T.L., and K.S.; project administration, K.J., T.L., and K.S.; funding acquisition, K.J., K.S., and T.L.

Funding: This research was funded by the APPL2.0 and APPLIII project within the Framework agreement in the field of invasive species in fruit growing and major pathologies (PROT. VZL_BZ 09.05.2018 0002552) and was co-funded by the Autonomous Province of Bozen/Bolzano, Italy and the South Tyrolean Apple Consortium.

Acknowledgments: We would like to thank Andreas Putti and Laura Russo of the Food Microbiology Laboratory at the Laimburg Research Centre for helpful and interesting discussions about yeast biology. Furthermore, we would like to thank Mirko Moser from the Fondazione Edmund Mach for discussing phytoplasma sequence data, Vicky Oberkofler for lab assistance, and Amy Kadison for her English proofreading.

Conflicts of Interest: The authors declare no conflict of interest.

Abbreviations

AP	Apple proliferation
AY-WB	Aster Yellow Witches' Broom
C. picta	*Cacopsylla picta*
FDA	Fluorescein diacetate
GFP	Green fluorescent protein
N.	*Nicotiana*
NES	Nuclear export signal
NLS	Nuclear localization signal
PI	Propidium iodide
P. mali	*Candidatus* Phytoplasma mali
PME2	Protein in Malus Expressed 2
qPCR	quantitative PCR
SNP	Single nucleotide polymorphism
Y2H	Yeast two-hybrid

References

1. Strauss, E. Phytoplasma research begins to bloom. *Science* **2009**, *325*, 388–390. [CrossRef]
2. Christensen, N.M.; Axelsen, K.B.; Nicolaisen, M.; Schulz, A. Phytoplasmas and their interactions with hosts. *Trends Plant Sci.* **2005**, *10*, 526–535. [CrossRef] [PubMed]
3. Hogenhout, S.A.; Oshima, K.; Ammar, E.-D.; Kakizawa, S.; Kingdom, H.N.; Namba, S. Phytoplasmas: Bacteria that manipulate plants and insects. *Mol. Plant Pathol.* **2008**, *9*, 403–423. [CrossRef] [PubMed]
4. Sugio, A.; MacLean, A.M.; Kingdom, H.N.; Grieve, V.M.; Manimekalai, R.; Hogenhout, S.A. Diverse targets of phytoplasma effectors: From plant development to defense against insects. *Annu. Rev. Phytopathol.* **2011**, *49*, 175–195. [CrossRef] [PubMed]
5. Bai, X.; Correa, V.R.; Toruño, T.Y.; Ammar, E.-D.; Kamoun, S.; Hogenhout, S.A. AY-WB phytoplasma secretes a protein that targets plant cell nuclei. *Mol. Plant Microbe Interact.* **2009**, *22*, 18–30. [CrossRef]
6. Lu, Y.-T.; Cheng, K.-T.; Jiang, S.-Y.; Yang, J.-Y. Post-translational cleavage and self-interaction of the phytoplasma effector SAP11. *Plant Signal. Behav.* **2014**, *9*, e28991. [CrossRef] [PubMed]
7. Lu, Y.-T.; Li, M.-Y.; Cheng, K.-T.; Tan, C.M.; Su, L.-W.; Lin, W.-Y.; Shih, H.-T.; Chiou, T.-J.; Yang, J.-Y. Transgenic plants that express the phytoplasma effector SAP11 show altered phosphate starvation and defense responses. *Plant Physiol.* **2014**, *164*, 1456–1469. [CrossRef]
8. Sugawara, K.; Honma, Y.; Komatsu, K.; Himeno, M.; Oshima, K.; Namba, S. The alteration of plant morphology by small peptides released from the proteolytic processing of the bacterial peptide TENGU. *Plant Physiol.* **2013**, *162*, 2005–2014. [CrossRef]

9. MacLean, A.M.; Sugio, A.; Makarova, O.V.; Findlay, K.C.; Grieve, V.M.; Tóth, R.; Nicolaisen, M.; Hogenhout, S.A. Phytoplasma effector SAP54 induces indeterminate leaf-like flower development in *Arabidopsis* plants. *Plant Physiol.* **2011**, *157*, 831–841. [CrossRef]

10. Sugio, A.; Kingdom, H.N.; MacLean, A.M.; Grieve, V.M.; Hogenhout, S.A. Phytoplasma protein effector SAP11 enhances insect vector reproduction by manipulating plant development and defense hormone biosynthesis. *Proc. Natl. Acad. Sci. USA* **2011**, *108*, E1254–E1263. [CrossRef]

11. Tomkins, M.; Kliot, A.; Marée, A.F.; Hogenhout, S.A. A multi-layered mechanistic modelling approach to understand how effector genes extend beyond phytoplasma to modulate plant hosts, insect vectors and the environment. *Curr. Opin. Plant Biol.* **2018**, *44*, 39–48. [CrossRef] [PubMed]

12. MacLean, A.M.; Orlovskis, Z.; Kowitwanich, K.; Zdziarska, A.M.; Angenent, G.C.; Immink, R.G.H.; Hogenhout, S.A. Phytoplasma effector SAP54 hijacks plant reproduction by degrading MADS-box proteins and promotes insect colonization in a RAD23-dependent manner. *PLoS Biol.* **2014**, *12*, e1001835. [CrossRef] [PubMed]

13. Janik, K.; Mithöfer, A.; Raffeiner, M.; Stellmach, H.; Hause, B.; Schlink, K.; Mithofer, A. An effector of apple proliferation phytoplasma targets TCP transcription factors-a generalized virulence strategy of phytoplasma? *Mol. Plant Pathol.* **2017**, *18*, 435–442. [CrossRef] [PubMed]

14. Strohmayer, A.; Moser, M.; Si-Ammour, A.; Krczal, G.; Boonrod, K. 'Candidatus Phytoplasma mali' genome encodes a protein that functions as a E3 Ubiquitin Ligase and could inhibit plant basal defense. *Mol. Plant Microbe Interact.* **2019**. [CrossRef] [PubMed]

15. Boonrod, K.; Munteanu, B.; Jarausch, B.; Jarausch, W.; Krczal, G. An immunodominant membrane protein (Imp) of 'Candidatus Phytoplasma mali' binds to plant actin. *Mol. Plant Microbe Interact.* **2012**, *25*, 889–895. [CrossRef] [PubMed]

16. Seemüller, E.; Kampmann, M.; Kiss, E.; Schneider, B. *HflB* gene-based phytopathogenic classification of 'Candidatus Phytoplasma mali' strains and evidence that strain composition determines virulence in multiply infected apple trees. *Mol. Plant Microbe Interact.* **2011**, *24*, 1258–1266. [CrossRef]

17. Schneider, B.; Sule, S.; Jelkmann, W.; Seemüller, E. Suppression of aggressive strains of 'Candidatus phytoplasma mali' by mild strains in *Catharanthus roseus* and *Nicotiana occidentalis* and indication of similar action in apple trees. *Phytopathology* **2014**, *104*, 453–461. [CrossRef] [PubMed]

18. Seemüller, E.; Zikeli, K.; Furch, A.C.U.; Wensing, A.; Jelkmann, W. Virulence of 'Candidatus Phytoplasma mali' strains is closely linked to conserved substitutions in AAA+ ATPase AP460 and their supposed effect on enzyme function. *Eur. J. Plant Pathol.* **2017**, *86*, 141. [CrossRef]

19. Kube, M.; Schneider, B.; Kuhl, H.; Dandekar, T.; Heitmann, K.; Migdoll, A.M.; Reinhardt, R.; Seemüller, E. The linear chromosome of the plant-pathogenic mycoplasma 'Candidatus Phytoplasma mali'. *BMC Genomics* **2008**, *9*, 306. [CrossRef]

20. Kube, M.; Mitrovic, J.; Duduk, B.; Rabus, R.; Seemüller, E. Current view on phytoplasma genomes and encoded metabolism. *Scientific World J.* **2012**, *2012*, 185942. [CrossRef]

21. Bendtsen, J.D.; Nielsen, H.; von Heijne, G.; Brunak, S. Improved prediction of signal peptides: SignalP 3.0. *J. Mol. Biol.* **2004**, *340*, 783–795. [CrossRef] [PubMed]

22. Krogh, A.; Larsson, B.; von Heijne, G.; Sonnhammer, E.L. Predicting transmembrane protein topology with a hidden Markov model: Application to complete genomes. *J. Mol. Biol.* **2001**, *305*, 567–580. [CrossRef] [PubMed]

23. Beckwith, J. The Sec-dependent pathway. *Res. Microbiol.* **2013**, *164*, 497–504. [CrossRef] [PubMed]

24. Kosugi, S.; Hasebe, M.; Tomita, M.; Yanagawa, H. Systematic identification of cell cycle-dependent yeast nucleocytoplasmic shuttling proteins by prediction of composite motifs. *Proc. Natl. Acad. Sci. USA* **2009**, *106*, 10171–10176. [CrossRef] [PubMed]

25. La Cour, T.; Kiemer, L.; Mølgaard, A.; Gupta, R.; Skriver, K.; Brunak, S. Analysis and prediction of leucine-rich nuclear export signals. *Protein Eng. Des. Sel.* **2004**, *17*, 527–536. [CrossRef] [PubMed]

26. Horton, P.; Park, K.-J.; Obayashi, T.; Fujita, N.; Harada, H.; Adams-Collier, C.J.; Nakai, K. WoLF PSORT: Protein localization predictor. *Nucleic Acids Res.* **2007**, *35*, W585–W587. [CrossRef] [PubMed]

27. Gasteiger, E.; Hoogland, C.; Gattiker, A.; Duvaud, S.; Wilkins, M.R.; Appel, R.D.; Bairoch, A. Protein identification and analysis tools on the ExPASy server. In *The Proteomics Protocols Handbook*; Walker, J.M., Ed.; Humana Press: Totowa, NJ, USA, 2005; pp. 571–607. ISBN 978-1-59259-890-8.

28. Journal of Bacteriology. Nomenclature: Genetic Nomenclature. Available online: https://jb.asm.org/content/nomenclature (accessed on 16 September 2019).

29. Luge, T.; Kube, M.; Freiwald, A.; Meierhofer, D.; Seemüller, E.; Sauer, S. Transcriptomics assisted proteomic analysis of *Nicotiana occidentalis* infected by *Candidatus* Phytoplasma mali strain AT. *Proteomics* **2014**, *14*, 1882–1889. [CrossRef] [PubMed]

30. Hogenhout, S.A.; Van der Hoorn, R.A.L.; Terauchi, R.; Kamoun, S. Emerging concepts in effector biology of plant-associated organisms. *Mol. Plant Microbe Interact.* **2009**, *22*, 115–122. [CrossRef] [PubMed]

31. Chaudhari, P.; Ahmed, B.; Joly, D.L.; Germain, H. Effector biology during biotrophic invasion of plant cells. *Virulence* **2014**, *5*, 703–709. [CrossRef] [PubMed]

32. Fu, X.; Liang, C.; Li, F.; Wang, L.; Wu, X.; Lu, A.; Xiao, G.; Zhang, G. The rules and functions of nucleocytoplasmic shuttling proteins. *Int. J. Mol. Sci.* **2018**, *19*, 1445. [CrossRef] [PubMed]

33. Corbett, A.H.; Silver, P.A. Nucleocytoplasmic transport of macromolecules. *Microbiol. Mol. Biol. Rev.* **1997**, *61*, 193–211. [PubMed]

34. Görlich, D.; Dabrowski, M.; Bischoff, F.R.; Kutay, U.; Bork, P.; Hartmann, E.; Prehn, S.; Izaurralde, E. A novel class of RanGTP binding proteins. *J. Cell Biol* **1997**, *138*, 65–80. [CrossRef] [PubMed]

35. Ward, B.; Medville, R.; Lazarowitz, S.-G.; Turgeon, R. The geminivirus BL1 movement protein is associated with endoplasmic reticulum-derived tubules in developing phloem cells. *J. Virol.* **1997**, *71*, 3726–3733. [PubMed]

36. Minato, N.; Himeno, M.; Hoshi, A.; Maejima, K.; Komatsu, K.; Takebayashi, Y.; Kasahara, H.; Yusa, A.; Yamaji, Y.; Oshima, K.; et al. The phytoplasmal virulence factor TENGU causes plant sterility by downregulating of the jasmonic acid and auxin pathways. *Sci. Rep.* **2014**, *4*, 7399. [CrossRef] [PubMed]

37. Kitazawa, Y.; Iwabuchi, N.; Himeno, M.; Sasano, M.; Koinuma, H.; Nijo, T.; Tomomitsu, T.; Yoshida, T.; Okano, Y.; Yoshikawa, N.; et al. Phytoplasma-conserved phyllogen proteins induce phyllody across the Plantae by degrading floral MADS domain proteins. *J. Exp. Bot.* **2017**, *68*, 2799–2811. [CrossRef] [PubMed]

38. Noueiry, A.O.; Lucas, W.J.; Gilbertson, R.L. Two proteins of a plant DNA virus coordinate nuclear and plasmodesmal transport. *Cell* **1994**, *76*, 925–932. [CrossRef]

39. Sanderfoot, A.A.; Ingham, D.J.; Lazarowitz, S.-G. A viral movement protein as a nuclear shuttle: The geminivirus BR1 movement protein contains domains essential for interaction with BL1 and nuclear localization. *Plant Physiol.* **1996**, *110*, 23–33. [CrossRef] [PubMed]

40. Sanderfoot, A.A.; Lazarowitz, S.-G. Cooperation in viral movement: The geminivirus BL1 movement protein interacts with BR1 and redirects it from the nucleus to the cell periphery. *The Plant Cell* **1995**, *7*, 1185–1194. [CrossRef] [PubMed]

41. Sanderfoot, A.A.; Lazarowitz, S.-G. Getting it together in plant virus movement: Cooperative interactions between bipartite geminivirus movement proteins. *Trends Cell Biol.* **1996**, *6*, 353–358. [CrossRef]

42. Fields, S.; Song, O.-K. A novel genetic system to detect protein–protein interactions. *Nature* **1989**, *340*, 245–246. [CrossRef]

43. Brückner, A.; Polge, C.; Lentze, N.; Auerbach, D.; Schlattner, U. Yeast two-hybrid, a powerful tool for systems biology. *Int. J. Mol. Sci.* **2009**, *10*, 2763–2788. [CrossRef] [PubMed]

44. Musetti, R.; Paolacci, A.; Ciaffi, M.; Tanzarella, O.A.; Polizzotto, R.; Tubaro, F.; Mizzau, M.; Ermacora, P.; Badiani, M.; Osler, R. Phloem cytochemical modification and gene expression following the recovery of apple plants from apple proliferation disease. *Phytopathology* **2010**, *100*, 390–399. [CrossRef] [PubMed]

45. Zimmermann, M.R.; Schneider, B.; Mithöfer, A.; Reichelt, M.; Seemüller, E.; Furch, A.C.U. Implications of *Candidatus Phytoplasma mali* infection on phloem function of apple trees. *Endocytobiosis Cell Res.* **2015**, *26*, 67–75.

46. Kay, S.; Hahn, S.; Marois, E.; Wieduwild, R.; Bonas, U. Detailed analysis of the DNA recognition motifs of the *Xanthomonas* type III effectors AvrBs3 and AvrBs3Deltarep16. *Plant J.* **2009**, *59*, 859–871. [CrossRef] [PubMed]

47. Win, J.; Chaparro-Garcia, A.; Belhaj, K.; Saunders, D.G.O.; Yoshida, K.; Dong, S.; Schornack, S.; Zipfel, C.; Robatzek, S.; Hogenhout, S.A.; et al. Effector biology of plant-associated organisms: Concepts and perspectives. *Cold Spring Harb. Symp. Quant. Biol.* **2012**, *77*, 235–247. [CrossRef] [PubMed]

48. Duan, Y.P.; Castañeda, A.; Zhao, G.; Erdos, G.; Gabriel, D.W. Expression of a single, host-specific, bacterial pathogenicity gene in plant cells elicits division, enlargement, and cell death. *Mol. Plant Microbe Interact.* **1999**, *12*, 556–560. [CrossRef]

49. Boch, J.; Scholze, H.; Schornack, S.; Landgraf, A.; Hahn, S.; Kay, S.; Lahaye, T.; Nickstadt, A.; Bonas, U. Breaking the code of DNA binding specificity of TAL-type III effectors. *Science* **2009**, *326*, 1509–1512. [CrossRef]

50. Domingues, M.N.; de Souza, T.A.; Cernadas, R.A.; de Oliveira, M.L.P.; Docena, C.; Farah, C.S.; Benedetti, C.E. The *Xanthomonas citri* effector protein PthA interacts with citrus proteins involved in nuclear transport, protein folding and ubiquitination associated with DNA repair. *Mol. Plant Pathol.* **2010**, *11*, 663–675. [CrossRef]

51. de Souza, T.A.; Soprano, A.S.; de Lira, N.P.V.; Quaresma, A.J.C.; Pauletti, B.-A.; Paes Leme, A.-F.; Benedetti, C.-E. The TAL effector PthA4 interacts with nuclear factors involved in RNA-dependent processes including a HMG protein that selectively binds poly(U) RNA. *PLoS ONE* **2012**, *7*, e32305. [CrossRef]

52. Kay, S.; Hahn, S.; Marois, E.; Hause, G.; Bonas, U. A bacterial effector acts as a plant transcription factor and induces a cell size regulator. *Science* **2007**, *318*, 648–651. [CrossRef]

53. Seemüller, E.; Schneider, B. Differences in virulence and genomic features of strains of 'Candidatus Phytoplasma mali', the apple proliferation agent. *Phytopathology* **2007**, *97*, 964–970. [CrossRef] [PubMed]

54. Seemüller, E.; Kampmann, M.; Kiss, E.; Schneider, B. Molecular differentiation of severe and mild strains of 'Candidatus Phytoplasma mali' and evidence that their interaction in multiply infected trees determines disease severity. *Bull. Insectology* **2011**, *64*, 163–164.

55. Rid, M.; Mesca, C.; Ayasse, M.; Gross, J. Apple proliferation phytoplasma influences the pattern of plant volatiles emitted depending on pathogen virulence. *Front. Ecol. Evol.* **2016**, *3*, 271. [CrossRef]

56. Taberlet, P.; Gielly, L.; Pautou, G.; Bouvet, J. Universal primers for amplification of three non-coding regions of chloroplast DNA. *Plant Mol. Biol.* **1991**, *17*, 1105–1109. [CrossRef] [PubMed]

57. Mittelberger, C.; Obkircher, L.; Oettl, S.; Oppedisano, T.; Pedrazzoli, F.; Panassiti, B.; Kerschbamer, C.; Anfora, G.; Janik, K. The insect vector *Cacopsylla picta* vertically transmits the bacterium 'Candidatus Phytoplasma mali' to its progeny. *Plant Pathol* **2017**, *66*, 1015–1021. [CrossRef]

58. Monti, M.; Martini, M.; Tedeschi, R. EvaGreen Real-time PCR protocol for specific 'Candidatus Phytoplasma mali' detection and quantification in insects. *Mol. Cell. Probes* **2013**, *27*, 129–136. [CrossRef]

59. Lampropoulos, A.; Sutikovic, Z.; Wenzl, C.; Maegele, I.; Lohmann, J.U.; Forner, J. GreenGate - A novel, versatile, and efficient cloning system for plant transgenesis. *PLoS ONE* **2013**, *8*, e83043. [CrossRef]

60. Janik, K.; Stellmach, H.; Mittelberger, C.; Hause, B. Characterization of phytoplasmal effector protein interaction with proteinaceous plant host targets using bimolecular fluorescence complementation (BiFC). In *Phytoplasmas: Methods and Protocols*; Musetti, R., Pagliari, L., Eds.; Humana Press: New York, NY, USA, 2019; pp. 321–331. ISBN 978-1-4939-8837-2.

61. Heslop-Harrison, J.; Heslop-Harrison, Y. Evaluation of pollen viability by enzymatically induced fluorescence; intracellular hydrolysis of fluorescein diacetate. *Stain Technol.* **1970**, *45*, 115–120. [CrossRef]

62. Janik, K.; Schlink, K. Unravelling the function of a bacterial effector from a non-cultivable plant pathogen using a yeast two-hybrid screen. *J. Vis. Exp.* **2017**, *119*, e55150. [CrossRef]

© 2019 by the authors. Licensee MDPI, Basel, Switzerland. This article is an open access article distributed under the terms and conditions of the Creative Commons Attribution (CC BY) license (http://creativecommons.org/licenses/by/4.0/).

International Journal of
Molecular Sciences

MDPI

Article

Proteomics Analysis of SsNsd1-Mediated Compound Appressoria Formation in *Sclerotinia sclerotiorum*

Jingtao Li [1,2,†], Xianghui Zhang [1,†], Le Li [1], Jinliang Liu [1], Yanhua Zhang [1,*] and Hongyu Pan [1,*]

[1] College of Plant Science, Jilin University, Changchun 130062, China; lijingtao789@126.com (J.L.);
    zhangxianghui@jlu.edu.cn (X.Z.); lile19890904@126.com (L.L.); jlliu@jlu.edu.cn (J.L.)

[2] College of Plant Health and Medicine, Qingdao Agricultural University, Qingdao 266109, China

* Correspondence: yh_zhang@jlu.edu.cn (Y.Z.); panhongyu@jlu.edu.cn (H.P.); Tel.: +86-431-8783-5712 (H.P.)

† These authors contributed equally to this work.

Received: 10 September 2018; Accepted: 24 September 2018; Published: 27 September 2018

Abstract: *Sclerotinia sclerotiorum* (Lib.) de Bary is a devastating necrotrophic fungal pathogen attacking a broad range of agricultural crops. In this study, although the transcript accumulation of SsNsd1, a GATA-type IVb transcription factor, was much lower during the vegetative hyphae stage, its mutants completely abolished the development of compound appressoria. To further elucidate how SsNsd1 influenced the appressorium formation, we conducted proteomics-based analysis of the wild-type and Δ*SsNsd1* mutant by two-dimensional electrophoresis (2-DE). A total number of 43 differentially expressed proteins (≥3-fold change) were observed. Of them, 77% were downregulated, whereas 14% were upregulated. Four protein spots fully disappeared in the mutants. Further, we evaluated these protein sequences by mass spectrometry analysis of the peptide mass and obtained functionally annotated 40 proteins, among which only 17 proteins (38%) were identified to have known functions including energy production, metabolism, protein fate, stress response, cellular organization, and cell growth and division. However, the remaining 23 proteins (56%) were characterized as hypothetical proteins among which four proteins (17%) were predicted to contain the signal peptides. In conclusion, the differentially expressed proteins identified in this study shed light on the Δ*SsNsd1* mutant-mediated appressorium deficiency and can be used in future investigations to better understand the signaling mechanisms of SsNsd1 in *S. sclerotiorum*.

Keywords: *Sclerotinia sclerotiorum*; SsNsd1; compound appressorium; two-dimensional electrophoresis; proteomics analysis; differential expression proteins

1. Introduction

Sclerotinia sclerotiorum (Lib.) de Bary is a destructive and hard-to-control plant necrotrophic fungal pathogen on a broad range of agricultural crops [1,2]. Developmentally, vegetative hyphae gathered together forming hardened, multicellular sclerotia enclosed by a melanized rind layer, which plays an important role in the development and pathogenesis of *S. sclerotiorum* [3,4]. Under suitable environmental conditions, sclerotia germinate to form vegetative hyphae or apothecia, and the latter release numerous ascospores that initiate new disease cycles [5]. Mycelia from sclerotia or ascospores can directly infect the plant tissues by forming compound appressoria (also known as infection cushions) from modified hyphae [2,4] or enter the plant tissue through open stomata by secreting oxalic acid [6]. Therefore, a better understanding of the developmental mechanism of appressorium is also critical to the control of this important plant disease.

The formation of compound appressoria in *S. sclerotiorum* has been reported to require a contact stimulus [7]. Prior to penetration, the tips of hyphae become swollen and extensively branched, and then form modified, multicellular, and melanin-rich compound appressoria [8,9]. The tip of compound appressorium could penetrate the host epidermis and form vesicles of bulbous [8]. Some events are consistent with this development process, such as the production and accumulation of oxalic acid (OA), cell wall-degrading enzymes (CWDE), and effector proteins, which contribute to *S. sclerotiorum* pathogenesis in myriad ways [9–14]. However, despite these important findings, the detailed molecular mechanism underpinning the development and formation of compound appressoria in *S. sclerotiorum* is still largely unclear.

In the past years, many genetic factors have already been characterized to be essential for appressoria development in *S. sclerotiorum*. The disruption of the oxalic acid biosynthesis gene (*Ssoah1*) promotes the compound appressorium development; however, the disruption of Sspks13 only eliminates the pigmentation of the compound appressorium without attenuating its infection and pathogenicity potential [2]. Significant accumulation of the oxalate decarboxylase (OxDC) gene *Ss-odc2* occurs during the compound appressorium development, and Δ*Ss-odc2* mutants were found to have less effective compound appressorium differentiation [11]. In addition, the secretory proteins Ss-Rhs1 and Ss-Caf1 were highly expressed during the hyphal infection process, whereas the silenced strains had decreased appressoria formation [9,15]. Furthermore, *Ss-ggt1* (γ-glutamyl transpeptidase gene), *sac1* (cAMP pathway adenylatecyclase gene), and *rgb1* (type-2A phosphoprotein phosphatase (PP2A) B regulatory subunit gene) have also been identified to be associated with the development of compound appressoria [16–18]. Recently, the type IV GATA zinc finger transcription factor SsNsd1, orthologous to the *Aspergillus nidulans* NsdD (never in sexual development) proteins and *Botrytis cinerea* BcLTF1 [19], was reported to regulate asexual–sexual development and appressoria formation [4]. Its knockout mutants were defective in the transition from hyphae to compound appressorium formation, resulting in a loss of infection-dependent pathogenicity on healthy hosts [4]. However, the signal pathway by which the SsNsd1 regulates the development and pathogenicity remains to be further elucidated.

Life sciences have been deeply influenced by the "omics" technologies in last decade, including genomics, transcriptomics, proteomics, and metabolomics, aiming at a global perspective on biological systems [20]. Proteomics strategies, such as the two-dimensional gel electrophoresis (2-DE) approaches, have been confirmed as efficient, rapid, and powerful means to identify proteins (or genes) followed by mass spectrometry, and matrix-assisted laser desorption/ionization (MALDI) [21]. Large-scale analyses of proteins by 2-DE have been conducted in a number of organisms, such as animals [22], plants [23], yeast [24], and fungi [25,26], which contributes considerably to our understanding of gene functions in the postgenomic era. However, the development and application of such methods in the filamentous plant-pathogenic fungus *S. sclerotiorum* have not yet been reported.

Modern agriculture faces a huge challenge in the prevention from the diseases caused by *S. sclerotiorum*. The transcription factor SsNsd1 was characterized to be essential for appressoria development in our previous study [4]. The possibility to control plant diseases by suppressing the compound appressorium formation would eliminate initial infections. Here, we used the SsNsd1 knockout mutant (Δ*SsNsd1*) to confirm the loss-of-function nature in compound appressorium development and conducted proteomics analysis by 2-DE. Using comparative proteomics analysis of the Δ*SsNsd1* mutant and the wild-type *S. sclerotiorum*, we attempted to identify the Δ*SsNsd1*-mediated differentially expressed proteins combined with peptide mass spectrometry analysis, which would contribute significantly to the SsNsd1-mediated compound appressorium formation.

2. Results

2.1. Phylogenetic Analysis of SsNsd1 and Other GATA-Type Proteins

In this study, similar proteins of SsNsd1 and other GATA-type proteins of *S. sclerotiorum* were searched from *Botrytis cinerea*, *Fusarium oxysporum*, *Magnaporthe oryzae*, and *Aspergillus oryzae* by BLAST. The homolog of SsNsd1 was obtained only from the *B. cinerea*. However, all other GATA-type proteins had homologs in *B. cinerea*, *F. oxysporum*, *M. oryzae*, and *A. oryzae*. Phylogenetic analysis of the putative amino acid sequence of these GATA-type proteins showed their genetic relationship in different fungi (Figure 1A). Coincidently, the GATA-type proteins of *S. sclerotiorum* were closely related to *B. cinerea*.

Besides, two forms of the type IV zinc finger motif (IVa and IVb) are also depicted in Figure 1A based on the residue loops. Most of the clades contained all five sequences from *S. sclerotiorum*, *B. cinerea*, *F. oxysporum*, *M. oryzae*, and *A. oryzae*. However, the SS1G_03775 and its homologs were separated as shown by the pink color clade. One branch was also separated from the clade of the SS1G_03252 and its homologs, which might be due to the poor similarity in different fungi.

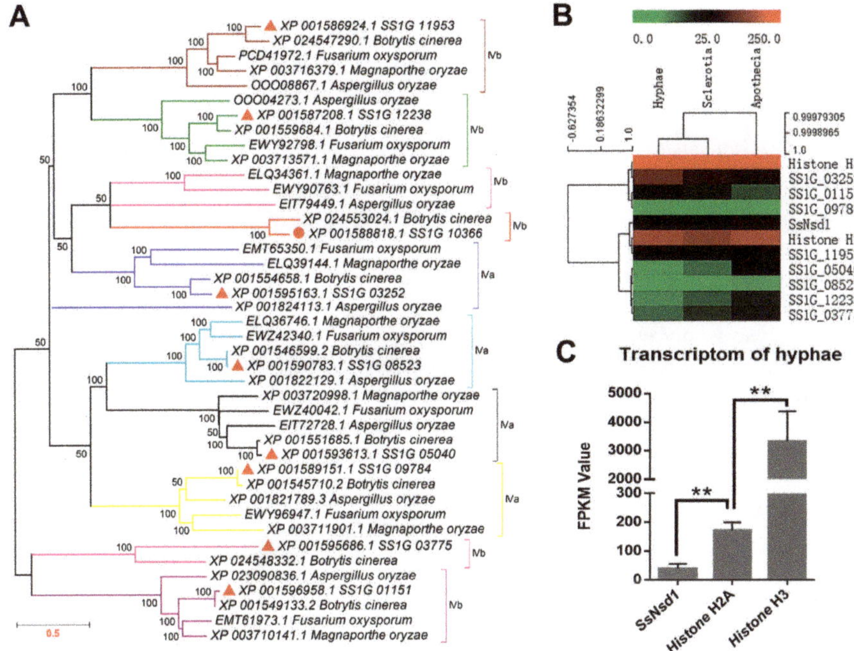

Figure 1. Phylogenetic analysis and transcription expression of SsNsd1 and other proteins containing GATA-type DNA domains in *S. sclerotiorum*. (**A**) Phylogenetic analysis of the amino acid sequences of SsNsd1 (SS1G_10366) and other GATA-type proteins in pathogenic fungi (*S. sclerotiorum*, *B. cinerea*, *F. oxysporum*, *M. oryzae*, and *A. oryzae*). A phylogenetic tree was generated by MAGE using the neighbor-joining method. The nine GATA-type proteins were separated by using a different branch color. (**B**) Hierarchical cluster of GATA-type genes and two histone genes in transcript abundance from three developments stages (hyphae, sclerotia, and apothecia) of *S. sclerotiorum*. Each gene is represented by a single row of colored boxes, and a single column indicates different development stages. The gene transcription abundance was evaluated by the fragments per kilobase of exon per million mapped fragments (FPKM) value. (**C**) The transcription level of *SsNsd1*, *Histone H2A*, and *Histone H3* genes at the hyphae developmental stage. The expression level of *SsNsd1* gene was significantly different from those of the histone genes ($n = 3$; ** $p < 0.01$).

2.2. Transcript Accumulation of SsNsd1 and Other GATA-Type Proteins

Digital gene expression (DGE) analysis based on FPKM values was performed using the transcriptomes during three the developmental stages of *S. sclerotiorum* (Figure 1B). The transcript accumulation in the GATA-type proteins was lower than those of the histone genes (*histone H3* and *histone H2A*). Only the SS1G_03252 protein showed a little higher transcript accumulation in the hyphae development stage, whereas most of the other GATA-type proteins displayed low contents with FPKM values ranging from 1.3 to 64 in all three development stages. Obviously, the SS1G_08523 and SS1G_09784 proteins were extremely low abundant with FPKM values under 10 in all three development stages.

As one of the GATA-type proteins, the varied expression patterns of SsNsd1 were previously examined and compared across developmental stages [4]. In this study, to further determine the regulation mechanism of the Ssnsd1 expression, we obtained its expression profile in the vegetative hyphae development stage (Figure 1C). The transcript accumulation of the *SsNsd1* gene was significantly lower than those of the histone genes (*histone H2A* or *histone H3*) during the vegetative mycelial growth prior to the compound appressorium formation. *SsNsd1* was expressed at the lowest level (FPKM value of 39.6), however, *Histone H2A* was expressed higher (FPKM value of 171.8). Moreover, *Histone H3* (FPKM value of 3337.3) displayed the highest expression level, indicating its predominant role, usually as a housekeeping gene. Overall, the transcript accumulation of SsNsd1 and other GATA-type proteins were exceedingly low in the development stages of *S. sclerotiorum*.

2.3. ΔSsNsd1 Mutant Suppressed Compound Appressorium Formation

Although SsNsd1 exhibited a low expression level during the hyphae development, it still played a crucial role in appressorium development (Figure 2). Phenotypically, the Δ*SsNsd1* mutant had inhibited normal production of pigmented compound appressoria from the vegetative hyphae, as established by paraffin film assays (Figure 2A). In determining whether Δ*SsNsd1* affected the compound appressorium formation or only the pigmentation, normal compound appressorium and penetration (invasive mycelium) were observed microscopically only on onion epidermal strips inoculated with the wild type (WT) strain (Figure 2B), but not on the Δ*SsNsd1* mutant strain. The WT strain could colonize onion cells, but no invasive mycelium was observed from Δ*SsNsd1* mutants (Figure 2C). Thus, SsNsd1 abolished the compound appressoria formation from the modified hyphae, resulting in the penetration-dependent loss of pathogenicity.

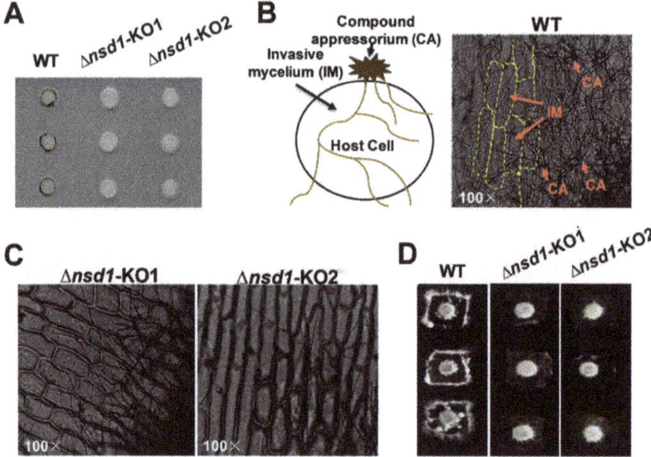

Figure 2. The defective compound appressoria of Δ*SsNsd1* mutant strain led to a loss of penetration into unwounded onion tissue. (**A**) Pigmented compound appressoria of wild type (WT) were observed on parafilm. Pictures were taken four days after transfer (DAT) of 5-mm-diameter mycelial plugs to parafilm. (**B**) Penetration assays with the WT on onion epidermal strips. Invasion mycelium (penetration) of WT strain on onion epidermal strips was observed by light microscopy two days after inoculation (DAI). (**C**) Penetration assays with the Δ*SsNsd1* strain on onion epidermal strips at 2 DAI. No invasion mycelium was observed on onion epidermal strips inoculated by Δ*SsNsd1* strain. (**D**) Penetration assays with the WT and the Δ*SsNsd1* mutant on onion epidermal strips at 4 DAI.

2.4. Diagram of the Identification of Differential Proteins

In this study on *S. sclerotiorum*, we applied the 2-DE technology. A diagram illustrating the approach of the determination of the proteomics changes is presented in Figure 3. In this diagram, the comprehensive protocol is described by individual steps of the application of this technique, i.e., sample preparation and solubilization, isoelectric focusing (IEF) in IPG strips, running SDS-PAGE gels, image analysis, differential spots identification and mass spectrometry (MS) analysis, and bioinformatic prediction.

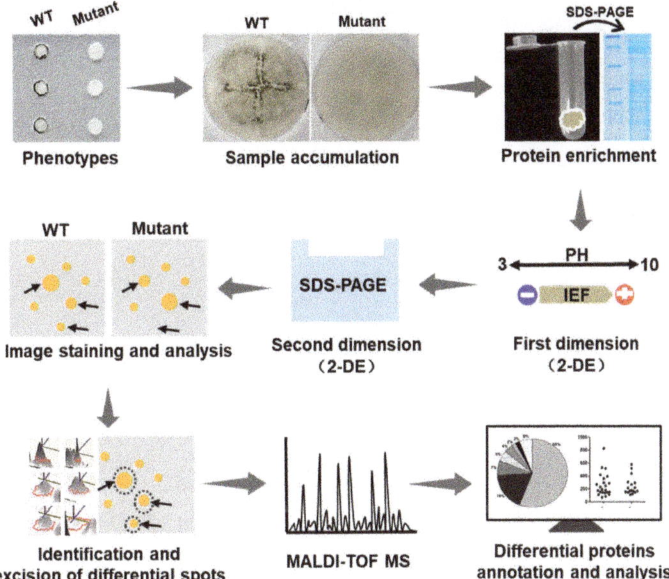

Figure 3. Schematic diagram illustrating the process of identifying the changes of proteomics between the Δ*SsNsd1* mutant and wild type (WT) during the compound appressorium formation. Generally, tissue samples with different phenotypes were subjected to protein extraction and SDS-PAGE test. Then the proteomics profiles were analyzed by two-dimensional gel electrophoresis (2-DE) to obtain the differential expression spots, which were further identified by mass spectrometry (MS) of the peptides and bioinformatics analyses (2-DE, two-dimensional gel electrophoresis; IEF, isoelectric focusing; SDS-PAGE, sodium dodecyl sulfate polyacrylamide gel electrophoresis).

2.5. Protein Extraction from Enriched Compound Appressoria

The cellophane induced the formation of abundant compound appressoria from the modified hyphae, which were observed macroscopically when the WT was cultured on cellophane (Figure 4A). No pigmentation was observed due to the lack of compound appressoria when Δ*SsNsd1* were cultured on cellophane (Figure 4A). Thus, this was an effective method to provide enriched compound appressoria tissue with simple sample collection, which was ideally suited for protein extraction.

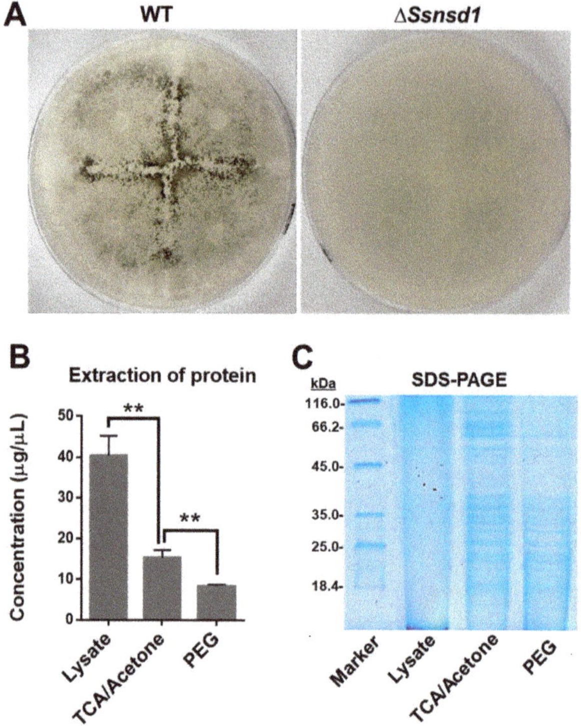

Figure 4. Protein extraction and SDS-PAGE test of accumulated compound appressoria tissue induced on cellophane. (**A**) Induction assay for compound appressoria development on potato dextrose agar (PDA) overlaid with cellophane at 3 DAI. (**B**) Quantification of protein concentration, which was enriched by three different extraction methods: lysate, trichloroacetic acid (TCA)/acetone precipitation, and polyethylene glycol PEG method (n = 3; ** $p < 0.01$). (**C**) Quality detection of proteomics by SDS-PAGE to optimize the method for protein extraction.

To optimize the best protein extraction method for 2-DE separation, we compared the concentration and quality of the extracted proteins using the three different methods. The highest concentration of protein was extracted by the use of the lysate as compared with those extracted by trichloroacetic acid (TCA)/acetone or polyethylene glycol (PEG) precipitation (Figure 4B); however, the quality of this protein, established by the SDS-PAGE test, was extremely poor (Figure 4C). In contrast, the protein extracted by further TCA/acetone or PEG precipitation exhibited clearer bands, but that extracted by TCA/acetone precipitation was at a higher concentration; the best result was obtained using the SDS-PAGE test (Figure 4C). Furthermore, the 2-DE clearly showed protein spots aggregation and transparent background (Figure 5). Thus, the protein extraction with further TCA-acetone precipitation could lay a foundation for a novel approach in comparative proteomics analysis by 2-DE technology in *S. sclerotiorum*.

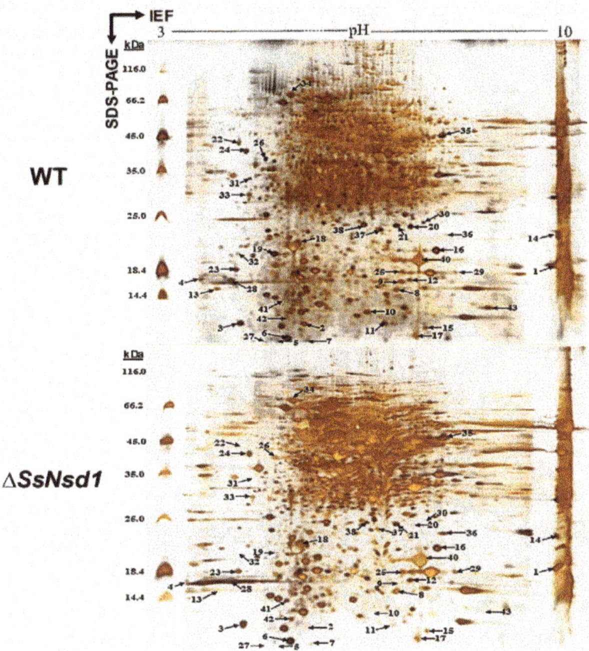

Figure 5. Silver nitrate-stained 2-DE gel image of the WT and Δ*SsNsd1* strains. The proteins were extracted from the accumulated compound appressoria tissue induced by cellophane. The numbers and arrows correspond to the identified differential expression proteins (≥3-fold change) for further MS analysis of the peptides.

2.6. Identification of the Differential Protein Spots by 2-DE Analysis

The proteins extracted from the WT and Δ*SsNsd1* culture on cellophane were separated by 2-DE (Figure 5). In this figure, 2-DE displayed well-visible protein spot aggregation and transparent background, while few nonspecific bands or foreign matters were observed. Using three biological replicates for both the wild-type and mutant strains, the clear 2-DE images were compared and analyzed by ImageMaster™ 2D Platinum 6.0 software. More than 2660 protein spots were detected reproducibly on each 2-DE gel image for the WT and Δ*SsNsd1* mutant, within the pH range of 4 to 10 and with relative molecular masses of 8 to 80 kDa. However, only 43 protein spots exhibited changes in the differential abundance (more than three-fold) between the WT and the Δ*SsNsd1* mutant, which are marked with arrows and numbers in Figure 5. Among these selected differential protein spots, 33 protein spots were downregulated, six proteins spots were upregulated, and four proteins spots disappeared in the Δ*SsNsd1* strain compared to the wild type (Figure 6A). Subsequently, all differential expression proteins were excised and subjected to MALDI-TOF analysis.

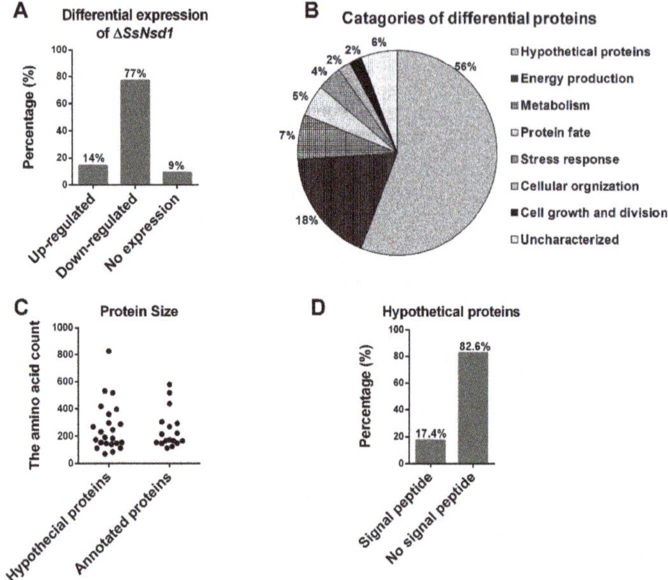

Figure 6. Analysis and evaluation of differentially expressed proteins. (**A**) Differential expression analysis of the identified protein spots in the Δ*SsNsd1* mutant compared to the WT strain. (**B**) Functional categories of differential expression proteins after MS identification and functional annotation. The percentage corresponds to the proportion of the annotated proteins in the classification. (**C**) The population distribution of the protein size evaluated by the amino acid count of the predicted proteins. (**D**) The signal peptide was predicted by running amino acid sequences of predicted hypothetical proteins on the SignalP server.

2.7. Prediction and Characteristics of the Identified Proteins

After prediction and functional annotation of these excised proteins, 40 proteins were identified. Of them, 17 proteins (38%) were predicted with the known functions (Table 1), and other 23 homologous to unnamed or predicted proteins were collectively designated as "hypothetical proteins", accounting for 56% (Table 2). However, three proteins (spots 7, 13, and 23) were not identified by MALDI for unknown reason, which were designated as "uncharacterized" (6%). The predicted known functional categories (38%) were further sorted into six functional categories, including energy production (18%), metabolisms (7%), protein fate (5%), stress responses (4%), cellular organization (2%), and cell growth and division (2%) (Figure 6B).

In addition to providing functional categories, the global view of the protein sizes was also evaluated by the numbers of the amino acids (Figure 6C). Based on the function annotation, they were divided into hypothetical and annotated proteins. Then, the distribution of their protein size was similar to that of 262 and 247 amino acids, respectively (Figure 6C). Furthermore, we predicted the signal peptide among these hypothetical proteins to obtain additional insights into the putative functions of the hypothetical proteins (Figure 6D, Table 2). Four proteins (17.4%) were predicted to contain the N-terminal signal peptide, which indicated they might be potential secretory proteins during the compound appressorium formation.

Table 1. List of the differential expression proteins identified as function known proteins.

Spot [a]	Protein Name [b]	Locus [c]	Accession No. [d]	Score [e]	PM [f]	Cov [g]
01↓	nucleoside diphosphate kinase	gi/156063126	XP_001597486.1	360	6	44
03↓	nuclear transport factor 2	gi/156052963	XP_001592408.1	364	4	34
04↓	60 s ribosomal protein L23	gi/154703604	EDO03343.1	85	6	47
08↓	ubiquitin-conjugating enzyme E2	gi/156042748	XP_001587931.1	141	5	48
09↓	nucleoside diphosphate kinase	gi/156063126	XP_001597485.1	130	8	74
12↓	peptidyl-prolyl cis-trans isomerase	gi/156054872	XP_001593362.1	532	9	53
16↓	peptidyl-prolyl cis-trans isomerase B	gi/156042834	XP_001587974.1	583	10	37
17↓	nucleoside diphosphate kinase	gi/156063126	XP_001597485.1	281	5	39
18↓	eukaryotic translation initiation factor 5A-1	gi/156063512	XP_001597678.1	97	2	44
25↓	peptidyl-prolyl cis-trans isomerase	gi/156054872	XP_001593362.1	620	12	57
26↓	elongation factor 1-beta	gi/156053087	XP_001592470.1	107	2	31
28↓	60 s ribosomal protein L23	gi/154703604	EDO03343.1	59	2	62
29↑	peptidyl-prolyl cis-trans isomerase	gi/156054872	XP_001593362.1	406	8	46
30↓	GTP-binding nuclear protein GSP1/Ran	gi/156057585	XP_001594716.1	215	7	47
32-	SCF complex subunit Skp1	gi/156065065	XP_001598454.1	603	13	69
35↓	citrate synthase, mitochondrial precursor	gi/156063018	XP_001597431.1	104	9	42
39↓	peptidyl-prolyl cis-trans isomerase	gi/156054872	XP_001593362.1	498	11	51

[a]: The number of identified protein spots was showed on the two-dimensional gel electrophoresis (2-DE) image; The identified function known proteins were shown in this table; [b]: Protein name searched by locus tag in NCBI result; [c]: Locus tag number in NCBI annotation; [d]: Accession number in NCBI; [e]: Mascot score (threshold score > 50); [f]: Peptide count; [g]: Percent sequence coverage (%); ↓: expression level of proteins was down-regulated in the mutant; ↑: expression level of protein was up-regulated in the mutant; -: protein spot was disappeared in the mutant.

Table 2. List of the differential expression proteins identified as hypothetical proteins.

Spot [a]	Protein Name [b]	Locus [c]	Accession No. [d]	Score [e]	PM [f]	SignalP [g]
02↓	hypothetical proteins	SS1G_05791	gi|156053886	412	6	No
05-	hypothetical proteins	SS1G_08534	gi|156049655	147	5	No
06↓	hypothetical proteins	SS1G_02967	gi|156061643	1090	16	Yes
10↓	hypothetical proteins	SS1G_13380	gi|156033320	329	7	No
11↓	hypothetical proteins	SS1G_10490	gi|156044772	136	5	No
14↑	hypothetical proteins	SS1G_05792	gi|156053888	404	6	No
15↓	hypothetical proteins	SS1G_03527	gi|156059030	58	3	No
19↓	hypothetical proteins	SS1G_03843	gi|156059662	545	10	No
20↓	hypothetical proteins	SS1G_06246	gi|156054796	143	4	No
21↓	hypothetical proteins	SS1G_06246	gi|156054796	658	10	No
22↓	hypothetical proteins	SS1G_11818	gi|156039363	1040	14	Yes
24↑	hypothetical proteins	SS1G_04923	gi|156055144	199	5	Yes
27-	hypothetical proteins	SS1G_09167	gi|156045782	46	7	No
31↓	hypothetical proteins	SS1G_02387	gi|156060497	237	5	No
33↓	hypothetical proteins	SS1G_06394	gi|156052457	389	4	No
34↓	hypothetical proteins	SS1G_02266	gi|156060255	1510	32	Yes
36↑	hypothetical proteins	SS1G_03994	gi|156056527	61	5	No
37↓	hypothetical proteins	SS1G_12818	gi|156036260	383	8	No
38↓	hypothetical proteins	SS1G_14285	gi|156031009	491	8	No
40↓	hypothetical proteins	SS1G_03527	gi|156059030	421	11	No
41↑	hypothetical proteins	SS1G_08260	gi|156049107	57	2	No
42↑	hypothetical proteins	SS1G_06959	gi|156051104	155	4	No
43-	hypothetical proteins	SS1G_06561	gi|156052787	186	3	No

[a]: The number of identified protein spots was showed on the 2-DE gel image; The hypothetical proteins were shown in this table; [b]: Protein name searched by locus tag in NCBI result; [c]: Locus tag number in NCBI annotation; [d]: Accession number in NCBI; [e]: Mascot score (threshold score > 50); [f]: Peptide count; [g]: Signal peptide predicted (Yes or No); ↓: expression level of protein was down-regulated in the mutant; ↑: expression level of protein was up-regulated in the mutant; -: protein spot was disappeared in the mutant.

2.8. Functional Analysis of Annotated Proteins

The 2-DE approach provided a powerful proteomic screening tool to identify the initial candidate differentially expressed proteins. We evaluated these predicted proteins after functional annotation and identified 17 proteins to be functionally known, representing 10 nonredundant unique proteins (Table 1). The fact that identical proteins were available in different protein spots might have been due to the protein modification or other unclear reasons as analyzed in the discussion section.

Here, we provide information concerning the analysis of the predicted functionally known proteins. Nucleoside diphosphate kinase (NDPK) (spots 1, 9, and 17) usually possesses kinase activity exerted by direct response to the G-protein signaling or indirect catalytic GDP–GTP exchange activity, which also plays a major role in the synthesis of nucleoside triphosphates [27–29]. The eukaryotic translation initiation factor 5A-1 (eIF5A-1; spot 18) is involved in the protein fate pathway [30], which activates the 60 s subunits combination, assists in the conformational changes of the 80 s subunits, and participates in the intracellular part; proteins associate with ribosomes cyclically during the elongation phase of the protein synthesis [31]. The elongation factor 1-β (spot 26) plays a central role in the elongation step in eukaryotic protein biosynthesis [32]. The 60 s ribosomal protein L23 (Spot 4) is usually involved in cell growth; its expression was decreased in the mutant (Table 1). Ribosomal protein is involved in regulating gene transcription, translation, and regulation of cell proliferation, differentiation, apoptosis, etc. [33].

The nuclear transport factor 2 (spot 3) mediates the nucleus introduction of GDP-bound RAN (ras-related nuclear) from the cytoplasm, which is of great significance in the cargo receptor-mediated nucleocytoplasmic transport [34]. The GTP-binding nuclear protein (spot 30) is also known as the GDP-bound RAN, which is involved in the nucleocytoplasmic transport processes, nuclear envelope formation, and mitotic spindle formation [35].

The predicted ubiquitin-conjugating (UBC) enzyme E2 (spot 8) belongs to the ubiquitin pathway enzymes, which are involved in protein degradation in eukaryotic cells [36]. The SCF (Skp1/Cul1/F-box) complex submit Skp1 (spot 32) is involved in the assembly of protein complex and joins in ubiquitin depending on the protein catabolism process [37]. Peptidyl-prolyl cis-trans isomerase (spots 12, 25, 16, and 39) was downregulated, whereas spot 29 was upregulated (Table 1). Peptidyl-prolyl *cis-trans* isomerase regulates the mitosis-related protease in the cell cycle by protein phosphorylation of the substrate proteins [38] or through other mechanisms such as the ubiquitin-mediated proteasomal degradation [39].

Citrate synthase (spot 35) is localized in the mitochondrial matrix and catalyzes the condensation reaction from acetyl coenzyme A (CoA) and oxaloacetate to form the six-carbon citrate [40]. Oxalic acid biogenesis is realized through the hydrolysis of oxaloacetate, which is a key pathogenicity factor accumulated during the compound appressorium development [10].

The functional analysis of the predicted differential proteins was mainly based on the evidence in mammalian, plant, or yeast cells. However, the proteins displayed their important role on the cell proliferation, differentiation, protein synthesis and degradation, protein transport and modification, etc., which might determine they function as a complex regulatory network during the compound appressorium formation in *S. sclerotiorum*.

3. Discussion

Novel strategies for prevention and control of the devastating plant pathogenic fungus *S. sclerotiorum* have been intensively investigated [5]. However, still no effective method has been discovered to control the diseases caused by this pathogen. Compound appressoria are formed unless penetration occurs directly via stomata, which could be one of the key targets for disease control. The GATA-family transcription factors are involved in several essential aspects of the life cycle of *M. oryzae*, especially in the regulation of appressorium development and sporulation [41]. In *S. sclerotiorum*, the GATA-type transcription factors SsSFH1 and SsNsd1 were recently reported to be involved in the development of compound appressoria [4,42]. Here, phylogenetic analysis was performed, and transcription accumulation of all the predicted GATA-type proteins was detected. Importantly, even the transcription accumulation of *SsNsd1* in the vegetative hyphae was significantly higher than that during the sclerotium and apothecium developmental stages [4]. The transcript accumulation remained at a much lower level even during vegetative mycelial growth, compared to that of the histone genes (Figure 1C). The *SsNsd1* gene knockout strain was defective in the development of appressorium, and no penetration was observed into unwounded onion epidermal cells (Figure 2), which confirmed the findings of a previous study [4]. In general, NsdD or its orthologous gene is involved in providing a regulatory balance between asexual and sexual development in ascomycete fungi (e.g., the development of perithecia, fruiting body, conidia, or sclerotia) [43–45]. In addition, NsdD is also involved in pathogenicity. The Δ*bclft1* mutants of *B. cinerea* exhibited only a postpenetration virulence defect without causing significant defects in the compound appressorium development; however, the *S. sclerotiorum* Δ*Ssnsd1* mutant was essentially reversed as established earlier [4]. Due to the particularly different infection defects between *B. cinerea* and *S. sclerotiorum*, in the present study, we focused on the compound appressorium deficiency phenotype and the biological role of SsNsd1, aiming to find the key target of appressorium formation-related genes that would enable the control of this pathogen.

The proteomics analysis as an evaluation of the final level of gene expression started out with techniques based on 2-DE and extended its reach by the use of MS-based techniques that have been increasingly employed in recent years [20]. Although alternative technologies, such as multidimensional protein identification technology (MudPIT), or arrays, have already emerged, thus far, there is no technology that matches 2-DE in its capability to realize routine parallel expression profiling of large mixtures [46]. Furthermore, 2-DE combined with the identification by MS is currently the major approach utilized in most of the undergoing proteome projects to develop a global understanding of the living cell [46]. Compared to the quantitative analysis based on MALDI-TOF techniques, LC–MS is currently in an early stage considering limitations, such as the availability of software, algorithms, etc. [20]. MALDI-TOF is already widely used in fungal proteome research, such as that in yeast [24], *A. fumigatus* [25], and *Cryomyces antarcticus* [26]. Therefore, we applied the 2-DE technology and MALDI-TOF mass spectrometry in this project to investigate the plant pathogenic fungus *S. sclerotiorum* using proteomics analysis for identification of differentially expressed proteins during the compound appressorium formation (Figure 3). Using suitable equipment and experienced laboratory personnel, this system approach can quickly perform the identification of functional proteins.

The 2-DE technology combined with IPGs has already conquered most limitations of carrier ampholyte-based 2-DE with in the respect of reproducibility, handling, resolution, and separation [47]. The efforts to develop the 2-DE technology further have been concentrated on improved solubilization/separation of hydrophobic proteins, show of low abundance, and more reliable quantitation by fluorescent dye technologies in recent years [46]. Despite the obvious advantages of the 2-DE technology, high quality proteins samples are always the bottleneck and precondition to the 2-DE project approach. As the 2-DE was firstly applied in *S. sclerotiorum*, we provided an optimal method for sample preparation after comparing three different pathways for protein extraction (Figure 4). By SDS-PAGE test, the protein extracted by further TCA/acetone precipitation and PEG methods has a better quality and could be used for 2-DE analysis. Both TCA/acetone and PEG methods are useful for minimizing protein degradation and removing interfering compounds, such as salt or polyphenols [48]. However, the amount of protein extracted by the PEG method is difficult to meet the requirements of 2-DE as established based on our results and those of a previous study [49]. In a comprehensive analysis, the TCA/acetone method displayed its advantages and was found to be the best method for protein extraction, which was consistent with the findings of a previous examination on another fungus, *A. fumigatus* [50]. Furthermore, in the 2-DE analysis, we achieved the separation of as many protein spots as needed on the gel, which was a prerequisite for the computerized analysis (Figure 5). In addition, a clear peptide mass spectrum was finally obtained by extraction of the different expressed protein spots, which laid the foundation of further research on SsNsd1-mediated differentially expressed proteins.

The development of the compound appressoria involves several distinct stages [9] and is tightly regulated by numerous genetic factors. In the present study, a total of 43 differentially expressed proteins were identified with significantly differential expression changes ($\geq$3-fold) by computer analysis (Figures 5 and 6). Most of them were downregulated, which indicated that the SsNsd1 transcript factor might positively regulate them. SsNsd1 might exert a reverse role in the signal pathway of the upregulated protein spots. By MS analysis of peptides and functional annotation, these functionally known proteins were predicted to be involved into energy production, metabolism, protein fate, stress response, cellular organization, and cell growth and division. However, attention had to be paid to the hypothetical proteins as they contained the signal peptide. The secretion and accumulation of effector proteins are usually coincident with the appressorium formation process, which contributes to *S. sclerotiorum* pathogenesis [9,15]. Therefore, these newly identified four proteins might have the effector protein role during the compound appressoria formation, but this notion needs to be further studied (Table 2). Overall, the differentially expressed proteins were finally obtained from the Δ*SsNsd1* mutant, which might play an important role during the compound appressorium formation. Furthermore, losing the capacity to produce compound appressorium could also lead to defective sclerotium development, which is a key factor in the disease cycle of *S. sclerotiorum*, such as the mutation of *Ss-ggt1* (a γ-glutamyltranspeptidase gene) [16], *sac1* (a cAMP pathway adenylatecyclase gene) [17], and *rgb1* (a type-2A phosphoprotein phosphatase (PP2A) B regulatory subunit gene) [18]. Therefore, these identified differential expression proteins were important gene resources involved in the development of *S. sclerotiorum*, which might be associated with the formation of both compound appressoria and sclerotia.

Last but not the least, a section on how to evaluate the 2-DE and the identified differential proteins is included here. After 2-DE, each protein could be theoretically resolved at a unique isoelectric point/molecular size coordinate [51]. Although hundreds of protein spots were also separated on the gel, omission of partial differential proteins can always occur due to unfavorable experimental factors, such as incomplete precipitation and/or dissolution of proteins [48], loss of sample during gel entry, inefficiency transfer of the protein from the first to the second dimension, loss of protein during staining [20], and truly absent spots from the samples [52]. In addition, some spots from 2-DE might result multiple protein identification, however, only the first identified protein with best protein score and most peptide counts was accepted for further study. Moreover, attention had also to be paid to the identical proteins (listed in Table 1), such as the nucleoside diphosphate kinase, peptidyl-prolyl cis-trans isomerase, and the 60 s ribosomal protein. Post-translational protein modifications affect the isoelectric point and, therefore, the focusing behavior of the protein in the first dimension [23], which could lead to the presence of identical proteins in different locations (i.e., spot 25 and spot 29; spot 4 and spot 28) (Table 1 and Figure 5). Post-translational modifications by fatty acid acylation, glycosylation, methylation, acetylation, or phosphorylation largely modulate the activity of most eukaryote proteins [53,54]. For example, certain signaling pathways were found to consist of series of phosphorylation and dephosphorylation events, which defined directionality and allowed different levels of feedback regulation [55]. In addition, the incomplete and insensitive separation can also lead to the appearance of identical proteins in different locations. Besides, attention is to be paid to the silver staining, as Coomassie brilliant blue was used in most of the 2-DE staining [56]. The silver staining has a low dynamic range, which has been criticized for the quantitative analyses of spots. However, the silver staining has very high sensitivity, which allows for a detection of very low protein amounts [20]; the improved and advanced image processing method could become feasible in better quantification of protein spots [57]. Besides, to obtain more data of the exact protein abundance, only the differential proteins with $\geq$3-fold changes were accepted for further study in this research. Therefore, the 2-DE technology of gradual optimization, further analysis of protein modifications, and other proteomic analysis methods are still needed to employ, which would present formidable challenges but generate indispensable insight into biological functions in *S. sclerotiorum*.

4. Materials and Methods

4.1. Fungal Strains and Culture Conditions

The wild-type (WT) *S. sclerotiorum* isolate 1980 and its derived mutant Δ*SsNsd1* were used in this study based on our previous reports [4]. The strains were routinely grown on potato dextrose agar (PDA) at normal room conditions. The WT and Δ*SsNsd1* stocks were stocked as dry sclerotia or as desiccated mycelia-colonized filter paper at −20 °C.

4.2. Phylogenetic Analysis of SsNsd1 and Other GATA-Type Proteins

From the genome of *S. sclerotiorum*, nine proteins are predicted to containing GATA-type DNA domains: SS1G_1036, SS1G_11953, SS1G_12238, SS1G_03252, SS1G_08523, SS1G_05040, SS1G_09784, SS1G_03775, and SS1G_01151 [4,42]. The BLASTX program at NCBI (http://www.ncbi.nlm.nih.gov/) was employed to search for the homologs of the sequence of the SsNsd1 (SS1G_1036) and other GATA-type proteins from pathogenic fungi (*Botrytis cinerea*, *Fusarium oxysporum*, *Magnaporthe oryzae*, and *Aspergillus oryzae*). The phylogenetic tree was generated using neighbor-joining method in MEGA5 [58]. Prediction of protein zinc finger domain was performed to classify the different categories. Most fungal GATA factors contain a single zinc finger domain, which can be divided into two distinct categories: the 17-residue loops ($CX_2CX_{17}CX_2C$; zinc finger type IVa) and the 18-residue loops ($CX_2CX_{18}CX_2C$; zinc finger type IVb) [59,60].

4.3. Digital Gene Expression of SsNsd1 and Other GATA-Type Proteins

The transcription accumulation of SsNsd1 from three different developmental stages (hyphae, sclerotia, and apothecia) has been characterized by qRT-PCR [4]. To further study the gene expression patterns of SsNsd1 prior to compound appressorium development, we profiled gene expression patterns in *S. sclerotiorum* from hyphae, sclerotia, and apothecia through RNA-seq approach. The transcription level of SsNsd1 and other GATA-type proteins were quantified using the fragments per kilobase of exon per million mapped fragments (FPKM) method [61]. The FPKM means were generated from three technical replicate samples. The FPKM values of *histone H3* (SS1G_09608.3) and *histone H2A* (SS1G_02052) were used as endogenous control for quantitative comparison with *SsNsd1* and other GATA-type protein genes. Hierarchical clustering was performed using the MeV program [62]. Clustering was based on the average of FPKM values.

4.4. Compound Appressorium Assays

Deficiency of compound appressoria was observed macroscopically by placing the freshly colonized mycelial agar plugs (5-mm diameter) on parafilm due to the presence of pigmented appressoria surrounding the agar plug [2]. Yellow onions were purchased from a local grocery store, and the onion epidermal strips were used for inoculation with a colonized PDA agar plug for observing the penetration of compound appressorium using light microscopy. For the enrichment of compound appressoria, colonized agar was cultured on PDA medium covered with cellophane and grown at a temperature of 22 to 25 °C as reported previously [2].

4.5. Two-Dimensional Gel Electrophoresis (2-DE) Strategy in This Study

The 2-DE combines isoelectric focusing separation based on isoelectric point of protein in the first dimension and sodium dodecyl sulfate (SDS) polyacrylamide gel electrophoresis (SDS-PAGE) to separate the complex mixtures of proteins according to the molecular size in the second dimension [23]. The Combined with identification by mass spectrometry (MS), 2-DE is currently the major method used in the majority of the ongoing proteome projects [46]. Besides, 2-DE gels are easy to handle and could be produced in a highly parallelized way. Furthermore, the corresponding software has also reached a level that allows for routine bioinformatic analysis. Meanwhile, in the presence of a suitable laboratory equipment and experienced personnel, analysis of samples can be theoretically completed through this approach with investments of time and efforts that are much smaller than those needed for the laboratory work [20]. Thus, these mature and coherent techniques are well-suited for comparative proteomics analysis of Δ*SsNsd1* mutant-mediated appressoria deficiency in *S. sclerotiorum*. In detail, tissues derived from two different strains, WT and Δ*SsNsd1*, were harvested, and the proteome was enriched and solubilized. The protein mixture was then applied to a "first dimension" gel strip that separated the proteins based on isoelectric focusing (IEF) in IPG strips. Next, the IPG strip was subjected to equilibration and running of multiple "second dimension" SDS–PAGE gels, where proteins were finally separated by their molecular charge and molecular size. After staining, the visual protein spots were recorded and analyzed by sophisticated software. Then, the differential protein spots were excised for MS analysis. Finally, the differential proteins were subjected to functional annotation and prediction analysis.

4.6. Protein Extraction and Optimization

The enriched compound appressoria on the cellophane were harvested and frozen in liquid nitrogen, and then ground to a fine powder for protein extraction. To optimize the method for protein extraction from *S. sclerotiorum* and separate the protein by 2-DE, during the compound appressoria production, the protein was extracted by direct lysate and further TCA/acetone or PEG precipitation methods. For the lysate method [63], 0.1 g powdered sample was suspended in 300 μL of precooled lysate (7 M urea, 2 M thiourea, 2% (*w/v*) 3-[(3-Cholamidopropyl)dimethylammonio]propanesulfonate (CHAPS), 20 mM Tris-HCl, and 20 mM dithiothreitol), then vortexed for 30 s and centrifuged at $15,000 \times g$ rpm for 10 min at 4 °C. The supernatant was further centrifuged at $15,000 \times g$ rpm for 30 min. Approximately 250 μL of the supernatant was taken as the crude protein solution and stored in a freezer at −80 °C. For further precipitation by trichloroacetic acid (TCA)/acetone [64], 80 μL of crude protein solution was suspended in 5 mL of cold TCA/acetone (10% TCA, 0.07% β-Mercaptoethanol (ME) in acetone) and mixed for 30 s, then precipitated at −20 °C for 2 h. Further, the precipitate was centrifuged at $15,000 \times g$ rpm for 15 min at 4 °C. Then, the supernatant was discarded, and the pellet washed three times with 800 μL of cold acetone and finally centrifuged at $15,000 \times g$ rpm for 15 min at 4 °C. Next, the pellet was desiccated using a vacuum dryer and stored at −80 °C. For the PEG precipitation [65], 80 μL of crude protein solution was suspended using 40% PEG solution for 30 s and then precipitated at −20 °C for 2 h. Further, the pellet was washed with 800 μL of cold acetone and desiccated, as described above. The final protein was resuspended in 80 μL of rehydration buffer (7 M urea, 2 M thiourea, 4% (*w/v*) CHAPS, 0.002% (*w/v*) bromophenol blue, 2% (*v/v*) Bio-Lyte, and 20 mM dithiothreitol). The concentration of the dissolved protein solution was determined according to the method of Bradford [66]. Then, the results were compared by routine SDS-PAGE test to optimize the method for protein extraction, and the extracted protein with the best quality was further separated by 2-DE for image analysis.

4.7. The 2-DE Assay and Image Analysis

Comparative 2-DE was performed using a Ettan™ IPGphor apparatus (GE Healthcare, Pittsburgh, PA, USA) for isoelectric focusing (IEF; first dimension), and an Ettan™ DALTsix (GE Healthcare) for the second dimension according to the manufacturer's instructions [48] and the protocol described in a previous report [56]. For IEF, 300 μg hydrated protein was loaded on immobilized pH gradient (IPG) strips (18 cm length, pH 3.0–10.0). The following IEF steps were used: 500 V for 1 h, 1000 V for 1 h, 4000 V for 1 h, 8000 V for 1 h with a linear gradient, holding at 8000 V until a total of at least 40,000 Vh was reached, then holding at 500 V for 20 h. After the strips were balanced twice, the second dimension was performed with 12% SDS-PAGE gel, and the total proteins were stained before further analysis. The stained gel image was captured with an Image Master LabScan (GE Healthcare), and the images were analyzed using the ImageMaster™ 2D Platinum 6.0 software (GE Healthcare) for spot detection, gel matching, and statistical analysis of the spots [56]. The image analysis remains one of the most labor-intensive parts of the 2-DE approach. In brief, triplicate images from three independent gels for the WT and mutant were obtained, while the normalization of the gels was carried out by the sum of the spot densities on each gel to compare the spots. The abundance of the individual protein spots was determined as vol.%. To identify the protein spots, the silver staining method was applied to the prepared gels. Silver nitrate was added to the solution before use, and then it was quickly admixed into the dyeing tray. The tray was then covered with an opaque cloth to reduce the decomposition of the silver nitrate utilized.

After visualization by staining and gel image analysis, protein spots with at least 3-fold spot volume ratio change ($p < 0.05$) were excised and subjected to mass spectrometry sequence analysis combined with database comparison. Statistical comparisons were conducted using the one-way ANOVA with the Tukey's HSD test.

4.8. MALDI-TOF Analysis and Prediction of Differential Proteins

In this study, the spots showing statistically significant changes were cut out from the preparative gels and washed twice with ultrapure water. Then, the protein spots were destained with 50% acetonitrile (ACN) in 25 mM NH_4HCO_3. After removing the destaining buffer, the gel pieces were lyophilized and rehydrated in 30 μL of 50 mM NH_4HCO_3 containing 50 ng trypsin (Promega, Madison, WI, USA) at 37 °C overnight. The supernatant of the resulting peptides was washed with 0.1% trifluoroacetic acid (TFA) in 67% ACN. Extracts were pooled and lyophilized for MS analysis. The MS spectra were obtained using an ABI 4800 MALDI-TOF/MS-MS Proteomics Analyzer (Applied Biosystems, Foster City, CA, USA) as previously described [56]. The positive ion reflector (2 kV accelerating voltage) with 1000 laser shots per spectrum and automatic data acquisition modes were used for data collection, and the TOF spectra were collected over the mass range within 800–4000 Da with a signal-to-noise ratio minimum set to 10 and a local noise window width of m/z 250. A maximum of 10 precursors per spot with a minimum signal/noise ratio of 50 were selected for data-dependent MS/MS analysis.

Then, the resultant MS and MS/MS spectra data were analyzed with the GPS Explorer software (Version 2.0, Applied Biosystems). The database search was performed on the Mascot server (http://www.matrixscience.com) by searching the NCBInr (nonredundant protein sequence) database of *S. sclerotiorum* (http://www.ncbi.nlm.nih.gov/) to identify the proteins. The other important parameters were set as follows trypsin cleavage, two missed cleavage allowed, carbamidomethylation set as fixed modification, oxidation of methionine allowed as variable modification, monoisotopic precursor mass, precursor ion mass tolerance set to ±100 ppm, and fragment mass tolerance set to ±0.5 Da. The protein was correctly identified if a sufficient number of peptides were matched with a high score to a protein in the database. Only the significant hits with a protein score of 100% and the highest peptide counts were recorded and analyzed. Then the predicted protein sequences that matched the sequences in the NCBInr database were further analyzed for functional category denomination [56,62]. Prediction of the signal peptide was done using the online SignalP 4.1 Server (http://www.cbs.dtu.dk/services/SignalP/).

4.9. Data Analysis

All graphs were exported by the GraphPad Prism 6 software (La Jolla, CA, USA). Statistical comparisons were done using the one-way ANOVA with the Tukey's HSD (Honestly Significant Difference) test in the PASW Statistics 18 (SPSS Inc., Chicago, IL, USA).

5. Conclusions

In this study, we combined TCA/acetone precipitation for protein extraction, 2-DE, and peptide mass analysis to develop a fast and simple method for studying the proteomics changes of Δ*SsNsd1* mutant during compound appressorium formation. In our approach the results from 2-DE gel analysis are put into a larger context by combining spot data with functional annotations to explore the SsNsd1-mediated compound appressoria formation. Visualizing results, such as differential expression, functional categories, and predicted effector proteins makes it possible to gain new insights from the data accumulated by the "omics" technologies. Thus, this system approach can be effectively used to identify important candidate proteins in response to the SsNsd1-mediated appressorium formation, but it will require subsequent, more detailed studies to determine the precise role of the differentially expressed proteins.

Author Contributions: J.L. (Jingtao Li), X.Z., and L.L. performed the experiments and analyzed the data. J.L. (Jingtao Li) wrote the manuscript. H.P. and Y.Z. conceived the study and provided funding. J.L. (Jinliang Liu) provided technical support. All authors commented on the manuscript.

Funding: This study was financially supported by the National Natural Science Foundation of China (31772108, 31471730, and 31271991) and the National Research and Development Program of China (2018YFD0201005).

Acknowledgments: We gratefully acknowledge Jeffrey A. Rollins (University of Florida) for donating the wild-type strain 1980 and the mutants. We also thank Gang Yu for manuscript proofreading and technical assistance.

Conflicts of Interest: The authors declare no conflict of interest.

References

1. Boland, G.J.; Hall, R. Index of plant hosts of *Sclerotinia sclerotiorum*. *Can. J. Plant Pathol.* **1994**, *16*, 93–108. [CrossRef]
2. Li, J.; Zhang, Y.; Zhang, Y.; Yu, P.-L.; Pan, H.; Rollins, J.A. Introduction of large sequence inserts by CRISPR-Cas9 to create pathogenicity mutants in the multinucleate filamentous pathogen *Sclerotinia sclerotiorum*. *mBio* **2018**, *9*, e00567-18. [CrossRef] [PubMed]
3. Bardin, S.D.; Huang, H.C. Research on biology and control of sclerotinia diseases in Canada. *Can. J. Plant Pathol.* **2001**, *23*, 88–98. [CrossRef]
4. Li, J.; Mu, W.; Veluchamy, S.; Liu, Y.; Zhang, Y.; Pan, H.; Rollins, J.A. The GATA-type IVb zinc-finger transcription factor SsNsd1 regulates asexual–sexual development and appressoria formation in *Sclerotinia sclerotiorum*. *Mol. Plant Pathol.* **2018**, *19*, 1679–1689. [CrossRef] [PubMed]
5. Bolton, M.D.; Thomma, B.P.H.J.; Nelson, B.D. *Sclerotinia sclerotiorum* (Lib.) de Bary: Biology and molecular traits of a cosmopolitan pathogen. *Mol. Plant Pathol.* **2006**, *7*, 1–16. [CrossRef] [PubMed]
6. Guimarães, R.L.; Stotz, H.U. Oxalate production by *Sclerotinia sclerotiorum* deregulates guard cells during infection. *Plant Physiol.* **2004**, *136*, 3703–3711. [CrossRef] [PubMed]
7. Garg, H.; Hua, L.; Sivasithamparam, K.; Kuo, J.; Barbetti, M.J. The infection processes of *Sclerotinia sclerotiorum* in cotyledon tissue of a resistant and a susceptible genotype of *Brassica napus*. *Ann. Bot.* **2010**, *106*, 897–908. [CrossRef] [PubMed]
8. Lumsden, R.D.; Dow, R.L. Histopathology of *Sclerotinia sclerotiorum* infection of bean. *Phytopathology* **1973**, *63*, 708. [CrossRef]
9. Xiao, X.; Xie, J.; Cheng, J.; Li, G.; Yi, X.; Jiang, D.; Fu, Y. Novel secretory protein Ss-Caf1 of the plant-pathogenic fungus *Sclerotinia sclerotiorum* is required for host penetration and normal sclerotial development. *Mol. Plant Microbe Interact.* **2014**, *27*, 40–55. [CrossRef] [PubMed]
10. Liang, X.; Liberti, D.; Li, M.; Kim, Y.-T.; Hutchens, A.; Wilson, R.; Rollins, J.A. Oxaloacetate acetylhydrolase gene mutants of *Sclerotinia sclerotiorum* do not accumulate oxalic acid, but do produce limited lesions on host plants. *Mol. Plant Pathol.* **2015**, *16*, 559–571. [CrossRef] [PubMed]
11. Liang, X.; Moomaw, E.W.; Rollins, J.A. Fungal oxalate decarboxylase activity contributes to *Sclerotinia sclerotiorum* early infection by affecting both compound appressoria development and function. *Mol. Plant Pathol.* **2015**, *16*, 825–836. [CrossRef] [PubMed]
12. Zuppini, A.; Navazio, L.; Sella, L.; Castiglioni, J.; Favaron, F.; Mariani, P. An endopolygalacturonase from *Sclerotinia sclerotiorum* induces calcium-mediated signaling and programmed cell death in soybean cells. *Mol. Plant Microbe Interact.* **2005**, *18*, 849–855. [CrossRef] [PubMed]

13. Yajima, W.; Liang, Y.; Kav, N.N.V. Gene disruption of an arabinofuranosidase/beta-xylosidase precursor decreases *Sclerotinia sclerotiorum* virulence on canola tissue. *Mol. Plant Microbe Interact.* **2009**, *22*, 783–789. [CrossRef] [PubMed]

14. Zhu, W.; Wei, W.; Fu, Y.; Cheng, J.; Xie, J.; Li, G.; Yi, X.; Kang, Z.; Dickman, M.B.; Jiang, D. A secretory protein of necrotrophic fungus *Sclerotinia sclerotiorum* that suppresses host resistance. *PLoS ONE* **2013**, *8*, e53901. [CrossRef] [PubMed]

15. Yu, Y.; Xiao, J.; Zhu, W.; Yang, Y.; Mei, J.; Bi, C.; Qian, W.; Qing, L.; Tan, W. Ss-Rhs1, a secretory Rhs repeat-containing protein, is required for the virulence of *Sclerotinia sclerotiorum*. *Mol. Plant Pathol.* **2017**, *18*, 1052–1061. [CrossRef] [PubMed]

16. Li, M.; Liang, X.; Rollins, J.A. *Sclerotinia sclerotiorum* γ-Glutamyl transpeptidase (Ss-Ggt1) is required for regulating glutathione accumulation and development of sclerotia and compound appressoria. *Mol. Plant Microbe Interact.* **2012**, *25*, 412–420. [CrossRef] [PubMed]

17. Jurick, W.M.; Rollins, J.A. Deletion of the adenylate cyclase (*sac1*) gene affects multiple developmental pathways and pathogenicity in *Sclerotinia sclerotiorum*. *Fungal Genet. Biol.* **2007**, *44*, 521–530. [CrossRef] [PubMed]

18. Erental, A.; Harel, A.; Yarden, O. Type 2A phosphoprotein phosphatase is required for asexual development and pathogenesis of *Sclerotinia sclerotiorum*. *Mol. Plant Microbe Interact.* **2007**, *20*, 944–954. [CrossRef] [PubMed]

19. Schumacher, J.; Simon, A.; Cohrs, K.C.; Viaud, M.; Tudzynski, P. The transcription factor BcLTF1 regulates virulence and light responses in the necrotrophic plant pathogen *Botrytis cinerea*. *PLoS Genet.* **2014**, *10*, e1004040. [CrossRef] [PubMed]

20. Berth, M.; Moser, F.M.; Kolbe, M.; Bernhardt, J. The state of the art in the analysis of two-dimensional gel electrophoresis images. *Appl. Microbiol. Biotechnol.* **2007**, *76*, 1223–1243. [CrossRef] [PubMed]

21. Pandey, A.; Mann, M. Proteomics to study genes and genomes. *Nature* **2000**, *405*, 837. [CrossRef] [PubMed]

22. Michael, F.; Peter, B.; Hanno, L.; Laura, S. The rat liver mitochondrial proteins. *Electrophoresis* **2002**, *23*, 311–328.

23. Mayer, K.; Albrecht, S.; Schaller, A. Targeted Analysis of Protein Phosphorylation by 2D Electrophoresis. In *Plant Phosphoproteomics: Methods and Protocols*; Schulze, W.X., Ed.; Springer: New York, NY, USA, 2015; pp. 167–176.

24. Makrantoni, V.; Antrobus, R.; Botting, C.H.; Coote, P.J. Rapid enrichment and analysis of yeast phosphoproteins using affinity chromatography, 2D-PAGE and peptide mass fingerprinting. *Yeast* **2005**, *22*, 401–414. [CrossRef] [PubMed]

25. Voedisch, M.; Scherlach, K.; Winkler, R.; Hertweck, C.; Horn, U.; Brakhage, A.A.; Kniemeyer, O. Characterisation of the hypoxic response of the human-pathogenic fungus *Aspergillus fumigatus* by 2D-gel electrophoresis. *Infection* **2009**, *37*, 19.

26. Zakharova, K.; Sterflinger, K.; Razzazi-Fazeli, E.; Noebauer, K.; Marzban, G. Global proteomics of the extremophile black fungus *Cryomyces antarcticus* using 2D-electrophoresis. *Nat. Sci.* **2014**, *6*, 978–995.

27. Dharmasiri, S.; Harrington, H.M.; Dharmasiri, N. Heat shock modulates phosphorylation status and activity of nucleoside diphosphate kinase in cultured sugarcane cells. *Plant Cell Rep.* **2010**, *29*, 1305–1314. [CrossRef] [PubMed]

28. Shen, Y.; Han, Y.-J.; Kim, J.-I.; Song, P.-S. Arabidopsis nucleoside diphosphate kinase-2 as a plant GTPase activating protein. *BMB Rep.* **2008**, *41*, 645–650. [CrossRef] [PubMed]

29. Hetmann, A.; Kowalczyk, S. Nucleoside diphosphate kinase isoforms regulated by phytochrome A isolated from oat coleoptiles. *Acta Biochim. Pol.* **2009**, *56*, 143–153. [PubMed]

30. Wang, X.-L.; Cai, H.-P.; Ge, J.-H.; Su, X.-F. Detection of eukaryotic translation initiation factor 4E and its clinical significance in hepatocellular carcinoma. *World J. Gastroenterol.* **2012**, *18*, 2540–2544. [CrossRef] [PubMed]

31. Xu, G.-D.; Shi, X.-B.; Sun, L.-B.; Zhou, Q.-Y.; Zheng, D.-W.; Shi, H.-S.; Che, Y.-L.; Wang, Z.-S.; Shao, G.-F. Down-regulation of eIF5A-2 prevents epithelial-mesenchymal transition in non-small-cell lung cancer cells. *J. Zhejiang Univ. Sci. B* **2013**, *14*, 460–467. [CrossRef] [PubMed]

32. Achilonu, I.; Elebo, N.; Hlabano, B.; Owen, G.R.; Papathanasopoulos, M.; Dirr, H.W. An update on the biophysical character of the human eukaryotic elongation factor 1 beta: Perspectives from interaction with elongation factor 1 gamma. *J. Mol. Recognit.* **2018**, *31*, e2708. [CrossRef] [PubMed]

33. Dai, M.-S.; Sun, X.-X.; Lu, H. Ribosomal protein L11 associates with c-Myc at 5 S rRNA and tRNA genes and regulates their expression. *J. Biol. Chem.* **2010**, *285*, 12587–12594. [CrossRef] [PubMed]

34. Cushman, I.; Bowman, B.R.; Sowa, M.E.; Lichtarge, O.; Quiocho, F.A.; Moore, M.S. Computational and biochemical identification of a nuclear pore complex binding site on the nuclear transport carrier NTF2. *J. Mol. Biol.* **2004**, *344*, 303–310. [CrossRef] [PubMed]

35. De Boor, S.; Knyphausen, P.; Kuhlmann, N.; Wroblowski, S.; Brenig, J.; Scislowski, L.; Baldus, L.; Nolte, H.; Krüger, M.; Lammers, M. Small GTP-binding protein Ran is regulated by posttranslational lysine acetylation. *Proc. Natl. Acad. Sci. USA* **2015**, *112*, E3679–E3688. [CrossRef] [PubMed]

36. Uben. Ubiquitin-conjugating Enzyme, E2. In *Encyclopedic Reference of Cancer*; Schwab, M., Ed.; Springer: Berlin/Heidelberg, Germany, 2001; p. 942.

37. Takayama, Y.; Mamnun, Y.M.; Trickey, M.; Dhut, S.; Masuda, F.; Yamano, H.; Toda, T.; Saitoh, S. Hsk1- and SCF^{Pof3}-dependent proteolysis of *S. pombe* Ams2 ensures histone homeostasis and centromere function. *Dev. Cell* **2010**, *18*, 385–396. [CrossRef] [PubMed]

38. Velazquez, H.A.; Hamelberg, D. Conformational selection in the recognition of phosphorylated substrates by the catalytic domain of human Pin1. *Biochemistry* **2011**, *50*, 9605–9615. [CrossRef] [PubMed]

39. Liou, Y.-C.; Zhou, X.Z.; Lu, K.P. Prolyl isomerase Pin1 as a molecular switch to determine the fate of phosphoproteins. *Trends Biochem. Sci.* **2011**, *36*, 501–514. [CrossRef] [PubMed]

40. Wiegand, A.; Remington, S.J. Citrate synthase: Structure, control, and mechanism. *Annu. Rev. Biophys. Biophys. Chem.* **1986**, *15*, 97–117. [CrossRef] [PubMed]

41. Quispe, C.F. GATA-Family Transcription Factors in *Magnaporthe oryzae*. Master's Thesis, University of Nebraska-Lincoln, Lincoln, NE, USA, 2011.

42. Liu, L.; Wang, Q.; Sun, Y.; Zhang, Y.; Zhang, X.; Liu, J.; Yu, G.; Pan, H. *Sssfh1*, a gene encoding a putative component of the RSC chromatin remodeling complex, is involved in hyphal growth, reactive oxygen species accumulation, and pathogenicity in *Sclerotinia sclerotiorum*. *Front. Microbiol.* **2018**, *9*. [CrossRef] [PubMed]

43. Tsitsigiannis, D.I.; Kowieski, T.M.; Zarnowski, R.; Keller, N.P. Endogenous lipogenic regulators of spore balance in *Aspergillus nidulans*. *Eukaryot. Cell* **2004**, *3*, 1398–1411. [CrossRef] [PubMed]

44. Han, K.-H.; Han, K.-Y.; Yu, J.-H.; Chae, K.-S.; Jahng, K.-Y.; Han, D.-M. The nsdD gene encodes a putative GATA-type transcription factor necessary for sexual development of *Aspergillus nidulans*. *Mol. Microbiol.* **2001**, *41*, 299–309. [CrossRef] [PubMed]

45. Lee, M.-K.; Kwon, N.-J.; Lee, I.-S.; Jung, S.; Kim, S.-C.; Yu, J.-H. Negative regulation and developmental competence in *Aspergillus*. *Sci. Rep.* **2016**, *6*, 28874. [CrossRef] [PubMed]

46. Weiss, W.; Görg, A. High-resolution two-dimensional electrophoresis. *Methods Mol. Biol.* **2009**, *564*, 13–32. [PubMed]

47. Weiss, W.; Görg, A. Two-dimensional electrophoresis for plant proteomics. *Methods Mol. Biol.* **2007**, *355*, 121–143. [PubMed]

48. Görg, A.; Weiss, W.; Dunn, M.J. Current two-dimensional electrophoresis technology for proteomics. *Proteomics* **2004**, *4*, 3665–3685. [CrossRef] [PubMed]

49. Görg, A.; Weiss, W. Two-Dimensional Electrophoresis with Immobilized pH Gradients. In *Proteome Research: Two-Dimensional Gel Electrophoresis and Identification Methods*; Rabilloud, T., Ed.; Springer: Berlin/Heidelberg, Germany, 2000; pp. 57–106.

50. Yu, Z.; Dan, H.; Wang, S. Comparison of methods for extraction of protein secreted from *Aspergillus fumigatus*. *Chin. J. Biol.* **2010**, *23*, 313–316.

51. Smith, R. Two-Dimensional Electrophoresis: An Overview. In *Two-Dimensional Electrophoresis Protocols*; Tyther, R., Sheehan, D., Eds.; Humana Press: Totowa, NJ, USA, 2009; pp. 2–17.

52. Pedreschi, R.; Hertog, M.L.; Carpentier, S.C.; Lammertyn, J.; Robben, J.; Noben, J.P.; Panis, B.; Swennen, R.; Nicolaï, B.M. Treatment of missing values for multivariate statistical analysis of gel-based proteomics data. *Proteomics* **2008**, *8*, 1371–1383. [CrossRef] [PubMed]

53. Boyle, W.J.; van der Geer, P.; Hunter, T. Phosphopeptide mapping and phosphoamino acid analysis by two-dimensional separation on thin-layer cellulose plates. *Methods Enzymol.* **1991**, *201*, 110–149. [PubMed]

54. Mann, M.; Jensen, O.N. Proteomic analysis of post-translational modifications. *Nat. Biotechnol.* **2003**, *21*, 255. [CrossRef] [PubMed]

55. Schulze, W.X. Proteomics approaches to understand protein phosphorylation in pathway modulation. *Curr. Opin. Plant Biol.* **2010**, *13*, 279–286. [CrossRef] [PubMed]

56. Zhang, X.; Fu, J.; Hiromasa, Y.; Pan, H.; Bai, G. Differentially expressed proteins associated with fusarium head blight resistance in wheat. *PLoS ONE* **2013**, *8*, e82079. [CrossRef] [PubMed]

57. Kishimoto, M.; Tatsumi, Y.; Tamesui, N.; Kumada, Y.; Horiuchi, J.I.; Okumura, K. Dynamic analysis of the silver staining gel image of 2-DE for protein spots segmentation. *J. Electrophor.* **2010**, *54*, 1–7. [CrossRef]

58. Li, J.; Yu, G.; Sun, X.; Zhang, X.; Liu, J.; Pan, H. AcEBP1, an ErbB3-Binding Protein (EBP1) from halophyte *Atriplex canescens*, negatively regulates cell growth and stress responses in Arabidopsis. *Plant Sci.* **2016**, *248*, 64–74. [CrossRef] [PubMed]

59. Teakle, G.R.; Gilmartin, P.M. Two forms of type IV zinc-finger motif and their kingdom-specific distribution between the flora, fauna and fungi. *Trends Biochem. Sci.* **1998**, *23*, 100–102. [CrossRef]

60. Reyes, J.C.; Muro-Pastor, M.I.; Florencio, F.J. The GATA family of transcription factors in Arabidopsis and rice. *Plant Physiol.* **2004**, *134*, 1718–1732. [CrossRef] [PubMed]

61. Yuan, J.-B.; Zhang, X.-J.; Liu, C.-Z.; Wei, J.-K.; Li, F.-H.; Xiang, J.-H. Horizontally transferred genes in the genome of pacific white shrimp, *Litopenaeus vannamei*. *BMC Evol. Biol.* **2013**, *13*, 165. [CrossRef] [PubMed]

62. Li, J.; Sun, X.; Yu, G.; Jia, C.; Liu, J.; Pan, H. Generation and analysis of expressed sequence tags (ESTs) from halophyte *Atriplex canescens* to explore salt-responsive related genes. *Int. J. Mol. Sci.* **2014**, *15*, 11172–11189. [CrossRef] [PubMed]

63. Papura, D.; Jacquot, E.; Dedryver, C.A.; Luche, S.; Riault, G.; Bossis, M.; Rabilloud, T. Two-dimensional electrophoresis of proteins discriminates aphid clones of Sitobion avenae differing in BYDV-PAV transmission. *Arch. Virol.* **2002**, *147*, 1881–1898. [CrossRef] [PubMed]

64. Méchin, V.; Damerval, C.; Zivy, M. Total protein extraction with TCA-acetone. *Methods Mol. Biol.* **2007**, *355*, 1–8. [PubMed]

65. Sim, S.-L.; He, T.; Tscheliessnig, A.; Mueller, M.; Tan, R.B.H.; Jungbauer, A. Protein precipitation by polyethylene glycol: A generalized model based on hydrodynamic radius. *J. Biotechnol.* **2012**, *157*, 315–319. [CrossRef] [PubMed]

66. Bradford, M.M. A rapid and sensitive method for the quantitation of microgram quantities of protein utilizing the principle of protein-dye binding. *Anal. Biochem.* **1976**, *72*, 248–254. [CrossRef]

 © 2018 by the authors. Licensee MDPI, Basel, Switzerland. This article is an open access article distributed under the terms and conditions of the Creative Commons Attribution (CC BY) license (http://creativecommons.org/licenses/by/4.0/).

International Journal of
Molecular Sciences

Article

Interspecies Outer Membrane Vesicles (OMVs) Modulate the Sensitivity of Pathogenic Bacteria and Pathogenic Yeasts to Cationic Peptides and Serum Complement

Justyna Roszkowiak [1], Paweł Jajor [2], Grzegorz Guła [1], Jerzy Gubernator [3], Andrzej Żak [4], Zuzanna Drulis-Kawa [1] and Daria Augustyniak [1,*]

1   Department of Pathogen Biology and Immunology, Institute of Genetics and Microbiology, University of Wroclaw, 51-148 Wroclaw, Poland; justyna.roszkowiak@uwr.edu.pl (J.R.); grzegorz.gula@uwr.edu.pl (G.G.); zuzanna.drulis-kawa@uwr.edu.pl (Z.D.-K.)
2   Department of Pharmacology and Toxicology, Faculty of Veterinary Medicine, Wroclaw University of Environmental and Life Sciences, 50-375 Wroclaw, Poland; pawel.jajor@upwr.edu.pl
3   Department of Lipids and Liposomes, Faculty of Biotechnology, University of Wroclaw, 50-383 Wroclaw, Poland; jerzy.gubernator@uwr.edu.pl
4   Ludwik Hirszfeld Institute of Immunology and Experimental Therapy, Polish Academy of Science, 53-114 Wroclaw, Poland; andrzej.zak@hirszfeld.pl
*   Correspondence: daria.augustyniak@uwr.edu.pl; Tel.: +48-71-375-6296

Received: 30 September 2019; Accepted: 5 November 2019; Published: 8 November 2019

Abstract: The virulence of bacterial outer membrane vesicles (OMVs) contributes to innate microbial defense. Limited data report their role in interspecies reactions. There are no data about the relevance of OMVs in bacterial-yeast communication. We hypothesized that model *Moraxella catarrhalis* OMVs may orchestrate the susceptibility of pathogenic bacteria and yeasts to cationic peptides (polymyxin B) and serum complement. Using growth kinetic curve and time-kill assay we found that OMVs protect *Candida albicans* against polymyxin B-dependent fungicidal action in combination with fluconazole. We showed that OMVs preserve the virulent filamentous phenotype of yeasts in the presence of both antifungal drugs. We demonstrated that bacteria including *Haemophilus influenza*, *Acinetobacter baumannii*, and *Pseudomonas aeruginosa* coincubated with OMVs are protected against membrane targeting agents. The high susceptibility of OMV-associated bacteria to polymyxin B excluded the direct way of protection, suggesting rather the fusion mechanisms. High-performance liquid chromatography-ultraviolet spectroscopy (HPLC-UV) and zeta-potential measurement revealed a high sequestration capacity (up to 95%) of OMVs against model cationic peptide accompanied by an increase in surface electrical charge. We presented the first experimental evidence that bacterial OMVs by sequestering of cationic peptides may protect pathogenic yeast against combined action of antifungal drugs. Our findings identify OMVs as important inter-kingdom players.

Keywords: outer membrane vesicles (OMVs); *Candida albicans*; antimicrobial peptides; complement; interspecies interactions; inter-kingdom protection; fungicidal activity; fluconazole; hyphae

---

1. Introduction

In the environment, different microbial populations, including bacteria and fungi, co-exist. Mixed microbial populations are also a common feature in many diseases. Hence, understanding which factors may be potentially important players in interspecies dynamics of growth and to what degree these dynamics are mediated by the host is very important. Among these factors are outer membrane vesicles (OMVs) of Gram-negative bacteria classified recently as secretion system type zero [1]. These proteoliposomal nanoparticles released from the cell play an important role in bacterial physiology and pathogenesis. During pathogenesis, they enhance cellular adherence, cause biofilm formation, and induce apoptosis or inflammation [2–4]. Furthermore, DNA-containing OMVs, through horizontal gene transfer, are important in interspecies communication as well as in host–pathogen interactions [5]. OMVs contribute also to innate bacterial defense through β-lactamase content or trapping and degrading the membrane-active peptides [6–10]. Accordingly, the involvement of OMVs in resistance to antibiotics with various modes of action is growing [11]. Both offensive and defensive functions of OMVs have been reported. In line with this, OMVs can deliver bactericidal toxins or enzymes to other bacteria. Lytic activities have been documented for OMVs derived from *Pseudomonas aeruginosa* or *Myxococcus xanthus* [12–14]. On the other hand, for example, *M. catarrhalis* OMVs carrying β-lactamases confer protection to *M. catarrhalis* and *Streptococcus pneumoniae* against β-lactam antibiotics [15]. Likewise, OMVs from β-lactam-resistant *E.coli* can protect β-lactam-susceptible *E.coli* and fully rescue them from β-lactam antibiotic-induced growth inhibition [16]. Several earlier studies have shown that OMVs are involved in the trapping of antimicrobials, thus providing protection among bacteria, essentially in intraspecies systems [6,8,17]. Nevertheless, limited data are available on the relevance of this phenomenon in various interspecies populations. Among mixed populations, the degree of OMV-dependent protection of individual partners can be completely different. One of the factors that may influence this is the physicochemical nature of the vesicle itself. It includes both the intrinsic antimicrobial binding capacity, which is the result of inherited or acquired resistance [18] but also the ability of vesicle, based on physicochemical properties, to interact with other target cells. There are several factors responsible for the latter phenomenon including electric charge and hydrophobicity of a cellular surface [19,20]. The metabolic activity of OMV-producing bacteria also seems to have a significant impact on this issue. For example, for *P. aeruginosa* it was reported that variation in physicochemical properties was dependent on growth phase (exponential versus stationary), influencing, therefore, the cell association activity [19]. All aforementioned characteristics of OMVs indicate the significant degree of selectivity in OMV–cell interactions. Therefore, considering variety of microbial populations, the question of which bacteria are and which are not protected against membrane-active agents is still very ambiguous. It is also of great importance to ask whether this protection can go beyond the protection against only bacteria, affecting pathogens from other kingdoms such as fungi. Hence, it is of interest to investigate the protective potency of OMVs in the light of their specificity to react with a target cell. In the present study, we used our well-characterized OMVs from *Moraxella catarrhalis* Mc6 [4,21] as model vesicles and characterized their protective potential against model membrane-active agents in various interspecies combinations. First, we examined the protective activity of OMVs in bacterial intra- and interspecies systems and mode of observed protection, showing by HPLC-UV, zeta potential measurement, and o-nitrophenyl-β-D-galactopyranoside ONPG-based permeabilization assay, highly effective sequestration. Next, we examined the protective potential of OMVs in the bacterial-yeast inter-kingdom system. As far as we know, our study is the first report that OMVs can protect *Candida albicans* against antifungals drugs. Finally, we investigated whether OMVs-dependent protection is a signature of only free vesicles or those associated with the cell.

2. Results

2.1. Mc6 OMVs Characteristics

As shown in transmission electron microscopy (TEM) image, the diameters of outer membrane vesicles from *M. catarrhalis* 6 had 30–200 nm (Figure 1a). The results of measurements of OMV particle size/zeta potential are shown in Figure 1b. The protein and lipooligosaccharide (LOS) components of OMVs are shown in Figure 1c,d, respectively. As we documented previously [21], the pivotal outer membrane proteins packaged in these vesicles were OmpCD, OmpE, UspA1 (ubiquitous surface protein A1, Hag/MID (*Moraxella* immunoglobulin D-binding protein), CopB, MhuA (hemoglobin-binding protein), TbpA (transferrin-binding protein A), TbpB (transferrin-binding protein B), LbpB (lactoferrin-binding protein B), OMP M35, and MipA (structural protein).

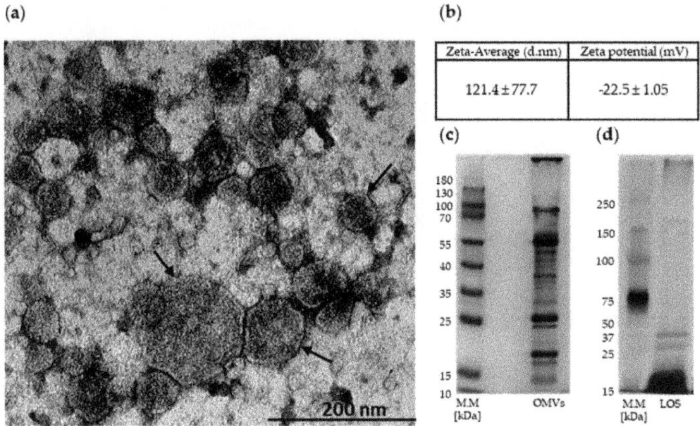

Figure 1. Physical characterization of outer membrane vesicles (OMVs) released from Mc6 cells: (**a**) TEM image of OMVs, vesicles are indicated by arrows (magnification, ×50,000); (**b**) the size distribution by volume and the zeta potential of vesicles, as assessed by the Zeta-sizer; each experiment was performed in triplicate; (**c**) the respresentative proteinogram of 12% SDS-PAGE electrophoresis of OMVs; the protein profiles were visualized using Coomassie staining; (**d**) the representative 10% SDS-PAGE electrophoresis of lipooligosaccharide (LOS)-OMVs visualized by silver staining.

2.2. *M. catarrhalis OMVs Passively Protect Cross-Pathogens against Polymyxin B-Dependent Killing*

Polymyxin B (PB) was used as a model of cationic peptide. In our study, using 4-h time-kill assay, the minimal bactericidal concentrations (MBC) of PB against 5×10^5–10^6 cfu/mL of prominent human pathogens, including nontypeable *Haemophilus influenzae*, *Pseudomonas aeruginosa*, and *Acinetobacter baumannii*, was in the range 0.5–5 µg/mL, and caused killing effect within 2 h (Figure 2a). When pathogenic bacteria were incubated with a bactericidal concentration of PB in combination with 20 µg/mL OMVs from Mc6, the bacteria showed active growth in contrast to the antibiotic alone, which remained at the level of control. Up to 10 times lower concentrations of OMVs had no effect or the effect was negligible. The amount of OMVs that was required to achieve complete protection against respiratory pathogens referred to 20 µg/mL. The higher concentrations of vesicles did not intensify the growth.

2.3. *OMVs Protect Serum-Sensitive Strains and Accelerate Growth of Serum-Resistant Strains against Complement*

To define the influence of OMVs on the survival of cross-pathogens in the presence of active serum (NHS), the 4 h complement bactericidal assays were performed (Figure 2b). Of the three studied cross-pathogens, only nontypeable *H. influenzae* appeared to be serum-sensitive. Two others, *A. baumannii* and *P. aeruginosa*, were serum-resistant. The resistant strains showed either 100% survival or only a slight decrease (within one order) after 4 h of incubation in 50% or 75% NHS. Nontypeable *H. influenzae* strains were sensitive to NHS-dependent killing with a reduction of cfu/mL in the range of six orders after 30 min of incubation (NTHi3) or 120 min of incubation (NTHi6) in the presence of 25% or 50% serum, respectively, compared to the initial inoculum. These strains co-incubated with NHS and 20 µg/mL of OMVs exhibited a time-dependent increase in survival, which was similar to growth of control incubated in the presence of heat-inactivated serum (HiNHS).

(a) (b)

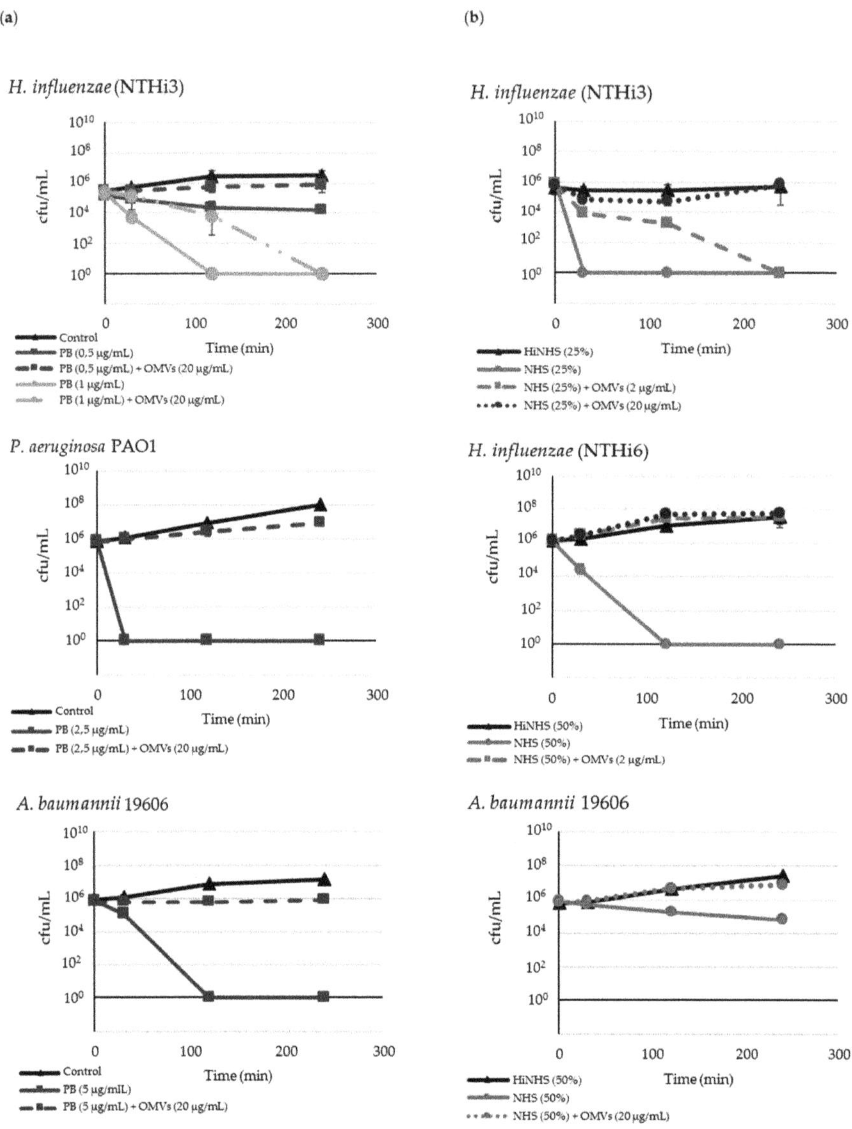

Figure 2. Mc6 OMVs protect bacteria from other species against polymxin B-dependent and human complement-dependent bactericidal activity: (**a**) Polymyxin B-dependent killing; (**b**) human serum (NHS) complement-dependent killing. Bacteria from log phase were incubated for 4 h in the presence of indicated concentrations of membrane-targeting agents alone or along with OMVs and plated in 0, 30, 120, and 240 min. Data are expressed as mean cfu/mL $\pm$ SD from three independent experiments performed in triplicate.

A. baumannii incubated only in the presence of active NHS was slightly killed by the lytic complement action within 4 h, whereas after parallel exposure to OMVs, its growth was at the level of HiNHS control. *P. aeruginosa* was highly resistant to 75% action of serum, thus co-incubation with OMVs did not change the growth dynamics (data not shown). When control bacteria were incubated in the presence of heat-inactivated NHS, neither killing nor other growth alterations were observed for any of the tested strains, indicating that observed lysis was complement-dependent. Overall, these findings indicate that OMVs efficiently protect serum-sensitive gram-negative pathogens against complement action while accelerating the growth of moderately serum-resistant strains.

2.4. Mc 6 OMVs Passively Protect Pathogenic Yeasts against Polymyxin B-Dependent Fungicidal Effect in Combination with Fluconazole

It has been previously reported that the antifungal effect of polymyxin B combined with fluconazole can be synergistic or potentiated [22]. We therefore initially examined the sensitivity of *Candida albicans* to combined action of both drugs. Our results showed, that for ~10^5 cfu/mL of this yeast, the MIC_{50} for fluconazole (FLC) and MIC_{100} for polymyxin B (PB) were, respectively, 1 µg/mL and 128 µg/mL. Next, based on checkerboard assay, we assessed the growth of *C. albicans* incubated with polymyxin B alone, fluconazole alone, or their combinations at various concentrations. The results revealed that polymyxin B at concentrations much lower than MIC (1/8 MIC and 1/16 MIC) exerts a potent antifungal effect against *C. albicans* when combined with 1 µg/mL (MIC_{50}) of FLC (Figure 3a).

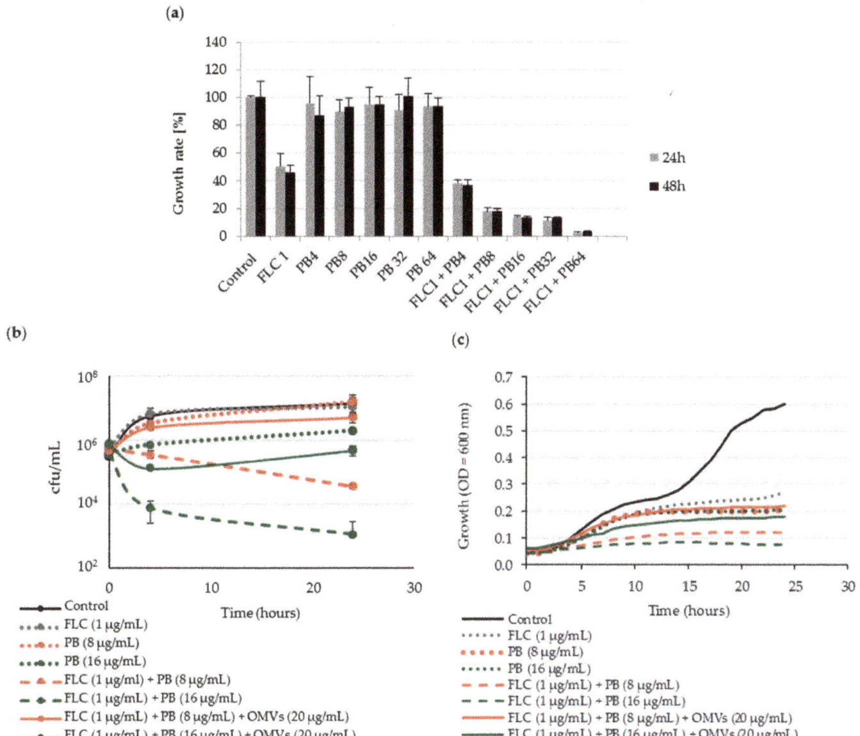

Figure 3. OMVs protect yeast against synergistic fungicidal activity of polymyxin B in combination with fluconazole: (**a**) Synergistic activities determined by checkerboard assay; (**b**) time-kill assay; (**c**) 24 h growth-curve kinetics for *C. albicans* incubated with drug alone and drug in combinations with or without OMVs. The kinetics were measured using Varioskan™ LUX reader with measurements at 1 h intervals. The data for experiments (a,b) show means ± SD from two independent experiments carried out in triplicate; the data for experiment (c) show mean values from two independent repetitions carried out in duplicate; for better readability, SDs were not included.

Next, using time-kill assay we showed that after 24 h of incubation, the combinations of FLC^1-PB^8 and FLC^1-PB^{16} (µg/mL) are fungicidal, which was expressed by one order and by two orders of decrease in viability, respectively (Figure 3b). In the presence of OMVs, this fungicidal action of both drugs was significantly abolished or weakened in comparison to action of a single drug and depending on PB concentration used. This protective effect of Mc6 OMVs against *C. albicans* was also confirmed in bacterial growth kinetics (Figure 3c). By estimating the profiles of *Candida* growth, as a result of every hour measurements carried out for 24 h at 37 °C, we have documented that the OMVs added to both aforementioned drug combinations abolished their inhibitory effect on *Candida* growth rate. Thus, the *Candida* growth in the presence of OMVs remained at the level for a single compound. Collectively, these results indicate that OMVs protect *C. albicans* from PB-dependent fungicidal effect in combination with fluconazole.

2.5. Mc6 OMVs Passively Enhance the Virulence of Pathogenic Yeasts Facilitating the Formation of Filaments

It was documented that azole drugs are necessary both during growth and induction step to have an effect on transition yeast-to hyphae in *C. albicans* [23]. The presence of FLC in the decreased range between 1–0.125 µg/mL both during growth and induction step caused considerable inhibition of hyphal growth (data not shown). Therefore, to determine the effects of fluconazole in combination with polymyxin B on hyphal growth, the yeasts were initially preincubacted overnight at 37 °C with 1/16 MIC of FLC (0.0625 µg/mL). During the 2 h induction of hyphal growth in the presence of 10% fetal bovine serum, 1/16 MIC of FLC was added together with 1/8, 1/16, and 1/32 MIC of PB in the presence of absence of 20 µg/mL of OMVs. Figure 4a,b shows that the number of hyphal cells, shown by the ratio of cells in yeast form compared to that with formed filaments after 2 h of incubation, decreases as the concentration of polymyxin B increases. Under the tested conditions, the combinations $FLC^{0.0625}$-PB^8 and $FLC^{0.0625}$-PB^{16} (µg/mL) showed almost complete inhibitory activity against the formation of hyphae. $FLC^{0.0625}$-PB^4 had also clear inhibitory effect on yeast-to-hyphae transition in comparison to FLC alone. The presence of 20 µg/mL OMVs in all tested drug combinations caused a significant increase in the percentage of hyphae-forming cells (Figure 4b), probably as a result of neutralizing PB-dependent potentiating effect of FLC. Furthermore, the increase in yeast-to-hyphal transition was accompanied by a significant extension of filaments (Figure 4c). In summary, these results indicate that in the presence of Mc6 OMVs, the inhibiting effect of FLC and PB, at concentrations significantly lower than MICs, on the formation of filaments is abolished. It shows that in the presence of OMVs, *C. albicans* may retain the virulent filamentous phenotype in the presence of both antifungal drugs, thus increasing its virulence.

2.6. OMV-Dependent Protection against PB and Complement is the Result of PB Sequestration on Free OMVs

To study how efficiently PB is neutralized by Mc6 OMVs, various biological and biochemical methods were used. Initially, a decrease of free PB was confirmed indirectly using *E.coli* ML35p mutant with constitutive β-galactosidase expression (Figure 5a).

In the experiment, different concentrations of Mc6 OMVs were introduced into the system containing *E. coli* ML-35p mutant at a concentration of ~10^6 cfu/mL and PB at a concentration of 5 µg/mL. By measuring the quantitative intracellular influx and hydrolysis of ONPG (β-galactosidase substrate), as a result of membrane permeabilization, we showed that OMV-dependent protection was dose-dependent and that PB was very quickly depleted from the environment in the presence of at least 20 µg/mL OMVs leading to a decrease in peptide activity (Figure 5a).

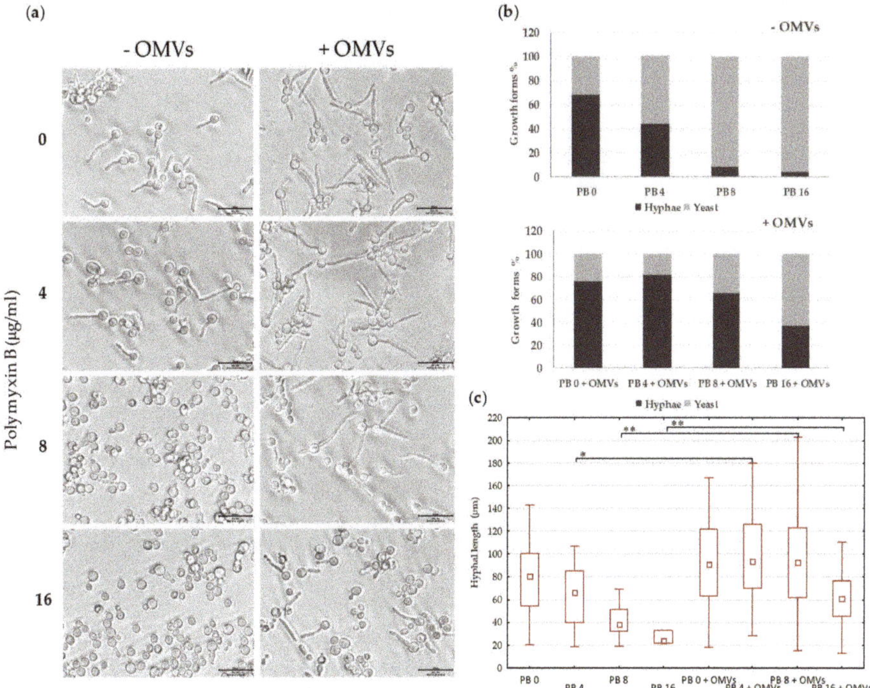

Figure 4. Effect of OMVs on filamentous growth of *C. albicans*: Yeast cells were preincubated overnight with 1/16 MIC of FLC (0.0625 µg/mL) without shaking and then, after washing, were incubated at OD = 0.4 in 0.4 mL of hyphae inducing medium (10% FBS in PBS) in 24-well microtiter plates with shaking (130 rpm) for 2 h at 37 °C in the presence of 0.0625 µg/mL FLC, indicated concentrations of PB, and with or without 20 µg/mL of OMVs. Three independent experiments were performed. (**a**) The filamentation was monitored under inverted microscope using 40× objective (Zeiss) and images were recorded. Scale bars = 100 µm. (**b**) The number of hyphal cells versus yeast cells was determined using ImageJ software. The pool of yeast cells contains yeasts and yeasts whose germ tubes did not exceed 10 µm. The data represent the mean values calculated for ≥ 200 cells for each of the eight tested options. (**c**) Box and Whisker plots of hyphal length: Minimal and maximal values, median, quartiles Q_1 and Q_3. Statistical analysis was performed by one-way ANOVA (* $p < 0.05$, ** $p < 0.01$).

Next, to directly confirm the role of OMVs in PB sequestration, we determined zeta potential, showing that the initially moderately negative zeta potential of intact OMVs was immediately and significantly neutralized by addition of at least 50 µg/mL of PB, in a concentration-dependent and diluent-dependent manner leading to membrane depolarization (Table 1). The presented results of alteration in Zeta potentials were similar after 30 min and 60 min incubation at 37 °C (data not shown). Finally, to quantitatively assess the magnitude of PB sequestration by OMVs, the residual free PB content remaining after ultrafiltration of OMV-PB complexes was measured using HPLC-UV. Figure 6 shows the chromatograms of recovered by ultrafiltration PB that was preincubated for 30 min either alone or with OMVs. The loss of PB following incubation with OMVs at 5 µg/mL and 20 µg/mL was almost 60% ($p < 0.001$) and over 96% ($p < 0.001$), respectively, in reference to standard PB (Table 2).

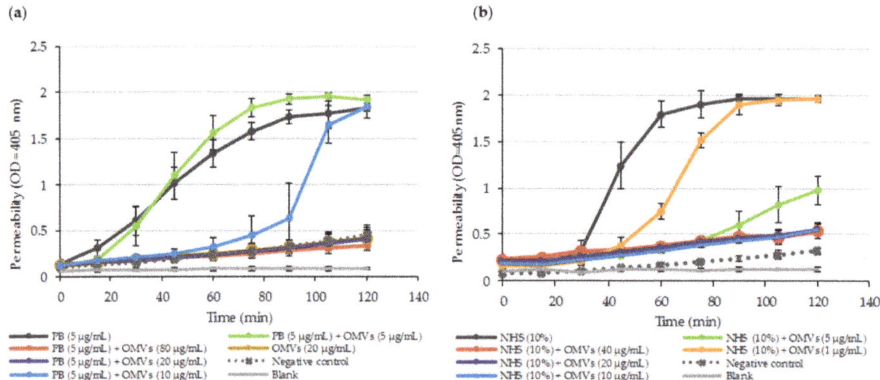

Figure 5. OMVs block the permeabilizing activity of membrane targeting agents. Change in *E. coli* ML-35p membrane permeability was assayed by a time-dependent influx of ONPG in the presence of membrane-targeting agents and different concentrations of Mc6 OMVs. Bacteria at log phase ($\sim10^6$ cfu/mL) were incubated on microplate at 37 °C for 120 min in the presence of indicated concentrations of OMVs together with (**a**) polymyxin B (PB) and (**b**) normal human serum (NHS). The absorbance was measured at indicated time points. The results are shown as mean ± SD from at least three independent experiments performed in duplicate.

Table 1. Alteration in Zeta potential of OMVs treated with polymyxin B.

PB Binding by OMVs in NaPB Buffer		PB Binding by OMVs in miliQ	
Treatment	Zeta Potential (mV)	Treatment	Zeta Potential (mV)
20 µg/mL OMVs (control)	−22.5 ± 1.05	20 µg/mL OMVs (control)	−24.7 ± 0.76
20 µg/mL OMVs + 5 µg/mL PB	−21.2 ± 1.06	20 µg/mL OMVs + 5 µg/mL PB	−25.9 ± 0.97
20 µg/mL OMVs + 50 µg/mL PB	−16.5 ± 0.76 *	20 µg/mL OMVs + 50 µg/mL PB	−9.6 ± 0.35 *
20 µg/mL OMVs + 250 µg/mL PB	−10.1 ± 0.46 *	20 µg/mL OMVs + 250 µg/mL PB	−1.17 ± 0.14 *

*-$p < 0.001$ in reference to control as determined by one-way ANOVA.

Collectively, our results demonstrate that PB is quickly and effectively depleted from the environment via OMV-dependent sequestration.

Similar experiments with *E.coli* ML35p mutant were performed with human active serum (NHS). Similar to PB, the strong permeabilizing potency of membrane attack complex (MAC), present in 10% human serum, against *E.coli* ML35p was considerably decreased or even dumped as the concentration of OMVs increased (Figure 5b). It suggests that by deposition of complement components on the OMV surface, vesicles trigger MAC formation away from target bacteria and thus protect them from MAC-mediated lysis. Accordingly, based on the quantitative formation of soluble non-proteolytic membrane attack complex (SC5b-9), we demonstrated that the Mc6 OMVs activate human serum complement in a concentration-dependent manner (Figure S1).

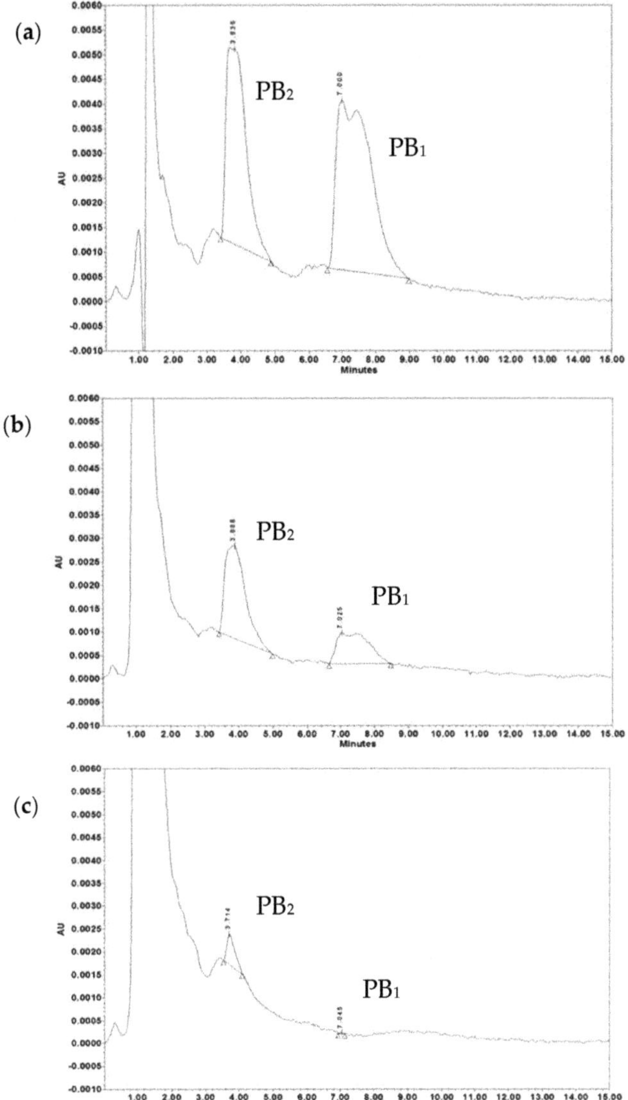

Figure 6. The HPLC spectra of polymyxin B incubated alone or along with OMVs. Chromatograms of (**a**) 50 µg/mL solution of polymyxin B, (**b**) 50 µg/mL of PB after treatment with 5 µg/mL of Mc6 OMVs, (**c**) 50 µg/mL of PB after treatment with 20 µg/mL of Mc6 OMVs. Samples were prepared as described in Materials and Method section. HPLC analyses were performed with Macherey–Nagel Nucleodur C18 Isis column. UV-detection at 215 nm. Double peak with Tr = 3.8 min corresponds mainly to polymyxin PB_2 and PB_3, and double peak with Tr= 7 min corresponds to polymyxin PB_1 and PB_1-I. The representative chromatograms are shown.

Table 2. Magnitude of PB sequestration on *M. catarrhalis* OMVs.

Treatment	PB$_2$ (µg/mL)	PB$_1$ (µg/mL)	PB$_2$ + PB$_1$ (µg/mL)
Standard PB	14.38 ± 0.48	17.45 ± 3.58	31.83 ± 3.47
PB after treatment with 5 µg/mL OMVs	7.76 ± 1.01	5.19 ± 1.05	13.67 ± 1.33 *
PB after treatment with 20 µg/mL OMVs	0.96 ± 0.42	0.06 ± 0.01	1.02 ± 0.42 *

The PB content was measured by HPLC-UV method as described in material and methods. The results are expressed as mean ± SD. * $p < 0.001$ in reference to standard PB as determined by one-way ANOVA.

2.7. OMVs Associated with Bacteria do not Protect against PB

To answer the question of whether Mc6 OMVs protect bacteria against the membrane-active agent by sequestration only indirectly, being far from the bacterial surface or through the direct shield of these cells, we determined the association capability of OMVs using flow cytometry. We used 4-times higher concentration of OMVs (80 µg/mL) to make sure that the entire cell population could be covered (Figure S2). We found that Mc6 OMVs incubated with studied strains strongly interacted within 30 min with the parental strain as well as with other strains from *Moraxella* species. After this time, more than 93% of measured events were fluorescent, indicating that FITC-labelled OMVs were associated with unlabeled bacteria. No increase in fluorescence was observed when Mc6 OMVs were incubated even up to 2 h with bacteria from other species indicating the complete lack of association (Figure 7a,b). These results indicate that OMVs released by *M. catarrhalis* associated only in intraspecies but not in interspecies tested systems, pointing to the specificity of this reaction. To verify whether the covering of *M. catarrhalis* with OMVs is another mode of protection against PB, 4 h time-kill assays (Figure 7c,d), as well as 24 h real-time bacterial growth kinetics (Figure 7e,f) were carried out on OMV-associated and not associated cells in the presence and absence of PB.

Not in the line with expectation, the results showed that for *M. catarrhalis* associated and not associated with OMVs, the lethal effect of PB occurred after 4 h (Figure 7c) and 2 h (Figure 7d), respectively. It indicates that the association does not ensure longer protection but only delays the bactericidal effect. The lack of protection was confirmed in growth kinetic experiments documenting that PB-dependent growth inhibition for OMV-associated *M. catarrhalis* and non-associated control was comparable and preserved for 24 h of incubation (Figure 7e). In the case of bacteria incubated in the presence of free OMVs and PB, the OMV-dependent protection is assured for ~7 h being on the level of control growth. Some modest and stable level of protection is maintained for 24 h (Figure 7f). Compared to the growth profiles of OMV-associated and non-associated *M. catarrhalis*, it was shown that the interaction with vesicles weakens the growth dynamics of the former. Overall, the results are evidence that OMVs associated with bacteria do not protect against PB, assuming of course that the bacterial cell is completely shielded by them. Consequently, it suggests the active fusion between OMVs and target cell membrane rather than only surface association.

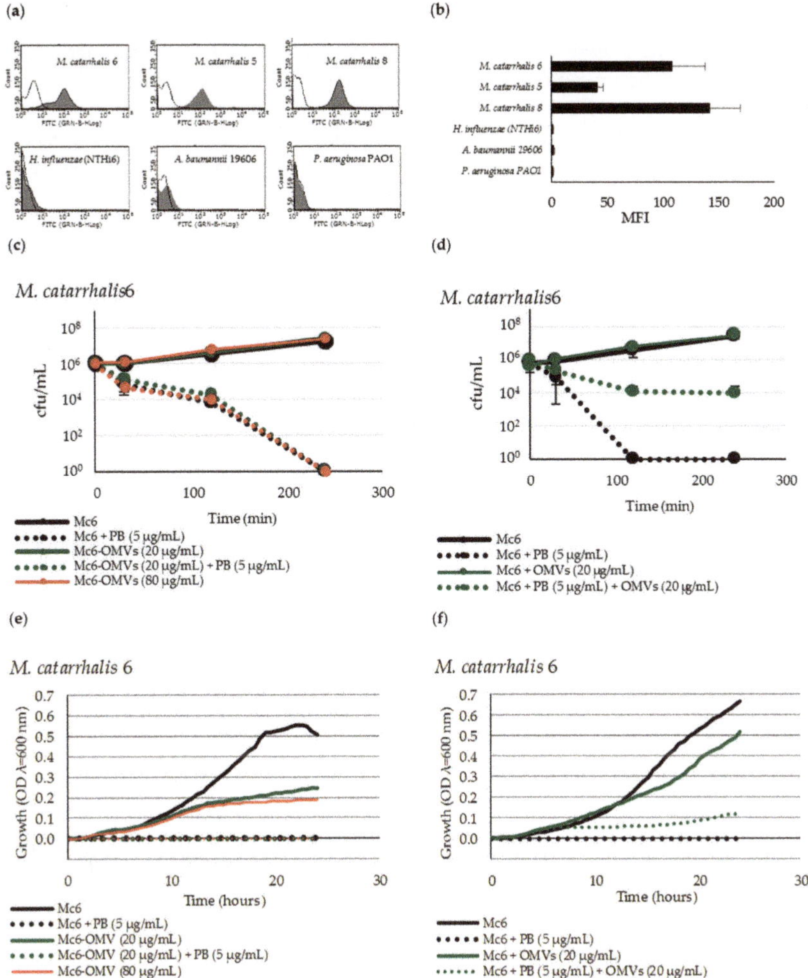

Figure 7. Mc6 OMVs highly associated with bacteria do not protect against polymyxin B-dependent killing. (**a**) Flow cytometry analysis of fluorescein isothiocyanate (FITC)-labelled OMVs associated with intraspecies bacteria (upper panel) and not associated with interspecies bacteria (lower panel), The fluorescence intensities of OMV-associated bacteria are shown as black histograms whereas control bacteria as dotted histograms. Representative plots are shown. (**b**) Quantification of bacteria associated with FITC-labelled OMVs. Data are expressed as mean fluorescent intensity (MFI) ± SD from three independent experiments performed in duplicates. (**c**) Bactericidal activity of polymyxin B against *M. catarrhalis* after association with OMVs. (**d**) Bactericidal activity of polymyxin B against *M. catarrhalis* incubated with free OMVs. (**e**) 24 h growth-curve kinetics for *M. catarrhalis* after association with OMVs. (**f**) 24 h growth-curve kinetics for *M. catarrhalis* incubated with free OMVs. The kinetics was measured using Varioskan™ LUX reader with measurements at 30 min intervals. In experiments (c,d) results are shown as mean ± SD from at least two independent experiments. In experiments (e,f) results are shown as mean from at least three independent experiments performed at duplicate and for better readability, SDs were not included.

3. Discussion

Bacteria and yeasts employ a variety of means to protect their envelopes against harmful environmental factors. Some of these factors cause membrane permeabilization, while others can inhibit the synthesis of cellular membrane components affecting the physical properties of the membrane [24,25]. The common strategy of bacteria to avoid membrane targeting factors is to overcome the negative charges of their surface envelope [24,26]. Alternatively, bacteria can release OMVs that act as extracellular decoys for some antimicrobials.

In this study, we report that OMVs may serve as a model of passive interspecies protection of prominent pathogenic bacteria but also pathogenic yeasts. While documenting the lethal effect, we showed that free OMVs block both the bactericidal action of peptide antibiotic polymyxin B and the lytic activity of complement. By quantitative measurement of alteration in electric charge on the OMV surface treated with PB as well as free PB determination by HPLC, we showed that 20 µg/mL of OMVs that had a significant protective effect against interspecies microorganisms caused almost complete depletion of this model cationic antimicrobial agent after co-incubation, thus indicating the immediate sequestration of the peptide on free OMVs. Accordingly, in similar experiments, we showed that OMVs inhibit serum lytic activity against prone pathogens, indicating that by deposition of complement components on the OMV surface, vesicles trigger MAC (SC5b-9) formation away from target bacteria and thus protect them from MAC-mediated lysis. The intensity of OMV-dependent complement activation in vitro was correlated with the number of free vesicles in the environment. Therefore, it is possible that any fluctuation in the number of released vesicles may positively or negatively influence complement activation of this potent innate mechanism.

Next, using the PB-dependent model, we investigated whether OMVs-dependent protection is a signature of only free vesicles or those associated with the cell. Using flow cytometry, we documented that the association of OMVs with bacterial cells may be very potent and species-specific, influencing the cell sensitivity to antimicrobial compound. Our vesicles were able to associate only with representatives of *M. catarrhalis* species while they did not work in interspecies systems. By evidencing the lethal effect of cell-associated OMVs exposed to polymyxin B, we showed that OMVs do not protect against AMPs this way, suggesting a fusion between OMVs and OM rather than only shielding. It is tempting to speculate that the specificity of interaction between OMVs and recipient cell is somehow pivotal in its sensitization at least to AMPs. We are therefore in agreement with Tashiro et al. that elucidating the selectivity in OMV-cell interaction is critical for an improved understanding of the outcome of this reaction. On the other hand, it has been documented that interactions with OMVs for some cross-pathogens can be specific, less specific, or not specific at all [27]. Thus, the OMV-dependent interplay may contribute to the generation of synergistic or antagonistic interactions between pathogens, resulting in more or less harmful outcomes for the host.

Next, we address the question of whether OMV-dependent trapping of cationic antimicrobials can protect pathogens from other kingdoms such as fungi of the species *C. albicans*. Polymyxins alone are effective against *C. albicans* only at relatively high concentrations [28]. Azoles including fluconazole are common antifungal drugs [29]. They affect the integrity of fungal membranes, altering their morphology and inhibiting growth [30,31]. It has been reported that quinolone and other antibiotics may augment the anti-candidal activity of azole and polyene agents [32]. Previous research also documented that PB in lower concentrations exerts a potent antifungal effect when combined with fluconazole [22,33]. Therefore, the elimination of the antibiotic from these systems by means of OMVs should decrease the fungus susceptibility to fluconazole. To test this hypothesis, we incubated *C. albicans* in the presence of both drugs and OMVs using FLC at MIC_{50} (1 µg/mL) and PB in the range much lower than MIC (1/8–1/16). To our knowledge, for the first time, we have provided evidence that bacterial OMVs can inhibit the fungicidal action of certain combined antifungal agents that are effective against pathogenic yeasts. Next, we documented that sub-MIC concentrations of FLC (1/16) and PB (in the range of 1/8–1/32) applied together weakened or almost completely inhibited the yeast-to-hyphae transition. Furthermore, in the presence of OMVs, the filamentous phenotype was considerably recovered and even strengthened despite exposure to both drugs. It is a very undesirable action of OMVs, since filamentation increases the virulence of *C. albicans*, which in hyphae form is more invasive and can attach in a higher number to epithelial cells than yeast and pseudohyphae forms [34,35]. Using analogies to the role of OMVs contributing to the sequestration of PB, it is tempting to speculate that the same mechanism of action exists in this case. Collectively, both of the aforementioned activities of bacterial vesicles seem to render *C. albicans* less vulnerable to destruction. The documented inter-kingdom OMV-based mutualistic relationship between bacteria and yeasts are, to our knowledge, a novel phenomenon. Therefore, its importance for more complex in vitro and in vivo conditions, as well as pathophysiology, remains to be clarified.

The complex *Candida*–bacteria interactions are not a rare occurrence and may have an important impact on the human disease by causing e.g., the faster biofilm growth or *Candida*-dependent induction of antimicrobial resistance of *Staphylococci* [36,37]. Sometimes, the results investigating cross-kingdom polymicrobial interactions are very contradictory even for the same microbial components. It was shown that direct or indirect (by released soluble molecules) contact of *P. aeruginosa* or *A. baumannii* with *C. albicans* or other fungi may lead to the killing of yeast [38,39]. It can also decrease fungal filamentation, biofilm formation, and conidia biomass [40]. The synergistic collaboration between the two pathogens was also documented. For example, the pre-colonization with *C. albicans*, which compromises the immune system, facilitates the emergence of *A. baumannii* pneumonia [41]. The interactions between microbes are not only affected by the specific combination of microorganisms, but also by the environment such as immunological milieu. Therefore, bacterial OMVs shape the behavior of neighboring microbes and the overall outcome of their interplay for the host. Due to the limited antifungal arsenal, the synergistic effect of PB and fluconazole or human broad-spectrum AMP lactoferrin with amphotericin B or fluconazole, which increases the activity of the antifungals against *Candida* spp, could be an alternative for treatment [22,42]. Likewise, AMPs such as defensins or gramicidin have shown to be a promising alternative to the current antimycotic and antibacterial therapies [43–45]. Furthermore, AMPs display a lower propensity to develop resistance than do conventional antibiotics [45]. In light of our research, however, these promising strategies should consider the unfavorable role of OMVs as a trap for host cationic peptides in mixed bacterial and bacterial–fungal infections. Our results may also have a meaning in medical microbiology. Because AMPs are already used in topical nasal antimicrobials in the treatment of nasal or paranasal cavity infections (sinusitis, maxillary, otitis media) [46,47], it is conceivable that abundantly produced OMVs, by very efficient AMP sequestration, may decrease the pharmacokinetics of these compounds.

Another important issue of our results is the number of OMVs needed to show biological activity. In general, the information about the number of OMVs produced in vivo is still very limited. Although OMV production in the course of infection has been documented [48], the magnitude of this production was not given. During the in vitro part of this study, however, the biologically active concentration of OMVs was 5 μg/mL per 10^3 cfu/mL. In our study, 20 μg/mL of OMVs was protective against 10^6 cfu/mL, indicating that OMV amounts used in our study are rational and may resemble the infectious condition.Furthermore, there is many data on how different stress factors may induce bacterial hypervesiculation incuding temperature stress [49], oxidative stress [50], hyperosmotic stress [49], or antibiotic stress [51]. So far, the correlation between OMVs production and pathophysiology of a specific disease was proven both in animal sepsis-like inflammation models [52–54] and in a patient with fatal meningococcal septicaemia [55]. Based on these aforementioned examples, the OMVs concentrations, which seems to be clinically relevant, are in the range 5–20 μg/mL.

Overall, our results on the protective and somehow deleterious for pathogens role of OMVs, in the bacteria–bacteria and bacteria–yeast interspecies systems, underline the enormous potential of these nanostructures as accelerating factors in case of various mixed infections. Furthermore, the OMV-dependent mode of actions may serve as a model of passive resistance of gram-negative bacteria not only to antimicrobials, but also to humoral defense components, which operate to disrupt cell membrane. Likewise, for dimorphic yeasts, the ability of OMVs to sequester membrane active compounds that augment the antifungal activity of azoles may have an important impact on *Candida* virulence. This work may serve as an important basis for further evaluation of OMVs-dependent interactions within pathogenic bacterial-fungal communities. Our results indicate that OMVs are important players in interspecies and cross-kingdom microbial interactions.

4. Materials and Methods

4.1. Materials

4.1.1. Reagents

BHI (Brain Heart Infusion, OXOID, Basingstoke, UK)); TSB (Tryptone Soya Broth, OXOID, Hampshire, England); TSA (Tryptone Soya Agar, OXOID, Hampshire, England), Bradford reagent (Protein Assay Dye Reagent Concentrate, Bio-Rad, München, Germany); β-nicotinoamide adenine dinucleotide hydrate (NAD, Sigma-Aldrich, Steinheim, Germany); fluconazole (FLC, Sigma-Aldrich, Poznan, Poland), hemin (Sigma-Aldrich, St. Louis, MO, USA); polymyxin B sulfate salt (PB, Sigma-Aldrich, Denmark); fluorescein isothiocyanate (FITC, ThermoScientific, Rockford, IL, USA); Hank's Buffer with Ca^{2+}, Mg^{2+} (HBSS, PAN Biotech, UK); RPMI 1640 (Lonza, Walkersville, MD, USA); *o*-nitrophenyl-β-D-galactopyranoside (ONPG, Sigma, Steinheim Germany), heat inactivated (56 °C, 1 h) FBS (Fetal bovine serum, Gibco Life Technologies, Grand Island, NY, USA).). HPLC chemicals: Acetonitrile (Sigma, München, Germany) for separation was HPLC far UV/gradient grade (J. T. Baker, Avantor™ Performance Material); 32 mM Na_2SO_4 solution for chromatographic usage was prepared with 4.5 g anhydrous sodium sulfate (POCh, Avantor™ Performance Material, Gliwice, Poland) and MiliQ (ultrapure water made with Simplicity UV Water Purification System, Merck Millipore, Saint-Quentin, France).

4.1.2. Microbial Strains and Growth Condition

The following microbial strains were used: *Moraxella catarrhalis* (Mc5, Mc6, Mc8), nontypeable *Haemophilus influenzae* (NTHi3, NTHi6), *Acinetobacter baumannii* (ATCC 19606), *Pseudomonas aeruginosa* (PAO1), *Candida albicans* (Ca1), mutant of *Escherichia coli* ML-35p, a lactose permease-deficient strain with constitutive cytoplasmic β-galactosidase. All strains were from the collection of our Institute. *M. catarrhalis* strains were grown on Columbia agar plates or BHI broth. NTHi strains were grown on chocolate agar plates or in BHI broth supplemented with hemin and NAD at final concentrations of 15 μg/mL each. *Moraxella* and *Haemophilus* strains were cultivated at 37 °C with 5% CO_2. *A. baumannii*; *E. coli ML-35p* and *P. aeruginosa* were routinely cultured in TSB medium at 37 °C. *C. albicans* was cultured in yeast extract-peptone-glucose (YPG) in 37 °C.

4.2. Methods

4.2.1. Outer Membrane Vesicles Isolation

Outer membrane vesicles (OMVs) isolation was performed as described previously [21]. The protein concentartions of purified OMVs preparations was determined by Bradford assay) and the quality of OMVs preparation was confirmed in 12% SDS-PAGE.

4.2.2. Time-Kill Assay

For testing PB or human serum (NHS) activity, the log-phase bacterial suspension (5×10^5–10^6 cfu/mL) was incubated with or without PB (in the range 0.5–5 μg/mL) or NHS (in the range 25%–75%) with the presence or absence of free OMVs (2 μg/mL or 20 μg/mL) in the final volume of 200 μL 1% medium (*w/v*). The experiments were performed from 0 to 240 min and the 10 μL aliquots of 10-time diluted bacterial suspensions were plated in triplicate on appropriate agar plates at 0, 30, 120, and 240 min time points. The colony counts and cfu/mL were calculated the next day. The controls for

NHS contained heat-inactivated serum (HiNHS), (56 °C, 30 min). The bactericidal activity of PB or NHS was expressed in each time point as cfu/mL in reference to cfu/mL in time 0. Analogous experiments with bacteria associated and non-associated with OMVs (20 µg/mL or 80 µg/mL) in the presence of PB (5 µg/mL) were performed. Analogous experiments for selected combinations of antifungals were carried out for *C. albicans* except that: (i) Initial inoculum was ~2×10^5–5×10^5 cfu/mL, (ii) tested antimicrobials were PB (8 µg/mL or 16 µg/mL) and fluconazole (1 µg/mL), (iii) the medium was 0.5% YPG (*w/v*), and (iv) incubation lasts 24 h. All microbicidal assays were performed at least two times in triplicate.

4.2.3. Growth Kinetics Assay

All growth kinetics experiments were performed on the flat-bottomed 96-well microplates (NUNC, Denmark) at 37 °C in volume 200 µL. Dynamic of growth was measured using Varioskan™ LUX reader with measurements at 30 min intervals for bacteria at 10^6 cfu/mL diluted (final concentration) in 1% BHI (*w/v*) and 60 min intervals for yeast at 10^5 cfu/mL (final concentration.) diluted in 0.5% YPG.

OMV-associated bacteria preparation: 0.5 mL of ~10^8 cfu/mL exponentially growing bacteria were washed by centrifugation with HBSS Ca^{2+} Mg^{2+} and resuspended in 100 µL OMVs (20 µg/mL or 80 µg/mL) in Eppendorf tube. Association was performed for 30 min at 37 °C with gentle mixing. Thereafter, all samples were washed with HBSS Ca^{2+}, Mg^2 by centrifugation (8000 rpm, 10 min, 4 °C) to remove free OMVs particles, diluted to 2×10^6 cfu/mL with HBSS Ca^{2+}, Mg^2, and used in growth kinetics assay.

4.2.4. Checkerboard Microdilution Assay

The microdilution assay was performed on flat-bottom microplate according to the CLSI (formerly NCCLS) standard [56] except that the initial inoculum for *C. albicans* was ~10^5 cfu/mL and cells were incubated at 37 °C without shaking. Synergy/growth potentiation was tested by the checkerboard method including a two-dimensional array of serial concentrations of both drugs. The fluconazole was used in concentrations 1–64 µg/mL whereas polymyxin B in concentrations 1–128 µg/mL. Wells without drugs or yeast inoculation were included as positive and negative controls, respectively. The MIC_{100} of polymyxin B and the MIC_{50} of fluconazole was defined as the lowest drug concentration that caused a decrease in absorbance of 100% and 50%, respectively, compared to control in drug-free medium.

4.2.5. Induction of Filamentation

Before the induction of filamentation, *C. albicans* cells were preincubated overnight in YPG supplemented with sub-MIC concentration of FLC at 37 °C without shaking. To induce hyphal transition, the suspensions of *C. albicans* (OD = 0.4) were treated with 10% heat inactivated FBS in PBS for 2 h at 37 °C in 24-well flat-bottom microplates (NUNC) in volume 0.4 mL, in the presence of sub-MIC concentrations of FLC (1/16) and PB (range 1/8–1/32) and with shaking (130 rpm). The samples were observed under inverted microscope Zeiss Axio Vert. A1 with objective Zeiss LD A-Plan (40 × /0.55 Ph1). The images wer recorded using Industrial Digital Camera 5.1 MP 1 /2.5″. The assessment of cell morphology and the length (µm) of hyphae was performed using ImageJ software.

4.2.6. Membrane Permeability Assay with ONPG

To assess the polymyxin B sequestration by *M. catarrhalis* OMVs in vitro, the time-dependent decrease of permeabilizing activity of this peptide against *E. coli* ML-35p with constitutive β-galactosidase activity was measured using the ONPG-mediated β-galactosidase microplate assay as previously described [57]. The final bacterial suspension (~10^6 cfu/mL) in 10 mM sodium phosphate NaPB (pH 7.4) were incubated in the flat-bottom 96-well microplates (NUNC, Denmark) with 5 µg/mL of PB in the presence of OMVs at concentration range 1–20 µg/mL and with 3 mM ONPG as β-galactosidase substrate in the final volume of 150 µL. Microplates were incubated at 37 °C for 1.5 h and optical densities at $\lambda = 405$ nm were measured every 15 min (spectrophotometer ASYS). All the assays were performed at least 3 times in duplicates.

4.2.7. Complement Activation

This method was described previously [58]. To assess complement activation by *M. catarrhalis* OMVs in vitro, the various concentrations of OMVs (1–20 µg/mL) were pretreated with active normal human serum (NHS) at a volume ratio 1:9 and the soluble terminal complement complex SC5b-9 was measured using ELISA kit according to manufacturer's instructions (Quidel Corporation, San Diego, USA). Unstimulated NHS served as negative controls.

4.2.8. OMV Association Assay and Flow Cytometry

OMVs labeling: Initially, OMVs (80 µg/mL) in PBS were concentrated using 50 kDa Vivaspin centrifugal concentrators (Amicon ultra, Millipore) at 14,000 $\times$ g for 10 min at 4 °C to remove PBS. The collected OMVs were reconstituted with 500 µL of 0.05 M carbonate/bicarbonate buffer (pH 9.5) and washed by centrifugation on vivaspin as described before. The collected OMVs at concentration 80 µg/mL were labeled with 1 mg/mL FITC at carbonate/bicarbonate buffer for 30 min at 37 °C with gentle mixing in the dark. The remaining fluorochrome was rinsed of 3 times with a 500 µL of cold carbonate/biscarbonate buffer each time, using 50 kDa Vivaspin. The final FITC-labelled OMVs were resuspended in Hank's buffer Ca^{2+}, Mg^{2+} at original concentration.

OMVs association with bacteria: 1 mL of each fresh bacterial culture corresponding to $OD_{550} = 0.2$ was centrifuged and subsequently washed with 1 mL of HBSS Ca^{2+}, Mg^{2+} (8000$\times$ g, 10 min, 4 °C). The pellet was resuspended in 200 µL of OMV-FITC conjugate (80 µg/mL OMVs) and incubated for 30 min at 37 °C with gentle mixing in the dark. Afterward, samples were washed with HBSS $Ca^{2+}Mg^{2}$ by centrifugation (8000$\times$ g, 10 min, 4 °C) to remove free OMVs particles and finally resuspended in 200 µL of HBSS Ca^{2+} Mg^{2}.

Flow cytometric analysis: To detect bacterial cells associated with FITC-labeled OMVs flow cytometry analysis was performed using GUAVA® easyCyte flow cytometer (Millipore, Seattle, WA, USA). Before analysis, the samples were diluted 1:100 to obtain approximately $1–5 \times 10^5$ cell/mL in HBSS Ca^{2+} Mg^{2+}. Fluorescence intensity of bacterial cells associated with OMVs was analyzed for green fluorescence in the FL1 channel, by collecting 5000 events. Data were expressed as mean fluorescence intensity (MFI). Data analysis was performed using InCyte Merck Guava software (Millipore, Hayward, CA, USA).

4.2.9. Estimation of Zeta Potential

The Zeta potential of OMVs was measured at room temperature (25 °C) by a Zetasizer Nano-ZS 90 (Malvern, UK). The instrument was equipped with a Helium–Neon laser (633 nm) as a source of light. The detection angle of Zetasizer at aqueous media was 173°. Considering the influence of factors such as conductivity (salt concentration) or pH of the solution on the Zeta potential, to minimize their influence, all Zeta potential measurements were performed with 10 mM NaPB buffer (pH 7.4) or in MiliQ.

4.2.10. HPLC-UV System and Method

HPLC chromatography of polymyxin B was performed as described previously [26,59]. The HPLC system consisted of a Water's 2695 Solvent Manager System with a built-in autosampler and 100 µL sample loop connected to Waters 2996 Photodiode Array Detector. Data collection and peaks integration were realized by computer with Water's Empower 3 Chromatography Data Software. The separation developed with the use of a Macherey-Nagel EC Nucleodur C18 Isis column (50 mm, 4.6 mm ID with 1.8 µm beads) with Hypersil Gold aQ 5 µm (10 mm $\times$ 4.6 mm ID) guard precolumn. The mobile phase consisted of 22% acetonitrile and 78% of 32 mM Na_2SO_4 in water (pH 3.2 achieved with H_2SO_4) in isocratic separation mode. Eluent flow 0.75 mL/min and detection realized with a UV detector at

215 nm. Column and sample temperature were respectively 30 °C and 5 °C; separation run time was 15 min. The injection volume was 50 μL. The calibration curve was in the range of 1.56 μg/mL to 100 μg/mL prepared with 10 mg/mL standard stock solution of polymyxin B sulfate salt. Each concentration was injected twice daily for precision into 3 or 2 independent samples, and between day eight samples. The linearity of the method was in a range from 6.25 to 100 μg/mL with 16% of the average coefficient of variation (CV%) for the sum of peaks, and with a coefficient of determination $R^2 = 0.997$. Since the major constituents of polymyxin B are B_1 and B_2 [59], (polymyxin B sulfate certificate of analysis, Sigma-Aldrich), these two components were analyzed altogether.

To quantify the results, the relative concentration was calculated as the mean PB concentration of sample/mean PB of control × 100%. Results were expressed as mean ± SD.

Sample preparation: 200 μL of reaction mixtures containing OMVs (5 μg/mL or 20 μg/mL) and PB (50 μg/mL) or PB alone (50 μg/mL) were incubated for 30 min at 37 °C with gentle mixing. The samples were ultrafiltrated using 50 kDa vivaspin centrifugal concentrators (Amicon ultra, Merck Millipore, Cork, Ireland) at 14 000× g for 15 min at RT to remove OMVs. The filtrate was collected and stored at −20 °C until use. Tested and control samples were processed identically.

4.2.11. LOS-OMVs Electrophoresis

The concentration of lipooligosaccharide (LOS) from Mc6 OMVs was determined based on the purpald assay [60]. Diluted samples were solubilized in Laemmli sample buffer and heated at 100 °C for 5 min. Proteinase K (20 mg mL^{-1}) was added per 20 mg of OMV proteins and incubated in a heating block at 60°C for 1 h. The presence of LOS in proteinase K-treated OMV samples was analyzed by dodecylosulfate gel electrophoresis. The 15 μL of samples were applied to the Glycine-SDS-PAGE (10%) gel path, corresponding to 5 μg LOS. Electrophoresis was carried during 2.5 h at a constant voltage of 80 V at 4 °C. The gel was fixed for 1 h at room temperature using a fixing solution (EtOH: Acetic Acid: MiliQ: 40:5:55); POCH, Gliwice, Poland) The fixed gel was stored overnight at 4 °C. MiliQ was exchanged for fresh before visualization using a modified Tsai and Frasch [61] silver staining protocol.

4.2.12. TEM

The OMVs preparation for TEM was described previously [4]. OMVs were visualized by standard negative staining using a formvar copper grid (Christine Gröpl Electronenmikroskopie, Tulln, Austria) and 2% (*w/v*) aqueous solution of uranyl acetate. The OMVs were imaged with a TEM operating at an acceleration voltage of 150 kV (Hitachi H-800, Japan).

4.2.13. Statistical Analysis

The data were expressed as the mean ± SD, and analyzed for the significant difference by one-way ANOVA or Kruskal–Wallis ANOVA rang using the Statistica (version 13.1) software (StatSoft, Krakow, Poland) Differences were considered statistically significant if $p < 0.05$.

Supplementary Materials: Supplementary materials can be found at http://www.mdpi.com/1422-0067/20/22/5577/s1.

Author Contributions: D.A., conceived the study; J.R., P.J., G.G., J.G., DA., performed experiments; D.A., J.R., P.J. and G.G. performed analysis and interpretation of data; A.Ż. performed TEM images; D.A. wrote the manuscript with input from J.R.; D.A., J.R., G.G. reviewed and edited the manuscript. D.A., wrote the revised version of the manuscript; Z.D.-K. and D.A. acquired funding. This work is a part of the PhD thesis by J.R.

Funding: This study was supported by research grants of University of Wroclaw (1016/S/IGiM/T-20).

Conflicts of Interest: The authors declare no conflict of interest.

Abbreviations

AMP	Antimicrobial peptide
FBS	Fetal bovine serum
FLC	Fluconazole
HiNHS	Heat-inactivated normal human serum
LOS	Lipooligosaccharide
MIC	Minimal inhibitory concentration
NHS	Normal human serum
OMV	Outer membrane vesicle
ONPG	o-nitrophenyl-β-D-galactopyranoside
PB	Polymyxin B

References

1. Guerrero-Mandujano, A.; Hernández-Cortez, C.; Ibarra, J.A.; Castro-Escarpulli, G. The outer membrane vesicles: Secretion system type zero. *Traffic* **2017**, *18*, 425–432. [CrossRef] [PubMed]
2. Dusane, D.H.; Zinjarde, S.S.; Venugopalan, V.P.; McLean, R.J.C.; Weber, M.M.; Rahman, P.K.S.M. Quorum sensing: Implications on rhamnolipid biosurfactant production. *Biotechnol. Genet. Eng. Rev.* **2010**, *27*, 159–184. [CrossRef] [PubMed]
3. Jan, A.T. Outer Membrane Vesicles (OMVs) of Gram-negative Bacteria: A Perspective Update. *Front. Microbiol.* **2017**, *8*, 1053. [CrossRef] [PubMed]
4. Augustyniak, D.; Roszkowiak, J.; Wiśniewska, I.; Skała, J.; Gorczyca, D.; Drulis-Kawa, Z. Neuropeptides SP and CGRP Diminish the *Moraxella catarrhalis* Outer Membrane Vesicle- (OMV-) Triggered Inflammatory Response of Human A549 Epithelial Cells and Neutrophils. *Mediat. Inflamm.* **2018**, *2018*, 1–15. [CrossRef]
5. Bitto, N.J.; Chapman, R.; Pidot, S.; Costin, A.; Lo, C.; Choi, J.; D'Cruze, T.; Reynolds, E.C.; Dashper, S.G.; Turnbull, L.; et al. Bacterial membrane vesicles transport their DNA cargo into host cells. *Sci. Rep.* **2017**, *7*. [CrossRef]
6. Kulkarni, H.M.; Nagaraj, R.; Jagannadham, M.V. Protective role of *E. coli* outer membrane vesicles against antibiotics. *Microbiol. Res.* **2015**, *181*, 1–7. [CrossRef]
7. Lee, J.; Lee, E.-Y.; Kim, S.-H.; Kim, D.-K.; Park, K.-S.; Kim, K.P.; Kim, Y.-K.; Roh, T.-Y.; Gho, Y.S. *Staphylococcus aureus* extracellular vesicles carry biologically active β-lactamase. *Antimicrob. Agents Chemother.* **2013**, *57*, 2589–2595. [CrossRef]
8. Manning, A.J.; Kuehn, M.J. Contribution of bacterial outer membrane vesicles to innate bacterial defense. *Bmc Microbiol.* **2011**, *11*, 258. [CrossRef]
9. Rumbo, C.; Fernández-Moreira, E.; Merino, M.; Poza, M.; Mendez, J.A.; Soares, N.C.; Mosquera, A.; Chaves, F.; Bou, G. Horizontal transfer of the OXA-24 carbapenemase gene via outer membrane vesicles: A new mechanism of dissemination of carbapenem resistance genes in *Acinetobacter baumannii*. *Antimicrob. Agents Chemother.* **2011**, *55*, 3084–3090. [CrossRef]
10. Chatterjee, S.; Mondal, A.; Mitra, S.; Basu, S. *Acinetobacter baumannii* transfers the *blaNDM-1* gene via outer membrane vesicles. *J. Antimicrob. Chemother.* **2017**, *72*, 2201–2207. [CrossRef]
11. Chattopadhyay, M.K.; Jaganandham, M.V. Vesicles-mediated resistance to antibiotics in bacteria. *Front. Microbiol.* **2015**, *6*, 758. [CrossRef] [PubMed]
12. Kadurugamuwa, J.L.; Mayer, A.; Messner, P.; Sára, M.; Sleytr, U.B.; Beveridge, T.J. S-layered *Aneurinibacillus* and *Bacillus spp.* are susceptible to the lytic action of *Pseudomonas aeruginosa* membrane vesicles. *J. Bacteriol.* **1998**, *180*, 2306–2311. [PubMed]
13. Evans, A.G.L.; Davey, H.M.; Cookson, A.; Currinn, H.; Cooke-Fox, G.; Stanczyk, P.J.; Whitworth, D.E. Predatory activity of *Myxococcus xanthus* outer-membrane vesicles and properties of their hydrolase cargo. *Microbiology* **2012**, *158*, 2742–2752. [CrossRef] [PubMed]
14. Tashiro, Y.; Yawata, Y.; Toyofuku, M.; Uchiyama, H.; Nomura, N. Interspecies Interaction between *Pseudomonas aeruginosa* and Other Microorganisms. *Microbe. Environ.* **2013**, *28*, 13–24. [CrossRef] [PubMed]
15. Schaar, V.; Nordström, T.; Mörgelin, M.; Riesbeck, K. *Moraxella catarrhalis* outer membrane vesicles carry β-lactamase and promote survival of *Streptococcus pneumoniae* and *Haemophilus influenzae* by inactivating amoxicillin. *Antimicrob. Agents Chemother.* **2011**, *55*, 3845–3853. [CrossRef]

16. Kim, S.W.; Park, S.B.; Im, S.P.; Lee, J.S.; Jung, J.W.; Gong, T.W.; Lazarte, J.M.S.; Kim, J.; Seo, J.-S.; Kim, J.-H.; et al. Outer membrane vesicles from β-lactam-resistant *Escherichia coli* enable the survival of β-lactam-susceptible *E. coli* in the presence of β-lactam antibiotics. *Sci. Rep.* **2018**, *8*, 5402. [CrossRef]

17. Kulkarni, H.M.; Jagannadham, M.V. Biogenesis and multifaceted roles of outer membrane vesicles from Gram-negative bacteria. *Microbiology* **2014**, *160*, 2109–2121. [CrossRef]

18. Joo, H.-S.; Fu, C.-I.; Otto, M. Bacterial strategies of resistance to antimicrobial peptides. *Philos. Trans. R. Soc. B Biol. Sci.* **2016**, *371*, 20150292. [CrossRef]

19. Tashiro, Y.; Ichikawa, S.; Shimizu, M.; Toyofuku, M.; Takaya, N.; Nakajima-Kambe, T.; Uchiyama, H.; Nomura, N. Variation of Physiochemical Properties and Cell Association Activity of Membrane Vesicles with Growth Phase in *Pseudomonas aeruginosa*. *Appl. Environ. Microbiol.* **2010**, *76*, 3732–3739. [CrossRef]

20. Hasegawa, Y.; Futamata, H.; Tashiro, Y. Complexities of cell-to-cell communication through membrane vesicles: Implications for selective interaction of membrane vesicles with microbial cells. *Front. Microbiol.* **2015**, *6*, 633. [CrossRef]

21. Augustyniak, D.; Seredyński, R.; McClean, S.; Roszkowiak, J.; Roszniowski, B.; Smith, D.L.; Drulis-Kawa, Z.; Mackiewicz, P. Virulence factors of *Moraxella catarrhalis* outer membrane vesicles are major targets for cross-reactive antibodies and have adapted during evolution. *Sci. Rep.* **2018**, *8*, 4955. [CrossRef] [PubMed]

22. Zhai, B.; Zhou, H.; Yang, L.; Zhang, J.; Jung, K.; Giam, C.Z.; Xiang, X.; Lin, X. Polymyxin B, in combination with fluconazole, exerts a potent fungicidal effect. *J. Antimicrob. Chemother.* **2010**, *65*, 931–938. [CrossRef] [PubMed]

23. Ha, K.C.; White, T.C. Effects of Azole Antifungal Drugs on the Transition from Yeast Cells to Hyphae in Susceptible and Resistant Isolates of the Pathogenic Yeast *Candida albicans*. *Antimicrob. agents Chemother.* **1999**, *43*, 763–768. [CrossRef] [PubMed]

24. Epand, R.M.; Walker, C.; Epand, R.F.; Magarvey, N.A. Molecular mechanisms of membrane targeting antibiotics. *Biochim. Biophys. Acta-Biomembr.* **2016**, *1858*, 980–987. [CrossRef] [PubMed]

25. Cowen, L.E.; Sanglard, D.; Howard, S.J.; Rogers, P.D.; Perlin, D.S. Mechanisms of antifungal drug resistance. *Cold Spring Harb. Perspect. Med.* **2015**, *5*. [CrossRef]

26. Andersson, D.I.; Hughes, D.; Kubicek-Sutherland, J.Z. Mechanisms and consequences of bacterial resistance to antimicrobial peptides. *Drug Resist. Updat.* **2016**, *26*, 43–57. [CrossRef]

27. Tashiro, Y.; Hasegawa, Y.; Shintani, M.; Takaki, K.; Ohkuma, M.; Kimbara, K.; Futamata, H. Interaction of Bacterial Membrane Vesicles with Specific Species and Their Potential for Delivery to Target Cells. *Front. Microbiol.* **2017**, *8*, 571. [CrossRef]

28. Yousfi, H.; Ranque, S.; Rolain, J.-M.; Bittar, F. In vitro polymyxin activity against clinical multidrug-resistant fungi. *Antimicrob. Resist. Infect. Control.* **2019**, *8*, 66. [CrossRef]

29. Scorzoni, L.; de Paula e Silva, A.C.A.; Marcos, C.M.; Assato, P.A.; de Melo, W.C.M.A.; de Oliveira, H.C.; Costa-Orlandi, C.B.; Mendes-Giannini, M.J.S.; Fusco-Almeida, A.M. Antifungal Therapy: New Advances in the Understanding and Treatment of Mycosis. *Front. Microbiol.* **2017**, *08*, 36. [CrossRef]

30. Kathiravan, M.K.; Salake, A.B.; Chothe, A.S.; Dudhe, P.B.; Watode, R.P.; Mukta, M.S.; Gadhwe, S. The biology and chemistry of antifungal agents: A review. *Bioorg. Med. Chem.* **2012**, *20*, 5678–5698. [CrossRef]

31. Tatsumi, Y.; Nagashima, M.; Shibanushi, T.; Iwata, A.; Kangawa, Y.; Inui, F.; Siu, W.J.J.; Pillai, R.; Nishiyama, Y. Mechanism of action of efinaconazole, a novel triazole antifungal agent. *Antimicrob. Agents Chemother.* **2013**, *57*, 2405–2409. [CrossRef] [PubMed]

32. Chen, S.C.A.; Lewis, R.E.; Kontoyiannis, D.P. Direct effects of non-antifungal agents used in cancer chemotherapy and organ transplantation on the development and virulence of *Candida* and *Aspergillus* species. *Virulence* **2011**, *2*, 280–295. [CrossRef] [PubMed]

33. Hsu, L.-H.; Wang, H.-F.; Sun, P.-L.; Hu, F.-R.; Chen, Y.-L. The antibiotic polymyxin B exhibits novel antifungal activity against *Fusarium* species. *Int. J. Antimicrob. Agents* **2017**, *49*, 740–748. [CrossRef] [PubMed]

34. Villar, C.C.; Kashleva, H.; Dongari-Bagtzoglou, A. Role of *Candida albicans* polymorphism in interactions with oral epithelial cells. *Oral Microbiol. Immunol.* **2004**, *19*, 262–269. [CrossRef] [PubMed]

35. Murzyn, A.; Krasowska, A.; Augustyniak, D.; Majkowska-Skrobek, G.; Łukaszewicz, M.; Dziadkowiec, D. The effect of *Saccharomyces boulardii* on *Candida albicans*-infected human intestinal cell lines Caco-2 and Intestin 407. *FEMS Microbiol. Lett.* **2010**, *310*, 17–23. [CrossRef]

36. Allison, D.L.; Willems, H.M.E.; Jayatilake, J.A.M.S.; Bruno, V.M.; Peters, B.M.; Shirtliff, M.E. *Candida*–Bacteria Interactions: Their Impact on Human Disease. *Microbiol. Spectr.* **2016**, *4*. [CrossRef]

37. Harriott, M.M.; Noverr, M.C. Importance of *Candida*-bacterial polymicrobial biofilms in disease. *Trends Microbiol.* **2011**, *19*, 557–563. [CrossRef]

38. Kerr, J.R.; Taylor, G.W.; Rutman, A.; Høiby, N.; Cole, P.J.; Wilson, R. *Pseudomonas aeruginosa* pyocyanin and 1-hydroxyphenazine inhibit fungal growth. *J. Clin. Pathol.* **1999**, *52*, 385–387. [CrossRef]

39. Gaddy, J.A.; Tomaras, A.P.; Actis, L.A. The *Acinetobacter baumannii* 19606 OmpA protein plays a role in biofilm formation on abiotic surfaces and in the interaction of this pathogen with eukaryotic cells. *Infect. Immun.* **2009**, *77*, 3150–3160. [CrossRef]

40. Krüger, W.; Vielreicher, S.; Kapitan, M.; Jacobsen, I.D.; Niemiec, M.J. Fungal-Bacterial Interactions in Health and Disease. *Pathogens* **2019**, *8*, 70. [CrossRef]

41. Roux, D.; Gaudry, S.; Khoy-Ear, L.; Aloulou, M.; Phillips-Houlbracq, M.; Bex, J.; Skurnik, D.; Denamur, E.; Monteiro, R.C.; Dreyfuss, D.; et al. Airway Fungal Colonization Compromises the Immune System Allowing Bacterial Pneumonia to Prevail. *Crit. Care Med.* **2013**, *41*, e191–e199. [CrossRef] [PubMed]

42. Venkatesh, M.P.; Rong, L. Human recombinant lactoferrin acts synergistically with antimicrobials commonly used in neonatal practice against coagulase-negative *staphylococci* and *Candida albicans* causing neonatal sepsis. *J. Med. Microbiol.* **2008**, *57*, 1113–1121. [CrossRef] [PubMed]

43. Silva, P.M.; GonÃ§alves, S.; Santos, N.C. Defensins: Antifungal lessons from eukaryotes. *Front. Microbiol.* **2014**, *5*, 97. [CrossRef] [PubMed]

44. Swidergall, M.; Ernst, J.F. Interplay between *Candida albicans* and the antimicrobial peptide armory. *Eukaryot. Cell* **2014**, *13*, 950–957. [CrossRef] [PubMed]

45. Li, J.; Koh, J.-J.; Liu, S.; Lakshminarayanan, R.; Verma, C.S.; Beuerman, R.W. Membrane Active Antimicrobial Peptides: Translating Mechanistic Insights to Design. *Front. Neurosci.* **2017**, *11*, 73. [CrossRef] [PubMed]

46. Goh, Y.H.; Goode, R.L. State of the Art Review: Current Status of Topical Nasal Antimicrobial Agents. *Laryngoscope* **2000**, *110*, 875–880. [CrossRef]

47. Ragioto, D.A.M.T.; Carrasco, L.D.M.; Carmona-Ribeiro, A.M. Novel gramicidin formulations in cationic lipid as broad-spectrum microbicidal agents. *Int. J. Nanomed.* **2014**, *9*, 3183–3192.

48. Tan, T.T.; Mörgelin, M.; Forsgren, A.; Riesbeck, K. *Haemophilus influenzae* Survival during Complement-Mediated Attacks Is Promoted by *Moraxella catarrhalis* Outer Membrane Vesicles. *J. Infect. Dis.* **2007**, *195*, 1661–1670. [CrossRef]

49. Baumgarten, T.; Sperling, S.; Seifert, J.; von Bergen, M.; Steiniger, F.; Wick, L.Y.; Heipieper, H.J. Membrane vesicle formation as a multiple-stress response mechanism enhances *Pseudomonas putida* DOT-T1E cell surface hydrophobicity and biofilm formation. *Appl. Environ. Microbiol.* **2012**, *78*, 6217–6224. [CrossRef]

50. MacDonald, I.A.; Kuehna, M.J. Stress-induced outer membrane vesicle production by *Pseudomonas aeruginosa*. *J. Bacteriol.* **2013**, *195*, 2971–2981. [CrossRef]

51. Yun, S.H.; Park, E.C.; Lee, S.Y.; Lee, H.; Choi, C.W.; Yi, Y.S.; Ro, H.J.; Lee, J.C.; Jun, S.; Kim, H.Y.; et al. Antibiotic treatment modulates protein components of cytotoxic outer membrane vesicles of multidrug-resistant clinical strain, *Acinetobacter baumannii* DU202. *Clin. Proteom.* **2018**, *15*, 28. [CrossRef] [PubMed]

52. Park, K.S.; Choi, K.H.; Kim, Y.S.; Hong, B.S.; Kim, O.Y.; Kim, J.H.; Yoon, C.M.; Koh, G.Y.; Kim, Y.K.; Gho, Y.S. Outer membrane vesicles derived from *Escherichia coli* induce systemic inflammatory response syndrome. *PLoS ONE* **2010**, *5*, e11334. [CrossRef] [PubMed]

53. Shah, B.; Sullivan, C.J.; Lonergan, N.E.; Stanley, S.; Soult, M.C.; Britt, L.D. Circulating bacterial membrane vesicles cause sepsis in rats. *Shock* **2012**, *37*, 621–628. [CrossRef] [PubMed]

54. Park, K.S.; Lee, J.; Jang, S.C.; Kim, S.R.; Jang, M.H.; Lötvall, J.; Kim, Y.K.; Gho, Y.S. Pulmonary inflammation induced by bacteria-free outer membrane vesicles from *Pseudomonas aeruginosa*. *Am. J. Respir. Cell Mol. Biol.* **2013**, *49*, 637–645. [CrossRef] [PubMed]

55. Namork, E.; Brandtzaeg, P. Fatal meningococcal septicaemia with "blebbing" meningococcus. *Lancet* **2002**, *360*, 1741. [CrossRef]

56. Wayne, P.A. Reference method for broth dilution antifungal susceptibility testing of yeasts, Approved standard-second edition. *Clsi Doc. M27-A2* **2002**.

57. Augustyniak, D.; Jankowski, A.; Mackiewicz, P.; Skowyra, A.; Gutowicz, J.; Drulis-Kawa, Z. Innate immune properties of selected human neuropeptides against *Moraxella catarrhalis* and nontypeable *Haemophilus influenzae*. *BMC Immunol.* **2012**, *13*, 24. [CrossRef]

58. Arabski, M.; Lisowska, H.; Lankoff, A.; Davydova, V.N.; Drulis-Kawa, Z.; Augustyniak, D.; Yermak, I.M.; Molinaro, A.; Kaca, W. The properties of chitosan complexes with smooth and rough forms of lipopolysaccharides on CHO-K1 cells. *Carbohydr. Polym.* **2013**, *97*, 284–292. [CrossRef]
59. Orwa, J.A.; Van Gerven, A.; Roets, E.; Hoogmartens, J. Liquid chromatography of polymyxin B sulphate. *J. Chromatogr. A* **2000**, *870*, 237–243. [CrossRef]
60. Inka, B. *Glycobiology Protocols*; Humana Press: Totowa, NJ, USA, 2006.
61. Tsai, C.-M.; Frasch, C.E. A sensitive silver stain for detecting lipopolysaccharides in polyacrylamide gels. *Anal. Biochem.* **1982**, *119*, 115–119. [CrossRef]

© 2019 by the authors. Licensee MDPI, Basel, Switzerland. This article is an open access article distributed under the terms and conditions of the Creative Commons Attribution (CC BY) license (http://creativecommons.org/licenses/by/4.0/).

International Journal of
Molecular Sciences

Article

Comparative Integrated Omics Analysis of the Hfq Regulon in *Bordetella pertussis*

Ana Dienstbier [1], Fabian Amman [2,3], Daniel Štipl [1], Denisa Petráčková [1] and Branislav Večerek [1,*]

[1] Laboratory of post-transcriptional control of gene expression, Institute of Microbiology v.v.i., 14220 Prague, Czech Republic; ana.sencilo@biomed.cas.cz (A.D.); daniel.stipl@biomed.cas.cz (D.S.); petrack@biomed.cas.cz (D.P.)

[2] Institute for Theoretical Chemistry, University of Vienna, Währinger Straße 17, A-1090 Vienna, Austria; fabian@tbi.univie.ac.at

[3] Division of Cell and Developmental Biology, Medical University of Vienna, Schwarzspanierstraße 17, A-1090 Vienna, Austria

* Correspondence: vecerek@biomed.cas.cz; Tel.: +420-241-062-507

Received: 5 June 2019; Accepted: 19 June 2019; Published: 24 June 2019

Abstract: *Bordetella pertussis* is a Gram-negative strictly human pathogen of the respiratory tract and the etiological agent of whooping cough (pertussis). Previously, we have shown that RNA chaperone Hfq is required for virulence of *B. pertussis*. Furthermore, microarray analysis revealed that a large number of genes are affected by the lack of Hfq. This study represents the first attempt to characterize the Hfq regulon in bacterial pathogen using an integrative omics approach. Gene expression profiles were analyzed by RNA-seq and protein amounts in cell-associated and cell-free fractions were determined by LC-MS/MS technique. Comparative analysis of transcriptomic and proteomic data revealed solid correlation ($r^2 = 0.4$) considering the role of Hfq in post-transcriptional control of gene expression. Importantly, our study confirms and further enlightens the role of Hfq in pathogenicity of *B. pertussis* as it shows that Δhfq strain displays strongly impaired secretion of substrates of Type III secretion system (T3SS) and substantially reduced resistance to serum killing. On the other hand, significantly increased production of proteins implicated in transport of important metabolites and essential nutrients observed in the mutant seems to compensate for the physiological defect introduced by the deletion of the *hfq* gene.

Keywords: *Bordetella pertussis*; Hfq; omics analysis; T3SS; serum resistance; solute-binding proteins

---

1. Introduction

Bordetella pertussis is a Gram-negative strictly human pathogen of the respiratory tract and the etiological agent of whooping cough (pertussis) [1]. This highly contagious disease is especially severe in infants and remains a major cause of infant mortality and morbidity worldwide, predominantly in developing countries [2]. Furthermore, pertussis incidence is currently on the rise in industrialized countries with highly vaccinated populations [3,4]. While there are several reasons for this phenomenon [5], there are two major disease-related factors contributing to recent increase in pertussis cases: short-lived immunity induced by current acellular vaccines and pathogen adaptation leading to escape from the immunity by antigenic variation [6–8]. The global reemergence of pertussis clearly suggests that we need to widen our understanding of the molecular mechanisms underlying the pathogenesis of *B. pertussis* [9,10]. In many pathogenic bacteria the RNA chaperone Hfq and small non-coding regulatory RNAs (sRNAs) emerged as critical players in posttranscriptional regulation of virulence and physiological fitness [11–13]. The Hfq protein forms ring-shaped hexamers that possess several RNA binding sites that allow for simultaneous interaction with both sRNA and mRNA molecules and stabilization of their interactions [14–16]. Besides its role in facilitation and stabilization of RNA duplexes, Hfq can actively remodel the structure of RNAs and also increase the stability of sRNAs [14,16,17].

Recently we have shown that Hfq is required for virulence of *B. pertussis* as the Δ*hfq* mutant was affected both in its ability to efficiently multiply and persist in mouse lungs as well as in its capacity to cause a lethal infection in mouse [18]. Furthermore, our global DNA microarray-based transcriptomic profiling of the *hfq* mutant suggested that Hfq protein significantly affects expression of more than 10% annotated genes [19]. Nevertheless, despite the high sensitivity, transcriptomic profiling does not capture post-transcriptional and post-translational modifications that affect the amounts of produced proteins. On the other hand, mass spectrometry-based proteomics lacks the sensitivity to detect low abundant proteins. Therefore, integrative analysis of both transcriptomic and proteomic datasets enables a more complete understanding of studied biological processes [20,21]. First studies based on such an approach revealed that the overlap between the outcomes of transcriptomic and proteomic analyses is not extensive irrespective of the organism [22–24]. This discrepancy was attributed in part to technological limitations of applied procedures and in part to inherent biological complexity of transcription and translation processes [25,26]. Especially, factors linked to translational efficiency, such as codon usage bias, strength and accessibility of ribosome binding site, secondary structure and stability of the transcript, and post-transcriptional activity of the regulatory proteins, contribute to poor correlation between determined transcript and protein levels [20,27–29]

Hfq is a key player in post-transcriptional control of gene expression in Gram-negative bacteria and therefore, its biological activities should in principle weaken the correlation between the gene expression and protein synthesis profiles. Recently, an integrative analysis of Hfq-specific transcriptomic and proteomic profiles based on high-throughput RNA-seq and LC-MS/MS technologies was performed in soil bacterium *Pseudomonas fluorescence*. It showed that such a multiomics approach allows for dissection of discrete contributions of Hfq to gene regulation at different levels [30]. Therefore, we were interested to perform such a comparative analysis in human pathogen to elucidate the Hfq-related variations at both transcriptomic and proteomic levels per se and also to decipher how the changes in gene expression profiles translate into protein production in *B. pertussis*. Our results indicate that considering the role of Hfq in the post-transcriptional control of gene expression, the correlation between transcriptome and proteome is relatively high. Furthermore, our data corroborate and further clarify the necessity of Hfq for physiological fitness and pathogenicity of *B. pertussis*.

2. Results

2.1. Identification of the Hfq Regulon by RNA-seq

Samples of total RNA isolated from biological triplicates of *B. pertussis* Tohama I strain and its isogenic Δ*hfq* strain cultures were analyzed by RNA-seq. RNA-seq analysis yielded on average 16 million reads, which were mapped to the *B. pertussis* genome. The comparison of global expression profiles showed that biological replicates of either wt or Δ*hfq* cells are highly uniform and thereby reproducible (Figure 1A). Principal component analysis (PCA) revealed that samples from wt strain and *hfq* mutant clustered separately along principal component 1 (94%) reflecting global changes in gene expression profiles resulting from deletion of the *hfq* gene (Figure 1B).

Differential expression (DE) analysis identified 653 significantly modulated *B. pertussis* genes (|log$_2$FC| > 1; q < 0.05) including 40 non-coding RNAs and 11 transfer RNA genes (Table S1). Among the DE genes, 281 genes were downregulated and 372 upregulated in the Δ*hfq* strain. Remarkably, 56 genes (8.3% of all DE genes) encoding the components of ATP-binding cassette transport system were significantly up- or downregulated in the mutant. In agreement with our previous microarray study, the expression of several genes within the type III secretion (T3SS) *bcs/btr* locus including *bsp22*, *bopN*, *bopB*, and *bopD* was substantially decreased in the mutant. Genes involved in iron–sulfur cluster protein biogenesis (*iscU*, *iscA*, *iscS*) were also highly downregulated in the *hfq* mutant. Among the genes displaying increased expression in the *hfq* mutant were those coding for ribosomal proteins, amino acid biosynthesis and transport, and, surprisingly, genes encoding pertussis toxin subunits and its secretion apparatus (*ptx/ptl* locus).

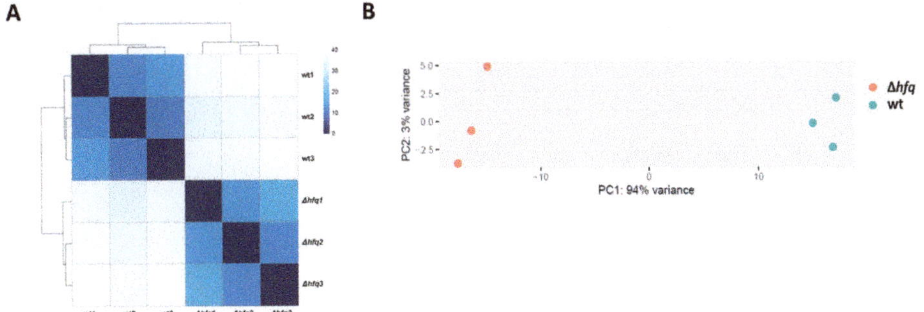

Figure 1. Clustering of transcriptomic data. (**A**) Heat map showing hierarchical clustering of the Euclidean sample-to-sample distance between transcriptomic profiles of wt and Δ*hfq* mutant. (**B**) Principal component analysis was applied to transcriptomic profiles of the wt strain (blue circles) and Δ*hfq* mutant (red circles). Each dot represents an independent biological replicate.

To get better insight into the functional profiles of Hfq-dependent genes we performed gene ontology (GO) enrichment analysis using all DE genes. GO term analysis revealed that genes belonging to several biological processes were significantly enriched in both up- and downregulated gene sets. As shown in Figure 2A, within the set of genes which were significantly upregulated in the *hfq* mutant, categories such as "Translation", "Regulation of transcription", and "Transmembrane transport" were highly enriched. On the other side, genes belonging to "Transmembrane transport", "Iron–sulfur cluster assembly", "Oxido-reduction process", "Pathogenesis", and "Protein secretion by the type III secretion system" terms were enriched among the transcripts which were significantly downregulated in the *hfq* mutant (Figure 2B). Apparently, GO term analysis of the DE genes recapitulated many of the observations from our previous studies [19].

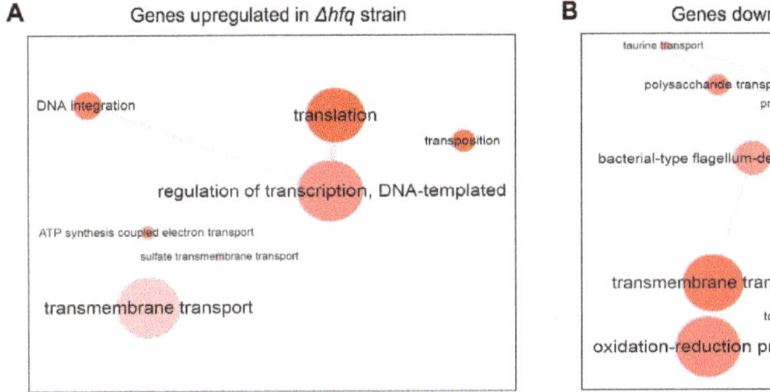

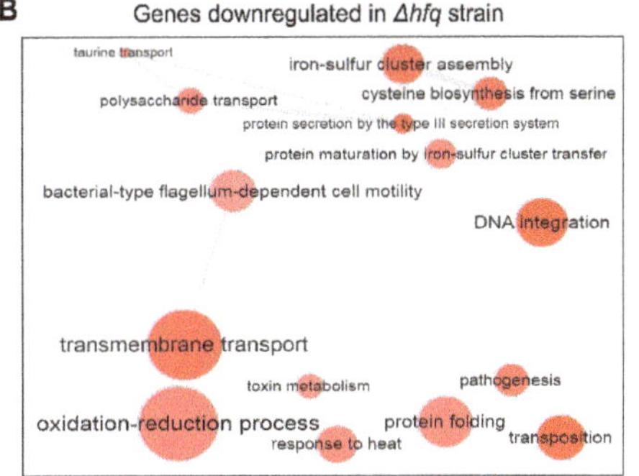

Figure 2. Gene ontology (GO) term analysis of genes significantly downregulated (**A**) or upregulated (**B**) in the Δ*hfq* mutant. Significantly enriched terms from the domain 'Biological processes' and their catenations are summarized and visualized by REVIGO as an interactive graph. Circle size encodes number of genes associated with respective category and red shades encode the significance level of the enrichment.

2.2. The Effect of Hfq on Proteome and Secretome Composition of B. pertussis

LC-MS/MS analysis of the *B. pertussis* proteome and secretome identified 1631 and 733 proteins, respectively, whose label-free quantification (LFQ) intensities passed our detection criteria. Principal component analysis of both datasets revealed that protein profiles identified in samples of the wt strain clustered separately from those of the Δ*hfq* strain (Figure 3).

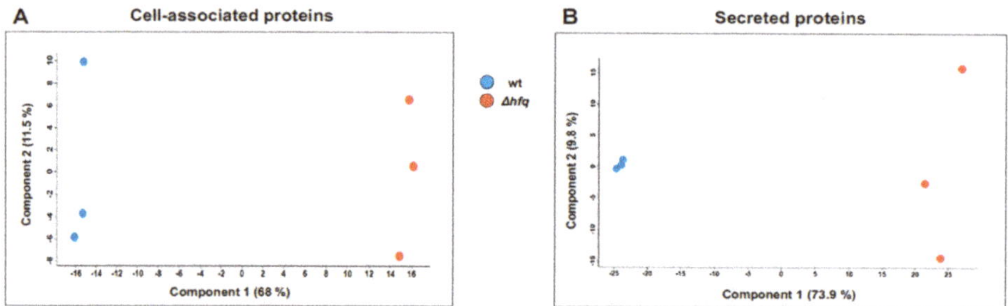

Figure 3. Principal component analysis (PCA) of proteomic samples. PCA was applied to protein profiles of the wt strain (blue circles) and Δ*hfq* mutant (red circles) determined in corresponding cell-associated (**A**) or secreted (**B**) protein fractions. Each dot represents an independent biological replicate.

For pellet proteins, the production of 489 proteins was found to be significantly modulated between wt and *hfq* mutant strains (Table S2). The abundance of 219 proteins was higher in the mutant including 19 "ON" proteins which were not detected in the wt strain. Among this set of proteins, GO terms such as "Cell cycle", "Peptidoglycan synthesis", and "Aromatic amino acid metabolism" were significantly enriched (Figure 4A). On the other hand, 270 proteins displayed significantly higher LFQ intensities in the wt strain including 10 "OFF" proteins which, in contrast to the wt samples, were not detected in the mutant. GO term analysis revealed that biological processes such as "Proteolysis", "Ion transport", "Pathogenesis", and "Protein secretion by the type III secretion system" were enriched among this group of proteins (Figure 4B).

As for the secreted fraction (Table S3), abundance of 114 proteins was higher in the mutant (including six "ON" proteins) and these proteins clustered into categories such as "Transmembrane transport", "Cell adhesion", and "Amino acid transport" (Figure 4C). On the other hand, 445 proteins displayed significantly higher LFQ intensities in the wt strain (including 136 "OFF" proteins). The GO term analysis of these differentially secreted proteins identified rather broad variety of processes including "Transmembrane transport", "Proteolysis", "Response to oxidative stress", "Protein secretion by the type III secretion system", and several processes linked to translation and amino acid biosynthesis (Figure 4D).

Among the proteins with highly increased abundance in the *hfq* mutant in both proteome and secretome datasets were prevalently members of various transporters including tripartite tricarboxylate transporters (TTT) (BP2066, BP3501, and BP1358), ABC transporters (BP2090, BP2352, and BP0663) and tripartite ATP-independent periplasmic transporters [31] (BP1487 and BP1489), lipoprotein BP2271, adhesin FhaS and all five pertussis toxin subunits. Of note, genes encoding these overproduced proteins also belonged to the set of the most upregulated genes in the mutant strain (Table 1).

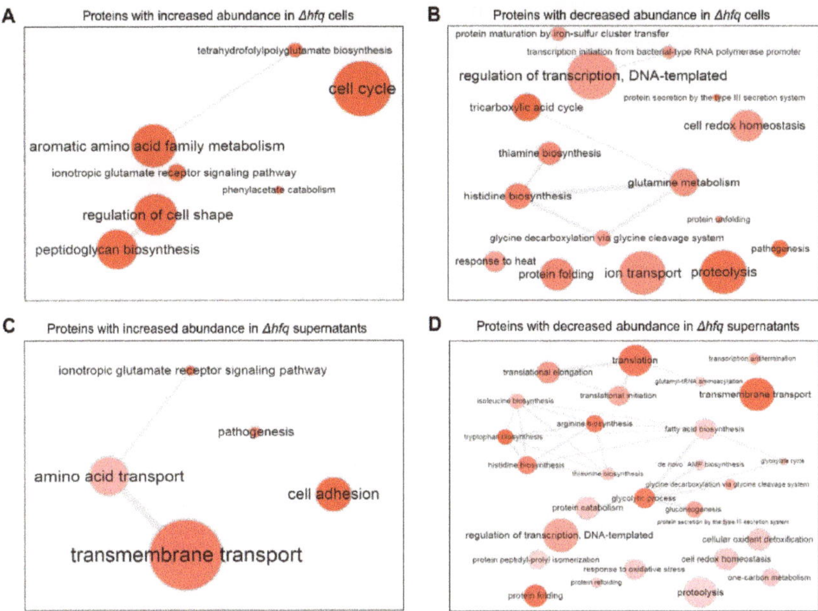

Figure 4. GO term enrichment analysis of genes either significantly upregulated (**A,C**) or downregulated (**B,D**) in the Δ*hfq* cells and Δ*hfq* culture supernatants, respectively. Biological processes significantly enriched for genes belonging to corresponding functional category are shown as an interactive graph. Significantly enriched terms from the domain "Biological processes" and their catenations are summarized and visualized by REVIGO. Circle size encodes number of genes associated with respective categories, red shades encode the significance level of the enrichment.

Table 1. List of genes showing consistently increased RNA and protein abundance in the *hfq* mutant.

Gene	RNA-seq[1]	Proteome[1]	Secretome[1]	Function
BP1358	4.18	3.93	2.78	TTT transporter
BP1487	4.34	2.92	1.65	TRAP transporter
BP1489	3.48	2.06	ND	TRAP transporter
BP2066	2.64	ON	8.80	TTT transporter
BP2090	1.92	ON	4.40	ABC transporter
BP2271	5.05	9.75	3.98	lipoprotein
BP2352	3.58	4.50	1.90	TRAP transporter
BP2667	2.11	6.47	5.72	adhesin FhaS
BP2692	1.51	2.10	1.90	ABC transporter
BP3501	3.36	4.86	ON	TTT transporter
BP3783	1.50	1.62	1.05	pertussis toxin subunit A
BP3784	1.80	1.14	1.04	pertussis toxin subunit B
BP3785	1.47	1.57	*0.81*	pertussis toxin subunit D
BP3786	1.14	ND	1.60	pertussis toxin subunit E
BP3787	1.49	1.76	1.26	pertussis toxin subunit C

[1] Log$_2$FC values of Δ*hfq*/wt comparison are shown for RNA-seq and proteomic analyses. Values which did not pass the statistical significance are shown in italics. ND: not determined in both strains in the respective analysis. ON: protein was not detected in the wt strain within the corresponding fraction.

A substantial number of proteins displayed significantly diminished levels in the *hfq* mutant in both proteomic analyses. The largest group of such proteins belonged to T3SS structural components and its secreted substrates. Among the 20 proteins displaying the most decreased abundance in the Δ*hfq* cells were nine T3SS-specific proteins. Likewise, in the supernatant fraction, amounts of Bsp22 and

BopD proteins were dramatically reduced ($\log_2$FC < −9) while BopC, Bcr4, BopB and BopN proteins could not be detected (Table 2).

Table 2. List of T3SS genes showing consistently decreased RNA and protein abundance in the *hfq* mutant.

Gene	Name	RNA-seq[1]	Proteome[1]	Secretome[1]
BP0500	*bopC*	*−0.69*	−1.19	OFF
BP2248	*bscJ*	−1.58	−2.16	ND
BP2250	*bcr4*	−1.81	−2.75	OFF
BP2251	*bcrH2*	−1.58	−1.44	ND
BP2252	*bopB*	−1.89	−2.23	OFF
BP2253	*bopD*	−1.85	−1.83	−9.95
BP2254	*bcrH1*	−1.26	*−0.96*	OFF
BP2256	*bsp22*	−1.22	−1.76	-9.13
BP2257	*bopN*	*−0.85*	−1.71	OFF

[1] $\log_2$FC values of Δ*hfq*/wt comparison are shown for RNA-seq and proteomic analyses. Values that did not pass the statistical significance are shown in italics. ND: not determined in both strains in the respective analysis. OFF: protein was not detected in the Δ*hfq* strain within the corresponding fraction.

Noticeably, one of the virulence factors displaying consistently reduced expression, production and secretion in the *hfq* mutant was the autotransporter Vag8. The Vag8 protein binds and recruits C1 esterase inhibitor and thereby inhibits complement activity and contributes to serum resistance [32–34]. Therefore, we asked whether the reduced production of Vag8 factor would compromise the capacity of the Δ*hfq* strain to evade complement-mediated killing. When compared to recent isolates, Tohama I strain exhibits high susceptibility to serum killing [34], nevertheless, in line with our assumption, the survival of the *hfq* mutant (0.06% ± 0.01%) was dramatically decreased when compared to the wt strain (1.36% ± 0.38%) (Figure 5).

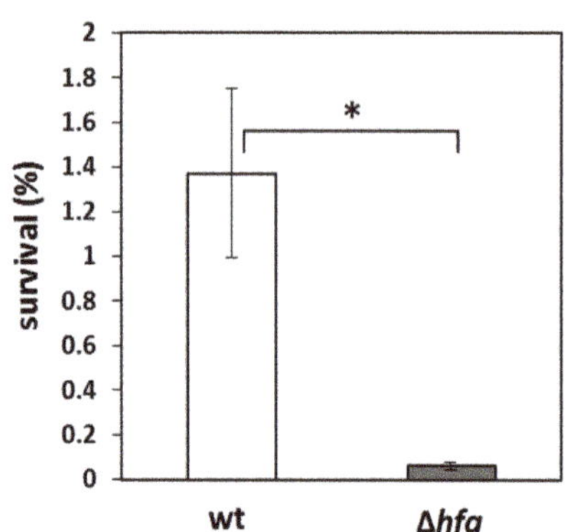

Figure 5. Serum killing assay of Δ*hfq* and Tohama I strains. Serum resistance is expressed as percentage of wt (white bar) and Δ*hfq* (grey bar) cells that survived upon incubation with 10% human serum when compared to controls (bacteria incubated with heat-inactivated serum). The error bars represent the standard deviation of the mean obtained from three biological replicates (*, *p* < 0.001). The result is representative of three independent experiments.

2.3. Correlation between Transcriptome, Proteome and Secretome Datasets

Considering previous reports, the correlation between RNA-seq and LC-MS/MS analysis of cell-associated proteins was relatively high ($r^2 = 0.40$, *p*-value 1.2×10^{-174}) with 148 proteins displaying same trend in abundance as corresponding genes in transcriptomic profiling (Figure 6A). As shown in Figure 6B, the concordance of RNA-seq data with secretome analysis was much less positive ($r^2 = 0.24$,

p-value 7.4×10^{-34}) with only 80 proteins showing similar trend between both analyses. Of note, among genes showing strong correlation with both proteomic datasets were those encoding the T3SS apparatus, pertussis toxin and its transport machinery, ABC, TRAP, and TTT transporters and other proteins involved in the primary metabolism. Apparently, highly modulated genes showed better correlation with abundance of corresponding proteins than those for which the expression was changed only slightly above the thresholds of significance.

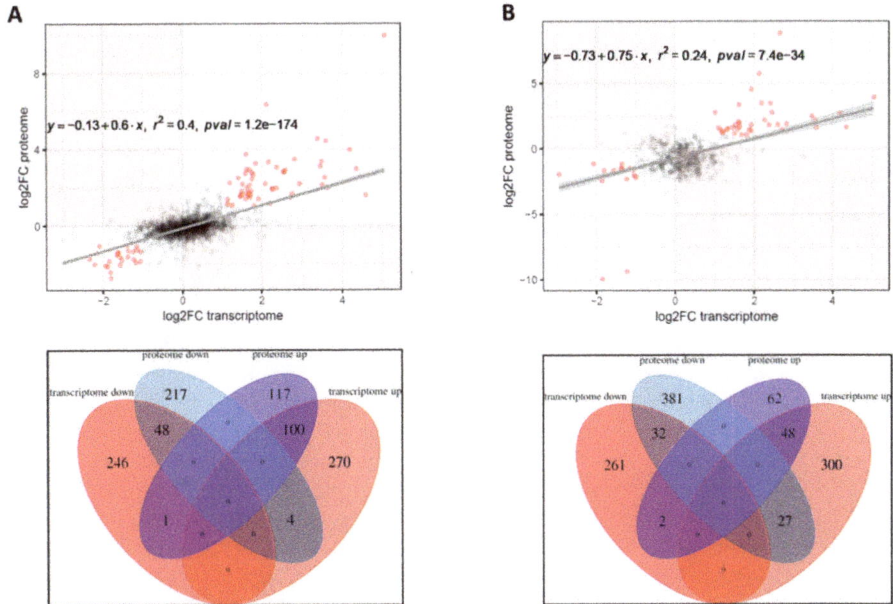

Figure 6. Correlation analysis between transcriptomic and proteomic datasets. (**A**) Scatterplots representing the pairwise comparisons of Δ*hfq*/wt log$_2$ ratios between transcriptome and either proteome (left) or secretome (right). Only the genes for which levels of corresponding proteins were reliably detected by label-free quantification were used for correlation analysis. Red dots depict genes which are significantly deregulated in both datasets (p-value < 0.05, |log$_2$FC| > 1). Line depicts the best fit as predicted by linear regression. (**B**) Venn diagrams showing the number of differentially expressed genes and proteins and the overlap between each dataset. Left: comparison of transcriptome and proteome. Right: comparison of transcriptome and secretome. Zero values indicate intersections which can not materialize (i.e., being up- and downregulated in the same dataset).

3. Discussion

In this study we present first integrative omics analysis of the Hfq regulon in the human pathogen. Our study had several objectives: (a) to corroborate the outcomes of our previous transcriptomic study, (b) identify novel targets of Hfq-specific regulatory activities using high-throughput omics techniques, and (c) compare and evaluate the general effects caused by an important post-transcriptional regulator at the level of transcriptome, proteome, and secretome. Compared to microarray profiling (368 protein coding differentially expressed (DE) genes), the differential expression RNA sequencing identified almost two-fold higher number of deregulated genes (602 protein coding DE genes) in the *hfq* mutant. This finding is not surprising considering the higher sensitivity and reproducibility of the RNA-seq method compared to DNA microarray technique [35,36].

Our data are in line with previous studies reporting modest correlation between transcriptomic and proteomic analyses. Nevertheless, considering the role of Hfq in the post-transcriptional control of gene expression, the correlation coefficient between transcriptome and proteome is relatively high when compared to other studies [22,23,37]. The comparison of RNA-seq with secretome analysis output yielded lower correlation values. This finding can possibly result from "contamination" of bacterial culture supernatants with abundant cytosolic proteins such as components of transcriptional and translational machineries. We speculate that these proteins were released from lysed cells during cultivation and sample preparation and therefore their levels do not correspond to changes in gene expression profiles between wt and Δ*hfq* strains.

Importantly, we corroborated several Hfq-specific effects on gene expression profiles which were seen in our previous microarray study [19]. We recapitulated the strong requirement of Hfq chaperone for T3SS functionality as the expression of T3SS genes and production as well as secretion of T3SS components were significantly reduced in the *hfq* mutant. Especially the differences observed in culture supernatants were enormous (more than two orders of magnitude). Several regulators were shown to play a role in control of T3SS activity in *B. pertussis*, including BvgAS two-component system and BtrAS regulatory circuit [38,39]. The response regulator BvgA activates the expression of an extracytoplasmic function sigma factor *btrS* (BP2234) as well as of *btrU*, *btrV*, and *btrW* genes [40]. While BtrS was shown to be required for efficient transcription of the *bsc* locus encoding the T3SS injectisome, BtrU, BtrV, and BtrW regulatory proteins encoded within the *btr* locus are required for secretion through the T3SS apparatus. Recently, a secreted antagonist of BtrS factor called BtrA (BP2233) exerting negative control over the expression of *Bordetella* T3SS genes was reported [39,41]. Secretion of the BtrA inhibitor reactivates BtrS and, consequently, activates the expression of the T3SS genes [39]. Nevertheless, we did not observe any significant changes in expression of the *btrAS* regulatory node. Moreover, BtrA protein could be detected only in the pellets and its levels were decreased in the mutant (log$_2$FC of −0.53). Apparently, the reduced expression, production and in particular secretion of T3SS components observed in the *hfq* mutant are independent of *btrAS* circuit. Relatively high LFQ intensities of T3SS secreted substrates detected in the wt strain were rather surprising. *B. pertussis* Tohama I represents a laboratory-adapted strain and was suggested to lose its ability to secrete T3SS components during long-term in vitro passaging [40,42,43]. Nevertheless, the capacity to secrete T3SS substrates in Tohama I can be regained upon contact with the host [19,43] or under nutrient limitation [44,45]. We did not use iron- or glutamate-limited media in our experiments and cells were collected in mid exponential phase of growth. Nevertheless, we cannot completely rule out the possibility that our cultures were partially nutrient-limited at the time of harvest. Of note, when compared to Hfq-specific effects at transcriptional and translational levels, the massive differences in protein abundances seen in culture supernatants suggest that Hfq is indirectly required for efficient secretion process through the T3SS apparatus.

In line with our previous reports, expression and production of autotransporter Vag8, a major player in complement evasion [32–34], was significantly reduced in the *hfq* mutant. In support, the *hfq* mutant displayed strongly reduced resistance to serum killing. Increased serum sensitivity of the *hfq* mutant was described also in *Neisseria meningitidis* [46]. We assume that this phenotype can be ascribed to reduced production of the Vag8 protein, as the amounts of BrkA, FhaB, and BapC factors reported to be involved in diversion of complement-mediated killing [47–49] were comparable (BrkA) or even higher in the mutant (FhaB). BapC autotransporter was not detected by proteomics and expression of *bapC* gene was increased in the mutant.

Similarly to several other *hfq*-deficient bacteria, the Δ*hfq* strain of *B. pertussis* displays growth deficit. Based on the results of our microarray study we hypothesized that *hfq* mutant of *B. pertussis* compensates the slower growth with increased production of translation machinery components and proteins involved in transport of nutrients [19]. In support, our current study reveals that the most upregulated genes and corresponding proteins found in the *hfq* mutant are represented predominantly by different types of transport proteins, namely, TTT, TRAP, and ABC transporter families. These solute-binding protein-dependent transporters allow uptake even at very low concentrations of ligands [50]. Interestingly, TTT family transporter genes called "Bug" genes (Bordetella uptake genes) are highly overrepresented in the *B. pertussis* genome as they encode 81 functional TTT proteins [51]. While ligands for majority of these proteins are unknown, crystal structures of BugD and BugE proteins identified their ligands as aspartate and glutamate, respectively [52,53]. Intriguingly, expression of *ptx/ptl* locus and, consequently, production and secretion of pertussis toxin subunits was significantly increased in the *hfq* mutant. In the light of reduced virulence of the mutant and the importance of this toxin for *B. pertussis* pathogenicity [54] it is rather surprising observation which may be conceived as compensatory response to the lack of Hfq.

With regard to observed high impact of *hfq* deletion on gene expression profiles it is of particular interest that abundance of at least 16 transcriptional regulators and five alternative sigma factors was significantly modulated in the *hfq* mutant. These results are in line with already described roles of Hfq in expression of alternative sigma factors [55–58] and suggest that similarly to other bacteria, a substantial part of the Hfq-specific effects seen in *B. pertussis* represents indirect regulation. For example, the expression and production of the iron transport repressor Fur is increased in the *hfq* mutant of *B. pertussis* and, consequently, the expression of several genes responsible for iron delivery was decreased in the mutant. One of the surprising results of this study was the relatively low impact of Hfq on abundance of non-coding RNAs. Recently we have identified small non-coding RNA RgtA that is involved in the regulation of the transport of glutamate, a key metabolite in the *B. pertussis* physiology and the abundance of which in the *hfq* mutant was strongly reduced [59]. Nevertheless, our data indicate that only 40 non-coding transcripts out of the recently identified 400 candidate sRNAs [60] changed their levels in the absence of Hfq. Similarly, integrative analysis of Hfq regulon in *P. fluorescence* identified only four ncRNAs out of 87 whose abundance was dependent on Hfq [30]. Thus, in the light of observed extensive changes in transcriptomic and proteomic profiles observed in *B. pertussis*, the relatively small impact on sRNA levels suggests that Hfq exerts some of its regulatory activities in the sRNA-independent fashion or does not substantially contribute to sRNA stability in *B. pertussis*.

Collectively, this study reveals that impact of Hfq on the gene and protein expression profiles in *B. pertussis* is very profound. The Hfq regulon is comprised of hundreds of genes/proteins making almost 20 % of its genome and covering broad variety of genes and their products involved in different cellular processes. Obviously, these pleiotropic effects associated with loss of Hfq in *B. pertussis* cannot be completely ascribed to its role in posttranscriptional circuits but instead may be related to other global regulators that are themselves targets of Hfq regulation such as transcriptional factors. We assume that several observed effects are linked to impaired growth of the mutant. Especially increased production of proteins implicated in transport of metabolites and essential elements seems to compensate for the physiological defect introduced by deletion of the *hfq* gene. Finally, our study corroborated and further clarified the necessity of Hfq for physiological fitness and pathogenicity of *B. pertussis*. It will be of our primary interest to characterize the exact mechanism rendering the production and secretion of T3SS components strongly dependent on Hfq. Furthermore, we are currently characterizing function of several identified Hfq-dependent sRNAs in the physiology of *B. pertussis*.

4. Materials and Methods

4.1. Bacterial Strains and Growth Conditions

The *Bordetella pertussis* Tohama I strain [61] and its isogenic *hfq* deletion mutant were grown on Bordet-Gengou agar (BGA) plates supplemented with 15% sheep blood for 3 to 4 days at 37 °C. For liquid cultures, bacteria were grown in Stainer–Scholte (SS) medium [62] supplemented with 0.1% cyclodextrin and 0.5% casamino acids (Difco) at 37 °C. To harvest samples for RNA and protein isolation, the *B. pertussis* cells were grown overnight in SS medium to mid exponential phase of growth ($OD_{600} \approx 1.0$). Three independent cultivation experiments were performed to collect three biological replicates for each of both strains for RNA and protein isolation.

4.2. RNA Isolation

Total RNA was isolated using TRI Reagent (Sigma, Darmstadt, Germany) according to manufacturer's protocol. Removal of DNA was achieved by treatment of samples with TURBO DNA-free kit (Thermo Fisher Scientific). RNA quality and quantity was determined by agarose gel electrophoresis and using the Nanodrop 2000 machine (Thermo, Carlsbad, CA, USA). Furthermore, the RNA quality was assessed at sequencing facility (Vienna Biocenter Core Facility, NGS unit) on an Agilent 2100 Bioanalyzer device. All samples displayed RNA integrity numbers higher than 9.

4.3. Library Preparation and Deep Sequencing

Ribosomal RNA was depleted with the Ribo-Zero rRNA Removal Kit for Bacteria (Illumina, San Diego, CA, USA). Libraries were prepared using NEBNext® Ultra™ II DNA Library Prep Kit for Illumina and sequenced on an Illumina HiSeq 2500 platform using HiSeqV4 chemistry with single-end 50-base-pair reads at the Vienna Biocenter Core Facilities Next Generation Sequencing unit. Reads were demultiplexed and quality trimming and adapter removal from the reads was performed using trimmomatic [63]. After quality control and adapter clipping, the reads were mapped to *B. pertussis* Tohama I reference genome using segemehl [64] with default parameters. Reads per gene counts were deduced with htseq-count with default parameters [65]. Differential gene expression analysis was performed with DESeq2 [66]. Genes with a |$\log_2$ fold change| > 1 and a q-value (p-value adjusted for multiple testing correction by the method of Benjamini and Hochberg [67]) < 0.05 were considered as significantly deregulated. RNA-seq data from the sequencing runs were deposited at the European Nucleotide Archive (ENA) under project accession number PRJEB32623.

4.4. Protein Isolation and Sample Preparation for Proteomics

Cultures of *B. pertussis* were pelleted by centrifugation (10,000× g, 4 °C, 10 min) to separate cell pellets and culture supernatants. Cells were resuspended in TEAB digestion buffer (100 mM Triethylammonium bicarbonate, pH 8.5, 2% sodium deoxycholate) and lysed by sonication. For analysis of supernatant fractions, supernatants were filtered through 0.22-µm filters and precipitated with 10% (w/v) trichloracetic acid (Sigma) overnight at 4 °C. Precipitated proteins were collected by centrifugation (14,000× g, 4 °C, 20 min), washed with 80% acetone (w/v) and finally dissolved in TEAB digestion buffer. Protein concentrations were determined using BCA protein assay kit (Thermo Fischer Scientific) and 20 µg of protein per sample were used for protein analysis. Cysteines were reduced with M Tris(2-carboxyethyl)phosphine (60 °C for 60 min) and blocked with 1M methyl methanethiosulfonate (10 min, room temperature). Samples were digested with trypsin (trypsin to protein ratio 1:20) at 37 °C overnight. Digestion of samples was stopped by addition of trifluoracetic acid (Sigma) to a final concentration of 1% (v/v). SDC was removed by extraction with ethylacetate [68] and peptides were desalted on C18 column (Michrom Bio, Auburn, CA, USA).

4.5. Label-Free Proteomic Analysis by LC-MS/MS

A nanoreversed phase column (EASY-Spray column, 50 cm × 75 µm ID, PepMap C18, 2 µm particles, 100 Å pore size) was used for LC-MS analysis. Mobile phase buffer A was composed of water and 0.1% formic acid. Mobile phase B was composed of acetonitrile and 0.1% formic acid. Samples were loaded onto the trap column (Acclaim PepMap300, C18, 5 µm, 300 Å wide pore, 300 µm × 5 mm) at a flow rate of 15 µL/min. Loading buffer was composed of water, 2% acetonitrile, and 0.1% trifluoroacetic acid. Peptides were eluted with gradient of B phase ranging from 4% to 35% over 60 min at a flow rate of 300 nL/min. Eluting peptide cations were converted to gas-phase ions by electrospray ionization and analyzed on a Thermo Orbitrap Fusion (Q-OT-qIT, Thermo Fischer). Survey scans of peptide precursors from 350 to 1400 m/z were performed at 12 resolution (at 200 m/z) with a 5 × 10^5 ion count target. Tandem MS (MS^2) was performed by isolation within 1.5-Th window with the quadrupole, HCD fragmentation with normalized collision energy of 30, and rapid scan MS analysis in the ion trap. The MS^2 ion count target value was set to 10^4 and the maximal injection time was 35 ms. Only those precursors with charge state 2–6 were sampled for MS^2. The dynamic exclusion duration was set to 45 s with a 10 ppm tolerance around the selected precursor and its isotopes. Monoisotopic precursor selection was turned on. The instrument was run in top speed mode with 2 s cycles [69].

Raw data were imported into MaxQuant software (version 1.5.3.8) [70] for identification and label-free quantification of proteins. The false discovery rate (FDR) was set to 1% for peptides and minimum specific length of seven amino acids. The Andromeda search engine [71] was used for the MS/MS spectra search against the Uniprot *Bordetella pertussis* database (downloaded on November 2016), containing 3258 entries. Enzyme specificity was set as C-terminal to Arg and Lys, also allowing cleavage at proline bonds and a maximum of two missed cleavages. Dithiomethylation of cysteine was selected as fixed modification and N- terminal protein acetylation and methionine oxidation as variable modifications. The "match between runs" feature of MaxQuant was used to transfer identifications to other LC-MS/MS runs based on their masses and retention time (maximum deviation 0.7 min) and this was also used in quantification experiments. Protein abundance was calculated from obtained label-free protein intensities using the MaxLFQ algorithm described recently [72]. Proteins with less than four MS/MS spectral counts were removed from the analysis. Statistics and data interpretation were performed using Perseus 1.6.1.3 software [73]. The normalized label free intensities were compared between wt and *hfq* mutant and each abundance ratio was tested for significance with two-group t-test (p-value < 0.05). The p-values were further adjusted for multiple testing correction to control the false discovery rate at cut off of 0.05 using the permutation test (number of randomization 250). Proteins with corrected p-value (q-value) < 0.05 were considered as significantly modulated. For downstream analyses (e.g., GO term enrichment) only proteins which were detected by at least two unique peptides in at least two of the three biological replicates were considered. Proteins for which label free intensities were not obtained in any of the replicates of either the wt or the Δhfq strain were considered as significantly modulated and defined as "ON/OFF". The proteomics data were deposited to the ProteomeXchange Consortium via the PRIDE [74] partner repository with the dataset identifier PXD013953.

4.6. GO Term Enrichment Analysis

To gain a comprehensive functional annotation of the reference genome, gene ontology (GO) terms per gene were deduced using blast2go [75]. For the GO term enrichment analysis significantly deregulated genes from the transcriptome and proteome analysis were split into up- and downregulated genes and each gene set was analyzed separately. Each GO term which is associated with more than one gene in the gene set was tested for enrichment in the gene set compared to the whole transcriptome, applying a Fisher's exact test. Afterwards, determined p-values were corrected for multiple testing by the method of Benjamini and Hochberg [67]. Enriched GO terms were further summarized and visualized by Revigo [76].

4.7. Transcriptome–Proteome Correlation Analyses

To correlate the effect of hfq gene deletion on the transcript and protein abundance globally, the $\log_2 FC$ of all genes the products of which were reliably detected (see Chapter 4.5) by the label-free quantification were compared. To this end, the 'lm' function from R was used to fit a linear model between these two datasets. Since the relative errors in $\log_2 FC$ measurements can be expected to be higher for genes with higher p-value, each data point was weighted in the course of model fitting by 1 - q-value where q-value represents the geometric mean of the q-values of the proteome and the transcriptome analysis.

4.8. Serum Killing Assay

Overnight-grown bacterial cultures were diluted in SS medium to 5×10^6 bacteria/ ml of culture and supplemented either with intact or heat-inactivated (56 °C, 30 min) 10% human serum (Sigma No).

Cells were incubated in parallel in the presence of both type of sera for 60 min at 37 °C in orbital incubator. Then the bactericidal activity was terminated by addition of 10 mM EDTA, serial dilutions of bacterial samples were plated onto BG agar and colony-forming units (CFU) were counted to assess bacterial survival. Survival was calculated as a percentage of CFU obtained from cultures treated with intact serum compared to CFUs from cultures treated with heat-inactivated serum (control, 100% survival). Mann–Whitney test was applied to assay the statistical significance of observed differences in sensitivity to serum killing.

Supplementary Materials: Supplementary materials can be found at http://www.mdpi.com/1422-0067/20/12/3073/s1.

Author Contributions: A.D. conceived and performed the experiments, analyzed and interpreted the data, and edited the manuscript; F.A. analyzed and interpreted the data and edited the manuscript; D.S. performed the experiment and analyzed the data; D.P. analyzed and interpreted the data and edited the manuscript; B.V. conceived the experiments, supervised the research project, and wrote the original draft of the manuscript. All authors approved the final draft.

Funding: This work was supported by grant 19-12338S (to B.V.) from the Czech Science Foundation (www.gacr.cz) and by funding from RVO61388971. This work was also supported by Mobility grant from Czech Academy of Sciences (MSM200201702) to A.D. Acquisition of the proteomic data was supported by the project BIOCEV— Biotechnology and Biomedicine Centre of the Academy of Sciences and Charles University (CZ.1.05/1.1.00/02.0109) from the European Regional Development Fund.

Acknowledgments: We thank J. Držmíšek for help with preparing samples for proteomic analysis and Rudy Antoine (Institute Pasteur, Lille) for helpful discussions. We are grateful to Karel Harant and Pavel Talacko from the Mass Spectrometry and Proteomics Service Laboratory, Faculty of Science, Charles University for performing the LC-MS/MS run.

Conflicts of Interest: The authors declare no conflict of interest. The funders had no role in the design of the study; in the collection, analyses, or interpretation of data; in the writing of the manuscript, or in the decision to publish the results.

Abbreviations

ABC	ATP-binding cassette
BGA	Bordet-Gengou agar
FDR	False discovery rate
TEAB	Triethylammonium bicarbonate
EDTA	Ethylenediamine tetra-acetic acid
CFU	Colony forming unit
DE	Differential expression
TTT	Tripartite tricarboxylate transporters
TRAP	Tripartite ATP-independent periplasmic transporters
PCA	Principal component analysis
GO	Gene ontology
LFQ	Label-free quantification
T3SS	Type 3 secretion system
FC	Fold change

References

1. Mattoo, S.; Cherry, J.D. Molecular pathogenesis, epidemiology, and clinical manifestations of respiratory infections due to bordetella pertussis and other bordetella subspecies. *Clin. Microbiol. Rev.* **2005**, *18*, 326–382. [CrossRef] [PubMed]
2. World Health Organization (WHO). Vaccine preventable deaths and the global immunization vision and strategy, 2006–2015. *MMWR Morb. Mortal. Wkly. Rep.* **2006**, *55*, 511–515.
3. Raguckas, S.E.; VandenBussche, H.L.; Jacobs, C.; Klepser, M.E. Pertussis resurgence: Diagnosis, treatment, prevention, and beyond. *Pharmacotherapy* **2007**, *27*, 41–52. [CrossRef] [PubMed]
4. Cherry, J.D. The present and future control of pertussis. *Clin. Infect. Dis.* **2010**, *51*, 663–667. [CrossRef] [PubMed]

5. Cherry, J.D. Pertussis: Challenges today and for the future. *PLoS Pathog.* **2013**, *9*, e1003418. [CrossRef] [PubMed]

6. Mooi, F.R.; Van Der Maas, N.A.; De Melker, H.E. Pertussis resurgence: Waning immunity and pathogen adaptation—Two sides of the same coin. *Epidemiol. Infect.* **2014**, *142*, 685–694. [CrossRef] [PubMed]

7. Sealey, K.L.; Belcher, T.; Preston, A. Bordetella pertussis epidemiology and evolution in the light of pertussis resurgence. *Infect. Genet. Evol.* **2016**, *40*, 136–143. [CrossRef]

8. Burdin, N.; Handy, L.K.; Plotkin, S.A. What is wrong with pertussis vaccine immunity? The problem of waning effectiveness of pertussis vaccines. *Cold Spring Harb. Perspect. Biol.* **2017**, *9*. [CrossRef]

9. Hewlett, E.L.; Burns, D.L.; Cotter, P.A.; Harvill, E.T.; Merkel, T.J.; Quinn, C.P.; Stibitz, E.S. Pertussis pathogenesis—What we know and what we don't know. *J. Infect. Dis.* **2014**, *209*, 982–985. [CrossRef]

10. The Periscope Consortium. Periscope: Road towards effective control of pertussis. *Lancet Infect. Dis.* **2019**, *19*, 179–186. [CrossRef]

11. Chao, Y.; Vogel, J. The role of hfq in bacterial pathogens. *Curr. Opin. Microbiol.* **2010**, *13*, 24–33. [CrossRef] [PubMed]

12. Feliciano, J.R.; Grilo, A.M.; Guerreiro, S.I.; Sousa, S.A.; Leitao, J.H. Hfq: A multifaceted rna chaperone involved in virulence. *Fut. Microbiol.* **2016**, *11*, 137–151. [CrossRef] [PubMed]

13. Papenfort, K.; Vogel, J. Regulatory rna in bacterial pathogens. *Cell Host Microbe* **2010**, *8*, 116–127. [CrossRef] [PubMed]

14. Vogel, J.; Luisi, B.F. Hfq and its constellation of rna. *Nat. Rev. Microbiol.* **2011**, *9*, 578–589. [CrossRef] [PubMed]

15. Vecerek, B.; Rajkowitsch, L.; Sonnleitner, E.; Schroeder, R.; Blasi, U. The c-terminal domain of escherichia coli hfq is required for regulation. *Nucleic Acids Res.* **2008**, *36*, 133–143. [CrossRef] [PubMed]

16. Updegrove, T.B.; Zhang, A.; Storz, G. Hfq: The flexible rna matchmaker. *Curr. Opin. Microbiol.* **2016**, *30*, 133–138. [CrossRef] [PubMed]

17. Brennan, R.G.; Link, T.M. Hfq structure, function and ligand binding. *Curr. Opin. Microbiol.* **2007**, *10*, 125–133. [CrossRef]

18. Bibova, I.; Skopova, K.; Masin, J.; Cerny, O.; Hot, D.; Sebo, P.; Vecerek, B. The rna chaperone hfq is required for virulence of bordetella pertussis. *Infect. Immun.* **2013**, *81*, 4081–4090. [CrossRef]

19. Bibova, I.; Hot, D.; Keidel, K.; Amman, F.; Slupek, S.; Cerny, O.; Gross, R.; Vecerek, B. Transcriptional profiling of bordetella pertussis reveals requirement of rna chaperone hfq for type iii secretion system functionality. *RNA Biol.* **2015**, *12*, 175–185. [CrossRef]

20. de Sousa Abreu, R.; Penalva, L.O.; Marcotte, E.M.; Vogel, C. Global signatures of protein and mrna expression levels. *Mol. Biosyst.* **2009**, *5*, 1512–1526. [CrossRef]

21. Kumar, D.; Bansal, G.; Narang, A.; Basak, T.; Abbas, T.; Dash, D. Integrating transcriptome and proteome profiling: Strategies and applications. *Proteomics* **2016**, *16*, 2533–2544. [CrossRef] [PubMed]

22. Gygi, S.P.; Rochon, Y.; Franza, B.R.; Aebersold, R. Correlation between protein and mrna abundance in yeast. *Mol. Cell Biol.* **1999**, *19*, 1720–1730. [CrossRef] [PubMed]

23. Greenbaum, D.; Colangelo, C.; Williams, K.; Gerstein, M. Comparing protein abundance and mrna expression levels on a genomic scale. *Genome Biol.* **2003**, *4*, 117. [CrossRef] [PubMed]

24. Ghazalpour, A.; Bennett, B.; Petyuk, V.A.; Orozco, L.; Hagopian, R.; Mungrue, I.N.; Farber, C.R.; Sinsheimer, J.; Kang, H.M.; Furlotte, N.; et al. Comparative analysis of proteome and transcriptome variation in mouse. *PLoS Genet.* **2011**, *7*, e1001393. [CrossRef] [PubMed]

25. Maier, T.; Guell, M.; Serrano, L. Correlation of mrna and protein in complex biological samples. *FEBS Lett.* **2009**, *583*, 3966–3973. [CrossRef] [PubMed]

26. Manzoni, C.; Kia, D.A.; Vandrovcova, J.; Hardy, J.; Wood, N.W.; Lewis, P.A.; Ferrari, R. Genome, transcriptome and proteome: The rise of omics data and their integration in biomedical sciences. *Brief Bioinform.* **2018**, *19*, 286–302. [CrossRef]

27. Nie, L.; Wu, G.; Zhang, W. Correlation of mrna expression and protein abundance affected by multiple sequence features related to translational efficiency in desulfovibrio vulgaris: A quantitative analysis. *Genetics* **2006**, *174*, 2229–2243. [CrossRef]

28. Arraiano, C.M.; Andrade, J.M.; Domingues, S.; Guinote, I.B.; Malecki, M.; Matos, R.G.; Moreira, R.N.; Pobre, V.; Reis, F.P.; Saramago, M.; et al. The critical role of rna processing and degradation in the control of gene expression. *FEMS Microbiol. Rev.* **2010**, *34*, 883–923. [CrossRef]

29. Kudla, G.; Murray, A.W.; Tollervey, D.; Plotkin, J.B. Coding-sequence determinants of gene expression in escherichia coli. *Science* **2009**, *324*, 255–258. [CrossRef]

30. Grenga, L.; Chandra, G.; Saalbach, G.; Galmozzi, C.V.; Kramer, G.; Malone, J.G. Analyzing the complex regulatory landscape of hfq—An integrative, multi-omics approach. *Front. Microbiol.* **2017**, *8*, 1784. [CrossRef]

31. Delgado-Ortega, M.; Marc, D.; Dupont, J.; Trapp, S.; Berri, M.; Meurens, F. Socs proteins in infectious diseases of mammals. *Vet. Immunol. Immunopathol.* **2013**, *151*, 1–19. [CrossRef] [PubMed]

32. Marr, N.; Shah, N.R.; Lee, R.; Kim, E.J.; Fernandez, R.C. Bordetella pertussis autotransporter vag8 binds human c1 esterase inhibitor and confers serum resistance. *PLoS ONE* **2011**, *6*, e20585. [CrossRef] [PubMed]

33. Hovingh, E.S.; van den Broek, B.; Kuipers, B.; Pinelli, E.; Rooijakkers, S.H.M.; Jongerius, I. Acquisition of c1 inhibitor by bordetella pertussis virulence associated gene 8 results in c2 and c4 consumption away from the bacterial surface. *PLoS Pathog.* **2017**, *13*, e1006531. [CrossRef] [PubMed]

34. Brookes, C.; Freire-Martin, I.; Cavell, B.; Alexander, F.; Taylor, S.; Persaud, R.; Fry, N.; Preston, A.; Diavatopoulos, D.; Gorringe, A. Bordetella pertussis isolates vary in their interactions with human complement components. *Emerg. Microbes Infect.* **2018**, *7*, 81. [CrossRef] [PubMed]

35. Marioni, J.C.; Mason, C.E.; Mane, S.M.; Stephens, M.; Gilad, Y. Rna-seq: An assessment of technical reproducibility and comparison with gene expression arrays. *Genome Res.* **2008**, *18*, 1509–1517. [CrossRef]

36. Zhao, S.; Fung-Leung, W.P.; Bittner, A.; Ngo, K.; Liu, X. Comparison of rna-seq and microarray in transcriptome profiling of activated t cells. *PLoS ONE* **2014**, *9*, e78644. [CrossRef] [PubMed]

37. Basler, M.; Linhartova, I.; Halada, P.; Novotna, J.; Bezouskova, S.; Osicka, R.; Weiser, J.; Vohradsky, J.; Sebo, P. The iron-regulated transcriptome and proteome of neisseria meningitidis serogroup c. *Proteomics* **2006**, *6*, 6194–6206. [CrossRef]

38. Yuk, M.H.; Harvill, E.T.; Miller, J.F. The bvgas virulence control system regulates type iii secretion in bordetella bronchiseptica. *Mol. Microbiol.* **1998**, *28*, 945–959. [CrossRef]

39. Ahuja, U.; Shokeen, B.; Cheng, N.; Cho, Y.; Blum, C.; Coppola, G.; Miller, J.F. Differential regulation of type iii secretion and virulence genes in bordetella pertussis and bordetella bronchiseptica by a secreted anti-sigma factor. *Proc. Natl. Acad. Sci. USA* **2016**, *113*, 2341–2348. [CrossRef]

40. Mattoo, S.; Yuk, M.H.; Huang, L.L.; Miller, J.F. Regulation of type iii secretion in bordetella. *Mol. Microbiol.* **2004**, *52*, 1201–1214. [CrossRef]

41. Kurushima, J.; Kuwae, A.; Abe, A. The type iii secreted protein bspr regulates the virulence genes in bordetella bronchiseptica. *PLoS ONE* **2012**, *7*, e38925. [CrossRef] [PubMed]

42. Fennelly, N.K.; Sisti, F.; Higgins, S.C.; Ross, P.J.; van der Heide, H.; Mooi, F.R.; Boyd, A.; Mills, K.H. Bordetella pertussis expresses a functional type iii secretion system that subverts protective innate and adaptive immune responses. *Infect. Immun.* **2008**, *76*, 1257–1266. [CrossRef]

43. Gaillard, M.E.; Bottero, D.; Castuma, C.E.; Basile, L.A.; Hozbor, D. Laboratory adaptation of bordetella pertussis is associated with the loss of type three secretion system functionality. *Infect. Immun.* **2011**, *79*, 3677–3682. [CrossRef] [PubMed]

44. Brickman, T.J.; Cummings, C.A.; Liew, S.Y.; Relman, D.A.; Armstrong, S.K. Transcriptional profiling of the iron starvation response in bordetella pertussis provides new insights into siderophore utilization and virulence gene expression. *J. Bacteriol.* **2011**, *193*, 4798–4812. [CrossRef] [PubMed]

45. Hanawa, T.; Kamachi, K.; Yonezawa, H.; Fukutomi, T.; Kawakami, H.; Kamiya, S. Glutamate limitation, bvgas activation, and (p)ppgpp regulate the expression of the bordetella pertussis type 3 secretion system. *J. Bacteriol.* **2016**, *198*, 343–351. [CrossRef] [PubMed]

46. Fantappie, L.; Metruccio, M.M.; Seib, K.L.; Oriente, F.; Cartocci, E.; Ferlicca, F.; Giuliani, M.M.; Scarlato, V.; Delany, I. The rna chaperone hfq is involved in stress response and virulence in neisseria meningitidis and is a pleiotropic regulator of protein expression. *Infect. Immun.* **2009**, *77*, 1842–1853. [CrossRef]

47. Barnes, M.G.; Weiss, A.A. Brka protein of bordetella pertussis inhibits the classical pathway of complement after c1 deposition. *Infect. Immun.* **2001**, *69*, 3067–3072. [CrossRef]

48. Berggard, K.; Johnsson, E.; Mooi, F.R.; Lindahl, G. Bordetella pertussis binds the human complement regulator c4bp: Role of filamentous hemagglutinin. *Infect. Immun.* **1997**, *65*, 3638–3643.

49. Noofeli, M.; Bokhari, H.; Blackburn, P.; Roberts, M.; Coote, J.G.; Parton, R. Bapc autotransporter protein is a virulence determinant of bordetella pertussis. *Microb. Pathog.* **2011**, *51*, 169–177. [CrossRef]

50. Rosa, L.T.; Bianconi, M.E.; Thomas, G.H.; Kelly, D.J. Tripartite atp-independent periplasmic (trap) transporters and tripartite tricarboxylate transporters (ttt): From uptake to pathogenicity. *Front. Cell Infect. Microbiol.* **2018**, *8*, 33. [CrossRef]

51. Antoine, R.; Jacob-Dubuisson, F.; Drobecq, H.; Willery, E.; Lesjean, S.; Locht, C. Overrepresentation of a gene family encoding extracytoplasmic solute receptors in bordetella. *J. Bacteriol.* **2003**, *185*, 1470–1474. [CrossRef] [PubMed]

52. Huvent, I.; Belrhali, H.; Antoine, R.; Bompard, C.; Locht, C.; Jacob-Dubuisson, F.; Villeret, V. Crystal structure of bordetella pertussis bugd solute receptor unveils the basis of ligand binding in a new family of periplasmic binding proteins. *J. Mol. Biol.* **2006**, *356*, 1014–1026. [CrossRef] [PubMed]

53. Huvent, I.; Belrhali, H.; Antoine, R.; Bompard, C.; Locht, C.; Jacob-Dubuisson, F.; Villeret, V. Structural analysis of bordetella pertussis buge solute receptor in a bound conformation. *Acta Crystallogr. D Biol. Crystallogr.* **2006**, *62*, 1375–1381. [CrossRef] [PubMed]

54. Carbonetti, N.H. Contribution of pertussis toxin to the pathogenesis of pertussis disease. *Pathog. Dis.* **2015**, *73*, ftv073. [CrossRef] [PubMed]

55. Brown, L.; Elliott, T. Efficient translation of the rpos sigma factor in salmonella typhimurium requires host factor i, an rna-binding protein encoded by the hfq gene. *J. Bacteriol.* **1996**, *178*, 3763–3770. [CrossRef] [PubMed]

56. Muffler, A.; Fischer, D.; Hengge-Aronis, R. The rna-binding protein hf-i, known as a host factor for phage qbeta rna replication, is essential for rpos translation in escherichia coli. *Genes Dev.* **1996**, *10*, 1143–1151. [CrossRef] [PubMed]

57. Ding, Y.; Davis, B.M.; Waldor, M.K. Hfq is essential for vibrio cholerae virulence and downregulates sigma expression. *Mol. Microbiol.* **2004**, *53*, 345–354. [CrossRef] [PubMed]

58. Figueroa-Bossi, N.; Lemire, S.; Maloriol, D.; Balbontin, R.; Casadesus, J.; Bossi, L. Loss of hfq activates the sigmae-dependent envelope stress response in salmonella enterica. *Mol. Microbiol.* **2006**, *62*, 838–852. [CrossRef] [PubMed]

59. Keidel, K.; Amman, F.; Bibova, I.; Drzmisek, J.; Benes, V.; Hot, D.; Vecerek, B. Signal transduction-dependent small regulatory rna is involved in glutamate metabolism of the human pathogen bordetella pertussis. *RNA* **2018**, *24*, 1530–1541. [CrossRef]

60. Amman, F.; D'Halluin, A.; Antoine, R.; Huot, L.; Bibova, I.; Keidel, K.; Slupek, S.; Bouquet, P.; Coutte, L.; Caboche, S.; et al. Primary transcriptome analysis reveals importance of is elements for the shaping of the transcriptional landscape of bordetella pertussis. *RNA Biol.* **2018**, *15*, 967–975. [CrossRef]

61. Kasuga, T.; Nakase, Y.; Ukishima, K.; Takatsu, K. Studies on haemophilis pertussis. Iii. Some properties of each phase of h. Pertussis. *Kitasato Arch. Exp. Med.* **1954**, *27*, 37–47. [PubMed]

62. Stainer, D.W.; Scholte, M.J. A simple chemically defined medium for the production of phase i bordetella pertussis. *J. Gen. Microbiol.* **1970**, *63*, 211–220. [CrossRef] [PubMed]

63. Bolger, A.M.; Lohse, M.; Usadel, B. Trimmomatic: A flexible trimmer for illumina sequence data. *Bioinformatics* **2014**, *30*, 2114–2120. [CrossRef] [PubMed]

64. Hoffmann, S.; Otto, C.; Kurtz, S.; Sharma, C.M.; Khaitovich, P.; Vogel, J.; Stadler, P.F.; Hackermuller, J. Fast mapping of short sequences with mismatches, insertions and deletions using index structures. *PLoS Comput. Biol.* **2009**, *5*, e1000502. [CrossRef] [PubMed]

65. Anders, S.; Pyl, P.T.; Huber, W. Htseq—A python framework to work with high-throughput sequencing data. *Bioinformatics* **2015**, *31*, 166–169. [CrossRef] [PubMed]

66. Love, M.I.; Huber, W.; Anders, S. Moderated estimation of fold change and dispersion for rna-seq data with deseq2. *Genome Biol.* **2014**, *15*, 550. [CrossRef] [PubMed]

67. Benjamini, Y.; Hochberg, Y. Controlling the false discovery rate: A practical and powerful approach to multiple testing. *J. R. Stat. Soc. B* **1995**, *57*, 289–300. [CrossRef]

68. Masuda, T.; Tomita, M.; Ishihama, Y. Phase transfer surfactant-aided trypsin digestion for membrane proteome analysis. *J. Proteome Res.* **2008**, *7*, 731–740. [CrossRef]

69. Hebert, A.S.; Richards, A.L.; Bailey, D.J.; Ulbrich, A.; Coughlin, E.E.; Westphall, M.S.; Coon, J.J. The one hour yeast proteome. *Mol. Cell Proteom.* **2014**, *13*, 339–347. [CrossRef]

70. Cox, J.; Mann, M. Maxquant enables high peptide identification rates, individualized p.P.B.-range mass accuracies and proteome-wide protein quantification. *Nat. Biotechnol.* **2008**, *26*, 1367–1372. [CrossRef]

71. Cox, J.; Neuhauser, N.; Michalski, A.; Scheltema, R.A.; Olsen, J.V.; Mann, M. Andromeda: A peptide search engine integrated into the maxquant environment. *J. Proteome Res.* **2011**, *10*, 1794–1805. [CrossRef] [PubMed]

72. Cox, J.; Hein, M.Y.; Luber, C.A.; Paron, I.; Nagaraj, N.; Mann, M. Accurate proteome-wide label-free quantification by delayed normalization and maximal peptide ratio extraction, termed maxlfq. *Mol. Cell Proteom.* **2014**, *13*, 2513–2526. [CrossRef] [PubMed]

73. Tyanova, S.; Temu, T.; Sinitcyn, P.; Carlson, A.; Hein, M.Y.; Geiger, T.; Mann, M.; Cox, J. The perseus computational platform for comprehensive analysis of (prote)omics data. *Nat. Methods* **2016**, *13*, 731–740. [CrossRef] [PubMed]

74. Perez-Riverol, Y.; Csordas, A.; Bai, J.; Bernal-Llinares, M.; Hewapathirana, S.; Kundu, D.J.; Inuganti, A.; Griss, J.; Mayer, G.; Eisenacher, M.; et al. The pride database and related tools and resources in 2019: Improving support for quantification data. *Nucleic Acids Res.* **2019**, *47*, D442–D450. [CrossRef]

75. Conesa, A.; Gotz, S.; Garcia-Gomez, J.M.; Terol, J.; Talon, M.; Robles, M. Blast2go: A universal tool for annotation, visualization and analysis in functional genomics research. *Bioinformatics* **2005**, *21*, 3674–3676. [CrossRef] [PubMed]

76. Supek, F.; Bosnjak, M.; Skunca, N.; Smuc, T. Revigo summarizes and visualizes long lists of gene ontology terms. *PLoS ONE* **2011**, *6*, e21800. [CrossRef] [PubMed]

© 2019 by the authors. Licensee MDPI, Basel, Switzerland. This article is an open access article distributed under the terms and conditions of the Creative Commons Attribution (CC BY) license (http://creativecommons.org/licenses/by/4.0/).

 International Journal of
Molecular Sciences

Brief Report

Targeting *Pseudomonas aeruginosa* in the Sputum of Primary Ciliary Dyskinesia Patients with a Combinatorial Strategy Having Antibacterial and Anti-Virulence Potential

Giuseppantonio Maisetta [1,*], Lucia Grassi [1], Semih Esin [1], Esingül Kaya [1], Andrea Morelli [2], Dario Puppi [2], Martina Piras [3], Federica Chiellini [2], Massimo Pifferi [3] and Giovanna Batoni [1]

[1] Department of Translational Research and New Technologies in Medicine and Surgery, University of Pisa, 56123 Pisa, Italy; lucia.grassi@med.unipi.it (L.G.); semih.esin@med.unipi.it (S.E.); esingulkaya@gmail.com (E.K.); giovanna.batoni@med.unipi.it (G.B.)

[2] Department of Chemistry and Industrial Chemistry, University of Pisa, 56124 Pisa, Italy; a.morelli@dcci.unipi.it (A.M.); d.puppi@dcci.unipi.it (D.P.); federica.chiellini@unipi.it (F.C.)

[3] Section of Pneumology and Allergology, Unit of Pediatrics, Pisa University Hospital, 56126 Pisa, Italy; martinaprs@gmail.com (M.P.); m.pifferi@med.unipi.it (M.P.)

* Correspondence: giuseppantonio.maisetta@dps.unipi.it; Tel.: +39-050-2213692; Fax: +39-050-2213711

Received: 2 December 2019; Accepted: 17 December 2019; Published: 20 December 2019

Abstract: In primary ciliary dyskinesia (PCD) patients, *Pseudomonas aeruginosa* is a major opportunistic pathogen, frequently involved in chronic infections of the lower airways. Infections by this bacterial species correlates with a worsening clinical prognosis and recalcitrance to currently available therapeutics. The antimicrobial peptide, lin-SB056-1, in combination with the cation chelator ethylenediaminetetraacetic acid (EDTA), was previously demonstrated to be bactericidal against *P. aeruginosa* in an artificial sputum medium. The purpose of this study was to validate the anti-*P. aeruginosa* activity of such a combination in PCD sputum and to evaluate the in vitro anti-virulence effects of EDTA. In combination with EDTA, lin-SB056-1 was able to significantly reduce the load of endogenous *P. aeruginosa* ex vivo in the sputum of PCD patients. In addition, EDTA markedly reduced the production of relevant bacterial virulence factors (e.g., pyocyanin, proteases, LasA) in vitro by two representative mucoid strains of *P. aeruginosa* isolated from the sputum of PCD patients. These results indicate that the lin-SB056-1/EDTA combination may exert a dual antimicrobial and anti-virulence action against *P. aeruginosa*, suggesting a therapeutic potential against chronic airway infections sustained by this bacterium.

Keywords: antimicrobial peptide; EDTA; *Pseudomonas aeruginosa*; primary ciliary dyskinesia; virulence factor; anti-virulence; sputum; chronic infection

---

1. Introduction

Primary ciliary dyskinesia (PCD) is an autosomal recessive disorder characterized by abnormal ciliary ultrastructure and function leading to impaired mucociliary clearance and recurrent respiratory infections [1]. Although *Haemophilus influenzae* is the pathogen most commonly isolated from patients with PCD until adolescence/early adulthood, in adult PCD patients, *P. aeruginosa* plays a major role, especially after the age of 30 [1]. Accordingly, a negative correlation between the abundance of *P. aeruginosa* in the airways of these patients and lung function has been reported [2,3]. The pathogenesis of *P. aeruginosa* infection is at least partially attributable to its ability to synthesize and secrete a number of virulence factors (e.g., pyoverdine, pyocyanin, proteases) and to form biofilms, in which bacterial cells are embedded in an alginate extracellular matrix [4]. Despite intensive antibiotic therapy, once the patients are stably colonized by *P. aeruginosa*, the eradication of the bacterium is rarely achieved [1,5]. Therefore, there is a critical need for novel antimicrobial drugs that can effectively lower *P. aeruginosa* load in the challenging environment of PCD lung.

Over the last decades, antimicrobial peptides (AMPs) have been intensively investigated as potential antibiotics against multidrug-resistant bacteria [6]. Most AMPs are cationic molecules with an amphipathic structure that selectively target bacterial membranes via electrostatic forces. In contrast to standard antibiotics, AMPs are generally effective against both quiescent and actively growing bacteria, display rapid killing kinetics, and demonstrate low propensity to select resistant mutants in vitro [7,8]. On the other hand, AMPs may display a reduction in their antibacterial potency in the presence of complex biological fluids such as sputum, plasma, or saliva due to the high concentration of salt found in these fluids and/or the presence of anionic proteins and host or bacterial proteases that may neutralize their activity [9,10].

As observed in cystic fibrosis (CF) lungs, the biofilm mode of growth of bacteria together with the lung mucus viscosity reduces the effectiveness of conventional antibiotic therapy in PCD patients [11]. Thus, the usage of adjuvants has been proposed to improve the diffusion of antimicrobials through the mucus and the biofilm matrix and facilitate the targeting of bacterial cells [12]. Previous studies have shown that the divalent cation chelator ethylenediaminetetraacetic acid (EDTA) can destabilize the biofilm structure by interfering with the ionic attractive forces among the biofilm matrix components [13,14]. EDTA is prescribed in a number of clinical conditions demonstrating high tolerability (up to 2 g once a week, intravenously injected) and efficacy [15]. Recently, we demonstrated that the optimized semi-synthetic antimicrobial peptide lin-SB056-1 in combination with EDTA is able to exert a synergistic bactericidal effect against *P. aeruginosa* in an artificial sputum medium resembling CF sputum [16]. Similarities and differences between CF and PDC sputum have been reported. For instance, while both diseases seem to be associated with a similar degree of airways neutrophilia, the concentration of interleukin-8 in sputum is higher in PCD than in CF patients, while neutrophil elastase activity is lower in PCD compared with CF [17]. In order to evaluate the therapeutic potential of the lin-SB056-1/EDTA combination in PCD, in this study, we evaluated its bactericidal activity ex vivo, against endogenous *P. aeruginosa* in the sputum from PCD patients. Importantly, the ex vivo sputum mimics, with good approximation, the lung environment, as it contains both host and bacterial components, including bronchial mucus contaminated by saliva, serum proteins, inflammatory mediators, desquamated epithelial cells, and pathogenic bacteria, as well as bacteria from the normal flora [18].

Previous reports have demonstrated the involvement of different cations (i.e., calcium, magnesium, and zinc) either in the regulation of gene expression or in the production and processing of virulence factors in *P. aeruginosa* [19,20]. Thus, herein, we also evaluated the ability of EDTA to reduce the production of relevant virulence factors of *P. aeruginosa* (e.g., pyoverdin, pyocyanin, proteases, biofilm production). Overall, the results obtained demonstrated that the lin-SB056-1/EDTA combination is able to significantly reduce *P. aeruginosa* load ex vivo and that EDTA is highly active in suppressing the production of relevant bacterial virulence factors, suggesting a dual antibacterial and anti-virulence potential of the combination.

2. Results

2.1. Killing Activity of lin-SB056-1 in Combination with Ethylenediaminetetraacetic Acid (EDTA) against Endogenous P. aeruginosa

P. aeruginosa strains were isolated from the sputum of six PCD patients known to be chronically infected with the bacterium. All the strains displayed a mucoid phenotype and different antibiotic susceptibility profiles (Supplementary Table S1).

Diluted sputum (5-fold) from each patient was incubated for 1.5 h with the peptide at 25 µg/mL, alone or in combination with EDTA (0.625 or 1.25 mM), and the colony forming unit (CFU) number of *P. aeruginosa* surviving the treatment was detected. During the incubation time, endogenous

P. aeruginosa did not grow in PCD sputum. While the peptide and EDTA were almost inactive when used alone, their combination exerted a significant synergistic killing effect against endogenous *P. aeruginosa*, although with different efficacy depending on the sputum sample (Figure 1). When compared to the corresponding controls, the reduction in CFU number caused by the combination ranged from approximately 0.3 Log-units (50% reduction, grey dot) to 3 Log-units (99.9% reduction, blue triangle) (Figure 1).

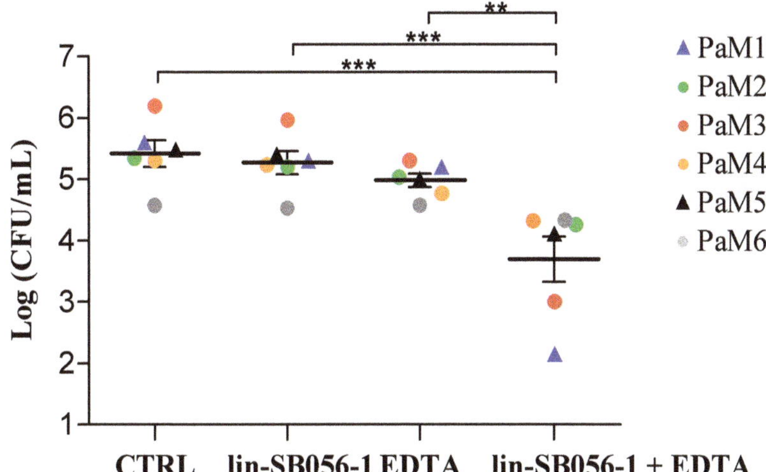

Figure 1. Antibacterial activity of peptide lin-SB056-1, ethylenediaminetetraacetic acid (EDTA), and both in combination against endogenous *P. aeruginosa* in primary ciliary dyskinesia (PCD) sputum. The effect of lin-SB056-1 and/or EDTA after 1.5 h of incubation in six diluted (1:5) sputum samples was assessed against endogenous *P. aeruginosa* strains (PaM1 to PaM6) by colony forming unit (CFU) counting. Lin-SB056-1 was tested at 25 μg/mL in combination with 0.625 mM EDTA against PaM1 and PaM5 strains (triangles), and with 1.25 mM EDTA against PaM2, PaM3, PaM4, and PaM6 strains (dots). Control (CTRL): bacteria incubated in diluted sputum only. Individual sputum samples are identified with different colors. Results represent the mean of 6 sputa done in duplicate. Error bars indicate the standard error of the mean. ** $p < 0.01$, *** $p < 0.001$ (one-way analysis of variance (ANOVA) followed by the Tukey–Kramer post-hoc test).

2.2. Effects of EDTA and lin-SB056-1 on Virulence Factors' Production by P. aeruginosa PaM1 and PaM5

Preliminary experiments indicated that PaM1and PaM5 strains are able to produce high levels of most of the virulence factors analyzed; therefore, these strains were selected to evaluate the effect of sub-inhibitory concentrations of EDTA on virulence factors' production. To this aim, we first evaluated the susceptibility of PaM1 and PaM5 strains to EDTA in liquid medium, in terms of minimum inhibitory concentration (MIC). A concentration of 1.25 mM EDTA was able to inhibit visible bacterial growth (MIC), while the concentrations of 0.075 and 0.15 mM were sub-inhibitory and therefore, were selected for the subsequent experiments.

Pyocyanin is a greenish pigment secreted by *P. aeruginosa* that enhances the inflammatory response and causes tissue damage in the host [21]. As shown in Figure 2a, EDTA at the concentrations of 0.075 and 0.15 mM, highly inhibited pyocyanin production by the PaM1 strain at 72 h (by 62% and 70%, respectively) as compared to the untreated cells. Regarding the PaM5 strain, which was a low pyocyanin producer (Figure 2b), EDTA at both concentrations caused a reduction of approximately 40% in the production of such pigment, although the difference did not reach statistical significance compared to the untreated cells.

P. aeruginosa produces and secretes a number of proteases, such as LasA, elastase B (LasB), protease IV, and alkaline protease, which are considered important virulence factors as they damage host tissues and interfere with host antibacterial defense mechanisms [22]. The total proteolytic activity of PaM1 was completely abolished in the presence of either 0.075 or 0.15 mM EDTA (Figure 2a). Similarly, EDTA significantly reduced the proteolytic activity of the PaM5 strain but only at the concentration of 0.15 mM (Figure 2b).

LasA is a zinc-dependent metalloprotease secreted by *P. aeruginosa*. It exhibits a staphylolytic activity, enhances the elastolytic activity of LasB in vivo, and induces shedding of syndecans, a family of cell surface heparan sulfate proteoglycans, from host cell surfaces [23]. LasA activity of both the PaM1 and PaM5 strains was significantly inhibited in the presence of 0.075 and 0.15 mM EDTA, with a reduction of approximately 70% and 80%, as compared to the untreated control, respectively (Figure 2a,b). Further experiments were undertaken in order to evaluate whether the reduction of LasA activity was ascribable to the inhibition of protease synthesis or, rather, to the inhibition of the enzyme activity due to zinc chelation by EDTA (Figure S1). To this aim, the assessment of the PaM1 staphylolytic activity was performed in the presence of exogenously added zinc (0.1 mM $ZnSO_4$). When $ZnSO_4$ was added directly in the enzyme assay, at the end of the incubation period (72 h), no significant increase in LasA activity was observed, suggesting that the low levels of LasA activity detected in the presence of EDTA were likely due to inhibition of protein synthesis and not to the chelation of the enzyme cofactor. In contrast, when $ZnSO_4$ was added at the beginning of the incubation of PaM1 with EDTA, LasA activity was restored (Supplementary Figure S1), indicating that the excess of zinc could overcome the inhibitory effect of EDTA. Overall, these data suggest that EDTA may act by interfering with the expression/procession of LasA protease by PaM1 strain rather than by sequestering the zinc cofactor.

Pyoverdin is a chelator involved in iron binding and cellular uptake in a low-iron environment [24]. EDTA at both concentrations tested did not reduce the level of pyoverdin in culture supernatants of both *P. aeruginosa* strains (Figure 2a,b).

Studies on mucoid *P. aeruginosa* isolates have shown that alginate plays a critical role in biofilm establishment and persistence by protecting bacteria against antibiotics and phagocytosis [25,26]. Although EDTA did not inhibit the production of alginate in culture supernatants of both strains (Figure 2a,b), it was able to significantly reduce the viscosity of PaM1 culture supernatants at both concentrations tested (Figure 2a).

Finally, EDTA was tested for its antibiofilm activity against the PaM1 strain. A reduction of 40% and 57% in PaM1 biofilm formation was observed in the presence of 0.075 and 0.15 mM EDTA, respectively (Figure 2a).

The impact of lin-SB056-1 on the production of virulence factors by PaM1 and PaM5 strains was also evaluated. As reported in Table S2, the peptide at sub-inhibiting concentrations did not reduce the production of any of the virulence factors analyzed for both bacterial strains tested.

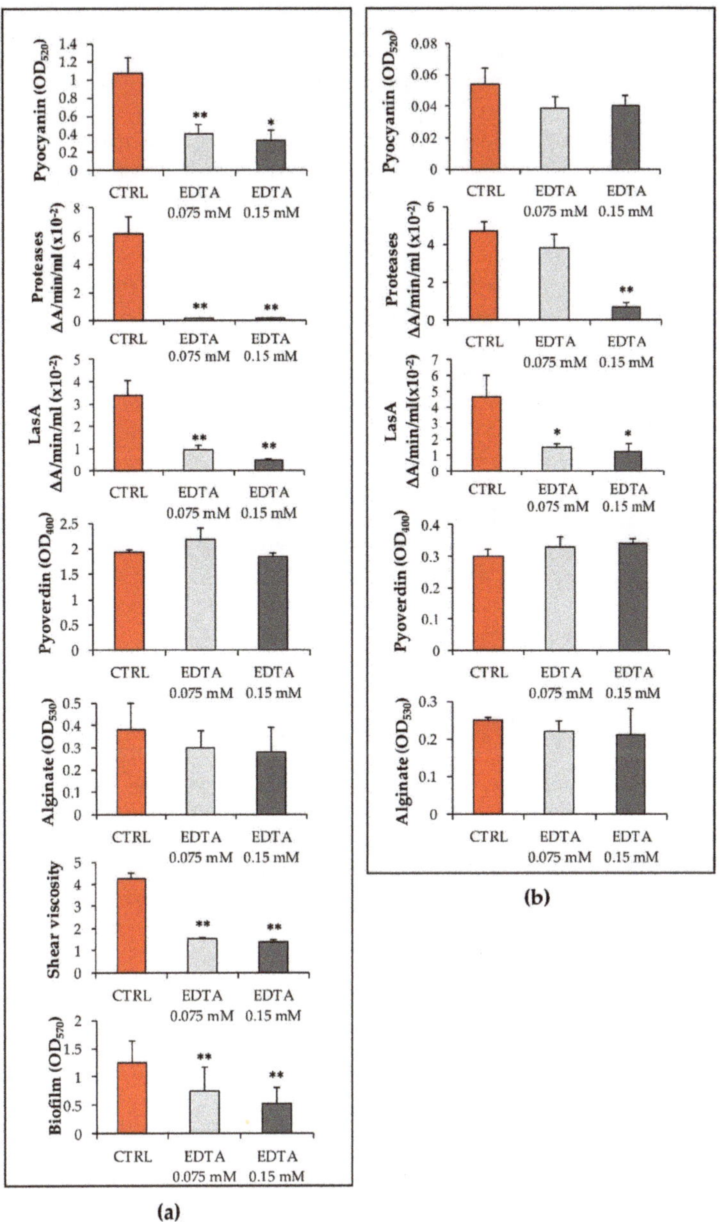

Figure 2. Effects of EDTA on virulence factor production by (**a**) PaM1 and (**b**) PaM5 strains. PaM1 and PaM5 cultures were incubated at 37 °C in the presence or absence of EDTA for 72 h. Following incubation, OD_{600} was measured prior the quantification of the virulence factors in culture supernatants (see the Materials and Methods Section for details). Values obtained were normalized by multiplying them by the ratio between OD_{600} of the control/OD_{600} of the corresponding EDTA-treated samples and reported as mean +/- SEM of three independent experiments. CTRL: bacteria incubated without EDTA; * $p < 0.05$, ** $p < 0.01$ (one-way ANOVA followed by the Tukey–Kramer post-hoc test).

3. Discussion

Similar to CF patients, eradication of chronic *P. aeruginosa* infection in PCD lungs is hardly obtained, and the reduction of bacterial density during chronic colonization or exacerbations is often the aim of the antimicrobial therapy [3]. In previous studies, we have shown that the combination of lin-SB056-1/EDTA possesses antimicrobial activity against *P. aeruginosa* in artificial sputum medium and prevents *P. aeruginosa* biofilm formation in an in vivo-like three-dimensional (3D) lung epithelial cell model [16,27]. Despite resembling the airway mucus, artificial sputum media normally behave like Newtonian fluids lacking many of the intramolecular interactions and covalent cross-links that give respiratory secretions their viscoelastic characteristic [28]. Furthermore, the genotype and physiological state of *P. aeruginosa* cells found in vivo may significantly differ from those of bacteria grown in laboratory media [29]. Hence, in this study, we sought to validate the anti-pseudomonal activity of the lin-SB056-1/EDTA combination in conditions more closely resembling the environment found in vivo. To this aim efficacy of the combination was tested ex vivo, against endogenous *P. aeruginosa* in the sputum of PCD patients. Differently from artificial sputum medium, patients' sputum contains host/bacterial components such as cell-derived factors, normal flora, inflammatory mediators, proteases, or peptidases that may exert an additional inhibitory effect on the peptide's activity. Nevertheless, herein, we showed that the combination of lin-SB056-1 and EDTA at sub-active and non-cytotoxic concentrations [16,27] determined a significant reduction in *P. aeruginosa* load in sputa of PCD chronically infected patients, despite certain differences in the level of reduction among different sputum samples being observed. Such differences are not surprising considering that inter-patient variables (e.g., clinical stage and severity of lung infection, bacterial load, sputum sample composition, and consistency) were not standardized in our experiments, in the attempt to mimic conditions found during the actual antimicrobial therapy. The mechanisms of the synergistic effect of EDTA on the peptide's activity might be multiple. Due to the chelation of divalent cations from their binding sites in lipopolysaccharide (LPS), EDTA may destabilize the bacterial outer membrane, thus increasing the permeability to lin-SB056-1 molecules and their interaction with the bacterial membranes. In addition, at least part of the synergistic effect observed in sputum could be ascribed to the ability of EDTA to reduce sputum viscosity, thus favoring peptide diffusion [16]. Finally, EDTA could also neutralize the inhibitory effect of sputum on the peptide's activity, sequestering cations that may interfere with the electrostatic interactions of the peptide with bacterial surface [9,10]. The possible use of EDTA in the treatment of pulmonary infections and its safety as an adjuvant has been highlighted in previous in vivo studies [30,31]. For instance, Liu and coworkers demonstrated, in the guinea pig model, that EDTA (30 mg/kg intraperitoneally injected) plus ciprofloxacin (4 µg/mL administered by inhalation) significantly reduced the *P. aeruginosa* CFU number per gram of lung tissue as compared to the single treatment groups [31].

Interestingly, the results obtained in this study clearly demonstrated that EDTA could not only favor the activity of lin-SB056-1 in ex vivo conditions, but could also reduce in vitro, at sub-inhibitory concentrations, the production of several *P. aeruginosa* virulence factors (i.e., pyocyanin, total protease and LasA), which are known to play a crucial role in the pathogenesis of *P. aeruginosa* infections. The action of EDTA as an anti-virulence molecule could be ascribed to its capacity of binding divalent cations, many of which are critical for the expression/processing of virulence factors of *P. aeruginosa*. In particular, the reduction of pyocyanin observed in the presence of EDTA is in line with previous reports demonstrating a positive correlation between calcium levels and the expression of proteins involved in the pathway of pyocyanin biosynthesis [32]. Analogously, the reduction of proteases by EDTA is in agreement with previous observations reporting that zinc ions are important for the efficient production and processing of different proteases, such as LasA, LasB, and protease IV [19,33].

On the contrary, EDTA did not inhibit pyoverdin production, in agreement with the observation that neither calcium nor magnesium enhances pyoverdin production [34,35]. Interestingly, although EDTA did not inhibit alginate production, it was able to markedly reduce the viscosity of culture supernatants of the PaM1 strain. It can be hypothesized that this effect may be due to the sequestration of calcium ions that are crucial for alginate cross-linking [36]. A similar mechanism may be involved in the ability of EDTA to reduce the formation of PaM1 biofilm, confirming previous observations in which EDTA significantly reduced biofilm formation by a mucoid strain of *P. aeruginosa* either in vitro or in a guinea-pig model of lung infection [31]. Overall, the ability of EDTA to reduce the accumulation and/or activity of important virulence factors might contribute to limit the pathogenicity of *P. aeruginosa*.

In conclusion, in the present study, we demonstrated that the lin-SB056-1/EDTA combination is able to significantly reduce *P. aeruginosa* load in PCD sputum, and that EDTA decreases the production of relevant virulence factors of mucoid *P. aeruginosa* in vitro. Such results suggest a dual antimicrobial and anti-virulence effect of the lin-SB056-1/EDTA combination and highlight the possible use of EDTA as an adjuvant in the treatment of chronic *P. aeruginosa* lung infections.

4. Materials and Methods

4.1. Sputum Collection and Treatment

The sputum samples were collected by spontaneous expectoration from six PCD patients following informed consent. The study was conducted in accordance with the Declaration of Helsinki, and the protocol was approved by the Ethics Committee of Pisa (Protocol number 62532, 11.06.2016). PCD patients included in this study (median age: 34 years) were chronically infected by *P. aeruginosa* and characterized by frequent relapses of infection. A volume of 0.5–1 mL of sputum was collected from each patient after interruption (at least 14 days) of the antibiotic therapy regimen. For easier handling of the samples, the dense and sticky sputa were diluted five-fold in sodium phosphate buffer 10 mM, pH 7.4 (SPB). Samples were plated on cetrimide (Sigma Aldrich, Saint Louis, MO, USA) and MacConkey (Oxoid Basingstoke, Hampshire, UK agar to confirm the presence of *P. aeruginosa* and assess the mucoid phenotype of the colonies, respectively.

4.2. Peptide and EDTA Solutions

Lin-SB056-1 peptide (KWKIRVRLSA-NH$_2$) was purchased from Peptide Protein Research, Ltd. (Fareham, UK) with a purity of 98%. EDTA (disodium salt) was obtained from Sigma-Aldrich. A stock solution of disodium-EDTA (0.5 M) was prepared in milli-Q water by adjusting the pH to 8.0 with NaOH (Sigma-Aldrich). The stock solution was then diluted in milli-Q water to obtain a working solution of 50 mM that was sterilized and stored at 4 °C.

4.3. Susceptibility Testing

Identification and susceptibility testing of *P. aeruginosa* strains isolated from sputum samples (PaM1, PaM2, PaM3, PaM4, PAM5, PaM6) were performed by MALDI-TOF (Bruker Daltonics, Bremen, Germany) and VITEK 2 automatic instruments (BioMerieux, Lyon, France), respectively (Table S1). Determination of minimum inhibitory concentration (MIC) of EDTA towards PaM1 and PaM5 strains was performed according to the standard microdilution method in Muller–Hinton broth (Oxoid) [37]. The MIC was defined as the lowest concentration of EDTA that completely inhibited visible growth of bacteria after 24 h of incubation.

4.4. Bactericidal Assay in Patients' Sputum

An aliquot of each PCD sputum was serially diluted and plated on selective cetrimide agar, to assess the CFU number of endogenous *P. aeruginosa* at time 0. After that, a volume of 90 µL of diluted (1:5) sputum of each patient was incubated with sub-bactericidal concentrations of peptide and EDTA, used alone or in combination, for 1.5 h at 37 °C. Following incubation, samples were serially diluted and plated on selective cetrimide agar for assessing the *P. aeruginosa* CFU number.

4.5. Assays for Evaluation of Virulence Factors in Culture Supernatants

Colonies of mucoid strains PaM1 and PaM5 grown on MacConkey agar were suspended in Luria Bertani (LB) broth (Sigma-Aldrich) to obtain an OD_{600} of 0.1. Cultures were incubated in the presence or in the absence of EDTA (0.075 or 0.15 mM) in static conditions at 37 °C for 72 h. Following incubation, the OD_{600} of the cultures was determined to account for bacterial density. After that, cultures were centrifuged at 10,000× *g* for 20 min at room temperature and culture supernatants were used for the quantification of virulence factors. The same protocol was followed to evaluate the eventual effects of linSB056-1 on the production of virulence factors. To this aim, the peptide was added to PaM1 and PaM5 cultures at the concentration of 6.25 µg/mL and 12.5 µg/mL that were sub-inhibitory for both bacterial strains.

Pyocyanin was extracted from cell-free supernatants with subsequent exposure to chloroform and 0.2 N hydrochloric acid (Sigma-Aldrich) and quantified at OD_{520} nm, as previously described [38].

Total proteolytic activity was determined using a modified skim milk assay [39]. Briefly, culture supernatants of PaM1 and PaM5 strains (0.5 mL) were incubated with 0.5 mL skim milk (Sigma-Aldrich) (1.25%) at 37 °C for 30 min and turbidity was measured at OD_{600} nm. The decrement in turbidity due to proteolytic activity was expressed as ΔA/min/mL.

Secreted LasA of *P. aeruginosa* has a staphylolytic activity, i.e., it causes a decrement in the OD_{600} of a culture of *Staphylococcus aureus*. LasA activity was assessed by evaluating the ability of cell-free supernatants from *P. aeruginosa* exposed or not exposed to EDTA to lyse boiled cells (intact) of *S. aureus* American Type Culture Collection (ATCC) 33591 and expressed as ΔA/min/mL [40]. Due to the role of zinc as a cofactor of LasA, in some experiments, the staphylolytic activity in the presence of EDTA was evaluated by adding 0.1 mM $ZnSO_4$ (Sigma-Aldrich) to the enzyme assay.

Pyoverdin was quantified by measuring the OD_{400} of cell-free supernatants [41].

The quantification of alginate was performed by carbazole-borate assay according to Heidari et al. [42]. Shear viscosity of culture supernatants was assessed by rheometric measurement at 25 °C, applying a shear stress of 1 Pa/s on 150 µl of supernatant using a gap between the rheometer plates of 52 µm (Rheometer Scientific RM500, Reologica Instruments AB, Lund, Sweden).

The value obtained for each virulence factor was multiplied by the ratio OD_{600} of the control/OD_{600} of the sample, to normalize for small differences in the culture densities between the controls and the EDTA-exposed samples after 72 h of incubation.

4.6. Biofilm Inhibition Assay

P. aeruginosa PaM1 grown in tryptone soy broth (TSB) for 48 h at 37 °C was diluted 1:20 in TSB supplemented with 0.25 mM $CaCl_2$. Bacterial suspensions were inoculated into flat-bottom polystyrene 96-well microplates (Corning Costar, Lowell, MA, USA) in the absence (negative control) or in the presence of EDTA at sub-inhibiting concentrations (0.075 and 0.15 mM). Microplates were incubated statically at 37 °C for 48 h and biofilm biomass was estimated by crystal violet (CV) (Sigma-Aldrich) staining assay, as previously described [43].

4.7. Statistical Analysis

Data reported in Figure 1 represent the mean of 6 experiments done in duplicate. Figure 2 and Supplementary Figure S1 depict the data obtained from three independent experiments, while Table S2 reports the mean of two independent experiments. Differences between mean values of groups were evaluated by one-way analysis of variance (ANOVA) followed by the Tukey–Kramer post-hoc test. A *p*-value < 0.05 was considered statistically significant. Data analysis was performed with GraphPad In Stat (GraphPad Software, La Jolla, CA, USA).

Supplementary Materials: Supplementary Materials can be found at http://www.mdpi.com/1422-0067/21/1/69/s1. Table S1: Patients' information, colony phenotype, and resistance profile of *P. aeruginosa* strains isolated from PDC sputum; Table S2: Effects of lin-SB056-1 on the production of virulence factors by PaM1 and PaM5 strains; Figure S1: Effect of exogenously added Zinc on LasA staphylolytic activity of PaM1 strain in the presence of EDTA.

Author Contributions: Conceptualization, G.M., S.E., M.P. (Massimo Pifferi), and G.B.; methodology, G.M., L.G., M.P. (Martina Piras), E.K., A.M., D.P.; validation, G.M., S.E., M.P. (Massimo Pifferi), and G.B.; formal analysis, G.M., L.G., M.P. (Martina Piras), E.K. and F.C.; writing—original draft preparation, G.M. and G.B.; writing—review and editing, L.G., S.E., E.K., F.C., and M.P. (Martina Piras), supervision, M.P. (Massimo Pifferi), S.E., and G.B.; funding acquisition, G.B. All authors have read and agreed to the published version of the manuscript.

Funding: This research was funded by the University of Pisa, grant number PRA 2017_18.

Conflicts of Interest: The authors declare no conflict of interest. The funders had no role in the design of the study; in the collection, analyses, or interpretation of data; in the writing of the manuscript, or in the decision to publish the results.

Abbreviations

ATCC	American Type Culture Collection
AMPs	Antimicrobial peptides
ANOVA	Analysis of variance
CF	Cystic fibrosis
CFU	Colony forming units
CV	Crystal violet
EDTA	Ethylenediaminetetraacetic acid
LB	Luria-Bertani broth
LPS	Lipopolysaccharide
MIC	Minimum Inhibitory Concentration
OD	Optical density
PCD	Primary Ciliary Dyskinesia
SPB	Sodium Phosphate Buffer
TSB	Tryptone Soy Broth

References

1. Wijers, C.D.M.; Chmiel, J.F.; Gaston, B.M. Bacterial infections in patients with primary ciliary dyskinesia: Comparison with cystic fibrosis. *Chron. Respir. Dis.* **2017**, *14*, 392–406. [CrossRef]

2. Cohen-Cymberknoh, M.; Weigert, N.; Gileles-Hillel, A.; Breuer, O.; Simanovsky, N.; Boon, M.; De Boeck, K.; Barbato, A.; Snijders, D.; Collura, M.; et al. Clinical impact of *Pseudomonas aeruginosa* colonization in patients with primary ciliary dyskinesia. *Respir. Med.* **2017**, *131*, 241–246. [CrossRef] [PubMed]

3. Crowley, S.; Holgersen, M.G.; Nielsen, K.G. Variation in treatment strategies for the eradication of *Pseudomonas aeruginosa* in primary ciliary dyskinesia across European centers. *Chron. Respir. Dis.* **2019**, *16*, 1–8. [CrossRef] [PubMed]

4. Bianconi, I.; Jeukens, J.; Freschi, L.; Alcalá-Franco, B.; Facchini, M.; Boyle, B.; Molinaro, A.; Kukavica-Ibrulj, I.; Tümmler, B.; Levesque, R.C.; et al. Comparative genomics and biological characterization of sequential *Pseudomonas aeruginosa* isolates from persistent airways infection. *BMC Genom.* **2015**, *16*, 1105–1118. [CrossRef]

5. Alanin Mikkel, C. Bacteriology and treatment of infections in the upper and lower airways in patients with primary ciliary dyskinesia: Addressing the paranasal sinuses. *Dan. Med. J.* **2017**, *64*, 5361–5378.

6. Haney, E.F.; Mansour, S.C.; Hancock, R.E. Antimicrobial Peptides: An introduction. *Methods Mol. Biol.* **2017**, *1548*, 3–22. [CrossRef]

7. Grassi, L.; Di Luca, M.; Maisetta, G.; Rinaldi, A.C.; Esin, S.; Trampuz, A.; Batoni, G. Generation of persister cells of *Pseudomonas aeruginosa* and *Staphylococcus aureus* by chemical treatment and evaluation of their susceptibility to membrane-targeting agents. *Front. Microbiol.* **2017**, *8*, 1917. [CrossRef]

8. Grassi, L.; Maisetta, G.; Esin, S.; Batoni, G. Combination strategies to enhance the efficacy of antimicrobial peptides against bacterial biofilms. *Front. Microbiol.* **2017**, *8*, 2409. [CrossRef]

9. Maisetta, G.; Di Luca, M.; Esin, S.; Florio, W.; Brancatisano, F.L.; Bottai, D.; Campa, M.; Batoni, G. Evaluation of the inhibitory effects of human serum components on bactericidal activity of human beta defensin 3. *Peptides* **2008**, *29*, 1–6. [CrossRef]

10. Batoni, G.; Maisetta, G.; Esin, S.; Campa, M. Human beta-defensin-3: A promising antimicrobial peptide. *Mini Rev. Med. Chem.* **2006**, *6*, 1063–1073. [CrossRef]

11. Sommer, L.M.; Alanin, M.C.; Marvig, R.L.; Nielsen, K.G.; Høiby, N.; von Buchwald, C.; Molin, S.; Johansen, H.K. Bacterial evolution in PCD and CF patients follows the same mutational steps. *Sci. Rep.* **2016**, *28*, 28732. [CrossRef] [PubMed]

12. Waters, V.; Smyth, A. Cystic fibrosis microbiology: Advances in antimicrobial therapy. *J. Cyst. Fibros.* **2015**, *14*, 551–560. [CrossRef] [PubMed]

13. Lefebvre, E.; Vighetto, C.; Di Martino, P.; Larreta Garde, V.; Seyer, D. Synergistic antibiofilm efficacy of various commercial antiseptics, enzymes and EDTA: A study of *Pseudomonas aeruginosa* and *Staphylococcus aureus* biofilms. *Int. J. Antimicrob. Agents* **2016**, *48*, 181–188. [CrossRef] [PubMed]

14. Percival, S.L.; Salisbury, A.M. The Efficacy of tetrasodium EDTA on biofilms. *Adv. Exp. Med. Biol.* **2018**, *1057*, 101–110. [CrossRef] [PubMed]

15. Ferrero, M.E. Rationale for the successful management of EDTA chelation therapy in human burden by toxic metals. *Biomed. Res. Int.* **2016**, *2016*, 8274504. [CrossRef]

16. Maisetta, G.; Grassi, L.; Esin, S.; Serra, I.; Scorciapino, M.A.; Rinaldi, A.C.; Batoni, G. The semi-synthetic peptide lin-SB056-1 in combination with EDTA exerts strong antimicrobial and antibiofilm activity against *Pseudomonas aeruginosa* in conditions mimicking cystic fibrosis sputum. *Int. J. Mol. Sci.* **2017**, *16*, 1994. [CrossRef]

17. Bush, A.; Payne, D.; Pike, S.; Jenkins, G.; Henke, M.O.; Rubin, B.K. Mucus properties in children with primary ciliary dyskinesia: Comparison with cystic fibrosis. *Chest* **2006**, *129*, 118–123. [CrossRef]

18. Kim, W.D. Lung mucus: A clinician's view. *Eur. Respir. J.* **1997**, *10*, 1914–1917. [CrossRef]

19. Olson, J.C.; Ohman, D.E. Efficient production and processing of elastase and LasA by *Pseudomonas aeruginosa* require zinc and calcium ions. *J. Bacteriol.* **1992**, *174*, 4140–4147. [CrossRef]

20. Guragain, M.; King, M.M.; Williamson, K.S.; Pérez-Osorio, A.C.; Akiyama, T.; Khanam, S.; Patrauchan, M.A.; Franklin, M.J. The *Pseudomonas aeruginosa* PAO1 two-component regulator CarSR regulates calcium homeostasis and calcium-induced virulence factor production through its regulatory targets CarO and CarP. *J. Bacteriol.* **2016**, *198*, 951–963. [CrossRef]

21. Rada, B.; Leto, T.L. Pyocyanin effects on respiratory epithelium: Relevance in *Pseudomonas aeruginosa* airway infections. *Trends Microbiol.* **2013**, *21*, 73–81. [CrossRef] [PubMed]

22. Tingpej, P.; Smith, L.; Rose, B.; Zhu, H.; Conibear, T.; Al Nassafi, K.; Manos, J.; Elkins, M.; Bye, P.; Willcox, M.; et al. Phenotypic characterization of clonal and nonclonal *Pseudomonas aeruginosa* strains isolated from lungs of adults with cystic fibrosis. *J. Clin. Microbiol.* **2007**, *45*, 1697–1704. [CrossRef] [PubMed]

23. Park, P.W.; Pier, G.B.; Preston, M.J.; Goldberger, O.; Fitzgerald, M.L.; Bernfield, M. Syndecan-1 shedding is enhanced by LasA, a secreted virulence factor of *Pseudomonas aeruginosa*. *J. Biol. Chem.* **2000**, *275*, 3057–3064. [CrossRef] [PubMed]

24. Edgar, R.J.; Hampton, G.E.; Garcia, G.P.C.; Maher, M.J.; Perugini, M.A.; Ackerley, D.F.; Lamont, I.L. Integrated activities of two alternative sigma factors coordinate iron acquisition and uptake by *Pseudomonas aeruginosa*. *Mol. Microbiol.* **2017**, *106*, 891–904. [CrossRef]

25. Hodges, N.A.; Gordon, C.A. Protection of *Pseudomonas aeruginosa* against ciprofloxacin and beta-lactams by homologous alginate. *Antimicrob. Agents Chemother.* **1991**, *35*, 2450–2452. [CrossRef] [PubMed]

26. Leid, J.G.; Willson, C.J.; Shirtliff, M.E.; Hassett, D.J.; Parsek, M.R.; Jeffers, A.K. The exopolysaccharide alginate protects *Pseudomonas aeruginosa* biofilm bacteria from IFN-gamma-mediated macrophage killing. *J. Immunol.* **2005**, *175*, 7512–7518. [CrossRef]

27. Grassi, L.; Batoni, G.; Ostyn, L.; Rigole, P.; Van den Bossche, S.; Rinaldi, A.C.; Maisetta, G.; Esin, S.; Coenye, T.; Crabbé, A. The antimicrobial peptide lin-SB056-1 and its dendrimeric derivative prevent *Pseudomonas aeruginosa* biofilm formation in physiologically relevant models of chronic infections. *Front. Microbiol.* **2019**, *10*, 198. [CrossRef]

28. Müller, A.; Wenzel, M.; Strahl, H.; Grein, F.; Saaki, T.N.V.; Kohl, B.; Siersma, T.; Bandow, J.E.; Sahl, H.G.; Schneider, T.; et al. Daptomycin inhibits cell envelope synthesis by interfering with fluid membrane microdomains. *Proc. Natl. Acad. Sci. USA* **2016**, *113*, 7077–7086. [CrossRef]

29. Sajjan, U.S.; Tran, L.T.; Sole, N.; Rovaldi, C.; Akiyama, A.; Friden, P.M.; Forstner, J.F.; Rothstein, D.M. P-113D, an antimicrobial peptide active against *Pseudomonas aeruginosa*, retains activity in the presence of sputum from cystic fibrosis patients. *Antimicrob. Agents Chemother.* **2001**, *45*, 3437–3444. [CrossRef]
30. Hachem, R.; Bahna, P.; Hanna, H.; Stephens, L.C.; Raad, I. EDTA as an adjunct antifungal agent for invasive pulmonary aspergillosis in a rodent model. *Antimicrob. Agents Chemother.* **2006**, *50*, 1823–1827. [CrossRef]
31. Liu, Z.; Lin, Y.; Lu, Q.; Li, F.; Yu, J.; Wang, Z.; He, Y.; Song, C. In vitro and in vivo activity of EDTA and antibacterial agents against the biofilm of mucoid *Pseudomonas aeruginosa*. *Infection* **2017**, *45*, 23–31. [CrossRef] [PubMed]
32. Sarkisova, S.; Patrauchan, M.A.; Berglund, D.; Nivens, D.E.; Franklin, M.J. Calcium-induced virulence factors associated with the extracellular matrix of mucoid *Pseudomonas aeruginosa* biofilms. *J. Bacteriol.* **2005**, *187*, 4327–4337. [CrossRef] [PubMed]
33. D'Orazio, M.; Mastropasqua, M.C.; Cerasi, M.; Pacello, F.; Consalvo, A.; Chirullo, B.; Mortensen, B.; Skaar, E.P.; Ciavardelli, D.; Pasquali, P.; et al. The capability of *Pseudomonas aeruginosa* to recruit zinc under conditions of limited metal availability is affected by inactivation of the ZnuABC transporter. *Metallomics* **2015**, *7*, 1023–1035. [CrossRef]
34. Marquart, M.E.; Dajcs, J.J.; Caballero, A.R.; Thibodeaux, B.A.; O'Callaghan, R.J. Calcium and magnesium enhance the production of *Pseudomonas aeruginosa* protease IV, a corneal virulence factor. *Med. Microbiol. Immunol.* **2005**, *194*, 39–45. [CrossRef] [PubMed]
35. Laux, D.C.; Corson, J.M.; Givskov, M.; Hentzer, M.; Møller, A.; Wosencroft, K.A.; Olson, J.C.; Krogfelt, K.A.; Goldberg, J.B.; Cohen, P.S. Lysophosphatidic acid inhibition of the accumulation of *Pseudomonas aeruginosa* PAO1 alginate, pyoverdin, elastase and LasA. *Microbiology* **2002**, *148*, 1709–1723. [CrossRef]
36. Körstgens, V.; Flemming, H.C.; Wingender, J.; Borchard, W. Influence of calcium ions on the mechanical properties of a model biofilm of mucoid *Pseudomonas aeruginosa*. *Water Sci. Technol.* **2001**, *43*, 49–57. [CrossRef]
37. European Committee for Antimicrobial Susceptibility Testing (EUCAST) of the European Society of Clinical Microbiology and Infectious Diseases (ESCMID). EUCAST definitive document E. Def 1.2, May 2000: Terminology relating to methods for the determination of susceptibility of bacteria to antimicrobial agents. *Clin. Microbiol. Infect.* **2000**, *6*, 503–508. [CrossRef]
38. Essar, D.W.; Eberly, L.; Hadero, A.; Crawford, IP. Identification and characterization of genes for a second anthranilate synthase in *Pseudomonas aeruginosa*: Interchangeability of the two anthranilate synthases and evolutionary implications. *J. Bacteriol.* **1990**, *172*, 884–900. [CrossRef]
39. El-Mowafy, S.A.; Abd, E.; Galil, K.H.; El-Messery, S.M.; Shaaban, M.I. Aspirin is an efficient inhibitor of quorum sensing, virulence and toxins in *Pseudomonas aeruginosa*. *Microb. Pathog.* **2014**, *74*, 25–32. [CrossRef]
40. Kessler, E.; Safrin, M.; Olson, J.C.; Ohman, D.E. Secreted LasA of *Pseudomonas aeruginosa* is a staphylolytic protease. *J. Biol. Chem.* **1993**, *268*, 7503–7508. [CrossRef]
41. Cox, C.D.; Adams, P. Siderophore activity of pyoverdin for *Pseudomonas aeruginosa*. *Infect. Immun.* **1985**, *48*, 130–138.
42. Heidari, A.; Noshiranzadeh, N.; Haghi, F.; Bikas, R. Inhibition of quorum sensing related virulence factors of *Pseudomonas aeruginosa* by pyridoxal lactohydrazone. *Microb. Pathogen.* **2017**, *112*, 103–110. [CrossRef] [PubMed]
43. Maisetta, G.; Batoni, G.; Caboni, P.; Esin, S.; Rinaldi, A.C.; Zucca, P. Tannin profile, antioxidant properties, and antimicrobial activity of extracts from two Mediterranean species of parasitic plant Cytinus. *BMC Complement. Altern. Med.* **2019**, *19*, 82–93. [CrossRef] [PubMed]

© 2019 by the authors. Licensee MDPI, Basel, Switzerland. This article is an open access article distributed under the terms and conditions of the Creative Commons Attribution (CC BY) license (http://creativecommons.org/licenses/by/4.0/).

International Journal of
Molecular Sciences

Review

From Gene to Protein—How Bacterial Virulence Factors Manipulate Host Gene Expression During Infection

Lea Denzer, Horst Schroten and Christian Schwerk *

Department of Pediatrics, Pediatric Infectious Diseases, Medical Faculty Mannheim, Heidelberg University, 68167 Mannheim, Germany; lea.denzer@medma.uni-heidelberg.de (L.D.); horst.schroten@umm.de (H.S.)
* Correspondence: christian.schwerk@medma.uni-heidelberg.de; Tel.: +49-621-383-3466

Received: 30 April 2020; Accepted: 20 May 2020; Published: 25 May 2020

Abstract: Bacteria evolved many strategies to survive and persist within host cells. Secretion of bacterial effectors enables bacteria not only to enter the host cell but also to manipulate host gene expression to circumvent clearance by the host immune response. Some effectors were also shown to evade the nucleus to manipulate epigenetic processes as well as transcription and mRNA procession and are therefore classified as nucleomodulins. Others were shown to interfere downstream with gene expression at the level of mRNA stability, favoring either mRNA stabilization or mRNA degradation, translation or protein stability, including mechanisms of protein activation and degradation. Finally, manipulation of innate immune signaling and nutrient supply creates a replicative niche that enables bacterial intracellular persistence and survival. In this review, we want to highlight the divergent strategies applied by intracellular bacteria to evade host immune responses through subversion of host gene expression via bacterial effectors. Since these virulence proteins mimic host cell enzymes or own novel enzymatic functions, characterizing their properties could help to understand the complex interactions between host and pathogen during infections. Additionally, these insights could propose potential targets for medical therapy.

Keywords: virulence factors; bacteria; host-pathogen interaction; gene expression; immune response; manipulation; inflammation; persistence; replicative niche

---

1. Introduction

Successful defence against extracellular and intracellular bacteria primarily relies on the ability of innate immune cells to sense present bacteria followed by activation of the adequate matching immune response. The identification of bacteria is enabled by a broad array of pathogen recognition receptors (PRRs), which recognize extensively conserved pathogen-associated molecular patterns (PAMPs) as nucleid acids, cell wall components and proteins from viruses, bacteria, fungi and parasites [1]. Toll like receptors (TLRs) and NOD-like receptors (NLRs) represent the two major classes of PRRs, acting at the cell surface or in the cytoplasm, respectively [2]. After activation, PRRs induce multiple signalling pathways aiming at the expression of proinflammatory cytokines, which regulate the innate immune response. The most prominent pathways involved are mediated by mitogen-activated protein kinases (MAPKs) or nuclear factor-κB (NF-κB), which are initiated by path-specific adapter proteins and transferred by downstream phosphorylation cascades [3]. Interestingly, all pathways can synergize to guarantee a specific and appropriate immune response for the present pathogen. This is enabled by the high diversity of PRRs, PAMPs and PRR adapter proteins and their numerous ways to be combined during innate immune response [1,4].

The immune response has to be tightly controlled to ensure a clearance of the bacteria but also to prevent tissue damage and necrosis as result of sepsis. There are several levels to influence the expression of inflammatory genes. A first level of interference is changing of the DNA's structure on the chromatin level. Epigenetic modulation enables remodelling of the chromatin to transfer heterochromatin into euchromatin allowing transcription or vice versa [5–7]. In addition, the affinity of promotors and other regulatory DNA sequences for RNA polymerases and transcription factors (TFs) can be influenced by cytosine or adenine methylation. To induce transcription, TFs and RNA polymerases are recruited to target genes, a step that represents another level to regulate gene expression. Only a minor portion (fewer than 2%) of genes is transcribed into mRNAs, instead the majority is transferred into so called non-coding RNAs (ncRNAs). The long ncRNAs (lncRNAs) as well as some classes of short ncRNAs are also involved in epigenetic regulations but its most studied group, the miRNAs (microRNA), are mostly involved in RNA destruction [8].

Another point for interference with gene expression is during processing of mRNAs, which includes 5′ capping, alternative splicing and polyadenylation [9,10]. Primary ncRNAs can be processed by the protein complex DICER (eukaryotic ribonucleases) to generate miRNAs, which can negatively regulate expression of its primary transcript. To guarantee proper cell function, mRNAs need to be degraded after a certain time frame; the RNA stability is, therefore, another switch to modify gene expression. Polyadenylation and 5′capping prolongs RNA stability but several enzymes are able to decap the 5′cap and to remove the polyadenyl tail of the RNA leaving an unprotected mRNA [11–13]. Nevertheless, these structures are crucial for translation initiation and can gain enough time for the mRNA to be translated. This represents a further step for interference with gene expression, as there is the need of several factors to induce and prolong translation [14]. Translation initialisation factors have to be recruited leading to ribosome assembly and binding of the first amino acid loaded tRNA (transfer RNA). To keep translation ongoing, elongation factors and ATP (Adenosine triphosphate) have to be present. The nascent protein chain then needs to fold into its physiological form to be active [15,16]. Therefore, protein-folding and activation catalysed by different chaperones is another important step during gene expression and is also tightly controlled. Finally, a last step to regulate gene expression is represented by the stability and degradation of, in some cases misfolded, proteins [17,18]. An overview of the steps during host gene expression targeted by bacterial pathogens, as well as the bacteria involved, is given in Figure 1.

Host cells fight bacteria with a proinflammatory cytokine response, lysosomal degradation, autophatic clearance and activation of the unfolded protein response, which in the end can lead to apoptosis [19]. Bacteria can hijack all these defence mechanisms by interfering with the host's gene expression at any level. For that purpose, bacteria express several virulence factors, the so-called effectors. These proteins are able to mimic host enzymes, thereby manipulating the host response following invasion favouring intracellular survival, persistence and spreading. Since proteins and enzymes of signal transduction pathways are involved in all defence mechanisms, they are a favoured target of effector proteins [19]. It is worth noting that other virulence proteins own novel enzymatic functions, which allow them to enter the nucleus and directly induce gene expression or repression. Therefore, these bacterial effectors are termed nucleomodulins [20,21]. In the following we will present an overview on the manipulation of host gene expression at different levels by nucleomodulins and other bacterial effectors.

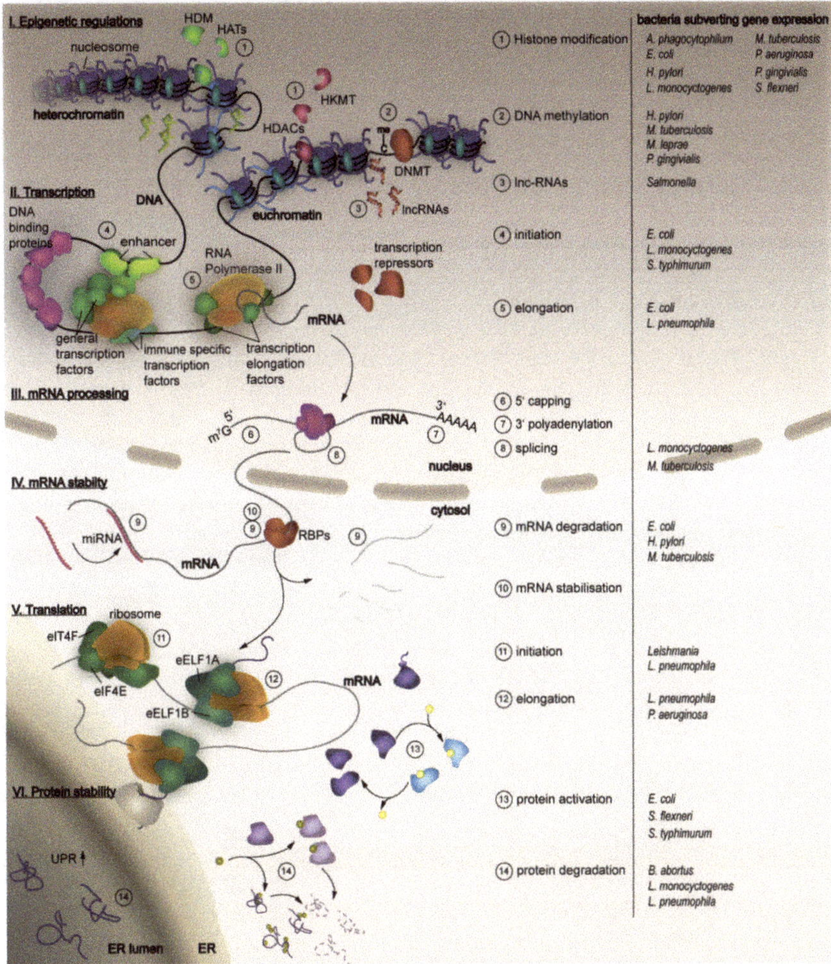

Figure 1. Steps of host gene expression manipulated by bacterial pathogens. The figure provides an overview over the main steps of gene expression that are indicated at the left side (I-VI). The numbers in the scheme highlight distinct characteristic processes that are part of each gene expression level and are listed in the legend at the right side. Different bacterial pathogens (indicated at the right) have been described to target the distinct steps and processes during host gene expression to their favor. For detailed information please refer to the text of this review.

2. Bacterial Virulence Factors Manipulating Host Gene Expression

2.1. Epigenetic Control of Gene Expression

The expression of genes is dependent on their accessibility for RNA polymerase II (RNA Pol II) and TFs. As approximately 147 base pairs of the DNA are wrapped around histone octamers build by the subunits H2A, H2B, H3 and H4 as well as the scaffold protein H1 to form the nucleosome, those sequences are protected from transcription [5–7]. Therefore, the packaging of the nucleosomes defines the chromatin state into euchromatin and heterochromatin enabling transcription or blocking it. In order to react properly to a certain stimulus, the chromatin state can be remodeled to give access to the required genes, a process called nuclear remodeling or histone modification. Enzymes posttranslationally modify the amino acids at the N-termini of the histone proteins (called histone tails) by acetylation, phosphorylation, methylation and ubiquitination in a reversible manner to modify the interaction between neighbored nucleosomes favoring an open or closed chromatin state [19,20,22]. Nucleosomes are then allowed to slide along the chromatin fiber in an ATP-dependent manner, to give access to the DNA sequence. This reveals the dual function of chromatin, to provide a natural scaffold and being part of an essential regulatory signaling network processing the incoming data to create a special transient biological output [23]. On top, established posttranslational modifications (PTMs) can be maintained beyond the initial signal and cell divisions inheriting cell type specific gene expression enabling cell lineage specification and cellular identity [23–25].

The enzymes responsible for the modulation of the histone tails are divided into "writers," which attach the chemical units, "readers," which recognize and translate them by recruitment of activating or repressing factors and "erasers," which remove the modifications. The resulting "epigenetic code" is highly dynamic, as each established modification influences the addition or removal of other modifications that in turn influence the own stability and persistence. Moreover, epigenetic mechanisms represent the missing link between more or less stable gene expression and the impact of environmental factors on gene expression that can also cause diseases as cancer [5,26,27]. Therefore, these enzymes represent a central role in the regulation of immune responses as alterations in their activity and expression profiles leading to global changes in the histone modification pattern have been detected as cause of several chronic immune diseases as asthma, chronic obstructive pulmonary disease, colitis, systemic lupus erythematosus and rheumatoid arthritis [28,29].

Additionally, the DNA can be methylated at cytosine or adenosine residues converting them into methyl-cytosine or methyl-adenosine to cause transcriptional repression [30]. Hypermethylation dominantly occurs at CpG islands, cytosine-guanine rich regions at promotor regions, disrupting TFs and RNA polymerase binding to DNA or recruiting other co-repressors. A hypermethylated gene, that was not methylated before is therefore, not suitable for transcription and with the recruitment of further silencing-factors, will finally be silenced. This kind of modification is thought to provide a stable gene silencing that can be inherited to the next generation of cells [31,32].

Moving away from the old definition of epigenetics as hereditable stable changes at chromatin and DNA without changing its sequence, modern opinion changed towards a highly dynamic and reversible mechanism of gene regulation also enabling short term adaptions to changing environments [30]. As consequence, regulation through ncRNAs are also included to the epigenetic regulatory repertoire that can be classified according to their length into short ncRNAs (<200 nucleotides), which include miRNAs or long ncRNAs (>200 nucleotides) [33].

2.1.1. Manipulation at the Level of Histone Modifications

After recognition of bacterial presence by PRRs, signaling cascades activate proinflammatory cytokine expression. To improve accessibility of TFs, such as NF-κB, to the promoters of inflammatory response genes, an activating histone modification as phosphorylation of Serine 10 on histone H3 (H3S10) is established, which itself is mediated by MAPK signalling. It has been shown, that the virulence factor LPS alone is able to induce a global increase of H3S10 leading to promotion of gene expression proving the high sensitivity of the immune reaction [34].

Recent studies revealed that bacteria directly interfere with a host's histone modifications to dampen the expression of proinflammatory cytokines by the secretion of effectors. Presence of *Listeria monocytogenes* induces phosphorylation of H3S10 but the bacterium is able to remove this activating phosphorylation within short time [35–38]. The secreted virulence factor Listeriolysin (LLO) mediates this mechanism and is also responsible for a global deacetylation of H3 and H4. Other bacteria, as *Clostridium perfringens* or *Streptococcus pneumoniae*, produce toxins, such as perfringolysin and pneumolysin, respectively, that belong to the same family as LLO and show also a similar effect on H3S10 phosphorylation [36]. The decreased levels of phosphorylated H3S10 and acetylated H4 at proinflammatory genes resulted in transcriptional downregulation thereby damping the immune response. As this observation is only dependent on the membrane-binding ability of LLO, it is most likely that LLO modulates the signal transduction to induce alterations in the histone modification pattern [36].

Like *L. monocytogenes*, *Shigella flexneri* is also able to inhibit H3S10 phosphorylation by secretion of phosphothreonine lyase effector OspF, which dephosphorylates MAPKs as p38 or ERK resulting in attenuated NF-κB binding at promotors of inflammatory genes [39]. Together with OspB, another effector of *Shigella*, OspF, interacts with the human retinoblastoma protein Rb that is capable of binding several chromatin-remodeling factors [40,41]. In this constellation, *Shigella* adjusts the chromatin structure at specific genes to downregulate host innate immunity.

L. monocytogenes owns another effector, which induces deacetylation on lysine 18 of histone H3 (H3K18). Thereby, Internalin B (InlB) activates the host histone deacetylase sirtuin 2 (SIRT 2), leading to repression of transcriptional start sites through occupation by SIRT 2 and following downregulation of the immune response, which could be attenuated by SIRT 2 inhibition [42]. The listerial virulence factor LntA enters the nucleus after infection of epithelial cells targeting the chromatin silencing complex component BAHD1. Together with heterochromatin protein 1 (HP1), methylated DNA-binding protein 1 (MBD1), histone deacetylases (HDAC1/2) and the KRAB-associated protein 1 (KAP1/TRIM28) that are involved in heterochromatin formation, BADH1 targets interferon-stimulated genes (ISG) for silencing by binding to their promotors [43,44]. This is inhibited by LntA, which is thought to promote chromatin-unwinding and as consequence upregulation of ISG by histone H3 acetylation. The exact mechanisms, how BAHD1 is recruited to its targets and how LntA interferes with this process has still to be investigated [21].

Another prominent histone modification is the methylation or demethylation of lysine residues, mediated by histone N-lysine methyltransferase (HKMT) or histone demethylases (HDM), respectively. Several bacteria express HKMT effectors, which enable them to directly interfere with host gene regulation as they are mimics of host chromatin modifiers. As there are many HKMT homologues in the repertoire of bacterial effectors described this mechanism seems to be a successful strategy to subvert host gene expression [45]. The nuclear effector (NUE), is secreted by *Chlamydia trachomatis* via a type III secretion system (T3SS) to enable its localization to the nucleus, where it might methylate H2B, H3 and H4. The homologous effectors RomA and LegAS4 secreted by *Legionella pneumophila Paris* and *L. pneumophila Philadelphia Lp02* strains, respectively, methylate H3 to alter host transcription but target different residues [45,46]. RomA represses global transcription by methylation of histone 3 lysine 14 (H3K14), a modification that is known to compete with the activating acetylation of H3K14 [46]. Contrary to RomA, LegAS4 increases transcription of ribosomal RNA genes (rRNA) through methylation of histone 3 lysine 4 (H3K4) but if this modification is mediated by LegAS4 alone it is not clear yet [45]. Interestingly all described bacterial methyltransferases own a conserved SET (Suppressor of variegation, Enhancer of zeste and Trithorax) domain, which uses a S-adenosyl-l-methionine (SAM) methyl donor to catalyze methyl group attachment to lysine residues [45,47]. One example is the effector BtSET, secreted by *Burkholderia thailandensis* that localizes to the nucleolus to methylate histone H3K4 promoting transcription of rRNA genes. Some effectors are capable of more unusual modifications, for example, the effector BaSET identified in *Bacillus anthracis* trimethylates histone H1 but none of the core histones. This effector represses the expression of NF-κB target genes after transient overexpression in mammalian cells and its deletion results in the loss of virulence [45–47].

Another modification, which differs from the known mechanisms of histone modification, is represented by dimethylation of histone 3 on arginine 42 (H3R42me2), a residue critical for DNA entry/exit from the nucleosome and not located at the histone N-termini. This modification is involved in the regulation of ROS (reactive oxygen species) production, which represents a crucial host defense mechanism against bacterial pathogens [48]. *Mycobacterium tuberculosis* represses genes involved in ROS production by secreting Rv1988, a methytransferase able to establish H3R42me2 to increase survival in host macrophages [48]. An overview of bacteria and their effectors that are secreted to induce histone modifications is given in Table 1.

Table 1. Histone modifications established by bacterial effectors.

Target	Modification	Bacterium	Effector	Mediator	Cellular Function	References
H3S10	de-phosphorylation	L. monocytogenes	LLO	unknown	Reduced expression of important immune regulators	[36]
H3S10	de-phosphorylation	S. flexneri	OspF	MAPK		[39,49,50]
H3S10	de-phosphorylation	Clostridium perfringens	perifringo-lysin	unknown	unknown	[36]
H3S10	de-phosphorylation	Streptococcus pneumoniae	pneumolysin	unknown	unknown	[36]
H3K18	deacetylation	L. monocytogenes	InlB	c-Met induced SIRT2 recruitment PI3K/AKT	Reduced expression of important immune regulators	[36]
H3K4	methylation	L. pneumophila Philadelphia LP02	LegAS4	direct	Transcriptional activation of ribosomal genes	[49]
H3K4	methylation	B. thaliandensis	BtSET	direct/NF-κB	Transcription of rRNA genes	[45,47]
H3K9	acetylation	M. tuberculosis	Rv3423.1	maybe direct		[51]
H3K14	methylation	L. pneumophila Paris	RomA	direct	Transcriptional repression	[46]
	acteylation	M. tuberculosis	Rv3423.1	maybe direct		[51]
H3K23	Global deacetylation	H. pylori		H3S10 dephosphorylation	Differential c-Jun and HSP70 expression	[52]
H1	trimethylation	B. anthracis	BaSET	direct/NF-κB	Transcriptional repression of NF-κB target genes	[47]
H2B, H3, H4	methylation	C. trachomatis	NUE	direct	Transcriptional repression	[53]
H3, H4	acetylation	P. gingivivalis	SCAFs, LPS	unknown	unknown	
H3	deacetylation	M. tuberculosis	unknown	HDAC1	Silencing of inteleukin-12β, suppression of T-helper1 response	[54]
H4	deacetylation	L. monocytogenes	LLO	H3S10 de-phosphory-lation	unknown	[36]
H4	deacetylation	M. tuberculosis	unknown	HDAC complex containing mammalian co-repressor Sin3A	Inhibition of interferon-γ-dependent HLA-DR gene expression	[55]

Influencing the expression of histone modifying enzymes is another possibility to affect histone modifications in favor of bacterial survival (see Table 2). Modulation of the histone deacetylase HDAC1 appears to be most targeted by pathogens, to manipulate the key acetylation system enabling protection against eradication. Infection with *Anaplasma phagocytophilum*, an intracellular pathogen causing human granulocytic anaplasmosis, causes upregulation of HDAC1 leading to a globally increased HDAC activity [56]. The recruitment of HDAC1 to AT-rich chromatin sites in promotors of host defense genes is mediated by the effector ankyrin A (AnkA) resulting in the reduction of histone H3 acetylation and the suppression of target genes such as *CYBB* that encodes Cytochrome b-245, beta polypeptide. As this element of the phagocyte NADPH oxidase is involved in the clearance of the pathogen by neutrophils, it is preferentially targeted [57–60]. Furthermore, AnkA functionally mimics SATB1, a protein able to bind AT-rich sequences distributed across distinct chromosomes at attachment regions of the nuclear matrix. Proteins with this ability are involved in nuclear matrix attachment, spatial organization of chromatin and large-scale transcriptional regulation [59,61–63]. AnkA could also perform as global organizer of the neutrophil genome, thereby acting locally (*cis*) and at a distance (*trans*) to a target gene. Moreover, pathogens as *Chlamydia psittaci* secrete nucleomodulins (SinC) that could act like AnkA and influence anchoring factors and lamins that control the dynamics of chromatin looping and organization, as the inner nuclear membrane proteins MAN1 and LAP1 [64].

Pseudomonas aeruginosa, an opportunistic pathogen that infects and colonizes inflamed airways and burn wounds, induces HDAC1 expression in human THP-1 monocytes with the help of a molecule usually used for quorum sensing, 2-aminoacetophenone [65,66]. This is followed by global histone H3K18 hypoacetylation and reduced expression of inflammatory cytokines and chemokines (e.g., TNF, IL-1b and MCP-1) resulting in dampened host defense against the bacterium.

Considering that this effect was also dampened by knockdown or inhibition of class I HDACs and the evidence that besides *A. phagocytophilum* and *P. aeruginosa* also *Porphyromonas gingivalis* modulates HDAC1 during infections, HDAC1 family members might play a central role in development of an epigenetic mediated tolerance against the pathogens [67]. In patients with chronic periodontitis, mRNA and protein levels of HDAC1 expression were globally increased compared to healthy individuals and colocalized with TNF expressing cells and tissues. Interestingly, epigenetic regulation mediated by *P. gingivalis* seems to be cell-type specific, since HDAC1 and HDAC2 are downregulated in gingival epithelial cells *in vitro*, while levels of acetylated histone H3 were increased in murine epithelial cells of the gingival tissue [68,69]. In addition, the host acetylation system is also often influenced by short chain fatty acids (SCAFs) produced by commensals or pathogenic bacteria as *P. gingivalis* (for recent reviews please refer to References [70] and [71]).

Table 2. Bacteria targeting histone modifying enzymes.

Target	Modification	Bacterium	Effector	Cellular function	References
HDAC1	induction	*A. phagocytophilum*	Ankyrin A	suppression of target genes as CYBB that encodes cytochrome b-245, beta polypeptide	[56]
HDAC1	induction	*P. aeruginosa*	2-Amnoacetophenone	reduced expression of inflammatory cytokines and chemokines	[65]
HDAC1	induction	*M. tuberculosis*	unknown	Silencing of inteleukin-12β, suppression of T helper1 response	[55]
HDAC2	repression	*P. gingivalis*	SCAFs	Activation of genes	[68,72–75]
P300	repression	*E. coli*	Proteinase NleC	decreased IL-8 production	[76]
HKMT	repression	*P. gingivalis*	SCAFs	Inhibition of heterochromatin marks	[73–75]

Another strategy followed by bacteria during host infection and manipulation of the epigenetic regulatory mechanisms is to proteolytically degrade histone acetyl transferase (HAT) family members.

One example of bacteria using this strategy are enteropathogenic and enterohaemorrhagic *Escherichia coli*, which secrete the effector protein NleC, a zinc-dependent metalloproteinase targeting intracellular signaling to dampen the host inflammatory response [76]. The protein specifically binds and degrades the host HAT p300 in infected cells leading to decreased IL-8 production, an effect that can be restored by p300 overexpression. Thus, HATs and HDACs can both be targeted by pathogenic bacteria to modulate epigenetics and inflammatory gene expression in their benefit.

2.1.2. How to Control Host DNA Methylation

DNA methylation is another way to control gene expression. There are several enzymes called DNA-(cytosine C5)-methyltransferases (DNMTs), which establish methyl residues to cytosine or adenosine residues, respectively [71]. In contrast, the removal of DNA methylation patterns is more complex, as the modified nucleotides or DNA sequences have to be exchanged by DNA-repair mechanisms or the methylation has to be oxidized to form 5-Hydroxymethylcytosine, which can be removed by enzymes [77]. DNA methylation patterns at promotors of tumor suppressor genes had already been discovered, when first hints pointed towards an influence of bacterial inflammation on mechanisms establishing DNA-methylation patterns after *Helicobacter pylori* infection. In this context, among others, genes associated with cell growth (*apc, p14 (ARF), p16 (INK4a)*), cell adherence (*cdh1, flnc, hand1, lox, hrasis, thbd, p14ARC*) and DNA-repair (*brca1, mgmt., hMLH1*) are influenced [52,78–80]. Similar observations of altered DNA-methylation patterns during inflammation were made following uropathogenic *E. coli, Campylobacter rectus* and *Mycobacterium leprae* infections [81–83]. Still, the questions if DNA-methylation is directly induced by bacteria or is a secondary reaction by the host due to persistent inflammations, as well as the underlying mechanisms, are not completely answered yet [84].

However, several *Mycoplasma* species to encode mammalian DNMTs like equivalents that target cytosine-phosphate-guanine (CpG) dinucleotides to establish methylation patterns in the bacterial genome [85–87]. Moreover, their expression in human cells results in their translocation to the nucleus, where they set up unusual methylation patterns on the host DNA. This was shown for the DNMTs Mhy1, Mhy2 and Mhy3 expressed by *Mycoplasma hyorhinis* in combination with up- and downregulation of certain genes resulting in activation of proliferation specific pathways, a process that might contribute to tumor progression [85,88].

Mycobacterium tuberculosis owns an effector called Rv2699 that can enter the nucleus of THP1 cells (a monocytic cell line derived from a patient with acute monocytic leukemia) and methylate cytosines outside CpG dinucleotides. Notably, Rv2699 prefers cytosine-phosphate-adenine or cytosine-phosphate-thymine sites to generate a type of methylation that is, with few exceptions, normally not present in mammalian adult differentiated cells [89,90]. However, non-CpG methylation could lead to a more stable type of modification that persists longer in the genome of infected nondividing macrophages, offering an advantage for *M tuberculosis* by establishing an intracellular environment for persistence [90]. A follow up study revealed that THP1 macrophages infected with *M. tuberculosis* strain H37Rv created genome-wide *de novo* methylation patterns at non-CpG dinucleotides that included hyper- and hypomethylated regions [90,91]. Additionally, clinical isolates infecting THP1 cells may downregulate IL-6 receptor expression by hypermethylation of CpG-dinucleotides at the promoter of the IL-6 receptor gene. Still, it has to be mentioned, that the observations of *M. tuberculosis* induced DNA-methylation patterns depend on the infected cell type.

Another interesting bacterial induced modification of gene expression is represented by differentiated Swann cells that adapt the phenotype of progenitor stem-like cells after *M. leprae* infection. This is probably induced by silencing of the *Sox10* gene after bacterial methylation [82]. In contrast to the decreased expression of Sox10, other genes involved in epithelial–mesenchymal transition (EMT) were demethylated and transcribed leading to the transformation of Swann cells into myofibers or smooth muscles in vitro and in vivo [92].

P. gingivalis was shown to increase the methylation of the TLR-2 promotor in gingival epithelial cells (GECs) reducing innate immunity activation and causing hyposensibility [69,93]. Besides, coinfection with *Filifactor alocis*, another pathogen associated with periodontitis is suggested to influence the whole cell transcriptome through impact on the nucleosome structure by reduced expression of H1 family members [73,74]. Other histone modifications induced by LPS or short chain fatty acids (SCFAs) produced by *P. gingivalis* are summarized in Tables 1 and 2.

Still, there is not much known about the relation of DNA-methylation and infection and the underlying causalities [71,84]. Considering that many of these modifications are observed in the context of cancer initiation and progression, further investigation may contribute to new therapeutic agents and cancer prophylaxis.

2.1.3. Regulation of Host Gene Expression via lncRNAs

The role of lncRNAs during modulation of gene expression has been discovered in the recent years. Similar to mRNAs, lncRNAs are transcribed by RNA polymerase II or III, followed by splicing, 5′capping and in some cases polyadenylation at the 3′end. Contrary to mRNAs, the expression of lncRNAs is much lower and in a cell-, tissue- and developmental stage-specific manner [94].

Dependent on of their position relative to the neighboring protein-coding gene, lncRNAs are classified as sense, antisense, bidirectional, intronic or intergenic and, despite their enormous number, they were previously considered as "dark matter" or "junk" in the genome [95]. *Au contraire*, lncRNAs are now respected as important physiological regulators during cell homeostasis, growth, differentiation and anti-viral responses [96–99]. In addition, gene imprinting, regulation of the p53 pathway, stem cell self-renewal and differentiation and DNA damage response were reported as lncRNA controlled mechanisms [100–103].

The functionality of lncRNAs is not restricted to the neighbored protein-coding gene (*in cis*), in contrast they are also able to act *in trans* to regulate gene expression across chromosomes. In this context, lncRNAs regulate different processes as chromatin remodeling, transcription and post-transcriptional regulation via their capacity as signals, decoys, guides and scaffolds [104,105]. Interestingly, another origin of lncRNAs is the expression of pseudogenes and gaining Influence over the expression of pseudogenes could, therefore, provide a possibility to control infectious responses [106].

Immune regulation through lncRNAs has already been known after viral infections but recent research indicates its involvement also whilst fighting bacteria [107]. In that context, 76 enhancer RNAs (eRNAs), 40 canonical lncRNAs, 65 antisense lncRNAs and 35 regions of bidirectional transcription are differentially expressed in human monocytes after LPS stimulation [108]. LPS stimulation alone induces a differential expression of about 27 lncRNAs leading to histone trimethylation or acetylation of neighboring genes after de-regulation, pointing towards their regulatory influence during the innate-immune response [109]. The observation, that 44% of total lncRNAs varied in their expression after *Salmonella* infection in HeLa cells could foster these results and substantiate them by a function in the early phase of infection as sensitive markers for pathogen activity [110]. In line with this, the lncRNA HOTAIR that contributes to transcriptional repression of HOX genes also promotes inflammation in mice cardiomyocytes by TNF-α production mediated through phosphorylation of p65 protein and NF-κB activation after LPS induced sepsis [111,112].

Long intergenic non-coding RNAs (lincRNAs) are a subtype of lncRNAs, as they are expressed from intergenic regions. In response to an LPS stimulus, bone-marrow dendritic cells expressed about 20 lincRNAs with the majority being dependent on NF-κB activity, including lincRNA-Cox2, which is also upregulated in bone marrow-derived macrophages following *L. monocytognes* infection [113,114]. Additionally, bacteria sabotage lncRNA activity, as BCG (attenuated strain *M. bovis* bacillus Calmette-Guérin BCG) infected macrophages repress the expression of 11 lncRNAs that are not dampened by infection with heat activated bacteria [115]. Still, possible subversion of lncRNA-mediated inflammatory regulation needs to be further investigated.

2.2. Bacterial Effectors Manipulating the Host Transcription Machinery

Proper RNA Pol II complex formation is essential for protein expression and tightly controlled by regulators, who are expressed by approximately 10% of all genes [116]. These are general or specific TFs, which serve as activators and repressors and determine specificity and efficiency of transcription at individual promotors [117,118]. Considering the large number of factors involved in transcriptional regulation, it is not surprising that bacteria target those regulators to drive transcription in their favor. For example, activator protein-1 (AP-1)–dependent gene transcription is inhibited by NleD, an effector of *E. coli* that cleaves and inactivates the MAPKs, JNK and p38. Also, the recently identified nucleomodulin OrfX secreted by *L. monocytogenes* that influences host transcription via its interaction with the Ring1 YY1-binding protein (RYBP), a multifunctional nuclear protein owning a zinc finger motif to interact with several TF components of the polycomb repressive complex 1 [119,120]. Moreover, RYBP promotes gene silencing and transcriptional repression of developmental genes, as it is part of the BCL6 corepressor (BCOR) complex [121,122]. In contrast, it is also involved in the activation of the Cdc6 promoter and mediates interaction of the TFs E2F and YY1 [123]. Furthermore, the TF p53 (a tumor suppression factor) is assumed to be stabilized by RYBP through binding of MDM2, an E3 ligase, preventing p53 from proteosomal degradation. As p53 controls intracellular levels of reactive oxygen (ROS) and nitrogen species (RNS) that are part of the immune defense of macrophages, OrfX targets RYBP for degradation to interrupt P53 activity promoting intracellular bacterial survival. Still, this model needs to be verified [120,124].

Salmonella Typhimurium inhibits the expression of NF-κB mediated genes by secretion of PipA, GogA and GtgA via its type II secretion system [125–128]. These proteins belong to the family of zinc metalloproteases and contain the short metal binding-motif HEXXH, which consists of two histidine residues coordinating the active-site zinc and a glutamate residue that is essential for catalytic activity [129]. They can cleave NF-κB TF subunits, including p65, RelB and cRel, thereby suppressing their ability to control the transcription of innate immune genes [125–128]. In addition, Jennings et al. predicted suppression of the transcriptional coactivator ribosomal protein S3 (RPS3) by GtgA family members, as it produces p65 (1–40) after cleavage of p65.

Enteropathogenic and enterohemorrhagic *E. coli* possess another zinc metalloprotease that also owns the HEXXH motif and is able to cleave p65, RelB and cRel, as well as NF-κB1 (p105/p50) and NF-κB2 (p100/p52) [126–128,130]. Following cleavage, the subunits are left inactive except for the N-terminus of p65, which prevents the nuclear import of the transcriptional coactivator ribosomal protein S3 (RPS3). This in turn inhibits the expression of a specific subset of NF-κB–dependent genes requiring RPS3 for their expression.

The non-pathogenic *E. coli* strain 83972, who is the agent causing persistent asymptomatic bacteriuria (ABU), suppresses host defense in the urinary tract by inhibition of RNA Pol II dependent transcription. While infections with pathogenic strains induce urinary tract infections, these bacteria create an asymptomatic carrier state that reminds of bacterial commensalism and protects patients against infection with more virulent strains. Therefore, therapeutic urinary tract inoculation with the ABU strain is a promising alternative to appease symptoms of therapy resistant, recurrent urinary tract infections [131–133].

Studies with patients and human cells treated with ABU strain 83972 revealed that 24 h after inoculation, over 60% of all genes were suppressed, including regulatory elements as transcriptional repressors, transcriptional activators, regulators of translation and chromatin or DNA organizing factors [134,135]. This phenomenon was observed for many genes of the innate immune response but about 22.5% of the effected genes are involved in Pol II transcription or in regulating Pol II–dependent pathways. After Ingenuity Pathway Analysis, a network incorporating *FOSB*, *HSPA6*, *RN7SK*, *RGS4* and *IFIT1*, inversely regulated genes that control Pol II for instance through TATA box–binding proteins (TBP) [136–138], appeared.

Another fascinating observation revealed that 50% of ABU strains lack virulence genes due to point mutations or deletions resulting in smaller genome sizes. Considering that ABU strains evolved from uropathogenic *E. coli*, this could be a hint for a reductive evolution creating a niche through active adaptation to the host environment [139]. Thereby, the ABU strain generates a commensal like state characterized by a well-balanced immune environment that finally protects the host from colonialization with more virulent strains and destructive immune activation [134,140–144].

Ambite et al. observed in a follow up study that obtaining one single virulence factor was enough to induce virulence of a non-virulent strain causing symptoms in the host, in contrast to the broad repertoire of virulence factors that are normally expressed by pathogens [145]. In this context, reconstitution of the *papG* adhesin gene recreated functional P-fimbriae leading to virulence of the avirulent ABU strain. Considering the high frequency of ABU strains carrying inactive papG genes, the loss of P-fimbriae might induce development of virulence attenuation and evolution towards commensalism [96,134].

L. pneumophila is another pathogen that induces global reprogramming of transcription, by interference with transcriptional elongation by Pol II. Its effector AnkH interacts with LARP7, a component of the 7SK small nuclear ribonucleoprotein (snRNP) complex involved in Pol II pausing. Thereby, the β-hairpin loop of the third ankyrin repeat of AnkH impairs LARP7 interaction with the other 7SK snRNP complex components resulting in promotion of gene wide transcriptional elongation [146]. The nucleomodulin SnlP expressed by *Legionella* also regulates RNA Pol II mediated transcription elongation by inhibition of SUPT5H that is part of the 5,6-Dichlorobenzimidazole 1-β-D-ribofuranoside (DRB, a selective inhibitor of transcriptional elongation by RNA pol II) sensitivity-inducing factor (DSIF) complex [147].

2.3. RNA Processing as Target during Infections

Following transcription the immature pre-mRNA is processed to mature mRNA, a process that includes 5′ capping, 3′polyadenylation and splicing and is essential for normal cell function [148]. Splicing of mRNAs can include omitting or retaining of exons to create a protein with altered structure, function and stability, a process called alternative splicing. In the human genome, more than 90% of all genes are adjusted by alternative splicing, which enables a variation and dynamism in the static genome as protein domains can be easily new combined to form isoforms with unique functional abilities [148,149]. The process is controlled by multi-molecular complexes that assemble at splice junctions, thereby evaluating splicing enhancer/silencer elements flanking splice junctions, which in their combination determine inclusion or exclusion of exons. Those elements are divided into *cis* elements including splicing enhancers and silencers and *trans* elements as snRNPs, hnRNPs, SR proteins (serine-arginine containing proteins) and several other accessory proteins [149,150]. Furthermore, the rate of transcription has a critical influence on alternative splicing, as a paused or decelerated RNA Pol II can use newly transcribed splice junctions that could have been skipped at higher translation speed [151–153].

Recent studies indicate that the host splicing machinery is targeted by pathogens to perturbate immune response. This has been extensively reported for viral infections but quite less is known about bacterial interference. Nevertheless, global alterations of splicing patterns were detected after infection of human macrophages with *M. tuberculosis*, *Salmonella* or *Listeria* [154–157]. More specifically, hnRNP M interacts with LLO leading to a hampered INF-γ response [158] and co-immunoprecipitation of splicing factors hnRNP U, hnRNP H, hnRNP A2/B1 isoform A2 and SRSF3 with the bacterial protein mtrA was shown in macrophages overexpressing specific secreted proteins that are infected with *M. tuberculosis* [159]. Another hint for the interaction of mycobacterial proteins and host splicing factors is the precipitation of host splicing proteins as SRSF2, SRRM2, SF1, HTATSF1, GCN1L1, CPSF6 and many more by the mycobacterial proteins EsxQ, Apa, Rv1827, LpqN, Rv2074 and Rv1816 [160].

Mycobacterium avium subsp. *paratuberculosis*, also induces alternative splicing of 46.2% of all genes, including two genes responsible for monocyte to macrophage differentiation-associated maturation and lysosome function. Since their splice variants lead to failure of macrophage maturation, bacterial intracellular persistence is improved in the early phase of infection by hampered clearance [161].

Furthermore, alternative splicing of RAB8B, a key regulator of phagosome maturation, induced by *M. tuberculosis* infected cells leads to the production of a truncated protein. The alternative splicing event results in nonsense-mediated decay of RAB8B mRNA resulting in lowered protein levels, that dampens phagosome maturation [156].

Analysis of RNAseq data revealed that specific genes are chosen by pathogens for the manipulation of alternative splicing. Indeed, most dominant isoforms of protein kinases produced end up with the loss of critical functional domains including kinase domain or protein–protein interaction domains like SH2, SH3 and PH domains [162]. Considering nonsense-mediated decay of RAB8B mRNA after *M. tuberculosis* infection, it is concluded that this mechanism describes the two strategies host and bacteria developed during their evolutionary concurrence [156]. In this theory, increase of transcription after infection represents the host response, whereas splicing into a truncated isoform, which is destinated for decay, exemplifies bacterial interference. The exact mechanism how bacteria manipulate alternative splicing is not clear yet. Possibly, they activate cryptic or weak splice sites in the host genome to alter the splicing pattern but this has still to be proven [149].

However, another aspect that needs to be considered is the high diversity with that individuals react to the same pathogenic agent. For example, only 5–10% of the individuals in tuberculosis endemic countries that had contact with *M. tuberculosis* develop disease, whereas the majority either eliminates the pathogen or controls it in a metabolically altered latent phase [163,164]. An explanation are single-nucleotide polymorphisms (SNPs) that disrupt splice-site consensus sequences in 15% of human disorders induced by inherited point mutations, whose influence induce strongly fluctuating pathological conditions after varying activation of disease associated genes [165–167]. This was already reported for diseases as diabetes and seems to be also true for infections, as several SNPs were identified that change the host susceptibility to *M. tuberculosis* infections for example, in the intronic region of human ASAP1 (dendritic cell migration). Another polymorphism in IL-7RA helps to protect against tuberculosis due to an impaired IL-7Ra splicing [168,169]. Therefore, alternative splicing gets into focus of possible medical therapy developing splicing inhibitors that are already tested for cancer [149,170,171]. Nevertheless, the knowledge about alternative splicing during bacterial infections and their interplay is very limited and deserves more attention, as this could give more insights in individual susceptibility and immunity.

2.4. The Advantage of Modulating Host RNA Stability and Degradation

The lifetime of mRNAs has a major impact on the amount of proteins that can be produced; the shorter an mRNA is present, the less it can be transcribed. Since the lifetime of an mRNA is dependent on its stability, there are mechanisms to increase resistance to degrading RNases [11–13]. First, the $5'm^7G$ cap and the $3'$poly-A tail that are established post transcriptionally at all mRNAs but especially the length of the poly A tail can vary between mRNAs, determining the duration of resistance against enzymatic degradation [9]. Besides, these structures are involved in virus clearance, as viral mRNAs lack these structures, what marks them for RNA degradation machinery [172]. In addition, the structure of the mRNA alone influences its stability, as hairpin-structures and other secondary structures are formed dependent on the sequence and therefore, increase the stability [11–13]. In the following paragraphs we want to highlight the regulation of mRNA stability and decay mediated by miRNAs during infections and how bacteria interfere with this part of the host immune defense.

Manipulation of miRNAs to Favor Bacterial Survival

Their physiological properties enable non-coding RNAs (ncRNAs) to base-pair with their targets and interfere with a twofold effect on gene expression—one single ncRNA can bind multiple targets, thereby influencing several pathways at once and one gene can be regulated by several ncRNA fine tuning gene expression [173]. miRNAs are involved in several cellular processes as cell proliferation or differentiation and after studies with human monocytes treated with LPS, miR-146 was identified as anti-inflammatory miRNA proving miRNA involvement in inflammation [174–176]. Indeed, subsequent studies revealed specific expression of miRNA sets including miR-155, miR-146, let-7 and miR-29 (see Table 3) due to infection with different bacterial pathogens regulated in a time dependent manner [177–181]. Another study with dendritic cells infected with six different bacteria demonstrated a core infectious response in a temporal manner including 49 miRNAs that were always expressed and may, therefore, play essential roles in infectious responses. Additionally expressed miRNAs might hint towards specific variability and signatures arising from the individual pathogens [182]. Interestingly, following infection, the proportion of miRNA variants, the so called isomiRs, varies, which is supposed to effect miRNA identity and regulatory potential but has not been proven yet [182].

The induction of miRNAs is often dependent on PRR and NF-κB pathway induction in response to bacterial stimuli as LPS. Interestingly, there is a dose dependent reaction to the stimulus, as a low dose activates miR-146 that acts as anti-inflammatory regulator promoting tolerance to low doses by targeting two members of the NF-κB pathway, TRAF6 (TNF Receptor-associated factor 6) and IRAK1 (IL-1R-associated kinase 1) [176,183]. In contrast, at high doses of LPS, TNF-α and Interferon β induce miR-155 via TAB2 to maintain the proinflammatory NF-κB activity fighting pathogens and exerting a negative feedback on the immune system. Therefore, both mi-RNAs protect the host from sepsis and overreaction [183,184] but in a dose dependant manner. miR-155 is also involved in T helper cell development or promoting autophagy by inhibition of the mTOR (mammalian/mechanistic target of rapamycin) pathway, indicating that it represents an important part of an efficient immune response [185–187]. Actually, upon *Citrobacter rodentium* or *L. monocytogenes* infection, miR-155 null mice showed slower clearance and an impaired CD8$^+$ T-cell response, respectively and miR-155 was identified as an essential factor during the vaccination process against *S.* Typhimurium [188–191].

Table 3. Activity of miRNAs in the host response. The arrows indicate changes of miRNA expression induced after bacterial infection that result in the described alterations..

miRNA	Target	Cellular Function/Cells Involved	Induced Changes During Infection	Described for Infection with	References
let-7B	TLR4	TLR sensing	↓ Promoted TLR sensing and NF-κB activity	several bacteria	[192]
let-7A, let-7D	IL-6, IL-10	Pro-inflammatory and anti-inflammatory cytokine	↓ Maintaining balanced immune response	several bacteria	[193]
let-7C	mTOR-pathway		↓ Modulation of T-cell activity	several bacteria	[194]
let-7I	SNAP23	Exosome release	↓ Antimicrobial response, cell to cell communication	several bacteria	[195]
miR-29	IFN-γ	Immune response to intracellular bacteria/NK-cells, CD4$^+$ and CD8$^+$ T-cells	↓ increased IFN-γ expression and bacterial clearance	*L. monocytogenes*, *Mycobacterium bovis* bacillus Calmette-Guérin (BCG)	[196]
miR-192, miR-378, miR-215, miR-148A, miR-200C, miR-200B	zinc finger E-box–binding homeobox ZEB1 and ZEB2 (transcriptional repressors of E-cadherin)	Intestinal homeostasis, gut transcriptome, tissue integrity, immunity and metabolism	↓	*L. monocytogenes*	[197,198]

Another miRNA family, the let-7 family, is repressed during infection or exposure to LPS, as Lin-28B expression is induced in a NF-κB dependent manner that blocks let-7 maturation [154,192,193,195,199,200]. Additionally, active repression of these miRNAs by bacteria has been reported in several studies (see Table 3). Many other bacteria induce miRNA expression and manipulate expression of these immune regulators in their favor (summarized in Table 4.)

Table 4. Bacteria manipulating host miRNA to circumvent immune response. Arrows indicate an up or downregulation and refer to the miRNA before.

Bacterium	miRNA	Cellular Function	miRNA Target	miRNA Expression Induced by	Bacterial Benefit	References
L. monocytogenes	miR-29 ↑	Expression of IFN-γ by NK cells	IFN-γ mRNA		unknown	[196]
Salmonella	miR-30c, miR-30e ↑	Host SUMOylation	Ubc9 (cellular E2 SUMO-conjugating enzyme)		Reduction of host SUMOylation	[201]
	miR-15 family ↓	Cell cycle	Cyclin D1	Inhibition of transcriptional factor E2F1 production	De-repression of cyclin D1,G1/S cell cycle transition, bacterial intracellular replication	[202]
	miR-128 ↑	Recruitment/ activation of macrophages	Macrophage colony-stimulating factor (M-CSF)	p53 signalling pathway	Impairment of M-CSF mediated macrophage recruitment	[203]
H. pylori	miR-21, miR-222 ↑		tumour suppressor RECK			[204,205]
	miR-30b ↑	Formation and maturation of autophago-somes	BECN1 and ATG12 transcripts		Autophagy interference, intracellular persistence and survival	[206]
M. tuberculosis	miR-132, miR-26a	Recruitment/ activation of macrophages	Tran-scriptional coactivator p 300		Downregulation of INF-γ signalling cascade, limitation of macrophage response to INF-γ signalling	[207]
	miR-125b, miR-155, miR-99b	Recruitment/ activation of macrophages	TNF-α		Reduction of TNF-α production, increase of SHIP1 transcription, reduced macrophage recruitment	[208,209]
	miR-155↑	Recruitment/ activation of macrophages and macrophage apoptosis	Repressing BACH1, SHIP1 and SOCS1, Rheb		Induction of heme oxygenase 1 expression, activation of serine/threonine kinase AKT	[199,210,211]
	let-7f↓	Inhibitor of NF-κB pathway	A20 (deubiquitinating enzyme)		favored bacterial survival in macrophages	[212]

In addition to *H. pylori* with miRNAs (see Table 4), there are two more interactions described, which depend on the effector CagA that activates NF-κB pathway. This effector induces an increased expression of miR-1289 that in turn leads to a decreased gastric acidity, as miR-1289 targets HKα, a component of the gastric H$^+$/K$^+$ ATPase [213]. Furthermore, Cag A induces cell cycle arrest at G1/S transition, which inhibits the renewal of the gastric epithelium, supporting *H. pylori* colonization [214]. In this context, miR372 and miR-373 expression is suppressed by Cag A, whereas miR-584 and miR-1290 expression is promoted. The latter target FOXA1, a negative regulator of the epithelial-mesenchymal transition, for inhibition, thereby favoring bacterial persistence and survival within the gastric epithelium [215]. Moreover, CagA suppresses miR320 and miR370 expression, who induce MCL1 (an anti-apoptotic gene) or downregulate FoxM1 expression, respectively. FoxM1 downregulation in turn activates p27^{K1P1} leading to cell cycle inhibition. Together, these factors decrease apoptosis and favor cell proliferation, which can lead to tumor development.

The *M. tuberculosis* effector ESAT-6 is also capable of manipulating miRNA expression to the benefit of the bacterium [212]. ESAT-6 downregulates let-7f expression in macrophages, leading to an enrichment of the deubiquitinating enzyme A20 that negatively regulates the NF-κB pathway. Furthermore, miR-155 expression is stimulated, resulting in BACH1 (a transcription regulator protein) and SHIP1 (SH-2 containing inositol 5′ polyphosphatase 1, a multifunctional protein expressed predominantly by hematopoietic cells) repression that in turn induces heme oxygenase 1 expression. Concurrently, serine/threonine kinase AKT is activated fostering bacterial dormancy and survival. This is subsidized by downregulation of SOCS1 (suppressor of cytokine signaling 1) and targeting of Rheb (Ras homolog enriched in brain), which is followed by macrophage apoptosis [199,210,211].

Evidence exists that bacteria are able to also produce their own regulatory RNAs that interfere with the host. In 28 bacterial genomes 68 possible candidate bacterial RNAs were found during an in silico search, which harbor secondary structures that could form miRNAs after host procession and be involved in 47 human diseases [216]. As an example, after *E. coli* ingestion *che-2* and *F42G9.6* gene expression was modulated and probably degraded in *Caenorhabditis elegans* by *E. coli* OxyS and DsrA ncRNAs [217]. Furthermore, *Mycobacterium marinum* expresses a pre-miRNA that associates with the host RISC complex in its mature, 23 nucleotide long form and could effectively downregulate its target mRNA when overexpressed [218].

It has been reported that exosomal transfer of host cell miRNAs is used to spread the host response to other cells and the ratios of miRNAs transported differ in a time and bacterial dependent manner. Hence, the ratios of miR-146a and miR-155 in exosomes can subsequently modulate host cell responses, favoring inflammation or recovery, respectively. The use of exosomes containing miRNAs could give rise for therapeutic possibilities to treat inflammation or to be used during vaccinations [219–226]. Moreover, exosomes containing miRNAs could be used for diagnosis, since they can be detected in many sample types (blood fluids, saliva, tears, urine, amniotic fluid, colostrum, breast milk, stool, etc.) and since there are unique patterns for each pathogen [219–221].

Regulation of gene expression through ncRNAs as miRNAs happens more immediately and flexibly than through transcriptional regulators [33]. The faster response is enabled by the cell and tissue type specific differentially regulated reservoir of ncRNAs, which also allows a precise fine-tuning of gene expression to organize immune defense and damage protection [33]. Taken together, investigation of the host-pathogen crosstalk with a focus on miRNAs and their usage and manipulation by bacteria provides new perspectives to fight bacterial mediated diseases.

2.5. Controlling Host Translation Improves Bacterial Persistance

Translation is a major regulator of gene expression and immune response. As many factors are needed to induce Ribosome association, start of translation and ongoing elongation, many ways exist to regulate or interfere with the translation machinery. Not surprising, inhibition of translation is a well-known strategy followed by bacteria to circumvent immune defense [227].

In most cases, eukaryotic translation is controlled during initiation, when ribosomes are recruited to the mRNA mediated by eukaryotic initiation factor 4F (eIF4F) that recognizes the 5′ cap structure with the help of its cap-binding subunit eIF4E [228]. The reversible association of this subunit with 4E-binding proteins (4E-BPs) inhibits the assembly of eIF4F and its release and activation are in turn mediated by the phosphorylation of the 4E-BPs [229–231]. This phosphorylation is induced by the serine/threonine kinase mTOR complex 1 (mTORC1), thereby requiring the protein Raptor for mTOR substrate binding whereas rapamycin binding inhibits phosphorylation and dissociation [232,233].

Legionella pneumophila (*L. pneumophila*) expresses five effectors, Lgt1, Lgt2, Lgt3, SidI and SidL, involved in global protein translation inhibition by interference with the eukaryotic elongation factors eELF1A and eELF1Bγ [227,234–236]. Moreover, *L. pneumophila* was also shown to negatively influence cap-dependent translation initiation mediated by ubiquitination of the mTOR pathway leading to suppression of the eukaryotic initiation factor 4E (eIF4E) through decreased eIF4E assembly into the translation initiation complex eIF4F [237].

Finally, the synthesis of IκB, an inhibitor of the NF-κB TF, is inhibited by *L. pneumophila*. This leads to a prolonged NF-κB activation resulting in the so-called effector-triggered response (ETR) including transcription of target genes, such as *Il23a* and *Csf2* that create a more pro-inflammatory state. Fascinatingly, mutants lacking effectors or the Dot/Icm type IV secretion system transferring them, still inhibit host translation via TLRs and NF-κB activation but not sufficient enough to fully induce ETR [234,235]. In addition, macrophages lacking TLR and Nod signaling still mediated MAPK signaling after exposure to the five *L. pneumophila* effectors that leads to inhibition of host translation. Therefore, translational inhibition does not exclusively rely on ETR but also on effector independent mechanisms that induce mTOR pathway downregulation and cytokine biasing [237]. In this context it is suggested that the host immune system senses not only for PAMPs but also for pathogen-encoded enzymatic activities that disrupt crucial cellular processes [227]. Interestingly, even if host translation is inhibited at the stage of initiation and elongation by *L. pneumophila*, there is still an inflammatory response detectable. The immune response is quite weak compared to its normal potential but few inflammatory cytokines, as IL-1α and IL-1β, circumvent translational inhibition by *L. pneumophila*, which is mediated by MyD88 signaling [227]. This demonstrates that the interference with cap-dependent host translation results in promotion of host defense, as highly abundant transcripts, which often encode proinflammatory cytokines, are favored for translation. Thereby, the bacterial benefit through blockage of host translation may consist of increased availability of amino acid nutrients beneath the dampened immune response, which is impeded by the host [234].

Translational inhibition is also known for *P. aeruginosa* infections in *C. elegans*, where Exotoxin A after its endocytosis into intestinal cells leads to suppression of elongation factor 2 (EF2), followed by selective translation of ZIP-2 and thus, induction of pathogen clearance [238,239]. As the 5′ UTR of *zip-2* contains several untranslated ORFs (uORFs), it was proposed that the uORFs could mediate selective translation. Moreover, inhibition of translation by pharmacological inhibitors also caused induction of various stress response genes including *Il6*, *Il23*, *Il1α* and *Il1β* [240–242]. As these cytokines are still expressed when the elongation machinery is attacked by *P. aeruginosa* with Exotoxin A, a consideration of similar functionalities of the 5′ UTR or the 3′ UTR of cytokine genes to uORFs was raised. However, the ADP-ribosylation of elongation factor 2 (EF2) in host cells also triggered a strong immune response that is supposed to be the result of a conserved surveillance mechanism in response to inhibition of the translation elongation machinery [236,243]. Thus, a set of elongation factors can be considered, that are resistant to modification by these effectors or are at least not targeted. Therefore, future research is needed to get more information about the underlying mechanisms and potential therapeutic targets [227].

2.6. Modification of Protein Degradation/Activity as Last Possibility to Evade Host Immune Defense?

2.6.1. Mechanisms to Interfere with Protein Degradation

To maintain cellular homeostasis, the quality of synthesized proteins must be controlled for a proper folding and products with quality issues must be degraded in a controlled manner. This quality control is taking place in the cytosol or the ER lumen, where chaperones and heat shock proteins catalyze and stabilize the protein folding [244]. In the case of physiological stress caused by DNA damage, chemical stimuli or pathogens, the ratio of misfolded or unfolded proteins in the ER lumen increases, causing additional stress. Then, the unfolded protein response (UPR), an evolutionary conserved signaling network, is activated resulting in downregulation of overall protein synthesis, except for chaperones and induction of ER associated protein degradation (ERAD [245]. The main kinases controlling the UPR, the inositol-requiring protein 1 (IRE1), PKR like ER kinase (PERK) and activating TF 6 (ATF6) are located inside the ER membranes. Their luminal domains bind the ER chaperone immunoglobulin binding protein (BiP) in unstressed conditions but in case of stress, BiP is released causing activation of the receptor proteins (IRE, PERKI and ATF6) [246]. Following BiP dissociation, oligomerization and autophosphorylation, the cytosolic RNase domain of IRE1 is activated targeting X-box-binding protein 1 mRNA (XBP1u) and transfers it into its spliced form (XPB1s). This enables transcription of the TF responsible for upregulation of UPR target genes fostering ERAD and enhances overall ER protein folding capacity [240–242]. The ER transmembrane kinase PERK1 also oligomerizes and autophosphorylates after activation, catalyzing the phosphorylation of the α-subunit of the eukaryotic initiation factor 2 (eIF2). This is followed by downregulation of global mRNA translation to reduce ER stress but favoring translation of some mRNAs as ATF4. ATF4 in turn induces several UPR target genes including C/EBP homologous protein (CHOP) [247,248]. In contrast to IRE1 and PERK, ATF6 translocates to the Golgi followed by its activation after proteolytical cleavage and activation of the b-ZIP TF to induce UPR target genes.

Since permanent ER-stress, which cannot be solved by UPR and ERAD, will finally induce apoptosis, intracellular bacteria have evolved strategies to interfere with those pathways [249–251]. Surprisingly, induction of UPR pathways can also promote bacterial replication, as bacterial effectors have been detected that induce UPR [19]. As result, it is difficult to refer an upregulation of UPR to bacteria using the increased ER folding capacity for their benefit or to the defense system of the host. In the case of *L. monocytogenes* infection, the effector LLO activates all three UPR pathways leading to induction of ER stress and reduction of bacterial survival. In contrast, the pharmacological block of UPR during infection reduced the intracellular replication of *Brucella melitensis* and *Brucella abortus* significantly [252].

B. melitensis and *B. abortus* both induce the IRE1 branch, a process often mediated by TLRs. The TLR adapter protein myeloid differentiation primary response gene 88 (MyD88) than mediates XBP1u splicing. Interestingly, the bacterial effector protein TcpB, secreted by *B. melitensis*, is able to induce the UPR target genes BiP, CHOP and ER DnaJ-like 4 (ERdj4) in a MyD88 independent manner; instead, the TcpB protein itself is required for UPR induction [253].

The induction of IRE1 by *B. abortus* is mediated by the secreted effector VceC after binding of BiP inside the ER lumen. This is followed by IRE1 dependent activation of Nod1/Nod2 innate immune signaling resulting in proinflammatory cytokine expression [254,255]. The ectopic expression of VceC leads to the structural reorganization of the ER and IL-6 production is stalled after infection with *B. abortus* vceC mutants unable to induce UPR. In mice infected with *B. abortus* vceC mutants necrosis was reduced and survival of the pups was increased [255]. Following infection with *B. abortus* WT, similar effects were observed after treatment with the general UPR inhibitor tauroursodeoxycholic acid, leading to the assumption that pharmacological UPR inhibition could be a promising treatment of *B. abortus* infections.

Several other bacteria are known to inhibit the UPR pathway, as it represents a major host defense mechanism involved in bacterial sensing mediated by TLR signaling. Some examples are summed up in Table 5 but unfortunately, the underlying mechanisms are rarely understood. In some cases, as for *L. pneumophila*, it is known that the observed downregulation of UPR is effector dependent, as mutants lacking functional T4SS were unable to induce those changes but the effectors and its targets have not been identified yet [256,257].

Table 5. Inhibition of unfolded protein response (UPR) by bacteria.

Effector	Bacterium	Target	Cellular Function	Manipulation	Reference
glucosyltransferase effector proteins	*L. pneumophila*	unknown	IRE1 branch of UPR	Inhibiting splicing of XBP1u mRNA	[256,257]
unknown	*L. pneumophila*	unknown	UPR	inhibit the translation of BiP and CHOP	[256–258]
unknown	*Simkania negevensis*	unknown	UPR	BiP induction (early), later on inhibition	[259]
unknown	*Simkania negevensis*	unknown	UPR	blocks the translocation of preexisting CHOP protein to the nucleus	[259]
unknown	*Simkania negevensis*	unknown	PERK branch of UPR	Reduced phosphorylated eIF2 levels	[259]
unknown	*L. monocytogenes*	unknown	PERK branch of UPR	Reduced phosphorylated eIF2 levels	[260]
unknown	*C. trachomatis*	unknown	PERK branch of UPR	Reduced phosphorylated eIF2 levels	[260]

Taken together, bacteria follow different strategies of interference with protein folding to ensure their intracellular survival. Manipulation of UPR is one strategy to achieve the best outcome for bacteria by either activation or inhibition of UPR. Activation may be induced to take advantage from increased protein folding capacity and lipid biosynthesis by host cells, whereas UPR blockage should hamper host defense, such as apoptosis or innate immunity [261]. In this context, further investigations are needed to understand the underlying mechanisms, how bacterial pathogens manipulate the UPR and which strategy is favored by the individual pathogens in a spatially and temporally manner.

2.6.2. Control of Protein Activity enables Bacteria to Direct Host Immune Reaction

As already mentioned above, the final level to regulate gene expression, for example, to avoid prolonged inflammatory response, is to control the activity of proteins followed by their degradation. The mechanisms to influence protein activity or to mark a protein for degradation are mediated via attachment of functional groups. To ensure a precise signal transduction, proteins are activated in most cases by addition of phosphate groups, which must be removed when the inducing stimulus is ending. The dephosphorylation of MAPKs and the resulting interruption of host signaling cascades, leading to reduced inflammation and an increase of bacterial replication inside the host, is a common bacterial strategy, for example, used by *S.* Typhimurium by secretion of the effector SpvC, a phosphothreonine lyase [262].

In addition, protein activity, subcellular localization and stability is not only regulated by ubiquitination or phosphorylation but also by reversible acetylation which is proposed for approximately 1700 proteins [263]. These include TFs, structural proteins and signal transduction regulators, indicating that reversible acetylation is critical for cell homeostasis [264]. As there are many examples for this kind of activation of proteins (e.g., histones) mentioned in the chapters above, the focus will here lie on the description of the signaling mechanism by ubiquitination. An ATP-dependent enzymatic cascade establishes covalent ubiquitin attachment to proteins, mediated by enzymes that activate ubiquitin (E1), conjugate ubiquitin (E2) and ligate ubiquitin (E3). Ubiquitin can be ubiquitinated at seven distinct lysine residues and is able to form linear or branched chains; the degree and the linkage determine, whether the substrate is degraded or associated with cell signaling [265]. The established ubiquitin modifications can be removed and modified by deubiquitinases (DUBs) to change the linkage pattern and, as consequence, the destination of the substrate.

Intracellular bacteria are able to mimic enzymes involved in ubiquitination processes, especially DUBs and E3 ubiquitin ligases, to modulate the ubiquitin pathway [266,267]. The SidE effector family (SidE, SdeA, SdeB and SdeC), secreted by intracellular bacteria as *L. pneumophila*, possesses domains conferring multiple enzymatic functions used for ubiquitination into a single effector without the requirement of ATP. Thereby, the mono-ADP-ribosyltransferase domain of SidE family members modifies host ubiquitin posttranslationally by attachment of phosphoribose on arginine 42 to generate ADP-ribosylated ubiquitin (ADPR-Ub) [268–271]. This intermediate is than cleaved by nucleotidase/phosphohydrolase/phosphodiesterase domains into AMP and phosphoribosylated ubiquitin (PR-Ub), which in the next step is attached to host proteins via a noncanonical serine-linked phosphodiester bond. The DUB domain (also found in SidE family members) than removes host ubiquitin modifications but not the SidE-mediated ubiquitination and reduces the level of ubiquitination on the surface of the *L. pneumophila*-containing vacuole (LCV) [271]. In addition, overexpression of SidE effector family members in mammalian or yeast cells generates a pool of PR-Ub that interfere with E1 and E2 enzymes hampering conventional host ubiquitination pathways. This results in interruption of mitophagy, immunity, as shown for TNF-induced NF-κB nuclear translocation, proteasomal degradation (for example of hypoxia inducing factor 1α) and other ubiquitin-regulated processes in the host [268,269]. Furthermore, ER structure and host membrane trafficking are modulated by SidE effector family members to enable LCV biogenesis. To ensure precise temporal control over the signal transduction mediated by ubiquitination, *L. pneumophila* secretes another DUB effector, SidJ, that is able to remove ubiquitin modifications established by the SidE effector family and the mammalian ubiquitination machinery [268,269,272].

S. flexneri also owns an effector with E3 ubiquitin ligase activity, IpaH9.8, that disrupts the NF-κB dependent pathway in the cytosol and impairs the activity of U2AF35, an mRNA splicing factor, leading to host inflammatory responses being suppressed [273–275]. The effector owns an N-terminal domain containing Leucine-Rich Repeats (LRR), also known as the LPX-domain, that is responsible for substrate recognition and a C-terminal E3 ubiquitin ligase domain, referred to as NEL domain (novel E3 ligase). Notably, the original structure of this domain differs from known eukaryotic E3 ligases, therefore, IpaH9.8 and its orthologues in other bacteria, for example, SspH 1 of *Salmonella enterica*, are part of a novel family of bacterial E3 ubiquitin ligases [276–281]. SspH 1 targets the host kinase PKN1 for proteasomal degradation, thereby functioning as ubiquitin ligase. In this context, it inhibits NF-κB dependent pro-inflammatory genes and regulates activation and function of neutrophils and macrophages as part of the androgen receptor pathway [278].

Enterohaemorrhagic *E. coli* express the T3SS effector protein NleG5-1, which contains a ubiquitin ligase U-box domain for ubiquitination and degradation of nuclear proteins. One target is a member of the mediator complex, MED15 that is a master regulator of RNA Pol II- dependent transcription [282]. The Ank-family expressed by *Orienta tsutsugamushi* (causative reagent of scrub typhus) is another protein family involved in ubiquitination and degradation, characterized by N-terminal Ank repeats and a C-terminal F-box-like domain termed as PRANC (pox protein repeats of ankyrin C terminus) motif. This family includes the proteins 1U5, 1A, 1B, 1E, 1F, 1U4 and 1U9 that interact with two members of multiprotein E3 ubiquitin ligase complexes, CULLIN-1 and SKP1 [283,284]. These proteins are supposed to act as mediators, as the ANK domain shall specifically bind target substrates, while the F-box recruits SKP1 promoting complex formation and finally inducing substrate degradation. This is also suggested for the degradation of the TF EF1α, probably induced via function of Ank 1U5 [283].

Especially the ubiquitin–proteasome system is a favored target of bacterial pathogens to manipulate the host cell cycle (summarized in Table 6). Two prominent targets are Skp1–Cullin1–F-Box protein (SCF, active throughout the cell cycle) and Anaphase-Promoting Complex/Cyclosome (APC/C, only regulatory active during Mitosis and late G1 phase), two classes of E3 ubiquitin ligase complexes inducing the degradation of cell cycle key regulators, thereby promoting its progression. A RING finger protein within both complexes enables binding to a scaffold cullin-like protein, the ubiquitin conjugating enzyme (E2) and distinct substrate-binding subunits. SCF are rated among the large family of Cullin-Ring E3 ubiquitin ligases (CRLs) as they are regulated via conjugation/deconjugation of the ubiquitin-like protein NEDD8 at the cullin subunit of SCF [285–289].

Table 6. Bacteria manipulating the host cell cycle.

Effector	Bacterium	Target	Cellular Function	Manipulation	Reference
VIRF	*Agrobacterium*	SCF	cell-cycle progression by targeting key regulators for rapid degradation	Promoting cell proliferation	[290,291]
cycle inhibiting factors (CIF)	*E. coli* (Cif_{Ec}) *Yersinia pseudotuberculosis* (Cif_{Yp}) *Burkolderia pseudomallei* (Cif_{Bp})	SCF CRL	arrest of the cell cycle at G1 and G2 phases	Inhibition of host cell proliferation	[292]
IPAB	*S. flexneri*	Mad2L2/MAD2B	inhibitor of the APC/Ccdh1	cell cycle arrest	[293]

Diverse animal pathogens, such as *E. coli*, *Yersinia pseudotuberculosis*, *Burkolderia pseudomallei* and *Photorhabdus* spp., express so-called Cycle inhibiting factors (Cif) that target SCFs and CRLs resulting in cell cycle arrest [292]. This effector expressed by *E. coli*, Cif_{Ec}, induces cell cycle arrest at G1/S and G2/M transitions by accumulation of cyclin dependent kinase inhibitors p27/$_{Kip1}$ and p21$_{Waf1/Cip1}$ and inhibition of key kinases. The appearing cytopathic effect is accompanied by cell enlargement and production of actin stress fibers.

Delay of mitotic progression was observed when IpaB, an effector of *S. flexneri*, induced unscheduled APC/C$_{cdh1}$ activation and degradation of its substrates [293]. IpaB has to localize to the nucleus during G2/M phase and has to bind Mad2L2/MAD2B, the mitotic spindle assembly checkpoint protein that inhibits APC/C$_{cdh1}$. The benefit of these mechanisms for the bacteria is not completely clarified yet but it is suggested that delay of cell renewal and cell turnover enables the bacteria to persist and further colonize the host tissues.

3. Conclusions

During the long coevolution of pathogenic bacteria and their host cells, several strategies developed on both sides to fight their counterpart and keep predominance. Thus, intracellular bacteria reach for the establishment of an intracellular niche that allows survival, replication and persistence. This state is achieved through disarming of the host immune defense while keeping it healthy enough to gain permanent nutrition excess. Bacterial effectors are crucial tools during the whole process, thereby targeting the host immune response at each level of gene expression. Even, if there are already many bacterial effectors reported mimicking host enzymes or featuring novel enzymatic functions, the complex interaction mechanisms and networks are still not completely understood. Moreover, it appears, that pathogenic bacteria can target different pathways simultaneously or one pathway with several effectors, thereby creating a species-specific modification pattern. Therefore, understanding the individual strategies gain new insights into the complex host-pathogen interactions during infections. Further investigation might unravel, which bacterial strategies are more efficient or where host cells already found strategies to circumvent attempts of bacterial manipulation. This highlights the importance of further research on bacterial subversion of the host immune response considering each level of gene expression, as new promising targets for successful bacterial clearance during medical therapy might emerge.

Author Contributions: Conceptualization, L.D. and C.S.; writing–original draft preparation, L.D.; writing–review and editing, H.S. and C.S. All authors have read and agreed to the published version of the manuscript.

Funding: This research received no external funding.

Conflicts of Interest: The authors declare no conflict of interest.

Abbreviations

ABU	Asymptomatic bacteriuria
AMP	Adenosine monophosphate
ADP	Adenosine diphosphate
ADPR-Ub	ADP-ribosylated ubiquitin
AnkA	Ankyrin A
APC/C	Anaphase-Promoting Complex/Cyclosome
ATF6	activating transcription factor 6
ATP	Adenosine triphosphate
BCG	*M. bovis* bacillus Calmette-Guérin
BiP	Immunoglobulin binding protein
CHOP	C/EBP homologous protein
Cif	Cycle inhibiting factors
CRLs	Cullin-Ring E3 ubiquitin ligases
CpG	Cytosine-phosphate-Guanosine
DNA	Desoxyribonucleidacid
DNMT	DNA-(cytosine C5)-methyltransferases
DUBs	Deubiquitinases
eELF	Eukaryotic elongation factor
EHEC	Enterohaemorrhagic *E. coli*
eIF	Eukaryotic initiation factor
ERAD	ER associated protein degradation
ERK	Extracellular signal-regulated kinases
ETR	effector-triggered response
GECs	gingival epithelial cells
HAT	Histone acetyl transferase
HDAC	Histone deacetylases
HKMT	Histone N-lysine methyltransferase
HvgA	Hypervirulent GBS adhesin
iagA	invasion associated gene A
Inl	Internalin
IL	Interleukin
IRAK1	IL-1R-associated kinase 1
IRE1	Inositol-requiring protein 1
ISG	Interferon-stimulated genes
JNK	c-JUN N-terminal kinases
lincRNAs	Long intergenic non-coding RNAs
LCV	*L. pneumophila*-containing vacuole
LLO	Listeriolysin O
lncRNA	Long non-coding RNA
LOS	Lipooligosaccharide
LOX 1	Lipoprotein receptor 1
LPS	Lipopolysaccharide
LRR	Leucine-Rich Repeats
MAPK	Mitogen activated protein kinase
MBD1	Methylated DNA-binding protein 1
mRNA	Messenger RNA
miRNA	MicroRNA
mTOR	Mammalian/mechanistic target of rapamycin
mTORC1	mTOR complex 1
NADPH	Nicotinamide adenine dinucleotide phosphate
ncRNA	Non-coding RNA
NF-κB	Nuclear factor κB
NO	Nitric oxide
NLR	NOD like receptor

NOD	Nucleotide-binding oligomerization domain
NUE	Nuclear effector
PAMPs	Pathogen-associated molecular patterns
PERK	PKR like ER kinase
PTM	Posttranslational modifications
PPRs	Pattern recognition receptors
PRANC	Pox protein repeats of ankyrin C terminus
PR-Ub	Phosphoribosylated ubiquitin
Rb	retinoblastoma protein
Rheb	Ras homolog enriched in brain
RNA	Ribonucleidacid
ROS	Reactive oxygen species
RPS3	Ribosomal protein S3
RNA Pol II	RNA Polymerase II
rRNA	Ribosomal RNA
RYBP	Ring1 YY1-binding protein
SAM	S-adenosyl-l-methionine
SCAFs	Short chain fatty acids
SCF	Skp1–Cullin1–F-Box protein
SET	Suppressor of variegation, Enhancer of zeste and Trithorax
SHIP1	SH-2 containing inositol 5′ polyphosphatase
SIRT 2	Sirtuin 2
SNPs	Single-nucleotide polymorphisms
snRNP	Small nuclear ribonucleoprotein
SR proteins	serine-arginine containing proteins
SOCS1	Suppressor of cytokine signaling 1
SSR	Serine-rich repeat
T3SS	Type three secretion system
TBP	TATA box–binding proteins
TF	Transcription factor
TLR	Toll like receptors
TNF	Tumor necrosis factor
TRAF6	TNF Receptor-associated factor 6
tRNA	Transfer RNA
Ub	Ubiquitin
uORFs	untranslated ORFs
UPR	unfolded protein response
XBP1	X-box-binding protein 1
4E-BPs	4E-binding proteins

Pathogens:

A. phagocytophilum	*Anaplasma phagocytophilum*
B. abortus	*Brucella abortus*
B. anthracis	*Bacillus anthracis*
B. melitensis	*Brucella melitensis*
B. thailandensis	*Burkholderia thailandensis*
C. perfringens	*Clostridium perfringens*
C. trachomatis	*Chlamydia trachomatis*
E. coli	*Escherichia coli*
H. pylori	*Helicobacter pylori*
L. monocytogenes	*Listeria monocytogenes*
L. pneumophila	*Legionella pneumophila*
M. leprae	*Mycobacterium leprae*
M. tuberculosis	*Mycobacterium tuberculosis*
P. aeruginosa	*Pseudomonas aeruginosa*
P. gingivialis	*Porphyromonas gingivalis*
S. pneumoniae	*Streptococcus pneumoniae*

S. flexneri	*Shigella flexneri*
Salmonella Typhimurium	*Salmonella enterica* subsp. *enterica* serotype Typhimurium

References

1. Akira, S.; Uematsu, S.; Takeuchi, O. Pathogen recognition and innate immunity. *Cell* **2006**, *124*, 783–801. [CrossRef] [PubMed]
2. Mogensen, T.H. Pathogen recognition and inflammatory signaling in innate immune defenses. *Clin. Microbiol. Rev.* **2009**, *22*, 240–273. [CrossRef] [PubMed]
3. Arthur, J.S.; Ley, S.C. Mitogen-activated protein kinases in innate immunity. *Nat. Rev. Immunol.* **2013**, *13*, 679–692. [CrossRef] [PubMed]
4. Dev, A.; Iyer, S.; Razani, B.; Cheng, G. NF-kappaB and innate immunity. *Curr. Top. Microbiol. Immunol.* **2011**, *349*, 115–143. [CrossRef] [PubMed]
5. Kouzarides, T. Chromatin modifications and their function. *Cell* **2007**, *128*, 693–705. [CrossRef] [PubMed]
6. Hargreaves, D.C.; Crabtree, G.R. ATP-dependent chromatin remodeling: Genetics, genomics and mechanisms. *Cell Res.* **2011**, *21*, 396–420. [CrossRef]
7. Luger, K.; Mader, A.W.; Richmond, R.K.; Sargent, D.F.; Richmond, T.J. Crystal structure of the nucleosome core particle at 2.8 A resolution. *Nature* **1997**, *389*, 251–260. [CrossRef]
8. Liu, W.; Ding, C. Roles of LncRNAs in Viral Infections. *Front. Cell. Infect. Microbiol.* **2017**, *7*, 205. [CrossRef]
9. Shatkin, A.J.; Manley, J.L. The ends of the affair: Capping and polyadenylation. *Nat. Struct. Biol.* **2000**, *7*, 838–842. [CrossRef]
10. Proudfoot, N.J.; Furger, A.; Dye, M.J. Integrating mRNA processing with transcription. *Cell* **2002**, *108*, 501–512. [CrossRef]
11. Moon, S.L.; Barnhart, M.D.; Wilusz, J. Inhibition and avoidance of mRNA degradation by RNA viruses. *Curr. Opin. Microbiol.* **2012**, *15*, 500–505. [CrossRef] [PubMed]
12. Moon, S.L.; Wilusz, J. Cytoplasmic viruses: Rage against the (cellular RNA decay) machine. *PLoS Pathog.* **2013**, *9*, e1003762. [CrossRef] [PubMed]
13. Decroly, E.; Ferron, F.; Lescar, J.; Canard, B. Conventional and unconventional mechanisms for capping viral mRNA. *Nat. Rev. Microbiol.* **2011**, *10*, 51–65. [CrossRef] [PubMed]
14. Kozak, M. Initiation of translation in prokaryotes and eukaryotes. *Gene* **1999**, *234*, 187–208. [CrossRef]
15. Dobson, C.M. Protein folding and misfolding. *Nature* **2003**, *426*, 884–890. [CrossRef]
16. Munoz, V.; Campos, L.A.; Sadqi, M. Limited cooperativity in protein folding. *Curr. Opin. Struct. Biol.* **2016**, *36*, 58–66. [CrossRef]
17. Tompa, P.; Prilusky, J.; Silman, I.; Sussman, J.L. Structural disorder serves as a weak signal for intracellular protein degradation. *Proteins* **2008**, *71*, 903–909. [CrossRef]
18. Sakamoto, K.M. Ubiquitin-dependent proteolysis: Its role in human diseases and the design of therapeutic strategies. *Mol. Genet. Metab.* **2002**, *77*, 44–56. [CrossRef]
19. Cornejo, E.; Schlaermann, P.; Mukherjee, S. How to rewire the host cell: A home improvement guide for intracellular bacteria. *J. Cell Biol.* **2017**, *216*, 3931–3948. [CrossRef]
20. Bierne, H.; Cossart, P. When bacteria target the nucleus: The emerging family of nucleomodulins. *Cell. Microbiol.* **2012**, *14*, 622–633. [CrossRef]
21. Bierne, H.; Pourpre, R. Bacterial Factors Targeting the Nucleus: The Growing Family of Nucleomodulins. *Toxins* **2020**, *12*, 220. [CrossRef] [PubMed]
22. Van Wolfswinkel, J.C.; Ketting, R.F. The role of small non-coding RNAs in genome stability and chromatin organization. *J. Cell Sci.* **2010**, *123*, 1825–1839. [CrossRef] [PubMed]
23. Lee, B.M.; Mahadevan, L.C. Stability of histone modifications across mammalian genomes: Implications for 'epigenetic' marking. *J. Cell. Biochem.* **2009**, *108*, 22–34. [CrossRef] [PubMed]
24. Schreiber, S.L.; Bernstein, B.E. Signaling network model of chromatin. *Cell* **2002**, *111*, 771–778. [CrossRef]
25. Smith, E.; Shilatifard, A. The chromatin signaling pathway: Diverse mechanisms of recruitment of histone-modifying enzymes and varied biological outcomes. *Mol. Cell* **2010**, *40*, 689–701. [CrossRef]
26. Cech, T.R.; Steitz, J.A. The noncoding RNA revolution-trashing old rules to forge new ones. *Cell* **2014**, *157*, 77–94. [CrossRef]
27. Mikeska, T.; Craig, J.M. DNA methylation biomarkers: Cancer and beyond. *Genes* **2014**, *5*, 821–864. [CrossRef]

28. Grabiec, A.M.; Tak, P.P.; Reedquist, K.A. Targeting histone deacetylase activity in rheumatoid arthritis and asthma as prototypes of inflammatory disease: Should we keep our HATs on? *Arthritis Res. Ther.* **2008**, *10*, 226. [CrossRef]

29. Zhang, Z.; Zhang, R. Epigenetics in autoimmune diseases: Pathogenesis and prospects for therapy. *Autoimmun. Rev.* **2015**, *14*, 854–863. [CrossRef]

30. Vaziri, F.; Tarashi, S.; Fateh, A.; Siadat, S.D. New insights of *Helicobacter pylori* host-pathogen interactions: The triangle of virulence factors, epigenetic modifications and non-coding RNAs. *World J. Clin. Cases* **2018**, *6*, 64–73. [CrossRef]

31. Jones, P.A.; Takai, D. The role of DNA methylation in mammalian epigenetics. *Science* **2001**, *293*, 1068–1070. [CrossRef] [PubMed]

32. Yoda, Y.; Takeshima, H.; Niwa, T.; Kim, J.G.; Ando, T.; Kushima, R.; Sugiyama, T.; Katai, H.; Noshiro, H.; Ushijima, T. Integrated analysis of cancer-related pathways affected by genetic and epigenetic alterations in gastric cancer. *Gastric Cancer* **2015**, *18*, 65–76. [CrossRef] [PubMed]

33. Duval, M.; Cossart, P.; Lebreton, A. Mammalian microRNAs and long noncoding RNAs in the host-bacterial pathogen crosstalk. *Semin. Cell Dev. Biol.* **2017**, *65*, 11–19. [CrossRef] [PubMed]

34. Saccani, S.; Pantano, S.; Natoli, G. p38-Dependent marking of inflammatory genes for increased NF-kappa B recruitment. *Nat. Immunol.* **2002**, *3*, 69–75. [CrossRef]

35. Hamon, M.A.; Cossart, P. Histone modifications and chromatin remodeling during bacterial infections. *Cell Host Microbe* **2008**, *4*, 100–109. [CrossRef]

36. Hamon, M.A.; Batsche, E.; Regnault, B.; Tham, T.N.; Seveau, S.; Muchardt, C.; Cossart, P. Histone modifications induced by a family of bacterial toxins. *Proc. Natl. Acad. Sci. USA* **2007**, *104*, 13467–13472. [CrossRef]

37. Schmeck, B.; Beermann, W.; van Laak, V.; Zahlten, J.; Opitz, B.; Witzenrath, M.; Hocke, A.C.; Chakraborty, T.; Kracht, M.; Rosseau, S.; et al. Intracellular bacteria differentially regulated endothelial cytokine release by MAPK-dependent histone modification. *J. Immunol.* **2005**, *175*, 2843–2850. [CrossRef]

38. Opitz, B.; Puschel, A.; Beermann, W.; Hocke, A.C.; Forster, S.; Schmeck, B.; van Laak, V.; Chakraborty, T.; Suttorp, N.; Hippenstiel, S. *Listeria monocytogenes* activated p38 MAPK and induced IL-8 secretion in a nucleotide-binding oligomerization domain 1-dependent manner in endothelial cells. *J. Immunol.* **2006**, *176*, 484–490. [CrossRef]

39. Arbibe, L.; Kim, D.W.; Batsche, E.; Pedron, T.; Mateescu, B.; Muchardt, C.; Parsot, C.; Sansonetti, P.J. An injected bacterial effector targets chromatin access for transcription factor NF-kappaB to alter transcription of host genes involved in immune responses. *Nat. Immunol.* **2007**, *8*, 47–56. [CrossRef]

40. Brennan, D.F.; Barford, D. Eliminylation: A post-translational modification catalyzed by phosphothreonine lyases. *Trends Biochem. Sci.* **2009**, *34*, 108–114. [CrossRef] [PubMed]

41. Zurawski, D.V.; Mumy, K.L.; Faherty, C.S.; McCormick, B.A.; Maurelli, A.T. *Shigella flexneri* type III secretion system effectors OspB and OspF target the nucleus to downregulate the host inflammatory response via interactions with retinoblastoma protein. *Mol. Microbiol.* **2009**, *71*, 350–368. [CrossRef] [PubMed]

42. Eskandarian, H.A.; Impens, F.; Nahori, M.A.; Soubigou, G.; Coppee, J.Y.; Cossart, P.; Hamon, M.A. A role for SIRT2-dependent histone H3K18 deacetylation in bacterial infection. *Science* **2013**, *341*, 1238858. [CrossRef] [PubMed]

43. Bierne, H.; Tham, T.N.; Batsche, E.; Dumay, A.; Leguillou, M.; Kerneis-Golsteyn, S.; Regnault, B.; Seeler, J.S.; Muchardt, C.; Feunteun, J.; et al. Human BAHD1 promotes heterochromatic gene silencing. *Proc. Natl. Acad. Sci. USA* **2009**, *106*, 13826–13831. [CrossRef] [PubMed]

44. Lebreton, A.; Lakisic, G.; Job, V.; Fritsch, L.; Tham, T.N.; Camejo, A.; Mattei, P.J.; Regnault, B.; Nahori, M.A.; Cabanes, D.; et al. A bacterial protein targets the BAHD1 chromatin complex to stimulate type III interferon response. *Science* **2011**, *331*, 1319–1321. [CrossRef] [PubMed]

45. Li, T.; Lu, Q.; Wang, G.; Xu, H.; Huang, H.; Cai, T.; Kan, B.; Ge, J.; Shao, F. SET-domain bacterial effectors target heterochromatin protein 1 to activate host rDNA transcription. *EMBO Rep.* **2013**, *14*, 733–740. [CrossRef]

46. Rolando, M.; Sanulli, S.; Rusniok, C.; Gomez-Valero, L.; Bertholet, C.; Sahr, T.; Margueron, R.; Buchrieser, C. *Legionella pneumophila* effector RomA uniquely modifies host chromatin to repress gene expression and promote intracellular bacterial replication. *Cell Host Microbe* **2013**, *13*, 395–405. [CrossRef]

47. Mujtaba, S.; Winer, B.Y.; Jaganathan, A.; Patel, J.; Sgobba, M.; Schuch, R.; Gupta, Y.K.; Haider, S.; Wang, R.; Fischetti, V.A. Anthrax SET protein: A potential virulence determinant that epigenetically represses NF-kappaB activation in infected macrophages. *J. Biol. Chem.* **2013**, *288*, 23458–23472. [CrossRef]

48. Yaseen, I.; Kaur, P.; Nandicoori, V.K.; Khosla, S. Mycobacteria modulate host epigenetic machinery by Rv1988 methylation of a non-tail arginine of histone H3. *Nat. Commun.* **2015**, *6*, 1–13. [CrossRef]

49. Li, H.; Xu, H.; Zhou, Y.; Zhang, J.; Long, C.; Li, S.; Chen, S.; Zhou, J.M.; Shao, F. The phosphothreonine lyase activity of a bacterial type III effector family. *Science* **2007**, *315*, 1000–1003. [CrossRef]

50. Zhu, Y.; Li, H.; Long, C.; Hu, L.; Xu, H.; Liu, L.; Chen, S.; Wang, D.C.; Shao, F. Structural insights into the enzymatic mechanism of the pathogenic MAPK phosphothreonine lyase. *Mol. Cell* **2007**, *28*, 899–913. [CrossRef]

51. Jose, L.; Ramachandran, R.; Bhagavat, R.; Gomez, R.L.; Chandran, A.; Raghunandanan, S.; Omkumar, R.V.; Chandra, N.; Mundayoor, S.; Kumar, R.A. Hypothetical protein Rv3423.1 of *Mycobacterium tuberculosis* is a histone acetyltransferase. *FEBS J.* **2016**, *283*, 265–281. [CrossRef] [PubMed]

52. Ding, S.Z.; Fischer, W.; Kaparakis-Liaskos, M.; Liechti, G.; Merrell, D.S.; Grant, P.A.; Ferrero, R.L.; Crowe, S.E.; Haas, R.; Hatakeyama, M.; et al. *Helicobacter pylori*-induced histone modification, associated gene expression in gastric epithelial cells and its implication in pathogenesis. *PLoS ONE* **2010**, *5*, e9875. [CrossRef] [PubMed]

53. Pennini, M.E.; Perrinet, S.; Dautry-Varsat, A.; Subtil, A. Histone methylation by NUE, a novel nuclear effector of the intracellular pathogen *Chlamydia trachomatis*. *PLoS Pathog.* **2010**, *6*, e1000995. [CrossRef] [PubMed]

54. Chandran, A.; Antony, C.; Jose, L.; Mundayoor, S.; Natarajan, K.; Kumar, R.A. *Mycobacterium tuberculosis* Infection Induces HDAC1-Mediated Suppression of IL-12B Gene Expression in Macrophages. *Front. Cell. Infect. Microbiol.* **2015**, *5*, 90. [CrossRef] [PubMed]

55. Wang, Y.; Curry, H.M.; Zwilling, B.S.; Lafuse, W.P. Mycobacteria inhibition of IFN-gamma induced HLA-DR gene expression by up-regulating histone deacetylation at the promoter region in human THP-1 monocytic cells. *J. Immunol.* **2005**, *174*, 5687–5694. [CrossRef]

56. Garcia-Garcia, J.C.; Barat, N.C.; Trembley, S.J.; Dumler, J.S. Epigenetic silencing of host cell defense genes enhances intracellular survival of the rickettsial pathogen *Anaplasma phagocytophilum*. *PLoS Pathog.* **2009**, *5*, e1000488. [CrossRef]

57. Garcia-Garcia, J.C.; Rennoll-Bankert, K.E.; Pelly, S.; Milstone, A.M.; Dumler, J.S. Silencing of host cell CYBB gene expression by the nuclear effector AnkA of the intracellular pathogen *Anaplasma phagocytophilum*. *Infect. Immun.* **2009**, *77*, 2385–2391. [CrossRef]

58. Rennoll-Bankert, K.E.; Dumler, J.S. Lessons from *Anaplasma phagocytophilum*: Chromatin remodeling by bacterial effectors. *Infect. Disord. Drug Targets* **2012**, *12*, 380–387. [CrossRef]

59. Rennoll-Bankert, K.E.; Garcia-Garcia, J.C.; Sinclair, S.H.; Dumler, J.S. Chromatin-bound bacterial effector ankyrin A recruits histone deacetylase 1 and modifies host gene expression. *Cell. Microbiol.* **2015**, *17*, 1640–1652. [CrossRef]

60. Park, J.; Kim, K.J.; Choi, K.S.; Grab, D.J.; Dumler, J.S. *Anaplasma phagocytophilum* AnkA binds to granulocyte DNA and nuclear proteins. *Cell. Microbiol.* **2004**, *6*, 743–751. [CrossRef]

61. Dumler, J.S.; Sinclair, S.H.; Pappas-Brown, V.; Shetty, A.C. Genome-Wide *Anaplasma phagocytophilum* AnkA-DNA Interactions Are Enriched in Intergenic Regions and Gene Promoters and Correlate with Infection-Induced Differential Gene Expression. *Front. Cell. Infect. Microbiol.* **2016**, *6*, 97. [CrossRef] [PubMed]

62. Kohwi-Shigematsu, T.; Kohwi, Y.; Takahashi, K.; Richards, H.W.; Ayers, S.D.; Han, H.J.; Cai, S. SATB1-mediated functional packaging of chromatin into loops. *Methods* **2012**, *58*, 243–254. [CrossRef] [PubMed]

63. Wang, T.Y.; Han, Z.M.; Chai, Y.R.; Zhang, J.H. A mini review of MAR-binding proteins. *Mol. Biol. Rep.* **2010**, *37*, 3553–3560. [CrossRef] [PubMed]

64. Mojica, S.A.; Hovis, K.M.; Frieman, M.B.; Tran, B.; Hsia, R.C.; Ravel, J.; Jenkins-Houk, C.; Wilson, K.L.; Bavoil, P.M. SINC, a type III secreted protein of *Chlamydia psittaci*, targets the inner nuclear membrane of infected cells and uninfected neighbors. *Mol. Biol. Cell* **2015**, *26*, 1918–1934. [CrossRef]

65. Bandyopadhaya, A.; Tsurumi, A.; Maura, D.; Jeffrey, K.L.; Rahme, L.G. A quorum-sensing signal promotes host tolerance training through HDAC1-mediated epigenetic reprogramming. *Nat. Microbiol.* **2016**, *1*, 1–9. [CrossRef]

66. Gajdacs, M. Carbapenem-Resistant But Cephalosporin-Susceptible *Pseudomonas aeruginosa* in Urinary Tract Infections: Opportunity for Colistin Sparing. *Antibiotics* **2020**, *9*, 153. [CrossRef]

67. Cantley, M.D.; Dharmapatni, A.A.; Algate, K.; Crotti, T.N.; Bartold, P.M.; Haynes, D.R. Class I and II histone deacetylase expression in human chronic periodontitis gingival tissue. *J. Periodontal Res.* **2016**, *51*, 143–151. [CrossRef]

68. Yin, L.; Chung, W.O. Epigenetic regulation of human beta-defensin 2 and CC chemokine ligand 20 expression in gingival epithelial cells in response to oral bacteria. *Mucosal Immunol.* **2011**, *4*, 409–419. [CrossRef]

69. Martins, M.D.; Jiao, Y.; Larsson, L.; Almeida, L.O.; Garaicoa-Pazmino, C.; Le, J.M.; Squarize, C.H.; Inohara, N.; Giannobile, W.V.; Castilho, R.M. Epigenetic Modifications of Histones in Periodontal Disease. *J. Dent. Res.* **2016**, *95*, 215–222. [CrossRef]

70. Grabiec, A.M.; Potempa, J. Epigenetic regulation in bacterial infections: Targeting histone deacetylases. *Crit. Rev. Microbiol.* **2018**, *44*, 336–350. [CrossRef]

71. Niller, H.H.; Masa, R.; Venkei, A.; Meszaros, S.; Minarovits, J. Pathogenic mechanisms of intracellular bacteria. *Curr. Opin. Infect. Dis.* **2017**, *30*, 309–315. [CrossRef] [PubMed]

72. Imai, K.; Inoue, H.; Tamura, M.; Cueno, M.E.; Inoue, H.; Takeichi, O.; Kusama, K.; Saito, I.; Ochiai, K. The periodontal pathogen *Porphyromonas gingivalis* induces the Epstein-Barr virus lytic switch transactivator ZEBRA by histone modification. *Biochimie* **2012**, *94*, 839–846. [CrossRef] [PubMed]

73. Aruni, A.W.; Mishra, A.; Dou, Y.; Chioma, O.; Hamilton, B.N.; Fletcher, H.M. Filifactor alocis–a new emerging periodontal pathogen. *Microbes Infect.* **2015**, *17*, 517–530. [CrossRef] [PubMed]

74. Aruni, A.W.; Zhang, K.; Dou, Y.; Fletcher, H. Proteome analysis of coinfection of epithelial cells with *Filifactor alocis* and *Porphyromonas gingivalis* shows modulation of pathogen and host regulatory pathways. *Infect. Immun.* **2014**, *82*, 3261–3274. [CrossRef] [PubMed]

75. Barros, S.P.; Offenbacher, S. Modifiable risk factors in periodontal disease: Epigenetic regulation of gene expression in the inflammatory response. *Periodontology 2000* **2014**, *64*, 95–110. [CrossRef]

76. Shames, S.R.; Bhavsar, A.P.; Croxen, M.A.; Law, R.J.; Mak, S.H.; Deng, W.; Li, Y.; Bidshari, R.; de Hoog, C.L.; Foster, L.J.; et al. The pathogenic *Escherichia coli* type III secreted protease NleC degrades the host acetyltransferase p300. *Cell. Microbiol.* **2011**, *13*, 1542–1557. [CrossRef] [PubMed]

77. Wu, S.C.; Zhang, Y. Active DNA demethylation: Many roads lead to Rome. *Nat. Rev. Mol. Cell Biol.* **2010**, *11*, 607–620. [CrossRef]

78. Nardone, G.; Compare, D.; De Colibus, P.; de Nucci, G.; Rocco, A. *Helicobacter pylori* and epigenetic mechanisms underlying gastric carcinogenesis. *Dig. Dis.* **2007**, *25*, 225–229. [CrossRef]

79. Santos, J.C.; Ribeiro, M.L. Epigenetic regulation of DNA repair machinery in *Helicobacter pylori*-induced gastric carcinogenesis. *World J. Gastroenterol.* **2015**, *21*, 9021–9037. [CrossRef]

80. Sitaraman, R. *Helicobacter pylori* DNA methyltransferases and the epigenetic field effect in cancerization. *Front. Microbiol* **2014**, *5*, 115. [CrossRef]

81. Bobetsis, Y.A.; Barros, S.P.; Lin, D.M.; Weidman, J.R.; Dolinoy, D.C.; Jirtle, R.L.; Boggess, K.A.; Beck, J.D.; Offenbacher, S. Bacterial infection promotes DNA hypermethylation. *J. Dent. Res.* **2007**, *86*, 169–174. [CrossRef] [PubMed]

82. Masaki, T.; Qu, J.; Cholewa-Waclaw, J.; Burr, K.; Raaum, R.; Rambukkana, A. Reprogramming adult Schwann cells to stem cell-like cells by leprosy bacilli promotes dissemination of infection. *Cell* **2013**, *152*, 51–67. [CrossRef] [PubMed]

83. Tolg, C.; Sabha, N.; Cortese, R.; Panchal, T.; Ahsan, A.; Soliman, A.; Aitken, K.J.; Petronis, A.; Bagli, D.J. Uropathogenic *E. coli* infection provokes epigenetic downregulation of CDKN2A (p16INK4A) in uroepithelial cells. *Lab. Investig.* **2011**, *91*, 825–836. [CrossRef] [PubMed]

84. Minarovits, J. Microbe-induced epigenetic alterations in host cells: The coming era of patho-epigenetics of microbial infections. A review. *Acta Microbiol. Immunol. Hung.* **2009**, *56*, 1–19. [CrossRef] [PubMed]

85. Chernov, A.V.; Reyes, L.; Peterson, S.; Strongin, A.Y. Depletion of CG-Specific Methylation in *Mycoplasma hyorhinis* Genomic DNA after Host Cell Invasion. *PLoS ONE* **2015**, *10*, e0142529. [CrossRef] [PubMed]

86. Luo, W.; Tu, A.H.; Cao, Z.; Yu, H.; Dybvig, K. Identification of an isoschizomer of the HhaI DNA methyltransferase in *Mycoplasma arthritidis*. *FEMS Microbiol. Lett.* **2009**, *290*, 195–198. [CrossRef]

87. Wojciechowski, M.; Czapinska, H.; Bochtler, M. CpG underrepresentation and the bacterial CpG-specific DNA methyltransferase M.MpeI. *Proc. Natl. Acad. Sci. USA* **2013**, *110*, 105–110. [CrossRef]

88. Chernov, A.V.; Reyes, L.; Xu, Z.; Gonzalez, B.; Golovko, G.; Peterson, S.; Perucho, M.; Fofanov, Y.; Strongin, A.Y. Mycoplasma CG- and GATC-specific DNA methyltransferases selectively and efficiently methylate the host genome and alter the epigenetic landscape in human cells. *Epigenetics* **2015**, *10*, 303–318. [CrossRef]

89. Pinney, S.E. Mammalian Non-CpG Methylation: Stem Cells and Beyond. *Biology* **2014**, *3*, 739–751. [CrossRef]
90. Sharma, G.; Upadhyay, S.; Srilalitha, M.; Nandicoori, V.K.; Khosla, S. The interaction of mycobacterial protein Rv2966c with host chromatin is mediated through non-CpG methylation and histone H3/H4 binding. *Nucleic Acids Res.* **2015**, *43*, 3922–3937. [CrossRef]
91. Sharma, G.; Sowpati, D.T.; Singh, P.; Khan, M.Z.; Ganji, R.; Upadhyay, S.; Banerjee, S.; Nandicoori, V.K.; Khosla, S. Genome-wide non-CpG methylation of the host genome during *M. tuberculosis* infection. *Sci. Rep.* **2016**, *6*, 25006. [CrossRef] [PubMed]
92. Hess, S.; Rambukkana, A. Bacterial-induced cell reprogramming to stem cell-like cells: New premise in host-pathogen interactions. *Curr. Opin. Microbiol.* **2015**, *23*, 179–188. [CrossRef] [PubMed]
93. Benakanakere, M.; Abdolhosseini, M.; Hosur, K.; Finoti, L.S.; Kinane, D.F. TLR2 promoter hypermethylation creates innate immune dysbiosis. *J. Dent. Res.* **2015**, *94*, 183–191. [CrossRef] [PubMed]
94. Djebali, S.; Davis, C.A.; Merkel, A.; Dobin, A.; Lassmann, T.; Mortazavi, A.; Tanzer, A.; Lagarde, J.; Lin, W.; Schlesinger, F.; et al. Landscape of transcription in human cells. *Nature* **2012**, *489*, 101–108. [CrossRef]
95. Doolittle, W.F. Is junk DNA bunk? A critique of ENCODE. *Proc. Natl. Acad. Sci. USA* **2013**, *110*, 5294–5300. [CrossRef]
96. Ambite, I.; Butler, D.S.C.; Stork, C.; Gronberg-Hernandez, J.; Koves, B.; Zdziarski, J.; Pinkner, J.; Hultgren, S.J.; Dobrindt, U.; Wullt, B.; et al. Fimbriae reprogram host gene expression—Divergent effects of P and type 1 fimbriae. *PLoS Pathog.* **2019**, *15*, e1007671. [CrossRef]
97. Fatica, A.; Bozzoni, I. Long non-coding RNAs: New players in cell differentiation and development. *Nat. Rev. Genet.* **2014**, *15*, 7–21. [CrossRef]
98. Hu, W.; Alvarez-Dominguez, J.R.; Lodish, H.F. Regulation of mammalian cell differentiation by long non-coding RNAs. *EMBO Rep.* **2012**, *13*, 971–983. [CrossRef]
99. Wapinski, O.; Chang, H.Y. Long noncoding RNAs and human disease. *Trends Cell Biol.* **2011**, *21*, 354–361. [CrossRef]
100. Latos, P.A.; Pauler, F.M.; Koerner, M.V.; Senergin, H.B.; Hudson, Q.J.; Stocsits, R.R.; Allhoff, W.; Stricker, S.H.; Klement, R.M.; Warczok, K.E.; et al. Airn transcriptional overlap but not its lncRNA products, induces imprinted Igf2r silencing. *Science* **2012**, *338*, 1469–1472. [CrossRef]
101. Liu, Q.; Huang, J.; Zhou, N.; Zhang, Z.; Zhang, A.; Lu, Z.; Wu, F.; Mo, Y.Y. LncRNA loc285194 is a p53-regulated tumor suppressor. *Nucleic Acids Res.* **2013**, *41*, 4976–4987. [CrossRef] [PubMed]
102. Sharma, V.; Khurana, S.; Kubben, N.; Abdelmohsen, K.; Oberdoerffer, P.; Gorospe, M.; Misteli, T. A BRCA1-interacting lncRNA regulates homologous recombination. *EMBO Rep.* **2015**, *16*, 1520–1534. [CrossRef] [PubMed]
103. Yang, Y.W.; Flynn, R.A.; Chen, Y.; Qu, K.; Wan, B.; Wang, K.C.; Lei, M.; Chang, H.Y. Essential role of lncRNA binding for WDR5 maintenance of active chromatin and embryonic stem cell pluripotency. *Elife* **2014**, *3*, e02046. [CrossRef] [PubMed]
104. Bonasio, R.; Shiekhattar, R. Regulation of transcription by long noncoding RNAs. *Annu. Rev. Genet.* **2014**, *48*, 433–455. [CrossRef]
105. Wang, K.C.; Chang, H.Y. Molecular mechanisms of long noncoding RNAs. *Mol. Cell* **2011**, *43*, 904–914. [CrossRef]
106. Rapicavoli, N.A.; Qu, K.; Zhang, J.; Mikhail, M.; Laberge, R.M.; Chang, H.Y. A mammalian pseudogene lncRNA at the interface of inflammation and anti-inflammatory therapeutics. *Elife* **2013**, *2*, e00762. [CrossRef]
107. Ding, Y.Z.; Zhang, Z.W.; Liu, Y.L.; Shi, C.X.; Zhang, J.; Zhang, Y.G. Relationship of long noncoding RNA and viruses. *Genomics* **2016**, *107*, 150–154. [CrossRef]
108. Ilott, N.E.; Heward, J.A.; Roux, B.; Tsitsiou, E.; Fenwick, P.S.; Lenzi, L.; Goodhead, I.; Hertz-Fowler, C.; Heger, A.; Hall, N.; et al. Long non-coding RNAs and enhancer RNAs regulate the lipopolysaccharide-induced inflammatory response in human monocytes. *Nat. Commun.* **2014**, *5*, 1–14. [CrossRef]
109. Mao, A.P.; Shen, J.; Zuo, Z. Expression and regulation of long noncoding RNAs in TLR4 signaling in mouse macrophages. *BMC Genomics* **2015**, *16*, 45. [CrossRef]
110. Westermann, A.J.; Forstner, K.U.; Amman, F.; Barquist, L.; Chao, Y.; Schulte, L.N.; Muller, L.; Reinhardt, R.; Stadler, P.F.; Vogel, J. Dual RNA-seq unveils noncoding RNA functions in host-pathogen interactions. *Nature* **2016**, *529*, 496–501. [CrossRef]

111. Rinn, J.L.; Kertesz, M.; Wang, J.K.; Squazzo, S.L.; Xu, X.; Brugmann, S.A.; Goodnough, L.H.; Helms, J.A.; Farnham, P.J.; Segal, E.; et al. Functional demarcation of active and silent chromatin domains in human HOX loci by noncoding RNAs. *Cell* **2007**, *129*, 1311–1323. [CrossRef] [PubMed]

112. Wu, H.; Liu, J.; Li, W.; Liu, G.; Li, Z. LncRNA-HOTAIR promotes TNF-alpha production in cardiomyocytes of LPS-induced sepsis mice by activating NF-kappaB pathway. *Biochem. Biophys. Res. Commun.* **2016**, *471*, 240–246. [CrossRef] [PubMed]

113. Carpenter, S.; Aiello, D.; Atianand, M.K.; Ricci, E.P.; Gandhi, P.; Hall, L.L.; Byron, M.; Monks, B.; Henry-Bezy, M.; Lawrence, J.B.; et al. A long noncoding RNA mediates both activation and repression of immune response genes. *Science* **2013**, *341*, 789–792. [CrossRef] [PubMed]

114. Guttman, M.; Amit, I.; Garber, M.; French, C.; Lin, M.F.; Feldser, D.; Huarte, M.; Zuk, O.; Carey, B.W.; Cassady, J.P.; et al. Chromatin signature reveals over a thousand highly conserved large non-coding RNAs in mammals. *Nature* **2009**, *458*, 223–227. [CrossRef]

115. Pawar, K.; Hanisch, C.; Palma Vera, S.E.; Einspanier, R.; Sharbati, S. Down regulated lncRNA MEG3 eliminates mycobacteria in macrophages via autophagy. *Sci. Rep.* **2016**, *6*, 1–13. [CrossRef]

116. Fuda, N.J.; Ardehali, M.B.; Lis, J.T. Defining mechanisms that regulate RNA polymerase II transcription in vivo. *Nature* **2009**, *461*, 186–192. [CrossRef]

117. Conaway, R.C.; Conaway, J.W. Function and regulation of the Mediator complex. *Curr. Opin. Genet. Dev.* **2011**, *21*, 225–230. [CrossRef]

118. Selth, L.A.; Sigurdsson, S.; Svejstrup, J.Q. Transcript Elongation by RNA Polymerase II. *Annu. Rev. Biochem.* **2010**, *79*, 271–293. [CrossRef]

119. Zhan, S.; Wang, T.; Ge, W.; Li, J. Multiple roles of Ring 1 and YY1 binding protein in physiology and disease. *J. Cell. Mol. Med.* **2018**, *22*, 2046–2054. [CrossRef]

120. Prokop, A.; Gouin, E.; Villiers, V.; Nahori, M.A.; Vincentelli, R.; Duval, M.; Cossart, P.; Dussurget, O. OrfX, a Nucleomodulin Required for *Listeria monocytogenes* Virulence. *mBio* **2017**, *8*. [CrossRef]

121. Garcia, E.; Marcos-Gutierrez, C.; del Mar Lorente, M.; Moreno, J.C.; Vidal, M. RYBP, a new repressor protein that interacts with components of the mammalian Polycomb complex and with the transcription factor YY1. *EMBO J.* **1999**, *18*, 3404–3418. [CrossRef] [PubMed]

122. Gearhart, M.D.; Corcoran, C.M.; Wamstad, J.A.; Bardwell, V.J. Polycomb group and SCF ubiquitin ligases are found in a novel BCOR complex that is recruited to BCL6 targets. *Mol. Cell. Biol.* **2006**, *26*, 6880–6889. [CrossRef] [PubMed]

123. Schlisio, S.; Halperin, T.; Vidal, M.; Nevins, J.R. Interaction of YY1 with E2Fs, mediated by RYBP, provides a mechanism for specificity of E2F function. *EMBO J.* **2002**, *21*, 5775–5786. [CrossRef]

124. Chen, D.; Zhang, J.; Li, M.; Rayburn, E.R.; Wang, H.; Zhang, R. RYBP stabilizes p53 by modulating MDM2. *EMBO Rep.* **2009**, *10*, 166–172. [CrossRef] [PubMed]

125. Sun, H.; Kamanova, J.; Lara-Tejero, M.; Galan, J.E. A Family of *Salmonella* Type III Secretion Effector Proteins Selectively Targets the NF-kappaB Signaling Pathway to Preserve Host Homeostasis. *PLoS Pathog.* **2016**, *12*, e1005484. [CrossRef] [PubMed]

126. Baruch, K.; Gur-Arie, L.; Nadler, C.; Koby, S.; Yerushalmi, G.; Ben-Neriah, Y.; Yogev, O.; Shaulian, E.; Guttman, C.; Zarivach, R.; et al. Metalloprotease type III effectors that specifically cleave JNK and NF-kappaB. *EMBO J.* **2011**, *30*, 221–231. [CrossRef]

127. Muhlen, S.; Ruchaud-Sparagano, M.H.; Kenny, B. Proteasome-independent degradation of canonical NFkappaB complex components by the NleC protein of pathogenic *Escherichia coli*. *J. Biol. Chem.* **2011**, *286*, 5100–5107. [CrossRef]

128. Yen, H.; Ooka, T.; Iguchi, A.; Hayashi, T.; Sugimoto, N.; Tobe, T. NleC, a type III secretion protease, compromises NF-kappaB activation by targeting p65/RelA. *PLoS Pathog.* **2010**, *6*, e1001231. [CrossRef]

129. Cerda-Costa, N.; Gomis-Ruth, F.X. Architecture and function of metallopeptidase catalytic domains. *Protein Sci.* **2014**, *23*, 123–144. [CrossRef]

130. Jennings, E.; Esposito, D.; Rittinger, K.; Thurston, T.L.M. Structure-function analyses of the bacterial zinc metalloprotease effector protein GtgA uncover key residues required for deactivating NF-kappaB. *J. Biol. Chem.* **2018**, *293*, 15316–15329. [CrossRef]

131. Agace, W.W.; Hedges, S.R.; Ceska, M.; Svanborg, C. Interleukin-8 and the neutrophil response to mucosal gram-negative infection. *J. Clin. Investig.* **1993**, *92*, 780–785. [CrossRef] [PubMed]

132. Sunden, F.; Hakansson, L.; Ljunggren, E.; Wullt, B. *Escherichia coli* 83972 bacteriuria protects against recurrent lower urinary tract infections in patients with incomplete bladder emptying. *J. Urol.* **2010**, *184*, 179–185. [CrossRef] [PubMed]

133. Wullt, B.; Bergsten, G.; Connell, H.; Rollano, P.; Gebretsadik, N.; Hull, R.; Svanborg, C. P fimbriae enhance the early establishment of *Escherichia coli* in the human urinary tract. *Mol. Microbiol.* **2000**, *38*, 456–464. [CrossRef] [PubMed]

134. Lutay, N.; Ambite, I.; Gronberg Hernandez, J.; Rydstrom, G.; Ragnarsdottir, B.; Puthia, M.; Nadeem, A.; Zhang, J.; Storm, P.; Dobrindt, U.; et al. Bacterial control of host gene expression through RNA polymerase II. *J. Clin. Investig.* **2013**, *123*, 2366–2379. [CrossRef] [PubMed]

135. Gajdacs, M.; Abrok, M.; Lazar, A.; Burian, K. Comparative Epidemiology and Resistance Trends of Common Urinary Pathogens in a Tertiary-Care Hospital: A 10-Year Surveillance Study. *Medicina* **2019**, *55*, 356. [CrossRef] [PubMed]

136. Egloff, S.; Van Herreweghe, E.; Kiss, T. Regulation of polymerase II transcription by 7SK snRNA: Two distinct RNA elements direct P-TEFb and HEXIM1 binding. *Mol. Cell. Biol.* **2006**, *26*, 630–642. [CrossRef] [PubMed]

137. Jeronimo, C.; Forget, D.; Bouchard, A.; Li, Q.; Chua, G.; Poitras, C.; Therien, C.; Bergeron, D.; Bourassa, S.; Greenblatt, J.; et al. Systematic analysis of the protein interaction network for the human transcription machinery reveals the identity of the 7SK capping enzyme. *Mol. Cell* **2007**, *27*, 262–274. [CrossRef]

138. Schaffar, G.; Breuer, P.; Boteva, R.; Behrends, C.; Tzvetkov, N.; Strippel, N.; Sakahira, H.; Siegers, K.; Hayer-Hartl, M.; Hartl, F.U. Cellular toxicity of polyglutamine expansion proteins: Mechanism of transcription factor deactivation. *Mol. Cell* **2004**, *15*, 95–105. [CrossRef]

139. Zdziarski, J.; Brzuszkiewicz, E.; Wullt, B.; Liesegang, H.; Biran, D.; Voigt, B.; Gronberg-Hernandez, J.; Ragnarsdottir, B.; Hecker, M.; Ron, E.Z.; et al. Host imprints on bacterial genomes—Rapid, divergent evolution in individual patients. *PLoS Pathog.* **2010**, *6*, e1001078. [CrossRef]

140. Hansson, S.; Jodal, U.; Lincoln, K.; Svanborg-Eden, C. Untreated asymptomatic bacteriuria in girls: II–Effect of phenoxymethylpenicillin and erythromycin given for intercurrent infections. *BMJ* **1989**, *298*, 856–859. [CrossRef]

141. Kunin, C.M.; White, L.V.; Hua, T.H. A reassessment of the importance of "low-count" bacteriuria in young women with acute urinary symptoms. *Ann. Intern. Med.* **1993**, *119*, 454–460. [CrossRef] [PubMed]

142. Lindberg, U.; Claesson, I.; Hanson, L.A.; Jodal, U. Asymptomatic bacteriuria in schoolgirls. I. Clinical and laboratory findings. *Acta Paediatr. Scand.* **1975**, *64*, 425–431. [CrossRef] [PubMed]

143. Nicolle, L.E. The Paradigm Shift to Non-Treatment of Asymptomatic Bacteriuria. *Pathogens* **2016**, *5*, 38. [CrossRef] [PubMed]

144. Darouiche, R.O.; Thornby, J.I.; Cerra-Stewart, C.; Donovan, W.H.; Hull, R.A. Bacterial interference for prevention of urinary tract infection: A prospective, randomized, placebo-controlled, double-blind pilot trial. *Clin. Infect. Dis.* **2005**, *41*, 1531–1534. [CrossRef]

145. Schreiber, H.L.t.; Conover, M.S.; Chou, W.C.; Hibbing, M.E.; Manson, A.L.; Dodson, K.W.; Hannan, T.J.; Roberts, P.L.; Stapleton, A.E.; Hooton, T.M.; et al. Bacterial virulence phenotypes of *Escherichia coli* and host susceptibility determine risk for urinary tract infections. *Sci. Transl. Med.* **2017**, *9*. [CrossRef]

146. Von Dwingelo, J.; Chung, I.Y.W.; Price, C.T.; Li, L.; Jones, S.; Cygler, M.; Abu Kwaik, Y. Interaction of the Ankyrin H Core Effector of *Legionella* with the Host LARP7 Component of the 7SK snRNP Complex. *mBio* **2019**, *10*. [CrossRef]

147. Schuelein, R.; Spencer, H.; Dagley, L.F.; Li, P.F.; Luo, L.; Stow, J.L.; Abraham, G.; Naderer, T.; Gomez-Valero, L.; Buchrieser, C.; et al. Targeting of RNA Polymerase II by a nuclear *Legionella pneumophila* Dot/Icm effector SnpL. *Cell. Microbiol.* **2018**, *20*, e12852. [CrossRef]

148. Wang, E.T.; Sandberg, R.; Luo, S.; Khrebtukova, I.; Zhang, L.; Mayr, C.; Kingsmore, S.F.; Schroth, G.P.; Burge, C.B. Alternative isoform regulation in human tissue transcriptomes. *Nature* **2008**, *456*, 470–476. [CrossRef]

149. Chauhan, K.; Kalam, H.; Dutt, R.; Kumar, D. RNA Splicing: A New Paradigm in Host-Pathogen Interactions. *J. Mol. Biol.* **2019**, *431*, 1565–1575. [CrossRef]

150. Black, D.L. Mechanisms of alternative pre-messenger RNA splicing. *Annu. Rev. Biochem.* **2003**, *72*, 291–336. [CrossRef]

151. Aslanzadeh, V.; Huang, Y.; Sanguinetti, G.; Beggs, J.D. Corrigendum: Transcription rate strongly affects splicing fidelity and cotranscriptionality in budding yeast. *Genome Res.* **2018**, *28*, 203–213. [CrossRef] [PubMed]

152. Chathoth, K.T.; Barrass, J.D.; Webb, S.; Beggs, J.D. A splicing-dependent transcriptional checkpoint associated with prespliceosome formation. *Mol. Cell* **2014**, *53*, 779–790. [CrossRef] [PubMed]

153. Davari, K.; Lichti, J.; Gallus, C.; Greulich, F.; Uhlenhaut, N.H.; Heinig, M.; Friedel, C.C.; Glasmacher, E. Rapid Genome-wide Recruitment of RNA Polymerase II Drives Transcription, Splicing and Translation Events during T Cell Responses. *Cell Rep.* **2017**, *19*, 643–654. [CrossRef] [PubMed]

154. Aartsma-Rus, A.; van Ommen, G.J. Less is more: Therapeutic exon skipping for Duchenne muscular dystrophy. *Lancet Neurol.* **2009**, *8*, 873–875. [CrossRef]

155. De Maio, F.A.; Risso, G.; Iglesias, N.G.; Shah, P.; Pozzi, B.; Gebhard, L.G.; Mammi, P.; Mancini, E.; Yanovsky, M.J.; Andino, R.; et al. The Dengue Virus NS5 Protein Intrudes in the Cellular Spliceosome and Modulates Splicing. *PLoS Pathog.* **2016**, *12*, e1005841. [CrossRef]

156. Kalam, H.; Fontana, M.F.; Kumar, D. Alternate splicing of transcripts shape macrophage response to *Mycobacterium tuberculosis* infection. *PLoS Pathog.* **2017**, *13*, e1006236. [CrossRef]

157. Pai, A.A.; Baharian, G.; Page Sabourin, A.; Brinkworth, J.F.; Nedelec, Y.; Foley, J.W.; Grenier, J.C.; Siddle, K.J.; Dumaine, A.; Yotova, V.; et al. Widespread Shortening of 3′ Untranslated Regions and Increased Exon Inclusion Are Evolutionarily Conserved Features of Innate Immune Responses to Infection. *PLoS Genet.* **2016**, *12*, e1006338. [CrossRef]

158. Luo, Z.; Li, Z.; Chen, K.; Liu, R.; Li, X.; Cao, H.; Zheng, S.J. Engagement of heterogeneous nuclear ribonucleoprotein M with listeriolysin O induces type I interferon expression and restricts *Listeria monocytogenes* growth in host cells. *Immunobiology* **2012**, *217*, 972–981. [CrossRef]

159. Agrawal, A.K.; Ranjan, R.; Chandra, S.; Rout, T.K.; Misra, A.; Reddy, T.J. Some proteins of *M. tuberculosis* that localise to the nucleus of THP-1-derived macrophages. *Tuberculosis* **2016**, *101*, 75–78. [CrossRef]

160. Penn, B.H.; Netter, Z.; Johnson, J.R.; Von Dollen, J.; Jang, G.M.; Johnson, T.; Ohol, Y.M.; Maher, C.; Bell, S.L.; Geiger, K.; et al. An Mtb-Human Protein-Protein Interaction Map Identifies a Switch between Host Antiviral and Antibacterial Responses. *Mol. Cell* **2018**, *71*, 637–648. [CrossRef]

161. Liang, G.; Malmuthuge, N.; Guan, Y.; Ren, Y.; Griebel, P.J.; Guan le, L. Altered microRNA expression and pre-mRNA splicing events reveal new mechanisms associated with early stage *Mycobacterium avium* subspecies paratuberculosis infection. *Sci. Rep.* **2016**, *6*, 1–12. [CrossRef] [PubMed]

162. Kalam, H.; Singh, K.; Chauhan, K.; Fontana, M.F.; Kumar, D. Alternate splicing of transcripts upon *Mycobacterium tuberculosis* infection impacts the expression of functional protein domains. *IUBMB Life* **2018**, *70*, 845–854. [CrossRef] [PubMed]

163. Russell, D.G.; Barry, C.E., 3rd; Flynn, J.L. Tuberculosis: What we don't know can and does, hurt us. *Science* **2010**, *328*, 852–856. [CrossRef] [PubMed]

164. Rioux, J.D.; Xavier, R.J.; Taylor, K.D.; Silverberg, M.S.; Goyette, P.; Huett, A.; Green, T.; Kuballa, P.; Barmada, M.M.; Datta, L.W.; et al. Genome-wide association study identifies new susceptibility loci for Crohn disease and implicates autophagy in disease pathogenesis. *Nat. Genet.* **2007**, *39*, 596–604. [CrossRef]

165. Baralle, D.; Lucassen, A.; Buratti, E. Missed threads. The impact of pre-mRNA splicing defects on clinical practice. *EMBO Rep.* **2009**, *10*, 810–816. [CrossRef]

166. Karambataki, M.; Malousi, A.; Tzimagiorgis, G.; Haitoglou, C.; Fragou, A.; Georgiou, E.; Papadopoulou, F.; Krassas, G.E.; Kouidou, S. Association of two synonymous splicing-associated CpG single nucleotide polymorphisms in calpain 10 and solute carrier family 2 member 2 with type 2 diabetes. *Biomed. Rep.* **2017**, *6*, 146–158. [CrossRef]

167. Soukarieh, O.; Gaildrat, P.; Hamieh, M.; Drouet, A.; Baert-Desurmont, S.; Frebourg, T.; Tosi, M.; Martins, A. Exonic Splicing Mutations Are More Prevalent than Currently Estimated and Can Be Predicted by Using In Silico Tools. *PLoS Genet.* **2016**, *12*, e1005756. [CrossRef]

168. Lundtoft, C.; Awuah, A.A.; Guler, A.; Harling, K.; Schaal, H.; Mayatepek, E.; Phillips, R.O.; Nausch, N.; Owusu-Dabo, E.; Jacobsen, M. An IL7RA exon 5 polymorphism is associated with impaired IL-7Ralpha splicing and protection against tuberculosis in Ghana. *Genes Immun.* **2019**, *20*, 514–519. [CrossRef]

169. Moller, M.; de Wit, E.; Hoal, E.G. Past, present and future directions in human genetic susceptibility to tuberculosis. *FEMS Immunol. Med. Microbiol.* **2010**, *58*, 3–26. [CrossRef]

170. Dlamini, Z.; Hull, R. Can the HIV-1 splicing machinery be targeted for drug discovery? *HIV AIDS* **2017**, *9*, 63–75. [CrossRef]

171. Bates, D.O.; Morris, J.C.; Oltean, S.; Donaldson, L.F. Pharmacology of Modulators of Alternative Splicing. *Pharm. Rev.* **2017**, *69*, 63–79. [CrossRef] [PubMed]

172. Molleston, J.M.; Cherry, S. Attacked from All Sides: RNA Decay in Antiviral Defense. *Viruses* **2017**, *9*, 2. [CrossRef] [PubMed]

173. Peterson, S.M.; Thompson, J.A.; Ufkin, M.L.; Sathyanarayana, P.; Liaw, L.; Congdon, C.B. Common features of microRNA target prediction tools. *Front. Genet.* **2014**, *5*, 23. [CrossRef] [PubMed]

174. O'Connell, R.M.; Rao, D.S.; Chaudhuri, A.A.; Baltimore, D. Physiological and pathological roles for microRNAs in the immune system. *Nat. Rev. Immunol.* **2010**, *10*, 111–122. [CrossRef] [PubMed]

175. Kedde, M.; Agami, R. Interplay between microRNAs and RNA-binding proteins determines developmental processes. *Cell Cycle* **2008**, *7*, 899–903. [CrossRef]

176. Taganov, K.D.; Boldin, M.P.; Chang, K.J.; Baltimore, D. NF-kappaB-dependent induction of microRNA miR-146, an inhibitor targeted to signaling proteins of innate immune responses. *Proc. Natl. Acad. Sci. USA* **2006**, *103*, 12481–12486. [CrossRef] [PubMed]

177. Jin, W.; Ibeagha-Awemu, E.M.; Liang, G.; Beaudoin, F.; Zhao, X.; Guan le, L. Transcriptome microRNA profiling of bovine mammary epithelial cells challenged with *Escherichia coli* or Staphylococcus aureus bacteria reveals pathogen directed microRNA expression profiles. *BMC Genomics* **2014**, *15*, 181. [CrossRef]

178. Zeng, F.R.; Tang, L.J.; He, Y.; Garcia, R.C. An update on the role of miRNA-155 in pathogenic microbial infections. *Microbes Infect.* **2015**, *17*, 613–621. [CrossRef]

179. Staedel, C.; Darfeuille, F. MicroRNAs and bacterial infection. *Cell. Microbiol.* **2013**, *15*, 1496–1507. [CrossRef]

180. Eulalio, A.; Schulte, L.; Vogel, J. The mammalian microRNA response to bacterial infections. *RNA Biol.* **2012**, *9*, 742–750. [CrossRef]

181. Maudet, C.; Mano, M.; Eulalio, A. MicroRNAs in the interaction between host and bacterial pathogens. *FEBS Lett.* **2014**, *588*, 4140–4147. [CrossRef] [PubMed]

182. Siddle, K.J.; Tailleux, L.; Deschamps, M.; Loh, Y.H.; Deluen, C.; Gicquel, B.; Antoniewski, C.; Barreiro, L.B.; Farinelli, L.; Quintana-Murci, L. bacterial infection drives the expression dynamics of microRNAs and their isomiRs. *PLoS Genet.* **2015**, *11*, e1005064. [CrossRef] [PubMed]

183. Schulte, L.N.; Westermann, A.J.; Vogel, J. Differential activation and functional specialization of miR-146 and miR-155 in innate immune sensing. *Nucleic Acids Res.* **2013**, *41*, 542–553. [CrossRef] [PubMed]

184. Tili, E.; Michaille, J.J.; Cimino, A.; Costinean, S.; Dumitru, C.D.; Adair, B.; Fabbri, M.; Alder, H.; Liu, C.G.; Calin, G.A.; et al. Modulation of miR-155 and miR-125b levels following lipopolysaccharide/TNF-alpha stimulation and their possible roles in regulating the response to endotoxin shock. *J. Immunol.* **2007**, *179*, 5082–5089. [CrossRef]

185. Hu, R.; Kagele, D.A.; Huffaker, T.B.; Runtsch, M.C.; Alexander, M.; Liu, J.; Bake, E.; Su, W.; Williams, M.A.; Rao, D.S.; et al. miR-155 promotes T follicular helper cell accumulation during chronic, low-grade inflammation. *Immunity* **2014**, *41*, 605–619. [CrossRef]

186. Wan, G.; Xie, W.; Liu, Z.; Xu, W.; Lao, Y.; Huang, N.; Cui, K.; Liao, M.; He, J.; Jiang, Y.; et al. Hypoxia-induced MIR155 is a potent autophagy inducer by targeting multiple players in the MTOR pathway. *Autophagy* **2014**, *10*, 70–79. [CrossRef]

187. Wang, J.; Yang, K.; Zhou, L.; Wu, Y.; Zhu, M.; Lai, X.; Chen, T.; Feng, L.; Li, M. MicroRNA-155 promotes autophagy to eliminate intracellular mycobacteria by targeting Rheb. *PLoS Pathog.* **2013**, *9*, e1003697. [CrossRef]

188. Clare, S.; John, V.; Walker, A.W.; Hill, J.L.; Abreu-Goodger, C.; Hale, C.; Goulding, D.; Lawley, T.D.; Mastroeni, P.; Frankel, G.; et al. Enhanced susceptibility to Citrobacter rodentium infection in microRNA-155-deficient mice. *Infect. Immun.* **2013**, *81*, 723–732. [CrossRef]

189. Lind, E.F.; Elford, A.R.; Ohashi, P.S. Micro-RNA 155 is required for optimal CD8+ T cell responses to acute viral and intracellular bacterial challenges. *J. Immunol.* **2013**, *190*, 1210–1216. [CrossRef]

190. Rodriguez, A.; Vigorito, E.; Clare, S.; Warren, M.V.; Couttet, P.; Soond, D.R.; van Dongen, S.; Grocock, R.J.; Das, P.P.; Miska, E.A.; et al. Requirement of bic/microRNA-155 for normal immune function. *Science* **2007**, *316*, 608–611. [CrossRef]

191. Imaizumi, T.; Tanaka, H.; Tajima, A.; Yokono, Y.; Matsumiya, T.; Yoshida, H.; Tsuruga, K.; Aizawa-Yashiro, T.; Hayakari, R.; Inoue, I.; et al. IFN-gamma and TNF-alpha synergistically induce microRNA-155 which regulates TAB2/IP-10 expression in human mesangial cells. *Am. J. Nephrol.* **2010**, *32*, 462–468. [CrossRef] [PubMed]

192. Teng, G.G.; Wang, W.H.; Dai, Y.; Wang, S.J.; Chu, Y.X.; Li, J. Let-7b is involved in the inflammation and immune responses associated with *Helicobacter pylori* infection by targeting Toll-like receptor 4. *PLoS ONE* **2013**, *8*, e56709. [CrossRef] [PubMed]

193. Schulte, L.N.; Eulalio, A.; Mollenkopf, H.J.; Reinhardt, R.; Vogel, J. Analysis of the host microRNA response to *Salmonella* uncovers the control of major cytokines by the let-7 family. *EMBO J.* **2011**, *30*, 1977–1989. [CrossRef] [PubMed]

194. Marcais, A.; Blevins, R.; Graumann, J.; Feytout, A.; Dharmalingam, G.; Carroll, T.; Amado, I.F.; Bruno, L.; Lee, K.; Walzer, T.; et al. microRNA-mediated regulation of mTOR complex components facilitates discrimination between activation and anergy in CD4 T cells. *J. Exp. Med.* **2014**, *211*, 2281–2295. [CrossRef]

195. Hu, G.; Gong, A.Y.; Roth, A.L.; Huang, B.Q.; Ward, H.D.; Zhu, G.; Larusso, N.F.; Hanson, N.D.; Chen, X.M. Release of luminal exosomes contributes to TLR4-mediated epithelial antimicrobial defense. *PLoS Pathog.* **2013**, *9*, e1003261. [CrossRef]

196. Ma, F.; Xu, S.; Liu, X.; Zhang, Q.; Xu, X.; Liu, M.; Hua, M.; Li, N.; Yao, H.; Cao, X. The microRNA miR-29 controls innate and adaptive immune responses to intracellular bacterial infection by targeting interferon-gamma. *Nat. Immunol.* **2011**, *12*, 861–869. [CrossRef]

197. Archambaud, C.; Nahori, M.A.; Soubigou, G.; Becavin, C.; Laval, L.; Lechat, P.; Smokvina, T.; Langella, P.; Lecuit, M.; Cossart, P. Impact of lactobacilli on orally acquired listeriosis. *Proc. Natl. Acad. Sci. USA* **2012**, *109*, 16684–16689. [CrossRef]

198. Archambaud, C.; Sismeiro, O.; Toedling, J.; Soubigou, G.; Becavin, C.; Lechat, P.; Lebreton, A.; Ciaudo, C.; Cossart, P. The intestinal microbiota interferes with the microRNA response upon oral *Listeria* infection. *mBio* **2013**, *4*, e00707-00713. [CrossRef]

199. Chen, X.M.; Splinter, P.L.; O'Hara, S.P.; LaRusso, N.F. A cellular micro-RNA, let-7i, regulates Toll-like receptor 4 expression and contributes to cholangiocyte immune responses against Cryptosporidium parvum infection. *J. Biol. Chem.* **2007**, *282*, 28929–28938. [CrossRef]

200. Izar, B.; Mannala, G.K.; Mraheil, M.A.; Chakraborty, T.; Hain, T. microRNA response to *Listeria monocytogenes* infection in epithelial cells. *Int. J. Mol. Sci.* **2012**, *13*, 1173–1185. [CrossRef]

201. Verma, S.; Mohapatra, G.; Ahmad, S.M.; Rana, S.; Jain, S.; Khalsa, J.K.; Srikanth, C.V. *Salmonella* Engages Host MicroRNAs to Modulate SUMOylation: A New Arsenal for Intracellular Survival. *Mol. Cell. Biol.* **2015**, *35*, 2932–2946. [CrossRef] [PubMed]

202. Maudet, C.; Mano, M.; Sunkavalli, U.; Sharan, M.; Giacca, M.; Forstner, K.U.; Eulalio, A. Functional high-throughput screening identifies the miR-15 microRNA family as cellular restriction factors for *Salmonella* infection. *Nat. Commun.* **2014**, *5*, 1–13. [CrossRef] [PubMed]

203. Zhang, T.; Yu, J.; Zhang, Y.; Li, L.; Chen, Y.; Li, D.; Liu, F.; Zhang, C.Y.; Gu, H.; Zen, K. *Salmonella* enterica serovar enteritidis modulates intestinal epithelial miR-128 levels to decrease macrophage recruitment via macrophage colony-stimulating factor. *J. Infect. Dis.* **2014**, *209*, 2000–2011. [CrossRef] [PubMed]

204. Li, N.; Tang, B.; Zhu, E.D.; Li, B.S.; Zhuang, Y.; Yu, S.; Lu, D.S.; Zou, Q.M.; Xiao, B.; Mao, X.H. Increased miR-222 in H. pylori-associated gastric cancer correlated with tumor progression by promoting cancer cell proliferation and targeting RECK. *FEBS Lett.* **2012**, *586*, 722–728. [CrossRef] [PubMed]

205. Zhang, Z.; Li, Z.; Gao, C.; Chen, P.; Chen, J.; Liu, W.; Xiao, S.; Lu, H. miR-21 plays a pivotal role in gastric cancer pathogenesis and progression. *Lab. Investig.* **2008**, *88*, 1358–1366. [CrossRef] [PubMed]

206. Tang, B.; Li, N.; Gu, J.; Zhuang, Y.; Li, Q.; Wang, H.G.; Fang, Y.; Yu, B.; Zhang, J.Y.; Xie, Q.H.; et al. Compromised autophagy by MIR30B benefits the intracellular survival of *Helicobacter pylori*. *Autophagy* **2012**, *8*, 1045–1057. [CrossRef]

207. Ni, B.; Rajaram, M.V.; Lafuse, W.P.; Landes, M.B.; Schlesinger, L.S. *Mycobacterium tuberculosis* decreases human macrophage IFN-gamma responsiveness through miR-132 and miR-26a. *J. Immunol.* **2014**, *193*, 4537–4547. [CrossRef]

208. Rajaram, M.V.; Ni, B.; Morris, J.D.; Brooks, M.N.; Carlson, T.K.; Bakthavachalu, B.; Schoenberg, D.R.; Torrelles, J.B.; Schlesinger, L.S. *Mycobacterium tuberculosis* lipomannan blocks TNF biosynthesis by regulating macrophage MAPK-activated protein kinase 2 (MK2) and microRNA miR-125b. *Proc. Natl. Acad. Sci. USA* **2011**, *108*, 17408–17413. [CrossRef]

209. Singh, Y.; Kaul, V.; Mehra, A.; Chatterjee, S.; Tousif, S.; Dwivedi, V.P.; Suar, M.; Van Kaer, L.; Bishai, W.R.; Das, G. *Mycobacterium tuberculosis* controls microRNA-99b (miR-99b) expression in infected murine dendritic cells to modulate host immunity. *J. Biol. Chem.* **2013**, *288*, 5056–5061. [CrossRef]

210. Kumar, R.; Halder, P.; Sahu, S.K.; Kumar, M.; Kumari, M.; Jana, K.; Ghosh, Z.; Sharma, P.; Kundu, M.; Basu, J. Identification of a novel role of ESAT-6-dependent miR-155 induction during infection of macrophages with *Mycobacterium tuberculosis*. *Cell. Microbiol.* **2012**, *14*, 1620–1631. [CrossRef]

211. Yang, S.; Li, F.; Jia, S.; Zhang, K.; Jiang, W.; Shang, Y.; Chang, K.; Deng, S.; Chen, M. Early secreted antigen ESAT-6 of *Mycobacterium tuberculosis* promotes apoptosis of macrophages via targeting the microRNA155-SOCS1 interaction. *Cell. Physiol. Biochem.* **2015**, *35*, 1276–1288. [CrossRef] [PubMed]

212. Kumar, M.; Sahu, S.K.; Kumar, R.; Subuddhi, A.; Maji, R.K.; Jana, K.; Gupta, P.; Raffetseder, J.; Lerm, M.; Ghosh, Z.; et al. MicroRNA let-7 modulates the immune response to *Mycobacterium tuberculosis* infection via control of A20, an inhibitor of the NF-kappaB pathway. *Cell Host Microbe* **2015**, *17*, 345–356. [CrossRef] [PubMed]

213. Zhang, Y.M.; Noto, J.M.; Hammond, C.E.; Barth, J.L.; Argraves, W.S.; Backert, S.; Peek, R.M., Jr.; Smolka, A.J. *Helicobacter pylori*-induced posttranscriptional regulation of H-K-ATPase alpha-subunit gene expression by miRNA. *Am. J. Physiol. Gastrointest. Liver Physiol.* **2014**, *306*, G606–G613. [CrossRef] [PubMed]

214. Belair, C.; Baud, J.; Chabas, S.; Sharma, C.M.; Vogel, J.; Staedel, C.; Darfeuille, F. *Helicobacter pylori* interferes with an embryonic stem cell micro RNA cluster to block cell cycle progression. *Silence* **2011**, *2*, 7. [CrossRef]

215. Zhu, Y.; Jiang, Q.; Lou, X.; Ji, X.; Wen, Z.; Wu, J.; Tao, H.; Jiang, T.; He, W.; Wang, C.; et al. MicroRNAs up-regulated by CagA of *Helicobacter pylori* induce intestinal metaplasia of gastric epithelial cells. *PLoS ONE* **2012**, *7*, e35147. [CrossRef]

216. Shmaryahu, A.; Carrasco, M.; Valenzuela, P.D. Prediction of bacterial microRNAs and possible targets in human cell transcriptome. *J. Microbiol.* **2014**, *52*, 482–489. [CrossRef]

217. Liu, H.; Wang, X.; Wang, H.D.; Wu, J.; Ren, J.; Meng, L.; Wu, Q.; Dong, H.; Wu, J.; Kao, T.Y.; et al. *Escherichia coli* noncoding RNAs can affect gene expression and physiology of Caenorhabditis elegans. *Nat. Commun.* **2012**, *3*, 1–11. [CrossRef]

218. Furuse, Y.; Finethy, R.; Saka, H.A.; Xet-Mull, A.M.; Sisk, D.M.; Smith, K.L.; Lee, S.; Coers, J.; Valdivia, R.H.; Tobin, D.M.; et al. Search for microRNAs expressed by intracellular bacterial pathogens in infected mammalian cells. *PLoS ONE* **2014**, *9*, e106434. [CrossRef]

219. Alexander, M.; Hu, R.; Runtsch, M.C.; Kagele, D.A.; Mosbruger, T.L.; Tolmachova, T.; Seabra, M.C.; Round, J.L.; Ward, D.M.; O'Connell, R.M. Exosome-delivered microRNAs modulate the inflammatory response to endotoxin. *Nat. Commun.* **2015**, *6*, 1–16. [CrossRef]

220. Liu, S.; da Cunha, A.P.; Rezende, R.M.; Cialic, R.; Wei, Z.; Bry, L.; Comstock, L.E.; Gandhi, R.; Weiner, H.L. The Host Shapes the Gut Microbiota via Fecal MicroRNA. *Cell Host Microbe* **2016**, *19*, 32–43. [CrossRef]

221. Weber, J.A.; Baxter, D.H.; Zhang, S.; Huang, D.Y.; Huang, K.H.; Lee, M.J.; Galas, D.J.; Wang, K. The microRNA spectrum in 12 body fluids. *Clin. Chem.* **2010**, *56*, 1733–1741. [CrossRef] [PubMed]

222. Wu, K.; Zhu, C.; Yao, Y.; Wang, X.; Song, J.; Zhai, J. MicroRNA-155-enhanced autophagy in human gastric epithelial cell in response to *Helicobacter pylori*. *Saudi J. Gastroenterol.* **2016**, *22*, 30–36. [CrossRef] [PubMed]

223. Benz, F.; Roy, S.; Trautwein, C.; Roderburg, C.; Luedde, T. Circulating MicroRNAs as Biomarkers for Sepsis. *Int. J. Mol. Sci.* **2016**, *17*, 78. [CrossRef] [PubMed]

224. Chakraborty, C.; Das, S. Profiling cell-free and circulating miRNA: A clinical diagnostic tool for different cancers. *Tumour Biol.* **2016**, *37*, 5705–5714. [CrossRef]

225. Guay, C.; Regazzi, R. Circulating microRNAs as novel biomarkers for diabetes mellitus. *Nat. Rev. Endocrinol.* **2013**, *9*, 513–521. [CrossRef]

226. Ueberberg, B.; Kohns, M.; Mayatepek, E.; Jacobsen, M. Are microRNAs suitable biomarkers of immunity to tuberculosis? *Mol. Cell. Pediatr.* **2014**, *1*, 8. [CrossRef]

227. Asrat, S.; Dugan, A.S.; Isberg, R.R. The frustrated host response to *Legionella pneumophila* is bypassed by MyD88-dependent translation of pro-inflammatory cytokines. *PLoS Pathog.* **2014**, *10*, e1004229. [CrossRef]

228. Gingras, A.C.; Raught, B.; Sonenberg, N. eIF4 initiation factors: Effectors of mRNA recruitment to ribosomes and regulators of translation. *Annu. Rev. Biochem.* **1999**, *68*, 913–963. [CrossRef]

229. Lin, T.A.; Kong, X.; Saltiel, A.R.; Blackshear, P.J.; Lawrence, J.C., Jr. Control of PHAS-I by insulin in 3T3-L1 adipocytes. Synthesis, degradation and phosphorylation by a rapamycin-sensitive and mitogen-activated protein kinase-independent pathway. *J. Biol. Chem.* **1995**, *270*, 18531–18538. [CrossRef]

230. Pause, A.; Belsham, G.J.; Gingras, A.C.; Donze, O.; Lin, T.A.; Lawrence, J.C., Jr.; Sonenberg, N. Insulin-dependent stimulation of protein synthesis by phosphorylation of a regulator of 5′-cap function. *Nature* **1994**, *371*, 762–767. [CrossRef]

231. Gingras, A.C.; Raught, B.; Gygi, S.P.; Niedzwiecka, A.; Miron, M.; Burley, S.K.; Polakiewicz, R.D.; Wyslouch-Cieszynska, A.; Aebersold, R.; Sonenberg, N. Hierarchical phosphorylation of the translation inhibitor 4E-BP1. *Genes Dev.* **2001**, *15*, 2852–2864. [CrossRef] [PubMed]

232. Hara, K.; Maruki, Y.; Long, X.; Yoshino, K.; Oshiro, N.; Hidayat, S.; Tokunaga, C.; Avruch, J.; Yonezawa, K. Raptor, a binding partner of target of rapamycin (TOR), mediates TOR action. *Cell* **2002**, *110*, 177–189. [CrossRef]

233. Beretta, L.; Gingras, A.C.; Svitkin, Y.V.; Hall, M.N.; Sonenberg, N. Rapamycin blocks the phosphorylation of 4E-BP1 and inhibits cap-dependent initiation of translation. *EMBO J.* **1996**, *15*, 658–664. [CrossRef]

234. Fontana, M.F.; Banga, S.; Barry, K.C.; Shen, X.; Tan, Y.; Luo, Z.Q.; Vance, R.E. Secreted bacterial effectors that inhibit host protein synthesis are critical for induction of the innate immune response to virulent *Legionella pneumophila*. *PLoS Pathog.* **2011**, *7*, e1001289. [CrossRef] [PubMed]

235. Fontana, M.F.; Shin, S.; Vance, R.E. Activation of host mitogen-activated protein kinases by secreted *Legionella pneumophila* effectors that inhibit host protein translation. *Infect. Immun.* **2012**, *80*, 3570–3575. [CrossRef] [PubMed]

236. Barry, K.C.; Fontana, M.F.; Portman, J.L.; Dugan, A.S.; Vance, R.E. IL-1alpha signaling initiates the inflammatory response to virulent *Legionella pneumophila* in vivo. *J. Immunol.* **2013**, *190*, 6329–6339. [CrossRef]

237. Ivanov, S.S.; Roy, C.R. Pathogen signatures activate a ubiquitination pathway that modulates the function of the metabolic checkpoint kinase mTOR. *Nat. Immunol.* **2013**, *14*, 1219–1228. [CrossRef]

238. Dunbar, T.L.; Yan, Z.; Balla, K.M.; Smelkinson, M.G.; Troemel, E.R. C. elegans detects pathogen-induced translational inhibition to activate immune signaling. *Cell Host Microbe* **2012**, *11*, 375–386. [CrossRef]

239. McEwan, D.L.; Kirienko, N.V.; Ausubel, F.M. Host translational inhibition by *Pseudomonas aeruginosa* Exotoxin A Triggers an immune response in Caenorhabditis elegans. *Cell Host Microbe* **2012**, *11*, 364–374. [CrossRef]

240. Cox, J.S.; Walter, P. A novel mechanism for regulating activity of a transcription factor that controls the unfolded protein response. *Cell* **1996**, *87*, 391–404. [CrossRef]

241. Yoshida, H.; Matsui, T.; Yamamoto, A.; Okada, T.; Mori, K. XBP1 mRNA is induced by ATF6 and spliced by IRE1 in response to ER stress to produce a highly active transcription factor. *Cell* **2001**, *107*, 881–891. [CrossRef]

242. Korennykh, A.V.; Egea, P.F.; Korostelev, A.A.; Finer-Moore, J.; Zhang, C.; Shokat, K.M.; Stroud, R.M.; Walter, P. The unfolded protein response signals through high-order assembly of Ire1. *Nature* **2009**, *457*, 687–693. [CrossRef] [PubMed]

243. Jorgensen, R.; Wang, Y.; Visschedyk, D.; Merrill, A.R. The nature and character of the transition state for the ADP-ribosyltransferase reaction. *EMBO Rep.* **2008**, *9*, 802–809. [CrossRef] [PubMed]

244. Schwarz, D.S.; Blower, M.D. The endoplasmic reticulum: Structure, function and response to cellular signaling. *Cell. Mol. Life Sci.* **2016**, *73*, 79–94. [CrossRef] [PubMed]

245. Walter, P.; Ron, D. The unfolded protein response: From stress pathway to homeostatic regulation. *Science* **2011**, *334*, 1081–1086. [CrossRef]

246. Gardner, B.M.; Pincus, D.; Gotthardt, K.; Gallagher, C.M.; Walter, P. Endoplasmic reticulum stress sensing in the unfolded protein response. *Cold Spring Harb. Perspect. Biol.* **2013**, *5*, a013169. [CrossRef]

247. Harding, H.P.; Zhang, Y.; Ron, D. Protein translation and folding are coupled by an endoplasmic-reticulum-resident kinase. *Nature* **1999**, *397*, 271–274. [CrossRef]

248. Harding, H.P.; Novoa, I.; Zhang, Y.; Zeng, H.; Wek, R.; Schapira, M.; Ron, D. Regulated translation initiation controls stress-induced gene expression in mammalian cells. *Mol. Cell* **2000**, *6*, 1099–1108. [CrossRef]

249. Adachi, Y.; Yamamoto, K.; Okada, T.; Yoshida, H.; Harada, A.; Mori, K. ATF6 is a transcription factor specializing in the regulation of quality control proteins in the endoplasmic reticulum. *Cell Struct. Funct.* **2008**, *33*, 75–89. [CrossRef]

250. Haze, K.; Yoshida, H.; Yanagi, H.; Yura, T.; Mori, K. Mammalian transcription factor ATF6 is synthesized as a transmembrane protein and activated by proteolysis in response to endoplasmic reticulum stress. *Mol. Biol. Cell* **1999**, *10*, 3787–3799. [CrossRef]

251. Iurlaro, R.; Munoz-Pinedo, C. Cell death induced by endoplasmic reticulum stress. *FEBS J.* **2016**, *283*, 2640–2652. [CrossRef] [PubMed]

252. Pillich, H.; Loose, M.; Zimmer, K.P.; Chakraborty, T. Activation of the unfolded protein response by *Listeria monocytogenes*. *Cell. Microbiol.* **2012**, *14*, 949–964. [CrossRef] [PubMed]

253. Smith, J.A.; Khan, M.; Magnani, D.D.; Harms, J.S.; Durward, M.; Radhakrishnan, G.K.; Liu, Y.P.; Splitter, G.A. Brucella induces an unfolded protein response via TcpB that supports intracellular replication in macrophages. *PLoS Pathog.* **2013**, *9*, e1003785. [CrossRef] [PubMed]

254. De Jong, M.F.; Starr, T.; Winter, M.G.; den Hartigh, A.B.; Child, R.; Knodler, L.A.; van Dijl, J.M.; Celli, J.; Tsolis, R.M. Sensing of bacterial type IV secretion via the unfolded protein response. *mBio* **2013**, *4*, e00418-00412. [CrossRef]

255. Keestra-Gounder, A.M.; Byndloss, M.X.; Seyffert, N.; Young, B.M.; Chavez-Arroyo, A.; Tsai, A.Y.; Cevallos, S.A.; Winter, M.G.; Pham, O.H.; Tiffany, C.R.; et al. NOD1 and NOD2 signalling links ER stress with inflammation. *Nature* **2016**, *532*, 394–397. [CrossRef]

256. Hempstead, A.D.; Isberg, R.R. Inhibition of host cell translation elongation by *Legionella pneumophila* blocks the host cell unfolded protein response. *Proc. Natl. Acad. Sci. USA* **2015**, *112*, E6790–E6797. [CrossRef]

257. Treacy-Abarca, S.; Mukherjee, S. *Legionella* suppresses the host unfolded protein response via multiple mechanisms. *Nat. Commun.* **2015**, *6*, 1–10. [CrossRef]

258. Starck, S.R.; Tsai, J.C.; Chen, K.; Shodiya, M.; Wang, L.; Yahiro, K.; Martins-Green, M.; Shastri, N.; Walter, P. Translation from the 5′ untranslated region shapes the integrated stress response. *Science* **2016**, *351*, aad3867. [CrossRef]

259. Mehlitz, A.; Karunakaran, K.; Herweg, J.A.; Krohne, G.; van de Linde, S.; Rieck, E.; Sauer, M.; Rudel, T. The chlamydial organism *Simkania negevensis* forms ER vacuole contact sites and inhibits ER-stress. *Cell. Microbiol.* **2014**, *16*, 1224–1243. [CrossRef]

260. Shrestha, N.; Bahnan, W.; Wiley, D.J.; Barber, G.; Fields, K.A.; Schesser, K. Eukaryotic initiation factor 2 (eIF2) signaling regulates proinflammatory cytokine expression and bacterial invasion. *J. Biol. Chem.* **2012**, *287*, 28738–28744. [CrossRef]

261. Celli, J.; Tsolis, R.M. Bacteria, the endoplasmic reticulum and the unfolded protein response: Friends or foes? *Nat. Rev. Microbiol.* **2015**, *13*, 71–82. [CrossRef] [PubMed]

262. Haneda, T.; Ishii, Y.; Shimizu, H.; Ohshima, K.; Iida, N.; Danbara, H.; Okada, N. *Salmonella* type III effector SpvC, a phosphothreonine lyase, contributes to reduction in inflammatory response during intestinal phase of infection. *Cell. Microbiol.* **2012**, *14*, 485–499. [CrossRef] [PubMed]

263. Kim, G.W.; Yang, X.J. Comprehensive lysine acetylomes emerging from bacteria to humans. *Trends Biochem. Sci.* **2011**, *36*, 211–220. [CrossRef] [PubMed]

264. Choudhary, C.; Kumar, C.; Gnad, F.; Nielsen, M.L.; Rehman, M.; Walther, T.C.; Olsen, J.V.; Mann, M. Lysine acetylation targets protein complexes and co-regulates major cellular functions. *Science* **2009**, *325*, 834–840. [CrossRef] [PubMed]

265. Haglund, K.; Dikic, I. Ubiquitylation and cell signaling. *EMBO J.* **2005**, *24*, 3353–3359. [CrossRef] [PubMed]

266. Yau, R.; Rape, M. The increasing complexity of the ubiquitin code. *Nat. Cell Biol.* **2016**, *18*, 579–586. [CrossRef]

267. Zhou, Y.; Zhu, Y. Diversity of bacterial manipulation of the host ubiquitin pathways. *Cell. Microbiol.* **2015**, *17*, 26–34. [CrossRef]

268. Bhogaraju, S.; Kalayil, S.; Liu, Y.; Bonn, F.; Colby, T.; Matic, I.; Dikic, I. Phosphoribosylation of Ubiquitin Promotes Serine Ubiquitination and Impairs Conventional Ubiquitination. *Cell* **2016**, *167*, 1636–1649. [CrossRef]

269. Kotewicz, K.M.; Ramabhadran, V.; Sjoblom, N.; Vogel, J.P.; Haenssler, E.; Zhang, M.; Behringer, J.; Scheck, R.A.; Isberg, R.R. A Single *Legionella* Effector Catalyzes a Multistep Ubiquitination Pathway to Rearrange Tubular Endoplasmic Reticulum for Replication. *Cell Host Microbe* **2017**, *21*, 169–181. [CrossRef]

270. Qiu, J.; Sheedlo, M.J.; Yu, K.; Tan, Y.; Nakayasu, E.S.; Das, C.; Liu, X.; Luo, Z.Q. Ubiquitination independent of E1 and E2 enzymes by bacterial effectors. *Nature* **2016**, *533*, 120–124. [CrossRef]

271. Sheedlo, M.J.; Qiu, J.; Tan, Y.; Paul, L.N.; Luo, Z.Q.; Das, C. Structural basis of substrate recognition by a bacterial deubiquitinase important for dynamics of phagosome ubiquitination. *Proc. Natl. Acad. Sci. USA* **2015**, *112*, 15090–15095. [CrossRef] [PubMed]

272. Qiu, J.; Yu, K.; Fei, X.; Liu, Y.; Nakayasu, E.S.; Piehowski, P.D.; Shaw, J.B.; Puvar, K.; Das, C.; Liu, X.; et al. A unique deubiquitinase that deconjugates phosphoribosyl-linked protein ubiquitination. *Cell Res.* **2017**, *27*, 865–881. [CrossRef] [PubMed]

273. Okuda, J.; Toyotome, T.; Kataoka, N.; Ohno, M.; Abe, H.; Shimura, Y.; Seyedarabi, A.; Pickersgill, R.; Sasakawa, C. *Shigella* effector IpaH9.8 binds to a splicing factor U2AF(35) to modulate host immune responses. *Biochem. Biophys. Res. Commun.* **2005**, *333*, 531–539. [CrossRef] [PubMed]

274. Seyedarabi, A.; Sullivan, J.A.; Sasakawa, C.; Pickersgill, R.W. A disulfide driven domain swap switches off the activity of *Shigella* IpaH9.8 E3 ligase. *FEBS Lett.* **2010**, *584*, 4163–4168. [CrossRef] [PubMed]

275. Toyotome, T.; Suzuki, T.; Kuwae, A.; Nonaka, T.; Fukuda, H.; Imajoh-Ohmi, S.; Toyofuku, T.; Hori, M.; Sasakawa, C. *Shigella* protein IpaH(9.8) is secreted from bacteria within mammalian cells and transported to the nucleus. *J. Biol. Chem.* **2001**, *276*, 32071–32079. [CrossRef]

276. Haraga, A.; Miller, S.I. A *Salmonella* type III secretion effector interacts with the mammalian serine/threonine protein kinase PKN1. *Cell. Microbiol.* **2006**, *8*, 837–846. [CrossRef]

277. Hicks, S.W.; Galan, J.E. Hijacking the host ubiquitin pathway: Structural strategies of bacterial E3 ubiquitin ligases. *Curr. Opin. Microbiol.* **2010**, *13*, 41–46. [CrossRef]

278. Keszei, A.F.; Tang, X.; McCormick, C.; Zeqiraj, E.; Rohde, J.R.; Tyers, M.; Sicheri, F. Structure of an SspH1-PKN1 complex reveals the basis for host substrate recognition and mechanism of activation for a bacterial E3 ubiquitin ligase. *Mol. Cell. Biol.* **2014**, *34*, 362–373. [CrossRef]

279. Lin, Y.H.; Machner, M.P. Exploitation of the host cell ubiquitin machinery by microbial effector proteins. *J. Cell Sci.* **2017**, *130*, 1985–1996. [CrossRef]

280. Norkowski, S.; Schmidt, M.A.; Ruter, C. The species-spanning family of LPX-motif harbouring effector proteins. *Cell. Microbiol.* **2018**, *20*, e12945. [CrossRef]

281. Rohde, J.R.; Breitkreutz, A.; Chenal, A.; Sansonetti, P.J.; Parsot, C. Type III secretion effectors of the IpaH family are E3 ubiquitin ligases. *Cell Host Microbe* **2007**, *1*, 77–83. [CrossRef] [PubMed]

282. Valleau, D.; Little, D.J.; Borek, D.; Skarina, T.; Quaile, A.T.; Di Leo, R.; Houliston, S.; Lemak, A.; Arrowsmith, C.H.; Coombes, B.K.; et al. Functional diversification of the NleG effector family in enterohemorrhagic *Escherichia coli*. *Proc. Natl. Acad. Sci. USA* **2018**, *115*, 10004–10009. [CrossRef] [PubMed]

283. Min, C.K.; Kwon, Y.J.; Ha, N.Y.; Cho, B.A.; Kim, J.M.; Kwon, E.K.; Kim, Y.S.; Choi, M.S.; Kim, I.S.; Cho, N.H. Multiple *Orientia tsutsugamushi* ankyrin repeat proteins interact with SCF1 ubiquitin ligase complex and eukaryotic elongation factor 1 alpha. *PLoS ONE* **2014**, *9*, e105652. [CrossRef] [PubMed]

284. Werden, S.J.; Lanchbury, J.; Shattuck, D.; Neff, C.; Dufford, M.; McFadden, G. The myxoma virus m-t5 ankyrin repeat host range protein is a novel adaptor that coordinately links the cellular signaling pathways mediated by Akt and Skp1 in virus-infected cells. *J. Virol.* **2009**, *83*, 12068–12083. [CrossRef]

285. Cui, J.; Yao, Q.; Li, S.; Ding, X.; Lu, Q.; Mao, H.; Liu, L.; Zheng, N.; Chen, S.; Shao, F. Glutamine deamidation and dysfunction of ubiquitin/NEDD8 induced by a bacterial effector family. *Science* **2010**, *329*, 1215–1218. [CrossRef]

286. El-Aouar Filho, R.A.; Nicolas, A.; De Paula Castro, T.L.; Deplanche, M.; De Carvalho Azevedo, V.A.; Goossens, P.L.; Taieb, F.; Lina, G.; Le Loir, Y.; Berkova, N. Heterogeneous Family of Cyclomodulins: Smart Weapons That Allow Bacteria to Hijack the Eukaryotic Cell Cycle and Promote Infections. *Front. Cell. Infect. Microbiol.* **2017**, *7*, 208. [CrossRef]

287. Morikawa, H.; Kim, M.; Mimuro, H.; Punginelli, C.; Koyama, T.; Nagai, S.; Miyawaki, A.; Iwai, K.; Sasakawa, C. The bacterial effector Cif interferes with SCF ubiquitin ligase function by inhibiting deneddylation of Cullin1. *Biochem. Biophys. Res. Commun.* **2010**, *401*, 268–274. [CrossRef]

288. Samba-Louaka, A.; Nougayrede, J.P.; Watrin, C.; Jubelin, G.; Oswald, E.; Taieb, F. Bacterial cyclomodulin Cif blocks the host cell cycle by stabilizing the cyclin-dependent kinase inhibitors p21 and p27. *Cell. Microbiol.* **2008**, *10*, 2496–2508. [CrossRef]

289. Taieb, F.; Nougayrede, J.P.; Oswald, E. Cycle inhibiting factors (cifs): Cyclomodulins that usurp the ubiquitin-dependent degradation pathway of host cells. *Toxins* **2011**, *3*, 356–368. [CrossRef]

290. Tzfira, T.; Vaidya, M.; Citovsky, V. Involvement of targeted proteolysis in plant genetic transformation by Agrobacterium. *Nature* **2004**, *431*, 87–92. [CrossRef]

291. Zaltsman, A.; Krichevsky, A.; Loyter, A.; Citovsky, V. Agrobacterium induces expression of a host F-box protein required for tumorigenicity. *Cell Host Microbe* **2010**, *7*, 197–209. [CrossRef] [PubMed]
292. Jubelin, G.; Chavez, C.V.; Taieb, F.; Banfield, M.J.; Samba-Louaka, A.; Nobe, R.; Nougayrede, J.P.; Zumbihl, R.; Givaudan, A.; Escoubas, J.M.; et al. Cycle inhibiting factors (CIFs) are a growing family of functional cyclomodulins present in invertebrate and mammal bacterial pathogens. *PLoS ONE* **2009**, *4*, e4855. [CrossRef] [PubMed]
293. Iwai, H.; Kim, M.; Yoshikawa, Y.; Ashida, H.; Ogawa, M.; Fujita, Y.; Muller, D.; Kirikae, T.; Jackson, P.K.; Kotani, S.; et al. A bacterial effector targets Mad2L2, an APC inhibitor, to modulate host cell cycling. *Cell* **2007**, *130*, 611–623. [CrossRef] [PubMed]

 © 2020 by the authors. Licensee MDPI, Basel, Switzerland. This article is an open access article distributed under the terms and conditions of the Creative Commons Attribution (CC BY) license (http://creativecommons.org/licenses/by/4.0/).

International Journal of
Molecular Sciences

Review

Bacterial Actin-Specific Endoproteases Grimelysin and Protealysin as Virulence Factors Contributing to the Invasive Activities of *Serratia*

Sofia Khaitlina *, Ekaterina Bozhokina, Olga Tsaplina and Tatiana Efremova

Institute of Cytology, Russian Academy of Sciences, 194064 St. Petersburg, Russia; bozhokina@yahoo.com (E.B.);
olga566@mail.ru (O.T.); tefi45@mail.ru (T.E.)
* Correspondence: skhspb@gmail.com

Received: 30 April 2020; Accepted: 2 June 2020; Published: 4 June 2020

Abstract: The article reviews the discovery, properties and functional activities of new bacterial enzymes, proteases grimelysin (ECP 32) of *Serratia grimesii* and protealysin of *Serratia proteamaculans*, characterized by both a highly specific "actinase" activity and their ability to stimulate bacterial invasion. Grimelysin cleaves the only polypeptide bond Gly42-Val43 in actin. This bond is not cleaved by any other proteases and leads to a reversible loss of actin polymerization. Similar properties were characteristic for another bacterial protease, protealysin. These properties made grimelysin and protealysin a unique tool to study the functional properties of actin. Furthermore, bacteria *Serratia grimesii* and *Serratia proteamaculans*, producing grimelysin and protealysin, invade eukaryotic cells, and the recombinant *Escherichia coli* expressing the grimelysin or protealysins gene become invasive. Participation of the cellular c-Src and RhoA/ROCK signaling pathways in the invasion of eukaryotic cells by *S. grimesii* was shown, and involvement of E-cadherin in the invasion has been suggested. Moreover, membrane vesicles produced by *S. grimesii* were found to contain grimelysin, penetrate into eukaryotic cells and increase the invasion of bacteria into eukaryotic cells. These data indicate that the protease is a virulence factor, and actin can be a target for the protease upon its translocation into the host cell.

Keywords: actin proteolysis; metalloproteinases; protease ECP 32; grimelysin; protealysin; bacterial invasion

---

1. Introduction

Invasion of opportunistic bacterial pathogenic into eukaryotic cells is a process of interaction of bacteria with eukaryotic cells. [1,2]. To invade host cells, invasive bacteria should initiate the host cell signaling system by the proteins that interact with cell receptors and modify cytoskeleton through direct interaction with actin and actin binding proteins. Bacterial effectors can mimic natural activators of small GTPases or directly stimulate the host signaling pathways [3,4]. On the other hand, efficiency of bacterial invasion depends on the physiological state of host cells and is determined by the processes associated with changes in the distribution of cell surface receptors and cytoskeleton rearrangements [5]. This implies the presence of specific bacterial virulence factors capable of interacting with or modifying eukaryotic cell receptors, as well as components of the signal transduction system and actin cytoskeleton.

Studying the mechanisms of actin polymerization, we discovered and characterized a new enzyme-bacterial protease that exhibited highly specific "actinase" activity [6]. The protease turned out to be an intracellular enzyme of bacteria originally identified by standard morphological, biochemical, and cultural properties as an atypical lactose-negative strain of *Escherichia coli* A2 [7]. The actin-specific protease isolated from these bacteria was identified as a single 32 kDa polypeptide, which gave the protease the name ECP 32 (Escherichia coli protease, 32 kDa) [7,8]. Protease ECP 32 cleaves actin at a single site between Gly-42 and Val-43 within the DNase I-binding loop on the top of the actin monomer [9,10] that is involved in extensive interactions with the neighboring subunits within the actin filament [11,12]. The high specificity of actin proteolysis made ECP 32 a unique tool to study actin properties and interactions [13–17]. On the other hand, the enzyme by itself turned out to be not unique.

Although ECP 32 has been described as an *E. coli* A2 enzyme the *N*-terminal amino acid sequence of ECP 32 AKTSSAGVVIRDIF could not be identified in published *Escherichia coli* genomic sequences [8,18]. Therefore, the systematic position of the ECP 32-producing bacterial strain was reinvestigated. Using about 50 biochemical reactions of the Vitek-2 system (bioMerrieux, Marcy l'Etoile, France) and partial sequencing of the 16S rRNA gene, the former *Escherichia coli* A2 strain was identified as *Serratia grimesii* A2 [18]. Then, the presumptive gene coding for the reference *S. grimesii* 30063 or *Serratia grimesii* A2 was cloned using published sequences of a similar protease protealysin identified in *S. proteamaculans* [19]. Amino acid sequences of two corresponding proteins from the *S. grimesii* A2 (former *E. coli* A2) and the reference strain of *S. grimesii* 30,063 were identical and contained the *N*-terminal 14 amino acids of protease ECP 32 as previously determined [8,18]. The same specific actin-hydrolyzing activity, characteristic of protease ECP 32 [9,10], was also revealed in bacterial extracts of the reference *S. grimesii* strain and *Escherichia coli* transformed by the presumptive gene encoding grimelysin (ECP32) in *S. grimesii* A2 [18]. Taken together, these data suggested that ECP 32 and grimelysin is the same enzyme named grimelysin [18], slightly different from protealysin [19]. While *S.grimesii* and *S. proteamaculans* belong to a cluster of bacteria within the *Serratia liquefaciens* group, a similar protease could be expected to be synthesized by other members of this group. Our preliminary data indicate that protease with the actinase activity is present in *Serratia marcescens*.

Along with the high similarity of the bacteria *S. grimesii* and *S. proteamaculans*, their proteases, grimelysin and protealysin, are highly homologous and differ by only 8 amino acid residues [18,19]. Moreover, similarly to the grimelysin-producing bacteria [20,21], *S. proteamaculans* 94 turned out to be one more bacterial strain where the actinase activity of metalloproteinase protealysin is coupled with bacterial invasion [21,22]. These results are consistent with the idea of the actin-specific metalloproteases being a factor that can be involved in bacterial invasion of eukaryotic cells. This paper describes the properties of grimelysin and protealysin in vitro and in vivo in the context of this idea. Our data indicate that the protease is a virulence factor and actin can be a target for the protease upon its translocation into the host cell.

2. Basic Properties and Substrate Specificity of Grimelysin and Protealysin

Grimelysin (ECP 32), discovered, purified and initially characterized as protease ECP 32 [6–8], was later shown to be identical to grimelysin [18]. Therefore, the properties of the enzyme identified for ECP 32 could be applied to grimelysin. However, here we retain the name grimelysin (ECP 32) and ECP-cleaved actin to comply with the published data where the protease was named ECP 32. Grimelysin (ECP 32), purified from a bacterial extract using sequential chromatography steps, is a single 32 kDa polypeptide, whose *N*-terminal sequence was determined to be AKTSSAGVVIRDIFL [8]. The optimum of the protease activity was observed in the range of pH 7–8 when actin and melittin were used as substrates [8,23]. The proteolytic activity increased with increasing ionic strength: in 50–100 mM NaCl the activity of grimelysin (ECP 32) towards melittin was shown to be nearly twice higher than in a low ionic strength solution [23,24]. It was also enhanced in the presence of millimolar ATP concentrations, though hydrolysis of melittin was not accompanied by ATP hydrolysis at a rate comparable with the cleavage rate. This implies that protease grimelysin (ECP 32) is not an ATP-dependent enzyme [23], which is important for the experiments involving actin because actin contains ATP as a tightly-bound nucleotide. The protease activity is inhibited by EDTA, EGTA, o-phenanthroline and zincone, and the EDTA-inactivated enzyme can be reactivated by cobalt, nickel and zinc ions [2,3]. Based on these data, grimelysin (ECP 32) was classified as a neutral metalloproteinase (EC 3.4.24) [8].

Limited proteolysis of skeletal muscle actin between Gly-42 and Val-43 [10] was observed at enzyme: substrate mass ratios of 1:25 to 1:3000 [8]. Two more sites, between Ala-29 and Val-30 and between Ser-33 and Ile-34, were cleaved by ECP 32 in heat- or EDTA-inactivated actin, apparently due to conformational changes around residues 28–34 buried in intact actin [8]. Besides actin, only melittin [18,19], histone H5, bacterial DNA-binding protein HU and chaperone DnaK [25] were found to be protease substrates. In agreement with this high substrate specificity, ECP 32 did not hydrolyze tropomyosin, troponin, α-actinin, casein, histone H2B, ovalbumin, bovine serum IgG, bovine serum albumin, bovine pancreatic ribonuclease A, trypsin, human heat shock protein HSP70, chicken egg lysozyme [7], insulin [24], DNAse I [9,13], gelsolin [14] and profilin [26]. The amino acid residues recognized by grimelysin (ECP 32) in actin and melittin are hydrophobic. This specificity is characteristic for thermolysin-like metalloproteinases [27]. However, high specificity of the enzyme seems to be determined predominantly by conformation at the actin cleavage site rather than its primary structure.

Grimelysin was obtained as a recombinant protein. This has been achieved by cloning the putative gene encoding grimelysin in *S. grimesii* A2 and in the reference *S. grimesii* 30063 [18] using published protealysin sequences identified in *S. proteamaculans* [19]. Grimelysin shared all properties characteristic for ECP 32 including a molecular weight of 32 kDa, an *N*-terminal 14 amino acid sequence, optimum activity in the range of pH 7–8 and inhibition with o-phenanthroline and EGTA [18].

Protealysin is a neutral zink-containing metalloprotease of *Serratia proteamaculans*. The protealysin gene was cloned from a genomic library of *S. proteamaculans* strain 94 isolated from spoiled meat. This protein was expressed in *Escherichia coli* and purified as described earlier [19]. Similarly to other thermolysin-like proteases [27,28], protealysin is synthesized as a precursor containing a propeptide of about 50 amino acids that is removed during formation of mature active protein [29]. The propeptide is much shorter than the propeptides of the thermolysin-like proteases and has no significant structural similarity to the propeptides of most thermolysin-like proteases [30–32]. A similar propeptide of 50 amino acids was also detected in the primary structure of the recombinant grimelysin. According to SDS-electrophoresis, recombinant proteins with or without propeptide had an apparent molecular weight of 37 and 32 kDa, respectively [19].

The molecular weight of the active recombinant protealysin 32 kDa and the *N*-terminal amino acid sequence AKTSTGGEVI are identical to those of grimelysin [8,19]. The optimal pH for azocasein hydrolysis is 7, and protealysin is completely inhibited by *o*-phenanthroline [19], i.e., has the same properties as grimelysin [8,18]. Protealysin and grimelysin (ECP 32) are also similar in their unique property of being able to digest actin specifically [8–10,22,33].

3. Specific Actinase Activity of Grimelysin (ECP 32) and Protealysin

The ECP 32-like limited proteolytic activity towards actin appears in bacterial lysates of *S. grimesii* and *S. proteamaculans* only at the late stationary phase of bacterial growth [7,18,22]. The bacterial lysates as well as the purified grimelysin (ECP 32) or recombinant grimelysin and protealysin cleave actin at a single site giving rise to two fragments of 36 and 8 kDa (Figure 1A) [10,13,22]. The fragments remain associated in the presence of the tightly bound calcium or magnesium cation needed to maintain the native actin conformation and dissociate after removal of the tightly bound cation with EDTA [13]. The *N*-terminal sequence of the 36 kDa actin fragment produced by ECP 32 was determined as Val-Met-Val-Gly-Met [10]. The same sequence was determined for the 36 kDa *N*-terminal actin fragment produced by the lysates of recombinant bacteria expressing the grimelysin or protealysin gene [18,22]. According to the amino acid sequence of actin [34], this peptide corresponds to the cleavage site between Gly-42 and Val-43 [10], which is not attacked by any known proteases. However, in the case of protealysin, this cleavage pattern was observed only at the protealysin/actin mass ratio of 1:50 or lower. At a higher protealysin to actin ratio, the 36 kDa fragment was further cleaved yielding two closely situated bands with an apparent molecular weight of 33 kDa [22]. The *N*-terminal sequences of these fragments are Leu-Lys-Tyr-Pro-Ile-Glu and Ile-Leu-Thr-Leu-Lys-Tyr, corresponding to the cleavage of the bonds Thr66–Leu67 and Gly63–Ile64, respectively [22]. The difference in activity of the purified bacterial grimelysin and recombinant protealysin may be due to their origin, both in terms of different bacteria and different purification protocols. Therefore, comparison of the actinase activities of recombinant grimelysin and protealysin would be of interest.

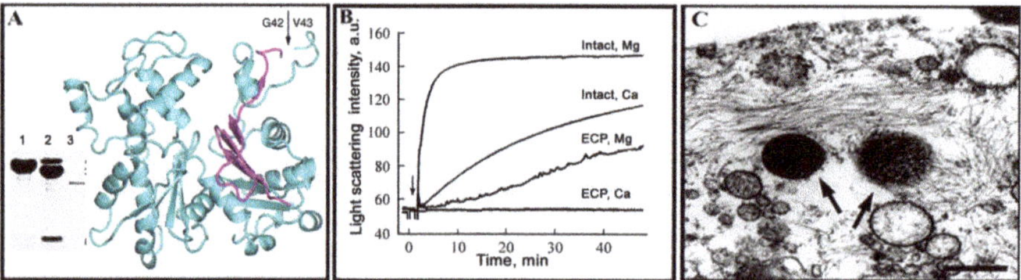

Figure 1. The actinase activity of grimelysin (ECP 32). (**A**) Three-dimensional structure of skeletal muscle actin cleaved with grimelysin (ECP 32) at a single site, Gly42-Val43, within the DNase I-binding loop [35]. Inset shows SDS electrophoresis of actin (lane 1), with the actin 36 and 8 kDa fragments produced by the grimelysin (ECP32) cleavage; the upper band is actin not completely cleaved in this experiment (lane 2) and isolated grimelysin (ECP 32) (lane 3) [8,10]. (**B**) Polymerization of ECP-cleaved skeletal muscle actin compared to that of intact actin. (**C**) Invasion of human larynx carcinoma Hep-2 cells by *Serratia grimesii* 30063. Intracellular bacteria are marked with arrows. Scale bar, 1 μm. Arrows indicate intracellular bacteria Reproduced from [13,21,35] with permissions from Elsevier Licences 4820990651384 and 1032953-1, 2020 (**A,B**) and from Wiley Licence 4821290007049, 2020 (**C**).

4. Specific Properties of Protease-Cleaved Actin

Although actin cleaved with grimelysin (ECP 32) between Gly42 and Val43 preserves its native conformation [10,13], the cleaved actin completely loses its ability to polymerize in the presence of Ca^{2+} [5,8] (Figure 1B). The ability of the cleaved actin to polymerize is partially restored upon substitution of the tightly bound Ca^{2+} with Mg^{2+} [13]. However, the degree of this polymerization is lower than that of intact actin, while the critical concentration of polymerization of ECP-Mg-actin and the exchange of subunits in the ECP-Mg-actin polymer increase 30 and 20–30 times, respectively (Figure 1B) [13]. This effect is due to a more open conformation of the cleaved actin monomers, which decreases the monomer nucleation step of cleaved actin polymerization or/and increases the actin filament dynamics, i.e., the dissociation/association kinetics of actin monomers at the filament ends [15]. The open conformation of the ECP-cleaved actin monomers is also preserved in the ECP-actin polymer [15]. Similar properties are inherent to actin cleaved with protealysin [36]. Thus, cleavage in globular actin of the only amino acid bond between Gly-42 and Val-43 causes local conformational changes that weaken the intermonomer contacts during actin polymerization and lead to enhanced polymer dynamics. As it turned out later, these properties correlated with the ability of the protease-producing bacteria to invade eukaryotic cells [20–22] (Figure 1C)

Similar instability arises if the protease cleaves the bond between Gly-42 and Val-43 in the subunits of the actin polymer. Incubation of filamentous actin (F-actin) with purified *S. proteamaculans* protealysin or with the lysates of the recombinant *Escherichia coli* producing protealysin showed that 20–40% of F-actin was digested at enzyme/actin mass ratios rising from 1:50 up to 1:5, respectively [33]. This process was accompanied by the increased steady-state ATPase activity (dynamics) of F-actin [33], which was also shown to be characteristic of the polymers forming from grimelysin (ECP 32)-cleaved actin [13]. The cleavage-produced increase in the dynamics of the modified filaments can be reversed with phalloidin [13], aluminium and sodium fluorides [22,36], as well as with actin-binding proteins gelsolin [14], myosin subfragment 1 [37] and tropomyosin [17], indicating that actin-binding proteins can restore integrity of actin filaments damaged by proteolysis.

This assumption was verified in vivo using microinjection of the purified protease grimelysin (ECP 32) preparation into *Amoeba proteus* cells, which revealed dynamics of the cytoskeleton structures by alterations in amoeba locomotion [38]. After injection of the protease solution, pseudopodia formation was ceased, and the cytoplasm motility slowed down and finally stopped [38]. Injected amoebae remained spread and immobile until the locomotion was slowly restored to the control level. No changes in the locomotion were observed when protease grimelysin (ECP 32) was injected in the presence of ECP 32-specific antibodies [38]. These results indicate that the protease-produced modifications of actin cytoskeleton can be reversible and might be used by bacteria to regulate their intracellular activity within eukaryotic and probably bacterial cells. Specifically, actin-like, bacterial proteins that form the bacterial cytoskeleton could be protease substrates.

5. Actin-Like Proteins of Bacteria

Many functions of bacterial cells are performed by filamentous structures similar to those of the cytoskeleton filaments of eukaryotic cells [39,40]. These structures are formed by globular proteins, which are regarded as actin homologues, actin-like proteins or bacterial actins [39] because three-dimensional structures of their monomers are similar to that of skeletal actin [39,40]. Also similarly to the eukaryotic globular actin, globular bacterial actin-like proteins assemble into two-stranded filaments consisting of two protofilaments coming together in various ways [39,40], although this sets them apart from the eukaryotic actin filaments assembled only in the form of two parallel strands. The most studied bacterial cytoskeleton proteins, MreB, ParM, MamK and Ftz, perform such functions as shape determination in rod-shaped bacteria, cell division, plasmid segregation and organelle positioning [39,40]. It would be tempting, therefore, to suggest that the actin-specific proteases, grimelysin and protealysin, could regulate these activities. However, all these bacterial proteins, including MreB, most closely related to actin [40–42] differ from eukaryotic actin due to the absence of the DNase I-binding loop inserted in eukaryotic actin subdomain 2 [41,42], where eukaryotic actin is cleaved by grimelysin or protealysin. Thus, in contrast to eukaryotic actin, bacterial actin-like proteins do not contain any specific cleavage site to be attacked by grimelysin or protealysin. This suggests that the specific actinase activity of grimelysin and protealysin against eukaryotic actin could be important if the protease is delivered by bacteria into eukaryotic cells.

6. Invasive Activity of Bacteria *Serratia grimesii* and *Serratia proteamaculans*

Capability of the actinase-producing bacteria to invade eukaryotic cells was first found upon incubation of human larynx carcinoma Hep-2 cells with bacteria *Serratia grimesii* A2 (*E. coli* A2) [20,43]. After 2 h of infection, bacteria were detected within the cells, mostly in vacuoles but also free in cytoplasm (Figure 1C). The invasion was accompanied by a change in the shape of the eukaryotic cells, disappearance of the actin stress fibers inside the cells and appearance of protrusions on the cell surface [43]. No changes were observed in the cells infected with the wild-type *Escherichia coli* CCM 5172 that do not produce actinases [43]. Correlation between synthesis of the specific actin hydrolyzing protease and the ability of bacteria to invade eukaryotic cells was also detected in *Shigella flexneri* mutant obtained by exposure of bacteria to furazolidone [44,45]. This treatment resulted in the formation of *Shigella flexneri* L forms whose revertants were not pathogenic but able to invade eukaryotic cells, while their extracts are capable of cleaving actin as the bacterial extracts containing grimelysin or protealysin (Figure 2) [44,45].

It was also shown that the transformed cells are more sensitive to invasion by actinase-producing bacteria than non-transformed or poorly transformed cells [20,46]. Specifically, the grimelysin (ECP 32)-producing bacteria invaded the transformed epithelial and fibroblasts cells A431, HeLa and 3T3-SV40, but they were not found in embryonic fibroblasts, primary human keratinocytes and in cells of poorly transformed 3T3 cell lines [20]. Later, a quantitative analysis of invasion confirmed rather a low susceptibility of 3T3 cells to *Serratia grimesii* invasion, compared to that of 3T3-SV40 cells [46]. The higher susceptibility of the immortal CaCo-2 and HeLa cells compared to their untransformed counterparts was also shown for invasion of *Listeria monocytogenes* [47]. Moreover, transfection of resistant *Listeria monocytogenes* cells by SV40 large T antigen was shown to induce highly transformed continuous cell lines with a susceptibility to bacteria phenotypes [48]. This difference may be produced by the different set of cell surface receptors contacting with bacteria in the "normal" and transformed cells, as is shown, for example, for the G-protein-coupled receptors [49].

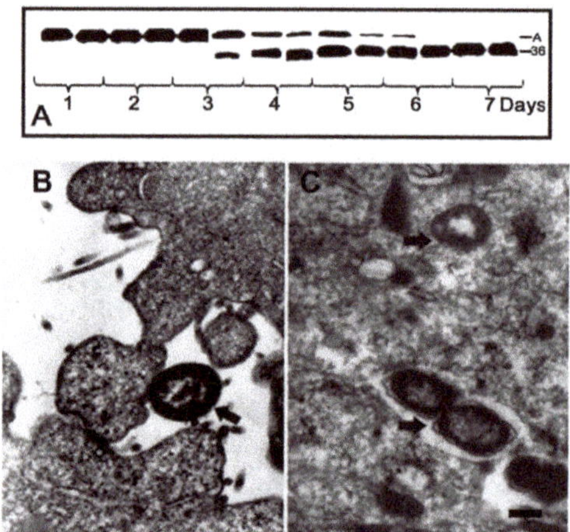

Figure 2. Apathogenic *Shigella flexneri* 5a2c mutant treated with furazolidone can invade eukaryotic cells. (**A**) Detection of the actinase activity in the lysates of *Shigella flexneri* 5a2c acquired upon the furazolidone treatment. Numbers indicate the time points of bacteria growth when their extracts were tested for the actinase activity. A, actin; 36, the 36 kDa actin fragment. (**B**) Incubation of human larynx carcinoma Hep-2 cells with the furazolidone-treated non-pathogenic *Shigella flexneri* 2a 4115. (**C**) Invasion of the pathogenic furazolidon-treated *Shigella flexneri* 5a2c into human larynx carcinoma Hep-2 cells [39]. Arrows indicate extracellular bacterium (**B**) and intracellular bacteria lying in vacuoles (**C**). Scale bar, 0.5 μm. Reproduced from [45] with permission from Springer Nature License 4823180887079, 2020.

To find out if grimelysin and protealysin are actively involved in the entry of *Serratia grimesii* and *Serratia proteamaculans* into host cells, human larynx carcinoma Hep-2 cells were infected with recombinant *Escherichia coli* expressing grimelysin or protealysin gene [21]. The results of this work showed that the extracts of the recombinant bacteria cleave actin at the only site corresponding to the cleavage with grimelysin or protealysin [21]. Recombinant *Escherichia coli* carrying the grimelysin or protealysin gene were found in the eukaryotic cells, both in vacuoles (Figure 3) and free in cytoplasm [21]. At the same time, no invasion was observed if the Hep-2 cells were incubated with the non-invasive *Escherichia coli*-carrying plasmids that did not contain the protease gene. Internalization of the non-invasive *Escherichia coli* was also not observed if protealysin was added to the culture medium [21]. These results showed the direct participation of grimelysin and protealysin in the invasion of the host cells by the protease-producing bacteria [21].

Using electron microscopy, two modes of *Serratia grimesii* invasion were revealed. In most cases, interaction of *Serratia grimesii* with eukaryotic HeLa M cells starts with formation of a tight contact between bacteria and the host cell, which is followed by bacteria internalization, apparently due to a specific interaction of bacterial adhesins with a specific cell surface receptor (Figure 4A,B) [50]. This process corresponds to the "zipper" mechanism of invasion involving activation of the signal system of the host cell followed by moderate cytoskeleton rearrangements [51–53]. However, in some cases, bacteria looked like they were trapped by filopodia (Figure 4C,D), probably induced by injected bacterial proteins triggering the bacterial uptake, as described in the "trigger mechanism" of invasion [54–57]. Recently, coexistence of both the trigger and zipper invasion mechanisms was postulated for *Salmonella* invasion [58–61]. These data provide a frame for revealing bacterial and cellular factors involved in *Serratia grimesii* and *Serratia proteamaculans* invasion.

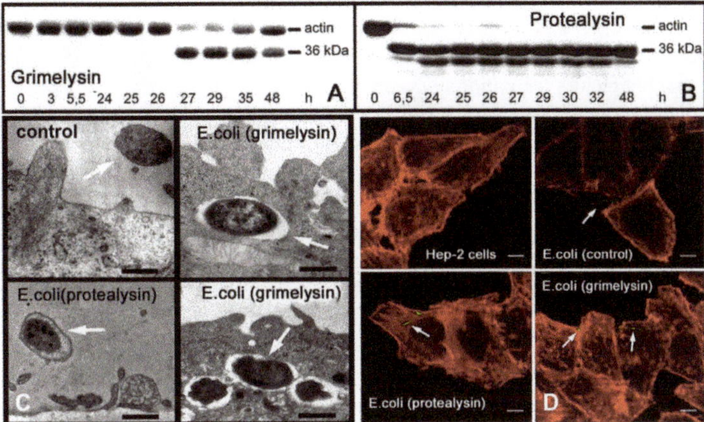

Figure 3. Actinase activity of recombinant *E. coli* transformed with the grimelysin or protealysin gene. (**A,B**) Detection of the actinase activity in the lysates of the recombinant *E. coli* expressing grimelysin or protealysin. Numbers indicate the time points of bacteria culturing when their extracts were tested for actinase activity [21]. (**C,D**) Invasion of human larynx carcinoma Hep-2 cells by recombinant *E. coli* expressing the grimelysin or protealysin gene, observed by electron (**C**) and confocal (**D**) microscopy [21]. Arrows indicate extracellular and internalized bacteria. By electron microscopy, bacteria are in vacuoles (**C**) that are not visible with confocal microscopy. Scale bars, 1 μm. (**D**) The samples were examined under a Leica TCS SL confocal scanning microscope using a dual argon ion (488 nm; green fluorescence) and helium/neon (543 nm; red fluorescence) laser system to visualize the FITC-stained bacteria and rhodamine phalloidin stained cytoskeleton, respectively. Scale bars, 1 μm (**C**) and 10 μm (**D**). Modified from [21] with permission from Wiley license 4827670496420, 2020.

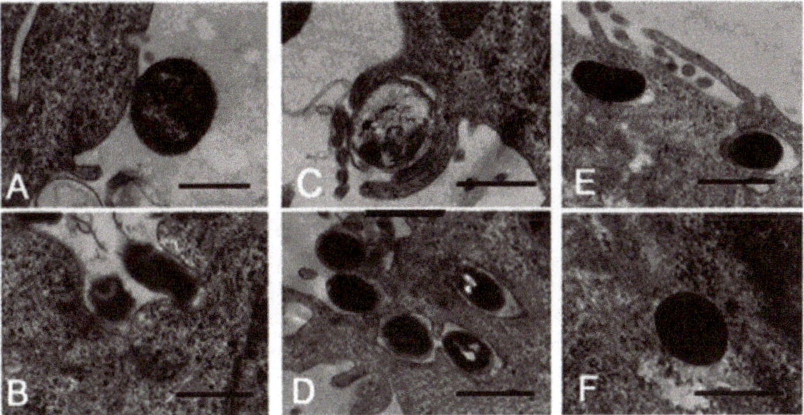

Figure 4. Electron microscopic identification of the initial steps in the invasion of eukaryotic cells by *Serratia grimesii*. (**A**) Tight contact of bacteria with HeLa M cells. (**B**) Extensive contact of bacteria with the surface of the host cell followed by the entry of bacteria into the host cell, which corresponds to the zipper mechanism of invasion. (**C,D**) Appearance of host cell filopodia, whose fusing traps bacterium and places it inside the vacuole, which corresponds to the trigger mechanism. (**E**) Bacteria live in the vacuoles. (**F**) Bacteria leave the vacuole to live free in the host cell cytoplasm. Scale bars, 1 μm. (Modified from [50] with permission from Pleiades Publishing, Ltd. 117342, 2020.).

7. Bacterial Virulence Factors Involved in *Serratia* Invasion

Transformation of the non-invasive *Escherichia coli* with the protealysin gene led to the appearance in the bacteria of the invasive activity [21]. However, inactivation of this gene did not abolish the ability of *Serratia proteamaculans* to invade eukaryotic cells [62]. Moreover, invasive activity of *Serratia proteamaculans* was five times higher than that of *Serratia grimesii* [63]. This indicates involvement of other factors regulating invasion.

Bacteria of *Serratia* genus are known to be facultative pathogens able to induce nosocomial infections or infections in immunocompromised patients [64–66]. A most pathogenic strain of *Serratia* genus is *Serratia marcescens*, pathogenicity of which is determined by a secreted pore-forming toxin (hemolysin) ShlA [67,68] and extracellular proteases [69]. Probing of *Serratia grimesii* and *Serratia proteamaculans* for these factors revealed the presence and high activities of pore-forming hemolysin ShlA and extracellular metalloprotease serralysin in *Serratia proteamaculans*. At the same time, in *Serratia grimesii*, the activity of the toxin ShlA was not detected, and the serralysin activity of the bacterial growth medium was very low [63]. It was also shown that iron depletion strongly enhanced the invasive activity of *S. proteamaculans*, increasing activities of hemolysin ShlA and serralysin, but did not affect these properties of *Serratia grimesii* [63]. These results showed that, along with protealysin, the invasive activity of *S. proteamaculans* is also determined by hemolysin and serralysin [63]. At the same time, grimelysin remains the only known virulent factor of *S. grimesii*.

The first step in bacterial invasion is the contact of bacteria with the surface of eukaryotic cells, often performed by bacterial outer membrane protein OmpX [70–72]. Indeed, transformation of *Escherichia coli* by a plasmid carrying the OmpX gene of *Serratia proteamaculans* caused a 3-fold increase in the adhesion of bacteria to the surface of eukaryotic cells Hep G2 and DF-2, without producing any effect on *Escherichia coli* invasion [73]. On the other hand, our preliminary data indicate that OmpX is a substrate for protealysin capable of enhancing the invasion of *Serratia proteamaculans* carrying the inactivated protealysin gene [73]. These data indicate that along with its direct participation in bacterial adhesion, OmpX could be involved in bacterial invasion, while, in turn, protealysin could regulate bacterial adhesion.

8. Cellular Factors Involved in *Serratia* Invasion

The penetration of bacteria into eukaryotic cells includes the contact of bacteria with eukaryotic cells, activation of the cell signaling system, reorganization of the cytoskeleton leading to bacterial uptake and spreading of the invaded bacteria within and between the cells [55,74,75]. The efficiency of this process is determined not only by the activity of bacterial virulent factors, but also by the sensitivity of eukaryotic cells, which, in turn, correlates with the degree of their transformation [20,47] and depends on the cell environment [76,77]. These factors were employed to find out whether bacterial invasion is sensitive to the cell type and the presence of antioxidant *N*-acethylcysteine (NAC) in the culture medium. The results of these experiments confirmed different effects of NAC on 3T3, 3T3-SV40 and HeLa cells [78]. Incubation of HeLa cells with NAC increased penetration of grimelysin-producing bacteria by 1.5–2 times for wild-type *Serratia grimesii* and by 3–3.5 times for recombinant *Escherichia coli* expressing the grimelysin gene [79]. These effects did not correlate with the cytoskeleton rearrangements but might be due to the NAC-induced upregulation of cell surface receptors playing a primary role in cell adhesion and cell–cell junctions [79].

Specifically, one of the cell surface receptors whose expression is regulated by *N*-acethylcysteine is E-cadherin, a transmembrane receptor known to be upregulated by NAC [77] and one of the two cell surface proteins that mediate adhesion and internalization of *Listeria monocytogenes* within epithelial cells [80,81]. Binding of bacteria internalin InlA to E-cadherin triggers the c-Src kinase-mediated phosphorylation of E-cadherin, followed by ubiquitination and E-cadherin internalization via the clathrin-mediated pathway [80–82]. In line with these data, incubation of 3T3 and 3T3-SV40 cells with NAC led both to increased expression of E-cadherin and increased sensitivity of these cells to invasion [46]. Co-localization of *S. grimesii* with E-cadherin of the 3T3 and 3T3-SV40 cells was also shown [46]. In addition, inhibitory analysis with the ROCK and c-Src kinase inhibitors revealed a correlation between c-Src and ROCK protein kinase activities, expression of E-cadherin and invasive activity of *Serratia grimesii* [83]. These data allow us to suggest that E-cadherin is at least one of the receptors involved in *Serratia grimesii* invasion.

On the other hand, participation of OmpX in *Serratia proteamaculans* invasion indicates that other cell surface receptors, including integrins and the epidermal growth factor receptor (EGFR), could be involved in this process, as it has been recently shown for the invasion of *Salmonella* [59–61]. In line with these data, infection of epithelial M-HeLa cells by bacteria *Serratia proteamaculans* led to the changes in localization of EGFR and fibronectin receptors α5-, β1-integrins. Accumulation of α5-, β1-integrins on the cell surface was accompanied by intensive attachment of bacteria to the sites of α5-integrin localization [84], indicating involvement of the outer membrane protein OmpX–fibronectin–integrin in the penetration of *Serratia proteamaculans* into eukaryotic cells.

9. Conclusions

Though *Serratia* are facultative pathogens able to cause nosocomial infections or infections in immunocompromised patients, hospital infection by *S. grimesii* or *S. proteamaculans* is low [64–66]. Consistent with this, the capability of *S. grimesii* to invade eukaryotic cells is also rather low, with only about 10% of the cells being invaded either by the wild-type or recombinant bacteria when cultured cells are infected in vitro [21]. This implies that production of grimelysin or protealysin does not make the bacteria pathogenic but rather provides them with an opportunity to be rendered pathogenic under specific conditions. Therefore, *Serratia grimesii* and *Serratia proteamaculans* invasion combines specific bacterial mechanisms to contact eukaryotic cells and the specificity of the cells to be invaded.

To penetrate eukaryotic cells, invasive bacteria should initiate their uptake by activating the signal transduction system and cytoskeleton rearrangements [1–4]. Thereby, pathogenic bacteria use specialized secretion systems to deliver bacterial virulence factors directly into the cytoplasm of the host cell to manipulate the intracellular pathways, thus following the trigger invasion mechanism [53–57,85]. In contrast, facultative bacterial pathogens use specific surface proteins that interact with the host cell receptors, according to the zipper invasion mechanism. These interactions activate receptor-mediated cell signaling cascades and lead to bacterial uptake [2,4,60,61,86]. Recent data showed however that the virulence components transported via the type III secretion system of pathogenic bacteria can be also translocated into the eukaryotic cells by the bacterial outer membrane vesicles (OMVs) [87]. Importantly, OMVs can be produced by opportunistic bacteria as well as by pathogenic ones [88–92], thus smoothing the difference between pathogenic bacteria and facultative bacterial pathogens.

Penetration of opportunistic bacteria *Serratia grimesii* and *S. proteamaculans* into eukaryotic cells [21] correlates with the presence in these bacteria of homologous metalloproteases grimelysin and protealysin, respectively [18,22]. Protealysin was found in the cytoplasm of 3T3-SV40 cells infected with *S. proteamaculans* or recombinant *Escherichia coli* expressing the protealysin gene [33], suggesting that protease can be transported into the cytoplasm of eukaryotic cells. Consistent with this suggestion, our preliminary data showed that the outer membrane vesicles produced by *Serratia grimesii* were able to transfer protease grimelysin into the cytoplasm of eukaryotic cells. This transfer enhanced the penetration of bacteria into eukaryotic cells if these cells were incubated with the vesicles prior to infection [93,94]. At least a part of this effect could be produced by the involvement of the OMV-delivered protease in the cytoskeleton dynamics. We have previously shown that the cleavage of actin with grimelysin (ECP32) enhanced turnover of actin monomers within actin filaments [15,38]. This, in turn, could contribute to the rearrangements of the actin cytoskeleton leading to bacteria internalization. The invasive activity of *Serratia* can also be regulated by activities of the intracellular factors, being, for example, diminished upon transfection of eukaryotic cells with anti-RhoA siRNA [83]. The cells transfected with anti-RhoA siRNA exhibited cell rounding, disassembly of actin cytoskeleton and formation of protrusions at the cell periphery [83], indicating an active involvement of the cytoskeleton in *S. grimesii* uptake.

Thus, the transition of opportunistic bacteria to the status of pathogenic is associated both with the activity of bacterial virulence factors and the sensitivity of eukaryotic cells to bacteria. In particular, the sensitivity of eukaryotic cells to bacterial invasion correlates with the degree of their transformation [20,46,47], the effects of antioxidants [78,79] and the activity of cell signaling components like c-SRC and ROCK kinases [83]. Further work is needed to integrate these individual cellular factors and bacterial virulence factors into the *Serratia* invasion mechanism.

Funding: The publication was partially supported by the "Director's Foundation of the Institute of Cytology".

Conflicts of Interest: The authors declare no conflicts of interest.

References

1. Bhavsar, A.P.; Guttman, J.A.; Finlay, B.B. Manipulation of host-cell pathways by bacterial pathogens. *Nature* **2007**, *449*, 827–834. [CrossRef]

2. Schnupf, P.; Sansonetti, P.J. Shigella Pathogenesis: New Insights through Advanced Methodologies. *Microbiol. Spectr.* **2019**, *7*, 15–39. [CrossRef] [PubMed]

3. Huang, Z.; Sutton, S.E.; Wallenfang, A.J.; Orchard, R.C.; Wu, X.; Feng, Y.; Chai, J.; Alto, N.M. Structural insights into host GTPase isoform selection by a family of bacterial GEF mimics. *Nat. Struct. Mol. Biol.* **2009**, *16*, 853–860. [CrossRef]

4. Pinaud, L.; Sansonetti, P.J.; Phalipon, A. Host Cell Targeting by Enteropathogenic Bacteria T3SS Effectors. *Trends Microbiol.* **2018**, *26*, 266–283. [CrossRef] [PubMed]

5. Rengarajan, M.; Hayer, A.; Theriot, J. A endothelial cells use a formin-dependent phagocytosis-like process to internalize the bacterium *Listeria monocytogenes*. *PLoS Pathog.* **2016**, *12*, e1005603. [CrossRef] [PubMed]

6. Mantulenko, V.B.; Khaitlina, S.Y..; Sheludko, N.S. High molecular weight proteolysis-resistant actin fragment. *Biochemistry* **1983**, *48*, 69–74.

7. Usmanova, A.M.; Khaitlina, S.Y. A specific actin-digesting protease from the bacterial strain *E. coli* A2. *Biochemistry* **1989**, *54*, 1074–1079.

8. Matveyev, V.V.; Usmanova, A.M.; Morozova, A.B.; Khaitlina, S.Y. Purification and characterization of the proteinase ECP-32 from *Escherichia coli* A2 strain. *Biochim. Biophys. Acta* **1996**, *1296*, 55–62. [CrossRef]

9. Khaitlina, S.Y..; Smirnova, T.D.; Usmanova, A.M. Limited proteolysis of actin by a specific bacterial protease. *FEBS Lett.* **1988**, *28*, 72–74. [CrossRef]

10. Khaitlina, S.Y..; Collins, J.H.; Kusnetsova, I.M.; Pershina, V.P.; Synakevich, I.G.; Turoverov, K.K.; Usmanova, A.M. Physico-chemical properties of actin cleaved with bacterial protease from *E. coli* A2 strain. *FEBS Lett.* **1991**, *279*, 49–51. [CrossRef]

11. Kabsch, W.; Mannherz, H.G.; Suck, D.; Pai, E.F.; Holmes, K.C. Atomic structure of the actin:DNase I complex. *Nature* **1990**, *347*, 37–44. [CrossRef] [PubMed]

12. Holmes, K.C.; Popp, D.; Gebhard, W.; Kabsch, W. Atomic model of the actin filament. *Nature* **1990**, *347*, 44–49. [CrossRef] [PubMed]

13. Khaitlina, S.Y..; Moraczewska, J.; Strzelecka-Golaszewska, H. The actin-actin interactions involving the N-terminal portion of the DNase I-binding loop are crucial for stabilization of the actin filament. *Eur. J. Biochem.* **1993**, *218*, 911–920. [CrossRef] [PubMed]

14. Khaitlina, S.; Hinssen, H. Conformational changes in actin Induced by Its Interaction with gelsolin. *Biophys. J.* **1997**, *73*, 929–937. [CrossRef]

15. Khaitlina, S.Y.; Strzelecka-Golaszewska, H. Role of the DNase-I-binding loop in dynamic properties of actin filament. *Biophys. J.* **2002**, *82*, 321–334. [CrossRef]

16. Moraczewska, J.; Gruszczynska-Biegala, J.; Redowicz, M.J.; Khaitlina, S.; Strzelecka-Golaszewska, H. The DNase-I binding loop of actin may play a role in the regulation of actin-myosin interaction by tropomyosin/troponin. *J. Biol. Chem.* **2004**, *279*, 31197–31204. [CrossRef]

17. Khaitlina, S.; Tsaplina, O.; Hinssen, H. Cooperative effects of tropomyosin on the dynamics of the actin filament. *Febs Lett.* **2017**, *591*, 1884–1891. [CrossRef]

18. Bozhokina, E.; Khaitlina, S.; Adam, T. Grimelysin, a novel metalloprotease from *Serratia grimesii*, is similar to ECP 32. *Biochem Biophys Res Commun.* **2006**, *367*, 888–892. [CrossRef]

19. Demidyuk, I.V.; Kalashnikov, A.E.; Gromova, T.Y.; Gasanov, E.V.; Safina, D.R.; Zabolotskaya, M.V.; Rudenskaya, G.N.; Kostrov, S.V. Cloning, sequencing, expression, and characterization of protealysin, a novel neutral proteinase from *Serratia proteamaculans* representing a new group of thermolysin-like proteases with short N-terminal region of precursor. *Protein Expr. Purif.* **2006**, *47*, 551–561. [CrossRef]
20. Efremova, T.; Ender, N.; Brudnaja, M.; Komissarchik, Y.; Khaitlina, S. Specific invasion of transformed cells by *Escherichia coli* A2 strain. *Cell Biol. Intern.* **2001**, *25*, 557–561. [CrossRef]
21. Bozhokina, E.S.; Tsaplina, O.A.; Efremova, T.N.; Kever, L.V.; Demidyuk, I.V.; Kostrov, S.V.; Adam, T.; Komissarchik, Y.Y.; Khaitlina, S.Y. Bacterial invasion of eukaryotic cells can be mediated by actin-hydrolysing metalloproteases grimelysin and protealysin. *Cell Biol Int.* **2011**, *35*, 111–118. [CrossRef] [PubMed]
22. Tsaplina, O.A.; Efremova, T.N.; Kever, L.V.; Komissarchik, Y.Y.; Demidyuk, I.V.; Kostrov, S.V.; Khaitlina, S.Y. Probing for actinase activity of protealysin. *Biochemistry* **2009**, *74*, 648–654. [CrossRef] [PubMed]
23. Kazanina, G.A.; Mirgorodskaia, E.P.; Mirgorodskaia, O.A.; Khaitlina, S.Y.. ECP 32 proteinase: Characteristics of the enzyme, study of specificity. *Bioorg. Khim.* **1995**, *21*, 761–766. [PubMed]
24. Mirgorodskaya, O.; Kazanina, G.; Mirgorodskaya, E.; Matveyev, V.; Thiede, B.; Khaitlina, S.Y.. Proteolytic cleavage of mellitin with the actin-digesting protease. *Protein Pept. Lett.* **1996**, *3*, 81–88.
25. Morozova, A.V.; Khaitlina, S.Y.; Malinin, A.Y. Heat Shock Protein DnaK—Substrate of actin-specific bacterial protease ECP 32. *Biochemistry* **2011**, *76*, 455–461. [CrossRef]
26. Khaitlina, S.; Lindberg, U. Dissociation of profilactin as a two-step process. *J. Muscle Res. Cell Motil.* **1995**, *16*, 188–189.
27. Rawlings, N.D.; Burrett, A.J. Evolutionary families of metallopeptidases. *Methods Enzym.* **1995**, *248*, 183–228. [PubMed]
28. Shinde, U.; Inouye, M. Intramolecular chaperones: Polypeptide extensions that modulate protein folding. *Semin. Cell Dev. Biol.* **2000**, *11*, 35–44. [CrossRef]
29. Gromova, T.Y.; Demidyuk, I.V.; Kozlovskiy, V.I.; Kuranova, I.P.; Kostrov, S.V. Processing of protealysin precursor. *Biochimie* **2009**, *91*, 639–645. [CrossRef]
30. Demidyuk, I.V.; Gasanov, E.V.; Safina, D.R.; Kostrov, S.V. Structural organization of precursors of thermolysin-like proteinases. *Protein J.* **2008**, *27*, 343–354. [CrossRef]
31. Demidyuk, I.V.; Gromova, T.Y.; Polyakov, K.M.; Melik-Adamyan, W.R.; Kuranova, I.P.; Kostrov, S.V. Crystal structure of the protealysin precursor: Insights into propeptide function. *J. Biol.Chem.* **2010**, *285*, 2003–2013. [CrossRef] [PubMed]
32. Demidyuk, I.V.; Shubin, A.V.; Gasanov, E.V.; Kostrov, S.V. Propeptides as modulators of functional activity of proteases. *Biomol. Concepts* **2010**, *1*, 305–322. [CrossRef]
33. Tsaplina, O.; Efremova, T.; Demidyuk, I.; Khaitlina, S. Filamentous actin is a substrate for protealysin, a metalloprotease of invasive Serratia proteamaculans. *Febs J.* **2012**, *279*, 264–274. [CrossRef] [PubMed]
34. Elzinga, M.; Collins, J.H.; Kuehl, W.M.; Adelstein, R.S. Complete amino-acid sequence of actin of rabbit skeletal muscle. *Proc. Natl. Acad. Sci. USA* **1973**, *70*, 2687–2691. [CrossRef] [PubMed]
35. Klenchin, V.A.; Khaitlina, S.Y.; Rayment, I. Crystal structure of polymerization-competent actin. *J. Mol. Biol.* **2006**, *362*, 140–150. [CrossRef]
36. Tsaplina, O.A.; Khaitlina, S.Y. Sodium fluoride as a nucleating factor for Mg-actin polymerization. *Biochem. Biophys. Res. Commun.* **2016**, *479*, 1746–1774. [CrossRef]
37. Wawro, B.; Khaitlina, S.Y.; Galinska-Rakoczy, A.; Strzelecka-Goaszewska, H. Role of DNase I-binding loop in myosin subfragment 1-induced actin polymerization. Implications to the polymerization mechanism. *Biophys. J.* **2005**, *88*, 2883–2896. [CrossRef]
38. Morozova, A.V.; Skovorodkin, I.N.; Khaitlina, S.Y.; Malinin, A.Y. Bacterial protease ECP 32 specifically hydrolyzing actin and its effect on cytoskeleton in vivo. *Biochemistry* **2001**, *66*, 83–90. [CrossRef]
39. Eun, Y.J.; Kapoor, M.; Hussain, S.; Garner, E.C. Bacterial filament systems: Towards understanding their emergent behavior and cellular functions. *J. Biol. Chem.* **2015**, *290*, 17181–17189. [CrossRef]
40. Gayathry, P. Bacterial actins and their interaction. *Curr. Top. Microbiol. Immunol.* **2017**, *399*, 22–1242.
41. Van den Ent, F.; Amos, L.; Löwe, J. Bacterial ancestry of actin and tubulin. *Curr Opin Microbiol.* **2001**, *4*, 634–638. [CrossRef]
42. Van den Ent, F.; Amos, L.A.; Löwe, J. Prokaryotic origin of the actin cytoskeleton. *Nature* **2001**, *413*, 39–44. [CrossRef] [PubMed]

43. Efremova, T.N.; Ender, N.A.; Brudnaia, M.S.; Komissarchik, I.I.; Khaĭtlina, S.I. Invasion of Escherichia coli A2 induces reorganization of actin microfilaments in Hep-2 cells. *Tsitologia* **1998**, *40*, 524–528.

44. Fedorova, Z.F.; Khaitlina, S.Y. Detection of actin-specific protease in the revertants *of Shigella flexnery* L. form. *Bull. Exp. Biol Med.* **1990**, *7*, 46–48.

45. Efremova, T.N.; Gruzdeva, I.G.; Matveev, I.V.; Bozhokina, E.S.; Komissarchuk, Y.Y.; Fedorova, Z.F.; Khaitlina, S.Y. Invasive characteristics of apathogenic *Shigella flexneri* 5a2c mutant obtained under the effect of furazolidone. *Bull. Exp. Biol. Med.* **2004**, *137*, 479–482. [CrossRef]

46. Ivlev, A.P.; Efremova, T.N.; Khaitlina, S.Y..; Bozhokina, E.S. Difference in susceptibility of 3T3 and 3T3-SV40 cells to invasion by opportunistic pathogens *Serratia grimesii*. *Cell Tissue Biol.* **2018**, *12*, 33–40. [CrossRef]

47. Velge, P.; Bottreau, E.; Kaeffer, B.; Pardon, P. Cell immortalization enhances *Listeria monocytogenes* invasion. *Med. Microbiol. Immunol.* **1994**, *183*, 145–158. [CrossRef] [PubMed]

48. Velge, P.; Kaeffer, B.; Bottreau, E.; Van Langendonck, N. The loss of contact inhibition and anchorage-dependent growth are key steps in the acquisition of *Listeria monocytogenes* susceptibility phenotype by non-phagocytic cells. *Biol. Cell.* **1995**, *85*, 55–66. [CrossRef] [PubMed]

49. Arakaki, A.K.S.; Pan, W.-A.; Trejo, J.A. GPCRs in cancer: Protease-activated receptors, endocytic adaptors and Signaling. *Int. J. Mol. Sci.* **2018**, *19*, 1886–1910. [CrossRef]

50. Bozhokina, E.S.; Kever, L.V.; Komissarchik, Y.Y.; Khaitlina, S.Y.; Efremova, T.N. Entry of facultative pathogen *Serratia grimesii* into Hela cells. Electron microscopic analysis. *Cell Tissue Biol.* **2016**, *10*, 60–68. [CrossRef]

51. Tiney, L.G.; Portnoy, D.A. Actin filaments and the growth, movement, and spread of the intracellular bacterial parasite, *Listeria monocytogenes*. *J. Cell Biol.* **1989**, *1091*, 597–608. [CrossRef]

52. Pizarro-Cerdá, J.; Kühbacher, A.; Cossart, P. Entry of *Listeria monocytogenes* in mammalian epithelial cells: An updated view. *Cold Spring Harb. Perspect. Med.* **2012**, *2*, a010009. [CrossRef]

53. Pizarro-Cerdá, J.; Cossart, P. *Listeria monocytogenes*: Cell biology of invasion and intracellular growth. *Microbiol Spectr.* **2018**, *6*, GPP3-00132018. [CrossRef] [PubMed]

54. Sansonetti, P.J. Molecular and cellular mechanisms of invasion of the intestinal barrier by enteric pathogens. The paradigm of *Shigella*. *Folia Microbiol.* **1998**, *43*, 239–246. [CrossRef] [PubMed]

55. Cossart, P.; Sansonetti, P.J. Bacterial invasion: The paradigms of enteroinvasive pathogens. *Science* **2004**, *304*, 242–248. [CrossRef]

56. Carayol, N.; Tran Van Nhieu, G. Tips and tricks about *Shigella* invasion of epithelial cells. *Curr Opin Microbiol.* **2013**, *16*, 32–37. [CrossRef]

57. Valencia-Gallardo, C.M.; Carayol, N.; Tran Van Nhieu, G. Cytoskeletal mechanics during *Shigella* invasion and dissemination in epithelial cells. *Cell Microbiol.* **2015**, *17*, 174–182. [CrossRef]

58. Rosselin, M.; Virlogeux-Payant, I.; Roy, C.; Bottreau, E.; Sizaret, P.Y.; Mijouin, L.; Germon, P.; Caron, E.; Velge, P.; Wiedemann, A. Rck of Salmonella enterica, subspecies enterica serovar enteritidis, mediates zipper-like internalization. *Cell Res.* **2010**, *20*, 647–664. [CrossRef]

59. Velge, P.; Wiedemann, A.; Rosselin, M.; Abed, N.; Boumart, Z.; Chausse, A.M.; Grepinet, O.; Namdari, F.; Roche, S.M.; Rossignol, A.; et al. Multiplicity of Salmonella entry mechanisms, a new paradigm for Salmonella pathogenesis. *Microbiol. Open* **2012**, *1*, 243–258. [CrossRef]

60. Boumart, Z.; Velge, P.; Wiedemann, A. Multiple invasion mechanisms and different intracellular behaviors: A new vision of Salmonella–host cell interaction. *Fems Microbiol Lett.* **2014**, *361*, 1–7. [CrossRef]

61. Mambu, J.; Virlogeux-Payant, I.; Holbert, S.; Grépinet, O.; Velge, P.; Wiedemann, A. An Updated view on the Rck Invasin of Salmonella: Still much to discover. *Front. Cell Infect. Microbiol.* **2017**, *7*, 500. [CrossRef] [PubMed]

62. Tsaplina, O. Cleavage of Ompx with protealysin can regulate *Serratia proteamaculans* invasion. In *Abstract Book of the 3rd International Conference SmartBio*; Vytautas Magnus University: Kaunas, Lithuania, 2019; p. 113.

63. Tsaplina, O.; Bozhokina, E.; Mardanova, A.; Khaitlina, S. Virulence factors contributing to invasive activities of Serratia grimesii and Serratia proteamaculans. *Arch. Microbiol.* **2015**, *197*, 481–488. [CrossRef]

64. Grimont, P.A.D.; Grimont, F.; Irino, K. Biochemical characterization of *Serratia liquefaciens* sensu stricto, *Serratia proteamaculans*, and *Serratia grimesii sp.*. *Curr. Microbiol.* **1982**, *7*, 69–74. [CrossRef]

65. Grimont, F.; Grimont, P.A.D. The genus Serratia. *Prokaryotes* **2006**, *6*, 219–244. [CrossRef]

66. Mahlen, S.D. *Serratia* infections: From military experiments to current practice. *Clin. Microbiol. Rev.* **2011**, *24*, 755–791. [CrossRef] [PubMed]

67. Hertle, R.; Schwarz, H. *Serratia marcescens* internalization and replication in human bladder epithelial cells. *BMC Infect. Dis.* **2004**, *4*, 16. [CrossRef]

68. Hertle, R. The family of *Serratia* type pore forming toxins. *Curr. Protein Pept. Sci.* **2005**, *6*, 313–325. [CrossRef]

69. Marty, K.B.; Williams, C.L.; Guynn, L.J.; Benedik, M.J.; Blanke, S.R. Characterization of a cytotoxic factor in culture filtrates of *Serratia marcescens*. *Infect. Immun.* **2002**, *70*, 1121–1128. [CrossRef]

70. Kolodziejek, A.M.; Sinclair, D.J.; Seo, K.S.; Schnider, D.R.; Deobald, C.F.; Rohde, H.N.; Viall, A.K.; Minnich, S.S.; Hovde, C.J.; Minnich, S.A.; et al. Phenotypic characterization of OmpX, an Ail homologue of Yersinia pestis KIM. *Microbiology* **2007**, *153*, 2941–2951. [CrossRef]

71. Kim, K.; Kim, K.P.; Choi, J.; Lim, J.A.; Lee, J.; Hwang, S.; Ryu, S. Outer membrane proteins A (OmpA) and X (OmpX) are essential for basolateral invasion of Cronobacter sakazakii. *Appl. Environ. Microbiol.* **2010**, *76*, 5188–5198. [CrossRef]

72. Meng, X.; Liu, X.; Zhang, L.; Hou, B.; Li, B.; Tan, C.; Li, Z.; Zhou, R.; Li, S. Virulence characteristics of extraintestinal pathogenic Escherichia coli deletion of gene encoding the outer membrane protein X. *J. Vet. Med. Sci.* **2016**, *78*, 1261–1267. [CrossRef] [PubMed]

73. Tsaplina, O.A. Participation of Serratia proteamaculans outer membrane protein (OmpX) in bacterial adhesion of eukaryotic cells. *Tsitologia (Rus).* **2018**, *50*, 817–820. [CrossRef]

74. Da Silva, C.V.; Cruz, L.; Araújo, N.S.; Angeloni, M.B.; Fonseca, B.B.; Gomes, A.O.; dos Carvalho, F.R.; Gonçalves, A.L.; Barbosa, B.F. A glance at Listeria and Salmonella cell invasion: Different strategies to promote host actin polymerization. *Int. J. Med. Microbiol.* **2012**, *302*, 19–32. [CrossRef] [PubMed]

75. Radoshevich, L.; Cossart, P. *Listeria monocytogenes*: Towards a complete picture of its physiology and pathogenesi. *Nat. Rev. Microbiol.* **2018**, *16*, 32–46. [CrossRef]

76. Parasassi, T.; Brunelli, R.; Costa, G.; De Spirito, M.; Krasnowska, E.; Lundeberg, T.; Pittaluga, E.; Ursini, F. Thiol redox transitions in cell signaling: A lesson from N-acetylcysteine. *Sci. World J.* **2010**, *10*, 1192–1202. [CrossRef]

77. Parasassi, T.; Brunelli, R.; Bracci-Laudiero, L.; Greco, G.; Gustafsson, A.C.; Krasnowska, E.K.; Lundeberg, J.; Lundeberg, T.; Pittaluga, E.; Romano, M.C.; et al. Differentiation of normal and cancer cells induced by sulfhydryl reduction: Biochemical and molecular mechanisms. *Cell Death Differ.* **2005**, *12*, 51128–51296. [CrossRef]

78. Gamalei, I.A.; Efremova, T.N.; Kirpichnikova, K.M.; Komissarchik, Y.Y.; Kever, L.V.; Polozov, Y.V.; Khaitlina, S.Y. Decreased sensitivity of transformed 3T3-SV40 cells treated with N-acetylcysteine to bacterial invasion. *Bull. Exp. Biol. Med.* **2006**, *142*, 90–93. [CrossRef]

79. Bozhokina, E.; Vakhromova, E.; Gamaley, I.; Khaitlina, S. N-Acetylcysteine increases susceptibility of HeLa cells to bacterial Invasion. *J. Cell. Biochem.* **2013**, *114*, 1568–1574. [CrossRef]

80. Bonazzi, M.; Veiga, E.; Pizarro-Cerda, J.; Cossart, P. Successive post-translational modifications of E-cadherin are required for InlA-mediated internalization of *Listeria monocytogenes*. *Cell Microbiol* **2008**, *10*, 2208–2222. [CrossRef]

81. Bonazzi, M.; Lecuit, M.; Cossart, P. *Listeria monocytogenes* internalin and E-cadherin: From structure to pathogenesis. *Cell Microbiol.* **2009**, *11*, 693–702. [CrossRef]

82. Ribet, D.; Cossart, P. How bacterial pathogens colonize their hosts and invade deeper tissues. *Microbes Infect.* **2015**, *17*, 173–183. [CrossRef] [PubMed]

83. Bozhokina, E.S.; Tsaplina, O.A.; Khaitlina, S.Y. The opposite effects of ROCK and Src kinase inhibitors on susceptibility of eukaryotic cells to invasion by bacteria *Serratia grimesii*. *Biochemistry* **2019**, *84*, 663–671. [CrossRef] [PubMed]

84. Tsaplina, A. Redistribution of the EGF receptor and α5-, β1-integins in response to infection of epithelial cells by Serratia proteamaculans. *Tsitologiya* **2020**, *62*, 349–355. [CrossRef]

85. Sasakawa, C. A new paradigm of bacteria-gut interplay brought through the study of *Shigella*. *Proc. Jpn. Acad. Ser. B.* **2010**, *86*, 229–243. [CrossRef]

86. Bahia, D.; Satoskar, A.R.; Dussurget, O. Editorial: Cell signaling in host–pathogen interactions: The host point of view. *Front. Immunol.* **2018**, *9*, 221. [CrossRef]

87. Kim, S.I.; Kim, S.; Kim, E.; Hwang, S.Y.; Yoon, H. Secretion of *Salmonella* pathogenicity island 1-encoded type III secretion system effectors by outer membrane vesicles in *Salmonella enterica* serovar Typhimurium. *Front. Microbiol.* **2018**, *23*, 2810. [CrossRef]

88. Chatterjee, D.; Chaudhuri, K. Association of cholera toxin with Vibrio cholerae outer membrane vesicles which are internalized by human intestinal epithelial cells. *FEBS Lett.* **2011**, *585*, 1357–1362. [CrossRef]

89. Kulp, A.; Kuehn, M.J. Biological functions and biogenesis of secreted bacterial outer membrane vesicles. *Annu. Rev. Microbiol.* **2010**, *64*, 163–184. [CrossRef]

90. Karthikeyan, R.; Gayathri, P.; Gunasekaran., P.; Lagannadham, M.V.; Rajendhran, J. Comprehensive proteomic analysis and pathogenic role of membrane vesicles of Listeria monocytogenes serotype 4b reveals proteins associated with virulence and their possible interaction with host. *Int. J. Med. Microbiol.* **2019**, *309*, 199–212. [CrossRef]

91. Rivera, J.; Cordero, R.J.B.; Nakouzi, A.S.; Frases, S.; Nicola, A.; Casadevall, A. Bacillus anthracis produces membrane-derived vesicles containing biologically active toxins. *Proc. Natl. Acad. Sci. USA* **2010**, *107*, 19002–19007. [CrossRef]

92. Vanhove, A.S.; Duperthuy, M.; Charrie're, G.M.; Le Roux, F.; Goudene'ge, D.; Gourbal, B. Outer membrane vesicles are vehicles for the delivery of *Vibrio tasmaniensis* virulence factors to oyster immune cells. *Environ. Microbiol.* **2015**, *17*, 1152–1165. [CrossRef] [PubMed]

93. Bozhokina, E.; Kever, L.; Komissarchik, Y. Secretion and internalization of outer membrane vesicles (OMV) from Serratia grimesii during bacterial invasion. In *Abstracts of the 2nd International Conference Smart Bio*; Vytautas Magnus University: Kaunas, Lithuania, 2018; p. 237.

94. Bozhokina, E. Outer membrane vesicles (OMV) of gram-negative bacteria Serratia grimesii: A role in bacterial invasion. In *Abstracts of the 3rd International Conference Smart Bio*; Vytautas Magnus University: Kaunas, Lithuania, 2019; p. 71.

© 2020 by the authors. Licensee MDPI, Basel, Switzerland. This article is an open access article distributed under the terms and conditions of the Creative Commons Attribution (CC BY) license (http://creativecommons.org/licenses/by/4.0/).

Review

Virulence Factors of Meningitis-Causing Bacteria: Enabling Brain Entry across the Blood–Brain Barrier

Rosanna Herold, Horst Schroten and Christian Schwerk *

Department of Pediatrics, Pediatric Infectious Diseases, Medical Faculty Mannheim, Heidelberg University, 68167 Mannheim, Germany; rosanna.herold@medma.uni-heidelberg.de (R.H.); horst.schroten@umm.de (H.S.)
* Correspondence: christian.schwerk@medma.uni-heidelberg.de; Tel.: +49-621-383-3466

Received: 26 September 2019; Accepted: 25 October 2019; Published: 29 October 2019

Abstract: Infections of the central nervous system (CNS) are still a major cause of morbidity and mortality worldwide. Traversal of the barriers protecting the brain by pathogens is a prerequisite for the development of meningitis. Bacteria have developed a variety of different strategies to cross these barriers and reach the CNS. To this end, they use a variety of different virulence factors that enable them to attach to and traverse these barriers. These virulence factors mediate adhesion to and invasion into host cells, intracellular survival, induction of host cell signaling and inflammatory response, and affect barrier function. While some of these mechanisms differ, others are shared by multiple pathogens. Further understanding of these processes, with special emphasis on the difference between the blood–brain barrier and the blood–cerebrospinal fluid barrier, as well as virulence factors used by the pathogens, is still needed.

Keywords: bacteria; blood–brain barrier; blood–cerebrospinal fluid barrier; meningitis; virulence factor

---

1. Introduction

Bacterial meningitis, as are bacterial encephalitis and meningoencephalitis, is an inflammatory disease of the central nervous system (CNS). It can be diagnosed by the presence of bacteria in the CNS. Despite advances in treatment, it is still a disease of global dimension, which can end fatally or leave long-term neurological sequelae in survivors [1].

The brain is well protected from invading pathogens by cellular barriers, with the two major barriers being the blood–brain barrier (BBB) and the blood–cerebrospinal fluid barrier (BCSFB) [2,3]. To reach the CNS, blood-borne bacteria must interact with and cross these barriers of the brain. During the course of infection attachment to the host cells is initiated, followed by the hijacking of several host cell pathways by the pathogens. This process can be used by the bacteria to enter host cells with subsequent intracellular survival and, for some pathogens, multiplication. Finally, after crossing into the brain, the bacteria elicit an immune response from the cells within the CNS that might contribute to the inflammatory events leading to disease, with potential disruption of barrier integrity [4].

The traversal or breach of these barriers by meningitis-causing pathogens is defined by a complex interplay between host cells and the pathogens, which use an array of virulence factors to facilitate this interaction, resulting in high morbidity and mortality [1]. These virulence factors are involved during protection in the bloodstream, such as the capsule of both gram-positive and gram-negative bacteria, mediate adhesion to and invasion into host cells, and are responsible for intracellular survival.

2. Barriers of the Central Nervous System

The barriers protecting the brain are essential for its function and the stability of its internal microenvironment [5]. Important morphological components found in most of these barriers are specialized intercellular tight junctions between the cells, which are the basis for the barrier property. Various transporters, such as ATP-binding cassette (ABC) transporters, as well as efflux pumps control the movement of molecules across these interfaces, making them another component of barrier function [6]. These mechanisms prevent substances such as toxins, pharmacologically active agents and pathogens from accessing the CNS [4,7,8].

2.1. Blood–Brain Barrier

The BBB is formed by brain microvascular endothelial cells (BMECs), astrocytes and pericytes. It presents a structural and functional barrier and acts as an interface between the CNS and the peripheral circulation [2,9]. It has an additional protective function, restricting the free movement of substances across and inhibiting the entry of pathogens and toxins into the CNS. By regulating the passage of molecules, it is furthermore responsible for maintaining the CNS homeostasis.

The structure of the BBB is defined by BMECs covering the inner surface of the capillaries by forming a continuous sheet of cells interconnected by tight junctions (TJs) [10]. Specific TJ proteins, including occludin and, at the BBB, claudin-5, and claudin-12, but not claudin-3, are responsible for limiting the paracellular diffusion of ion and solutes across the barrier [11,12]. BMECs are characterized by less cytoplasmic vesicles, a larger number of mitochondria and a high amount of intercellular junctions, resulting in a high transendothelial resistance and low paracellular flux [13]. Transporter systems of the BBB include solute carrier-mediated transport, receptor-mediated transport, as well as active efflux and ion transport [14]. The BMECs are supported in shaping the BBB by associated cells such as astrocytes and pericytes [10]. The mechanisms of microbial CNS invasion have been elucidated by the use of in vivo and in vitro studies. To model infections of the CNS, mice and rats are most commonly used. To model the BBB in in vitro cell culture systems, primary human, rodent, and bovine brain microvascular endothelial cells have been used. Extensive studies of the human BBB have been facilitated by the availability of immortalized brain endothelial cells such as human brain microvascular endothelial cells (HBMECs) and hCMEC/D3 [1].

2.2. Blood–Cerebrospinal Fluid Barrier

Another major barrier protecting the brain is the BCSFB, which can be separated into a barrier to the inner CSF in the ventricles at the choroid plexus (CP) and barriers to the outer CSF located at the arachnoidea and blood vessels present in the subarachnoidal space [15]. Importantly, the choroidal BCSFB is essential in the trafficking of immune cells into the CNS [3,16]. The epithelial cells displaying polarity and forming tight junctions, which specifically express claudin-1, -2, -3, and -11, are the morphological correlates of the BCSFB at the CP [3,11,12]. The presence of specific transporter systems, such as ABC transporters, and low pinocytotic activity regulating the crossing of substances are key characteristics of the CP [17,18]. Furthermore, the BCSFB is crucial for the protection of the CNS from pathogens [19].

To cause diseases of the CNS bacterial pathogens need to overcome these barriers [1,17]. During this process, several different steps are influenced by bacterial virulence factors. To analyze the role of the CP during infectious diseases of the CNS, animal models can be utilized. Important tools for studying the BCSFB in vitro are primary models of CP epithelial cells, such as primary porcine CP epithelial cells (PCPEC). Another functional model of the BCSFB is a human CP epithelial papilloma cell line (HIBCPP cells) which displays apical/basolateral polarity and a good barrier function [17].

3. Stages during the Pathogenesis of BACTERIAL Meningitis

Bacterial meningitis can be caused by a variety of different gram-positive and gram-negative pathogens. There are mechanisms and steps of disease progression that are similar among them as shown in Figure 1. Colonization of the mucosal surfaces is often the first step of disease progression, followed by different strategies of dissemination into the bloodstream [4]. Once the pathogens have entered the bloodstream and successfully counteracted innate and adaptive immune defenses, they have to translocate across the barriers of the brain to invade the CNS and cause severe inflammation. Attachment to the brain endothelium or the epithelial cells of the choroid plexus and subsequent invasion mark the initial step of translocation into the CNS [4].

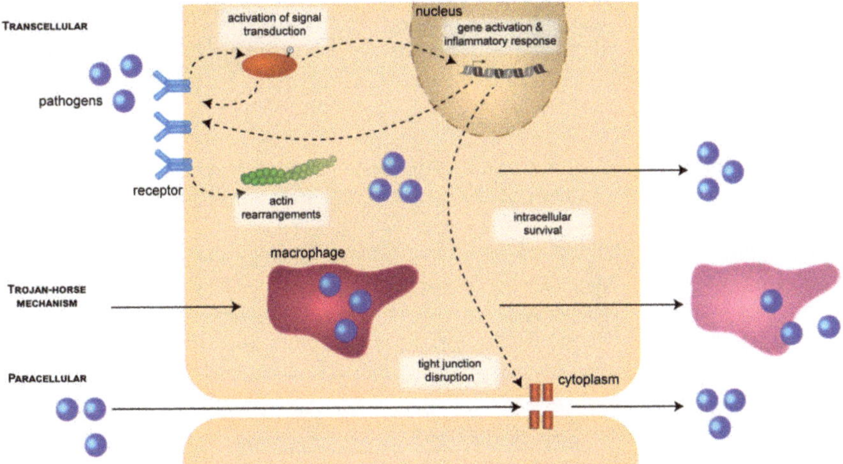

Figure 1. CNS entry pathways and stages during the pathogenesis of bacterial meningitis. Bacterial pathogens can cross the BBB or BCSFB paracellularly between neighboring cells, in a "Trojan horse" fashion inside infected host macrophages, or transcellularly by invading epithelial or endothelial cells. Cellular entry can be launched by the zipper mechanism involving binding to host cell receptors or by the trigger mechanism (see main text for details). During this process, activation of signal transduction pathways can cause initiation of actin rearrangements. Activation of signal transduction can also be triggered during the transcellular pathway or the "Trojan horse" mechanism, but is not indicated in the figure for reasons of clarity. Once in the cytoplasm, the pathogens need to survive inside the cells for further disease progress. Activation of host genes causes an inflammatory response that can lead to disruption of tight junctions.

3.1. Attachment and Invasion

To facilitate adhesion and invasion of the barriers protecting the brain, a threshold level of bacteremia has been shown to be required [20–22], which is correlated with the severity of infection and likeliness of developing meningitis [23]. However, direct invasion from neighboring infected tissues can occur as well.

Bacterial adhesion to host cell surfaces is a complex process. It involves multiple adhesion molecules of the pathogen interacting with a variety of target receptors. These interactions, which can involve several adhesins of one microbe, can occur in a sequential manner. Hereby, the initial interactions can trigger the expression of further host receptors, which are then targeted by other bacterial adhesins [24].

Many pathogens have been shown to bind to extracellular matrix proteins to facilitate initial attachment to the host cells. Furthermore, binding of bacterial adhesins to specific host cell receptors can in turn induce different signal transduction pathways, resulting in tight attachment or internalization of the bacteria into the host cells [4]. The use of pili or fibrils for invasion of HBMECs was observed for a multitude of meningitis-causing pathogens [8], making them highly important virulence factors for invasion of the CNS.

Endocytosis of pathogens into non-phagocytotic cells is initiated by one of two mechanisms: the "zipper" and the "trigger mechanism". Some pathogens express surface proteins capable of interacting with transmembrane receptors on the hosts cells which are connected to the cytoskeleton.

The "zipper mechanism", in particular, is defined by an interaction of a bacterial ligand and a host-specific membrane receptor initiating signaling events that lead to internalization of the pathogen through endocytosis [25]. The "trigger mechanism" is a micropinocytosis-related process that involves formation of actin-rich membrane ruffles. These are formed by localized changes in actin dynamics and membrane remodeling triggered by delivery of active effectors, which can be injected by the needle-like structure of a type three secretion system (T3SS), into the cytosol of host cells initiating signaling cascades [26,27].

3.1.1. CNS Entry Routes

Meningitis-causing pathogens most commonly cross host barriers in a transcellular or paracellular manner [2]. These processes are associated with protein interactions between pathogens and the host's cells. Transcellular traversal is characterized by pathogens crossing the barrier cells without evidence of TJ disruption or traversal between cells [2]. This is accomplished by intracellular invasion of the barrier cells and exploitation of signaling pathways. Paracellular traversal, on the other hand, involves penetration of pathogens between the host's cells and can occur with and without permanent disruption of TJs [2,4]. Furthermore, the release of bacterial toxins can lead to disruption of barrier function and promote paracellular traversal. Another means of entry is the "Trojan-horse" mechanism, which describes penetration of the barrier by transmigration within infected phagocytes [2]. It has been suggested that the infected phagocytes adhere to the luminal side of brain capillaries. This can occur with and without the activation of BMECs and is followed by either transcellular or paracellular traversal of the BBB [28].

3.1.2. Signal-Transduction Mechanisms and Cytoskeletal Rearrangements

Meningitis-causing bacterial pathogens have been shown to utilize host cell signaling molecules to facilitate infection. The mechanisms deployed can vary between the different pathogens, as well as the host tissues.

To both enter and leave the host cells, several pathogens have developed mechanisms to use the actin polymerization machinery of the host [29,30]. Pathogens not only use this process to spread throughout the cells, but have also developed mechanisms to subvert these regulatory mechanisms [31]. Furthermore, meningitis-causing pathogens use different signal-transduction mechanisms that result in rearrangements of the actin cytoskeleton [2]. This allows the pathogens to initiate attachment and entry of host cells, movement within and among cells as well as vacuole formation. This complex process of actin cytoskeletal remodeling can involve many factors, such as Rho-family GTPases and a variety of actin-binding proteins [30,32].

The innate immune system of the host can be triggered by various molecules which are characteristic for the bacteria. These pathogen-associated molecular patterns (PAMPs) are then recognized by eukaryotic pattern recognition receptors (PRRs), which in turn induce signaling cascades such as the nuclear factor κB (NF-κB) and mitogen-activated protein kinase (MAPK) pathways [33,34]. Activation of these signaling cascades triggers pro-inflammatory responses like the up-regulation of cytokines [35].

3.2. Intracellular Survival

3.2.1. Multiplication and Intracellular Survival

Extracellular multiplication is the most common process for propagation by pathogenic bacteria. Furthermore, replication and persistence inside host cells has been demonstrated for a variety of meningitis-causing pathogens as well [36]. Entry into the host's cells and replication within these protects the pathogens from clearance by the complement system and circulating antibodies. However, these pathogens have to overcome several cellular defense mechanisms such as the upregulation and secretion of neutrophil-specific factors in HBMECs. This response by the BBB is assumed to serve to recognize pathogens resulting in their clearance. However, overactivation of the cellular response through continued exposure to the pathogens could result in increased inflammation and compromised barrier integrity [8].

3.2.2. Disruption of Barrier Integrity and Inflammatory Response

Characteristically, bacterial meningitis is accompanied by a severe inflammatory response leading to neuronal damage. Activation of the transcription factor NF-κB upon invasion of the brain barrier tissues results in high levels of inflammatory cytokines in the blood and CSF [37]. This proinflammatory response can be triggered by bacterial cell wall components. Examples would include the lipopolysaccharide (LPS) for gram-negative bacteria, and lipoteichoic acid (LTA) for gram-positive bacteria. Furthermore, increased permeability of the barriers can be triggered by both bacterial toxins, as well as the initiation of host inflammatory pathways in response to the infection [38]. Tissue damage during bacterial meningitis arises from the initiated inflammatory cascade involving cytokines and chemokines, as well as proteolytic enzymes and oxidants [4]. Consequences of the release of these inflammatory substances, caused by multiplication of pathogens in the CNS, are damage of neurons and edema [39]. The hosts' immune response is therefore unable to embank infection of the CNS, and may even contribute to adverse events during bacterial meningitis [4].

The release of proinflammatory molecules not only increases permeability of the BBB but also attracts leukocytes to the CNS [40]. Cell death can be caused by cytokines, reactive oxygen species, reactive nitrogen species and matrix metalloproteinases [41–45].

4. Roles of Bacterial Virulence Factors During Invasion Through the Barriers of the CNS

To be able to enter the CNS, bacterial pathogens use several virulence factors, which are involved in the different steps of pathogenesis. Of major importance is the capsule of both gram-positive and gram-negative bacteria. The capsular polysaccharide has a protective function in bloodstream survival [46]. However, it was observed to attenuate invasion of the BBB and BCSFB [47–49]. This could result from electrostatic repulsion or from the masking of bacterial surface structures [8]. The necessity of the capsule for the pathogens survival in the blood but simultaneous hindrance of invasion of the host's tissues indicates a need for the regulation of capsule expression [50].

Besides the capsule, a multitude of further virulence factors are used by gram-positive and gram-negative bacteria [51]. These include, among others, adhesins and internalins, pore-forming toxins, and factors involved in intracellular movement as well as cell-to cell spread and are summarized in Figure 2.

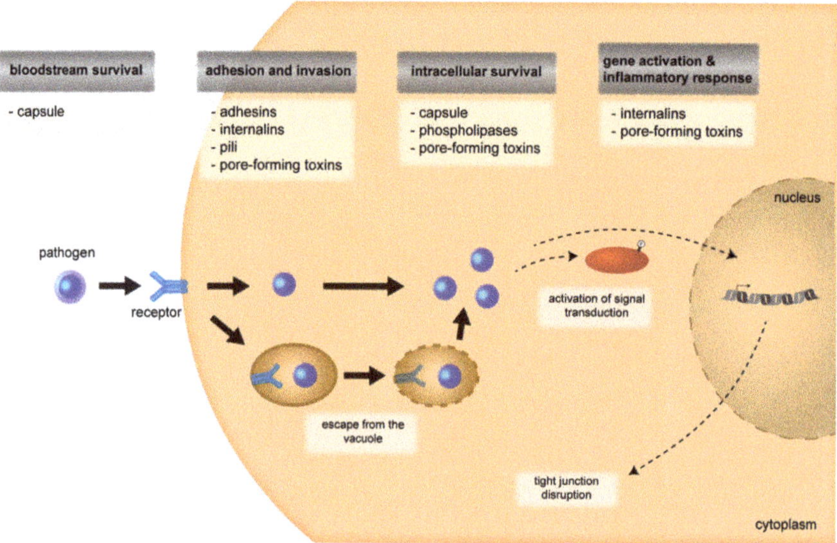

Figure 2. Multiple virulence factors are involved in the different steps of pathogenesis during bacterial meningitis. Expression of a capsule can support bloodstream survival of both gram-positive and gram-negative bacteria. It has been described that down-regulation of capsule expression occurs during adhesion to and invasion into host cells, which is mediated by adhesins, internalins, pili, and pore-forming toxins. Pore-forming toxins can also be involved during escape from vacuoles inside of host cells, as well as intracellular survival. These steps are further supported by pore-forming toxins and the capsule. Several virulence factors, including internalins and pore-forming toxins, activate host cell signal transduction and mediate gene activation causing an inflammatory response.

In this review, we will focus particularly on virulence factors, for which evidence for an involvement during brain entry across the blood–brain barriers has been proposed. A summary of these virulence factors for gram-positive and gram-negative bacteria is given in Table 1 at the end of this section.

4.1. Gram-Positive Bacteria

4.1.1. Listeria monocytogenes

Listeria monocytogenes (*L. monocytogenes*) is a facultative intracellular gram-positive bacterium. It can traverse several physiological barriers and finally enter the brain via the BBB or BCSFB, especially in immunocompromised individuals [52]. It is ingested through highly contaminated food by the host. Once ingested, *L. monocytogenes* traverses the intestinal epithelial barrier into the lamina propria followed by dissemination of the pathogen via the lymph and blood [53]. *L. monocytogenes* has multiple target organs, including the liver and spleen, and can enter the CNS across the barriers of the brain [53]. In addition to direct traversal of the BBB and BCSFB via the transcellular route, transportation across the BBB within leukocytes and retrograde migration within axons of cranial nerves have been described [54,55].

L. monocytogenes can enter non-phagocytotic cells by hijacking the host's receptor-mediated endocytosis machinery using the zipper mechanism. The two major invasion proteins of *L. monocytogenes* are internalin (InlA) and InlB, which bind to eukaryotic cell membrane members E-cadherin and tyrosine kinase receptor protein Met, respectively. These interactions induce receptor-mediated endocytosis of the pathogen. *L. monocytogenes* has been demonstrated to use one or both internalins to mediate invasion of the BBB and BCSFB [25,56,57]. A recent study has further demonstrated the importance of the bacterial surface protein InlF, showing that interaction with surface vimentin was required for an optimal colonialization of the brain [58].

The MAPK signaling cascade is activated during the invasion of *L. monocytogenes* [35,59,60]. In a model system of the BCSFB consisting of choroid plexus epithelial cells, the requirement of MAPK activation for listerial entry was demonstrated. Both extracellular signal-regulated kinases (ERK) 1 and 2 and p38 inhibition resulted in decreased bacterial invasion into this model system suggesting their involvement in the pathogens traversal of the BCSFB [34].

It was previously described that ubiquitination of E-cadherin and Met leads to the recruitment of the clathrin-mediated endocytosis machinery. This in turn results in the polymerization of the actin cytoskeleton. During this process, dynamin recruits several factors that result in two waves of actin rearrangements and subsequently result in the entry of the pathogen inside of vacuoles [61–63]. Accordingly, an in vitro study using a model of the BCSFB based on HIBCPP cells, revealed that *L. monocytogenes* invasion is inhibited if dynamin-mediated endocytosis is blocked [34].

Another essential virulence factor of *L. monocytogenes* is the pore-forming cytolysin Listeriolysin O (LLO). Activation of the NF-κB signaling pathway by LLO was reported in the human embryonic kidney HEK-293 cell line [64], as well as MAPK signaling [65,66]. It is secreted by *L. monocytogenes* and promotes the pathogens intracellular survival. After entering the host cell, lysis of the vacuole is initiated through LLO and the bacterial phospholipases PlcA and PlcB, and followed by intracellular spread in the cytoplasm [61].

Once *L. monocytogenes* has reached the cytoplasm of the host´s cells, it has been demonstrated to move around and enter neighboring cells using actin comet tails and membrane protrusions to facilitate its spread [61,67]. This F-actin-based intracellular motility is dependent on the expression of another essential listerial virulence factor, ActA [68].

Activation of the NF-κB signaling pathway is, as previously described, achieved through LLO. Another mechanism involving NF-κB is its activation by InlC, which is secreted intracellularly. It can directly interact with the subunit of the IκB kinase complex, IKKα. By phosphorylating IκB, this complex is critical for the activation of NF-κB, a major regulator of innate immune response. InlC was shown to impair phosphorylation of IκB, thereby scaling down the hosts immune response [69], and is also involved in cell-to-cell spread [70].

4.1.2. *Streptococcus suis*

Streptococcus suis (*S. suis*) is a zoonotic gram-positive bacterium and one of the most important porcine bacterial pathogens. Serotype 2 of *S. suis* has been described to be a major cause of meningitis, especially in South and East Asia [71]. To reach the CNS, *S. suis* has to colonize the host and traverse epithelial barriers in order to reach the bloodstream, where it needs to survive. *S. suis* has been demonstrated to cross the BBB and the BCFSB in human in vitro models as well as in porcine models [48,71–73].

The presence of a capsule is essential for survival in the bloodstream. However, it was demonstrated to attenuate invasion for *S. suis* in epithelial cells [48,72,74]. A link between capsule expression and carbohydrate metabolism has been described, indicating adaptation of *S. suis* to different environments. High concentrations of nutrients, as found in the bloodstream, coincided with high expression of the capsule, whereas in the CNS, which is low in nutrients, expression was reduced [50,75]. Attachment of *S. suis* to BMECs has been demonstrated in human and porcine in vitro models of the BBB [76,77]. Invasion has been reported in porcine models but at very low rates [73,78]. During the adhesion process, in these in vitro model systems, the capsule had no effect on adherence [76]. In contrast, in porcine and human models of the BCSFB, both attachment and significant invasion of *S. suis* strains were demonstrated. The use of unencapsulated mutants further increased invasion rates, indicating a role of the capsule and regulation of its expression [48,72]. Other important virulence factors for invasion of the barriers of the brain are cell wall components such as lipoteichoic acid, LPXTG-anchored proteins as well as lipoproteins [79].

Also, enolase has been identified as a virulence factor of *S. suis*. Previously thought to act only as a glycolytic enzyme, this protein, with a highly conserved sequence, has been implicated in the invasion process of various pathogens. For *S. suis*, enolase has been shown to increase BBB permeability as well as promoting the release of interleukin IL-8 [80]. Transmigration might further be promoted by the thiol-activated cytolysin suilysin, which induces pore formation in membranes. Furthermore, the use of bacterial mutants lacking suilysin has shown that it is not essential for invasion of the host [81]. It has, however, been demonstrated to promote association with epithelial cells, making it another major virulence factor of the pathogen [82].

The upregulation of different cytokines and chemokines in response to *S. suis* infection in BMECs has been reported. Exemplary is the induction of IL-6 and IL-8, stimulated by *S. suis* [83]. An inflammatory response was also described in a porcine model of the BCSFB. Here, induction of tumor necrosis factor (TNF) α and matrix metalloproteinase (MMP)-3 gene expression were described. This was paralleled by rearrangements of the tight junction proteins ZO-1, occludin and claudin-1, and loss of actin at the apical cell pole as well as the induction of stress fiber formation at the basolateral side of the barrier [84]. The expression of TNFα after stimulation by *S. suis* in porcine choroid plexus epithelial cells further promoted adhesion and transmigration of polymorphonuclear neutrophils (PMN) through the barrier, which is a critical step during bacterial meningitis. Interestingly, some PMNs contained internalized *S. suis*, indicating the possibility of the pathogen exploiting the Trojan-horse mechanism [85].

4.1.3. *Streptococcus pneumoniae*

Streptococcus pneumoniae (*S. pneumoniae*) is a gram-positive pathogen which is the major cause of bacterial meningitis in the developing world [86]. Close to 30% of individuals carry *S. pneumoniae* asymptomatically. Nasopharyngeal colonization is followed by systemic invasion and access to the bloodstream. Invasion of the CNS via the barriers of the brain is the major cause for meningitis [87]. Furthermore, olfactory neuron invasion was observed to be an entry route for *S. pneumoniae* [88]. To facilitate invasion of the CNS, *S. pneumoniae* utilizes several virulence factors such as the pneumococcal capsule and surface proteins as well as secreted proteins.

Bacterial interactions with the host and the subsequent development of bacterial meningitis are promoted by a high level of bacteremia [4]. Accordingly, attachment of *S. pneumoniae* to the choroid plexus in an in vivo mouse model was observed only during late stages of infection with high levels of bacteremia [87]. Further studies of the interaction with the BCSFB would be needed to determine if *S. pneumoniae* can use it as entry gate to the CNS.

The capacity of *S. pneumoniae* to invade host tissues is majorly determined by its capsule. Similar to other pathogens capable of causing bacterial meningitis, survival in the bloodstream is dependent on maximum capsule expression [89,90], whereas attachment to host tissues is hindered by its presence [91]. As the binding of various pneumococcal surface proteins is hindered by the capsule, altered expression of the capsule through quorum sensing and phase variation has been described [92,93].

Pneumococci have been demonstrated to use multiple virulence factors to initiate attachment to HBMECs in several in vitro studies. The platelet endothelial cell adhesion molecule (PECAM-1) has been implicated as receptor for neuraminidase A (Nad A) in pneumococcal attachment [94,95]. Importantly, NadA has been described as an important virulence factor, anchored in the cell wall of *S. pneumoniae* that can cleave sialic acid of the host substrates. It has been implicated in triggering the transforming growth factor-β (TGF-β) signaling cascade during the interaction with the BBB, resulting in decreased barrier integrity and an increase in invasion [96]. Another important receptor involved in pneumococcal BBB invasion is the polymeric immunoglobulin receptor (pIgR) as the major adhesin of the pneumococcal pilus-1, RrgA, was observed to bind to it as well as PECAM-1. Human platelet-activating factor receptor (PAFR) was demonstrated to be further involved in attachment, together with the before mentioned PECAM-1 and pIgR [97]. Additionally, a second type of pneumococcal pili has been described to mediate adhesion to host cells [98]. Furthermore, choline binding protein PspC was shown to bind only to pIgR [97], and interaction with the laminin receptor by *S. pneumoniae* is initiated by choline-binding protein A (CbpA) [99]. A proteome-based approach in a mouse meningitis model, addressing adaptive capabilities of the pathogens to a defined host compartment, has highlighted a crucial role for two highly expressed pneumococcal proteins; ComDE, a regulatory two-component system, and AliB, a substrate-binding protein of an oligopeptide transporter, in pneumococcal meningitis [100].

A major virulence factor of *S. pneumoniae* is pneumolysin, a pore-forming toxin which acts in a cholesterol-dependent manner. Pneumolysin has recently been demonstrated to induce the expression of CERB-binding protein (CBP), a coactivator of transcription. This results in the release of TNF-α and IL-6, which in turn lead to increased permeability of the BBB through increased apoptosis of the cells both in vivo and in vitro [101]. It has furthermore been implicated in paracellular traversal of the BBB by *S. pneumoniae* as a result of reduced barrier integrity [102].

In addition to the disruption of the BBB by pneumolysin, generation of H_2O_2 through α-glycerophosphate oxidase (GlpO) was observed to have a cytotoxic effect on HBMECs [103]. A study using rat brain tissues investigated the effect of *S. pneumoniae* infection on nucleotide-binding oligomerization domain 2 (NOD2) and inflammatory factors, suggesting that NOD2 may hold a role in the activation of inflammatory pathways and subsequent BBB damage [104].

A study by Coutinho et al. analyzed the CSF of patients with pneumococcal meningitis, detecting high levels of pro-inflammatory cytokines such as TNF-α, IL-1β, and IL-6 and anti-inflammatory cytokines IL-10 and TGF-β. Furthermore, chemokines IL-8, MIP-1a and MCP-1 were detected [45]. Furthermore, an in vivo study in neonatal Wistar rats demonstrated an increase in cytokines prior to BBB breakdown after induction of pneumococcal meningitis [105]. Entry of the CNS by the pneumococci is followed by rapid multiplication. Components released by the pathogen during this process are then recognized by PPRs resulting in a strong inflammatory response and subsequent BBB impairment [4].

4.1.4. Group B streptococcus

Group B streptococcus (GBS, *Streptococcus agalactiae*) is a β-haemolytic, gram-positive pathogen and the leading cause of meningitis in human neonates [106]. Classification of GBS strains is done by sequence type based on an allelic profile of seven loci [107]. The development of GBS meningitis is dependent on bloodstream survival and development of a high level of bacteremia of the pathogen, followed by the disruption of the BBB or possibly the BCSFB. This is followed by the multiplication of GBS in the CNS, culminating in severe inflammation and neural damage [4]. The necessity of high-level bacteremia indicates that bloodstream survival is an important virulence factor of the pathogen along with the sialylated GBS capsular polysaccharide [108]. Additionally, reduction of capsule expression by GBS was demonstrated to increase virulence and intracellular persistence [49]

Direct interaction of GBS with the BBB endothelium has been demonstrated, resulting in traversal of the barrier and subsequent infection of the CNS [49,109]. Both direct invasion of the BBB and/or brain invasion as a direct result of increased permeability of the barrier have been observed [4]. The use of in vitro models has further shown transcellular crossing of the pathogen [109]. To this end, a variety of virulence factors have been described. One of these factors is a surface anchored novel protein specific for the GBS ST-17 clone, which is associated with meningitis in infants after the first week of life, called hypervirulent GBS adhesin (HvgA) and is required for GBS hypervirulence. Increased adherence of strains expressing HvgA was detected for intestinal epithelial cells, choroid plexus epithelial cells, and microvascular endothelial cells of the BBB [110]. Another necessary determinant of the GBS interaction with the BBB is the expression of cell-wall anchored pili [111]. Interestingly, these pili displayed similar function in adhesion and invasion of GBS, given that the role of pili has been best described for gram-negative bacteria. Two proteins involved in the formation of the pili, encoded by *pilA* and *pilB*, were shown to facilitate the interaction with the BBB, wherein PilA is promoting attachment of GBS and PilB is mediating internalization of the bacterium [112]. Interaction of PilA with collagen promotes its interaction with the $\alpha_2\beta_1$ integrin, initiating the integrin signaling machinery [38]. Interaction with HBMECs is further enabled by the GBS fibronectin-binding protein streptococcal fibronectin binding protein A (SfbA) [113]. Additionally, GBS serine-rich repeat (Srr) glycoprotein was suggested to both promote bloodstream survival, facilitated by the adherence to fibrinogen, as well as adhesion to HBMECs [114]. An antigen I/II family adhesin, BspC, was recently demonstrated to interact with host cell vimentin during the pathogenesis of GBS meningitis, thereby promoting adherence of the pathogen in vitro as well as contributing to the development of GBS meningitis in vivo [115].

During invasion of type III GBS in HBMECs, tyrosine phosphorylation of the focal adhesion kinase (FAK) was demonstrated. Not only was the phosphorylation of the FAK required for invasion of GBS, it further induced association with PI3-kinase and paxillin, an actin filament adaptor protein [116]. BBB penetration of GBS was shown to also involve the *invasion associated gene A (iagA)*. Mice challenged with a mutant version of this gene, encoding a glycosyltransferase homolog, developed bacteremia comparable to the wild type (WT) mice but had significantly lower mortality. In addition, the *IagA* gene encodes an enzyme, the glycolipid diglucosyldiacylglycerol, which functions as a cell membrane anchor for LTA, indicating that proper LTA anchoring is necessary for invasion of GBS into the BBB [117].

Similarly to *S. pneumoniae*, GBS can secrete a pore-forming toxin to disrupt barrier function in infected BMECs [109]. In addition, a further consequence of the interaction of PilA with $\alpha_2\beta_1$ integrin is the activation of host chemokine expression as well as neutrophil recruitment, which was correlated with increased permeability of the BBB [38]. Furthermore, GBS hyaluronidase HylB was demonstrated to induce BBB opening in a dose-dependent manner [118]. Another factor influencing BBB dysfunction was described during infection of induced pluripotent stem cell-derived brain endothelial cells with GBS, which resulted in the inhibition of P-glycoprotein, an important efflux transporter for the maintenance of brain homeostasis [119]. Barrier disruption and subsequent bacterial passage was further demonstrated to be promoted by reduced expression of tight junction components ZO-1, Claudin-5 and Occludin in HBMECs. This was facilitated by the induction of transcriptional repressor Snail1 and sufficient to promote tight junction disruption. This process, which was shown to be dependent on ERK 1/2 MAPK signaling as well as bacterial cell wall components, marks another mechanism of BBB disruption by GBS [120].

4.2. Gram-Negative Bacteria

4.2.1. Escherichia coli

E. coli, a gram-negative bacillary organism, is a common cause of meningitis and still an important cause of mortality and morbidity throughout the world. Circulating *E. coli* have been shown to traverse the BBB and the BCSFB as a result of hematogenous spread [22,121]. Expression of the K1 capsule and O-lipopolysaccharide are critical determinants of *E. coli* meningitis, especially in neonates [122–124]. Several factors have been demonstrated to influence *E. coli* invasion of the CNS such as a high level of bacteremia for the invasion of the blood–brain barrier, as well as a variety of virulence factors that initiate binding and translocation into the CNS [2].

The first step in *E. coli* invasion of the CNS is attachment to the cells of the brain barriers. The two major virulence factors associated with attachment to the blood–brain barrier are type 1 fimbriae and OmpA [125,126]. The virulence factor IbeA has been associated with the subsequent invasion process [127], as well as cytotoxic necrotizing factor 1 (CNF1) [128]. Deletion of *ompA* and *ibeA* reduced infection of choroid plexus epithelial cells in a human model of the BCSFB, whereas deletion of *fimH* enhanced invasion but simultaneously decreased adhesion of *E. coli* strains in the same model [121]. In an in vitro model of the BBB, Ecgp was identified as receptor for OmpA [129].

To promote internalization, *E. coli* induces rearrangements of the actin cytoskeleton. In HBMECs, this has been demonstrated to trigger a zipper-like mechanism that envelops the bacterium and initiates internalization into the cell. This process is dependent on both the actin cytoskeleton as well as microtubules [4]. Induction of tyrosine phosphorylation of FAK and cytoskeletal proteins by *E. coli* was demonstrated in an in vitro model of brain endothelial cells [130]. In addition, phosphatidylinositol

3-kinase (PI3K) interaction with FAK and PI3-kinase signaling was shown to be necessary for successful invasion of *E. coli* in HBMECs [131]. In turn, PI3K activates phospholipase PLCγ, resulting in increased Ca^{2+} levels in the cells [132]. The cell adhesion molecule ICAM-1 was selectively upregulated during invasion of *E. coli* in brain endothelial cells through the interaction of OmpA and its receptor Ecgp. This upregulation was dependent on the previously described PI3K signaling pathways, as well as protein kinase C (PKC)-α and NK-κB signaling [133].

Traversal of *E. coli* across the BBB has been studied extensively using HBMECs as in vitro models [22]. Transmission electron microscopy revealed the pathogen crossing the HBMECs in membrane-bound vacuoles without intracellular replication [134]. Infection of HBMECs by *E. coli* K1 was further demonstrated to activate caveolin-1, resulting in the uptake of the pathogen via the caveolae. Furthermore, caveolin-1 interacts with phosphorylated protein kinase Cα at the site of *E. coli* attachment [135]. To promote intracellular survival within the vacuoles, the *E. coli* K1 capsule is essential. It can modulate the maturation of the vacuoles and prevent fusion with lysosomes, thereby enabling traversal of live bacteria to the CNS [136]. There are gene clusters, essential for the production of the precursors that are the basis of the K1 capsule, called the *neuDB* genes. They were identified as an essential virulence factor promoting intracellular survival in HBMEC [122,136]. Furthermore, *neuDB* was also shown to be of importance during infection of an in vitro model of the BCSFB [121].

The breakdown of barrier function and neuroinflammation are considered key mechanisms in the invasion of the brain by pathogenic *E. coli*. In brain endothelial cells, upregulation of platelet-derived growth factor-B (PDGF-B) and ICAM-1 was demonstrated after infection with *E. coli* [137]. Further in vivo and in vitro studies suggest an involvement of PDGF-B in BBB permeability, mediating breakdown of tight junction proteins. Upregulation of ICAM-1 on the other hand was shown to initiate the inflammatory response of the CNS, mediating neutrophils or monocyte recruitment during infection [138]. The activation of PKC–α and its association with vascular-endothelial cadherins at the TJs of HBMECs resulted in increased cellular permeability and decrease in transendothelial electrical resistance by releasing β-catenin from the junctions. Notably, only *E. coli* strains expressing *ompA* could induce this increase in barrier permeability [139]. The infection of HBMECs with *E. coli* was further demonstrated to promote the production of nitric oxide (NO) by activating inducible nitric oxide synthase and, as a result, displayed enhanced invasion rates and increased permeability of HBMEC monolayers [140]. NO production was hypothesized to additionally be triggered by the modulation of pterin synthesis, which is involved in cell differentiation, pain modulation as well as mRNA stability. Infection of HBMECs was demonstrated to induce the rate-limiting enzyme in pterin synthesis, guanosine triphosphate cyclohydrolase (GCH1), indicating its role in the invasion process. GCH1 further interacts with Ecgp96, the receptor for OmpA [141]. These findings indicate an essential role for NO during *E. coli* invasion of the BBB. The response of HBMECs to meningitic and non-meningitic *E. coli* infections highlighted the role for macrophage migration inhibitory factor (MIF) during infection. MIF, a proinflammatory cytokine that has been described as a major factor during infection and septic shock, was demonstrated to have a role in BBB damage, as evidenced by the induction of a significant decrease in ZO-1 and occludin, as well as inflammation [142].

A further study described the role of *E. coli* K1 virulence factor *cglD* in polymorphonuclear leukocyte transendothelial migration [37]. A follow up analysis demonstrated a contribution of *cglD* to NF-κB pathway activation in HBMECs, promoting PMN adhesion and transendothelial migration across the BBB [143].

4.2.2. *Neisseria meningitidis*

Neisseria meningitidis (*N. meningitidis*) is a human-specific gram-negative bacterium. It can colonize the nasopharynx extracellularly and is often non-pathogenic and commensal. Some strains can cause life-threatening diseases such as meningitis. To reach the barriers protecting the CNS, *N. meningitidis* has to overcome the mucosal epithelium and enter the bloodstream. To survive in the bloodstream and subsequently enter the CNS, *N. meningitidis* utilizes different virulence factors. These protect the bacterium from being killed by the hosts complement system or other effectors. They include the polysaccharide capsule and other surface structures, such as pili and other adhesins [144]. Factor H-binding protein is another important factor for bloodstream survival and evading the host innate immune system by binding factor H, which is a negative regulator of complement activation and alternative pathway and is bound to the surface of the pathogen [145,146]. *N. meningitidis* has been proposed to cross both the BBB and the BCSFB to the inner and the outer CSF [48,147–149].

The polysaccharide capsule of *N. meningitidis* is a major contributor to meningococcal disease and has been described as its main virulence factor. It can undergo genetic regulation and has the capability to mask the function of non-pilus adhesins [150]. While the capsule is essential for bloodstream survival, adhesion and invasion of host tissues are inhibited by the capsule [151]. Attenuated invasion was further described for capsulated strains of *N. meningitidis* in an in vitro model of the BCSFB [48]. The loss of the bacterial capsule for members of group B and group C meningococcal strains lead to increased uptake into HBMECs [152], and capsule and pili of *N. meningitidis* are downregulated upon contact with epithelial cells [153].

Adherence of *N. meningitidis* to host cells is facilitated by different virulence factors such as pili and surface exposed proteins like Opa and Opc, which further contribute to meningococcal disease. The type IV pili are a crucial adhesin expressed by *N. meningitidis*. They are involved in the attachment of capsulated virulence strains to host surfaces, extending from the bacterial surface through the capsule [154]. These long filamentous structures, made up of the major pilin protein and three minor pilins, promote adhesion of *N. meningitidis* to human endothelial cells via their interaction with CD147 [155]. In addition to the major pilin, type IV pili adhesion is also dependent on the expression of the outer membrane protein PilC [156–158]. Although affinity of pilin monomers to CD147 is weak, their assembly into the type IV pili as well as high expression of CD147 on the BBB supports the interaction [155]. Another group of adhesin-like structures, termed minor adhesion proteins, is expressed at low levels in vitro but may be of importance in in vivo environments which have been shown to alter the transcriptome of *N. meningitidis* [159].

While host, as well as tissue specificity, is determined by the pili and Opa proteins, invasion is facilitated by both Opa and Opc [160]. Opa proteins have been shown to bind to the carcinoembryonic antigen-related cellular adhesion molecule (CEACAM) receptor family, as well as the extracellular matrix proteins fibronectin and/or vitronectin [161,162]. An increase in adhesion and entry into HBMECs has been reported for *N. meningitidis* expressing Opc. This effect is facilitated by binding of Opc to serum vitronectin or fibronectin [163]. Also, activation of the acid sphingomyelidase/ceramide system, which involves clustering of ErbB2, an important receptor involved in bacterial uptake, by Opc-expressing *N. meningitidis* played a major role in determining the pathogens invasiveness [164]. Furthermore, the bacterial adhesins PilQ and PorA were shown to bind to the laminin receptor, thereby initiating contact with the BBB [99].

Attachment and invasion of human cells by three hypervirulent serogroup B strains of *N. meningitidis* is furthermore promoted by the *Neisseria* adhesin A (NadA), a phase-variable, surface-exposed protein [165–167]. For its interaction with the host, different membrane proteins have been suggested. NadA was shown to target human β1 integrin subunits [150]. Also, a genome-wide microarray analysis provided evidence for an interaction between NadA and the endothelial low-density oxidized lipoprotein receptor 1 (LOX-1) [168]. Following initial binding to the cell surface, the formation of microcolonies was observed, which in turn promoted the formation of specific molecular complexes called "cortical plaques". These structures contain accumulated ezrin, moesin, tyrosine-phosphorylated proteins, ICAM-1/-2, CD44 and epidermal growth factor receptors (EGFR) as well as localized polymerized cortical actin, resulting in major reorganization of host cell morphology. Changes in cell surface morphology are a prerequisite for bacterial uptake and the formation of microvilli-like cellular projections which protect the microcolonies from shear stress of the bloodstream [169].

Following attachment, *N. meningitidis* induces host cell signaling events to facilitate invasion into host tissues. These include the recruitment of ezrin as well as the activation of Src kinase and cortactin [170–172]. Furthermore, the FAK is essential for integrin-mediated internalization of *N. meningitidis* in HBMECs, enabling endocytosis through the interplay between FAK, Src and cortactin. This is facilitated by *N. meningitidis* using the integrin signaling pathways to mediate signaling from activated integrins upon attachment to the cytoskeleton [173]. These events promote the reorganization of the actin cytoskeleton resulting in the formation of membrane protrusions that take up the pathogens by surrounding them for internalization [144]. Genes involved in cytoskeletal reorganization were shown to be differentially expressed after infection of HBMECs [174]. Analysis of whole-cell lysates of human endothelial and epithelial cells have revealed an interaction of the Opc protein with alpha actinin, a modulator of various signaling pathways and cytoskeletal functions [175].

Disruption of the BBB integrity was observed in in vitro studies as a consequence of opening intracellular junctions [176]. Furthermore, cell detachment is initiated through activation of MMP 8, promoting cleavage of the TJ protein occludin [177]. Alterations of intracellular junctions were observed in a human brain microvascular endothelial cell line, potentially opening up a paracellular route of crossing the BBB into the CNS through recruitment of the Par3/Par6/PKCζ polarity complex [178].

Meningococcal disease is accompanied by an acute inflammatory response [179]. The transcription factor (TF) NF-κB, associated with the release of proinflammatory cytokines and chemokines during the inflammatory response, was shown to be active in an in vitro CP epithelial model after infection with *N. meningitidis*. Its activation was believed to arise from heterodimerization of TLR2 and TLR6 [33]. In HBMEC the p38 MAPK had an impact on the release of IL-6 and IL-8, whereas the c-JUN N-terminal kinases 1 and 2 (JNK1 and JNK2) were important for invasion [180].

4.2.3. *Haemophilus influenzae*

Haemophilus influenzae (*H. influenzae*) is a gram-negative bacterium capable of colonizing the upper respiratory tract. The makeup of the polysaccharide capsule determines the pathogens classification into serotypes a to f, of which *H. influenzae* serotype b (*Hib*) is responsible for the most severe infections, such as meningitis, especially in children under the age of 5 years [181]. Although there is widespread vaccination against *Hib*, some populations remain vulnerable and the vaccines do not protect against other serotypes [182]. Furthermore, in areas where Hib vaccines are used, nontypeable *H. influenzae* (NTHi) strains can now be the most common cause of meningitis [183]. Similar to the before described pathogens, *H. influenzae* has to cross the epithelial barrier of the upper respiratory tract and, after dissemination and survival in the bloodstream, cross the barriers of the brain to enter the CNS [184]. Traversal of both the BBB and BCSFB has been demonstrated for *H. influenzae* [47,51].

Attachment to host cells is enabled by the bacterial capsule and fimbriae which are subject to reversible phase variation [185]. *H. influenzae* shares a common strategy to enter endothelial cells with *S. pneumoniae* and *N. meningitidis*, which involves binding to the PAFR, mediated by phosphorylcholine (ChoP) [186,187]. This interaction results in the entry of the pathogens into the BBB through activation of β-arrestin–mediated uptake [188]. Binding of PAFR by lipooligosaccharide (LOS) glycoforms containing ChoP was also demonstrated during invasion NTHi. This binding resulted in the activation of host cell signaling by coupling with pertussis toxin-sensitive (PTX) heterotrimeric G protein complexes and invasion of the pathogen. It was further suggested that this mechanism was more efficient than micropinocytosis [187]. Binding to laminin receptor, another shared mechanism of these pathogens, is initiated by OmpP2, facilitating the interaction with the brain endothelium by *H. influenzae* [99].

In an in vitro model of the BCSFB, *Hib* as well as clinical isolates of *Hib* and *H. influenzae* serotype f (*Hif*) were shown to adhere and invade the HIBCPP cells as intracellular bacteria. Both fimbriae and the capsule lead to attenuated invasion [47]. Also, a study using *H. influenzae* serotype a (*Hia*) and a co-culture of HBMECs and pericytes demonstrated an activation of stimulated A_{2A} and A_{2B} adenosine receptors after infection. This in turn led to the release of Vascular Endothelial Growth Factor (VEGF) by the pericytes leading to pericyte detachment and endothelial cell proliferation resulting in overall BBB impairment [189].

Using rat models of meningitis, a dose-dependent increase of BBB permeability was observed after inoculation with *Hib* LPS [190]. Later studies showed that the permeability of the BBB was also increased in rats after inoculation with *H. influenzae* outer membrane vesicles (OMV), suggesting a role for these vesicles in transporting *Hib* LPS to the CSF during meningitis [191]. Furthermore, it was shown that zyxin, a cytoskeletal protein implicated in the protection of TJs in the BBB, is critical for the integrity of the BBB and, as a consequence, for protecting against invading pathogens such as *H. influenzae* [192]. Overall, the inflammatory response of patients during the infection with *H. influenzae* is determined by several virulence factors including the capsule, adhesion proteins, pili and outer membrane proteins as well as LPS and IgA1 protease [193].

Table 1. Evidence for the involvement of gram-positive and gram-negative bacteria during brain entry at the BBB and BCSFB.

Pathogen	Entry Mechanisms		Major Virulence Factors	
Gram-Positive	**BBB**	**BCSFB**	**BBB**	**BCSFB**
L. monocytogenes	Transcellular route [54] "Trojan horse" mechanism within leukocytes [54] Retrograde migration within axons of cranial nerves [54]	Transcellular route [57]	Major invasion protein InlB inducing receptor-mediated endocytosis [25] Bacterial surface protein InlF interacting with surface vimentin [58] Pore-forming cytolysin LLO inducing signaling pathways (NF-κB, MAPK) [64,66] ActA promoting F-actin-based intracellular motility [68]	Major invasion proteins InlA and InlB inducing receptor-mediated endocytosis [34,57]
S. suis	Invasion at low rates in porcine models [73,78].	Invasion demonstrated for porcine and human in vitro models [48,72] Possibly "Trojan-horse" mechanism [85]	Enolase increasing BBB permeability [80] Suilysin inducing pore formation in membranes [81]	Regulation of capsule expression [48,72]
S. pneumoniae	Translocation across BBB in vivo and in vitro [87,194]	Only attachment observed in an in vivo mouse model during late stages of infection with high levels of bacteremia [87].	Altered expression of the capsule for attachment [92,93] Interaction with BBB through NadA [96] Pore-forming toxin pneumolysin [101]	
GBS	Traversal of BBB in vivo and in vitro [49,109,195]		Expression of cell-wall anchored pili [111] PilA: promoting attachment of GBS [112] PilB: mediating internalization [112]	

Table 1. *Cont.*

Pathogen	Entry Mechanisms		Major Virulence Factors	
Gram-Negative	BBB	BCSFB	BBB	BCSFB
E. coli	Traversal of BBB in vivo and in vitro [22]	Traversal of BCSFB in vitro [121]	Attachent facilitated by type 1 fimbriae and OmpA [125,126] Invasion induced by IbeA [127] and CNF1 [128] Intracellular survival promoted by the *E. coli* K1 capsule [136]	Role of fimH during adhesion [121] Involvement of OmpA, FimH and IbeA in invasion [121]
N. meningitidi	Traversal of BBB in vivo and in vitro [2,196]	Traversal of BCSFB in vitro of choroid plexus epithelial cells [48] Invasion of outer BCSFB in induced pluripotent stem cell-derived brain endothelial cells [149]	Protective function of the polysaccharide capsule during bloodstream survival but attenuated tissue invasion [151] Adherence through pili and surface exposed proteins [154] Invasion is facilitated by Opa and Opc [160]	Capsule attenuates invasion in vitro [48]
H. influenzae	Traversal of BBB in vitro [51]	Traversal of BCSFB in vitro [47]	Entry via binding of PAFR [186,187] Attachment facilitated by binding of the laminin receptor [99]	Capsule and fimbriae attenuate invasion [47] Invasion if *H. influenzae* was observed as intracellular bacterium [47]

5. Conclusions

A variety of studies have focused on how pathogens cross the blood-CNS barriers, mostly focusing on the BBB. Although significant progress has been made in identifying mechanisms of host–pathogen interactions during bacterial meningitis, additional efforts towards identifying bacterial and host cell targets are needed. The diversity of mechanisms used by these pathogens presents the need for further research. To this end, the identification of common mechanisms used by multiple pathogens will be of great significance and further assist the development of effective therapies. Of high importance is the distinction between the mechanisms the pathogens use for crossing the BCSFB and the BBB. Pathogens exploiting epitopes of both barriers would present interesting targets for the development of therapeutics.

Author Contributions: Conceptualization, R.H. and C.S.; writing—original draft preparation, R.H.; writing—review and editing, H.S. and C.S.

Funding: This research received no external funding.

Acknowledgments: We acknowledge financial support by Deutsche Forschungsgemeinschaft within the funding programme Open Access Publishing, by the Baden-Württemberg Ministry of Science, Research and the Arts and by Ruprecht-Karls-Universität Heidelberg.

Conflicts of Interest: The authors declare no conflict of interest

Abbreviations

ABC	ATP-binding cassette
BBB	Blood–brain barrier
BCSFB	Blood–cerebrospinal fluid barrier
BMEC	Brain microvascular endothelial cells
CBP	CERB-binding protein
CbpA	Choline-binding protein A
CECAM	Carcinoembryonic antigen-related cellular adhesion molecule
ChoP	Phosphorylcholine
CNF1	Cytotoxic necrotizing factor 1
CNS	Central nervous system
CP	Choroid plexus
EGFR	Epidermal growth factor receptors
ERK	Extracellular signal-regulated kinases
FAK	Focal adhesion kinase
GCH1	Guanosine triphosphate cyclohydrolase
GlpO	α-glycerophosphate oxidase
HBMECs	Human brain microvascular endothelial cells
HIBCPP	Human choroid plexus epithelial papilloma
HvgA	Hypervirulent GBS adhesin
iagA	*invasion associated gene A*
Inl	Internalin
JNK	c-JUN N-terminal kinases
LLO	Listeriolysin O
LOS	Lipooligosaccharide
LOX 1	Lipoprotein receptor 1
LPS	Lipopolysaccharide
LTA	Lipoteichoic acid
MAPK	Mitogen activated protein kinase
MIF	Macrophage migration inhibitory factor
MMP	Matrix metalloproteinase

Nad A	Neuraminidase A
NadA	*Neisseria* adhesin A
NF-κB	Nuclear factor κB
NO	Nitric oxide
NOD2	Nucleotide-binding oligomerization domain 2
OMV	Outer membrane vesicles
PAFR	Platelet-activating factor receptor
PAMPs	Pathogen-associated molecular patterns
PCPEC	Primary porcine CP epithelial cells
PDGF-B	Platelet-derived growth factor-B
PECAM	Platelet endothelial cell adhesion molecule
PI3K	Phosphatidylinositol 3-kinase
PKC	Protein kinase C
plgR	Polymeric immunoglobulin receptor
PMN	Polymorphnuclear neutrophils
PPRs	Pattern recognition receptors
PTX	Pertussis toxin-sensitive
Ssr	Serine-rich repeat
T3SS	Type three secretion system
TF	Transcription factor
TGF-β	Transforming growth factor-β
TJs	Tight junctions
TNF	Tumor necrosis factor
VEGF	Vascular Endothelial Growth Factor
GBS	*Group B streptococcus, Streptococcus agalactiae*
E. coli	*Escherichia coli*
H. influenzae	*Haemophilus influenzae*
Hia	*H. influenzae* serotype a
Hib	*H. influenzae* serotype b
Hif	*H. influenzae* serotype f
L. monocytogenes	*Listeria monocytogenes*
N. meningitidis	*Neisseria meningitidis*
NTHi	nontypeable *H. influenzae*
S. pneumoniae	*Streptococcus pneumoniae*
S. suis	*Streptococcus suis*

References

1. Dando, S.J.; Mackay-Sim, A.; Norton, R.; Currie, B.J.; St John, J.A.; Ekberg, J.A.K.; Batzloff, M.; Ulett, G.C.; Beacham, I.R. Pathogens penetrating the central nervous system: Infection pathways and the cellular and molecular mechanisms of invasion. *Clin. Microbiol. Rev.* **2014**, *27*, 691–726. [CrossRef] [PubMed]
2. Kim, K.S. Mechanisms of microbial traversal of the blood-brain barrier. *Nat. Rev. Microbiol.* **2008**, *6*, 625–634. [CrossRef] [PubMed]
3. Wolburg, H.; Paulus, W. Choroid plexus: Biology and pathology. *Acta Neuropathol.* **2010**, *119*, 75–88.
4. Doran, K.S.; Fulde, M.; Gratz, N.; Kim, B.J.; Nau, R.; Prasadarao, N.; Schubert-Unkmeir, A.; Tuomanen, E.I.; Valentin-Weigand, P. Host-pathogen interactions in bacterial meningitis. *Acta Neuropathol.* **2016**, *131*, 185–209. [CrossRef] [PubMed]
5. Marques, F.; Sousa, J.C.; Sousa, N.; Palha, J.A. Blood-brain-barriers in aging and in alzheimer's disease. *Mol. Neurodegener.* **2013**, *8*, 38.
6. Saunders, N.R.; Habgood, M.D.; Mollgard, K.; Dziegielewska, K.M. The biological significance of brain barrier mechanisms: Help or hindrance in drug delivery to the central nervous system? *F1000Res.* **2016**, *5*. [CrossRef]
7. Lauer, A.N.; Tenenbaum, T.; Schroten, H.; Schwerk, C. The diverse cellular responses of the choroid plexus during infection of the central nervous system. *Am. J. Physiol. Cell Physiol.* **2018**, *314*, C152–C165.

8. Van Sorge, N.M.; Doran, K.S. Defense at the border: The blood-brain barrier versus bacterial foreigners. *Future Microbiol.* **2012**, *7*, 383–394.

9. Daneman, R.; Prat, A. The blood-brain barrier. *Cold Spring Harb. Perspect. Biol.* **2015**, *7*. [CrossRef]

10. Abbott, N.J.; Patabendige, A.A.K.; Dolman, D.E.M.; Yusof, S.R.; Begley, D.J. Structure and function of the blood-brain barrier. *Neurobiol. Dis.* **2010**, *37*, 13–25. [CrossRef]

11. Dias, M.C.; Coisne, C.; Lazarevic, I.; Baden, P.; Hata, M.; Iwamoto, N.; Francisco, D.M.F.; Vanlandewijck, M.; He, L.Q.; Baier, F.A.; et al. Claudin-3-deficient c57bl/6j mice display intact brain barriers. *Sci. Rep.* **2019**, *9*, 203. [CrossRef] [PubMed]

12. Tietz, S.; Engelhardt, B. Brain barriers: Crosstalk between complex tight junctions and adherens junctions. *J. Cell Biol.* **2015**, *209*, 493–506. [CrossRef] [PubMed]

13. Hawkins, R.A.; O'Kane, R.L.; Simpson, I.A.; Vina, J.R. Structure of the blood-brain barrier and its role in the transport of amino acids. *J. Nutr.* **2006**, *136*, 218s–226s. [CrossRef] [PubMed]

14. Sweeney, M.D.; Zhao, Z.; Montagne, A.; Nelson, A.R.; Zlokovic, B.V. Blood-brain barrier: From physiology to disease and back. *Physiol. Rev.* **2019**, *99*, 21–78. [CrossRef]

15. Weller, R.O.; Sharp, M.M.; Christodoulides, M.; Carare, R.O.; Mollgard, K. The meninges as barriers and facilitators for the movement of fluid, cells and pathogens related to the rodent and human cns. *Acta Neuropathol.* **2018**, *135*, 363–385. [CrossRef]

16. Liddelow, S.A. Development of the choroid plexus and blood-csf barrier. *Front. Neurosci.* **2015**, *9*. [CrossRef]

17. Schwerk, C.; Tenenbaum, T.; Kim, K.S.; Schroten, H. The choroid plexus-a multi-role player during infectious diseases of the cns. *Front. Cell. Neurosci.* **2015**, *9*. [CrossRef]

18. Bernd, A.; Ott, M.; Ishikawa, H.; Schroten, H.; Schwerk, C.; Fricker, G. Characterization of efflux transport proteins of the human choroid plexus papilloma cell line hibcpp, a functional in vitro model of the blood-cerebrospinal fluid barrier. *Pharm. Res.* **2015**, *32*, 2973–2982. [CrossRef]

19. Ransohoff, R.M.; Engelhardt, B. The anatomical and cellular basis of immune surveillance in the central nervous system. *Nat. Rev. Immunol.* **2012**, *12*, 623–635. [CrossRef]

20. Sullivan, T.D.; Lascolea, L.J.; Neter, E. Relationship between the magnitude of bacteremia in children and the clinical-disease. *Pediatrics* **1982**, *69*, 699–702.

21. Dietzman, D.E.; Fischer, G.W.; Schoenknecht, F.D. Neonatal escherichia coli septicemia–bacterial counts in blood. *J. Pediatr.* **1974**, *85*, 128–130. [CrossRef]

22. Kim, K.S. Human meningitis-associated escherichia coli. *Ecosal Plus* **2016**, *7*. [CrossRef] [PubMed]

23. Bell, L.M.; Alpert, G.; Campos, J.M.; Plotkin, S.A. Routine quantitative blood cultures in children with haemophilus influenzae or streptococcus pneumoniae bacteremia. *Pediatrics* **1985**, *76*, 901–904. [PubMed]

24. Virji, M. Ins and outs of microbial adhesion. *Top. Curr. Chem.* **2009**, *288*, 139–156. [PubMed]

25. Cossart, P.; Helenius, A. Endocytosis of viruses and bacteria. *Cold Spring Harb. Perspect. Biol.* **2014**, *6*, a016972. [CrossRef] [PubMed]

26. Cossart, P.; Sansonetti, P.J. Bacterial invasion: The paradigms of enteroinvasive pathogens. *Science* **2004**, *304*, 242–248. [CrossRef] [PubMed]

27. Mota, L.J.; Cornelis, G.R. The bacterial injection kit: Type iii secretion systems. *Ann. Med.* **2005**, *37*, 234–249. [CrossRef] [PubMed]

28. Santiago-Tirado, F.H.; Doering, T.L. False friends: Phagocytes as trojan horses in microbial brain infections. *PLoS Path.* **2017**, *13*, e1006680. [CrossRef]

29. Selbach, M.; Backert, S. Cortactin: An achilles' heel of the actin cytoskeleton targeted by pathogens. *Trends Microbiol.* **2005**, *13*, 181–189. [CrossRef]

30. Stradal, T.E.B.; Schelhaas, M. Actin dynamics in host-pathogen interaction. *Febs Lett.* **2018**, *592*, 3658–3669. [CrossRef]

31. Frischknecht, F.; Way, M. Surfing pathogens and the lessons learned for actin polymerization. *Trends Cell Biol.* **2001**, *11*, 30–38. [CrossRef]

32. Lamason, R.L.; Welch, M.D. Actin-based motility and cell-to-cell spread of bacterial pathogens. *Curr. Opin. Microbiol.* **2017**, *35*, 48–57. [CrossRef] [PubMed]

33. Borkowski, J.; Li, L.; Steinmann, U.; Quednau, N.; Stump-Guthier, C.; Weiss, C.; Findeisen, P.; Gretz, N.; Ishikawa, H.; Tenenbaum, T.; et al. Neisseria meningitidis elicits a pro-inflammatory response involving ikappabzeta in a human blood-cerebrospinal fluid barrier model. *J. Neuroinflammation* **2014**, *11*, 163. [CrossRef] [PubMed]

34. Dinner, S.; Kaltschmidt, J.; Stump-Guthier, C.; Hetjens, S.; Ishikawa, H.; Tenenbaum, T.; Schroten, H.; Schwerk, C. Mitogen-activated protein kinases are required for effective infection of human choroid plexus epithelial cells by listeria monocytogenes. *Microbes Infect.* **2017**, *19*, 18–33. [CrossRef] [PubMed]

35. Krachler, A.M.; Woolery, A.R.; Orth, K. Manipulation of kinase signaling by bacterial pathogens. *J. Cell Biol.* **2011**, *195*, 1083–1092. [CrossRef] [PubMed]

36. Pizarro-Cerda, J.; Cossart, P. Bacterial adhesion and entry into host cells. *Cell* **2006**, *124*, 715–727. [CrossRef] [PubMed]

37. Wang, S.; Peng, L.; Gai, Z.; Zhang, L.; Jong, A.; Cao, H.; Huang, S.H. Pathogenic triad in bacterial meningitis: Pathogen invasion, nf-kappab activation, and leukocyte transmigration that occur at the blood-brain barrier. *Front. Microbiol* **2016**, *7*, 148. [CrossRef]

38. Banerjee, A.; Kim, B.J.; Carmona, E.M.; Cutting, A.S.; Gurney, M.A.; Carlos, C.; Feuer, R.; Prasadarao, N.V.; Doran, K.S. Bacterial pili exploit integrin machinery to promote immune activation and efficient blood-brain barrier penetration. *Nat. Commun.* **2011**, *2*, 462. [CrossRef]

39. Koedel, U.; Scheld, W.M.; Pfister, H.W. Pathogenesis and pathophysiology of pneumococcal meningitis. *Lancet Infect. Dis.* **2002**, *2*, 721–736. [CrossRef]

40. Kim, S.Y.; Buckwalter, M.; Soreq, H.; Vezzani, A.; Kaufer, D. Blood-brain barrier dysfunction-induced inflammatory signaling in brain pathology and epileptogenesis. *Epilepsia* **2012**, *53*, 37–44. [CrossRef]

41. Haberl, R.L.; Anneser, F.; Kodel, U.; Pfister, H.W. Is nitric-oxide involved as a mediator of cerebrovascular changes in the early phase of experimental pneumococcal meningitis. *Neurol. Res.* **1994**, *16*, 108–112. [CrossRef] [PubMed]

42. Leppert, D.; Leib, S.L.; Grygar, C.; Miller, K.M.; Schaad, U.B.; Hollander, G.A. Matrix metalloproteinase (mmp)-8 and mmp-9 in cerebrospinal fluid during bacterial meningitis: Association with blood-brain barrier damage and neurological sequelae. *Clin. Infect. Dis.* **2000**, *31*, 80–84. [CrossRef] [PubMed]

43. Zwijnenburg, P.J.; de Bie, H.M.; Roord, J.J.; van der Poll, T.; van Furth, A.M. Chemotactic activity of cxcl5 in cerebrospinal fluid of children with bacterial meningitis. *J. Neuroimmunol.* **2003**, *145*, 148–153. [CrossRef] [PubMed]

44. Azuma, H.; Tsuda, N.; Sasaki, K.; Okuno, A. Clinical significance of cytokine measurement for detection of meningitis. *J. Pediatr* **1997**, *131*, 463–465. [CrossRef]

45. Coutinho, L.G.; Grandgirard, D.; Leib, S.L.; Agnez-Lima, L.F. Cerebrospinal-fluid cytokine and chemokine profile in patients with pneumococcal and meningococcal meningitis. *BMC Infect. Dis.* **2013**, *13*, 326. [CrossRef]

46. Cress, B.F.; Englaender, J.A.; He, W.Q.; Kasper, D.; Linhardt, R.J.; Koffas, M.A.G. Masquerading microbial pathogens: Capsular polysaccharides mimic host-tissue molecules. *Fems Microbiol. Rev.* **2014**, *38*, 660–697. [CrossRef]

47. Hauser, S.; Wegele, C.; Stump-Guthier, C.; Borkowski, J.; Weiss, C.; Rohde, M.; Ishikawa, H.; Schroten, H.; Schwerk, C.; Adam, R. Capsule and fimbriae modulate the invasion of haemophilus influenzae in a human blood-cerebrospinal fluid barrier model. *Int. J. Med. Microbiol.* **2018**, *308*, 829–839. [CrossRef]

48. Schwerk, C.; Papandreou, T.; Schuhmann, D.; Nickol, L.; Borkowski, J.; Steinmann, U.; Quednau, N.; Stump, C.; Weiss, C.; Berger, J.; et al. Polar invasion and translocation of neisseria meningitidis and streptococcus suis in a novel human model of the blood-cerebrospinal fluid barrier. *PLoS ONE* **2012**, *7*, e30069. [CrossRef]

49. Gendrin, C.; Merillat, S.; Vornhagen, J.; Coleman, M.; Armistead, B.; Ngo, L.; Aggarwal, A.; Quach, P.; Berrigan, J.; Rajagopal, L. Diminished capsule exacerbates virulence, blood-brain barrier penetration, intracellular persistence, and antibiotic evasion of hyperhemolytic group b streptococci. *J. Infect. Dis.* **2018**, *217*, 1128–1138. [CrossRef]

50. Wu, Z.F.; Wu, C.Y.; Shao, J.; Zhu, Z.Z.; Wang, W.X.; Zhang, W.W.; Tang, M.; Pei, N.; Fan, H.J.; Li, J.G.; et al. The streptococcus suis transcriptional landscape reveals adaptation mechanisms in pig blood and cerebrospinal fluid. *RNA* **2014**, *20*, 882–898. [CrossRef]

51. Al-Obaidi, M.M.J.; Desa, M.N.M. Mechanisms of blood brain barrier disruption by different types of bacteria, and bacterial-host interactions facilitate the bacterial pathogen invading the brain. *Cell. Mol. Neurobiol.* **2018**, *38*, 1349–1368. [CrossRef] [PubMed]

52. Disson, O.; Lecuit, M. Targeting of the central nervous system by listeria monocytogenes. *Virulence* **2012**, *3*, 213–221. [CrossRef] [PubMed]

53. Radoshevich, L.; Cossart, P. Listeria monocytogenes: Towards a complete picture of its physiology and pathogenesis. *Nat. Rev. Microbiol.* **2018**, *16*, 32–46. [CrossRef] [PubMed]

54. Drevets, D.A.; Leenen, P.J.M.; Greenfield, R.A. Invasion of the central nervous system by intracellular bacteria. *Clin. Microbiol. Rev.* **2004**, *17*, 323–347. [CrossRef] [PubMed]

55. Pagelow, D.; Chhatbar, C.; Beineke, A.; Liu, X.K.; Nerlich, A.; van Vorst, K.; Rohde, M.; Kalinke, U.; Forster, R.; Halle, S.; et al. The olfactory epithelium as a port of entry in neonatal neurolisteriosis. *Nat. Commun.* **2018**, *9*, 4269. [CrossRef] [PubMed]

56. Greiffenberg, L.; Goebel, W.; Kim, K.S.; Daniels, J.; Kuhn, M. Interaction of listeria monocytogenes with human brain microvascular endothelial cells: An electron microscopic study. *Infect. Immun.* **2000**, *68*, 3275–3279. [CrossRef]

57. Grundler, T.; Quednau, N.; Stump, C.; Orian-Rousseau, V.; Ishikawa, H.; Wolburg, H.; Schroten, H.; Tenenbaum, T.; Schwerk, C. The surface proteins inla and inlb are interdependently required for polar basolateral invasion by listeria monocytogenes in a human model of the blood-cerebrospinal fluid barrier. *Microbes Infect.* **2013**, *15*, 291–301. [CrossRef]

58. Ghosh, P.; Halvorsen, E.M.; Ammendolia, D.A.; Mor-Vaknin, N.; O'Riordan, M.X.D.; Brumell, J.H.; Markovitz, D.M.; Higgins, D.E. Invasion of the brain by listeria monocytogenes is mediated by inlf and host cell vimentin. *MBio* **2018**, *9*. [CrossRef]

59. Tang, P.; Rosenshine, I.; Finlay, B.B. Listeria-monocytogenes, an invasive bacterium, stimulates map kinase upon attachment to epithelial-cells. *Mol. Biol. Cell* **1994**, *5*, 455–464. [CrossRef]

60. Tang, P.; Sutherland, C.L.; Gold, M.R.; Finlay, B.B. Listeria monocytogenes invasion of epithelial cells requires the mek-1/erk-2 mitogen-activated protein kinase pathway. *Infect. Immun.* **1998**, *66*, 1106–1112.

61. Pizarro-Cerda, J.; Kuhbacher, A.; Cossart, P. Entry of listeria monocytogenes in mammalian epithelial cells: An updated view. *Cold Spring Harb. Perspect. Med.* **2012**, *2*, a010009. [CrossRef] [PubMed]

62. Veiga, E.; Cossart, P. Listeria hijacks the clathrin-dependent endocytic machinery to invade mammalian cells. *Nat. Cell Biol.* **2005**, *7*. [CrossRef] [PubMed]

63. Veiga, E.; Guttman, J.A.; Bonazzi, M.; Boucrot, E.; Toledo-Arana, A.; Lin, A.E.; Enninga, J.; Pizarro-Cerda, J.; Finlay, B.B.; Kirchhausen, T.; et al. Invasive and adherent bacterial pathogens co-opt host clathrin for infection. *Cell Host Microbe* **2007**, *2*, 340–351. [CrossRef] [PubMed]

64. Kayal, S.; Lilienbaum, A.; Join-Lambert, O.; Li, X.X.; Israel, A.; Berche, P. Listeriolysin o secreted by listeria monocytogenes induces nf-kappa b signalling by activating the i kappa b kinase complex. *Mol. Microbiol.* **2002**, *44*, 1407–1419. [CrossRef]

65. Tang, P.; Rosenshine, I.; Cossart, P.; Finlay, B.B. Listeriolysin o activates mitogen-activated protein kinase in eucaryotic cells. *Infect. Immun.* **1996**, *64*, 2359–2361.

66. Weiglein, I.; Goebel, W.; Troppmair, J.; Rapp, U.R.; Demuth, A.; Kuhn, M. Listeria monocytogenes infection of hela cells results in listeriolysin o-mediated transient activation of the raf-mek-map kinase pathway. *Fems Microbiol. Lett.* **1997**, *148*, 189–195. [CrossRef]

67. Lambrechts, A.; Gevaert, K.; Cossart, P.; Vandekerckhove, J.; Van Troys, M. Listeria comet tails: The actin-based motility machinery at work. *Trends Cell Biol.* **2008**, *18*, 220–227. [CrossRef]

68. Kocks, C.; Gouin, E.; Tabouret, M.; Berche, P.; Ohayon, H.; Cossart, P. L-monocytogenes-induced actin assembly requires the acta gene-product, a surface protein. *Cell* **1992**, *68*, 521–531. [CrossRef]

69. Gouin, E.; Adib-Conquy, M.; Balestrino, D.; Nahori, M.A.; Villiers, V.; Colland, F.; Dramsi, S.; Dussurget, O.; Cossart, P. The listeria monocytogenes inlc protein interferes with innate immune responses by targeting the i kappa b kinase subunit ikk alpha. *Proc. Natl. Acad. Sci. USA* **2010**, *107*, 17333–17338. [CrossRef]

70. Rajabian, T.; Gavicherla, B.; Heisig, M.; Muller-Altrock, S.; Goebel, W.; Gray-Owen, S.D.; Ireton, K. The bacterial virulence factor inlc perturbs apical cell junctions and promotes cell-to-cell spread of listeria. *Nat. Cell Biol.* **2009**, *11*, 1212–1218. [CrossRef]

71. Gottschalk, M.; Xu, J.G.; Calzas, C.; Segura, M. Streptococcus suis: A new emerging or an old neglected zoonotic pathogen? *Future Microbiol.* **2010**, *5*, 371–391. [CrossRef] [PubMed]

72. Tenenbaum, T.; Papandreou, T.; Gellrich, D.; Friedrichs, U.; Seibt, A.; Adam, R.; Wewer, C.; Galla, H.J.; Schwerk, C.; Schroten, H. Polar bacterial invasion and translocation of streptococcus suis across the blood-cerebrospinal fluid barrier in vitro. *Cell. Microbiol.* **2009**, *11*, 323–336. [CrossRef] [PubMed]

73. Vanier, G.; Segura, M.; Gottschalk, M. Characterization of the invasion of porcine endothelial cells by streptococcus suis serotype 2. *Can. J. Vet. Res. Rev. Can. De Rech. Vet.* **2007**, *71*, 81–89.

74. Benga, L.; Goethe, R.; Rohde, M.; Valentin-Weigand, P. Non-encapsulated strains reveal novel insights in invasion and survival of streptococcus suis in epithelial cells. *Cell. Microbiol.* **2004**, *6*, 867–881. [CrossRef]

75. Willenborg, J.; Fulde, M.; de Greeff, A.; Rohde, M.; Smith, H.E.; Valentin-Weigand, P.; Goethe, R. Role of glucose and ccpa in capsule expression and virulence of streptococcus suis. *Microbiology* **2011**, *157*, 1823–1833. [CrossRef]

76. Charland, N.; Nizet, V.; Rubens, C.E.; Kim, K.S.; Lacouture, S.; Gottschalk, M. Streptococcus suis serotype 2 interactions with human brain microvascular endothelial cells. *Infect. Immun.* **2000**, *68*, 637–643. [CrossRef]

77. Benga, L.; Friedl, P.; Valentin-Weigand, P. Adherence of *streptococcus suis* to porcine endothelial cells. *J. Vet. Med. Ser. B-Infect. Dis. Vet. Public Health* **2005**, *52*, 392–395. [CrossRef]

78. Vanier, G.; Segura, M.; Friedl, P.; Lacouture, S.; Gottschalk, M. Invasion of porcine brain microvascular endothelial cells by streptococcus suis serotype 2. *Infect. Immun.* **2004**, *72*, 1441–1449. [CrossRef]

79. Fittipaldi, N.; Segura, M.; Grenier, D.; Gottschalk, M. Virulence factors involved in the pathogenesis of the infection caused by the swine pathogen and zoonotic agent streptococcus suis. *Future Microbiol.* **2012**, *7*, 259–279. [CrossRef]

80. Sun, Y.; Li, N.; Zhang, J.; Liu, H.; Liu, J.; Xia, X.; Sun, C.; Feng, X.; Gu, J.; Du, C.; et al. Enolase of streptococcus suis serotype 2 enhances blood-brain barrier permeability by inducing il-8 release. *Inflammation* **2016**, *39*, 718–726. [CrossRef]

81. Lun, S.C.; Perez-Casal, J.; Connor, W.; Willson, P.J. Role of suilysin in pathogenesis of streptococcus suis capsular serotype 2. *Microb. Pathog.* **2003**, *34*, 27–37. [CrossRef]

82. Seitz, M.; Baums, C.G.; Neis, C.; Benga, L.; Fulde, M.; Rohde, M.; Goethe, R.; Valentin-Weigand, P. Subcytolytic effects of suilysin on interaction of streptococcus suis with epithelial cells. *Vet. Microbiol.* **2013**, *167*, 584–591. [CrossRef] [PubMed]

83. Vadeboncoeur, N.; Segura, M.; Al-Numani, D.; Vanier, G.; Gottschalk, M. Pro-inflammatory cytokine and chemokine release by human brain microvascular endothelial cells stimulated by streptococcus suis serotype 2. *Fems Immunol. Med. Microbiol.* **2003**, *35*, 49–58. [CrossRef] [PubMed]

84. Tenenbauma, T.; Matalon, D.; Adam, R.; Seibt, A.; Wewer, C.; Schwerk, C.; Galla, H.J.; Schrotena, H. Dexamethasone prevents alteration of tight junction-associated proteins and barrier function in porcine choroid plexus epithelial cells after infection with streptococcus suis in vitro. *Brain Res.* **2008**, *1229*, 1–17. [CrossRef] [PubMed]

85. Wewer, C.; Seibt, A.; Wolburg, H.; Greune, L.; Schmidt, M.A.; Berger, J.; Galla, H.J.; Quitsch, U.; Schwerk, C.; Schroten, H.; et al. Transcellular migration of neutrophil granulocytes through the blood-cerebrospinal fluid barrier after infection with streptococcus suis. *J. Neuroinflammation* **2011**, *8*, 51. [CrossRef] [PubMed]

86. Scarborough, M.; Thwaites, G.E. The diagnosis and management of acute bacterial meningitis in resource-poor settings. *Lancet Neurol.* **2008**, *7*, 637–648. [CrossRef]

87. Iovino, F.; Orihuela, C.J.; Moorlag, H.E.; Molema, G.; Bijlsma, J.J.E. Interactions between blood-borne streptococcus pneumoniae and the blood-brain barrier preceding meningitis. *PLoS ONE* **2013**, *8*, e68408. [CrossRef]

88. Van Ginkel, F.W.; McGhee, J.R.; Watt, J.M.; Campos-Torres, A.; Parish, L.A.; Briles, D.E. Pneumococcal carriage results in ganglioside-mediated olfactory tissue infection (vol 100, pg 14363, 2003). *Proc. Natl. Acad. Sci. USA* **2004**, *101*, 6834.

89. Wartha, F.; Beiter, K.; Albiger, B.; Fernebro, J.; Zychlinsky, A.; Normark, S.; Henriques-Normark, B. Capsule and d-alanylated lipoteichoic acids protect streptococcus pneumoniae against neutrophil extracellular traps. *Cell. Microbiol.* **2007**, *9*, 1162–1171. [CrossRef]

90. Middleton, D.R.; Paschall, A.V.; Duke, J.A.; Avci, F.Y. Enzymatic hydrolysis of pneumococcal capsular polysaccharide renders the bacterium vulnerable to host defense. *Infect. Immun.* **2018**, *86*. [CrossRef]

91. Keller, L.E.; Jones, C.V.; Thornton, J.A.; Sanders, M.E.; Swiatlo, E.; Nahm, M.H.; Park, I.H.; McDaniel, L.S. Pspk of streptococcus pneumoniae increases adherence to epithelial cells and enhances nasopharyngeal colonization. *Infect. Immun.* **2013**, *81*, 173–181. [CrossRef] [PubMed]

92. Li-Korotky, H.S.; Lo, C.Y.; Banks, J.M. Interaction of pneumococcal phase variation, host and pressure/gas composition: Virulence expression of nana, hyla, pspa and cbpa in simulated otitis media. *Microb. Pathog.* **2010**, *49*, 204–210. [CrossRef] [PubMed]

93. Shainheit, M.G.; Muie, M.; Camilli, A. The core promoter of the capsule operon of streptococcus pneumoniae is necessary for colonization and invasive disease. *Infect. Immun.* **2014**, *82*, 694–705. [CrossRef] [PubMed]

94. Iovino, F.; Molema, G.; Bijlsma, J.J.E. Platelet endothelial cell adhesion molecule-1, a putative receptor for the adhesion of streptococcus pneumoniae to the vascular endothelium of the blood-brain barrier. *Infect. Immun.* **2014**, *82*, 3555–3566. [CrossRef]

95. Uchiyama, S.; Carlin, A.F.; Khosravi, A.; Weiman, S.; Banerjee, A.; Quach, D.; Hightower, G.; Mitchell, T.J.; Doran, K.S.; Nizet, V. The surface-anchored nana protein promotes pneumococcal brain endothelial cell invasion. *J. Exp. Med.* **2009**, *206*, 1845–1852. [CrossRef] [PubMed]

96. Gratz, N.; Loh, L.N.; Mann, B.; Gao, G.; Carter, R.; Rosch, J.; Tuomanen, E.I. Pneumococcal neuraminidase activates tgf-beta signalling. *Microbiology* **2017**, *163*, 1198–1207.

97. Iovino, F.; Engelen-Lee, J.Y.; Brouwer, M.; van de Beek, D.; van der Ende, A.; Seron, M.V.; Mellroth, P.; Muschiol, S.; Bergstrand, J.; Widengren, J.; et al. Pigr and pec am-1 bind to pneumococcal adhesins rrga and pspc mediating bacterial brain invasion. *J. Exp. Med.* **2017**, *214*, 1619–1630. [CrossRef]

98. Bagnoli, F.; Moschioni, M.; Donati, C.; Dimitrovska, V.; Ferlenghi, I.; Facciotti, C.; Muzzi, A.; Giusti, F.; Emolo, C.; Sinisi, A.; et al. A second pilus type in streptococcus pneumoniae is prevalent in emerging serotypes and mediates adhesion to host cells. *J. Bacteriol.* **2008**, *190*, 5480–5492. [CrossRef]

99. Orihuela, C.J.; Mahdavi, J.; Thornton, J.; Mann, B.; Wooldridge, K.G.; Abouseada, N.; Oldfield, N.J.; Self, T.; Ala'Aldeen, D.A.; Tuomanen, E.I. Laminin receptor initiates bacterial contact with the blood brain barrier in experimental meningitis models. *J. Clin. Invest.* **2009**, *119*, 1638–1646. [CrossRef]

100. Schmidt, F.; Kakar, N.; Meyer, T.C.; Depke, M.; Masouris, I.; Burchhardt, G.; Gomez-Mejia, A.; Dhople, V.; Havarstein, L.S.; Sun, Z.; et al. In vivo proteomics identifies the competence regulon and alib oligopeptide transporter as pathogenic factors in pneumococcal meningitis. *PLoS Path.* **2019**, *15*, e1007987. [CrossRef]

101. Chen, J.Q.; Li, N.N.; Wang, B.W.; Liu, X.F.; Liu, J.L.; Chang, Q. Upregulation of cbp by ply can cause permeability of blood-brain barrier to increase meningitis. *J. Biochem. Mol. Toxicol.* **2019**, e22333. [CrossRef] [PubMed]

102. Zysk, G.; Schneider-Wald, B.K.; Hwang, J.H.; Bejo, L.; Kim, K.S.; Mitchell, T.J.; Hakenbeck, R.; Heinz, H.P. Pneumolysin is the main inducer of cytotoxicity to brain microvascular endothelial cells caused by streptococcus pneumoniae. *Infect. Immun.* **2001**, *69*, 845–852. [CrossRef] [PubMed]

103. Mandi, L.K.; Wang, H.; Van der Hoek, M.B.; Paton, J.C.; Ogunniyi, A.D. Identification of a novel pneumococcal vaccine antigen preferentially expressed during meningitis in mice. *J. Clin. Invest.* **2012**, *122*, 2208–2220.

104. Wang, Y.; Liu, X.; Liu, Q. Nod2 expression in streptococcus pneumoniae meningitis and its influence on the blood-brain barrier. *Can. J. Infect. Dis Med. Microbiol* **2018**, *2018*. [CrossRef] [PubMed]

105. Barichello, T.; Fagundes, G.D.; Generoso, J.S.; Moreira, A.P.; Costa, C.S.; Zanatta, J.R.; Simoes, L.R.; Petronilho, F.; Dal-Pizzol, F.; Vilela, M.C.; et al. Brain-blood barrier breakdown and pro-inflammatory mediators in neonate rats submitted meningitis by streptococcus pneumoniae. *Brain Res.* **2012**, *1471*, 162–168. [CrossRef]

106. Brouwer, M.C.; Tunkel, A.R.; van de Beek, D. Epidemiology, diagnosis, and antimicrobial treatment of acute bacterial meningitis. *Clin. Microbiol. Rev.* **2010**, *23*, 467–492. [CrossRef]

107. Jones, N.; Bohnsack, J.F.; Takahashi, S.; Oliver, K.A.; Chan, M.S.; Kunst, F.; Glaser, P.; Rusniok, C.; Crook, D.W.M.; Harding, R.M.; et al. Multilocus sequence typing system for group b streptococcus. *J. Clin. Microbiol.* **2003**, *41*, 2530–2536. [CrossRef]

108. Maisey, H.C.; Doran, K.S.; Nizet, V. Recent advances in understanding the molecular basis of group b streptococcus virulence. *Expert Rev. Mol. Med.* **2008**, *10*, e27. [CrossRef]

109. Nizet, V.; Kim, K.S.; Stins, M.; Jonas, M.; Chi, E.Y.; Nguyen, D.; Rubens, C.E. Invasion of brain microvascular endothelial cells by group b streptococci. *Infect. Immun.* **1997**, *65*, 5074–5081.

110. Tazi, A.; Disson, O.; Bellais, S.; Bouaboud, A.; Dmytruk, N.; Dramsi, S.; Mistou, M.Y.; Khun, H.; Mechler, C.; Tardieux, I.; et al. The surface protein hvga mediates group b streptococcus hypervirulence and meningeal tropism in neonates. *J. Exp. Med.* **2010**, *207*, 2313–2322. [CrossRef]

111. Lauer, P.; Rinaudo, C.D.; Soriani, M.; Margarit, I.; Maione, D.; Rosini, R.; Taddei, A.R.; Mora, M.; Rappuoli, R.; Grandi, G.; et al. Genome analysis reveals pili in group b streptococcus. *Science* **2005**, *309*, 105. [CrossRef] [PubMed]

112. Maisey, H.C.; Hensler, M.; Nizet, V.; Doran, K.S. Group b streptococcal pilus proteins contribute to adherence to and invasion of brain microvascular endothelial cells. *J. Bacteriol.* **2007**, *189*, 1464–1467. [CrossRef] [PubMed]

113. Mu, R.; Kim, B.J.; Paco, C.; Del Rosario, Y.; Courtney, H.S.; Doran, K.S. Identification of a group b streptococcal fibronectin binding protein, sfba, that contributes to invasion of brain endothelium and development of meningitis. *Infect. Immun.* **2014**, *82*, 2276–2286. [CrossRef] [PubMed]

114. Seo, H.S.; Mu, R.; Kim, B.J.; Doran, K.S.; Sullam, P.M. Binding of glycoprotein srr1 of streptococcus agalactiae to fibrinogen promotes attachment to brain endothelium and the development of meningitis. *PLoS Path.* **2012**, *8*, e1002947. [CrossRef] [PubMed]

115. Deng, L.W.; Spencer, B.L.; Holmes, J.A.; Mu, R.; Rego, S.; Weston, T.A.; Hu, Y.; Sanches, G.F.; Yoon, S.; Park, N.; et al. The group b streptococcal surface antigen i/ii protein, bspc, interacts with host vimentin to promote adherence to brain endothelium and inflammation during the pathogenesis of meningitis. *PLoS Path.* **2019**, *15*, e1007848. [CrossRef]

116. Shin, S.; Maneesh, P.S.; Lee, J.S.; Romer, L.H.; Kim, K.S. Focal adhesion kinase is involved in type iii group b streptococcal invasion of human brain microvascular endothelial cells. *Microb. Pathog.* **2006**, *41*, 168–173. [CrossRef]

117. Doran, K.S.; Engelson, E.J.; Khosravi, A.; Maisey, H.C.; Fedtke, I.; Equils, O.; Michelsen, K.S.; Arditi, M.; Peschel, A.; Nizet, V. Blood-brain barrier invasion by group b streptococcus depends upon proper cell-surface anchoring of lipoteichoic acid. *J. Clin. Invest.* **2005**, *115*, 2499–2507. [CrossRef]

118. Luo, S.; Cao, Q.; Ma, K.; Wang, Z.F.; Liu, G.J.; Lu, C.P.; Liu, Y.J. Quantitative assessment of the blood-brain barrier opening caused by streptococcus agalactiae hyaluronidase in a balb/c mouse model. *Sci Rep.-Uk* **2017**, *7*, 13529. [CrossRef]

119. Kim, B.J.; McDonagh, M.A.; Deng, L.; Gastfriend, B.D.; Schubert-Unkmeir, A.; Doran, K.S.; Shusta, E.V. Streptococcus agalactiae disrupts p-glycoprotein function in brain endothelial cells. *Fluids Barriers Cns* **2019**, *16*, 1–10. [CrossRef]

120. Kim, B.J.; Hancock, B.M.; Bermudez, A.; Del Cid, N.; Reyes, E.; van Sorge, N.M.; Lauth, X.; Smurthwaite, C.A.; Hilton, B.J.; Stotland, A.; et al. Bacterial induction of snail1 contributes to blood-brain barrier disruption. *J. Clin. Invest.* **2015**, *125*, 2473–2483. [CrossRef]

121. Rose, R.; Hauser, S.; Stump-Guthier, C.; Weiss, C.; Rohde, M.; Kim, K.S.; Ishikawa, H.; Schroten, H.; Schwerk, C.; Adam, R. Virulence factor-dependent basolateral invasion of choroid plexus epithelial cells by pathogenic *Escherichia coli* in vitro. *Fems Microbiol. Lett.* **2018**, *365*. [CrossRef] [PubMed]

122. Kim, K.S.; Itabashi, H.; Gemski, P.; Sadoff, J.; Warren, R.L.; Cross, A.S. The k1-capsule is the critical determinant in the development of escherichia-coli meningitis in the rat. *J. Clin. Invest.* **1992**, *90*, 897–905. [CrossRef] [PubMed]

123. Cross, A.S.; Kim, K.S.; Wright, D.C.; Sadoff, J.C.; Gemski, P. Role of lipopolysaccharide and capsule in the serum resistance of bacteremic strains of escherichia-coli. *J. Infect. Dis.* **1986**, *154*, 497–503. [CrossRef] [PubMed]

124. Logue, C.M.; Doetkott, C.; Mangiamele, P.; Wannemuehler, Y.M.; Johnson, T.J.; Tivendale, K.A.; Li, G.W.; Sherwood, J.S.; Nolan, L.K. Genotypic and phenotypic traits that distinguish neonatal meningitis-associated escherichia coli from fecal e. Coli isolates of healthy human hosts. *Appl. Environ. Microbiol.* **2012**, *78*, 5824–5830. [CrossRef]

125. Teng, C.H.; Cai, M.; Shin, S.; Xie, Y.; Kim, K.J.; Khan, N.A.; Di Cello, F.; Kim, K.S. Escherichia coli k1 rs218 interacts with human brain microvascular endothelial cells via type 1 fimbria bacteria in the fimbriated state. *Infect. Immun.* **2005**, *73*, 2923–2931. [CrossRef]

126. Khan, N.A.; Shin, S.; Chung, J.W.; Kim, K.J.; Elliott, S.; Wang, Y.; Kim, K.S. Outer membrane protein a and cytotoxic necrotizing factor-1 use diverse signaling mechanisms for escherichia coli k1 invasion of human brain microvascular endothelial cells. *Microb. Pathog.* **2003**, *35*, 35–42. [CrossRef]

127. Huang, S.H.; Wan, Z.S.; Chen, Y.H.; Jong, A.Y.; Kim, K.S. Further characterization of escherichia coli brain microvascular endothelial cell invasion gene ibea by deletion, complementation, and protein expression. *J. Infect. Dis.* **2001**, *183*, 1071–1078. [CrossRef]

128. Wang, M.H.; Kim, K.S. Cytotoxic necrotizing factor 1 contributes to escherichia coli meningitis. *Toxins* **2013**, *5*, 2270–2280. [CrossRef]

129. Prasadarao, N.V. Identification of escherichia coli outer membrane protein a receptor on human brain microvascular endothelial cells. *Infect. Immun.* **2002**, *70*, 4556–4563. [CrossRef]

130. Reddy, M.A.; Wass, C.A.; Kim, K.S.; Schlaepfer, D.D.; Prasadarao, N.V. Involvement of focal adhesion kinase in escherichia coli invasion of human brain microvascular endothelial cells. *Infect. Immun.* **2000**, *68*, 6423–6430. [CrossRef]

131. Reddy, M.A.; Prasadarao, N.V.; Wass, C.A.; Kim, K.S. Phosphatidylinositol 3-kinase activation and interaction with focal adhesion kinase in escherichia coli k1 invasion of human brain microvascular endothelial cells. *J. Biol. Chem.* **2000**, *275*, 36769–36774. [CrossRef] [PubMed]

132. Sukumaran, S.K.; McNamara, G.; Prasadarao, N.V. Escherichia coli k-1 interaction with human brain micro-vascular endothelial cells triggers phospholipase c-gamma1 activation downstream of phosphatidylinositol 3-kinase. *J. Biol. Chem.* **2003**, *278*, 45753–45762. [CrossRef] [PubMed]

133. Selvaraj, S.K.; Periandythevar, P.; Prasadarao, N.V. Outer membrane protein a of escherichia coli k1 selectively enhances the expression of intercellular adhesion molecule-1 in brain microvascular endothelial cells. *Microb. Infect.* **2007**, *9*, 547–557. [CrossRef] [PubMed]

134. Prasadarao, N.V.; Wass, C.A.; Stins, M.F.; Shimada, H.; Kim, K.S. Outer membrane protein a-promoted actin condensation of brain microvascular endothelial cells is required for escherichia coli invasion. *Infect. Immun.* **1999**, *67*, 5775–5783.

135. Sukumaran, S.K.; Quon, M.J.; Prasadarao, N.V. Escherichia coli k1 internalization via caveolae requires caveolin-1 and protein kinase c alpha interaction in human brain microvascular endothelial cells. *J. Biol. Chem.* **2002**, *277*, 50716–50724. [CrossRef]

136. Kim, K.J.; Elliott, S.J.; Di Cello, F.; Stins, M.F.; Kim, K.S. The k1 capsule modulates trafficking of e-coli-containing vacuoles and enhances intracellular bacterial survival in human brain microvascular endothelial cells. *Cell. Microbiol.* **2003**, *5*, 245–252. [CrossRef]

137. Yang, R.C.; Huang, F.; Fu, J.Y.; Dou, B.B.; Xu, B.J.; Miao, L.; Liu, W.T.; Yang, X.P.; Tan, C.; Chen, H.C.; et al. Differential transcription profiles of long non-coding rnas in primary human brain microvascular endothelial cells in response to meningitic escherichia coli. *Sci. Rep.* **2016**, *6*, 38903. [CrossRef]

138. Yang, R.C.; Qu, X.Y.; Xiao, S.Y.; Li, L.; Xu, B.J.; Fu, J.Y.; Lv, Y.J.; Amjad, N.; Tan, C.; Kim, K.S.; et al. Meningitic escherichia coli-induced upregulation of pdgf-b and icam-1 aggravates blood-brain barrier disruption and neuroinflammatory response. *J. Neuroinflamm.* **2019**, *16*, 101. [CrossRef]

139. Sukumaran, S.K.; Prasadarao, N.V. Escherichia coli k1 invasion increases human brain microvascular endothelial cell monolayer permeability by disassembling vascular-endothelial cadherins at tight junctions. *J. Infect. Dis.* **2003**, *188*, 1295–1309. [CrossRef]

140. Mittal, R.; Gonzalez-Gomez, I.; Goth, K.A.; Prasadarao, N.V. Inhibition of inducible nitric oxide controls pathogen load and brain damage by enhancing phagocytosis of escherichia coli k1 in neonatal meningitis. *Am. J. Pathol.* **2010**, *176*, 1292–1305. [CrossRef]

141. Shanmuganathan, M.V.; Krishnan, S.; Fu, X.W.; Prasadarao, N.V. Attenuation of biopterin synthesis prevents escherichia coli k1 invasion of brain endothelial cells and the development of meningitis in newborn mice. *J. Infect. Dis.* **2013**, *207*, 61–71. [CrossRef] [PubMed]

142. Liu, W.T.; Lv, Y.J.; Yang, R.C.; Fu, J.Y.; Liu, L.; Wang, H.; Cao, Q.; Tan, C.; Chen, H.C.; Wang, X.R. New insights into meningitic escherichia coli infection of brain microvascular endothelial cells from quantitative proteomics analysis. *J. Neuroinflamm.* **2018**, *15*, 291. [CrossRef] [PubMed]

143. Zhang, K.; Shi, M.J.; Niu, Z.; Chen, X.; Wei, J.Y.; Miao, Z.W.; Zhao, W.D.; Chen, Y.H. Activation of brain endothelium by escherichia coli k1 virulence factor cgld promotes polymorphonuclear leukocyte transendothelial migration. *Med. Microbiol. Immunol.* **2019**, *208*, 59–68. [CrossRef] [PubMed]

144. Coureuil, M.; Lecuyer, H.; Bourdoulous, S.; Nassif, X. A journey into the brain: Insight into how bacterial pathogens cross blood-brain barriers. *Nat. Rev. Microbiol.* **2017**, *15*, 149–159. [CrossRef]

145. Schneider, M.C.; Exley, R.M.; Chan, H.; Feavers, I.; Kang, Y.H.; Sim, R.B.; Tang, C.M. Functional significance of factor h binding to neisseria meningitidis. *J. Immunol.* **2006**, *176*, 7566–7575. [CrossRef]

146. Seib, K.L.; Scarselli, M.; Comanducci, M.; Toneatto, D.; Masignani, V. Neisseria meningitidis factor h-binding protein fhbp: A key virulence factor and vaccine antigen. *Expert Rev. Vaccines* **2015**, *14*, 841–859. [CrossRef]

147. Kim, B.J.; Shusta, E.V.; Doran, K.S. Past and current perspectives in modeling bacteria and blood brain barrier interactions. *Front. Microbiol.* **2019**, *10*, 1336. [CrossRef]

148. Stephens, D.S. Biology and pathogenesis of the evolutionarily successful, obligate human bacterium neisseria meningitidis. *Vaccine* **2009**, *27*, B71–B77. [CrossRef]

149. Gomes, S.F.M.; Westermann, A.J.; Sauerwein, T.; Hertlein, T.; Forstner, K.U.; Ohlsen, K.; Metzger, M.; Shusta, E.V.; Kim, B.J.; Appelt-Menzel, A.; et al. Induced pluripotent stem cell-derived brain endothelial cells as a cellular model to study neisseria meningitidis infection. *Front. Microbiol.* **2019**, *10*. [CrossRef]

150. Nagele, V.; Heesemann, J.; Schielke, S.; Jimenez-Soto, L.F.; Kurzai, O.; Ackermann, N. Neisseria meningitidis adhesin nada targets beta1 integrins: Functional similarity to yersinia invasin. *J. Biol. Chem.* **2011**, *286*, 20536–20546. [CrossRef]

151. Pizza, M.; Rappuoli, R. Neisseria meningitidis: Pathogenesis and immunity. *Curr. Opin. Microbiol.* **2015**, *23*, 68–72. [CrossRef] [PubMed]

152. Unkmeir, A.; Latsch, K.; Dietrich, G.; Wintermeyer, E.; Schinke, B.; Schwender, S.; Kim, K.S.; Eigenthaler, M.; Frosch, M. Fibronectin mediates opc-dependent internalization of neisseria meningitidis in human brain microvascular endothelial cells. *Mol. Microbiol.* **2002**, *46*, 933–946. [CrossRef] [PubMed]

153. Deghmane, A.E.; Giorgini, D.; Larribe, M.; Alonso, J.M.; Taha, M.K. Down-regulation of pili and capsule of neisseria meningitidis upon contact with epithelial cells is mediated by crga regulatory protein. *Mol. Microbiol.* **2002**, *43*, 1555–1564. [CrossRef] [PubMed]

154. Berry, J.L.; Pelicic, V. Exceptionally widespread nanomachines composed of type iv pilins: The prokaryotic swiss army knives. *Fems Microbiol. Rev.* **2015**, *39*, 134–154. [CrossRef]

155. Bernard, S.C.; Simpson, N.; Join-Lambert, O.; Federici, C.; Laran-Chich, M.P.; Maissa, N.; Bouzinba-Segard, H.; Morand, P.C.; Chretien, F.; Taouji, S.; et al. Pathogenic neisseria meningitidis utilizes cd147 for vascular colonization. *Nat. Med.* **2014**, *20*, 725–731. [CrossRef]

156. Rudel, T.; Scheurerpflug, I.; Meyer, T.F. Neisseria pilc protein identified as type-4 pilus tip-located adhesin. *Nature* **1995**, *373*, 357–359. [CrossRef]

157. Nassif, X.; Beretti, J.L.; Lowy, J.; Stenberg, P.; O'Gaora, P.; Pfeifer, J.; Normark, S.; So, M. Roles of pilin and pilc in adhesion of neisseria meningitidis to human epithelial and endothelial cells. *Proc. Natl. Acad. Sci. USA* **1994**, *91*, 3769–3773. [CrossRef]

158. Morand, P.C.; Tattevin, P.; Eugene, E.; Beretti, J.L.; Nassif, X. The adhesive property of the type iv pilus-associated component pilc1 of pathogenic neisseria is supported by the conformational structure of the n-terminal part of the molecule. *Mol. Microbiol.* **2001**, *40*, 846–856. [CrossRef]

159. Virji, M. Pathogenic neisseriae: Surface modulation, pathogenesis and infection control. *Nat. Rev. Microbiol.* **2009**, *7*, 274–286. [CrossRef]

160. Carbonnelle, E.; Hill, D.J.; Morand, P.; Griffiths, N.J.; Bourdoulous, S.; Murillo, I.; Nassif, X.; Virji, M. Meningococcal interactions with the host. *Vaccine* **2009**, *27*, B78–B89. [CrossRef]

161. Virji, M.; Makepeace, K.; Ferguson, D.J.; Achtman, M.; Moxon, E.R. Meningococcal opa and opc proteins: Their role in colonization and invasion of human epithelial and endothelial cells. *Mol. Microbiol.* **1993**, *10*, 499–510. [CrossRef] [PubMed]

162. Gray-Owen, S.D. Neisserial opa proteins: Impact on colonization, dissemination and immunity. *Scand. J. Infect. Dis.* **2003**, *35*, 614–618. [CrossRef] [PubMed]

163. Sa, E.C.C.; Griffiths, N.J.; Virji, M. Neisseria meningitidis opc invasin binds to the sulphated tyrosines of activated vitronectin to attach to and invade human brain endothelial cells. *PLoS Pathog* **2010**, *6*, e1000911.

164. Simonis, A.; Hebling, S.; Gulbins, E.; Schneider-Schaulies, S.; Schubert-Unkmeir, A. Differential activation of acid sphingomyelinase and ceramide release determines invasiveness of neisseria meningitidis into brain endothelial cells. *PLoS Path.* **2014**, *10*, e1004160. [CrossRef] [PubMed]

165. Metruccio, M.M.; Pigozzi, E.; Roncarati, D.; Berlanda Scorza, F.; Norais, N.; Hill, S.A.; Scarlato, V.; Delany, I. A novel phase variation mechanism in the meningococcus driven by a ligand-responsive repressor and differential spacing of distal promoter elements. *PLoS Pathog* **2009**, *5*, e1000710. [CrossRef] [PubMed]

166. Comanducci, M.; Bambini, S.; Caugant, D.A.; Mora, M.; Brunelli, B.; Capecchi, B.; Ciucchi, L.; Rappuoli, R.; Pizza, M. Nada diversity and carriage in neisseria meningitidis. *Infect. Immun.* **2004**, *72*, 4217–4223. [CrossRef] [PubMed]

167. Capecchi, B.; Adu-Bobie, J.; Di Marcello, F.; Ciucchi, L.; Masignani, V.; Taddei, A.; Rappuoli, R.; Pizza, M.; Arico, B. Neisseria meningitidis nada is a new invasin which promotes bacterial adhesion to and penetration into human epithelial cells. *Mol. Microbiol.* **2005**, *55*, 687–698. [CrossRef]

168. Scietti, L.; Sampieri, K.; Pinzuti, I.; Bartolini, E.; Benucci, B.; Liguori, A.; Haag, A.F.; Lo Surdo, P.; Pansegrau, W.; Nardi-Dei, V.; et al. Exploring host-pathogen interactions through genome wide protein microarray analysis. *Sci. Rep.* **2016**, *6*, 27996. [CrossRef]

169. Simonis, A.; Schubert-Unkmeir, A. Interactions of meningococcal virulence factors with endothelial cells at the human blood-cerebrospinal fluid barrier and their role in pathogenicity. *Febs Lett.* **2016**, *590*, 3854–3867. [CrossRef]

170. Lambotin, M.; Hoffmann, I.; Laran-Chich, M.P.; Nassif, X.; Couraud, P.O.; Bourdoulous, S. Invasion of endothelial cells by neisseria meningitidis requires cortactin recruitment by a phosphoinositide-3-kinase/rac1 signalling pathway triggered by the lipo-oligosaccharide. *J. Cell Sci.* **2005**, *118*, 3805–3816. [CrossRef]

171. Soyer, M.; Charles-Orszag, A.; Lagache, T.; Machata, S.; Imhaus, A.F.; Dumont, A.; Millien, C.; Olivo-Marin, J.C.; Dumenil, G. Early sequence of events triggered by the interaction of neisseria meningitidis with endothelial cells. *Cell. Microbiol.* **2014**, *16*, 878–895. [CrossRef] [PubMed]

172. Takahashi, H.; Watanabe, H.; Kim, K.S.; Yokoyama, S.; Yanagisawa, T. The meningococcal cysteine transport system plays a crucial role in neisseria meningitidis survival in human brain microvascular endothelial cells. *Mbio* **2018**, *9*. [CrossRef] [PubMed]

173. Slanina, H.; Hebling, S.; Hauck, C.R.; Schubert-Unkmeir, A. Cell invasion by neisseria meningitidis requires a functional interplay between the focal adhesion kinase, src and cortactin. *PLoS ONE* **2012**, *7*, e39613. [CrossRef] [PubMed]

174. Schubert-Unkmeir, A.; Sokolova, O.; Panzner, U.; Eigenthaler, M.; Frosch, M. Gene expression pattern in human brain endothelial cells in response to neisseria meningitidis. *Infect. Immun.* **2007**, *75*, 899–914. [CrossRef] [PubMed]

175. Sa, E.C.C.; Griffiths, N.J.; Murillo, I.; Virji, M. Neisseria meningitidis opc invasin binds to the cytoskeletal protein alpha-actinin. *Cell. Microbiol.* **2009**, *11*, 389–405. [CrossRef]

176. Coureuil, M.; Lecuyer, H.; Scott, M.G.H.; Boularan, C.; Enslen, H.; Soyer, M.; Mikaty, G.; Bourdoulous, S.; Nassif, X.; Marullo, S. Meningococcus hijacks a beta 2-adrenoceptor/beta-arrestin pathway to cross brain microvasculature endothelium. *Cell* **2010**, *143*, 1149–1160. [CrossRef]

177. Schubert-Unkmeir, A.; Konrad, C.; Slanina, H.; Czapek, F.; Hebling, S.; Frosch, M. Neisseria meningitidis induces brain microvascular endothelial cell detachment from the matrix and cleavage of occludin: A role for mmp-8. *PLoS Path.* **2010**, *6*, e1000874. [CrossRef]

178. Coureuil, M.; Mikaty, G.; Miller, F.; Lecuyer, H.; Bernard, C.; Bourdoulous, S.; Dumenil, G.; Mege, R.M.; Weksler, B.B.; Romero, I.A.; et al. Meningococcal type iv pili recruit the polarity complex to cross the brain endothelium. *Science* **2009**, *325*, 83–87. [CrossRef]

179. Waage, A.; Brandtzaeg, P.; Halstensen, A.; Kierulf, P.; Espevik, T. The complex pattern of cytokines in serum from patients with meningococcal septic shock-association between interleukin-6, interleukin-1, and fatal outcome. *J. Exp. Med.* **1989**, *169*, 333–338. [CrossRef]

180. Sokolova, O.; Heppel, N.; Jagerhuber, R.; Kim, K.S.; Frosch, M.; Eigenthaler, M.; Schubert-Unkmeir, A. Interaction of neisseria meningitidis with human brain microvascular endothelial cells: Role of map- and tyrosine kinases in invasion and inflammatory cytokine release. *Cell. Microbiol.* **2004**, *6*, 1153–1166. [CrossRef]

181. Watt, J.P.; Wolfson, L.J.; O'Brien, K.L.; Henkle, E.; Deloria-Knoll, M.; McCall, N.; Lee, E.; Levine, O.S.; Hajjeh, R.; Mulholland, K.; et al. Burden of disease caused by haemophilus influenzae type b in children younger than 5 years: Global estimates. *Lancet* **2009**, *374*, 903–911. [CrossRef]

182. Zarei, A.E.; Almehdar, H.A.; Redwan, E.M. Hib vaccines: Past, present, and future perspectives. *J. Immunol. Res.* **2016**. [CrossRef] [PubMed]

183. Bakaletz, L.O.; Novotny, L.A. Nontypeable haemophilus influenzae (nthi). *Trends Microbiol.* **2018**, *26*, 727–728. [CrossRef] [PubMed]

184. Doran, K.S.; Banerjee, A.; Disson, O.; Lecuit, M. Concepts and mechanisms: Crossing host barriers. *Cold Spring Harb. Perspect. Med.* **2013**, *3*, a010090. [CrossRef]

185. Van Ham, S.M.; van Alphen, L.; Mooi, F.R.; van Putten, J.P. Phase variation of h. Influenzae fimbriae: Transcriptional control of two divergent genes through a variable combined promoter region. *Cell* **1993**, *73*, 1187–1196. [CrossRef]

186. Cundell, D.R.; Gerard, N.P.; Gerard, C.; Idanpaan-Heikkila, I.; Tuomanen, E.I. Streptococcus pneumoniae anchor to activated human cells by the receptor for platelet-activating factor. *Nature* **1995**, *377*, 435–438. [CrossRef]

187. Swords, W.E.; Ketterer, M.R.; Shao, J.; Campbell, C.A.; Weiser, J.N.; Apicella, M.A. Binding of the non-typeable haemophilus influenzae lipooligosaccharide to the paf receptor initiates host cell signalling. *Cell. Microbiol.* **2001**, *3*, 525–536. [CrossRef]

188. Radin, J.N.; Orihuela, C.J.; Murti, G.; Guglielmo, C.; Murray, P.J.; Tuomanen, E.I. Beta-arrestin 1 participates in platelet-activating factor receptor-mediated endocytosis of streptococcus pneumoniae. *Infect. Immun.* **2005**, *73*, 7827–7835. [CrossRef]

189. Caporarello, N.; Olivieri, M.; Cristaldi, M.; Scalia, M.; Toscano, M.A.; Genovese, C.; Addamo, A.; Salmeri, M.; Lupo, G.; Anfuso, C.D. Blood-brain barrier in a haemophilus influenzae type a in vitro infection: Role of adenosine receptors a(2a) and a(2b). *Mol. Neurobiol.* **2018**, *55*, 5321–5336. [CrossRef]

190. Wispelwey, B.; Lesse, A.J.; Hansen, E.J.; Scheld, W.M. Haemophilus influenzae lipopolysaccharide-induced blood brain barrier permeability during experimental meningitis in the rat. *J. Clin. Invest.* **1988**, *82*, 1339–1346. [CrossRef]

191. Wispelwey, B.; Hansen, E.J.; Scheld, W.M. Haemophilus influenzae outer membrane vesicle-induced blood-brain barrier permeability during experimental meningitis. *Infect. Immun.* **1989**, *57*, 2559–2562. [PubMed]

192. Parisi, D.N.; Martinez, L.R. Intracellular haemophilus influenzae invades the brain: Is zyxin a critical blood brain barrier component regulated by tnf-alpha? *Virulence* **2014**, *5*, 645–647. [CrossRef] [PubMed]

193. Kostyanev, T.S.; Sechanova, L.P. Virulence factors and mechanisms of antibiotic resistance of haemophilus influenzae. *Folia Med.* **2012**, *54*, 19–23. [CrossRef]

194. Yau, B.; Hunt, N.H.; Mitchell, A.J.; Too, L.K. Bloodbrain barrier pathology and cns outcomes in streptococcus pneumoniae meningitis. *Int. J. Mol. Sci.* **2018**, *19*, 3555. [CrossRef] [PubMed]

195. Ferrieri, P.; Burke, B.; Nelson, J. Production of bacteremia and meningitis in infant rats with group-b streptococcal serotypes. *Infect. Immun.* **1980**, *27*, 1023–1032.

196. Nassif, X.; Bourdoulous, S.; Eugene, E.; Couraud, P.O. How do extracellular pathogens cross the blood-brain barrier? *Trends Microbiol.* **2002**, *10*. [CrossRef]

 © 2019 by the authors. Licensee MDPI, Basel, Switzerland. This article is an open access article distributed under the terms and conditions of the Creative Commons Attribution (CC BY) license (http://creativecommons.org/licenses/by/4.0/).

International Journal of
Molecular Sciences

Review

Coagulase-Negative Staphylococci Pathogenomics

Xavier Argemi [1,2,*], Yves Hansmann [1,2], Kevin Prola [2] and Gilles Prévost [2]

[1] Service des Maladies Infectieuses et Tropicales, Hôpitaux Universitaires, 67000 Strasbourg, France; yves.hansmann@chru-strasbourg.fr

[2] Université de Strasbourg, CHRU Strasbourg, Fédération de Médecine Translationnelle de Strasbourg, EA 7290, Virulence Bactérienne Précoce, F-67000 Strasbourg, France; kevin.prola@etu.unistra.fr (K.P.); prevost@unistra.fr (G.P.)

* Correspondence: xavier_argemi@hotmail.com or xavier_argemi@chru-strasbourg.fr; Tel.: +33-369550545; Fax: +33-369551836

Received: 22 January 2019; Accepted: 7 March 2019; Published: 11 March 2019

Abstract: Coagulase-negative Staphylococci (CoNS) are skin commensal bacteria. Besides their role in maintaining homeostasis, CoNS have emerged as major pathogens in nosocomial settings. Several studies have investigated the molecular basis for this emergence and identified multiple putative virulence factors with regards to *Staphylococcus aureus* pathogenicity. In the last decade, numerous CoNS whole-genome sequences have been released, leading to the identification of numerous putative virulence factors. Koch's postulates and the molecular rendition of these postulates, established by Stanley Falkow in 1988, do not explain the microbial pathogenicity of CoNS. However, whole-genome sequence data has shed new light on CoNS pathogenicity. In this review, we analyzed the contribution of genomics in defining CoNS virulence, focusing on the most frequent and pathogenic CoNS species: *S. epidermidis*, *S. haemolyticus*, *S. saprophyticus*, *S. capitis*, and *S. lugdunensis*.

Keywords: pathogenomics; coagulase-negative staphylococci; virulence factors; whole genome sequencing

---

1. Introduction

Bacterial virulence is a complex concept that must be considered from the clinical, molecular, and genomic perspectives. Clinically, the virulence of a pathogen, of a specific species or even of a clonal strain, can refer to its inherent capacity to provoke specific clinical manifestations that can be linked to the production of a virulence factor, often a protein. Microbiologically, Robert Koch postulates illustrated bacterial virulence, which Stanley Falkow redefined in the late 1980s to provide a molecular version of it [1]. Molecularly, the virulence factor is found in the pathogenic strains of a species and not in non-pathogenic ones. Secondly, the inactivation of the encoding gene in an animal model should attenuate the virulence of the strain. Thirdly, the reintroduction/reactivation of the gene should restore virulence. Nevertheless, in the past 20 years, neither Koch's nor Falkow's hypotheses have been able to completely define virulence factors. The development of rapid and cost-effective genome sequencing technologies has provided the opportunity, among others, to discover a wide range of genes that could be linked to bacterial pathogenicity. This ability to determine the pathogenic capacities of a bacterium from its genome is known as pathogenomics [2]. Virulence factor databases allow for fast and easy identification of putative virulence factors in whole-genome sequence data based only on sequence homologies. The Virulence Factor Database was released in 2005 and is an example of a regularly-updated database, of which a fourth version was released in 2018 [3]. Other updated databases exist, such as PHI-base and Victors [4,5]. Thus, some virulence factors retain the definition of Stanley Falkow, with a direct causative link with bacterial pathogenicity and some possible clinical manifestations. Conversely, some other factors, such as those that are produced by human commensal bacteria, do not act as virulence factors under normal conditions, but these factors will take part in bacterial pathogenicity under specific conditions, such as during cutaneous barrier breach. Staphylococci are examples of such dual characterizations of virulence factors.

Staphylococci comprise *Staphylococcus aureus*, which is a human commensal bacterium that colonizes nearly 30% of the human skin and mucosae, and coagulase-negative staphylococci (CoNS), which include species such as *Staphylococcus epidermidis*, which colonizes nearly all human skin [6]. *S. aureus* is usually considered separately from CoNS and as a virulent species. Indeed, it produces a wide range of toxins which can act as virulence factors leading to specific clinical manifestations with a direct causative link. For example, the Panton-Valentine leucocidin (LukSF-PV) is a bicomponent toxin that may lead to necrotizing pneumonia [7]. *S. aureus* produces other toxins such as epidermolysins (ETA-B-D) and superantigens, e.g., toxic shock syndrome toxin (TSST-1), that lead to specific clinical diseases and turn some particular strains into pathogens that fully meet the Koch and Falkow postulates [8,9]. *S. aureus* might also produce numerous adhesion factors, capsule, and biofilm-associated proteins, hemolysins, and immune evasion proteins whose main purpose is to persist on human skin as a commensal and probably beneficial bacterium [10]. Many of those proteins are cell wall-anchored and comprise a group of covalently-attached proteins with specific motifs that bind to the extracellular matrix and are usually identified as microbial surface components recognizing adhesive matrix molecules (MSCRAMMs) [11]. Many other proteins are cell-wall-attached but display distinct functions such as iron-regulated surface proteins, and some genes encode for putative MSCRAMMs but without any molecular and structural confirmation, and can be identified more generally as surface adhesins, a terminology used in this review. Those adhesion factors play a crucial role in the pathogenicity *of S. aureus*. The occurrence of a cutaneous barrier breach will allow its penetration in a new environment, where those factors will behave as real virulence factors, allowing its adherence, persistence, and multiplication in a normally sterile environment. Whole-genome sequence analysis of *S. aureus* has revealed the existence of a very large repertoire of such toxins, adhesion molecules, and biofilm-associated genes, sometimes without any molecular confirmation but based on sequence homologies that lead to the characterization of *S. aureus* as a major pathogenic species [12].

CoNS are often considered as simple commensal bacteria due to their rarity in clinical pathology and the absence of virulence factors like those produced by *S. aureus*. However, the emergence of nosocomial infections with CoNS has led clinicians and researchers to reconsider the status of these bacteria [13]. The publication of several CoNS genomes has been an essential tool to better understand CoNS pathogenicity. The identification of virulence factors, such as toxins with clinical impacts, remains exceptional or even controversial, and it appears that CoNS have a very large repertoire of genes that encode for adhesion factors, biofilm production, hemolysins, exoenzymes, and superantigens. Most CoNS also have similar regulatory systems to *S. aureus*, e.g., the *agr* system, for which homologs have been identified in numerous CoNS species [13]. The cutaneous barrier breach is a critical step to turning the CoNS species into pathogens, turning factors that are mainly implicated in the bacterial life cycle on the skin into virulence factors leading to pathological manifestations. In this situation, once again, the concept of virulence factors does not apply to the definitions provided by Robert Koch and Stanley Falkow. Considering the great species diversity of CoNS, *S. epidermidis* is the most frequent CoNS implicated in clinical diseases, and it remains the best-studied species at a molecular and genomic level [14]. Nevertheless, none of the virulence factors that were discovered could clearly categorized a specific strain as pathogenic or commensal, even if the strains belonging to the clonal type ST2 have been over-represented in the specific context of *S. epidermidis* neonatal sepsis [15]. ST2 clonal type always bears the insertion sequence *IS256* and *ica* genes, which are implicated in biofilm production, but the direct correlation of those loci with *S. epidermidis* invasiveness has never been confirmed [16]. Some other species also came up as significant pathogens, such as *S. lugdunensis* that has been widely investigated due to the severity and the specificity of clinical manifestations of infection due to this species [17]. *S. haemolyticus*, *S. saprophyticus*, and *S. capitis* are three other species for which the rate and the type of infection led to molecular and genomic investigations in the search of virulence factors [18,19]. *S. caprae* is a CoNS that usually belongs to the animal skin flora, but it can also colonize human skin. This CoNS has been described in veterinary infections (mastitis principally) and in human osteoarticular infections. Interestingly, this CoNS belongs to a group of CoNS that is comprised of *S. epidermidis*, *S. capitis*, and *S. saccharolyticus* [13,20–22].

The identification of virulence factors in staphylococci has resulted in the identification of mobile genetic elements. Indeed, the majority of *S. aureus* virulence genes are located on mobile genetic elements such as pathogenicity islands, plasmids, and phages, which represent up to 25% of the *S. aureus* genome and play crucial roles in host adaptation and the modulation of virulence [23–25]. Several bioinformatic tools have been developed to identify such genomic regions based on comparative genomics [26–28]. Thus, the identification of putative virulence factors can be driven by the identification of these genomic elements, but it has appeared that the presence of such mobile genetic elements is very common in CoNS, but that they do not bear any virulence factors which are similar to *S. aureus*. More generally, this large repertoire of mobile genetic elements in staphylococci species explains why, despite a limited core genome, those species usually displayed an open pan genome with a very high number of existing genes, sometimes shared by a very limited number of clones [29–31]. Worthy of note is the fact that *S. lugdunensis* has a closed pan genome which contains several barriers to horizontal gene transfers, which could explain this unique genomic characteristic amongst CoNS [32].

In this review, we aimed to describe the occurrence of virulence factors, or more accurately, of pathogenicity factors, based on whole-genome sequence data. We focused this systematic review on the major CoNS pathogens for which such data have been published: *S. epidermidis*, *S. lugdunensis*, *S. saprophyticus*, *S. haemolyticus*, *S. caprae*, and *S. capitis*. A comprehensive review of all CoNS pathogenicity factors has been described elsewhere and is covered in references [13,33–39].

2. Research Method and Results

This systematic review was based on PubMed published articles in the English language. We used the following MeSH terms: "Virulence", "Virulence factors", "Pathogenicity", "Whole genome sequencing", and "Genomics". In particular, we searched for the following species: "*Staphylococcus epidermidis*", "*Staphylococcus lugdunensis*", "*Staphylococcus capitis*", "*Staphylococcus caprae*", "*Staphylococcus haemolyticus*", and "*Staphylococcus saprophyticus*". We excluded studies that were not based on whole-genome data, and those for which genome sequences were not deposited in public databases (e.g., GenBank, European Nucleotide Archive).

We identified eight studies for *S. epidermidis* [29,30,40–45], four for *S. capitis* [46–49], three for *S. lugdunensis* [32,50,51], two for *S. haemolyticus* [52,53], and one for *S. caprae* [54] and *S. saprophyticus* [55].

2.1. Staphylococcus epidermidis

S. epidermidis is by far the most prevalent CoNS in microbiological samples and the primary cause of CoNS-related infections, particularly in nosocomial settings [56]. This species has been widely studied at a molecular level and has provided the largest amount of data regarding the presence of virulence factors that might be identified as pathogenicity factors [16]. Biofilm formation is the main path by which *S. epidermidis* colonizes and infects prosthetic and medical devices as reviewed by Otto et al. [57]. The biofilm life cycle is a complex process which implicates MSCRAMMs, the proliferation of the exopolysaccharide matrix, and finally, biofilm dispersion, mostly due to phenol-soluble modulin (PSM) peptides [57].

Nearly 500 genomes have been made available on GenBank, mainly draft genomes, but also 12 complete genomes. *S. epidermidis* whole-genome-based studies to determine virulence determinants are detailed in Table 1. In 2003, Zhang et al. provided the first genome-based analysis of *S. epidermidis* virulence [40]. The authors provided the first sequence of the reference strain ATCC 12228, a non-biofilm forming, non-infection associated strain, and they found several putative pathogenicity factors, including multiple MSCRAMMs encoding genes and putative exoenzymes and toxins such as metalloproteases and Delta/Beta-hemolysins. Interestingly, a revised version of this genome was released in 2017 which was determined using PacBio (PacBio, Pacific Biosciences, Menlo Park, CA, USA) long read sequencing; it provided a slightly different genome from the one released by Zhang et al., and established a new reference genome for comparative genomic studies [58]. In 2005, Gill et al. published a new complete genome from *S. epidermidis* strain RP62a and performed comparative genomic analyses with four *S. aureus* genomes, and included the genome from *S. epidermidis* ATCC 12228 [41]. As expected, the authors found multiple virulence factors in the *S. aureus* genome. Nearly one-half of those factors were carried by seven pathogenicity islands that were not found on *S. epidermidis*. Nevertheless, some other mobile genetic elements were identified in this species (one genomic island and two integrated plasmids), some bearing PSM and putative MSCRAMMs. Some other mobile genetic elements were found in *S. epidermidis* genomes such as prophages, insertion sequences, and *SSCmec-like* cassettes. The authors also identified several other putative virulence factors such as proteases (serine and cysteine protease), lipases, and hemolysins (*Beta/Delta haemolysin*) loci. The presence of mobile genetic elements and virulence factors led the authors to consider that horizontal gene transfer between staphylococci had to be considered as a source of variability and, in this case, as a cause of their pathogenicity.

Conlan et al. provided in 2012 the first large comparative genomic analysis of 30 *S. epidermidis* whole genomes, showing that *S. epidermidis* had an open pan genome, as expected, but more interestingly, that whole-genome-based phylogenetic trees could distinguish between commensal and pathogenic strains. The presence of the *formate dehydrogenase* gene (*fdh*) could explain this observation [42]. Meric et al. found in a second study based on whole-genome comparative analyses that *S. epidermidis* and *S. aureus*, in hospital settings, share some genes involved in pathogenicity, suggesting the existence of horizontal gene transfer between these species [29]. The added value of whole-genome analysis was confirmed by Virginia Post et al., who compared whole-genome sequences of 104 *S. epidermidis* isolates from patients with orthopedic-device-related infections. This study found a correlation between patient outcome and some loci [30]. Patients with multiple surgeries due to treatment failure were more likely to be infected with the biofilm-associated gene *bhp*, the antiseptic resistance gene *qacA*, the cassette genes *ccrA* and *ccrB*, and the *IS256-like* transposase gene. This whole-genome-based study identified, for the first time and directly from whole-genome data, loci that might be involved in *S. epidermidis* pathogenicity based on well-documented clinical data.

While such studies have yielded novel and sometimes revolutionary insights into *S. epidermidis* virulence, the presence of such loci does not provide any information regarding their expression, regulation, and their impact on clinical implications. Yet, whole-genome data can be used as a complementary tool when a putative virulence factor is identified, sometimes starting from clinical considerations. In 2013, Fournier et al. provided observations of a patient with *S. epidermidis* endocarditis, and the authors found several clues that could explain the virulence of this strain [43]. Besides known virulence factors such as MSCRAMMs, exoenzymes, and hemolysins, the authors identified a previously unreported prophage, a new toxin/antitoxin module, and a complete *icaABCD* operon, which was usually not observed in non-pathogenic strains. As observed by Conlan et al., this pathogenic strain also lacked the *fdh* gene.

The existence of enterotoxins and pathogenicity islands in *S. epidermidis* has been controversial, despite observations by Madhusoodanan et al. of such a genetic element in the clinical strain *S. epidermidis* FRI909, which carried two functional enterotoxin genes *sec3* and *sell* [59]. This unique observation was initially considered as a possible genetic "accident". In 2016, we described two strains isolated from septic shock patients that produced a staphylococcal enterotoxin C with homology to *S. aureus* enterotoxin C3 [60]. By using whole-genome data, we thereafter provided evidence that the *sec3* from our strains was located on a pathogenicity island very similar to the one described by Madhusoodanan et al., named SePI-1/SeCI-1 [44]. We also identified several plasmids carrying resistance genes and sharing homologies with *S. aureus*. This result confirmed that some mobile genetic elements from *S. epidermidis* might come from *S. aureus*, with the transfer of virulence factors.

Table 1. *Staphylococcus epidermidis* whole-genome-based studies to determine virulence determinants.

Date	Ref	Type of Analysis	Genomes	Strain ID	Accession Number	Strain Origin		Putative Genetic Determinants of Pathogenicity
2003	[40]	Pathogenicity factor search	1	ATCC 12228	NZ_CP022247	Skin swab	pSE-12228-01 pSE-12228-02 pSE-12228-03 pSE-12228-04 pSE-12228-05 pSE-12228-06	Plasmids
							sar, agr	Regulation systems
							ssp	Serine protease
							geh	Lipase
							clp, sep	Clp protease, metalloprotease, nuclease
							hld	Delta haemolysin
							atl	Autolysin
							sdr, embp, ebps	Adhesins/MSCRAMMs
2005	[41]	Pathogenicity factor search	1	RP62a	NC_002976.3	Catheter-associated infection	*clp, ssp, nuc*	Clp protease, serine and cysteine protease, nuclease
							lip, geh	Lipases
							hlb, hld	Haemolysins
							psm	Phenol-soluble modulins
							sdr, ses, ebh, ebp, fbe	Adhesins/MSCRAMMs
							νSEγ	Genomic island carrying PSMs genes
							νSE1, νSE2	Integrated plasmids carrying cadmium resistance gene and putative MSCRAMMs
							cap	Capsule synthesis
		Comparative genomics with 4 *S. aureus* genomes	2	Complete genomes from the study and from GenBank	-	-	-	*S. epidermidis* and *S. aureus* present syntenic genomes and horizontal gene exchange probably happen between those species in both directions, possibly mobilizing virulence factors from *S. aureus* to *S. epidermidis*.
2012	[42]	Comparative genomics between commensal and pathogenic *S. epidermidis*	30	Complete genomes from the study	-	-	-	*S. epidermidis* has an open pan genome, the variable genome comprises mobile genetic elements, the *fdh* gene distinguishes commensal from pathogenic bacteria.

Table 1. *Cont.*

Date	Ref	Type of Analysis	Genomes	Strain ID	Accession Number	Strain Origin	Putative Genetic Determinants of Pathogenicity	
2013	[43]	Pathogenicity factor search	1	12142587	GCA_000304575.1	Endocarditis	*ica*	Intercellular adhesion factors
							fbe, sdr, ebp, ebh	Adhesins/MSCRAMMs
							dtl, mpr, vra, apr	*Resistance to antimicrobial peptides*
							ssp	Serine and cysteine protease
							lip, geh	Lipases
							psmβ	Phenol-soluble modulin
							hlb	Beta-haemolysin
2015	[29]	Comparative genomics with 241 *S. aureus* genomes	83	Complete genomes from GenBank	-	-	-	*S. epidermidis* and *S. aureus* share a core genome of 1478 genes, and if homologous recombinations are rare between both species, interspecies transfer of mobile genetic elements might happen and shape the genome of both species regarding the environmental conditions.
2017	[30]	Comparative genomics between *S. epidermidis* strains after osteoarticular infections	104	Complete genomes from the study	-	-	-	Some genes from *S. epidermidis* were associated with bad outcome in patients: *bhp* (biofilm formation), *qacA* (antiseptics resistance), *ccrA-ccrB* (cassette chromosome recombinase), IS256-like (transposase).
2018	[44]	Comparative genomics between 3 *S. epidermidis* genomes	3	Complete genomes from the study and from GenBank	-	-	-	*S. epidermidis* pathogenic strains that produce a C-3 like enterotoxin bear a SEC-coding sequence located on a composite pathogenicity island *SePI-1/SeCI-1* that suggest the existence of HGT from *S. aureus* to *S. epidermidis*.
2018	[45]	Pathogenicity factor search	1	G6_2	ERR387168	Environmental strain	P1 to P6	Plasmids
							-	Presence of multiple genes implicated in drug resistance
							atl	Autolysin
							ebh, ebp, sdr	Adhesins/MSCRAMMs
							ssp, nuc	Serine and cysteine protease, nuclease
							lip, geh	Lipases
							ica	Intercellular adhesion factors
							hlb, hld	Haemolysin

Recently, Xu et al. performed a whole-genome analysis of a multi-resistant *S. epidermidis* strain isolated in the environment, i.e., in a hotel room, and identified multiple loci dedicated to antibiotic resistance but also to numerous virulence genes, as described previously [45].

S. epidermidis is a major cause of nosocomial infections within other CoNS. The existence of several factors that modulate its commensal life cycle on the skin also turns this species into a pathogen when entering sterile sites. Whole-genome studies have provided specific and unique data regarding the general evolution of its genome and interactions with other staphylococci, such as *S. aureus*. However, whole-genome data also appear as crucial and unique complementary analyses when a specific strain is isolated clinically, or when a putative virulence factor is molecularly identified.

2.2. Staphylococcus lugdunensis

S. lugdunensis is considered a significant pathogen in human infection [17]. Clinically, its virulence is most likely lower than that of *S. aureus*. Clinical observations have emphasized the frequency and severity of skin and soft tissue infections, endocarditis, and osteoarticular infections [61–63]. Microbiologically, this CoNS produces a coagulase, which is a common feature with *S. aureus* [64]. Its pathogenicity has been noted by several authors, and multiple virulence factors have been identified molecularly, including adhesion molecules like a fibrinogen-binding protein; cytotoxins such as δ-like-hemolysin; a complete biofilm synthesis system, including an *agr* regulating operon; and as seen in *S. aureus*, a large system dedicated to iron metabolism [65–67]. Interestingly, Heilbronner et al. showed that duplication of the *isd* locus provided a selective advantage to *S. lugdunensis* in the case of iron limitation, possibly improving bacteria survival and pathogenicity. Recently, we identified and purified a novel metalloprotease named lugdulysin that could be implicated in osteoarticular infections [61].

S. lugdunensis whole-genome-based studies to determine virulence determinants are listed in Table 2. The first complete genome of *S. lugdunensis* was published by Tse et al. in 2010 [68]. Since then, 17 complete genomes have been published, and eight additional draft genomes are available in GenBank. Nevertheless, the first exploitation of a complete genome was provided by Heilbronner et al. in 2011 [50]. The authors published the complete genome of strain N920143 and identified several putative virulence factors based on comparative genomics and homologies searches. They identified, for the first time, a complete operon implicated in iron metabolism (*isd*: *isdB-J-C-K-E-F-G*), several genes coding for MSCRAMMs such as fibrinogen and Von Willebrand adhesion factors, a *streptolysin S-like toxin*, an *IcaABCD* operon for biofilm synthesis, *agr* regulation genes, and even an *esx* locus with homology to the esx-ESAT 6 secretion systems that are implicated in the secretion of virulence factors in *Mycobacterium tuberculosis*. Due to the frequent localization of virulence genes on mobile genetic elements in *S. aureus*, in 2017 we also published the complete genome of seven strains of *S. lugdunensis* and identified multiple mobile genetic elements [51]. We identified a very unusual number of plasmids and prophages by using computational methods, including four new prophages and five plasmids, in which three were previously described in other CoNS and one in *S. aureus*. This study suggested that horizontal gene exchange could occur between staphylococci, including *S. aureus*. Descriptions of plasmids and prophages in CoNS remain rare, but, interestingly, they have been performed mainly in *S. epidermidis* and *S. haemolyticus*.

Table 2. *Staphylococcus lugdunensis* whole-genome-based studies to determine virulence determinants.

Date	Ref	Type of Analysis	Genomes	Strain ID	Accession Number	Strain Origin	Putative Genetic Determinants of Pathogenicity	
2011	[50]	Pathogenicity factor search	1	N920143	NC_017353.1	Human breast abscess	*isd*	Iron-regulated surface determinants
							sls, vWbl, fbl	MSCRAMMs
							sst	Iron-regulated ferric siderophore uptake system
							agr	Regulation system
							ess	6 toxin secretion system
							ica	Intercellular adhesion factors
							cap	Capsule synthesis
							mprf/dlt	Immune evasion
							CRISPR	Clustered regularly interspaced short palindromic repeats sequence
							φSL1	Prophage
2017	[51]	Mobile genetic elements search	7	C_33 VISLISI_21 VISLISI_22 VISLISI_25 VISLISI_27 VISLISI_33 VISLISI_37	NZ_CP020768.1 NZ_CP020762.1 NZ_CP020764.1 CP020763.1 NZ_CP020735.1 NZ_CP020769.1 CP020761.1	Skin swab Bacteremia Endocarditis Knee Prosthesis Knee prosthesis Liver abscess Endocarditis	pVISLISI_1 to pVISLISI_5	Plasmids
							φSL2 to φSL4	Prophages
2018	[32]	Comparative genomics with 15 *S. aureus* and 13 *S. epidermidis* genomes	15	All complete genome from GenBank	-	-	-	Identification of a closed pan genome in *S. lugdunensis* and multiple barriers to horizontal gene transfer (restriction-modification, CRISPR/Cas, and toxin/antitoxin systems).

S. lugdunensis is characterized by its unusual antimicrobial susceptibility, even if some rare strains are oxacillin-resistant. In 2017, Chang et al. fully characterized, for the first time, two novel variants of staphylococcal cassette chromosome *mec* elements in two oxacillin-resistant *S. lugdunensis* strains by using whole-genome sequence data [69]. The authors also identified several homologies with *S. aureus*, *S. haemolyticus*, and *S. epidermidis* cassette chromosome regions. If this result suggested once again the existence of horizontal gene exchanges between staphylococci, including *S. aureus*, the very unusual antimicrobial susceptibly of *S. lugdunensis* remained unexplained in comparison to other CoNS, which displayed a very high rate of resistance, particularly in nosocomial settings [19]. We recently published a comparative genomic analysis of *S. lugdunensis* whole-genome with *S. aureus* and *S. epidermidis* [32]. *S. epidermidis* and *S. aureus* are characterized by an open pan genome, which means that those species display a virtually unlimited number of new genes when adding different genomes, even if they conserve a limited number of genes through the strains called the core genome. Conversely, *S. lugdunensis* display a closed pan genome, and all published genomes were very similar, with only a very limited number of new genes among the strains. This unexpected characteristic could explain the very well conserved antimicrobial susceptibility of *S. lugdunensis* that rarely acquires resistance genes. We also found that this observation could rely on the presence of numerous barriers to horizontal gene transfer in this species, including restriction-modification, CRISPR/Cas, and toxin/antitoxin systems.

S. lugdunensis is a very particular CoNS and might be closer to *S. aureus* than other CoNS in terms of virulence. Whole-genome sequence studies have emphasized its originality on a genomic scale, including the presence of multiple unusual factors implicated in its pathogenicity and that are uncommon for CoNS, like its iron metabolism system. Overall, its genome encodes for several factors that take part in its survival on the skin as a commensal bacterium, such as adhesion proteins and biofilm capacities. Once again, those factors probably act as virulence factors once the cutaneous barrier is broken. Comparative genomic studies have also revealed that it displays a unique profile in terms of genomic plasticity, which could explain some of its microbiological characteristics, such as its highly-conserved antimicrobial susceptibility.

2.3. Staphylococcus capitis

S. capitis is the third CoNS that has been described in clinical infections and for which whole-genome studies focusing on its virulence are available [13]. *S. capitis* infections have been described in various clinical situations implicating biofilm production as endocarditis, catheter-related bacteremia, and prosthetic joint infections [70,71]. This species also causes neonatal sepsis in neonatal intensive care units [72,73].

S. capitis whole-genome-based studies to determine virulence determinants are listed in Table 3. The first draft genome of this species was released in 2009, and the first complete genome was published in 2015 by Cameron et al. to determine the genetic determinants that could support its pathogenicity [46]. The authors identified several loci that could encode for virulence factors and performed a comparative genomic analysis with the *S. epidermidis* genome. They identified putative virulence regulators (as *agr* homologs), biofilm production loci, genes encoding for exoenzymes (as metalloproteases and hemolysins), PSMs, and MSCRAMMs. This genome-scale analysis led to a functional analysis of biofilm and PSMs production.

Table 3. *Staphylococcus capitis* whole-genome-based studies to determine virulence determinants.

Date	Ref	Type of Analysis	Genomes	Strain ID	Accession Number	Strain Origin	Putative Genetic Determinants of Pathogenicity	
2015	[46]	Pathogenicity factor search	1	AYP1020	NZ_CP007601.1	Bacteremia	pAYP1020	Plasmid
							-	Prophage
							-	Insertion sequence
							agr, sar, sae, arl, rot, sigB	Regulation system
							ica	Intercellular adhesion factors
							cap	Capsule synthesis
							hlb	Haemolysin
							clp, ssp, sep, htr, spl	Clp protease, cysteine and serine proteases, metalloprotease
							lip, geh	Lipase
							psm	Phenol-soluble modulins
							fbe, aap, ebh, ses, ebp, bhp, sdr	Adhesins/MSCRAMMs
							atl	Autolysin
2014	[47]	Drug resistance genes	1	LNZR-1	NZ_JGYJ00000000.1	Bacteremia	-	Presence of multiple genes implicated in drug resistance
2016	[48]	Pathogenicity factor search	4	NRCS-A CR01 CR03 CR04 CR05	NZ_CBUB000000000.1 NZ_CVUF00000000.1 NZ_CTEM00000000.1 NZ_CTEO00000000.1	Bacteremia	-	Plasmids
							IS256, IS272, IS431mec-like	Insertion sequences
							hsdMSR	Type Restriction/Modification system
							-	Presence of multiple genes implicated in drug resistance
							atl	Autolysin
							ebh, ebp, sdr	Adhesins/MSCRAMMs
							ica	Intercellular adhesion factors
							clp, ssp, nuc	Clp protease, cysteine and serine protease, nuclease
							cap	Capsule synthesis
							hlb, hld	Haemolysin
							psm	Phenol-soluble modulins
							sar, rot, mgr	Regulation system
2017	[49]	Pathogenicity factor search	1	TE8	NZ_JMGB00000000.1	Cutaneous swab	*Epidermicin-like* and *Gallidermin* gene clusters	Antimicrobial peptides
							psm	Phenol-soluble modulins with antimicrobial activities

At a larger scale, it appears that some clonal strains might be implicated in the context of neonatal sepsis as suggested by Butin et al., who observed the worldwide endemicity in 17 countries of a multidrug-resistant strain by using pulsed-field gel electrophoresis patterns but also whole-genome sequence data [74]. Multidrug resistance has emerged as a crucial issue in *S. capitis* infections, with nearly half of the strains being oxacillin-resistant in hospital settings; occasionally, some strains with vancomycin reduced susceptibility have been identified [72,73]. Li et al. provided a complete whole-genome analysis of a multi-resistant strain causing bacteremia, and found a very high and unusual number of genes implicated in such a resistant profile [47]. Recently, Simoes et al. published the first whole-genome sequence of *S. capitis* strain NRCS-A, which is a multi-resistant clone implicated in neonatal infections worldwide, by using PacBio (PacBio, Pacific Biosciences, Menlo Park, CA, USA) long read sequencing, and provided comparative genomic analysis with three other NRCS-A sequenced clones coming from different countries [48]. The authors identified, as expected, multiple resistant genes, but interestingly, some of them were located on mobile genetic elements. The authors identified several virulence genes which had been described previously by Cameron et al., but more specifically, they identified which of them was clone-specific as a putative restriction-modification system and a nisin resistance gene, which could explain the persistence of NRCS-A clones in patients' gut microbiota. In 2017, Kumar et al. used whole-genome sequence data to reveal the existence of four loci encoding for antimicrobial peptides [49]. The authors were then able to perform a functional analysis to confirm the functionality (antimicrobial activity against Gram-positive bacteria, including *S. aureus*) of those genes, and formulate the hypothesis that such antimicrobial peptides could explain the adaptation and the persistence of *S. capitis* on the human skin.

S. capitis is considered a serious pathogen, particularly in neonatal settings, and whole-genome studies have provided useful and unique data regarding the spread of its antimicrobial resistance, but have also emphasized its ability to colonize and persist on human skin, which is directly linked to its pathogenicity regarding catheter-associated infections.

2.4. Staphylococcus caprae

S. caprae is a normal component of the animal skin flora and can cause mastitis, principally in goats [13]. It also belongs to the human skin flora and is an observed cause of osteoarticular infections and also material-associated infections and bacteremia [20]. Until recently, only six draft genomes of *S. caprae* were available from humans and goats. *S. caprae* whole-genome-based studies to determine virulence determinants are listed in Table 4. In 2018, Watanabe et al. published a comparative genomic analysis of three human strains by performing a whole-genome assembly which leads to three complete chromosomes [54]. The authors compared the conserved genome parts of *S. epidermidis*, *S. capitis*, and *S. caprae*, as all previous phylogeny studies found all three species belonged to the same group (*S. epidermidis* group) [22]. They found that all three species shared a biofilm- and capsule-associated loci, and they also shared PSM genes. They also identified several putative MSCRAMMs genes. We observed that the three species belonged to the *S. epidermidis* group and shared very common features for skin resident bacteria. The similarity in clinical presentation of infection caused by these species in human disease has been illuminated.

Table 4. *Staphylococcus caprae, haemolyticus,* and *saprophyticus* whole-genome-based studies to determine virulence determinants.

Date	Ref	Type of Analysis	Genomes	Strain ID	Genome Accession number	Strain Origin		Putative Genetic Determinants of Pathogenicity
						S. capare		
2018	[54]	Pathogenicity factor search	3	JMUB145 JMUB590 JMUB898	AP018585 AP018586 AP018587	Bacteremia Nasal swab Bacteremia	*cap*	Capsule synthesis
							clp, esp	Clp protease
							psm	Phenol-soluble modulins
							ses, bhp, sdr, clfB-like, fbe, embp	Adhesins/MSCRAMMs
							geh, lip	Lipases
							atl	Autolysin
						S. haemolyticus		
2005	[52]	Pathogenicity factor search	1	JCSC1435	NC_007168.1	Human	pSHaeA pSHaeB	Plasmids
							φSh1	Prophages
							-	Presence of multiple genes implicated in drug resistance
							sdr, ebp	Adhesins/SCRAMMs
							spl, clp, nuc	Clp protease, serine and cysteine proteases, nuclease
							cap	Capsule synthesis
							atl	Autolysin
							hla	Alpha-haemolysin
2016	[53]	Pathogenicity factor search	1	S167	NZ_CP013911.1	-	pSH108	Plasmid
							agr, sar, arl	Regulation systems
							atl	Autolysin
						S. saprophyticus		
2005	[55]	Pathogenicity factor search	1	ATCC15305	NC_007350.1	Urine	pSSP1 pSSP2	Plasmids
							pro, put, opu, aqp, nha	Transport system related to urine environement
							cap	Capsule synthesis
							uafA	Adhesin specialized in urinary tract cells adherence
							ure	Urease activity

2.5. Staphylococcus haemolyticus

S. haemolyticus is a commensal bacterium but is also a frequent nosocomial pathogen that has been described mainly in catheter-related bacteremia, with one of the highest degrees of methicillin resistance among CoNS [19,75,76]. *S. haemolyticus* whole-genome-based studies to determine virulence determinants are listed in Table 4. *S. haemolyticus* was one of the first CoNS with a complete genome published by Takeuchi et al. in 2005 [52]. The authors identified several putative virulence factors in this pioneering study, such as hemolysins and bacterial capsule. Until now, no publication has completely explored the presence of virulence genes in this species, warranting future studies. Some previous studies have described the presence of cytotoxins, enterotoxins, and PSMs at the molecular level, which is exceptional among CoNS [77,78]. This species is also a biofilm-producing species, but interestingly, in 2016, Hong et al. published the first complete sequence of *S. haemolyticus* by using PacBio (PacBio, Pacific Biosciences, Menlo Park, CA, USA) long read sequencing and found no *ica* operon, but an *agr* regulation system homolog, suggesting that alternative mechanisms were implicated in biofilm formation [53]. Thus, *S. haemolyticus* is a significant pathogen in humans, but it appears that a complete whole-genome analysis of those virulence factors is lacking, whereas, several complete genomes are already available with this species bearing very unusual virulence factors for a CoNS.

2.6. Staphylococcus saprophyticus

S. saprophyticus is a commensal CoNS and can lead to lower urinary tract infections in young women [79,80]. Colonization of the genital area and the gut has been linked to this atypical type of infection, but most importantly, it was shown in vitro that this CoNS has an unusual capacity to adhere to urothelial cells and to produce urease. Nevertheless, those characteristics are not unique in CoNS and do not fully explain the capacity of this CoNS to cause urinary tract infections. *S. saprophyticus* whole-genome-based studies to determine virulence determinants are listed in Table 4. In 2005, Kuroda et al. established the first complete genome of *S. saprophyticus*, and until now, only six complete genomes for this species have been made available in GenBank [55]. The authors identified three factors that could explain the specificity of *S. saprophyticus* pathogenicity by using whole-genome data and a comparative genomic approach. They identified a novel—and unique among CoNS—cell wall-anchored protein, UafA, associated at a molecular level with a high capacity to adhere to cells from the urinary tract. They also identified a unique uro-adaptative transport system and urease production as the two other factors that could be linked to the specific pathogenicity of *S. saprophyticus*. Since 2005, whole-genome virulence studies have been lacking, but adherence and persistence onto the urinary tract have been confirmed as the main factors that could be linked to pathogenicity [81–83]. Overall, *S. saprophyticus*, lacks many of the adhesion proteins and other virulence factors that have been identified in CoNS from the *S. epidermidis* group, *S. caprae*, and *S. lugdunensis*, which probably explains the differences that are observed at a clinical level.

2.7. Phylogenetic Relationship among Staphylococcus Species

In 2012, Lamers et al. proposed a refined classification of staphylococci using a combination of Bayesian and maximum likelihood analysis of multilocus data based on four loci (non coding 16S rRNA, *dnaJ*, *rpoB*, and *tuf*) [22]. The authors identified six species groups and 15 cluster groups among staphylococci, a classification that is considered as a current standard for staphylococci phylogenetic classification [13]. Interestingly, *S. lugdunensis* appeared in a unique cluster group, a particularity that could be link to the very low allelic polymorphism of this species and the existence of a closed pangenome, a unique characteristic among CoNS [32,84]. In addition, *S. epidermidis*, *S. capare*, and *S. capitis* which are described here as pathogenic species in nosocomial settings belong to the same cluster group, along with *S. saccharolyticus*. *S. saprophyticus* and *S. haemolyticus* belong to distinct cluster groups in this classification. This classification does not clarify the mechanisms by which staphylococci acquire genes, including virulence factors, but only emphasize the uniqueness of *S. lugdunensis*, even if more generally, all staphylococci have strong barriers preventing horizontal gene transfers, making their transformation extremely difficult even in vitro [85].

3. Conclusions

The emergence of nosocomial infections with CoNS has illuminated the role of numerous virulence factors. In the context of infections and breaches of the skin barrier, these factors allow virulent adherence, persistence, and multiplication of CoNS. The use of whole-genome sequence data has evidenced the multiplicity of such factors in CoNS. It would be more appropriate to characterize these elements as pathogenicity factors, even if some of them can appear exceptionally as virulence factors that may lead to specific clinical conditions, as seen with *S. epidermidis*-producing strains.

The increasing data that are now available at the clinical, molecular, and genomic levels even make possible the development of innovative approaches to characterize bacterial pathogenicity. Deneke et al. proposed such a novel approach by using a machine learning workflow that determines the pathogenicity of a novel bacterial species based on genomic data only, an eventuality that is not rare with the increasing use of metagenomic analyses in various environments and microbiota [86].

Nevertheless, if sequencing technologies have become cheaper and whole-genome sequence data easier to produce, they will remain as complementary tools to molecular and clinical studies that have to be coordinated in a structured framework [87]. Good practices and quality controls have become crucial to ensure the quality of the released genomes and the validity of analyses [88,89].

Author Contributions: Conceptualization: X.A. and G.P.; Methodology: X.A. and G.P.; Analysis of the studies: X.A. and Y.H.; Writing–Original Draft Preparation: X.A. and K.P.; Writing–Review and Editing: G.P. and Y.H.; Supervision: G.P.

Funding: No funding.

Acknowledgments: The authors would like to thank Enago (www.enago.com) for the English language review.

Conflicts of Interest: The authors declare no conflict of interest.

References

1. Falkow, S. Molecular Koch's postulates applied to microbial pathogenicity. *Rev. Infect. Dis.* **1988**, *10* (Suppl. 2), S274–S276. [CrossRef]
2. Pallen, M.J.; Wren, B.W. Bacterial pathogenomics. *Nature* **2007**, *449*, 835–842. [CrossRef] [PubMed]
3. Liu, B.; Zheng, D.; Jin, Q.; Chen, L.; Yang, J. VFDB 2019: A comparative pathogenomic platform with an interactive web interface. *Nucleic Acids Res.* **2019**, *47*, D687–D692. [CrossRef] [PubMed]
4. Urban, M.; Cuzick, A.; Rutherford, K.; Irvine, A.; Pedro, H.; Pant, R.; Sadanadan, V.; Khamari, L.; Billal, S.; Mohanty, S.; et al. PHI-base: A new interface and further additions for the multi-species pathogen-host interactions database. *Nucleic Acids Res.* **2017**, *45*, D604–D610. [CrossRef] [PubMed]
5. Sayers, S.; Li, L.; Ong, E.; Deng, S.; Fu, G.; Lin, Y.; Yang, B.; Zhang, S.; Fa, Z.; Zhao, B.; et al. Victors: A web-based knowledge base of virulence factors in human and animal pathogens. *Nucleic Acids Res.* **2019**, *47*, D693–D700. [CrossRef] [PubMed]
6. Grice, E.A.; Segre, J.A. The skin microbiome. *Nat. Rev. Microbiol.* **2011**, *9*, 244–253. [CrossRef] [PubMed]
7. Shallcross, L.J.; Fragaszy, E.; Johnson, A.M.; Hayward, A.C. The role of the Panton-Valentine leucocidin toxin in staphylococcal disease: A systematic review and meta-analysis. *Lancet Infect. Dis.* **2013**, *13*, 43–54. [CrossRef]
8. Spaulding, A.R.; Salgado-Pabón, W.; Kohler, P.L.; Horswill, A.R.; Leung, D.Y.M.; Schlievert, P.M. Staphylococcal and streptococcal superantigen exotoxins. *Clin. Microbiol. Rev.* **2013**, *26*, 422–447. [CrossRef] [PubMed]
9. Lacey, K.A.; Geoghegan, J.A.; McLoughlin, R.M. The Role of Staphylococcus aureus Virulence Factors in Skin Infection and Their Potential as Vaccine Antigens. *Pathogens* **2016**, *5*, 22. [CrossRef] [PubMed]
10. Otto, M. Staphylococcus aureus toxins. *Curr. Opin. Microbiol.* **2014**, *17*, 32–37. [CrossRef] [PubMed]

11. Foster, T.J.; Geoghegan, J.A.; Ganesh, V.K.; Höök, M. Adhesion, invasion and evasion: The many functions of the surface proteins of Staphylococcus aureus. *Nat. Rev. Microbiol.* **2014**, *12*, 49–62. [CrossRef] [PubMed]

12. Strauß, L.; Ruffing, U.; Abdulla, S.; Alabi, A.; Akulenko, R.; Garrine, M.; Germann, A.; Grobusch, M.P.; Helms, V.; Herrmann, M.; et al. Detecting *Staphylococcus aureus* Virulence and Resistance Genes: A Comparison of Whole-Genome Sequencing and DNA Microarray Technology. *J. Clin. Microbiol.* **2016**, *54*, 1008–1016. [CrossRef] [PubMed]

13. Becker, K.; Heilmann, C.; Peters, G. Coagulase-negative staphylococci. *Clin. Microbiol. Rev.* **2014**, *27*, 870–926. [CrossRef] [PubMed]

14. Otto, M. *Staphylococcus epidermidis*—The "accidental" pathogen. *Nat. Rev. Microbiol.* **2009**, *7*, 555–567. [CrossRef] [PubMed]

15. Dong, Y.; Speer, C.P.; Glaser, K. Beyond sepsis: *Staphylococcus epidermidis* is an underestimated but significant contributor to neonatal morbidity. *Virulence* **2018**, *9*, 621–633. [CrossRef] [PubMed]

16. Otto, M. Molecular basis of *Staphylococcus epidermidis* infections. *Semin. Immunopathol.* **2011**, *34*, 201–214. [CrossRef] [PubMed]

17. Argemi, X.; Hansmann, Y.; Riegel, P.; Prévost, G. Is *Staphylococcus lugdunensis* Significant in Clinical Samples? *J. Clin. Microbiol.* **2017**, *55*, 3167–3174. [CrossRef] [PubMed]

18. Soumya, K.R.; Philip, S.; Sugathan, S.; Mathew, J.; Radhakrishnan, E.K. Virulence factors associated with Coagulase Negative Staphylococci isolated from human infections. *3 Biotech.* **2017**, *7*, 140. [CrossRef] [PubMed]

19. Argemi, X.; Riegel, P.; Lavigne, T.; Lefebvre, N.; Grandpré, N.; Hansmann, Y.; Jaulhac, B.; Prévost, G.; Schramm, F. Implementation of Matrix-Assisted Laser Desorption Ionization-Time of Flight Mass Spectrometry in Routine Clinical Laboratories Improves Identification of Coagulase-Negative Staphylococci and Reveals the Pathogenic Role of *Staphylococcus lugdunensis*. *J. Clin. Microbiol.* **2015**, *53*, 2030–2036. [CrossRef] [PubMed]

20. D'Ersu, J.; Aubin, G.G.; Mercier, P.; Nicollet, P.; Bémer, P.; Corvec, S. Characterization of *Staphylococcus caprae* Clinical Isolates Involved in Human Bone and Joint Infections, Compared with Goat Mastitis Isolates. *J. Clin. Microbiol.* **2016**, *54*, 106–113. [CrossRef] [PubMed]

21. Seng, P.; Barbe, M.; Pinelli, P.O.; Gouriet, F.; Drancourt, M.; Minebois, A.; Cellier, N.; Lechiche, C.; Asencio, G.; Lavigne, J.P.; et al. *Staphylococcus caprae* bone and joint infections: A re-emerging infection? *Clin. Microbiol. Infect. Off. Publ. Eur. Soc. Clin. Microbiol. Infect. Dis.* **2014**, *20*, O1052–O1058. [CrossRef] [PubMed]

22. Lamers, R.P.; Muthukrishnan, G.; Castoe, T.A.; Tafur, S.; Cole, A.M.; Parkinson, C.L. Phylogenetic relationships among *Staphylococcus* species and refinement of cluster groups based on multilocus data. *BMC Evol. Biol.* **2012**, *12*, 171. [CrossRef] [PubMed]

23. McCarthy, A.J.; Loeffler, A.; Witney, A.A.; Gould, K.A.; Lloyd, D.H.; Lindsay, J.A. Extensive horizontal gene transfer during *Staphylococcus aureus* co-colonization in vivo. *Genome Biol. Evol.* **2014**, *6*, 2697–2708. [CrossRef] [PubMed]

24. Lindsay, J.A. *Staphylococcus aureus* genomics and the impact of horizontal gene transfer. *Int. J. Med. Microbiol.* **2014**, *304*, 103–109. [CrossRef] [PubMed]

25. Malachowa, N.; DeLeo, F.R. Mobile genetic elements of *Staphylococcus aureus*. *Cell. Mol. Life Sci.* **2010**, *67*, 3057–3071. [CrossRef] [PubMed]

26. Che, D.; Hasan, M.S.; Chen, B. Identifying Pathogenicity Islands in Bacterial Pathogenomics Using Computational Approaches. *Pathogens* **2014**, *3*, 36–56. [CrossRef] [PubMed]

27. Arndt, D.; Grant, J.R.; Marcu, A.; Sajed, T.; Pon, A.; Liang, Y.; Wishart, D.S. PHASTER: A better, faster version of the PHAST phage search tool. *Nucleic Acids Res.* **2016**, *44*, W16–W21. [CrossRef] [PubMed]

28. Dhillon, B.K.; Laird, M.R.; Shay, J.A.; Winsor, G.L.; Lo, R.; Nizam, F.; Pereira, S.K.; Waglechner, N.; McArthur, A.G.; Langille, M.G.I.; et al. IslandViewer 3: More flexible, interactive genomic island discovery, visualization and analysis. *Nucleic Acids Res.* **2015**, *43*, W104–W108. [CrossRef] [PubMed]

29. Méric, G.; Miragaia, M.; de Been, M.; Yahara, K.; Pascoe, B.; Mageiros, L.; Mikhail, J.; Harris, L.G.; Wilkinson, T.S.; Rolo, J.; et al. Ecological Overlap and Horizontal Gene Transfer in Staphylococcus aureus and *Staphylococcus epidermidis*. *Genome Biol. Evol.* **2015**, *7*, 1313–1328. [CrossRef] [PubMed]

30. Post, V.; Harris, L.G.; Morgenstern, M.; Mageiros, L.; Hitchings, M.D.; Méric, G.; Pascoe, B.; Sheppard, S.K.; Richards, R.G.; Moriarty, T.F. A comparative genomics study of *Staphylococcus epidermidis* from orthopedic device-related infections correlated with patient outcome. *J. Clin. Microbiol.* **2017**, *55*, 3089–3103. [CrossRef] [PubMed]

31. Bosi, E.; Monk, J.M.; Aziz, R.K.; Fondi, M.; Nizet, V.; Palsson, B.Ø. Comparative genome-scale modelling of *Staphylococcus aureus* strains identifies strain-specific metabolic capabilities linked to pathogenicity. *Proc. Natl. Acad. Sci. USA* **2016**, *113*, E3801–E3809. [CrossRef] [PubMed]

32. Argemi, X.; Matelska, D.; Ginalski, K.; Riegel, P.; Hansmann, Y.; Bloom, J.; Pestel-Caron, M.; Dahyot, S.; Lebeurre, J.; Prévost, G. Comparative genomic analysis of *Staphylococcus lugdunensis* shows a closed pan-genome and multiple barriers to horizontal gene transfer. *BMC Genomics* **2018**, *19*, 621. [CrossRef] [PubMed]

33. Natsis, N.E.; Cohen, P.R. Coagulase-Negative *Staphylococcus* Skin and Soft Tissue Infections. *Am. J. Clin. Dermatol.* **2018**, *19*, 671–677. [CrossRef] [PubMed]

34. Heilmann, C.; Ziebhur, W.; Becker, K. Are coagulase-negative staphylococci virulent? *Clin. Microbiol. Infect.* **2018**. [Epub ahead of print]. [CrossRef] [PubMed]

35. Rogers, K.L.; Fey, P.D.; Rupp, M.E. Coagulase-Negative Staphylococcal Infections. *Infect. Dis. Clin. N. Am.* **2009**, *23*, 73–98. [CrossRef] [PubMed]

36. Piette, A.; Verschraegen, G. Role of coagulase-negative staphylococci in human disease. *Vet. Microbiol.* **2009**, *134*, 45–54. [CrossRef] [PubMed]

37. Longauerova, A. Coagulase negative staphylococci and their participation in pathogenesis of human infections. *Bratisl. Lekárske Listy* **2006**, *107*, 448–452.

38. Otto, M. Virulence factors of the coagulase-negative staphylococci. *Front. Biosci. J. Virtual Libr.* **2004**, *9*, 841–863. [CrossRef]

39. Von Eiff, C.; Peters, G.; Heilmann, C. Pathogenesis of infections due to coagulasenegative staphylococci. *Lancet Infect. Dis.* **2002**, *2*, 677–685. [CrossRef]

40. Zhang, Y.-Q.; Ren, S.-X.; Li, H.-L.; Wang, Y.-X.; Fu, G.; Yang, J.; Qin, Z.-Q.; Miao, Y.-G.; Wang, W.-Y.; Chen, R.-S.; et al. Genome-based analysis of virulence genes in a non-biofilm-forming *Staphylococcus epidermidis* strain (ATCC 12228). *Mol. Microbiol.* **2003**, *49*, 1577–1593. [CrossRef] [PubMed]

41. Gill, S.R.; Fouts, D.E.; Archer, G.L.; Mongodin, E.F.; Deboy, R.T.; Ravel, J.; Paulsen, I.T.; Kolonay, J.F.; Brinkac, L.; Beanan, M.; et al. Insights on evolution of virulence and resistance from the complete genome analysis of an early methicillin-resistant *Staphylococcus aureus* strain and a biofilm-producing methicillin-resistant *Staphylococcus epidermidis* strain. *J. Bacteriol.* **2005**, *187*, 2426–2438. [CrossRef] [PubMed]

42. Conlan, S.; Mijares, L.A.; NISC Comparative Sequencing Program; Becker, J.; Blakesley, R.W.; Bouffard, G.G.; Brooks, S.; Coleman, H.; Gupta, J.; Gurson, N.; et al. *Staphylococcus epidermidis* pan-genome sequence analysis reveals diversity of skin commensal and hospital infection-associated isolates. *Genome Biol.* **2012**, *13*, R64. [PubMed]

43. Fournier, P.-E.; Gouriet, F.; Gimenez, G.; Robert, C.; Raoult, D. Deciphering genomic virulence traits of a *Staphylococcus epidermidis* strain causing native-valve endocarditis. *J. Clin. Microbiol.* **2013**, *51*, 1617–1621. [CrossRef] [PubMed]

44. Argemi, X.; Nanoukon, C.; Affolabi, D.; Keller, D.; Hansmann, Y.; Riegel, P.; Baba-Moussa, L.; Prévost, G. Comparative Genomics and Identification of an Enterotoxin-Bearing Pathogenicity Island, SEPI-1/SECI-1, in *Staphylococcus epidermidis* Pathogenic Strains. *Toxins* **2018**, *10*, 93. [CrossRef] [PubMed]

45. Xu, Z.; Misra, R.; Jamrozy, D.; Paterson, G.K.; Cutler, R.R.; Holmes, M.A.; Gharbia, S.; Mkrtchyan, H.V. Whole Genome Sequence and Comparative Genomics Analysis of Multi-drug Resistant Environmental *Staphylococcus epidermidis* ST59. *G3 GenesGenomesGenetics* **2018**, *8*, 2225–2230. [CrossRef] [PubMed]

46. Cameron, D.R.; Jiang, J.-H.; Hassan, K.A.; Elbourne, L.D.H.; Tuck, K.L.; Paulsen, I.T.; Peleg, A.Y. Insights on virulence from the complete genome of *Staphylococcus capitis*. *Front. Microbiol.* **2015**, *6*, 980. [CrossRef] [PubMed]

47. Li, X.; Lei, M.; Song, Y.; Gong, K.; Li, L.; Liang, H.; Jiang, X. Whole genome sequence and comparative genomic analysis of multidrug-resistant *Staphylococcus capitis* subsp. urealyticus strain LNZR-1. *Gut Pathog.* **2014**, *6*, 45. [CrossRef] [PubMed]

48. Simões, P.M.; Lemriss, H.; Dumont, Y.; Lemriss, S.; Rasigade, J.-P.; Assant-Trouillet, S.; Ibrahimi, A.; El Kabbaj, S.; Butin, M.; Laurent, F. Single-Molecule Sequencing (PacBio) of the *Staphylococcus capitis* NRCS-A Clone Reveals the Basis of Multidrug Resistance and Adaptation to the Neonatal Intensive Care Unit Environment. *Front. Microbiol.* **2016**, *7*, 1991. [CrossRef] [PubMed]

49. Kumar, R.; Jangir, P.K.; Das, J.; Taneja, B.; Sharma, R. Genome Analysis of *Staphylococcus capitis* TE8 Reveals Repertoire of Antimicrobial Peptides and Adaptation Strategies for Growth on Human Skin. *Sci. Rep.* **2017**, *7*, 10447. [CrossRef] [PubMed]

50. Heilbronner, S.; Holden, M.T.G.; van Tonder, A.; Geoghegan, J.A.; Foster, T.J.; Parkhill, J.; Bentley, S.D. Genome sequence of *Staphylococcus lugdunensis* N920143 allows identification of putative colonization and virulence factors: *Staphylococcus lugdunensis* genome sequence. *FEMS Microbiol. Lett.* **2011**, *322*, 60–67. [CrossRef] [PubMed]

51. Argemi, X.; Martin, V.; Loux, V.; Dahyot, S.; Lebeurre, J.; Guffroy, A.; Martin, M.; Velay, A.; Keller, D.; Riegel, P.; et al. Whole-Genome Sequencing of Seven Strains of *Staphylococcus lugdunensis* Allows Identification of Mobile Genetic Elements. *Genome Biol. Evol.* **2017**, *9*. [CrossRef]

52. Takeuchi, F.; Watanabe, S.; Baba, T.; Yuzawa, H.; Ito, T.; Morimoto, Y.; Kuroda, M.; Cui, L.; Takahashi, M.; Ankai, A.; et al. Whole-Genome Sequencing of *Staphylococcus haemolyticus* Uncovers the Extreme Plasticity of Its Genome and the Evolution of Human-Colonizing Staphylococcal Species. *J. Bacteriol.* **2005**, *187*, 7292–7308. [CrossRef] [PubMed]

53. Hong, J.; Kim, J.; Kim, B.-Y.; Park, J.-W.; Ryu, J.-G.; Roh, E. Complete Genome Sequence of Biofilm-Forming Strain *Staphylococcus haemolyticus S167*. *Genome Announc.* **2016**, *4*, e00567-16. [CrossRef] [PubMed]

54. Watanabe, S.; Aiba, Y.; Tan, X.-E.; Li, F.-Y.; Boonsiri, T.; Thitiananpakorn, K.; Cui, B.; Sato'o, Y.; Kiga, K.; Sasahara, T.; et al. Complete genome sequencing of three human clinical isolates of *Staphylococcus caprae* reveals virulence factors similar to those of S. epidermidis and S. capitis. *BMC Genomics* **2018**, *19*, 810. [CrossRef] [PubMed]

55. Kuroda, M.; Yamashita, A.; Hirakawa, H.; Kumano, M.; Morikawa, K.; Higashide, M.; Maruyama, A.; Inose, Y.; Matoba, K.; Toh, H.; et al. Whole genome sequence of *Staphylococcus saprophyticus* reveals the pathogenesis of uncomplicated urinary tract infection. *Proc. Natl. Acad. Sci. USA* **2005**, *102*, 13272–13277. [CrossRef] [PubMed]

56. Wisplinghoff, H.; Bischoff, T.; Tallent, S.M.; Seifert, H.; Wenzel, R.P.; Edmond, M.B. Nosocomial bloodstream infections in US hospitals: analysis of 24,179 cases from a prospective nationwide surveillance study. *Clin. Infect. Dis. Off. Publ. Infect. Dis. Soc. Am.* **2004**, *39*, 309–317. [CrossRef] [PubMed]

57. Le, K.Y.; Park, M.D.; Otto, M. Immune Evasion Mechanisms of *Staphylococcus epidermidis* Biofilm Infection. *Front. Microbiol.* **2018**, *9*, 359. [CrossRef] [PubMed]

58. MacLea, K.S.; Trachtenberg, A.M. Complete Genome Sequence of *Staphylococcus epidermidis* ATCC 12228 Chromosome and Plasmids, Generated by Long-Read Sequencing. *Genome Announc.* **2017**, *5*, e00954-17. [CrossRef] [PubMed]

59. Madhusoodanan, J.; Seo, K.S.; Remortel, B.; Park, J.Y.; Hwang, S.Y.; Fox, L.K.; Park, Y.H.; Deobald, C.F.; Wang, D.; Liu, S.; et al. An Enterotoxin-Bearing Pathogenicity Island in *Staphylococcus epidermidis*. *J. Bacteriol.* **2011**, *193*, 1854–1862. [CrossRef] [PubMed]

60. Nanoukon, C.; Argemi, X.; Sogbo, F.; Orekan, J.; Keller, D.; Affolabi, D.; Schramm, F.; Riegel, P.; Baba-Moussa, L.; Prévost, G. Pathogenic features of clinically significant coagulase-negative staphylococci in hospital and community infections in Benin. *Int. J. Med. Microbiol.* **2017**, *307*, 75–82. [CrossRef] [PubMed]

61. Argemi, X.; Prévost, G.; Riegel, P.; Keller, D.; Meyer, N.; Baldeyrou, M.; Douiri, N.; Lefebvre, N.; Meghit, K.; Ronde Oustau, C.; et al. VISLISI trial, a prospective clinical study allowing identification of a new metalloprotease and putative virulence factor from *Staphylococcus lugdunensis*. *Clin. Microbiol. Infect. Off. Publ. Eur. Soc. Clin. Microbiol. Infect. Dis.* **2017**, *23*, e1–e334. [CrossRef] [PubMed]

62. Sabe, M.A.; Shrestha, N.K.; Gordon, S.; Menon, V. *Staphylococcus lugdunensis*: A rare but destructive cause of coagulase-negative staphylococcus infective endocarditis. *Eur. Heart J. Acute Cardiovasc. Care* **2014**, *3*, 275–280. [CrossRef] [PubMed]

63. Bocher, S.; Tonning, B.; Skov, R.L.; Prag, J. *Staphylococcus lugdunensis*, a Common Cause of Skin and Soft Tissue Infections in the Community. *J. Clin. Microbiol.* **2009**, *47*, 946–950. [CrossRef] [PubMed]

64. Freney, J.; Brun, Y.; Bes, M.; Meugnier, H.; Grimont, F.; Grimont, P.A.D.; Nervi, C.; Fleurette, J. *Staphylococcus lugdunensis* sp. nov. and *Staphylococcus schleiferi* sp. nov., Two Species from Human Clinical Specimens. *Int. J. Syst. Bacteriol.* **1988**, *38*, 168–172. [CrossRef]

65. Zapotoczna, M.; Heilbronner, S.; Speziale, P.; Foster, T.J. Iron-Regulated Surface Determinant (Isd) Proteins of *Staphylococcus lugdunensis*. *J. Bacteriol.* **2012**, *194*, 6453–6467. [CrossRef] [PubMed]

66. Szabados, F.; Nowotny, Y.; Marlinghaus, L.; Korte, M.; Neumann, S.; Kaase, M.; Gatermann, S.G. Occurrence of genes of putative fibrinogen binding proteins and hemolysins, as well as of their phenotypic correlates in isolates of *S. lugdunensis* of different origins. *BMC Res. Notes* **2011**, *4*, 113. [CrossRef] [PubMed]

67. Piroth, L.; Que, Y.-A.; Widmer, E.; Panchaud, A.; Piu, S.; Entenza, J.M.; Moreillon, P. The Fibrinogen- and Fibronectin-Binding Domains of *Staphylococcus aureus* Fibronectin-Binding Protein A Synergistically Promote Endothelial Invasion and Experimental Endocarditis. *Infect. Immun.* **2008**, *76*, 3824–3831. [CrossRef] [PubMed]

68. Tse, H.; Tsoi, H.W.; Leung, S.P.; Lau, S.K.P.; Woo, P.C.Y.; Yuen, K.Y. Complete Genome Sequence of *Staphylococcus lugdunensis* Strain HKU09-01. *J. Bacteriol.* **2010**, *192*, 1471–1472. [CrossRef] [PubMed]

69. Chang, S.-C.; Lee, M.-H.; Yeh, C.-F.; Liu, T.-P.; Lin, J.-F.; Ho, C.-M.; Lu, J.-J. Characterization of two novel variants of staphylococcal cassette chromosome mec elements in oxacillin-resistant *Staphylococcus lugdunensis*. *J. Antimicrob. Chemother.* **2017**, *72*, 3258–3262. [CrossRef] [PubMed]

70. Cui, B.; Smooker, P.M.; Rouch, D.A.; Daley, A.J.; Deighton, M.A. Differences between two clinical *Staphylococcus capitis* subspecies as revealed by biofilm, antibiotic resistance, and pulsed-field gel electrophoresis profiling. *J. Clin. Microbiol.* **2013**, *51*, 9–14. [CrossRef] [PubMed]

71. Tevell, S.; Hellmark, B.; Nilsdotter-Augustinsson, Å.; Söderquist, B. *Staphylococcus capitis* isolated from prosthetic joint infections. *Eur. J. Clin. Microbiol. Infect. Dis.* **2017**, *36*, 115–122. [CrossRef] [PubMed]

72. Van Der Zwet, W.C.; Debets-Ossenkopp, Y.J.; Reinders, E.; Kapi, M.; Savelkoul, P.H.M.; Van Elburg, R.M.; Hiramatsu, K.; Vandenbroucke-Grauls, C.M.J.E. Nosocomial spread of a *Staphylococcus capitis* strain with heteroresistance to vancomycin in a neonatal intensive care unit. *J. Clin. Microbiol.* **2002**, *40*, 2520–2525. [CrossRef] [PubMed]

73. Rasigade, J.-P.; Raulin, O.; Picaud, J.-C.; Tellini, C.; Bes, M.; Grando, J.; Ben Saïd, M.; Claris, O.; Etienne, J.; Tigaud, S.; et al. Methicillin-Resistant *Staphylococcus capitis* with Reduced Vancomycin Susceptibility Causes Late-Onset Sepsis in Intensive Care Neonates. *PLoS ONE* **2012**, *7*, e31548. [CrossRef] [PubMed]

74. Butin, M.; Martins-Simões, P.; Rasigade, J.-P.; Picaud, J.-C.; Laurent, F. Worldwide Endemicity of a Multidrug-Resistant *Staphylococcus capitis* Clone Involved in Neonatal Sepsis. *Emerg. Infect. Dis.* **2017**, *23*, 538–539. [CrossRef] [PubMed]

75. Barros, E.M.; Ceotto, H.; Bastos, M.C.F.; Dos Santos, K.R.N.; Giambiagi-Demarval, M. *Staphylococcus haemolyticus* as an important hospital pathogen and carrier of methicillin resistance genes. *J. Clin. Microbiol.* **2012**, *50*, 166–168. [CrossRef] [PubMed]

76. Czekaj, T.; Ciszewski, M.; Szewczyk, E.M. *Staphylococcus haemolyticus*—An emerging threat in the twilight of the antibiotics age. *Microbiol. Read. Engl.* **2015**, *161*, 2061–2068. [CrossRef] [PubMed]

77. Pinheiro, L.; Brito, C.I.; de Oliveira, A.; Martins, P.Y.F.; Pereira, V.C.; da Cunha, M.d.L.R.d.S. *Staphylococcus epidermidis* and *Staphylococcus haemolyticus*: Molecular Detection of Cytotoxin and Enterotoxin Genes. *Toxins* **2015**, *7*, 3688–3699. [PubMed]

78. Da, F.; Joo, H.-S.; Cheung, G.Y.C.; Villaruz, A.E.; Rohde, H.; Luo, X.; Otto, M. Phenol-Soluble Modulin Toxins of *Staphylococcus haemolyticus*. *Front. Cell. Infect. Microbiol.* **2017**, *7*, 206. [CrossRef] [PubMed]

79. Raz, R.; Colodner, R.; Kunin, C.M. Who Are You—*Staphylococcus saprophyticus*? *Clin. Infect. Dis.* **2005**, *40*, 896–898. [CrossRef] [PubMed]

80. Hayami, H.; Takahashi, S.; Ishikawa, K.; Yasuda, M.; Yamamoto, S.; Uehara, S.; Hamasuna, R.; Matsumoto, T.; Minamitani, S.; Watanabe, A.; et al. Nationwide surveillance of bacterial pathogens from patients with acute uncomplicated cystitis conducted by the Japanese surveillance committee during 2009 and 2010: Antimicrobial susceptibility of *Escherichia coli* and *Staphylococcus saprophyticus*. *J. Infect. Chemother. Off. J. Jpn. Soc. Chemother.* **2013**, *19*, 393–403. [CrossRef] [PubMed]

81. De Paiva-Santos, W.; de Sousa, V.S.; Giambiagi-deMarval, M. Occurrence of virulence-associated genes among *Staphylococcus saprophyticus* isolated from different sources. *Microb. Pathog.* **2018**, *119*, 9–11. [CrossRef] [PubMed]

82. King, N.P.; Sakinç, T.; Ben Zakour, N.L.; Totsika, M.; Heras, B.; Simerska, P.; Shepherd, M.; Gatermann, S.G.; Beatson, S.A.; Schembri, M.A. Characterisation of a cell wall-anchored protein of *Staphylococcus saprophyticus* associated with linoleic acid resistance. *BMC Microbiol.* **2012**, *12*, 8. [CrossRef] [PubMed]

83. Kline, K.A.; Ingersoll, M.A.; Nielsen, H.V.; Sakinc, T.; Henriques-Normark, B.; Gatermann, S.; Caparon, M.G.; Hultgren, S.J. Characterization of a novel murine model of *Staphylococcus saprophyticus* urinary tract infection reveals roles for Ssp and SdrI in virulence. *Infect. Immun.* **2010**, *78*, 1943–1951. [CrossRef] [PubMed]

84. Dahyot, S.; Lebeurre, J.; Argemi, X.; François, P.; Lemée, L.; Prévost, G.; Pestel-Caron, M. Multiple-Locus Variable Number Tandem Repeat Analysis (MLVA) and Tandem Repeat Sequence Typing (TRST), helpful tools for subtyping *Staphylococcus lugdunensis*. *Sci. Rep.* **2018**, *8*, 11669. [CrossRef] [PubMed]

85. Monk, I.R.; Shah, I.M.; Xu, M.; Tan, M.-W.; Foster, T.J. Transforming the untransformable: Application of direct transformation to manipulate genetically *Staphylococcus aureus* and *Staphylococcus epidermidis*. *mBio* **2012**, *3*, e00277-11. [CrossRef] [PubMed]

86. Deneke, C.; Rentzsch, R.; Renard, B.Y. PaPrBaG: A machine learning approach for the detection of novel pathogens from NGS data. *Sci. Rep.* **2017**, *7*, 39194. [CrossRef] [PubMed]

87. Priest, N.K.; Rudkin, J.K.; Feil, E.J.; van den Elsen, J.M.H.; Cheung, A.; Peacock, S.J.; Laabei, M.; Lucks, D.A.; Recker, M.; Massey, R.C. From genotype to phenotype: Can systems biology be used to predict *Staphylococcus aureus* virulence? *Nat. Rev. Microbiol.* **2012**, *10*, 791–797. [CrossRef] [PubMed]

88. Utturkar, S.M.; Klingeman, D.M.; Land, M.L.; Schadt, C.W.; Doktycz, M.J.; Pelletier, D.A.; Brown, S.D. Evaluation and validation of de novo and hybrid assembly techniques to derive high-quality genome sequences. *Bioinform. Oxf. Engl.* **2014**, *30*, 2709–2716. [CrossRef] [PubMed]

89. Arnold, C.; Edwards, K.; Desai, M.; Platt, S.; Green, J.; Conway, D. Setup, Validation, and Quality Control of a Centralized Whole-Genome-Sequencing Laboratory: Lessons Learned. *J. Clin. Microbiol.* **2018**, *56*, e00261-18. [CrossRef] [PubMed]

 © 2019 by the authors. Licensee MDPI, Basel, Switzerland. This article is an open access article distributed under the terms and conditions of the Creative Commons Attribution (CC BY) license (http://creativecommons.org/licenses/by/4.0/).

Review

Pathogenicity and Virulence of *Trueperella pyogenes*: A Review

Magdalena Rzewuska *, Ewelina Kwiecień, Dorota Chrobak-Chmiel, Magdalena Kizerwetter-Świda, Ilona Stefańska and Małgorzata Gieryńska

Department of Preclinical Sciences, Faculty of Veterinary Medicine, Warsaw University of Life Sciences, Ciszewskiego 8 St., 02-786 Warsaw, Poland; ewelina1708@gmail.com (E.K.); dorota.chrobak@wp.pl (D.C.-C.); magdakiz@wp.pl (M.K.-Ś.); i.stefanska@gmail.com (I.S.); malgorzata_gierynska@sggw.pl (M.G.)

* Correspondence: magdalena_rzewuska@sggw.pl; Tel.: +48-225936032

Received: 16 May 2019; Accepted: 31 May 2019; Published: 4 June 2019

Abstract: Bacteria from the species *Trueperella pyogenes* are a part of the biota of skin and mucous membranes of the upper respiratory, gastrointestinal, or urogenital tracts of animals, but also, opportunistic pathogens. *T. pyogenes* causes a variety of purulent infections, such as metritis, mastitis, pneumonia, and abscesses, which, in livestock breeding, generate significant economic losses. Although this species has been known for a long time, many questions concerning the mechanisms of infection pathogenesis, as well as reservoirs and routes of transmission of bacteria, remain poorly understood. Pyolysin is a major known virulence factor of *T. pyogenes* that belongs to the family of cholesterol-dependent cytolysins. Its cytolytic activity is associated with transmembrane pore formation. Other putative virulence factors, including neuraminidases, extracellular matrix-binding proteins, fimbriae, and biofilm formation ability, contribute to the adhesion and colonization of the host tissues. However, data about the pathogen–host interactions that may be involved in the development of *T. pyogenes* infection are still limited. The aim of this review is to present the current knowledge about the pathogenic potential and virulence of *T. pyogenes*.

Keywords: *Trueperella pyogenes*; virulence; pyolysin; infection; pathogenicity; immune response; *Actinomycetales*

---

1. Introduction

The species *Trueperella pyogenes* [1], previously classified as *Arcanobacterium pyogenes* [2], *Actinomyces pyogenes* [3,4], and formerly as *Corynebacterium pyogenes* [5], belongs to the family *Actinomycetaceae*, in the order *Actinomycetales* of the class *Actinobacteria*, the so-called actinomycetes [6]. A dendrogram representing the phylogenetic relationship between this bacterium and some other pathogenic *Actinomycetales* is shown in Figure 1. *T. pyogenes* is a Gram-positive, pleomorphic, non-spore-forming, non-motile, non-capsulated, facultatively anaerobic rod, which is characterized by a fermentative metabolism and strong proteolytic activity [4]. Its growth requirements are not excessive, but media enriched with blood or serum need to be used for the culture. The preliminary recognition of *T. pyogenes* isolates is based on the cell morphology; the features of colonies, which are surrounded by a zone of beta-haemolysis on blood agar; and a negative catalase assay. Then, the biochemical properties can be tested for species determination [7]. Sometimes, additional bacteriological methods other than the conventional ones are necessary for the differentiation and appropriate identification of isolates. New techniques, such as loop-mediated isothermal amplification (LAMP) assay, matrix-assisted laser desorption/ionization time-of-flight (MALDI-TOF) mass spectrometry, Fourier transform infrared (FT-IR) spectroscopy, or 16S rRNA gene sequencing may be useful for the diagnostics of *T. pyogenes* infections [8–13]. Those methods enable the recognition of the closely related taxa of the order *Actinomycetales*, and sometimes the reclassification of some of them [14].

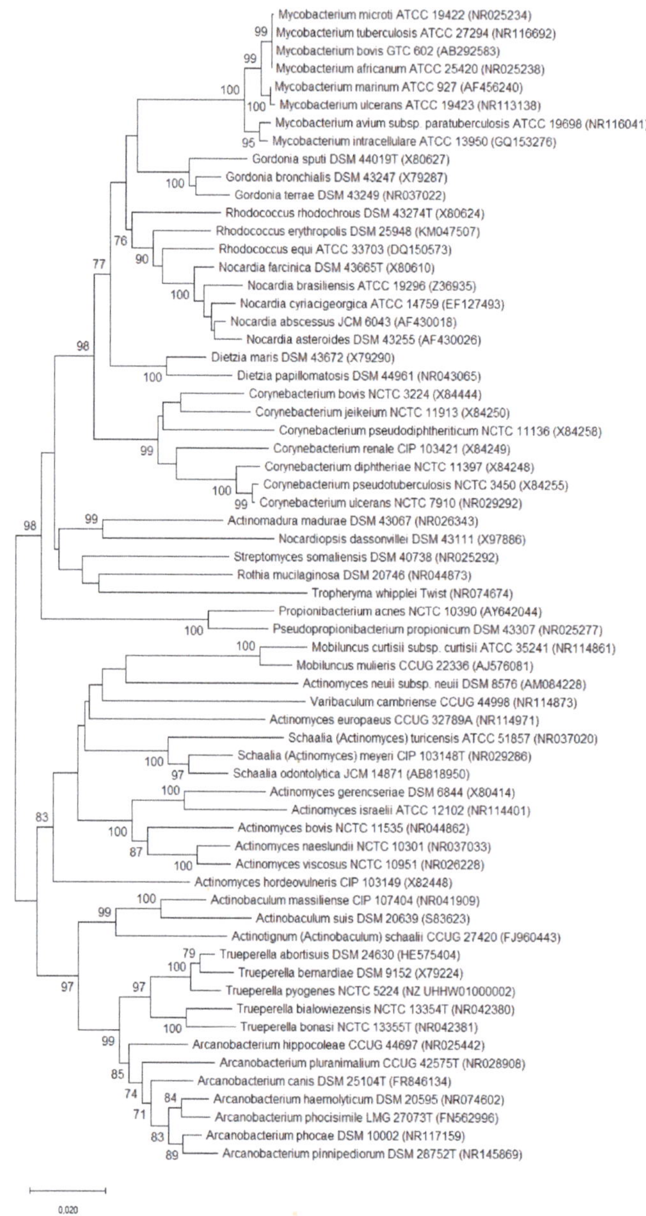

Figure 1. Neighbor-joining [15,16] phylogenetic tree based on 16S rRNA gene sequences (a total of 1422 positions in the final dataset) of *Trueperella pyogenes*, other *Trueperella* species, and related taxa. Only bootstrap values of 70%, based on 1000 replicates, are shown next to the branches [17]. The scale bar represents the number of substitutions per site. The analysis involved 64 nucleotide sequences derived from the GenBank® database, and the evolutionary relationships were calculated in MEGA X [18].

T. pyogenes is considered to be a part of the biota of skin and mucous membranes of the upper respiratory and urogenital tracts of animals [19–21]. Moreover, this bacterium was also isolated from the wall of bovine rumen and swine stomachs, as a gastrointestinal microbiota [22,23], and from the udders of clinically healthy cows [24]. However, *T. pyogenes* is also an important opportunistic pathogen. This species, like other well-known actinomycetes, such as *Corynebacterium pseudotuberculosis* or *Rhodococcus equi*, is an etiological agent of common suppurative infections in animals. In the case of *C. pseudotuberculosis*, the highest susceptibility to the infection is observed in small ruminants, and abscesses are located mainly in the lymph nodes [25,26]. *R. equi* is a cause of pyogenic infections, mostly in horses, and lesions are found mainly in the respiratory tract [27,28]. Whereas *T. pyogenes* is pathogenic for a variety of animal species, and purulent or necrotic lesions may occur in different host tissues. Interestingly, there were no relationships found between the virulence gene profiles of *T. pyogenes* strains and their origin, a type of infection and a host [13,21,29–31]. However, Ashrafi Tamai et al. [32] reported a significant association between the virulence genotypes and clonal types of *T. pyogenes* isolates, and the severity of the clinical symptoms in postpartum cows with metritis.

Although *T. pyogenes* is a bacterium that has been known of for a long time, many questions concerning the mechanisms of infection pathogenesis, as well as its reservoirs and routes of transmission, still remain poorly understood. The aim of this review is to present the current knowledge about the pathogenicity and virulence potential of this opportunistic animal pathogen.

2. Pathogenicity

T. pyogenes infections occur in both domestic and wild animals worldwide, but are rare in humans. The prevalence of *T. pyogenes* isolation may differ, depending on a host species and a geographic region. The majority of published data concerns *T. pyogenes* infections in food animals and comes from Europe [21,33–38], China [39–43], Japan [44,45], Brazil [30,46], and the United States [47–50]. In livestock, the diseases caused by *T. pyogenes* generate significant economic losses, mainly in cattle and swine breeding, causing a reduction of meat and milk yield, as well as decreased reproductive efficiency and sometimes the necessity to cull diseased animals. The clinical course of these suppurative infections may be severe, with different mortality rates, which increase in the case of misdiagnosis or inappropriate treatment. Beta-lactams, tetracyclines, and macrolides are the antibiotics most often used to treat *T. pyogenes* infections. A question of the therapy efficacy is especially important, as the antimicrobial resistance in *T. pyogenes* becomes an emerging problem because of the common use of these drugs in agriculture [29,32,34,39,40,42,44,47,49,51,52].

Generally, the infections caused by *T. pyogenes* have an opportunistic nature, in which adverse environmental and host-related factors play a relevant role in the disease establishment [45,50,53,54]. Nevertheless, the risk factors of the infection development are sometimes difficult to estimate. Curiously, there are no obvious differences observed between the virulence genotypes of commensal and clinical *T. pyogenes* isolates, although, in some investigations, a gene encoding one of the virulence determinants, fimbria A, was found more frequently in the isolates obtained from infected cows than from healthy ones [49,55]. However, the in vitro study of Ibrahim et al. [56] showed significant differences in the expression level of eight known virulence genes in *T. pyogenes* isolated from the uterus of a cow with clinical endometritis, and the isolate from the uterus of a healthy cow. This indicates the importance of the regulatory mechanisms of the virulence gene expression in the infection development, which can be also demonstrated in an example of the pyolysin gene expression [57]. In other study, a relationship between the clonal types of isolates and their origin, from a diseased or healthy host, was noted [21]. All of these findings suggest that other unknown bacterial factors may also contribute to the establishment and development of the *T. pyogenes* infection.

T. pyogenes may cause infection as a primary etiological agent, but more frequently, this species is involved in polymicrobial diseases, such as mastitis [45], uterine infections [55], interdigital phlegmon [58], or liver abscesses [59]. This bacterium may be recovered from a mix infection of various bacterial species, but especially frequently with Gram-negative anaerobes, such as *Fusobacterium necrophorum*, *Bacteroides* spp., or *Peptoniphilus* (formerly *Peptostreptococcus*) *indolicus*. In these cases, purulent and necrotic lesions are usually observed, leading to systemic signs and resulting in animal death. A particularly strong synergistic interaction occurs between *T. pyogenes* and *F. necrophorum* [60].

It consists in the stimulation of *F. necrophorum* growth by a diffusible and heat-labile product of *T. pyogenes*, which probably decreases the oxygen pressure and the oxidation-reduction potential in a site of infection, generating conditions advantageous for this anaerobe [61]. On the other hand, a leukotoxin produced by *F. necrophorum* protects *T. pyogenes* against phagocytosis, because of its ability to lysis leukocytes or to induce their apoptosis, depending on its concentration [59]. Moreover, lactic acid, which is a metabolic product of *T. pyogenes*, can be used by *F. necrophorum* as an energy substrate. Other bacteria, especially *Escherichia coli*, are also often associated with *T. pyogenes* co-infections, mostly postpartum uterine infections [55,62,63]. However, the synergistic effect between both of these bacteria is not evident. The findings of Zhao et al. [64] indicated that *N*-acyl homoserine lactones from *E. coli* and *Pseudomonas aeruginosa*, which act as the quorum-sensing (QS) signal molecules, can inhibit the growth and virulence of *T. pyogenes* in vitro. The recent in vivo study of Huang et al. [65], conducted in a mouse model, confirmed these observations.

2.1. T. pyogenes Infections in Animals

In livestock, *T. pyogenes* infections occur mainly in cattle, swine, sheep, and goats, rarely in horses or birds, and are often associated with heavy economic losses.

In cattle, *T. pyogenes* mainly causes infections of the reproductive tract and the mammary gland, as well as pneumonia and liver abscessation. The most prevalent diseases in dairy cows related with this bacterium are metritis and endometritis, which may develop in a clinical form in about 23–52% of animals after parturition [29,32,38,39,66]. *T. pyogenes* together with many other bacteria comprises a vaginal biota of healthy cows [67], and may also colonize and persist in the uterus of dairy cows with normal puerperium [21,38]. However, as an opportunistic pathogen, this bacterium can invade the distant parts of the reproductive tract, especially after parturition, when the protective epithelium of the endometrium is disrupted, and it can also increase the influx of inflammatory cells in these tissues [50,68]. The presence of *T. pyogenes* in the endometrium is correlated with the damage of the tissue because of the cytolytic activity of the pyolysin against the endometrial stromal cells, which are particularly sensitive to this cholesterol-dependent toxin [68,69]. The ability of *T. pyogenes* to produce inflammatory lesions in the endometrium was confirmed by the findings of Lima et al. [70]. They noted that after the intrauterine infusion of *T. pyogenes* suspension containing 10^9 colony-forming units/mL, moderate to severe endometrial inflammation developed in the studied cows, and additionally, premature luteolysis was observed in some animals. The uterine inflammation, usually later postpartum, associated with *T. pyogenes*, has a form of subclinical or clinical endometritis, and may reduce the reproductive performance and milk yield [63,71,72]. Bonnett et al. [73] noted that cows with *T. pyogenes* infection took significantly longer to conceive. Moreover, the study of Boer et al. [63] demonstrated that *T. pyogenes* isolation at day 21 postpartum was associated with the subsequent diagnosis of purulent vaginal discharge, and this observation was confirmed by Sheldon et al. [74]. *T. pyogenes* can also have a lethal impact on the oviductal epithelial cells, as it was demonstrated by Mesgaran et al. [75] in the in vitro study. In general, in the cows infected by *T. pyogenes*, an increased prevalence of clinical endometritis is observed [50]. Frequently, uterine disorders are from co-infections with *T. pyogenes* and other bacteria, such as *E. coli*, *Streptococcus* spp., *Staphylococcus* spp., *Fusobacterium* spp., *Prevotella* spp., and *Clostridium* spp. [39,55,62,71,76–78]. Such polymicrobial uterine infections, especially those with anaerobes, result in an increased purulent secretion and higher severity of lesions [76,79]. The problems regarding parturition, the subsequent negative energy balance, or hyperketonemia are considered to be important risk factors for these diseases [63,80].

Another important and common *T. pyogenes* infection in cattle is mastitis, which may affect lactating and dry cows, as well as heifers [30,37,43,45,46,81–84]. *T. pyogenes* is well known as one of the crucial agents of polymicrobial infection, called summer mastitis, which occurs mainly in pastured cows during the summer, and is associated with pathogen transmission by an insect, *Hydrotaea irritans* [85,86]. However, Madsen et al. [87] demonstrated no differences in the rates of *T. pyogenes* isolation from the mastitis cases in stabled and pastured cattle. Similarly, Ribeiro et al. [46] and Ishiyama et al. [45] did not observe a seasonality of the *T. pyogenes* mastitis occurrence during their long-term survey of the disease. Moreover, it was reported that *T. pyogenes* alone can cause clinical mastitis called pyogenes mastitis, even with a high severity of symptoms [45,85]. The mammary gland inflammation caused by this bacterium is characterized by severe pyogenic lesions in the mammary tissue, and malodorous and purulent milk, especially in case of co-infection with anaerobes, decreasing the milk yield and the low recovery rate [45]. The anaerobes most often involved in mastitis together with *T. pyogenes* are *P. indolicus*, *F. necrophorum*, and *Prevotella melaninogenica* (formerly *Bacteroides melaninogenicus*). All of these bacteria can colonize the mucous membranes and skin of clinically healthy cattle, but *P. indolicus* and *T. pyogenes* were the most frequently isolated from a teat skin [88]. Regardless of the many investigations on *T. pyogenes* mastitis, little is still known about the factors involved in the establishment and persistence of that infection.

T. pyogenes also contributes to many other disorders in cattle, among them, liver abscesses and interdigital phlegmon have a more significant economic impact. Those infections are mixed with anaerobes, mainly *F. necrophorum*, characterized by the synergistic interaction of both of the bacteria mentioned previously [59]. Liver abscesses related to *T. pyogenes* infection occur mostly in feedlot cattle, with a varying frequency from 2% to 80% [59,89–91]. The infection is associated with *T. pyogenes* presence in the rumen wall, in which the primary lesions can form and then bacteria can penetrate by a hepatic portal venous system to a liver, where abscesses form as secondary infection foci [60]. In the case of abscesses rupturing, systemic signs can develop, resulting in animal death. The development of liver abscesses is probably associated with ruminal acute or subacute acidosis, which can induce primary damage in the protective surface of the rumen wall.

Bovine interdigital phlegmon incidence in dairy cows is usually 2–5% per lactation, but some cases of outbreaks were also reported [35,58,92,93]. The role of *T. pyogenes* in the pathogenesis of this polymicrobial disease is not well defined. Kontturi et al. [58] reported that this bacterium is rather a secondary pathogen associated with the healing stage of the infection, while Bay et al. [94] did not detect *T. pyogenes* at all in the interdigital phlegmon investigation using 16S rRNA gene sequencing.

Moreover, *T. pyogenes* may cause a variety of other purulent infections in cattle such as pneumonia, encephalitis, pyelonephritis and kidney abscesses, lymphadenitis, endocarditis, and abscesses of various localization [30,31,36,44,46,95,96]. This pathogen was also isolated from cases of septicaemia and abortion [36,44,46,97].

In swine, *T. pyogenes* is a common agent of pneumonia, pleuritis, endocarditis, osteoarthritis, polyarthritis, mastitis, reproductive tract infections, and septicaemia [13,29,35,44,98–102]. Abscesses—superficial, muscular, or located in different organs—occur frequently, and may lead to the development of systemic purulent infection and inflammation of lungs, liver, kidneys, muscles, bones, joints, or other tissues [13,46,103]. In many cases, these are infection mixed with different microorganisms, as is observed in *T. pyogenes* infections in cattle. Diseases of swine associated with *T. pyogenes* are an emerging clinical, epidemiological, and economic problem, because they usually result in the necessity for the elimination of infected animals from a herd, and the discard of carcasses with suppurative lesions at slaughterhouses [103]. It has been reported that a number of factors can predispose to the development of these disorders, among other viral infections, that cause immunosuppression, for example with porcine reproductive and respiratory syndrome virus (PRRSV) [104].

In small ruminants, *T. pyogenes* is mostly a cause of abscesses formation in different tissues and localized in various parts of the body, including bone marrow and foot (footrot) abscesses [29,31,105–110]. Those lesions differ from that observed in the case of caseous lymphadenitis caused by *C. pseudotuberculosis* [111]. Moreover, *T. pyogenes* can be associated with purulent, mainly polymicrobial disorders in sheep and goats, such as pneumonia, lymphadenitis, arthritis, osteomyelitis, reproductive tract infections, mastitis, and septicaemia [30,35,36,46,112–114].

In other domestic animals, infections related to *T. pyogenes* are rare. It may be connected with the fact that this bacterium does not belong to their normal biota. Therefore, there are only a few data on the occurrence of *T. pyogenes* infections in companion animals. The first reported cases referred to otitis externa in a cat and cystitis in a dog [115], lung abscess in a dog [116], and feline pyothorax [117]. Moreover, *T. pyogenes* was isolated from cases of wound infection, abscesses, vaginitis, pneumonia, and encephalitis in dogs [30,35,36,46]. Recently, an interesting case of *T. pyogenes* and *Brucella abortus* co-infection in a cat and a dog was described by Wareth et al. [118]. Both animals lived on a dairy cattle farm, where cases of abortion and mastitis in cows were noted. The bacteria were isolated in a mixed culture from the uterine discharge of a bitch after abortion and a cat with pyometra.

Infections related to *T. pyogenes* were noted sporadically in horses, and included single cases of metritis, orchitis, mastitis, septicaemia, umbilical infection in foals, abscesses, and wound infection [30,36,46,119].

Furthermore, two cases of suppurative disorders in rabbits caused by *T. pyogenes* were published. In the first case, this bacterium was isolated from lung lesions, and in the second one, from necrotic foci in liver, spleen, lung, and brain [36,120].

The incidence of *T. pyogenes* infections in birds seems to be very low. Some cases of clinical lameness and osteomyelitis in turkeys were reported [121,122]. A unique case of liver abscesses in pigeons was described by Priya et al. [123].

In wildlife, *T. pyogenes* may be associated with many types of purulent disorders occurring in free-living and captive animals. *T. pyogenes* infections were reported the most frequently in ruminants and other herbivores, in which the bacteria were also found as a resident microbiota of the skin and mucous membranes of respiratory and urogenital tracts [20,54,124].

In the United States and Canada, *T. pyogenes* is an emerging pathogen of captive or free-ranging white-tailed deer (*Odocoileus virginianus*), which cause intracranial abscessation—suppurative meningoencephalitis disease complex (the intracranial abscess disease) that occurs with a variable frequency, depending on animal populations [54,124–127]. *T. pyogenes* may be also involved in pneumonia and necrobacillosis occurring in this animal species [128,129].

Furthermore, *T. pyogenes* was also isolated as an etiological agent of chronic purulent infections, including keratoconjunctivitis, brain and foot abscesses, in other cervids, such as Key deer (*Odocoileus virginianus clavium*) [130], fallow deer (*Dama dama*) [53], roebuck (*Capreolus capreolus*) [131], red deer (*Cervus elaphus*) [132], and mule deer (*Odocoileus hemionus*) [133]. *T. pyogenes* is a prevalent cause of variably located abscesses in forest musk deer (*Moschus berezovskii*), noted mainly in farm animals [40,134].

In European bison (*Bison bonasus*), *T. pyogenes* is associated with abscesses of the liver, spleen, lungs, lymph nodes, skin, and especially with urogenital tract infections [135]. In female bison, abscesses of various localizations were observed. In male bison, *T. pyogenes* is considered to be one of the etiological agents of a chronic necrotizing and ulcerative inflammation of the prepuce and penis (balanoposthitis) [135].

Additionally, *T. pyogenes* infections were reported in antelopes [29,136,137], an okapi (*Okapia Johnstoni*) [138], bison (*Bison bison*) [139], Bighorn sheep (*Ovis canadensis*) [140], a chamois (*Rupicapra pyrenaica*) [112], camels [141–144], an elk [36], and reindeer [145,146]. Moreover, some other sporadic cases of infectious diseases associated with *T. pyogenes* were described in Grey Slender lorises (*Loris lydekkerianus nordicus*) [147,148], macaws, and elephants [48], as well as in reptiles, a bearded dragon, and a gecko [149].

2.2. T. pyogenes Infections in Humans

Infections caused by *T. pyogenes* in humans are sporadic, mostly occur in immunosuppressed patients, and are connected with occupational exposure, especially relating to contact with farm animals and their environment [150]. Published data concerning *T. pyogenes* infections in humans are limited, and include, among others, reports on endocarditis [151–155], endemic leg ulcers [156,157], pneumonia [158], arthritis [159,160], sepsis [161], and various purulent lesions and abscesses [162]. As *T. pyogenes* was never demonstrated as a commensal microbiota of humans, those infections should be considered as zoonotic diseases [48]. Although, a probability of animal-to-human transmission of this pathogen has not been well estimated and confirmed.

3. Reservoir, Transmission, and Routes of Infection

Most actinomycetes are widespread in the natural environment, being found in various ecological niches as saprophytes, but they may also constitute a commensal biota of humans and animals, and some of them can be opportunistic pathogens [163,164]. Curiously, there is no published characterization of environmental *T. pyogenes* isolates, for example from soil, which might be considered non-pathogenic, as it is known, for instance, in case of environmental strains of *R. equi*, a typical soil opportunistic pathogen [165]. Although there are no data on *T. pyogenes* ability to replicate in soil or water, certainly, it can persist for some period in such environmental conditions. For example, *C. pseudotuberculosis*, the related actinomycete, can survive for several weeks in the environment [166]. Considering the nutritional requirements of *T. pyogenes*, it should be assumed that the bacterium can replicate only in the environment rich in peptides, fermentable carbohydrates, inositol, and hemin, and at the temperature range of 20 to 40 °C [4]. Therefore, it seems that a main reservoir and source of this bacterium are animals of various species.

Little is known about the dissemination of *T. pyogenes* infections, and about the transmission of the pathogen. It is suggested that the majority of infections have an endogenous character, as the bacteria are a common component of the skin and mucous membrane biota [48]. However, the possibility of exogenous infections should also be considered, because bacteria may be transmitted, for instance, by contaminated husbandry utensils and equipment, or directly from animal to animal [30,46]. The natural environment contaminated by *T. pyogenes* is suggested to be an important source of the bacteria, especially in the case of mammary or foot disorders in domestic and wild animals [46,53,109]. In addition, climate conditions, such as a high humidity and mild temperature, are factors favouring the infection occurrence [53]. In the case of summer mastitis, *T. pyogenes*, exceptionally, may be transmitted between animals by biting flies [86]. However, a possibility of tissue penetration by these bacteria through a micro-trauma caused by ectoparasites, such as ticks, should be also considered. The preliminary results of the study on ticks collected from the skin of European bison indicated a potential contribution of those arthropods to *T. pyogenes* transmission [167].

The observations from different studies indicate a potential threat of *T. pyogenes* transmission from wild to domestic animals, or the other way around [53,135]. It is very possible, regarding the frequent co-occurrence of wildlife species and livestock animals on the same agricultural areas, such as pastures and meadows, which can serve as a reservoir for this pathogen.

The data on the genetic relationship of *T. pyogenes* isolates occurring in a particular host population, and on the dissemination of strains amongst animals and the environment are lacking. The results of a few epidemiological studies on bovine and swine isolates are available [10,13,21,32,148]. In some of them, an association between the clonal types and the development of clinical infection by a strain was shown [21,32,148], while in other investigations, such a correlation was not found [10,13]. A variety of molecular methods, characterized by various discriminatory powers, were used for *T. pyogenes* differentiation, which enabled the phylogenetic analysis of strains to be performed, for example the 16S rRNA gene sequencing [135,149], single enzyme amplified fragments length polymorphism (SE-AFLP) [13], BOX-PCR [21,32], the superoxide dismutase A gene (*sodA*) sequencing [36], random amplification polymorphic DNA (RAPD-PCR), and multilocus sequence analysis (MLSA) [148]. The usefulness of some of them was proved by Nagib et al. [148] during the study on *T. pyogenes* isolates from lorises, which showed a close relationship, indicating their clonal origin. Unfortunately, the gold standard method used for genomic fingerprinting, pulse field gel electrophoresis (PFGE), was never reported as a technique applied for *T. pyogenes* phylogenetic investigation.

The routes of *T. pyogenes* infections are frequently difficult to establish. The *T. pyogenes* infections develop mainly as a consequence of the mechanical injuries of skin and mucous membranes [48]. Wounds and abrasions of skin are common routes of infection, for example, udder injuries in the case of *T. pyogenes* mastitis, or hoof injuries in the case of interdigital phlegmon. In the case of mucous membranes damage, such as in metritis, the bacteria can invade deeper tissues after endometrial epithelium, disruption during parturition. Another described route of infection is associated with improperly performed surgical procedures, such as castration or intramuscular injection [103]. Bacteria can also invade tissues through micro-injuries of the skin caused by some arthropoda, like flies or ticks, as it has been mentioned above. In deer, behaviour may influence the pathogen transmission and susceptibility to cranial/intracranial abscess infection, which occur more frequently in adult males [125,127]. Interestingly, Belser et al. [54] noted a higher prevalence of *T. pyogenes* along the forehead of the males of white-tailed deer than the females, when the rate of the pathogen carriage on the nasopharyngeal mucosa was the same for both sexes. Therefore, it is suggested that the disease development may be a result of cuts or abrasions arising from antler rubbing on trees or from sparring between males.

4. Pathogenesis of *T. pyogenes* Infection

The pathogenesis of microbial infection is determined by a variety of bacterial and host-related factors. In the case of *T. pyogenes*, the pathogenicity is attributed to the determinants, which induce the formation of abscesses, empyemas, and pyogranulomatous lesions. However, pathogen–host interactions in *T. pyogenes* infection are still poorly understood. Some known and putative bacterial factors contribute to the development of *T. pyogenes* infection, but their role in the infection pathogenesis remains insufficiently explained. Moreover, many researchers highlighted a high phenotypic and genotypic diversity among the studied *T. pyogenes* isolates of various origins, also taking into consideration virulence genotypes [13,21,29–31]. This finding indicates the genetic heterogeneity in the studied *T. pyogenes* populations. On the other hand, Cohen et al. [124] detected significantly more virulence genes in *T. pyogenes* isolates from populations of white-tailed deer males, in which a high incidence of the intracranial abscess disease was observed, compared with those detected in the isolates from apparently healthy animals in populations in which the disease was not noted. This finding suggests that the essential differences in the pathogenic potential may exist among *T. pyogenes* strains, depending on their origin.

4.1. T. pyogenes Virulence

Only few virulence factors in *T. pyogenes* are recognized to date (Figure 2). They include pyolysin (PLO), the only known toxin of this bacterium; some adhesive factors, such as fimbriae neuraminidases and extracellular matrix-binding proteins; different exoenzymes, such as serine proteases with gelatinase and caseinase activity, or DNAses; as well as the ability to invade host cells and to create biofilm formation [48]. The significance of some of them in the pathogenicity is not clear enough. In general, profiles of genes encoding particular virulence determinants are not correlated with a type of infection and host species. However, a few cases of association between the presence of some determinants and a type of infection were reported [49,55].

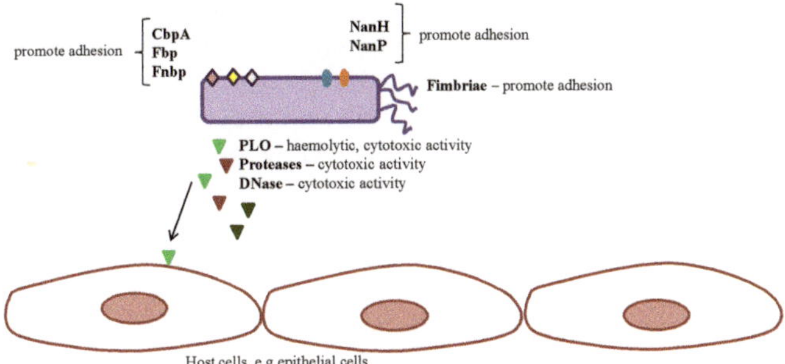

Figure 2. Schematic diagram of known virulence factors of *T. pyogenes*. Abbreviations: CpbA—collagen-binding protein; Fbp—fibrinogen-binding protein; Fnbp—fibronectin-binding protein; NanH—neuraminidase H; NanP—neuraminidase P; PLO—pyolysin.

4.1.1. Pyolysin

PLO is considered to be both a major virulence factor of *T. pyogenes* and a host-protective antigen [168–170], and, until now, the gene (*plo*) encoding this protein was detected in all wild-type *T. pyogenes* strains. This is an exotoxin belonging to the cholesterol-dependent (also called cholesterol-binding) cytolysins (CDCs), which are a family of 51- to 60-kDa single-chain proteins produced by many species of Gram-positive bacteria, for example *Streptococcus pneumoniae* (pneumolysin), *Streptococcus suis* (suilysin), *Streptococcus pyogenes* (streptolysin O), *Paenibacillus alvei* (alveolysin), *Clostridium perfringens* (perfringolysin O), or *Listeria monocytogenes* (listeriolysin O). [171,172]. PLO is the most divergent protein of this group, presenting only 38% to 45 % identity and 58% to 64% similarity to other toxins from the CDC family, when the conserved core of the CDCs (corresponding to amino acid fragment 38–500 of perfringolysin O) is taken into consideration during the analysis [172]. The highest identity of PLO is shown to suilysin, novyilysin, botulinolysin B, and tetanolysin O, which range from 45% to 44% [172]. This protein was purified and characterized for the first time by Ding and Lämmler in 1996 [173], however, its toxic and haemolytic activities were previously reported by Lovell [174]. As a member of the CDC family, PLO displays a cytotoxic effect on a variety of host cells, for example, erythrocytes, polymorphonuclear neutrophils (PMNs), macrophages, epithelial cells, fibroblasts, and endometrial stromal cells. [68,168,171,172]. The cytolytic activity of PLO is associated with its ability to bind to the plasma membrane and to form transmembrane pores, which is a common feature of CDCs [168,172]. The in vitro study showed that PLO is able to cause the lysis of human, sheep, horse, rabbit, and guinea pig erythrocytes [4]. Specific antibodies against PLO completely neutralize its haemolytic activity, suggesting that PLO is the sole haemolysin produced by *T. pyogenes* [168].

The significance of PLO as a primary virulence factor of *T. pyogenes* was confirmed in many in vitro and in vivo experiments, with use of PLO-deficient mutants or recombinant proteins [168,175,176]. PLO has a lethal effect on mice and rabbits, and a dermonecrotic effect on guinea pigs after intravenous and intraperitoneal injection [168,174,177–179]. Jost et al. [168] observed a loss of haemolytic activity in the PLO-1 mutant strain as a result of the insertional inactivation of the plo gene, as well as its 1.8-log$_{10}$ reduction in virulence for mice in comparison to the wild-type strain. They also, for the first time, documented the importance of PLO for the in vivo survival of bacteria, studying the effect of the co-challenge of a wild-type strain, and a mutant strain with the inactivated *plo* gene in a mouse model. Similarly, Zhao et al. [179] demonstrated that the *T. pyogenes* strain with an in vitro higher expression of the *plo* gene was more virulent for mice in the in vivo study than the strain with a lower *plo* expression, designated as avirulent.

Billington et al. [180] cloned and sequenced the *plo* gene encoding PLO, which was located in an open reading frame (ORF) of 1605 bp in the *T. pyogenes* chromosom. The *plo* gene sequence was also described by Ikegami et al. [181], for which the deduced pyolysin sequence showed a 99.4% similarity and a 97.6% identity to the PLO sequence previously reported by Billington et al. [180]. A consensus ribosome binding site, two promoter sequences (similar to the *E. coli* σ^{70} promoter), and three direct repeats (DR1 to D3 with sequence ATTTTTG(C)TGG) were found upstream of the *plo* gene, and a transcriptional terminator region was found downstream of this gene [180]. Moreover, Rudnick et al. [170] found that the *plo* gene, along with the mentioned sequences and ORF (*orf121*), encoded a 13.4 kDa protein of an unknown function, form a genomic islet of 2.7 kb, characterized by a reduced content of G+C (50.2%). This islet, flanked by two housekeeping genes, *smc* and *ftsY* (62.5% G+C), is located in the *T. pyogenes* chromosome (Figure 3) [170]. Interestingly, the codon usage of *plo* and *orf121* differ from that of the genes flanking the islet. The genes *smc* and *ftsY* are common in many other bacteria, mainly Gram-positive, and are essential for *T. pyogenes* viability [170]. Downstream of the *ftsY* gene are the *ffh* gene, and then the ORF, *orf353*. The *ftsY* and *ffh* genes encode the FtsY and Ffh proteins, respectively, which have a high similarity to the signal recognition particles from the other bacterial species [170]. The *ftsY*, *ffh*, and *orf353* genes in *T. pyogenes* are located in close proximity, and form an operon-like arrangement that is unusual in most other bacteria. It seems that, as in other species, these genes are needed for bacterial growth.

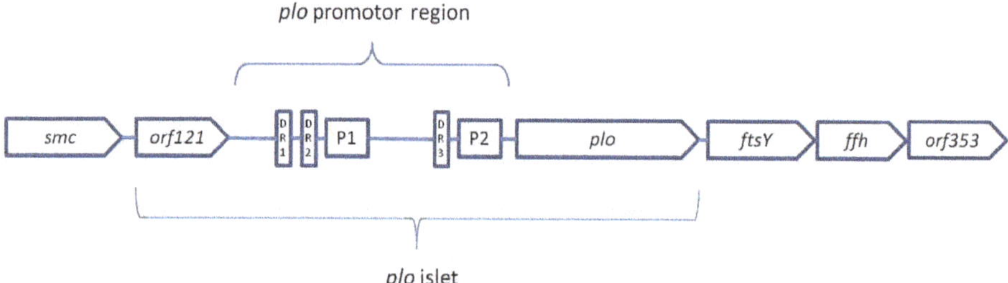

Figure 3. Schematic presentation of the *T. pyogenes* chromosomal region containing the *plo* gene and the surrounding genes. P1 and P2 indicate the positions of two promoters; DR 1–3 indicate the positions of three direct repeats. The scale is not designated.

The differences between the islet and the flanking genes suggest a possibility of horizontal transfer of the *plo* gene, though any integrase or transposon sequences were not found in this chromosomal region. Taking into consideration the fact that the *plo* gene is present in all wild-type *T. pyogenes* strains, including commensal isolates, it seems probable that *plo* was inserted into the intergenic region between *smc* and *ftsY* by homologous recombination; it has to be retained, as both of these genes are obligatorily required for bacterial growth. In addition, it was confirmed by Rudnick et al. [170] that the *plo* islet is a conserved region of the *T. pyogenes* chromosome found in all of the studied isolates of various origins.

T. pyogenes is an opportunistic pathogen that, being a part of the host microbiota, has no damaging effect on its organism. Therefore, it is suggested that the regulation of the PLO expression may be critical for the establishment of the status of this bacterium as a pathogen or a commensal. Rudnick et al. [170] reported that the *plo* gene expression is not regulated in vitro by *orf121* located upstream of this gene. In their subsequent investigation, two promoters, P1 and P2, located in the *plo* islet, were identified as the regulatory sequences controlling the transcription of *plo* [57]. It was observed that the in vitro haemolytic activity of *T. pyogenes* was in growth phase depending on a peak in the early stationary phase [57,173]. It indicated that *plo* is up-regulated during that phase, and indeed a significant increase in the *plo* specific mRNA was noted. Thus, the suggestion that the PLO expression is controlled at the transcriptional level was confirmed [57]. It was also shown that the P2 promoter is predominant and highly active during the stationary phase of the in vitro growth of bacteria, while the P1 activity is weak. Besides the P1 and P2 promoters, direct repeats (DRs) located in the *plo* promoter region may be involved in the regulation of the *plo* gene transcription. The study of Rudnick et al. [57]

showed that DRs, especially DR3, may function as binding sites for a soluble *T. pyogenes* factor being a transcriptional regulator. However, regarding AT-rich sequences of DRs, which can easily bend, it was supposed that DRs may also act as activators of transcription, by promoting RNA polymerase binding to the promoter.

The predicted, on the basis of the *plo* gene sequence, length of the PLO molecule is 534 amino acids (aa), and then, this protein should have a molecular mass of 57.9 kDa [180]. A signal peptidase cleavage site was found between 27 and 28 aa, indicating that the mature PLO molecule has a weight of 55.1 kDa. According to the Funk et al. [182] study, PLO is an oxygen-stable protein, heat-labile (destroyed at 56 and 100 °C), destroyed by pH 3 and 11, and id sensitive to treatment of protease, trypsin, and amylase.

In contrast to the majority of CDCs, the PLO activity is insensitive to thiols or other reducing agents, in other words, PLO does not require thiol activation [180]. This difference is associated with the divergence within the amino acid sequence of undecapeptide (491–501 aa) of PLO. The undecapeptide is a highly conserved region of each CDC, located near the C terminus of the protein. In some CDCs, including PLO, amino acid substitutions occur in this region [180,183]. In case of PLO undecapeptide (EATGLAWDPWW), a unique cysteine residue (C_{491}), responsible for the thiol-activated nature of CDCs, is replaced with an alanine (A_{492}). This substitution has no essential effect on the haemolytic activity of PLO, but is related to the oxygen stability of this cytotoxin [184]. Moreover, other changes in the PLO undecapeptide were noted, such as the insertion of a proline residue (P_{499}) and the deletion of an arginine residue at the end of the CDCs undecapeptide. It seems that these changes have important effects on the conformation and charge of the undecapeptide [180]. The proline insertion has the greatest effect on the PLO activity, as the deletion of this amino acid resulted in a lack of haemolytic activity [183]. The same effect on the PLO haemolytic activity was observed in the case of the removal or substitution of any of the three tryptophan residues [183]. Therefore, as it was previously demonstrated by Billington et al. [180], the amino acid sequence of the PLO undecapeptide is required for full cytolytic activity.

The cytolytic activity of PLO, like other CDCs, is suppressed by free cholesterol in a concentration of 1μg/mL [180]. However, the presence of cholesterol in the target plasma membrane is absolutely necessary for pore formation by CDCs. The role of the undecapeptide in the initial plasma membrane binding seems to be crucial, in particular, the first tryptophan residue (W_{497}) is important. Although, Billington et al. [183] suggested that other fragments of PLO might also participate in recognition and binding to the host cell membranes.

The known secondary and tertiary structures of some CDCs shows a significant similarity [172]. A crystal structure of the PLO molecule presumably is homologous to the structure of other CDCs. The tertiary structure model of PLO, based on the perfringolysin O (PFO)—PLO sequence alignment—was proposed by Pokrajec et al. [185]. The monomeric PLO molecule is rich in β-sheet elements and consists of four domains (D1 to D4), of which D2 and D3 are packed against each other, while D1 is located at the *N*-terminal and D4 at the C-terminal end of the molecule (Figure 4).

Domain 1 contains a-helix and β-sheet elements. The precise role of D1 in the pore-forming process remains undefined. Zhang et al. [186], studying an effect of the replacement of aspartic acid (D_{238}) in this domain with arginine, showed an important role of D1 in maintaining the pore-forming activity of PLO. The investigation of Imaizumi et al. [177] indicated that domain 1, especially the region of 55–74 amino acids, is required for the haemolytic activity of PLO. Yan et al. [187] also confirmed the importance of this region of domain 1, as they demonstrated that the substitution of each amino acid from 58 to 62 in the PLO molecule, particularly isoleucine 61, resulted in the complete loss of its haemolytic activity. Moreover, six B-cell linear epitopes in D1 were identified by Yang et al. [188].

Domain 2 of the molecule forms a long, curved single layer of anti-parallel β-sheet, while the structure of domain 3 is more complex and consists of five strands of anti-parallel β-sheet surrounded by helical layers and short α-helices at either side of the central β-sheet. Domains 2 and 3 are involved in the monomer oligomerization and insertion into the plasma membrane. The separation of D2 and

D3, as well as the structural collapse of D2, are required during these processes [189]. The conversion of the D3 short α-helices to two transmembrane β-hairpines starts the molecule insertion into the bilayer, and is a crucial step of the transmembrane β-barrel formation [172].

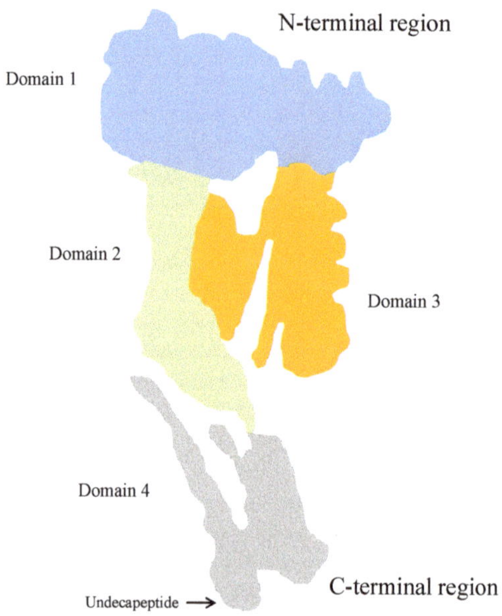

Figure 4. Simplified model of the pyolysin molecule, based on the perfringolysin O-structure.

Domain 4 of the PLO molecule is the only contiguous domain characterized by a β-sandwich structure, connected by a single peptide with the rest of the protein. As it has been mentioned above, the conserved undecapeptide sequence, rich in tryptophan residues, is located at the tip of this domain. This part of the PLO is responsible for host plasma membrane recognition and cholesterol binding [185,190]. However, Pokrajac et al. [185] also demonstrated the ability of D4 to self-oligomerization, but not pore formation. Their findings indicate a greater functional role of this PLO domain in cytolysin activity, compared with the previously described CDCs.

The molecular mechanism of pore-forming by PLO and other CDCs seems to be very similar, and is best described for PFO [172,185]. PLO molecules are secreted to the extracellular environment, probably via the general secretion system, as in the most CDCs [172]. In a form of water-soluble monomers, they bind to the cholesterol-containing areas in the plasma membrane of the eukaryotic host cells [183]. Binding to cholesterol by the PLO domain 4 initiates the oligomerization of the monomers and the formation of incomplete arc-shaped structures, called th pre-pore. Subsequently, the PLO oligomers are inserted into the bilayer by the transmembrane β-hairpins of domain 3, which leads to the formation of a large transmembrane β-barrel pore, which is <50 nm in diameter and protrudes <7 nm above the plasma membrane surface [191]. As a consequence, this process leads to th eleakage of ions and other cytoplasmatic molecules, and finally to the lysis of the host cell.

It seems that a predominant assignment of the cytolytic activity of PLO is to enable access to free iron and other growth factors essential for the bacterial replication contained in the host cells. Furthermore, the ability to lysis phagocytes protects the bacteria against the host immune response. On the other hand, at lower concentrations, PLO can modulate the host immune response.

Despite a wide knowledge about the molecular structure and function of PLO, the precise role of this toxin in the pathogenesis of particular *T. pyogenes* infections, characterized by different course and clinical manifestations, remains unclear.

4.1.2. Fimbriae

The occurrence of fimbriae in Gram-positive bacteria is a rarity. *T. pyogenes* is just one of the exceptions [48]. *T. pyogenes* fimbriae have a filamentous structure, at 200–700 nm in length and 2.5–4.5 nm in width, and probably occur in a limited number on bacterial cells (less than 10 per cell) [48,192]. In *T. pyogenes*, five fimbriae were described, FimA, FimB, FimC, FimE, and FimG [48,192]. The presence of other fimbrial subunits can be deduced on the base of the *T. pyogenes* genome sequences, for example FimJ [192,193]. FimA, 45.7 kDa protein, contains a signal peptide, an E box sequence, and is encoded by the *fimA* gene [48]. FimB, 90.5 kDa protein, has a similar structure, but additionally contains a cell wall-sorting domain and a fibronectin-binding domain, and is encoded by the *fimB* gene [48]. Both genes, *fimA* and *fimB*, are located in a fimbrial gene operon, together with the *srtA* gene encoding a putative sortase. The other fimbriae, FimC, FimE, and FimG, are encoded by the genes *fimC*, *fimE*, and *fimG*, respectively.

The expression of the fimbriae in vitro is poor; therefore, it is difficult to characterize their properties, thus only their genes or mRNAs are better known. Liu et al. [192] investigated the in vitro expression of FimA, FimC, and FimE. However, even though different culture conditions were used, only the FimE expression could be detected. Zhao et al. [179] studied the expression of the virulence factors of *T. pyogenes*, including FimA and FimC, in vitro and in vivo on a mouse model. Their results showed that in in vivo conditions, the *fimA* gene was earlier expressed than *fimC*, which suggests that FimA is a dominating fimbria in *T. pyogenes*. This is in accordance with the findings of many studies on fimbrial genes distribution that demonstrated a high prevalence of *fimA* among various isolates, whereas the genes of other fimbriae were detected with different frequencies [13,21,29–31].

A role of fimbriae in the *T. pyogenes* infection is associated with the bacterial adherence to host cells, but the precise mechanisms of this interaction are not well understood. Taking into consideration that *T. pyogenes* can be a commensal as well as a pathogen, it seems that fimbriae are equally important for bacteria in both of these forms. Likewise, in the case of infections caused by many other bacteria, *T. pyogenes* fimbriae are required for cell adhesion and the colonization of host tissues [49,55]. Liu et al. [192], based on their study results, speculated that the poor expression of FimA can limit the formation of other fimbriae, and that the up-regulation of this fimbria production may be associated with *T. pyogenes* pathogenicity.

4.1.3. Extracellular Matrix-Binding Proteins

T. pyogenes also expresses other proteins, which determine the adhesive properties, and which are associated with the cell wall of the bacteria. These proteins probably also promote adhesion and enhance colonisation through the ability to bind extracellular matrix (ECM) compounds [48]. In the case of Gram-positive bacteria, these proteins belong to the microbial surface components recognizing the adhesive matrix molecules (MSCRAMM) family [194]. Firstly, *T. pyogenes* shows a fibrinogen-binding activity, however, it was inhibited by proteases [195]. In addition, the 20 kDa fibronectin-binding protein associated with the *T. pyogenes* cell wall was also detected, but not well characterized [48].

Collagen-binding protein A (CbpA) represents the MSCRAMM family, and is one of the better known extracellular matrix-binding proteins produced by *T. pyogenes* [48,194,196]. This 121.9 kDa surface protein is encoded by the chromosomal *cbpA* gene, and displays a 50.4% similarity to Cna, the collagen adhesin of *Staphylococcus aureus*. CbpA binds almost all types of collagen, but not other ECM. However, Pietrocola et al. [194] reported that CbpA can bind fibronectin, but in a distinct subsite than that recognized by collagen. This protein is involved in the adherence of bacteria to epithelial and fibroblast cell lines [196]. It may be supposed that the *T. pyogenes* strain, which produces CbpA, has higher potential to colonize collagen-rich tissues. Although, the *cbpA* gene was found in many isolates of various origins, with different frequencies [13,21,29–31].

4.1.4. Neuraminidases

Neuraminidases (sialidases) are one of the most important enzymes of the sialic acid catabolism. Neuraminidases cleave the terminal sialic acid residues from complex glycoproteins, glycolipids, and carbohydrates from the host cell receptors. Neuraminidases are produced by many species of Gram-negative and Gram-positive bacteria or viruses [197]. Bacterial neuraminidases are the main factors for promoting adhesion by exposing the cryptic host cell receptors. In addition, neuraminidases can promote tissue colonisation by reducing mucus viscosity. These compounds also weaken the host's immune response by exposing IgA particles, making them more susceptible to bacterial proteases [48,197–199]. Bacterial neuraminidases are secreted as extracellular enzymes or are anchored to the cell wall. While in Gram-negative bacteria, neuraminidases are usually secreted, neuraminidases of the Gram-positive bacteria are commonly in a cell-associated form [200–204].

The first note about the 50 kDa extracellular neuraminidase of *T. pyogenes* was reported by Schaufuss and Lämmler [205]. Afterwards, two neuraminidases expressed by *T. pyogenes* were better characterized [206,207]. Neuraminidase H (NanH) and neuraminidase P (NanP) are 107 kDa and 186,8 kDa proteins, respectively. These neuraminidases are attached to the *T. pyogenes* cell wall, which results from the presence of specific anchoring motifs in their molecules (LPxTG—for NanH; LAWTG—for NanP). NanH and NanP proteins contain the conserved catalytic RIP/RLP motif (Arg-Ile/Leu-Pro) and five copies of the Asp box motif (Ser-X-Asp-X-Gly-X-Thr-Trp), commonly occurring in the bacterial neuraminidases [206,207]. NanH is the most similar to the neuraminidase from *Actinomyces viscosus* (61.8% similarity and 31.2% identity) [206]. NanP shows a similarity to the neuraminidase from *Micromonospora viridifaciens* (61.6% similarity and 45.3% identity) [207]. However, some similarity between NanH and NanP (53.8% similarity and 38.8% identity) was also demonstrated [48].

T. pyogenes strains can produce neuraminidases H and P, encoded by the *nan*H and *nan*P genes, respectively [206,207]. The occurrence of these genes is characteristic of the majority of *T. pyogenes* isolates from cattle [21,29–31,36,206–208] and swine [29,30,36]. However, the *nan*H and *nan*P genes were found at high rates in the bovine isolates studied by Zastempowska and Lassa [37]. The genes encoding neuraminidases are detected with various frequencies in the *T. pyogenes* isolates from small ruminants, such as goats and sheep [29–31,36], and in the forest musk deer isolates [134], while the presence of these genes among the isolates from European bison is much less common [29,135]. In several studies, the *nan*P gene was detected more frequently than the *nan*H gene in the tested isolates [30,135,208]. However, Rogovskyy et al. [31] reported that most of *T. pyogenes* isolates from small ruminants carried mainly the *nan*H gene.

T. pyogenes neuraminidases play an important role in adhesion to the host cells, as it was shown in the study with epithelial cells [207]. Moreover, *T. pyogenes* neuraminidases are probably necessary for the colonization of host tissues during the early phase of infection, but they are not essential during further stages of a disease development [48]. The different prevalence of the genes encoding neuraminidases in the *T. pyogenes* isolates from various origins, suggests that other adhesion factors may be also produced by this species.

4.1.5. Biofilm

The ability to perform biofilm formation is a well-known feature of microorganisms, which promotes their resistance to many disadvantageous factors, for instance antimicrobials, as well as increases the adherence properties and enables better protection against the immune response of the host [48]. *T. pyogenes*, as with many other bacteria that are resident on the mucous membranes of a host, is able to perform biofilm formation. Biofilm production was noted in a majority of the studied *T. pyogenes* isolates from different animal species and various types of infections [43,134]. Biofilm formation by *T. pyogenes* is controlled by a two-component regulatory system PloS/PloR, where PloR up-regulates the expression of biofilm [64,179,209]. In the case of polymicrobial infection, a biofilm formation by *T. pyogenes* may be inhibited by the QS molecules produced by other bacteria, for example

E. coli [64,210]. As in some cases biofilm production may promote the infection development, it seems important to find a therapeutical option for reducing this bacterial property. Da Silva Duarte et al. [211] evaluated the use of the *E. coli* phage UFV13 for the disruption of the *T. pyogenes* biofilm, and they obtained a significant decrease of biofilm formation caused by this phage. This finding indicates a necessity of further studies on this issue.

4.1.6. Regulation of Virulence Factor Expression

Jost and Billington [48] suggested the existence of a two-component system comprising a sensor histidine kinase (HK) and response regulator (RR) proteins in *T. pyogenes*. These signalling systems are common among bacteria, and they enable the detection of an environmental or cellular signal that leads to an appropriate cellular response [212,213]. Zhao et al. [64] determined a LuxR-type two-component regulatory system in *T. pyogenes*, named PloS/PloR, using a comparative transcriptome analysis between two *T. pyogenes* isolates from musk deer with different haemolytic activities. The results of the transcriptome analysis showed the presence of ten typical DNA-binding response regulators (RRs) in the *T. pyogenes* genome. However, a protein crystal structure analysis revealed that only one among the identified DNA-binding RRs had a structure similar to that described previously by Jost and Billington [48]. This RR, called PloR, had a CheY-type receiver domain and a C-terminal LuxR-type HTH DNA-binding domain, while a cognate sensor HK was named PloS. Zhao et al. [64] demonstrated that the expression level of the *ploR* gene is correlated with the expression level of the *plo* gene, indicating PloR as an activator of *plo*, which up-regulates PLO production. On the other hand, it was observed that PloR can down-regulate the expression of *T. pyogenes* proteases [48]. In addition, it seems that PloR is able to up-regulate the biofilm formation and probably the expression of the fimbriae in *T. pyogenes*.

The function of the two-component system in *T. pyogenes* was also investigated in vitro using the *N*-acyl homoserine lactones of *P. aeruginosa* and *E. coli* as QS signal molecules, which can bind to the upstream sensor HK, PloS, and via the PloS/PloR system, may regulate the virulence of bacteria by the inhibition of their growth and biofilm production [64]. Huang et al. [65], studying an effect of *N*-acyl homoserine lactones on the *T. pyogenes* virulence in a mouse model, confirmed that those QS molecules inhibited the expression of *plo*, *ploR*, and *ploS*, and increased the survival rate of mice. In conclusion, the function of the two-component system PloS/PloR and the QS system seems to have the essential significance for the modulation of the *T. pyogenes* virulence.

4.2. Induction of Host Defense Mechanisms by T. pyogenes

The pathogenicity of *T. pyogenes* is assigned mainly to PLO, a cholesterol-dependent cytolysin, which promotes cell lysis and is responsible for altering the host cytokine profile. This toxin also exerts cytolytic, dermonecrotic, and lethal effects on numerous types of cells, including immunocompetent and non-immune cells [176,186]. It is responsible for the complement cascade induction; modulation of the host cytokine profile; inhibition of respiratory burst; and bactericidal activities in neutrophils, monocytes, and macrophages. These features lead to the modulation of the immune response and to evasion from a host's immune mechanisms. Because of the presence of neuraminidases NanH and NanP, *T. pyogenes* is able to adhere to the alimentary and respiratory epithelial mucosa, increasing the mucus viscosity that facilitates bacterial colonization and diminishes the host immune response, including damage to a first line of defence—destruction of IgA by bacterial proteases. Those serine proteases, as it was mentioned above, are involved in the invasion and destruction of tissues [48,196].

For many years, the assessment of PLO involvement in the development of the disease was conducted using in vitro (established cell lines or primary cell cultures) and in vivo (mouse and large animal) models [48,68,69,75,168,186].

One of the first antibacterial mechanisms of immune response is inflammation, resulting in the recruitment of neutrophils, monocytes, and macrophages to a site of infection. The involvement of those cell populations leads to the phagocytosis and killing of bacteria, clearance of infection, and stimulation of adaptive immune response. Still, *T. pyogenes* interactions with host phagocytic cells strongly influence the outcome of the disease. Using macrophage established cell line J774A.1, it was clearly shown that this pathogen was easily phagocytosed, and could survive within macrophages up to 72 h, with the survival rate diminishing over time. Probably, this is caused by the presence of antibiotics in a medium used for culturing infected macrophages. Once bacteria exit the infected cell, they are killed by the antibiotic present in the medium. Similar results were obtained when the macrophage cell line RAW264.7 was infected with either *T. pyogenes* ATCC 19411 strain or *T. pyogenes* European bison isolate. In this study, phagosome formation, intracellular bacteria survival, and changes in the mitochondrial network and distribution were established at 2, 6, and 24 h post infection (pi) [214]. The formation of phagosomes was noticed as early as 2 h pi, and they were still present within the cells up to 24 h pi, with bacteria either within the phagosome or in the cytoplasm of the infected cells, and the recovery of viable bacteria was possible up to 24 h pi (Figure 5). However, the mitochondrial network morphology and distribution was not visibly altered in the RAW264.7 cells infected with either the ATCC 19411 strain or European bison isolate of *T. pyogenes* when compared with the uninfected control cells. Further investigations are required in order to elucidate the mechanism of *T. pyogenes* pathogenesis. Nonetheless, these data are in agreement with the observations done on bovine polymorphonuclear neutrophils (PMNs) infected with *T. pyogenes*, in which increased phagocytosis was noticed. As this pathogen can survive within phagocytic cells, increased phagocytosis is strongly connected with bacterial dissemination throughout the host [195]. Studies done in a mouse model clearly showed that PLO is responsible for peritonitis [168].

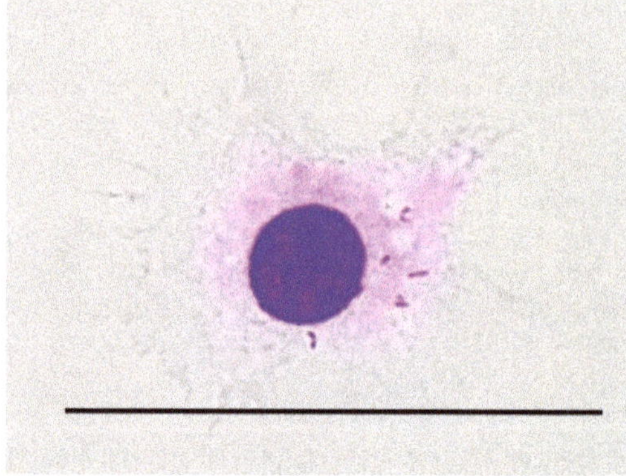

Figure 5. The macrophage of the RAW264.7 cell line infection with *T. pyogenes*. Staining using the Giemsa–MayGrünwald method; scale bar = 20 μm (photo courtesy of Anna Hupało-Sikorska, BSc).

One of the common, most recognized infections in dairy and beef cattle are uterine infections caused by *T. pyogenes*. They result from the recruitment of neutrophils, from monocytes, and from the induction of a proinflammatory response, accompanied with formation of mucopurulent discharges within the infected uterus. This was confirmed on a large animal model. The uterine endometrium of healthy cows was scratched with a cytobrush (damage to the epithelial layer), and subsequently, an intrauterine administration of 5×10^8 cfu of *T. pyogenes* was performed. Mucopurulent uterine discharge was evident in the challenged animals, and was associated with the isolation of this pathogen [68].

However, the ability of cows to defend against *T. pyogenes* infection depends on the interaction between the invading pathogen and the host's innate immunity. An application of bovine endometrial cell culture in an in vitro model elucidated the PLO involvement in tissue damage and the development of the endometrium pathology of the postpartum period. Recombinant PLO (rPLO) and bacteria-free filtrates (BFF) of *T. pyogenes* had a similar haemolytic activity on the target cells. Still, the stromal cells and epithelial cells expressed a different level of susceptibility concerning treatment with those factors. The endometrial stromal cells were markedly more sensitive than the epithelial cells. This haemolytic effect was abrogated when the endometrial cells were treated with BFF derived from *T. pyogenes* with a deleted *plo* gene (*T. pyogenesΔplo*), and when anti-PLO antibodies were applied. The difference in susceptibility or resistance to PLO between the stromal and epithelial cells can be explained by a lower level of cholesterol expression in the epithelial cells. It has biological implications to PLO-mediated uterine tissue damage. The columnar epithelium of the endometrium establishes a first line of defence by forming a physical barrier between the uterine lumen and the underlying stroma. As it expresses less sensitivity to PLO, it protects the tissue from colonization by *T. pyogenes*. Once this barrier is disrupted, the bacteria can colonize the endometrium, and the development of the disease is induced. However, PLO did not induce inflammation, as the treatment of bovine endometrial culture cells with rPLO did not result in increased levels of IL-1β, IL-6, or CXCL8 in the culture supernatant. Nevertheless, during clinical infection, *T. pyogenes* often causes tissue damage in association with other species of bacteria, expressing together with *T. pyogenes* pathogens, the associated molecular patterns (PAMPs) and damage associated molecular patterns (DAMPs) responsible for the induction of inflammatory response.

The described sensitivity of stromal cells to PLO-mediated cytolysis is one of the explanations for how *T. pyogenes* acts as an opportunistic pathogen, responsible for endometrial damage once the protective epithelium is lost after parturition [68].

It has to be remembered that the development of a disease depends on the ability of an infectious agent to adhere, to colonize a host's target tissue, and develop mechanism(s) that are responsible for avoiding a host's defence and the involvement of its innate immunity. In the case of *T. pyogenes*, bovine uterine infections also interplay, as follows: the host–virulent strain of *T. pyogenes* or host–opportunistic strain of *T. pyogenes*, should be considered. The model presented by Ibrahim et al. [56] considers all of these factors. They focused on in vitro investigating the interactions between a bovine endometrial epithelial cell culture with either the *T. pyogenes* strain isolated from the uterus of a postpartum cow developing clinical endometritis, or *T. pyogenes* collected from the uterus of a healthy cow. The assessment of the presence of strain-specific factors participating in the development of the bovine clinical endometritis was also performed. The outcome of the analysis on a genetic level revealed that both strains presented genes encoding virulence factors; however, there were differences between the expression levels of *plo*, *cbpA*, *nanH*, *nanP*, *fimA*, *fimC*, *fimE*, and *fimG* genes, favouring the virulent strain. The proteins encoded by those genes participate in adhesion to the host cells [48,207], so a virulent strain had a higher ability to attach to the epithelium, because of the elevated level of the cell-wall associated molecules. Both strains expressed a cytolytic effect on cultured cells, which means both strains secreted PLO, a major virulence factor. However, this effect was abrogated when the bovine endometrial epithelial cells were treated with heat inactivated bacteria or bacteria-free filtrates. These findings are in line with the results obtained by Amos et al. [68], showing that BFF and rPLO were not responsible for the stimulation of a proinflammatory response in endometrial cells.

To mimic the interplay between the host target tissue and *T. pyogenes* in the chronic proinflammatory environment, endometrial endothelial cells treated with bacteria were co-cultured in the presence/absence of peripheral blood mononuclear cells (PBMCs). The presence of PBMCs did not influence the viability of the endometrial epithelial cells more than 16 h, upon the treatment with live bacteria, irrespective of a strain. However, the early detection (4 h) of genes encoding proinflammatory mediators (*PTGS2*, *CXCL3*, *IL6*, and *CXCL8*) was noticed in the endometrial endothelial cells treated with a virulent strain of *T. pyogenes*, in the presence of PBMCs. This in vitro interaction suggests that the leukocytes found within the endometrium may be responsible for sensitizing the epithelial cells to the initial bacterial infection through the enhancement of uterine innate immunity induction (Figure 6). These data suggest that the development of endometritis in dairy cows after parturition may be caused by a specific characteristic of certain strains of *T. pyogenes*. Moreover, the presence of immune cells can be the reason for the amplification of the proinflammatory response in the endometrial epithelial tissue towards *T. pyogenes* pathogenic strains.

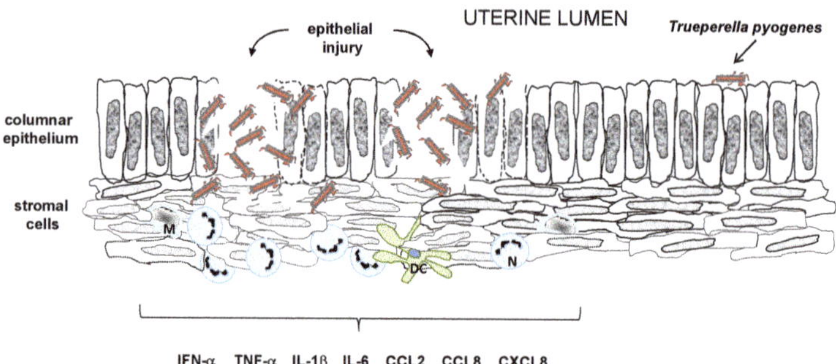

Figure 6. Schematic presentation of the pathogenesis of uterine infection with *T. pyogenes*. Recruitment of monocytes (M), neutrophils (N), dendritic cells (DC), and other proinflammatory cells—secretion of proinflammatory mediators.

The research of Zhang et al. [186] indicated that the inflammation-inducing effect of PLO depends on its cytolytic activity, but not on PAMPs activity. These findings are in contrast to the results of Amos et al. [68] and Ibrahim et al. [56].

Studies done on bovine oviductal epithelial cells (BOECs), in an in vitro model, suggested a specific mechanism employed by *T. pyogenes* to avoid an immune response. In the course of *T. pyogenes* infection, there was no induction of any clear proinflammatory response in the BOECs at either the transcriptional or protein level. The BOECs co-cultured with *T. pyogenes* remained viable during the first 24 h of incubation when treated with MOI 0.01. However, a higher MOI caused the death of those cells within 24 h of co-culturing. What is interesting, is that the BOECs co-cultured with *T. pyogenes* expressed a similar level of *IL1A*, *IL1B*, and *TNFA* mRNA as the control cells did. Similar results were obtained when the level of mRNA concerning *CXCL8*, *CXCL1/2*, and *PTGS2* was assessed. Although, at the same time, the BOECs co-cultured with *Bacillus pumilus* responded to stimulation, with an increased level of cytokines and chemokines mRNA. It seems that the lack of responsiveness in the BOECs induced by *T. pyogenes* is one of the escape routes from immune surveillance developed by this pathogen, as the expression of the genes encoding the proinflammatory mediators was similar to the control levels [75].

5. *T. pyogenes* Genome

The first complete sequence of the *T. pyogenes* genome was described in 2014 by Machado and Bicalho [215]. However, to date, 19 genome assemblies of this species are available in the GenBank nucleotide database (NCBI). Complete genome sequences were reported for 10 strains, isolated as follows: four from swine, three from cattle, one from goat, one from forest musk deer, and one from water buffalo. Additionally, the vast majority of the sequenced genome of *T. pyogenes* were derived from strains, which were isolated from animals in China. Other available genome assemblies are scaffolds ($n = 3$) and contigs ($n = 6$). The main features of complete *T. pyogenes* genomes are summarized in Table 1.

The *T. pyogenes* genome was found to consist of a single circular chromosome comprising from 2.25 to 2.43 mega base pairs, depending on the strain. This species is characterized by high G+C% content. Moreover, the differences in the rate of G+C between the *T. pyogenes* strains are not significant (59.4–59.8%). tRNA is the most abundant RNA type in these bacteria. The share of rRNA in the whole nucleic acid content ranges from three to nine. In addition, there are also several non-coding RNAs (one or three, depending on the strain). The *T. pyogenes* genomes are indicated by the relatively high numbers of pseudogenes (from 27 to 191).

Table 1. General features of the complete *T. pyogenes* genomes available in the GenBank nucleotide database (NCBI).

Strain Designation	Strain Origin	Size (Mb)	GC%	Number of				GenBank Accession nr
				Genes	CDS	RNA	Pseudogenes	
TP4	swine	2.43	59.4	2202	2102	59	41	CP033905.1
TP3	swine	2.38	59.4	2156	2058	58	40	CP033904.1
TP2	cattle	2.25	59.7	2023	1938	58	27	CP033903.1
TP1	cattle	2.33	59.8	2125	1993	58	74	CP033902.1
TP-2849	swine	2.38	59.4	2158	2063	58	37	CP029004.1
TP4479	swine	2.38	59.4	2153	2058	58	37	CP029001.1
Arash114	water buffalo	2.34	59.5	2145	2054	56	35	CP028833.1
2012CQ-ZSH	goat	2.30	59.7	2050	1806	53	191	CP012649.1
TP8	forest musk deer	2.27	59.6	2091	2001	50	40	CP007003.1
TP6375	cattle	2.34	59.5	2082	1984	53	45	CP007519.1

The *T. pyogenes* genome encodes many genes, for which the functions were assigned by the homologies to known proteins. Apart from the virulence-associated genes, the genome of this species can contain several antibiotic resistance genes. Furthermore, the *luxS* gene regulating the formation of biofilms was also found in the genome sequence [209]. In addition, the presence of a member of the *T4virus* (vB_EcoM-UFV13) indicates the efficiency in controlling the biofilm formation by *T. pyogenes* [216]. Moreover, the *T. pyogenes* genome contains four housekeeping genes (*metG, tuf, gyrA,* and *fusA*). The *tuf* and *fusA* genes encode the translation elongation factors Tu and G, respectively. Two other genes confer the DNA gyrase subunit A (*gyrA*) and methionine—tRNA ligase (*metG*). These genes were successfully used in the multilocus sequence analysis (MLSA) of this species [148]. The genetic analysis showed that *T. pyogenes* had a more complex system of amino acid and lipid metabolism, and more genes involved in the pathogenicity than are found in *Arcanobacterium hemolyticum*. These differences may affect the ability to cause infections in different host types by *T. pyogenes* isolates. Moreover, all of the groups of the family *Actinomycetaceae* have phosphotransferase systems that are probably essential for the colonization of a wide range of hosts and for the initiation of abscess formation. Some genes were lost or acquired as a result of lateral evolution, which helps in the adaptation of bacteria to a new environment [6]. The *T. pyogenes* species was reclassified based on 16S rRNA gene sequences, the menaquinone structure, and the phospholipid composition from the genus *Arcanobacterium* to the genus *Trueperella*, together with four other species, such as *Trueperella abortisuis, Trueperella bernardiae, Trueperella bonsai,* and *Trueperella bialowiezense*. Furthermore, 16S rRNA gene sequence similarities between strains from the genus *Trueperella* ranged from 95.3% to 98.6% [1]. There was also a draft genome sequence of one human *T. bernardiae* isolate deposited in the NCBI [217].

The genomes of *T. pyogenes* also include plasmids. Until now, two plasmids have been described for this species [178,218]. First, a native pAP1 (U83788) is a circular plasmid of 2439 bp, containing three open reading frames (ORFs). Plasmid pAP1 encodes the *rep* gene, which is required for the initiation of replication. Two other genes, *orf112* and *orf129*, encode the hypothetical proteins of an unknown function. However, pAP1 does not include any antibiotic resistance genes [218]. The other, pAP2 (AY255627), is a circular plasmid of 9304 bp, and it contains seven ORFs. In pAP2, *repA* is probably required for the replication of this plasmid. Moreover, there are other genes, namely: *tetR*(33), which encodes the repressor protein and *gcrY*, and *orf95* with unknown functions. In addition, pAP2 is characterized by the presence of two genes encoding antibiotic resistance determinants—*erm*(X) and *tetA*(33). In *T. pyogenes, erm*(X) determines the resistance to macrolide antibiotics, while *tetA*(33) is associated with low-level tetracycline resistance [178]. Furthermore, plasmid pAP2 contains the insertion sequence IS*6100*, which is found either in Gram-positive or in Gram-negative bacteria [219].

The knowledge of the complete genome sequence may allow for the identification of new genes that may contribute to the pathogenicity and antibiotic resistance of *T. pyogenes*. In addition, genome sequencing can help to understand the basics of the metabolism and evolution of the bacterial species.

6. Immunoprotection and Perspectives

The antimicrobial treatment of *T. pyogenes* infections is often ineffective, because of the increasing resistance of bacteria or the limited distribution of drugs to the site of infection, for example, to abscesses. Hence, vaccination should be considered as a primary method of *T. pyogenes* infection prevention. It seems that some of the virulence factors of *T. pyogenes* could be promising candidates for vaccine antigens.

Most studies concerning the stimulation of protective immunity against *T. pyogenes* infection were done on a mouse model, and only a few were done using a ruminant or swine model. The first trials to protect ruminants against infection with *T. pyogenes* were focused on animal treatment with whole cells of *T. pyogenes* or a culture supernatant, and were mostly unsuccessful; however, some therapeutic effects on mastitis development were observed [48,220,221]. In swine, the vaccination of pregnant sows with an autovaccine was an attempt to control the losses of newborn animals. Studies done by Kostro et al. [222] revealed that the administration of *T. pyogenes* cells treated with a phenol solution to pregnant sows, six and three weeks before anticipated delivery, significantly increased the number of CD4$^+$ T cells, CD8$^+$ T cells, and CD25$^+$ T cells, as well as enhanced the levels of IFN-γ, TNF-α, and IL-10 in a colostrum and milk [222]. The recent trials were concentrated on the protection of heifers against post partum uterine diseases. Pregnant heifers were vaccinated using different routes of immunization (subcutaneously and intravaginally), and different compositions of the vaccine, as follows: (i) inactivated bacterial whole cells (*E. coli*, *T. pyogenes*, and *F. necrophorum*) together with recombinant proteins, FimH, PLO, and leukotoxin (LKT)); (ii) recombinant proteins FimH, PLO, and LKT only; (iii) inactivated bacterial whole cells (*E. coli*, *T. pyogenes*, and *F. necrophorum*) only [223]. The obtained results demonstrated that subcutaneous vaccination significantly decreased the incidents of puerperal metritis, but intravaginal vaccination failed in preventing the disease.

Mice used for the assessment of protective immunity induction against infection with *T. pyogenes* provide a reliable model that can serve for further comprehensive studies on other animal species. The administration of formalin-inactivated recombinant PLO (rPLO) induced the protection of mice against intraperitoneal challenge with 10^8 of *T. pyogenes* cells, and PLO-specific antibodies were detected in the sera of the immunized mice [168,180].

The current trend in vaccine design is concentrated on the development of such a product that will induce immunity against multiple pathogens. Using a mouse model, Zhang et al. [224] and Hu et al. [176] induced specific immunity that protected animals against challenge with *T. pyogenes* and *C. perfringens*. However, their approach to solve the problem concerning the stimulation of protective immunity was quite different. Hu and colleagues [176] designed a chimeric protein called rPC-PD4, which was composed of *C. perfringens* truncated regions of C-domain of phospholipase C (rPLC), and D4 domain of *T. pyogenes* PLO (rPLO), and was encoded by chimeric genes incorporated into a plasmid vector. The production of specific antibodies, and the presence and level of proinflammatory cytokines (TNF-α, IL-1β), chemokines (CXCL8), and IL-10 in sera of immunized mice were assessed; however, only partial immunity was observed after challenge with *T. pyogenes* or *C. perfringens*. Zhang's group [224] concentrated not only on PLC and PLO as major immunogens, but also on the involvement of formaldehyde inactivated cultures of *T. pyogenes* and *C. perfringens* as vaccine antigens, combined with aluminium hydroxide gel. The outcome of their proposal is promising, as the survival rate of immunized mice after challenge with *T. pyogenes* and *C. perfringens* was 80% and 100%, respectively. These studies [176,224] present a very interesting approach for the development of a new, safe vaccine against polymicrobial infections.

Nonetheless, the latest trials to stimulate immunoprotection against *T. pyogens* infection are the most promising so far. The studies of Huang's group [225,226] done on a mouse model, concentrated on genetic immunization. Two pathways to stimulate protective immunity against *T. pyogenes* infection were used. The first one applied the modification of the cytokine environment of immunized animals together with delivering the gene encoding PLO [225], the second one focused on immunization with a DNA vaccine containing genes encoding four different virulence factors of *T. pyogenes*, and simultaneous vaccine protection against destruction by the host environment [226].

In the first study, concurrent immunization with two plasmids carrying the *plo* gene and the *IL1β* gene, respectively, resulted in the protection of mice from intraperitoneal challenge, with 3.7×10^8 cfu of *T. pyogenes* [225]. PLO-specific antibodies were detected in the sera of immunized mice with a higher titer of the IgG2a subtype than the IgG1 subtype. The T-cell profile analysis indicated an increased level of CD8$^+$ T-cells and CD4$^+$ T-cells. These results correlated with the detection and assessment of the cytokine levels evaluated in the supernatant collected from the spleen lymphocytes culture (increased levels of IFN-γ, and IL-2, and slightly diminished IL-4). In this model, IL-1β enhanced the immunogenicity of the vaccine by influencing the activity of the macrophages, the major population of the phagocytic cell, as well as antigen presenting cells (APCs), and also the B-cells, which act as APCs during early phases of immune response development, later transforming into plasma cells secreting antigen-specific antibodies, hence humoral immunity.

In the second research of Huang's group [226], the *plo*, *cbpA*, *fimA*, and *nanH* genes were paralleled as a single chimeric gene with an attached CpG ODN 1826 motif, so as to overcome low immunogenicity and susceptibility to host endonucleases. Finally, the chimeric gene was incorporated into the expression plasmid vector, and then this construct was encapsulated in chitosan nanoparticles to protect the DNA. The chimeric gene DNA vaccine introduced intramuscularly initiated production of antibodies specific for the target epitopes of PLO, CbpA, NanH, and FimA in mice. Moreover, the level of IgG2a was higher than the IgG1 in the sera of immunized mice, suggesting that immunization with this vaccine mainly primed the Th1 profile.

Compared with the subunit and conventional vaccines, DNA vaccines express some advantages as future stimulants of immunity. They are easy to manufacture, stable, and confer potential safety. Moreover, many genes encoding specific antigen proteins can be inserted in a carrier genome, and once introduced to the host, stimulate protective immunity against a variety of infections. In this context, the studies done by Huang et al. [225,226] are very important, not only in the induction of effective protection against *T. pyogenes* infection. They also proved that the formulated chitosan-CpG ODN nanoparticles are stable in eukaryotic cells, are able to protect the chimeric gene in the encapsulated plasmid from degradation, promote its expression, and also participate in the induction and enhancement of the immune response, so they could serve as a safe and efficient drug release carrier system.

7. Conclusions

T. pyogenes has significant importance as an opportunistic animal pathogen causing a variety of purulent infections. These infections pose a huge economic problem in livestock breeding. Therefore, it is crucial to understand the mechanisms of pathogenicity of this bacterium and the routes of its transmission for the development of an effective prevention strategy.

Although some of virulence determinants in *T. pyogenes* are well recognized, little is known about the dissemination of infections. It is interesting, there is still limited data about the pathogen–host interactions that may be involved in the development of *T. pyogenes* infection. Especially, the determination of factors that have a key role in transformation from a commensal to a pathogenic bacterium is needed. Although it seems that animals are the main reservoir of this bacterium, the epidemiological relationships between the particular isolates remain unclear.

Author Contributions: Conceptualization, M.R.; writing (original draft preparation), M.R., E.K., D.C.-C., M.K.-Ś., I.S., and M.G.; writing (review), M.R., E.K., and M.G.

Funding: This work was supported by KNOW (the Leading National Research Centre) Scientific Consortium "Healthy Animal–Safe Food", decision of Polish Ministry of Science and Higher Education, no. 05-1/KNOW2/2015.

Acknowledgments: The authors thank Dorota Rzewuska for the technical support during editing the manuscript.

Conflicts of Interest: The authors declare no conflict of interest.

References

1. Yassin, A.F.; Hupfer, H.; Siering, C.; Schumann, P. Comparative chemotaxonomic and phylogenetic studies on the genus *Arcanobacterium* Collins et al. 1982 emend. Lehnen et al. 2006: Proposal for *Trueperella* gen. nov. and emended description of the genus *Arcanobacterium*. *Int. J. Syst. Evol. Microbiol.* **2011**, *61*, 1265–1274. [CrossRef] [PubMed]
2. Ramos, C.P.; Foster, G.; Collins, M.D. Phylogenetic analysis of the genus *Actinomyces* based on 16S rRNA gene sequences: Description of *Arcanobacterium phocae* sp. nov., *Arcanobacterium bernardiae* comb. nov., and *Arcanobacterium pyogenes* comb. nov. *Int. J. Syst. Bacteriol.* **1997**, *47*, 46–53. [CrossRef] [PubMed]
3. Collins, M.D.; Jones, D.; Kroppenstedt, R.M.; Schleifer, K.H. Chemical studies as a guide to the classification of *Corynebacterium pyogenes* and *Corynebacterium haemolyticum*. *J. Gen. Microbiol.* **1982**, *128*, 335–341. [CrossRef] [PubMed]
4. Reddy, C.A.; Cornell, C.P.; Fraga, A.M. Transfer of *Corynebacterium pyogenes* (Glage) Eberson to the genus *Actinomyces* as *Actinomyces pyogenes* (Glage) comb. nov. *Int. J. Syst. Bacteriol.* **1982**, *32*, 419–429. [CrossRef]
5. Rogosa, M.; Cummins, C.S.; Lelliott, R.A.; Keddie, R.M. Coryneform group of bacteria. In *Bergey's Manual of Determinative Bacteriology*, 8th ed.; Buchanan, R.E., Gibbons, N.E., Eds.; The Williams and Wilkins Company: Baltimore, MD, USA, 1974; pp. 599–632.
6. Zhao, K.; Li, W.; Kang, C.; Du, L.; Huang, T.; Zhang, X.; Wu, M.; Yue, B. Phylogenomics and Evolutionary Dynamics of the Family *Actinomycetaceae*. *Genome Biol. Evol.* **2014**, *6*, 2625–2633. [CrossRef] [PubMed]
7. Ding, H.; Lämmler, C. Evaluation of the API Coryne test system for identification of *Actinomyces pyogenes*. *Zentralbl. Veterinarmed. B* **1992**, *39*, 273–276. [CrossRef]
8. Hijazin, M.; Hassan, A.A.; Alber, J.; Lämmler, C.; Timke, M.; Kostrzewa, M.; Prenger-Bernighoff, E.; Zschöck, M. Evaluation of Matrix-Assisted Laser Desorption Ionization-Time of Flight Mass Spectrometry (MALDI-TOF MS) for species identification of bacteria of genera *Arcanobacterium* and *Trueperella*. *Vet. Microbiol.* **2012**, *157*, 243–245. [CrossRef]
9. Zhang, W.; Meng, X.; Wang, J. Sensitive and rapid detection of *Trueperella pyogenes* using loop-mediated isothermal amplification method. *J. Microbiol. Methods* **2013**, *93*, 124–126. [CrossRef]
10. Nagib, S.; Rau, J.; Sammra, O.; Lämmler, C.; Schlez, K.; Zschöck, M.; Prenger-Berninghoff, E.; Klein, G.; Abdulmawjood, A. Identification of *Trueperella pyogenes* isolated from bovine mastitis by Fourier transform infrared spectroscopy. *PLoS ONE* **2014**, *9*, e10465. [CrossRef]
11. Randall, L.P.; Lemma, F.; Koylass, M.; Rogers, J.; Ayling, R.D.; Worth, D.; Klita, M.; Steventon, A.; Line, K.; Wragg, P.; et al. Evaluation of MALDI-ToF as a method for the identification of bacteria in the veterinary diagnostic laboratory. *Res. Vet. Sci.* **2015**, *101*, 42–49. [CrossRef]
12. Abdulmawjood, A.; Wickhorst, J.; Hashim, O.; Sammra, O.; Hassan, A.A.; Alssahen, M.; Lämmler, C.; Prenger-Berninghoff, E.; Klein, G. Application of a loop-mediated isothermal amplification (LAMP) assay for molecular identification of *Trueperella pyogenes* isolated from various origins. *Mol. Cell Probes* **2016**, *30*, 205–210. [CrossRef] [PubMed]
13. Moreno, L.Z.; Matajira, C.E.C.; da Costa, B.L.P.; Ferreira, T.S.P.; Silva, G.F.R.; Dutra, M.C.; Gomes, V.T.M.; Silva, A.P.S.; Christ, A.P.G.; Sato, M.I.Z.; et al. Characterization of porcine *Trueperella pyogenes* by matrix-assisted laser desorption ionization-time of flight mass spectrometry (MALDI-TOF MS), molecular typing and antimicrobial susceptibility profiling in Sao Paulo State. *Comp. Immunol. Microbiol. Infect. Dis.* **2017**, *51*, 49–53. [CrossRef] [PubMed]
14. Rowlinson, M.C.; Bruckner, D.A.; Hinnebusch, C.; Nielsen, K.; Deville, J.G. Clearance of *Cellulosimicrobium cellulans* Bacteriemia in a Child without Central Venous Catheter Removal. *J. Clin. Microbiol.* **2006**, *44*, 2650–2654. [CrossRef] [PubMed]
15. Saitou, N.; Nei, M. The neighbor-joiningmethod: A newmethod for reconstructingphylogenetictrees. *Mol. Biol. Evol.* **1987**, *4*, 406–425. [PubMed]
16. Tamura, K.; Nei, M.; Kumar, S. Prospects for inferringverylargephylogenies by using the neighbor-joiningmethod. *Proc. Natl. Acad. Sci. USA* **2004**, *101*, 11030–11035. [CrossRef] [PubMed]
17. Felsenstein, J. Confidencelimits on phylogenies: Anapproachusing the bootstrap. *Evolution* **1985**, *39*, 783–791. [CrossRef] [PubMed]
18. Kumar, S.; Stecher, G.; Li, M.; Knyaz, C.; Tamura, K. MEGA X: MolecularEvolutionary Genetics Analysis acrosscomputingplatforms. *Mol. Biol. Evol.* **2018**, *35*, 1547–1549. [CrossRef]

19. Queen, C.; Ward, A.C.; Hunter, D.L. Bacteria isolated from nasal and tonsillar samples of clinically healthy Rocky Mountain bighorn and domestic sheep. *J. Wildl. Dis.* **1994**, *30*, 1–7. [CrossRef] [PubMed]

20. Rzewuska, M.; Osińska, B.; Bielecki, W.; Skrzypczak, M.; Stefańska, I.; Binek, M. Microflora of urogenital tract in European bison (*Bison bonasus*). In *Health Threats for the Eurpean Bison Particularly in Free-Roaming Populations in Poland*; Kita, J., Anusz, K., Eds.; Warsaw University of Life Sciences: Warsaw, Poland, 2006; pp. 27–28.

21. Silva, E.; Gaivão, M.; Leitão, S.; Jost, B.H.; Carneiro, C.; Vilela, C.L.; Lopes da Costa, L.; Mateus, L. Genomic characterization of *Arcanobacterium pyogenes* isolates recovered from the uterus of dairy cows with normal puerperium or clinical metritis. *Vet. Microbiol.* **2008**, *132*, 111–118. [CrossRef] [PubMed]

22. Narayanan, S.; Nagaraja, T.G.; Wallace, N.; Staats, J.; Chengappa, M.M.; Oberst, R.D. Biochemical and ribotypic comparison of *Actinomyces pyogenes* and *A. pyogenes*-like organisms from liver abscesses, ruminal wall, and ruminal contents of cattle. *Am. J. Vet. Res.* **1998**, *59*, 271–276. [PubMed]

23. Jost, B.H.; Post, K.W.; Songer, J.G.; Billington, S.J. Isolation of *Arcanobacterium pyogenes* from the porcine gastric mucosa. *Vet. Res. Commun.* **2002**, *26*, 419–425. [CrossRef] [PubMed]

24. Spittel, S.; Hoedemaker, M. Mastitis diagnosis in dairy cows using PathoProof real-time polymerase chain reaction assay in comparison with conventional bacterial culture in a Northern German field study. *Berl. Munchener Tierarztliche Wochenschr.* **2012**, *125*, 494–502.

25. Williamson, L.H. Caseous lymphadenitis in small ruminants. *Vet. Clin. N. Am. Food Anim. Pract.* **2001**, *17*, 359–371. [CrossRef]

26. Stefańska, I.; Gieryńska, M.; Rzewuska, M.; Binek, M. Survival of *Corynebacterium pseudotuberculosis* within macrophages and induction of phagocytes death. *Pol. J. Vet. Sci.* **2010**, *13*, 143–149. [PubMed]

27. Muscatello, G. *Rhodococcus equi* pneumonia in the foal—Part 1: Pathogenesis and epidemiology. *Vet. J.* **2012**, *192*, 20–26. [CrossRef] [PubMed]

28. Witkowski, L.; Rzewuska, M.; Takai, S.; Chrobak-Chmiel, D.; Kizerwetter-Świda, M.; Feret, M.; Gawryś, M.; Witkowski, M.; Kita, J. Molecular characterization of *Rhodococcus equi* isolates from horses in Poland: pVapA characteristics and plasmid new variant, 85-kb type V. *BMC Vet. Res.* **2017**, *13*, 35. [CrossRef] [PubMed]

29. Rzewuska, M.; Czopowicz, M.; Gawryś, M.; Markowska-Daniel, I.; Bielecki, W. Relationships between antimicrobial resistance, distribution of virulence factor genes and the origin of *Trueperella pyogenes* isolated from domestic animals and European bison (*Bison bonasus*). *Microb. Pathog.* **2016**, *96*, 35–41. [CrossRef]

30. Risseti, R.M.; Zastempowska, E.; Twarużek, M.; Lassa, H.; Pantoja, J.C.F.; de Vargas, A.P.C.; Guerra, S.T.; Bolanõs, C.A.D.; de Paula, C.L.; Alves, A.C.; et al. Virulence markers associated with *Trueperella pyogenes* infections in livestock and companion animals. *Lett. Appl. Microbiol.* **2017**, *65*, 125–132. [CrossRef] [PubMed]

31. Rogovskyy, A.S.; Lawhon, S.; Kuczmanski, K.; Gillis, D.C.; Wu, J.; Hurley, H.; Rogovska, Y.V.; Konganti, K.; Yang, C.Y.; Duncan, K. Phenotypic and genotypic characteristics of *Trueperella pyogenes* isolated from ruminants. *J. Vet. Diagn. Investig.* **2018**, *30*, 348–353. [CrossRef]

32. Ashrafi Tamai, I.; Mohammadzadeh, A.; Zahraei Salehi, T.; Mahmoodi, P. Genomic characterisation, detection of genes encoding virulence factors and evaluation of antibiotic resistance of *Trueperella pyogenes* isolated from cattle with clinical metritis. *Antonie Van Leeuwenhoek* **2018**, *111*, 2441–2453. [CrossRef] [PubMed]

33. Guérin-Faublée, V.; Flandrois, J.P.; Broye, E.; Tupin, F.; Richard, Y. *Actinomyces pyogenes*: Susceptibility of 103 clinical animal isolates to 22 antimicrobial agents. *Vet. Res.* **1993**, *24*, 251–259. [PubMed]

34. Werckenthin, C.; Alesík, E.; Grobbel, M.; Lübke-Becker, A.; Schwarz, S.; Wieler, L.H.; Wallmann, J. Antimicrobial susceptibility of *Pseudomonas aeruginosa* from dogs and cats as well as *Arcanobacterium pyogenes* from cattle and swine as determined in the BfT-GermVet monitoring program 2004–2006. *Berl. Munchener Tierarztliche Wochenschr.* **2007**, *120*, 412–422.

35. Urumova, V.; Lyutzkanov, M.; Tsachev, I.; Marutsov, P.; Zhelev, G. Investigations on the involvement of *Arcanobacterium pyogenes* in various infections in productive and companion animals and sensitivity of isolates to antibacterials. *Revue Méd. Vét.* **2009**, *160*, 582–585.

36. Hijazin, M.; Ülbegi-Mohyla, H.; Alber, J.; Lämmler, C.; Hassan, A.A.; Abdulmawjood, A.; Prenger-Berninghoff, E.; Weiss, R.; Zschöck, M. Molecular identification and further characterization of *Arcanobacterium pyogenes* isolated from bovine mastitis and from various other origins. *J. Dairy Sci.* **2011**, *94*, 1813–1819. [CrossRef] [PubMed]

37. Zastempowska, E.; Lassa, H. Genotypic characterization and evaluation of an antibiotic resistance of *Trueperella pyogenes* (*Arcanobacterium pyogenes*) isolated from milk of dairy cows with clinical mastitis. *Vet. Microbiol.* **2012**, *161*, 153–158. [CrossRef] [PubMed]

38. Pohl, A.; Lübke-Becker, A.; Heuwieser, W. Minimum inhibitory concentrations of frequently used antibiotics against *Escherichia coli* and *Trueperella pyogenes* isolated from uteri of postpartum dairy cows. *J. Dairy Sci.* **2018**, *101*, 1355–1364. [CrossRef] [PubMed]

39. Liu, M.; Wu, C.; Liu, Y.; Zhao, J.; Yang, Y.; Shen, J. Identification, susceptibility, and detection of integron-gene cassettes of *Arcanobacterium pyogenes* in bovine endometritis. *J. Dairy Sci.* **2009**, *92*, 3659–3666. [CrossRef] [PubMed]

40. Zhao, K.L.; Liu, Y.; Zhang, X.Y.; Palahati, P.; Wang, H.N.; Yue, B.S. Detection and characterization of antibiotic resistance genes in *Arcanobacterium pyogenes* strains from abscesses of forest musk deer. *J. Med. Microbiol.* **2011**, *60*, 1820–1826. [CrossRef] [PubMed]

41. Zhang, D.X.; Tian, K.; Han, L.M.; Wang, Q.X.; Liu, Y.C.; Tian, C.L.; Liu, M.C. Resistance to β-lactam antibiotic may influence nanH gene expression in Trueperella pyogenes isolated from bovine endometritis. *Microb. Pathog.* **2014**, *71–72*, 20–24. [CrossRef] [PubMed]

42. Zhang, D.; Zhao, J.; Wang, Q.; Liu, Y.; Tian, C.; Zhao, Y.; Yu, L.; Liu, M. *Trueperella pyogenes* isolated from dairy cows with endometritis in Inner Mongolia, China: Tetracycline susceptibility and tetracycline-resistance gene distribution. *Microb. Pathog.* **2017**, *105*, 51–56. [CrossRef] [PubMed]

43. Alkasir, R.; Wang, J.; Gao, J.; Ali, T.; Zhang, L.; Szenci, O.; Bajcsy, Á.C.; Han, B. Properties and antimicrobial susceptibility of *Trueperella pyogenes* isolated from bovine mastitis in China. *Acta Vet. Hung.* **2016**, *64*, 1–12. [CrossRef] [PubMed]

44. Yoshimura, H.; Kojima, A.; Ishimaru, M. Antimicrobial susceptibility of *Arcanobacterium pyogenes* isolated from cattle and pigs. *J. Vet. Med. B Infect. Dis. Vet. Public Health* **2000**, *47*, 139–143. [CrossRef] [PubMed]

45. Ishiyama, D.; Mizomoto, T.; Ueda, C.; Takagi, N.; Shimizu, N.; Matsuura, Y.; Makuuchi, Y.; Watanabe, A.; Shinozuka, Y.; Kawai, K. Factors affecting the incidence and outcome of *Trueperella pyogenes* mastitis in cows. *J. Vet. Med. Sci.* **2017**, *79*, 626–631. [CrossRef] [PubMed]

46. Ribeiro, M.G.; Risseti, R.M.; Bolanõs, C.A.D.; Caffaro, K.A.; de Morais, A.C.B.; Lara, G.H.B.; Zamprogna, T.O.; Paes, A.C.; Listoni, F.J.P.; Franco, M.M.J. *Trueperella pyogenes* multispecies infections in domestic animals: A retrospective study of 144 cases (2002 to 2012). *Vet. Q.* **2015**, *35*, 82–87. [CrossRef]

47. Trinh, H.T.; Billington, S.J.; Field, A.C.; Songer, J.G.; Jost, B.H. Susceptibility of *Arcanobacterium pyogenes* from different sources to tetracycline, macrolide and lincosamide antimicrobial agents. *Vet. Microbiol.* **2002**, *85*, 353–359. [CrossRef]

48. Jost, B.H.; Billington, S.J. *Arcanobacterium pyogenes*: Molecular pathogenesis of an animal opportunist. *Antonie Van Leeuwenhoek* **2005**, *88*, 87–102. [CrossRef] [PubMed]

49. Santos, T.M.; Caixeta, L.S.; Machado, V.S.; Rauf, A.K.; Gilbert, R.O.; Bicalho, R.C. Antimicrobial resistance and presence of virulence factor genes in *Arcanobacterium pyogenes* isolated from the uterus of postpartum dairy cows. *Vet. Microbiol.* **2010**, *145*, 84–89. [CrossRef]

50. Bicalho, M.L.; Lima, F.S.; Machado, V.S.; Meira, E.B.; Ganda, E.K.; Foditsch, C.; Bicalho, R.C.; Gilbert, R.O. Associations among *Trueperella pyogenes*, endometritis diagnosis, and pregnancy outcomes in dairy cows. *Theriogenology* **2016**, *85*, 267–274. [CrossRef]

51. Markowska-Daniel, I.; Urbaniak, K.; Stępniewska, K.; Pejsak, Z. Antibiotic susceptibility of bacteria isolated from respiratory tract of pigs in Poland between 2004 and 2008. *Pol. J. Vet. Sci.* **2010**, *13*, 29–36.

52. Feßler, A.T.; Schwarz, S. Antimicrobial Resistance in *Corynebacterium* spp., *Arcanobacterium* spp., and *Trueperella pyogenes*. *Microbiol. Spectr.* **2017**, *5*. [CrossRef]

53. Lavín, S.; Ruiz-Bascarán, M.; Marco, I.; Abarca, M.L.; Crespo, M.J.; Franch, J. Foot infections associated with *Arcanobacterium pyogenes* in free-living fallow deer (*Dama dama*). *J. Wildl. Dis.* **2004**, *40*, 607–611. [CrossRef] [PubMed]

54. Belser, E.H.; Cohen, B.S.; Keeler, S.P.; Killmaster, C.H.; Bowers, J.W.; Miller, K.V. Epethelial presence of *Trueperella pyogenes* predicts site-level presence of cranial abscess disease in white-tailed deer (*Odocoileus virginianus*). *PLoS ONE* **2015**, *10*, e0120028. [CrossRef] [PubMed]

55. Bicalho, M.L.; Machado, V.S.; Oikonomou, G.; Gilbert, R.O.; Bicalho, R.C. Association between virulence factors of *Escherichia coli*, *Fusobacterium necrophorum*, and *Arcanobacterium pyogenes* and uterine diseases of dairy cows. *Vet. Microbiol.* **2012**, *157*, 125–131. [CrossRef] [PubMed]

56. Ibrahim, M.; Peter, S.; Wagener, K.; Drillich, M.; Ehling-Schulz, M.; Einspanier, R.; Gabler, C. Bovine Endometrial Epithelial Cells Scale Their Pro-inflammatory Response In vitro to Pathogenic *Trueperella pyogenes* Isolated from the Bovine Uterus in a Strain-Specific Manner. *Front. Cell. Infect. Microbiol.* **2017**, *7*, 264. [CrossRef] [PubMed]

57. Rudnick, S.T.; Jost, B.H.; Billington, S.J. Transcriptional regulation of pyolysin production in the animal pathogen, *Arcanobacterium pyogenes*. *Vet. Microbiol.* **2008**, *132*, 96–104. [CrossRef] [PubMed]

58. Kontturi, M.; Junni, R.; Simojoki, H.; Malinen, E.; Seuna, E.; Klitgaard, K.; Kujala-Wirth, M.; Soveri, T.; Pelkonen, S. Bacterial species associated with interdigital phlegmon outbreaks in Finnish dairy herds. *BMC Vet. Res.* **2019**, *29*, 44. [CrossRef] [PubMed]

59. Tadepalli, S.; Narayanan, S.K.; Stewart, G.C.; Chengappa, M.M.; Nagaraja, T.G. *Fusobacterium necrophorum*: A ruminal bacterium that invades liver to cause abscesses in cattle. *Anaerobe* **2009**, *15*, 36–43. [CrossRef] [PubMed]

60. Scanlan, C.M.; Hathcock, T.L. Bovine rumenitis-liver abscess complex: A bacteriological review. *Cornell Vet.* **1983**, *73*, 288–297. [PubMed]

61. Roberts, D.S. The pathogenic synergy of *Fusiformis necrophorus* and *Corynebacterium pyogenes*. II. The response of *F. necrophorus* to a filterable product of *C. pyogenes*. *Br. J. Exp. Pathol.* **1967**, *48*, 674–679.

62. Azawi, O.I.; Al-Abidy, H.F.; Ali, A.J. Pathological and bacteriological studies of hydrosalpinx in buffaloes. *Reprod. Domest. Anim.* **2010**, *45*, 416–420. [CrossRef]

63. de Boer, M.; Heuer, C.; Hussein, H.; McDougall, S. Minimum inhibitory concentrations of selected antimicrobials against *Escherichia coli* and *Trueperella pyogenes* of bovine uterine origin. *J. Dairy Sci.* **2015**, *98*, 4427–4438. [CrossRef] [PubMed]

64. Zhao, K.; Li, W.; Huang, T.; Song, X.; Zhang, X.; Yue, B. Comparative transcriptome analysis of *Trueperella pyogenes* reveals a novel antimicrobial strategy. *Arch. Microbiol.* **2017**, *199*, 649–655. [CrossRef] [PubMed]

65. Huang, T.; Song, X.; Zhao, K.; Jing, J.; Shen, Y.; Zhang, X.; Yue, B. Quorum-sensing molecules N-acyl homoserine lactones inhibit *Trueperella pyogenes* infection in mouse model. *Vet. Microbiol.* **2018**, *213*, 89–94. [CrossRef] [PubMed]

66. Malinowski, E.; Lassa, H.; Markiewicz, H.; Kaptur, M.; Nadolny, M.; Niewitecki, W.; Ziętara, J. Sensitivity to antibiotics of *Arcanobacterium pyogenes* and *Escherichia coli* from the uteri of cows with metritis/endometritis. *Vet. J.* **2011**, *187*, 234–238. [CrossRef] [PubMed]

67. Zambrano-Nava, S.; Boscán-Ocando, J.; Nava, J. Normal bacterial flora from vaginas of Criollo Limonero cows. *Trop. Anim. Health Prod.* **2011**, *43*, 291–294. [CrossRef] [PubMed]

68. Amos, M.R.; Healey, G.D.; Goldstone, R.J.; Mahan, S.M.; Düvel, A.; Schuberth, H.J.; Sandra, O.; Zieger, P.; Dieuzy-Labaye, I.; Smith, D.G.; et al. Differential endometrial cell sensitivity to a cholesterol-dependent cytolysin links *Trueperella pyogenes* to uterine disease in cattle. *Biol. Reprod.* **2014**, *90*, 54. [CrossRef]

69. Griffin, S.; Healey, G.D.; Sheldon, I.M. Isoprenoids increase bovine endometrial stromal cell tolerance to the cholesterol-dependent cytolysin from *Trueperella pyogenes*. *Biol. Reprod.* **2018**, *99*, 749–760. [CrossRef]

70. Lima, F.S.; Greco, L.F.; Bisinotto, R.S.; Ribeiro, E.S.; Martinez, N.M.; Thatcher, W.W.; Santos, J.E.; Reinhard, M.K.; Galvão, K.N. Effects of intrauterine infusion of *Trueperella pyogenes* on endometrial mRNA expression of proinflammatory cytokines and luteolytic cascade genes and their association with luteal life span in dairy cows. *Theriogenology* **2015**, *84*, 1263–1272. [CrossRef]

71. Wagener, K.; Grunert, T.; Prunner, I.; Ehling-Schulz, M.; Drillich, M. Dynamics of uterine infections with *Escherichia coli*, *Streptococcus uberis* and *Trueperella pyogenes* in post-partum dairy cows and their association with clinical endometritis. *Vet. J.* **2014**, *202*, 527–532. [CrossRef]

72. Carneiro, L.C.; Cronin, J.G.; Sheldon, I.M. Mechanisms linking bacterial infections of the bovine endometrium to disease and infertility. *Reprod. Biol.* **2016**, *16*, 1–7. [CrossRef]

73. Bonnett, B.N.; Martin, S.W.; Meek, A.H. Associations of clinical findings, bacteriological and histological results of endometrial biopsy with reproductive performance of postpartum dairy cows. *Prev. Vet. Med.* **1993**, *15*, 205–220. [CrossRef]

74. Sheldon, I.M.; Lewis, G.S.; LeBlanc, S.; Gilbert, R.O. Defining postpartum uterine disease in cattle. *Theriogenology* **2006**, *65*, 1516–1530. [CrossRef] [PubMed]

75. Danesh Mesgaran, S.; Gärtner, M.A.; Wagener, K.; Drillich, M.; Ehling-Schulz, M.; Einspanier, R.; Gabler, C. Different inflammatory responses of bovine oviductal epithelial cells in vitro to bacterial species with distinct pathogenicity characteristics and passage number. *Theriogenology* **2018**, *106*, 237–246. [CrossRef] [PubMed]

76. Bonnett, B.N.; Martin, S.W.; Gannon, V.P.; Miller, R.B.; Etherington, W.G. Endometrial biopsy in Holstein-Friesian dairy cows. III. Bacteriological analysis and correlations with histological findings. *Can. J. Vet. Res.* **1991**, *55*, 168–173. [PubMed]

77. Sheldon, I.M.; Cronin, J.; Goetze, L.; Donofrio, G.; Schuberth, H.J. Defining postpartum uterine disease and the mechanisms of infection and immunity in the female reproductive tract in cattle. *Biol. Reprod.* **2009**, *81*, 1025–1032. [CrossRef] [PubMed]

78. Jaureguiberry, M.; Madoz, L.V.; Giuliodori, M.J.; Wagener, K.; Prunner, I.; Grunert, T.; Ehling-Schulz, M.; Drillich, M.; de la Sota, R.L. Identification of *Escherichia coli* and *Trueperella pyogenes* isolated from the uterus of dairy cows using routine bacteriological testing and Fourier transform infrared spectroscopy. *Acta Vet. Scand.* **2016**, *58*, 81. [CrossRef] [PubMed]

79. Sheldon, I.M.; Williams, E.J.; Miller, A.N.; Nash, D.M.; Herath, S. Uterine diseases in cattle after parturition. *Vet. J.* **2008**, *176*, 115–121. [CrossRef]

80. Prunner, I.; Wagener, K.; Pothmann, H.; Ehling-Schulz, M.; Drillich, M. Risk factors for uterine diseases on small- and medium-sized dairy farms determined by clinical, bacteriological, and cytological examinations. *Theriogenology* **2014**, *82*, 857–865. [CrossRef]

81. Seno, N.; Azuma, R. A study on heifer mastitis in Japan and its causative microorganisms. *Natl. Inst. Anim. Health Q. (Tokyo)* **1983**, *23*, 82–91.

82. Quinn, A.K.; Vermunt, J.J.; Twiss, D.P. *Arcanobacterium pyogenes* mastitis in a 18-month-old heifer. *N. Z. Vet. J.* **2002**, *50*, 167–168. [CrossRef]

83. Ericsson Unnerstad, H.; Lindberg, A.; Persson Waller, K.; Ekman, T.; Artursson, K.; Nilsson-Ost, M.; Bengtsson, B. Microbial aetiology of acute clinical mastitis and agent-specific risk factors. *Vet. Microbiol.* **2009**, *137*, 90–97. [CrossRef] [PubMed]

84. Hertl, J.A.; Schukken, Y.H.; Welcome, F.L.; Tauer, L.W.; Gröhn, Y.T. Effects of pathogen-specific clinical mastitis on probability of conception in Holstein dairy cows. *J. Dairy Sci.* **2014**, *97*, 6942–6954. [CrossRef] [PubMed]

85. Hillerton, J.E.; Bramley, A.J.; Watson, C.A. The epidemiology of summer mastitis: A survey of clinical cases. *Br. Vet. J.* **1987**, *143*, 520–530. [CrossRef]

86. Chirico, J.; Jonsson, P.; Kjellberg, S.; Thomas, G. Summer mastitis experimentally induced by *Hydrotaea irritans* exposed to bacteria. *Med. Vet. Entomol.* **1997**, *11*, 187–192. [CrossRef] [PubMed]

87. Madsen, M.; Aalbaek, B.; Hansen, J.W. Comparative bacteriological studies on summer mastitis in grazing cattle and pyogenes mastitis in stabled cattle in Denmark. *Vet. Microbiol.* **1992**, *32*, 81–88. [CrossRef]

88. Madsen, M.; Høi Sørensen, G.; Aalbaek, B.; Hansen, J.W.; Bjørn, H. Summer mastitis in heifers: Studies on the seasonal occurrence of *Actinomyces pyogenes*, *Peptostreptococcus indolicus* and *Bacteroidaceae* in clinically healthy cattle in Denmark. *Vet. Microbiol.* **1992**, *30*, 243–255. [CrossRef]

89. Nagaraja, T.G.; Chengappa, M.M. Liver abscesses in feedlot cattle: A review. *J. Anim. Sci.* **1998**, *76*, 287–298. [CrossRef]

90. Nagaraja, T.G.; Lechtenberg, K.F. Liver abscesses in feedlot cattle. *Vet. Clin. N. Am. Food Anim. Pract.* **2007**, *23*, 351–369. [CrossRef]

91. Amachawadi, R.G.; Nagaraja, T.G. Liver abscesses in cattle: A review of incidence in Holsteins and of bacteriology and vaccine approaches to control in feedlot cattle. *J. Anim. Sci.* **2016**, *94*, 1620–1632. [CrossRef]

92. Oberbauer, A.M.; Berry, S.L.; Belanger, J.M.; McGoldrick, R.M.; Pinos-Rodriquez, J.M.; Famula, T.R. Determining the heritable component of dairy cattle foot lesions. *J. Dairy Sci.* **2013**, *96*, 605–613. [CrossRef]

93. Kontturi, M.; Kujala, M.; Junni, R.; Malinen, E.; Seuna, E.; Pelkonen, S.; Soveri, T.; Simojoki, H. Survey of interdigital phlegmon outbreaks and their risk factors in free stall dairy herds in Finland. *Acta Vet. Scand.* **2017**, *59*, 46. [CrossRef] [PubMed]

94. Bay, V.; Griffiths, B.; Carter, S.; Evans, N.J.; Lenzi, L.; Bicalho, R.C.; Oikonomou, G. 16S rRNA amplicon sequencing reveals a polymicrobial nature of complicated claw horn disruption lesions and interdigital phlegmon in dairy cattle. *Sci. Rep.* **2018**, *8*, 15529. [CrossRef] [PubMed]

95. Power, H.T.; Rebhun, W.C. Bacterial endocarditis in adult dairy cattle. *J. Am. Vet. Med. Assoc.* **1983**, *182*, 806–808. [PubMed]

96. Ertas, H.B.; Kilic, A.; Özbey, G.; Muz, A. Isolation of *Arcanobacterium* (*Actinomyces*) *pyogenes* from abscessed cattle kidney and identification by PCR, Turk. *J. Vet. Anim. Sci.* **2005**, *29*, 455–459.

97. Hinton, M. *Corynebacterium pyogenes* and bovine abortion. *J. Hyg. Camb.* **1974**, *72*, 365–368. [CrossRef]

98. Turner, G.V. A microbiological study of polyarthritis in slaughter pigs. *J. S. Afr. Vet. Assoc.* **1982**, *53*, 99–101. [PubMed]

99. Høie, S.; Falk, K.; Lium, B.M. An abattoir survey of pneumonia and pleuritis in slaughter weight swine from 9 selected herds. IV. Bacteriological findings in chronic pneumonic lesions. *Acta Vet. Scand.* **1991**, *32*, 395–402.

100. Hariharan, H.; MacDonald, J.; Carnat, B.; Bryenton, J.; Heaney, S. An investigation of bacterial causes of arthritis in slaughter hogs. *J. Vet. Diagn. Investig.* **1992**, *4*, 28–30. [CrossRef]

101. Iwamatsu, S.; Morio, A.; Watanabe, K.; Ikeo, T.; Yamashita, T.; Yoshino, H. Suppurative cerebral lesions caused by infection with *Actinomyces pyogenes* in breeding swine. *J. Jpn. Vet. Med. Assoc.* **1986**, *39*, 238–242. [CrossRef]

102. Dong, W.L.; Kong, L.C.; Wang, Y.; Gou, C.L.; Xu, B.; Ma, H.X.; Gao, Y.H. Aminoglycoside resistance of *Trueperella pyogenes* isolated from pigs in China. *J. Vet. Med. Sci.* **2017**, *79*, 1836–1839. [CrossRef]

103. Jarosz, Ł.S.; Gradzki, Z.; Kalinowski, M. Trueperella pyogenes infections in swine: Clinical course and pathology. *Pol. J. Vet. Sci.* **2014**, *17*, 395–404. [CrossRef] [PubMed]

104. Pejsak, Z.; Markowska-Daniel, I.; Samorek, M.; Truszczyński, M. Wykrywanie i ocena właściwości krajowych izolatów *Arcanobacterium pyogenes* wyosobnionych od świń. *Med. Weter.* **2006**, *62*, 781–784.

105. Tadayon, R.A.; Cheema, A.H.; Muhammed, S.I. Microorganisms associated with abscesses of sheep and goats in the south of Iran. *Am. J. Vet. Res.* **1980**, *41*, 798–802. [PubMed]

106. Al Dughaym, A.M. Isolation of *Serratia, Arcanobacterium* and *Burkholderia* species from visceral and cutaneous abscesses in four emaciated ewes. *Vet. Rec.* **2004**, *155*, 425–426. [CrossRef] [PubMed]

107. Lin, C.C.; Chen, T.H.; Shyu, C.L.; Su, N.Y.; Chan, J.P. Disseminated abscessation complicated with bone marrow abscess caused by *Arcanobacterium pyogenes* in a goat. *J. Vet. Med. Sci.* **2010**, *72*, 1089–1092. [CrossRef]

108. Barbour, E.K.; Itani, H.H.; Shaib, H.A.; Saade, M.F.; Sleiman, F.T.; Nour, A.A.; Harakeh, S. Koch's postulate of *Arcanobacterium pyogenes* and its immunogenicity in local and imported Saanen goats. *Vet. Ital.* **2010**, *46*, 319–327.

109. Raadsma, H.W.; Egerton, J.R. A review of footrot in sheep: Aetiology, risk factors and control methods. *Livest. Sci.* **2013**, *156*, 106–114. [CrossRef]

110. Wani, A.H.; Verma, S.; Sharma, M.; Wani, A. Infectious lameness among migratory sheep and goats in north-west India, with particular focus on anaerobes. *Rev. Sci. Tech.* **2015**, *34*, 855–867. [CrossRef]

111. Stefańska, I.; Rzewuska, M.; Binek, M. *Corynebacterium pseudotuberculosis*—Zakażenia u zwierząt. *Postęp. Mikrobiol.* **2007**, *46*, 101–112.

112. Lavin, S.; Marco, I.; Franch, J.; Abarca, L. Report of a case of pyogenic arthritis associated with *Actinomyces pyogenes* in a chamois (*Rupicapra pyrenaica*). *Zentralbl. Veterinarmed. B* **1998**, *45*, 251–253. [CrossRef]

113. Seifi, H.A.; Saifzadeh, S.; Farshid, A.A.; Rad, M.; Farrokhi, F. Mandibular pyogranulomatous osteomyelitis in a Sannen goat. *J. Vet. Med. A Physiol. Pathol. Clin. Med.* **2003**, *50*, 219–221. [CrossRef] [PubMed]

114. Kidanemariam, A.; Gouws, J.; van Vuuren, M.; Gummow, B. Ulcerative balanitis and vulvitis of Dorper sheep in South Africa: A study on its aetiology and clinical features. *J. S. Afr. Vet. Assoc.* **2005**, *76*, 197–203. [CrossRef] [PubMed]

115. Billington, S.J.; Post, K.W.; Jost, B.H. Isolation of *Arcanobacterium* (*Actinomyces*) *pyogenes* from cases of feline otitis externa and canine cystitis. *J. Vet. Diagn. Investig.* **2002**, *14*, 159–162. [CrossRef] [PubMed]

116. Hesselink, J.W.; van den Tweel, J.G. Hypertrophic osteopathy in a dog with a chronic lung abscess. *J. Am. Vet. Med. Assoc.* **1990**, *196*, 760–762.

117. Dickie, C.W. Feline pyothorax caused by a *Borrelia*-like organism and *Corynebacterium pyogenes*. *J. Am. Vet. Med. Assoc.* **1979**, *174*, 516–517. [PubMed]

118. Wareth, G.; El-Diasty, M.; Melzer, F.; Murugaiyan, J.; Abdulmawjood, A.; Sprague, L.D.; Neubauer, H. *Trueperella pyogenes* and *Brucella abortus* coinfection in a dog and a cat on a dairy farm in Egypt with recurrent cases of mastitis and abortion. *Vet. Med. Int.* **2018**, *2018*, 2056436. [CrossRef] [PubMed]

119. Rampacci, E.; Passamonti, F.; Bottinelli, M.; Stefanetti, V.; Cercone, M.; Nannarone, S.; Gialletti, R.; Beccati, F.; Coletti, M.; Pepe, M. Umbilical infections in foals: Microbiological investigation and management. *Vet. Rec.* **2017**, *180*, 543. [CrossRef]

120. Shahbazfar, A.A.; Kolahian, S.; Ashrafi Helan, J.; Mohammadpour, H. Multi abscessation with multinodular abscesses in a New Zealand white rabbit (*Oryctolagus cuniculus*) following *Arcanobacterium pyogenes* infection. *Revue Méd. Vét.* **2013**, *164*, 23–26.

121. Barbour, E.K.; Brinton, M.K.; Caputa, A.; Johnson, J.B.; Poss, P.E. Characteristics of *Actinomyces pyogenes* involved in lameness of male turkeys in north-central United States. *Avian Dis.* **1991**, *35*, 192–196. [CrossRef] [PubMed]

122. Brinton, M.K.; Schellberg, L.C.; Johnson, J.B.; Frank, R.K.; Halvorson, D.A.; Newman, J.A. Description of osteomyelitis lesions associated with *Actinomyces pyogenes* infection in the proximal tibia of adult male turkeys. *Avian Dis.* **1993**, *37*, 259–262. [CrossRef]

123. Priya, P.M.; Abhinay, G.; Ambily, R.; Siju, J. A Unique Case of *Trueperella Pyogenes* Causing Hepatic Abscesses in Pigeons. *Int. J. Vet. Med. Res. Rep.* **2013**, 1–3. [CrossRef]

124. Cohen, B.S.; Belser, E.H.; Keeler, S.P.; Yabsley, M.J.; Miller, K.V. A Headache from Our Past? Intracranial Abscess Disease, Virulence Factors of *Trueperella Pyogenes*, And A Legacy of Translocating White-Tailed Deer (*Odocoileus Virginianus*). *J. Wildl. Dis.* **2018**, *54*, 671–679. [CrossRef]

125. Karns, G.R.; Lancia, R.A.; Deperno, C.S.; Conner, M.C.; Stoskopf, M.K. Intracranial abscessation as a natural mortality factor for adult male white-tailed deer (*Odocoileus virginianus*) in Kent County, Maryland, USA. *J. Wildl. Dis.* **2009**, *45*, 196–200. [CrossRef]

126. Turner, M.M.; Deperno, C.S.; Conner, M.C.; Eyler, T.B.; Lancia, R.A.; Klaver, R.W.; Stoskopf, M.K. Habitat, wildlife, and one health: *Arcanobacterium pyogenes* in Maryland and Upper Eastern Shore white-tailed deer populations. *Infect. Ecol. Epidemiol.* **2013**, *3*, 19175. [CrossRef]

127. Cohen, B.S.; Belser, E.H.; Keeler, S.P.; Yabsley, M.J.; Miller, K.V. Isolation and genotypic characterization of *Trueperella* (*Arcanobacterium*) *pyogenes* recovered from active cranial abscess infections of male white-tailed deer (*Odocoileus virginianus*). *J. Zoo Wildl. Med.* **2015**, *46*, 62–67. [CrossRef]

128. Chirino-Trejo, M.; Woodbury, M.R.; Huang, F. Antibiotic sensitivity and biochemical characterization of *Fusobacterium* spp. and *Arcanobacterium pyogenes* isolated from farmed white-tailed deer (*Odocoileus virginianus*) with necrobacillosis. *J. Zoo Wildl. Med.* **2003**, *34*, 262–268. [CrossRef]

129. Tell, L.A.; Brooks, J.W.; Lintner, V.; Matthews, T.; Kariyawasam, S. Antimicrobial susceptibility of *Arcanobacterium pyogenes* isolated from the lungs of white-tailed deer (*Odocoileus virginianus*) with pneumonia. *J. Vet. Diagn. Investig.* **2011**, *23*, 1009–1013. [CrossRef]

130. Nettles, V.F.; Quist, C.F.; Lopez, R.R.; Wilmers, T.J.; Frank, P.; Roberts, W.; Chitwood, S.; Davidson, W.R. Morbidity and mortality factors in key deer (*Odocoileus virginianus clavium*). *J. Wildl. Dis.* **2002**, *38*, 685–692. [CrossRef]

131. Wickhorst, J.P.; Hassan, A.A.; Sheet, O.H.; Eisenberg, T.; Sammra, O.; Alssahen, M.; Lämmler, C.; Prenger-Berninghoff, E.; Zschöck, M.; Timke, M.; et al. *Trueperella pyogenes* isolated from a brain abscess of an adult roebuck (*Capreolus capreolus*). *Folia Microbiol. (Praha)* **2018**, *63*, 17–22. [CrossRef]

132. Badger, S.B. Infectious epiphysitis and valvular endocarditis in a red deer (*Cervus elaphus*). *N. Z. Vet. J.* **1982**, *30*, 17–18. [CrossRef]

133. Muñoz Gutiérrez, J.F.; Sondgeroth, K.S.; Williams, E.S.; Montgomery, D.L.; Creekmore, T.E.; Miller, M.M. Infectious keratoconjunctivitis in free-ranging mule deer in Wyoming: A retrospective study and identification of a novel alphaherpesvirus. *J. Vet. Diagn. Investig.* **2018**, *30*, 663–670. [CrossRef]

134. Zhao, K.; Tian, Y.; Yue, B.; Wang, H.; Zhang, X. Virulence determinants and biofilm production among *Trueperella pyogenes* recovered from abscesses of captive forest musk deer. *Arch. Microbiol.* **2013**, *195*, 203–209. [CrossRef]

135. Rzewuska, M.; Stefańska, I.; Osińska, B.; Kizerwetter-Świda, M.; Chrobak, D.; Kaba, J.; Bielecki, W. Phenotypic characteristics and virulence genotypes of *Trueperella* (*Arcanobacterium*) *pyogenes* strains isolated from European bison (*Bison bonasus*). *Vet. Microbiol.* **2012**, *160*, 69–76. [CrossRef]

136. Portas, T.J.; Bryant, B.R. Morbidity and mortality associated with *Arcanobacterium pyogenes* in a group of captive blackbuck (*Antilope cervicapra*). *J. Zoo Wildl. Med.* **2005**, *36*, 286–289. [CrossRef]

137. Twomey, D.F.; Boon, J.D.; Sayers, G.; Schock, A. *Arcanobacterium pyogenes* septicemia in a southern pudu (*Pudu puda*) following uterine prolapse. *J. Zoo Wildl. Med.* **2010**, *41*, 158–160. [CrossRef]

138. Franzen, D.; Lamberski, N.; Zuba, J.; Richardson, G.L.; Fischer, A.T.; Rantanen, N.W. Diagnosis and Medical and Surgical Management of Chronic Infectious Fibrinous Pleuritis in an Okapi (*Okapia Johnstoni*). *J. Zoo Wildl. Med.* **2015**, *46*, 427–430. [CrossRef]

139. Zarnke, R.L.; Schlater, L.K. Abscesses in a free-ranging bison in Alaska. *J. Wildl. Dis.* **1984**, *20*, 151–152.

140. Besser, T.E.; Frances Cassirer, E.; Highland, M.A.; Wolff, P.; Justice-Allen, A.; Mansfield, K.; Davis, M.A.; Foreyt, W. Bighorn sheep pneumonia: Sorting out the cause of a polymicrobial disease. *Prev. Vet. Med.* **2013**, *108*, 85–93. [CrossRef]

141. Moustafa, A.M. First observation of camel (*Camelus dromedarius*) lymphadenitis in Libya. A case report. *Rev. Elev. Med. Vet. Pays. Trop.* **1994**, *47*, 313–314.

142. Wareth, G.; Murugaiyan, J.; Khater, D.F.; Moustafa, S.A. Subclinical pulmonary pathogenic infection in camels slaughtered in Cairo, Egypt. *J. Infect. Dev. Ctries.* **2014**, *8*, 909–913. [CrossRef]

143. Ali, A.; Derar, R.; Al-Sobayil, F.; Al-Hawas, A.; Hassanein, K. A retrospective study on clinical findings of 7300 cases (2007–2014) of barren female dromedaries. *Theriogenology* **2015**, *84*, 452–456. [CrossRef]

144. Tarazi, Y.H.; Al-Ani, F.K. An outbreak of dermatophilosis and caseous lymphadenitis mixed infection in camels (*Camelus dromedaries*) in Jordan. *J. Infect. Dev. Ctries.* **2016**, *10*, 506–511. [CrossRef]

145. Aschfalk, A.; Josefsen, T.D.; Steingass, H.; Müller, W.; Goethe, R. Crowding and winter emergency feeding as predisposing factors for kerato-conjunctivitis in semi-domesticated reindeer in Norway. *Dtsch. Tierarztl. Wochenschr.* **2003**, *110*, 295–298.

146. Tryland, M.; das Neves, C.G.; Sunde, M.; Mørk, T. Cervid herpesvirus 2, the primary agent in an outbreak of infectious keratoconjunctivitis in semidomesticated reindeer. *J. Clin. Microbiol.* **2009**, *47*, 3707–3713. [CrossRef]

147. Eisenberg, T.; Nagib, S.; Hijazin, M.; Alber, J.; Lämmler, C.; Hassan, A.A.; Timke, M.; Kostrzewa, M.; Prenger-Berninghoff, E.; Schauerte, N.; et al. *Trueperella pyogenes* as cause of a facial abscess in a grey slender loris (*Loris lydekkerianus nordicus*)—A case report. *Berl. Munchener Tierarztliche Wochenschr.* **2012**, *125*, 407–410.

148. Nagib, S.; Glaeser, S.P.; Eisenberg, T.; Sammra, O.; Lämmler, C.; Kämpfer, P.; Schauerte, N.; Geiger, C.; Kaim, U.; Prenger-Berninghoff, E.; et al. Fatal infection in three Grey Slender Lorises (*Loris lydekkerianus nordicus*) caused by clonally related *Trueperella pyogenes*. *BMC Vet. Res.* **2017**, *13*, 273. [CrossRef]

149. Ülbegi-Mohyla, H.; Hijazin, M.; Alber, J.; Lämmler, C.; Hassan, A.A.; Abdulmawjood, A.; Prenger-Berninghoff, E.; Weiss, R.; Zschöck, M. Identification of *Arcanobacteriumpyogenes* isolated by post mortem examinations of a bearded dragon and a gecko by phenotypic and genotypic properties. *J. Vet. Sci.* **2010**, *11*, 265–267. [CrossRef]

150. Funke, G.; von Graevenitz, A.; Clarridge, J.E.; Bernard, K.A. Clinical microbiology of coryneform bacteria. *Clin. Microbiol. Rev.* **1997**, *10*, 125–159. [CrossRef]

151. Jootar, P.; Gherunpong, V.; Saitanu, K. *Corynebacterium pyogenes* endocarditis. Report of a case with necropsy and review of the literature. *J. Med. Assoc. Thail.* **1978**, *61*, 596–601.

152. Kavitha, K.; Latha, R.; Udayashankar, C.; Jayanthi, K.; Oudeacoumar, P. Three cases of *Arcanobacterium pyogenes*-associated soft tissue infection. *J. Med. Microbiol.* **2010**, *59*, 736–739. [CrossRef]

153. Plamondon, M.; Martinez, G.; Raynal, L.; Touchette, M.; Valiquette, L. A fatal case of *Arcanobacterium pyogenes* endocarditis in a man with no identified animal contact: Case report and review of the literature. *Eur. J. Clin. Microbiol. Infect. Dis.* **2007**, *26*, 663–666. [CrossRef]

154. Chesdachai, S.; Larbcharoensub, N.; Chansoon, T.; Chalermsanyakorn, P.; Santanirand, P.; Chotiprasitsakul, D.; Ratanakorn, D.; Boonbaichaiyapruck, S. *Arcanobacterium pyogenes* endocarditis: A case report and literature review. *Southeast. Asian. J. Trop. Med. Public Health* **2014**, *45*, 142–148.

155. Andy Tang, S.O.; Leong, T.S.; Ruixin, T.; Chua, H.H.; Chew, L.P. Thrombotic thrombocytopenic purpura-like syndrome associated with *Arcanobacterium pyogenes* endocarditis in a post-transplant patient: A case report. *Med. J. Malaysia.* **2018**, *73*, 344–346.

156. Kotrajaras, R.; Tagami, H. *Corynebacterium pyogenes*. Its pathogenic mechanism in epidemic leg ulcers in Thailand. *Int. J. Dermatol.* **1987**, *26*, 45–50. [CrossRef]

157. Drancourt, M.; Oulès, O.; Bouche, V.; Peloux, Y. Two cases of *Actinomyces pyogenes* infection in humans. *Eur. J. Clin. Microbiol. Infect. Dis.* **1993**, *12*, 55–57. [CrossRef] [PubMed]

158. Hermida Amejeiras, A.; Romero Jung, P.; Cabarcos Ortiz De Barrón, A.; Treviño Castallo, M. One case of pneumonia with *Arcanobacterium pyogenes*. *An. Med. Int.* **2004**, *21*, 334–336.

159. Nicholson, P.; Kiely, P.; Street, J.; Mahalingum, K. Septic arthritis due to *Actinomyces pyogenes*. *Injury* **1998**, *29*, 640–642. [CrossRef]

160. Lynch, M.; O'Leary, J.; Murnaghan, D.; Cryan, B. *Actinomyces pyogenes* septic arthritis in a diabetic farmer. *J. Infect.* **1998**, *37*, 71–73. [CrossRef]

161. Levy, C.E.; Pedro, R.J.; von Nowakonski, A.; Holanda, L.M.; Brocchi, M.; Ramo, M.C. *Arcanobacterium pyogenes* sepsis in farmer, Brazil. *Emerg. Infect. Dis.* **2009**, *15*, 1131–1132. [CrossRef]

162. Gahrn-Hansen, B.; Frederiksen, W. Human infections with *Actinomyces pyogenes* (*Corynebacterium pyogenes*). *Diagn. Microbiol. Infect. Dis.* **1992**, *15*, 349–354. [CrossRef]

163. Sullivan, D.C.; Chapman, S.W. Bacteria that masquerade as fungi: Actinomycosis/nocardia. *Proc. Am. Thorac. Soc.* **2010**, *7*, 216–221. [CrossRef] [PubMed]

164. Mochon, A.B.; Sussland, D.; Saubolle, M.A. Aerobic Actinomycetes of Clinical Significance. *Microbiol. Spectr.* **2016**, *4*. [CrossRef]

165. Prescott, J.F. *Rhodococcus equi*: An animal and human pathogen. *Clin. Microbiol. Rev.* **1991**, *4*, 20–34. [CrossRef] [PubMed]

166. Dorella, F.A.; Pacheco, L.G.; Oliveira, S.C.; Miyoshi, A.; Azevedo, V. *Corynebacterium pseudotuberculosis*: Microbiology, biochemical properties, pathogenesis and molecular studies of virulence. *Vet. Res.* **2006**, *37*, 201–218. [CrossRef] [PubMed]

167. Rzewuska, M.; Rodo, A.; Bielecki, W. Dermacentor reticulatus ticks as possible vectors of *Trueperella pyogenes* infection in European bison (*Bison bonasus*): Preliminary studies. *J. Comp. Pathol.* **2016**, *154*, 120. [CrossRef]

168. Jost, B.H.; Songer, J.G.; Billington, S.J. An *Arcanobacterium* (*Actinomyces*) *pyogenes* mutant deficient in production of the pore-forming cytolysin pyolysin has reduced virulence. *Infect. Immun.* **1999**, *67*, 1723–1728. [PubMed]

169. Billington, S.J.; Songer, J.G.; Jost, B.H. Molecular characterization of the pore-forming toxin, pyolysin, a major virulence determinant of *Arcanobacterium pyogenes*. *Vet. Microbiol.* **2001**, *82*, 261–274. [CrossRef]

170. Rudnick, S.T.; Jost, B.H.; Songer, J.G.; Billington, S.J. The gene encoding pyolysin, the pore-forming toxin of *Arcanobacterium pyogenes*, resides within a genomic islet flanked by essential genes. *FEMS Microbiol. Lett.* **2003**, *225*, 241–247. [CrossRef]

171. Alouf, J.E. Cholesterol-binding cytolytic protein toxins. *Int. J. Med. Microbiol.* **2000**, *290*, 351–356. [CrossRef]

172. Heuck, A.P.; Moe, P.C.; Johnson, B.B. The cholesterol-dependent cytolysin family of gram-positive bacterial toxins. *Subcell. Biochem.* **2010**, *51*, 551–577. [PubMed]

173. Ding, H.; Lämmler, C. Purification and further characterization of a haemolysin of *Actinomyces pyogenes*. *Zentralbl. Veterinarmed. B* **1996**, *43*, 179–188. [CrossRef] [PubMed]

174. Lovell, R. Further studies on the toxin of *Corynebacterium pyogenes*. *J. Pathol. Bacteriol.* **1944**, *56*, 525–529. [CrossRef]

175. Kume, T.; Tainaka, M.; Saito, H.; Hiruma, M.; Nishio, S.; Kashiwazaki, M.; Mitani, K.; Nakajima, Y. Research on experimental *Corynebacterium pyogenes* infections in pigs. *Kitasato Arch. Exp. Med.* **1983**, *56*, 119–135. [PubMed]

176. Hu, Y.; Zhang, W.; Bao, J.; Wu, Y.; Yan, M.; Xiao, Y.; Yang, L.; Zhang, Y.; Wang, J. A chimeric protein composed of the binding domains of *Clostridium perfringens* phospholipase C and *Trueperella pyogenes* pyolysin induces partial immunoprotection in a mouse model. *Res. Vet. Sci.* **2016**, *107*, 106–115. [CrossRef] [PubMed]

177. Imaizumi, K.; Serizawa, A.; Hashimoto, N.; Kaidoh, T.; Takeuchi, S. Analysis of the functional domains of *Arcanobacterium pyogenes* pyolysin using monoclonal antibodies. *Vet. Microbiol.* **2001**, *81*, 235–242. [CrossRef]

178. Jost, B.H.; Field, A.C.; Trinh, H.T.; Songer, J.G.; Billington, S.J. Tylosin resistance in *Arcanobacterium pyogenes* is encoded by an *erm* X determinant. *Antimicrob. Agents Chemother.* **2003**, *47*, 3519–3524. [CrossRef] [PubMed]

179. Zhao, K.; Liu, M.; Zhang, X.; Wang, H.; Yue, B. In vitro and in vivo expression of virulence genes in *Trueperella pyogenes* based on a mouse model. *Vet. Microbiol.* **2013**, *163*, 344–350. [CrossRef]

180. Billington, S.J.; Jost, B.H.; Cuevas, W.A. The *Arcanobacterium* (*Actinomyces*) *pyogenes* hemolysin, pyolysin, is a novel member of the thiol-activated cytolysin family. *J. Bacteriol.* **1997**, *179*, 6100–6106. [CrossRef]

181. Ikegami, M.; Hashimoto, N.; Kaidoh, T.; Sekizaki, T.; Takeuchi, S. Genetic and biochemical properties of a hemolysin (pyolysin) produced by a swine isolate of *Arcanobacterium* (*Actinomyces*) *pyogenes*. *Microbiol. Immunol.* **2000**, *44*, 1–7. [CrossRef]

182. Funk, P.G.; Staats, J.J.; Howe, M.; Nagaraja, T.G.; Chengappa, M.M. Identification and partial characterization of an *Actinomyces pyogenes* hemolysin. *Vet. Microbiol.* **1996**, *50*, 129–142. [CrossRef]

183. Billington, S.J.; Esmay, P.A.; Songer, J.G.; Jost, B.H. Identification and role in virulence of putative iron acquisition genes from *Corynebacterium pseudotuberculosis*. *FEMS Microbiol. Lett.* **2002**, *208*, 41–45. [CrossRef] [PubMed]

184. Imaizumi, K.; Matsunaga, K.; Higuchi, H.; Kaidoh, T.; Takeuchi, S. Effect of amino acid substitutions in the epitope regions of pyolysin from *Arcanobacterium pyogenes*. *Vet. Microbiol.* **2003**, *91*, 205–213. [CrossRef]

185. Pokrajac, L.; Harris, J.R.; Sarraf, N.; Palmer, M. Oligomerization and hemolytic properties of the C-terminal domain of pyolysin, a cholesterol-dependent cytolysin. *Biochem. Cell Biol.* **2013**, *91*, 59–66. [CrossRef] [PubMed]

186. Zhang, W.; Wang, H.; Wang, B.; Zhang, Y.; Hu, Y.; Ma, B.; Wang, J. Replacing the 238th aspartic acid with an arginine impaired the oligomerization activity and inflammation-inducing property of pyolysin. *Virulence* **2018**, *9*, 1112–1125. [CrossRef] [PubMed]

187. Yan, M.; Hu, Y.; Bao, J.; Xiao, Y.; Zhang, Y.; Yang, L.; Wang, J.; Zhang, W. Isoleucine 61 is important for the hemolytic activity of pyolysin of *Trueperella pyogenes*. *Vet. Microbiol.* **2016**, *182*, 196–201. [CrossRef] [PubMed]

188. Yang, L.; Zhang, Y.; Wang, H.; Ma, B.; Xu, L.; Wang, J.; Zhang, W. Identification of B-cell linear epitopes in domains 1-3 of pyolysin of *Trueperella pyogenes* using polyclonal antibodies. *Vet. Microbiol.* **2017**, *210*, 24–31. [CrossRef] [PubMed]

189. Pokrajac, L.; Baik, C.; Harris, J.R.; Sarraf, N.S.; Palmer, M. Partial oligomerization of pyolysin induced by a disulfide-tethered mutant. *Biochem. Cell Biol.* **2012**, *90*, 709–717. [CrossRef]

190. Harris, J.R.; Lewis, R.J.; Baik, C.; Pokrajac, L.; Billington, S.J.; Palmer, M. Cholesterol microcrystals and cochleate cylinders: Attachment of pyolysin oligomers and domain 4. *J. Struct. Biol.* **2011**, *173*, 38–45. [CrossRef]

191. Preta, G.; Jankunec, M.; Heinrich, F.; Griffin, S.; Sheldon, I.M.; Valincius, G. Tethered bilayer membranes as a complementary tool for functional and structural studies: The pyolysin case. *Biochim. Biophys. Acta* **2016**, *1858*, 2070–2080. [CrossRef]

192. Liu, M.; Wang, B.; Liang, H.; Ma, B.; Wang, J.; Zhang, W. Determination of the expression of three fimbrial subunit proteins in cultured *Trueperella pyogenes*. *Acta Vet. Scand.* **2018**, *60*, 53. [CrossRef]

193. Bisinotto, R.S.; Filho, J.C.O.; Narbus, C.; Machado, V.S.; Murray, E.; Bicalho, R.C. Identification of fimbrial subunits in the genome of *Trueperella pyogenes* and association between serum antibodies against fimbrial proteins and uterine conditions in dairy cows. *J. Dairy Sci.* **2016**, *99*, 3765–3776. [CrossRef] [PubMed]

194. Pietrocola, G.; Valtulina, V.; Rindi, S.; Jost, B.H.; Speziale, P. Functional and structural properties of CbpA, a collagen-binding protein from *Arcanobacterium pyogenes*. *Microbiology* **2007**, *153*, 3380–3389. [CrossRef] [PubMed]

195. Lämmler, C.; Ding, H. Characterization of Fibrinogen-binding Properties of *Actinomyces pyogenes*. *J. Vet. Med. B* **1994**, *41*, 588–596. [CrossRef]

196. Esmay, P.A.; Billington, S.J.; Link, M.A.; Songer, J.G.; Jost, B.H. The *Arcanobacterium pyogenes* collagen-binding protein, CbpA, promotes adhesion to host cells. *Infect. Immun.* **2003**, *71*, 4368–4374. [CrossRef]

197. Taylor, G. Sialidases: Structures, biological significance and therapeutic potential. *Curr. Opin. Struct. Biol.* **1996**, *6*, 830–837. [CrossRef]

198. Pettigrew, M.M.; Fennie, K.P.; York, M.P.; Daniels, J.; Ghaffar, F. Variation in the Presence of Neuraminidase Genes among *Streptococcus pneumoniae* Isolates with Identical Sequence Types. *Infect. Immun.* **2006**, *74*, 3360–3365. [CrossRef]

199. Moriyama, T.; Barksdale, L. Neuraminidase of *Corynebacterium diphtheriae*. *J. Bacteriol.* **1967**, *9*, 1565–1581.

200. Soong, G.; Muir, A.; Gomez, M.I.; Waks, J.; Reddy, B.; Planet, P.; Singh, P.K.; Kanetko, Y.; Wolfgang, M.C.; Hsiao, Y.S.; et al. Bacterial neuraminidase facilitates mucosal infection by participating in biofilm production. *J. Clin. Investig.* **2006**, *116*, 2297–2305. [CrossRef]

201. Moncla, B.J.; Braham, P.; Hillier, S.L. Sialidase (neuraminidase) activity among gram-negative anaerobic and capnophilic bacteria. *J. Clin. Microbiol.* **1990**, *28*, 422–425.

202. Crennell, S.J.; Garman, E.F.; Philippon, C.; Vasella, A.; Laver, W.G.; Vim, E.R.; Taylor, G.L. The structures of *Salmonella typhimurium* LT2 neuraminidase and its complexes with three inhibitors at high resolution. *J. Mol. Biol.* **1996**, *259*, 264–280. [CrossRef]

203. Ada, G.L.; French, E.L.; Lind, P.E. Purification and properties of neuraminidase from *Vibrio cholera*. *J. Gen. Microbiol.* **1961**, *24*, 409–425. [CrossRef] [PubMed]

204. Godoy, V.G.; Dallas, M.M.; Russo, T.A.; Malamy, M.H. A role for *Bacteroides fragilis* neuraminidase in bacterial growth in two model systems. *Infect. Immun.* **1993**, *61*, 4415–4426. [PubMed]

205. Schaufuss, P.; Lämmler, C. Characterization of Extracellular Neuraminidase produced by *Actinomyces pyogenes*. *Zentralblatt. Bakteriol.* **1989**, *271*, 28–35. [CrossRef]

206. Jost, B.H.; Songer, J.G.; Billington, S.J. Cloning, expression, and characterization of a neuraminidase gene from *Arcanobacterium pyogenes*. *Infect. Immun.* **2001**, *69*, 4430–4437. [CrossRef] [PubMed]

207. Jost, B.H.; Songer, J.G.; Billington, S.J. Identification of a second *Arcanobacterium pyogenes* neuraminidase and involvement of neuraminidase activity in host cell adhesion. *Infect. Immun.* **2002**, *70*, 1106–1112. [CrossRef] [PubMed]

208. Ozturk, D.; Turutoglu, H.; Pehlivanoglu, F.; Guler, L. Virulence Genes, Biofilm Production and Antibiotic Susceptibility in *Trueperella pyogenes* isolated from Cattle. *Isr. J. Vet. Med.* **2016**, *71*, 36–42.

209. Zhang, S.; Qiu, J.; Yang, R.; Shen, K.; Xu, G.; Fu, L. Complete genome sequence of *Trueperella pyogenes*, isolated from infected farmland goats. *Genome Announc.* **2016**, *4*, e01421-16. [CrossRef]

210. Kasimanickam, V.R.; Owen, K.; Kasimanickam, R.K. Detection of genes encoding multidrug resistance and biofilm virulence factor in uterine pathogenic bacteria in postpartum dairy cows. *Theriogenology* **2016**, *85*, 173–179. [CrossRef]

211. da Silva Duarte, V.; Dias, R.S.; Kropinski, A.M.; da Silva Xavier, A.; Ferro, C.G.; Vidigal, P.M.P.; da Silva, C.C.; de Paula, S.O. A T4virus prevents biofilm formation by *Trueperella pyogenes*. *Vet. Microbiol.* **2018**, *218*, 45–51. [CrossRef]

212. Zschiedrich, C.P.; Keidel, V.; Szurmant, H. Molecular Mechanisms of Two-Component Signal Transduction. *J. Mol. Biol.* **2016**, *428*, 3752–3775. [CrossRef]

213. Herrou, J.; Crosson, S.; Fiebig, A. Structure and function of HWE/HisKA2-family sensor histidine kinases. *Curr. Opin. Microbiol.* **2017**, *36*, 47–54. [CrossRef] [PubMed]

214. Hupało-Sikorska, A.; Gregorczyk, K.P.; Rzewuska, M.; Szulc-Dąbrowska, L.; Struzik, J.; Gieryńska, M. Involvement of macrophages during Trueperella pyogenes infection. In *Proceedings of XVIII Conference DIAGMOL 2017—Molecular Biology in Diagnostics of Infectious Diseases and Biotechnology*; Warsaw University of Life Sciences: Warsaw, Poland, 2017; pp. 45–46.

215. Machado, V.S.; Bicalho, R.C. Complete genome sequence of *Trueperella pyogenes*, an important opportunistic pathogen of livestock. *Genome Announc.* **2014**, *2*, e00400-14. [CrossRef] [PubMed]

216. da Silva Duarte, V.; Dias, R.S.; Kropinski, A.M.; Vidigal, P.M.P.; Sousa, F.O.; da Silva Xavier, A.; da Silva, C.C.; de Paula, S.O. Complete genome sequence of vB_EcoM_UFV13, a new bacteriophage able to disrupt *Trueperella pyogenes* biofilm. *Genome Announc.* **2016**, *4*, e01292-16.

217. Bernier, A.M.; Bernard, K. Draft genome sequence of *Trueperella bernardiae* LCDC 89-0504T, isolated from a human blood culture. *Genome Announc.* **2016**, *4*, e01634-15. [PubMed]

218. Billington, S.J.; Jost, B.H.; Songer, J.G. The *Arcanobacterium (Actinomyces) pyogenes* plasmid pAP1 is a member of the pIJ101/pJV1 family of rolling circle replication plasmids. *J. Bacteriol.* **1998**, *180*, 3233–3236.

219. Tauch, A.; Götker, S.; Pühler, A.; Kalinowski, J.; Thierbach, G. The 27.8-kb R-plasmid pTET3 from *Corynebacterium glutamicum* encodes the aminoglycoside adenyltransferase gene cassette *aad*A9 and the regulated tetracycline efflux system Tet 33 flanked by active copies of the widespread insertion sequence *IS6100*. *Plasmid* **2002**, *48*, 117–129. [CrossRef]

220. Lovell, R.; Foggie, A.; Pearson, J.K. Field trials with *Corynebacterium pyogenes* alum-precipitated toxoid. *J. Comp. Pathol.* **1950**, *60*, 225–229. [CrossRef]

221. Hunter, P.; van der Lugt, J.J.; Gouws, J.J.; Onderstepoort, J. Failure of an *Actinomyces pyogenes* vaccine to protect sheep against an intravenous challenge. *Vet. Res.* **1990**, *57*, 239–241.

222. Kostro, K.; Lisiecka, U.; Żmuda, A.; Niemczuk, K.; Stojecki, K.; Taszkun, I. Selected parameters of immune response in pregnant sows vaccinated against *Trueperella pyogenes*. *Bull. Vet. Inst. Pulawy* **2014**, *58*, 17–21. [CrossRef]

223. Machado, V.S.; Bicalho, M.L.; Meira Junior, E.B.; Rossi, R.; Ribeiro, B.L.; Lima, S.; Santos, T.; Kussler, A.; Foditsch, C.; Ganda, E.K.; et al. Subcutaneous immunization with inactivated bacterial components and purified protein of *Escherichia coli*, *Fusobacterium necrophorum* and *Trueperella pyogenes* prevents puerperal metritis in Holstein dairy cows. *PLoS ONE* **2014**, *9*, e91734. [CrossRef]

224. Zhang, W.; Wang, P.; Wang, B.; Ma, B.; Wang, J. A combined *Clostridium perfringens/Trueperella pyogenes* inactivated vaccine induces complete immunoprotection in a mouse model. *Biologicals* **2017**, *47*, 1–10. [CrossRef] [PubMed]

225. Huang, T.; Zhao, K.; Zhang, Z.; Tang, C.; Zhang, X.; Yue, B. DNA vaccination based on pyolysin co-immunized with IL-1β enhances host antibacterial immunity against *Trueperella pyogenes* infection. *Vaccine* **2016**, *34*, 3469–3477. [CrossRef] [PubMed]

226. Huang, T.; Song, X.; Jing, J.; Zhao, K.; Shen, Y.; Zhang, X.; Yue, B. Chitosan-DNA nanoparticles enhanced the immunogenicity of multivalent DNA vaccination on mice against *Trueperella pyogenes* infection. *J. Nanobiotechnol.* **2018**, *16*, 8. [CrossRef] [PubMed]

© 2019 by the authors. Licensee MDPI, Basel, Switzerland. This article is an open access article distributed under the terms and conditions of the Creative Commons Attribution (CC BY) license (http://creativecommons.org/licenses/by/4.0/).

International Journal of
Molecular Sciences

Review

Divergent Approaches to Virulence in *C. albicans* and *C. glabrata*: Two Sides of the Same Coin

Mónica Galocha [1,2]**, Pedro Pais** [1,2]**, Mafalda Cavalheiro** [1,2]**, Diana Pereira** [1,2]**, Romeu Viana** [1,2] **and Miguel C. Teixeira** [1,2,]*****

[1] Department of Bioengineering, Instituto Superior Técnico, Universidade de Lisboa, 1049-001 Lisboa, Portugal; monicagalocha@gmail.com (M.G.); pedrohpais@tecnico.ulisboa.pt (P.P.); mafalda.cavalheiro@tecnico.ulisboa.pt (M.C.); diana.pereira@tecnico.ulisboa.pt (D.P.); romeuviana@outlook.com (R.V.)

[2] iBB-Institute for Bioengineering and Biosciences, Biological Sciences Research Group, Instituto Superior Técnico, University of Lisbon, Av. Rovisco Pais, 1049-001 Lisboa, Portugal

***** Correspondence: mnpct@tecnico.ulisboa.pt; Tel.: +351-218417772; Fax: +351-218419199

Received: 10 April 2019; Accepted: 8 May 2019; Published: 11 May 2019

Abstract: *Candida albicans* and *Candida glabrata* are the two most prevalent etiologic agents of candidiasis worldwide. Although both are recognized as pathogenic, their choice of virulence traits is highly divergent. Indeed, it appears that these different approaches to fungal virulence may be equally successful in causing human candidiasis. In this review, the virulence mechanisms employed by *C. albicans* and *C. glabrata* are analyzed, with emphasis on the differences between the two systems. Pathogenesis features considered in this paper include dimorphic growth, secreted enzymes and signaling molecules, and stress resistance mechanisms. The consequences of these traits in tissue invasion, biofilm formation, immune system evasion, and macrophage escape, in a species dependent manner, are discussed. This review highlights the observation that *C. albicans* and *C. glabrata* follow different paths leading to a similar outcome. It also highlights the lack of knowledge on some of the specific mechanisms underlying *C. glabrata* pathogenesis, which deserve future scrutiny.

Keywords: *Candida*; host-pathogen interaction; virulence; biofilm formation; morphology; immune evasion

---

1. Introduction

Infections caused by fungi affect millions of people worldwide, with the overall mortality rate estimated to be roughly 1,350,000 deaths per year [1]. Among pathogenic fungi, *Candida* species are responsible for the most common invasive fungal disease in developed countries—the candidiasis [2]. *Candida* species live as commensals on mucosal surfaces where they are constituents of the normal microbiota of the oral cavity and gastrointestinal and vaginal tracts. However, they can opportunistically become pathogenic under suitable conditions, such as host-disrupted microbiota or immunocompromised hosts, being responsible for clinical manifestations from mucocutaneous overgrowth to bloodstream infections [3–5]. Of the various *Candida* species, *Candida albicans* and *Candida glabrata* not only account for 60% of *Candida* species present in the human body, but also constitute the most prevalent of the pathogenic *Candida* species, being responsible for more than 400,000 life-threatening infections worldwide every year [3,6].

C. albicans and *C. glabrata* are the two most common pathogenic yeasts of humans, yet they are phylogenetically, genetically, and phenotypically very different. On one hand, *C. albicans* diploid genome carries several gene families that are associated with virulence [7]. These include the ALS (agglutinin-like sequence) adhesins, required for host adhesion, secreted aspartyl proteases (SAPs) and phospholipases, which allow for the degradation of host barriers and the invasion of surrounding tissue, and proteins involved in oligopeptide and iron transfer [8–10]. On the other hand, *C. glabrata* mechanisms of tissue invasion are mostly unknown, although it is hypothesized to possibly occur by endocytosis induction of host cells [11]. Its haploid genome encodes a large group of glycosylphosphatidylinositol (GPI)-anchored cell wall proteins, such as the adhesins from the *EPA* gene family, implicated in fungus–host interactions or biofilm formation, and a family of aspartic proteases (yapsins) which are mainly associated with cell wall remodeling and possible immune evasion [12,13]. Moreover, key virulence attributes of *C. albicans*, which are known to be the basis of its pathogenicity, are absent in *C. glabrata* [14]. Switching from yeast to hyphal growth not only allows consistent biofilm production but also enables *C. albicans* to be highly invasive and escape macrophage engulfment [15–19]. Nevertheless, both species are known to use biofilms to colonize the surface of several medical devices based on different materials [20]. Unlike *C. albicans*, it has been demonstrated that *C. glabrata* lets itself be taken up by macrophages, where it persists and divides for long periods of time eventually leading to cell lysis due to fungal load [21,22]. It has the ability of detoxifying oxidative radical species and disrupting normal phagosomal maturation, leading to the inhibition of phagolysosome formation and phagosome acidification [21,23].

The interaction between *Candida* and its host cells is characterized by a complex interplay between the expression of fungal virulence factors and the host immune system, and the presence of other microorganisms affects this interplay. This review aims to explore and compare the remarkably distinct paths toward virulence trailed by the two most common causative agents of candidiasis worldwide. On one hand, *C. albicans* is known for its ability to evade host defenses and form bulk biofilms due to its ability to undergo filamentous growth, while on the other hand, *C. glabrata* is an unusually stress-tolerant organism able to survive and replicate inside the immune system cells. Despite having such distinct virulence features, *C. glabrata* and *C. albicans* are frequently co-isolated [11].

2. Host Damage and Invasion

There is a variety of defense mechanisms through which the human host is able to prevent invasion by pathogenic microorganisms, such as *C. albicans* and *C. glabrata*. These mechanisms consist not only of physical but also of chemical barriers. For instance, epithelial cells, which in most cases are the first line of contact between host and pathogen, function as the prime physical barrier restraining *Candida* from invasion of the underlying tissue. On one hand, these cells are interconnected through "tight junctions" preventing the entry of microorganisms into interepithelial space and eventually into the bloodstream [24,25]. On the other hand, some types of epithelial cells, such as those in the intestinal or vaginal epithelium, are able to produce a mucus layer by secreting mucins [24,26]. This layer impairs *Candida* invasion by preventing contact with the epithelium surface. Likewise, in the oral cavity the flow of saliva plays an important role as both a physical and a chemical barrier as it not only prevents the adhesion to mucosa and dental surfaces but also contains several antimicrobial agents that impair the contact of *Candida* with the oral epithelium [27,28]. Another chemical barrier against *Candida* establishment is the presence of gastric acid and bile in the digestive system which creates a harsh environment for fungal growth. Nevertheless, these human pathogens are known to have a remarkable ability to adapt to these adverse conditions and proliferate.

C. albicans relies on two distinct invasion mechanisms to gain entry into host cells: (i) induced endocytosis and (ii) active penetration of hyphal forms through physical forces of hyphae production associated with lytic enzyme secretion [29] (Figure 1). Nonetheless, depending on the host cell, these invasion mechanisms are thought to be exploited to a different extent. For instance, it was demonstrated that while invasion into oral epithelial cells occurs via both routes, invasion into intestinal epithelial cells occurs only via active penetration under normal conditions [29,30].

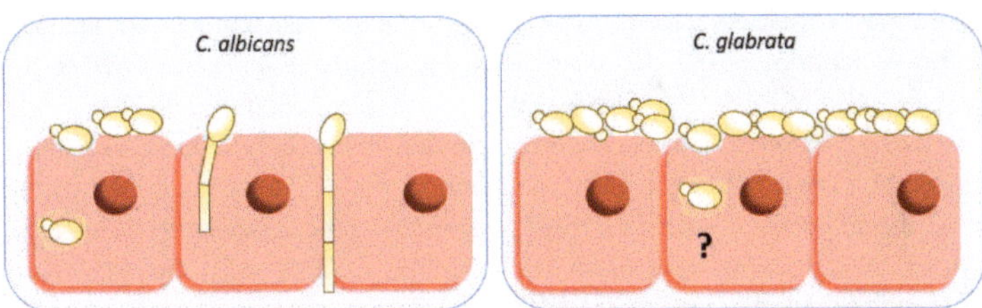

Figure 1. Schematic representation of *C. albicans* and *C. glabrata* host damage and invasion. *C. albicans* enter host cells through induced endocytosis or by active penetration (inter- and intra-cellular) of hyphal forms associated with the release of hydrolytic enzymes, resulting in the damage of cells and loss of epithelial integrity. Induced endocytosis of host cells is thought to be the mechanism behind *C. glabrata* tissue invasion.

Living within host cells is a profitable strategy since it enables cells to deal with the host immune system and antimicrobial treatment, and there are plenty of nutrients available and an absence of competition with other commensal microorganisms. Additionally, dissemination into deeper tissues and ultimately into the bloodstream is easier from within the cell. Fungal invasion via induced endocytosis is dependent on dynamic microfilaments of the host. In *C. albicans*, the GPI-anchored hypha-associated protein Als3 interacts with mammalian cadherins, mimicking the establishment of adherence junctions. This process leads to rearrangements in the actin cytoskeleton that ultimately lead to fungal cell internalization [31]. Similarly, Ssa1, a member of the HPS70 heat shock protein family, was reported to play the same role as Als3, being essential for maximal host cell damage and subsequently fungal cell endocytosis [32]. However, *EED1* was the first fungal gene reported as being required for epithelial escape and interepithelial dissemination and not for initial invasion into epithelial cells [33]. Moreover, it is thought that the contact with the epithelium, among several other stimuli, is a highly potent inducer of *C. albicans* filamentation. In turn, the contact between hyphal cells and epithelial cells induces host defense mechanisms such as the formation of epithelial cell protrusions surrounding the hyphae, and membrane ruffling, which is characteristic for induced endocytosis [34].

The ability to switch between yeast and hyphal growth forms is one of the most discussed and best-investigated virulence attributes of *C. albicans*. This morphology switch is activated by well-established kinase-based signal transduction pathways and is triggered by diverse host environmental cues, including temperature, pH, serum, and CO_2 and is linked to several steps during host invasion [35,36]. The extracellular signals are transmitted via Ras to both protein kinase A and the MAP kinase cascade [37–39], inducing hyphal differentiation through the activation of a number of transcription factors such as Efg1 [40] and Ume6 [41]. The transcriptional repressor Nrg1 is inactivated and removed from the hyphal-specific gene promoters, thereby allowing the induction of hyphal morphogenesis [42,43].

Invasion into epithelial cells via active penetration relies on a combination of physical pressure employed by the growing hyphae and the secretion of hydrolytic enzymes. Moreover, hyphal cells are capable of directional growth in response to contact with a solid surface (thigmotropism) which enables *C. albicans* to specifically identify and invade intercellular junctions, thereby damaging the epithelium compact structure [44]. Interestingly, it was very recently discovered that *C. albicans* release hydroxyphenylacetic acid (HPA) during hyphal growth [45]. Its production seems to occur through the same pathway and the same precursors as tyrosol, which is able to stimulate hypha induction in *C. albicans* [46], therefore their biological functions are likely to be the same. Additionally, along with the active penetration by hyphal cells, there is also the secretion of hydrolytic enzymes, such as SAPs and phospholipases, which can digest epithelial cell surface components enabling the entrance into or between host cells [47]. SAPs are the best-characterized members of the *C. albicans* hydrolytic enzymes, as is well reviewed by Hube and Naglik [48]. Interestingly, these tissue damaging enzymes not only have distinct optimal pH requirements but also are growth stage and infection site related [47]. For instance, *SAP1–SAP3* are yeast growth-associated and have optimum activity at lower pH values, while *SAP4–SAP6* are hyphal growth-associated and have optimum activity at higher pH values [49]. Similarly, it was demonstrated that *SAP1*, *SAP3*, and *SAP8* are preferentially expressed in vaginal, rather than oral, *C. albicans* infections [47]. One possible explanation for the existence of a significant number of different *SAP* genes may be the necessity for specific and optimized proteinases during the different stages of an infection [48]. Additionally, the cytolytic peptide toxin of *C. albicans* candidalysin, encoded by the hypha-associated gene *ECE1*, was recently found to be essential for damage of enterocytes and is a key factor in subsequent fungal translocation, suggesting that transcellular translocation of *C. albicans* through intestinal layers is mediated by candidalysin [25]. Moreover, phytase activity in *C. albicans* was demonstrated to be important for virulence. Phytate is a major storage form of phosphorus in plants and is abundant in the human diet and intestinal tract [50]. Recently, it was reported that decreased phytase activity leads to a reduced ability to form hyphae, attenuated in vitro adhesion, and reduced ability to penetrate human epithelium [51].

While the transition from yeast to hyphae has been extensively studied in *C. albicans*, the switch from hyphae to yeast still remains poorly understood [35]. Nevertheless, both yeast and hyphae forms are found in infection sites, which suggests that both forms are implicit in the infection process. Interestingly, it was demonstrated that depending on the infected organ, one or the other morphology predominates [52]. It is thought that the yeast form is important for dissemination upon infection, whereas hyphae forms are more relevant to attachment, host invasion, and tissue damage [53].

Unlike *C. albicans*, and despite the existence of some reports demonstrating that *C. glabrata* forms pseudohyphae [54,55], the pathogenicity of *C. glabrata* seems to be independent of morphology. The most common route for this pathogen to reach the bloodstream is through the iatrogenic breach of natural barriers, such as the use of catheters, trauma, or surgery.

In 2000, Csank and Haynes [54] reported for the first time that *C. glabrata* can undergo morphological change and grow as a pseudohyphae on solid nitrogen starvation media. This invasive growth mode could be a possible mechanism of host invasion, however, this phenomenon has not yet been reported in vivo. Despite lacking this prime virulence feature, this opportunistic pathogen is still able to reach the human bloodstream and cause infection. In some cases, *C. glabrata* can involuntarily reach the bloodstream through nosocomial conditions, namely surgery, catheter, parenteral nutrition, and burn injury [56]. However, even when these external factors are abrogated, *C. glabrata* is able to invade the host and colonize different tissues, as shown in an intragastrointestinal mouse model of infection [57] and in a chorioallantoic membrane chicken embryo model of infection [58]. Thus, *C. glabrata* must have other invasion mechanisms.

Co-infection with other microorganisms may be a possible explanation to the invasive capacity of *C. glabrata* since this yeast is often co-isolated in infections with *C. albicans* [59–61] or even other pathogens such as *Clostridium difficile* [62]. In the co-infected environment, *C. glabrata* cells may exploit the tissue invasion and destruction caused by *C. albicans* to access nutrients and reach the bloodstream. In fact, Tati et al. (2016) [63] demonstrated that when mice are infected with *C. glabrata* alone, oropharyngeal candidiasis is negligible, however, when co-infected with *C. albicans*, an increased colonization by *C. glabrata* was observed. This effect was attributed to the binding of *C. glabrata* to *C. albicans* hyphae [63] and similar results were reported by Alves et al. (2014) [64] using a reconstituted human vaginal epithelium. Nonetheless, the intracytoplasmic presence of *C. glabrata* was detected in oral epithelial cells [65] and vaginal epithelial cells [66] and it has been shown that when endocytosis is inhibited, the internalization of *C. glabrata* is prevented [65]. This suggests that induced endocytosis by host cells could be the most likely mechanism of *C. glabrata* internalization (Figure 1).

As referred to before, the tissue/cell damaging ability of *C. glabrata* is lower compared to *C. albicans*. In *C. albicans*, secreted hydrolytic enzymes are considered to be important destructive factors that damage host tissues, providing nutrients for its propagation. However, the production of these hydrolytic enzymes is very low or even null in *C. glabrata*, wherefore its importance for virulence does not seem to be as relevant as it is in *C. albicans* [67–73]. The proteinase enzyme is responsible for protein degradation resulting in tissue invasion. Among the proteases, SAPs are considered crucial for the pathogenicity of *C. albicans* [74]. Although *C. glabrata* does not express SAPs [75], its genome contains 11 non-secreted GPI-linked aspartyl proteases (*YPS* genes), which are surface-exposed aspartic proteases required for virulence, known as yapsins [13]. These yapsins are important for cell wall maintenance, remodeling, cell to cell interactions, and resistance to cell wall stress, however, their direct link with virulence is not very well characterized yet. Otherwise, phospholipases promote the destruction of cell membrane phospholipids, causing cell damage and lysis which allows a greater invasive capacity [76]. Some of the *C. glabrata* strains are able to produce these enzymes [67,68,70,71,77], which appear to play a role in *C. glabrata*-associated persistent candidemia [9]. However, phospholipase production in *C. glabrata* is lower than in *C. albicans*, and in some cases inexistent [69,72,73], therefore its relation to *C. glabrata* virulence is not clear and needs further analysis.

3. Adhesion and Biofilm Formation

The ability to infect and prevail in the human host is related to different pathogenesis factors, of which biofilm formation excels [19,78–80]. *Candida* species ability to form biofilms on medical devices increases mortality rates associated with infections, while often forcing the treatment to include the removal of the medical device [81]. A lot of efforts have been put into understanding the molecular basis of *Candida* species biofilm formation [20].

Adhesion is one of the most relevant and advantageous capacities of the yeast cell wall. It allows cells to colonize mucosal surfaces and prevail in a nutritional environment, being the first critical step for biofilm formation, which serves as a shield against adverse conditions, as well as a highly drug-resistant reservoir of infective cells [82,83]. *C. albicans* and *C. glabrata* pathogenesis has been strongly related to adhesion, which is considered a crucial virulence factor in these yeasts [84–86]. In this regard, both *Candida* species are able to not only attach to mammalian host cells (epithelial, endothelial, and immune cells) but also to other microbes (bacteria and other *Candida* species) and abiotic surfaces, such as medical devices [85,87]. Several studies have tried to understand the nature of adherence to plastic surfaces. For instance, cell surface hydrophobicity (CSH) seems to have a positive correlation with adhesion in both *C. albicans* and *C. glabrata* species, thus, adhesion is mediated by van der Waals forces. Moreover, compared to *C. albicans*, the relative CSH of *C. glabrata* has been shown to be significantly higher [88–90].

Both *Candida* species have a set of proteins that enable attachment, known as adhesins. The most important *C. albicans* adhesins are agglutinin-like sequence (Als) proteins (Als1–7 and Als9) [91] and hypha-associated GPI-linked protein (Hwp1), known to be required for adhesion and virulence in vivo, and also being associated to biofilm formation through interaction with Als1 and Als3 adhesins [92–94]. Other adhesins required for adhesion and biofilm formation include Hwp2 [95] and Eap1 [96–98]. *C. glabrata* expresses a large group of adhesins, belonging to the epithelial adhesin (Epa) family, which is encoded by 17 to 23 genes, depending on the strain [78]. Among these, Epa1 is a major virulence player in *C. glabrata*, and mediates 95% of in vitro adhesion to epithelial cells [99]. This adhesin is highly heterogeneous among *C. glabrata* clinical isolates, being an important virulence factor [100]. Epa6 and Epa7 are involved in kidney and bladder colonization in vivo and boost biofilm formation [101–104]. Transcriptomic and proteomic studies have revealed that besides *EPA* genes, *C. glabrata* holds other biofilm-related adhesin families, such as Pwp (encoded by seven members *PWP1–7*), Aed (*AED1* and *AED2*), and Awp (encoded by 12 members *AWP1–6* and *AWP8–13*), which are usually found in significantly higher numbers in clinical isolates, consistently with an important role in pathogenesis [12,105–107].

The successful pathogenicity of these yeasts relies on its flexibility, which allows for adaptation and proliferation under both nutrient-rich and nutrient-poor conditions. Several studies have reported the importance of host and antifungal selective pressure on virulence traits as adhesion [108,109]. A recent study conducted by Vale-Silva et al. (2017) [110] used the PacBio technology to compare the genomes of two sequential *C. glabrata* clinical isolates and observed a significant increase in the number of adhesin-encoding genes (101 and 107) when compared to the CBS138 genome (63), despite the limited variation between the two studied isolates. The same authors, along with Ni et al. (2018) [111], further linked the increased expression of the adhesin gene *EPA1* with gain-of-function (GOF) mutations in the *PDR1* gene in drug-resistant clinical isolates [112]. *EPA1* has been strongly related to adhesion of *C. glabrata* cells to mammalian epithelial cells both in vitro [99] and in vivo [113]. Furthermore, Salazar et al. (2018) [114] also observed a GOF mutation in the *PDR1* gene, which led to changes in the transcriptome when compared to the CBS138 strain. Among the genes identified as having the highest number of non-synonymous SNPs, there were several genes encoding adhesins and, agreeing with Vale-Silva et al. (2013) [115], the number of adhesin-expressed genes varied when compared to other GOF *PDR1* mutations [114]. This reinforces the idea that antifungal treatment deploys a tight selective pressure which results in changes at the genomic and transcriptional levels, particularly affecting adhesin-encoding genes [116]. The transcription factor Cst6 was found to also play a role in *C. glabrata* adhesion and biofilm formation, negatively regulating the expression of *EPA6* [103]. In *C. albicans*, a transcriptional regulatory network comprising nine regulators (Bcr1, Brg1, Efg1, Flo8, Gal4, Ndt80, Rob1, Rfx2, and Tec1) was identified in in vitro and in vivo studies as underlying the biofilm formation phenomenon in this pathogenic yeast [117–121].

Hyphae formation, which is exclusive to *C. albicans*, when compared to *C. glabrata* is also an important trait in biofilm development. Various studies have described hyphae as exhibiting improved adhesion to the human epithelium, with these cells displaying increased expression of Als1, Als3, and Hwp1 [122–125]. Interestingly, Tati et al. (2016) [63] characterized the co-colonization of *C. glabrata* and *C. albicans* in a murine model of oropharyngeal candidiasis (OPC) and demonstrated that this is driven by specific adhesins in both species. Namely, the *C. albicans* Als3 and Als1 adhesins are crucial for in vitro binding of *C. glabrata* cells to *C. albicans* hyphae and for further in vivo establishment of infection. Considering *C. glabrata* cells, incubation with *C. albicans* hyphae led to the overexpression of the adhesins *EPA8*, *EPA19*, *AWP2*, *AWP7*, and *CAGL0F00181* [63].

After adhesion, *Candida* species are known to be able to form a 3D-structure of cells embedded in a gel-like matrix [20,126,127]. In order to achieve this, surface-adhered cells begin to adhere to other *Candida* cells, initiating the formation of discrete colonies, which correspond to an early phase of biofilm formation. At this point, the intermediate phase begins with the cellular production and secretion of important molecules, known as extracellular polymeric substances (EPS), that will constitute the extracellular matrix of the biofilm, protecting the cells and ensuring a more developed 3D-structure. The final structure is reached after the maturation phase, where more cells and the matrix are originated. Mature biofilms might also suffer the detachment of some cells that can spread to form new biofilms on other niches of the host, a process called the dispersal phase [20,126,128,129]. Although this process is true for every *Candida* species, there are differences between *C. albicans* and *C. glabrata* biofilms, regarding their dimensions and structure, cell morphology, EPS produced and secreted, response to environmental cues, and resistance to antifungal drugs (Figure 2).

A very clear difference between *C. albicans* and *C. glabrata* mature biofilms is the dimension and total biomass of each biofilm. *C. glabrata* in vivo biofilm formation leads to a thickness of 75–90 $\pm$ 5 μm, which is half of the normal thickness of *C. albicans* biofilms [130], with much less biomass in the end of biofilm formation compared to *C. albicans* [131]. The organization of the biofilm structure also differs between these two species. *C. albicans* biofilms are arranged in a three-dimensional structure with different morphologies and empty spaces between cells [132], where microchannels are formed [133]. On the other hand, *C. glabrata* biofilms are thinner, but display a higher density of cells, tightly packed together [132]. Although *C. glabrata*'s biofilms are composed by yeast cells only [131,132], the same is not true for *C. albicans* biofilms, where different morphologies arise. *C. albicans* mature biofilms are composed by a dense network of pseudohyphae, hyphae, and yeast cells [134]. This filamentation process in biofilm formation is controlled by the transcription factor Efg1, without which *C. albicans* only forms scarce monolayers of elongated yeast cells on polyurethane catheters and polystyrene [135,136].

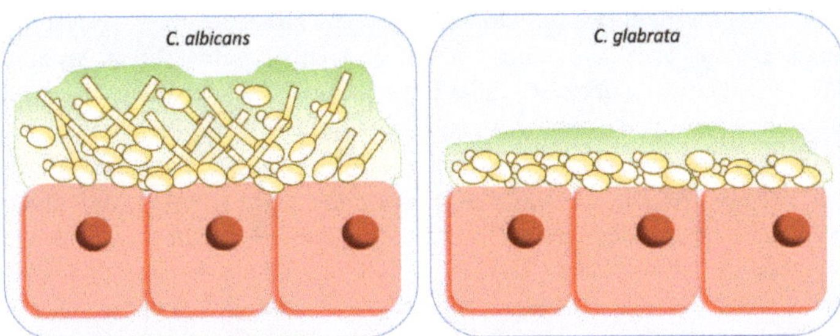

Figure 2. Schematic representation of *C. albicans* and *C. glabrata* biofilm formation. *C. albicans* forms thicker biofilms, with much more biomass in the end of biofilm formation and produces more extracellular matrix than *C. glabrata*. *C. albicans* mature biofilms are composed by a dense network of pseudohyphae, hyphae, and yeast cells, whereas *C. glabrata* biofilms are composed by compact yeast cells only, forming a thin but dense biofilm.

Although *C. albicans* produces more extracellular matrix than *C. glabrata* [137], the main components of the matrix, proteins, and carbohydrates, are the same in both biofilms [131,138–140]. Interestingly, *C. glabrata* has a very high content of proteins and carbohydrates, which is five times higher than that found on the biofilms of other non-*albicans Candida* species [131]. *C. albicans'* matrix is also composed by other lipids (mainly neutral glycerolipids, polar glycerolipids, and a small percentage of sphingolipids) [139], phosphorus, and uronic acid [140]. Biofilms of both species also have a small content of extracellular DNA [131,138–140].

Interestingly, *C. albicans* and *C. glabrata* also behave differently on different surfaces when it comes to initiating biofilm formation. Cleary, each species has a different propensity to form biofilm on a given surface. For instance, *C. albicans* is known to adhere better to latex and silicone elastomer while showing less biofilm formation on polyvinyl chloride, polyurethane, or 100% silicone [134]. *C. albicans* also adheres and forms biofilm on different surfaces of denture base materials, having higher biofilm formation on the surface of alloy and lower biofilm formation on methacrylate-based denture material [141]. Moreover, polyetherurethane treated with 6% of polyethylene oxide was found to reduce the metabolic activity of cells and the total biomass of *C. albicans* biofilms [142]. On the other hand, while all other pathogenic *Candida* species have greater biofilm formation on Teflon, *C. glabrata* prefers polyvinyl chloride to form biofilm [143]. Interestingly, other components of the environment might alter the ability of *Candida* species to adhere to a given surface. For instance, the presence of saliva has been shown to decrease the ability of *C. albicans* to form biofilm in vitro [144,145].

Depending on the antifungal drug, *C. albicans* and *C. glabrata* biofilms might be able to resist the therapeutic action of the drug. For instance, Choi and colleagues (2007) [146] measured the in vitro susceptibilities of biofilms of *C. glabrata* and *C. albicans* bloodstream isolates, showing that both biofilms were resistant to fluconazole and only moderately resistant to amphotericin B, while exposure to 0.25 to 1 µg/mL of caspofungin and micafungin lead to an 80% reduction of the biofilms. Moreover, voriconazole is also able to reduce *C. albicans* and *C. glabrata* biofilms, being present at 0.25 mg/L or being used as a surface coating at different concentrations [147]. Nevertheless, when growing on an RPMI 1640 medium, *C. glabrata* mature biofilms have shown to be less susceptible to caspofungin and anidulafungin than *C. albicans* mature biofilms on a polystyrene surface [148], showing that each species biofilms might resist differently to the same antifungal drug. Although reacting differently to antifungal drug exposure, the strategy to achieve resistance seems to be very similar between the two species. The mechanisms known to underlie resistance to antifungal drugs in *Candida* biofilms are believed to be related to alterations in the metabolic activity, the role of the extracellular matrix as a barrier for diffusion, the role of its EPS components, and the presence of persister cells within the biofilm [20,149]. Both species suffer the upregulation of drug efflux pump-encoding genes [138,150], as well as seem to rely on the β-1,3-glucans present on the extracellular matrix [151,152].

Biofilms also allow for the survival of *Candida* species by protecting them against the host immune system. The extracellular matrix is essential for the protection against the action of neutrophils by inhibiting the release of neutrophil extracellular traps (NETs) [153]. Moreover, the β-glucans present on the matrix avoid the activation of neutrophils, actually inhibiting the reactive oxygen response, thereby being an important distracting mechanism to evade the innate immune system [154]. Biofilms are also believed to resist well the action of the innate immune system due to its heterogenicity, given the different types of cells and different metabolic activity the cells might present on biofilms [155].

The combination of *C. albicans* and *C. glabrata* to form biofilm has been well described and the two species are usually found together in niches of candidiasis patients [60]. A recent study has shown that a ratio of *C. albicans* to *C. glabrata* of 1:3, significantly increases the total biofilm biomass comparatively to a *C. albicans* monoculture or ratio of *C. albicans* to *C. glabrata* of 1:1. This co-culture biofilm exhibited a high heterogenicity with *C. albicans* hyphae and *C. glabrata* cells clustered together in a 3D-structure. Interestingly, an upregulation of *HWP1* and *ALS3* genes is observed in this mixed-species biofilm, as well as an increased resistance to caspofungin [3]. *C. albicans* and *C. glabrata* are also known to form biofilms with bacteria from different host niches, usually relying on quorum-sensing mechanisms for the establishment of an interaction [20,155]. All the possible interactions between species increase the complexity of this vast field, pointing out the big clinical challenge of biofilm formation.

4. Host Immune System Evasion

Throughout infection, when the first line of defense has been breached by invasion into deeper tissues, *Candida* pathogens have to cope with cells of the host innate immune system. Interaction with the host immune system, and the ability to overcome it, is one of the main virulence features for fungal pathogens.

At early stages of infection, upon an interaction between *Candida* pathogens and epithelial cells, the former are recognized as invasive microorganisms by Pattern Recognition Receptors (PRRs) localized at host epithelial cell surfaces. PRRs interact with Pathogen Associated Molecular Patterns (PAMPs), such as β-1,3-glucan or chitin, present on microbial cells, thereby inducing a host response [156,157]. Epithelial cells, that are part of the innate immunity, not only secrete antimicrobial peptides, such as β-defensins and LL-37 [158–160], to try to control fungal infection, but also release proinflammatory mediators, such as chemokines and cytokines, triggering the recruitment of phagocytic cells, such as neutrophils, macrophages, and dendritic cells. These innate immune system cells also have PRRs in their surfaces, such as the C-type lectin receptor Dectin-1 [161], allowing the recognition of the invading pathogens and thereby inducing phagocytosis [162]. After phagocytosis, dendritic cells are responsible for the link between innate and adaptative antifungal immunity, presenting the *Candida*-specific antigens to naïve T-helper cells [163]. Therefore, in an immunocompetent host, this host–*Candida* interaction ultimately leads to the elimination of the pathogen. Otherwise, in immunocompromised individuals, a persistent infection, such as chronic mucocutaneous candidiasis, candidemia and/or persistent visceral candidiasis might be established.

Immune interaction can be translated in distinct spectrums, from avoidance of recognition by host immune cells to escaping or surviving immune attack. Masking PAMPs on the cell wall to avoid recognition, macrophage activation, and consequent phagocytosis is a common strategy of fungal pathogens during interaction with immune cells [164]. Generally, yeasts' cell wall is composed of a carbohydrate-rich inner layer and a protein-rich outer layer of heavily mannosylated proteins and phospholipomannan [165]. The outer layer acts as a shield of immunostimulatory components of the inner layer, such as β-1,3-glucan or chitin, playing a key role in protection and evasion from immune recognition [166]. β-1,3-glucan is the main polysaccharide present in the cell wall of *C. albicans*, *C. glabrata*, and other pathogenic *Candida* species, and is a key PAMP recognized by the host immune system [167]. Recognition of β-1,3-glucan by Dectin-1 receptor prompts phagocytosis by macrophages and neutrophils [161].

In order to avoid immune recognition, *C. albicans* resorts to cell wall remodeling, effectively masking β-glucan from the cell surface. The first reported case of active PAMP masking by *Candida* species was reported by Ballou et al. (2016) [167]. *C. albicans* was shown to mask β-glucan in response to lactate [167], which is a relevant physiological metabolite present in *Candida* niches, such as the vaginal tract and blood, or produced by the host microbiota [168], with which *Candida* interact. Lactate-mediated β-glucan masking is modulated by a signaling pathway associated with the G-protein coupled receptor Gpr1 and the transcription factor Crz1 [167]. This pathway reduces *C. albicans* uptake by macrophages and decreases the inflammatory response (TNFα and MIP1α) and neutrophil recruitment [167]. The participation of lactate was also observed later in low oxygen environments [169]. *C. albicans* was found to mask β-glucan upon oxygen deprivation, hindering recognition by Dectin-1 of polymorphonuclear leukocytes (PMNs). This was seen to modulate PMN responses, crippling phagocytosis, action of extracellular DNA traps, and reactive oxygen species (ROS) production [169]. Interestingly, β-glucan masking was prolonged by the build-up of lactate levels produced by PMNs [169]. Later, another study reported that hypoxia promotes β-glucan masking in *C. albicans* [170]. Hypoxia-induced masking is dependent on mitochondrial function and cAMP-protein kinase A (PKA) signaling, leading to reduced macrophage phagocytosis and cytokine (IL-10, RANTES, and TNF-α) production [170].

Changes in carbon source result in cell wall modifications with correspondent changes in virulence and immune properties [108,171]. As mentioned before, the cell wall is a complex structure with not only glucan, but also mannans, phosphomannans, and chitin [157], although distinct *Candida* species display different glucan exposure and mannan complexity [166]. β-glucans and chitin are located in the inner-most layer, while mannans are present in the outer layer [172–174]. Because of such structure, mannan plays an important role in reducing immunogenic exposure of β-glucan [175], but coordinated chitin and glucan exposure has also been reported to occur in *C. albicans* [176–178]. Moreover, cell wall structure and mannans affect virulence in different ways in *C. albicans* and *C. glabrata* [179–181]. Recently, mannan structure was found to affect glucan exposure in both *C. albicans* and *C. glabrata*, albeit in distinct ways. Deletion of mannosyltransferase family genes was associated with loss of negatively-charged acid-labile mannan and less efficient glucan masking in *C. albicans* (e.g., Δcgmnn2), while in *C. glabrata* increased glucan exposure density was associated with mutants displaying shorter backbones (e.g., Δcgmnn1 and Δcganp1) [166]. Previously, another study had shown that the β-1,6-mannosyltransferase encoded by *C. albicans MNN10* is required for backbone synthesis and influences immune recognition [182]. Absence of Mnn10 results in reduced *C. albicans* virulence, enhanced antifungal immunity by T helper cells, and increased recruitment of monocytes and neutrophils. Reinforcing the notion of a complex interplay among cell wall polysaccharides, mannosyltransferase activity was also associated with β-1,3-glucan masking from Dectin-1 recognition and modulatory action of cytokine production by macrophages [182]. Another study showed how *C. albicans* cell wall responds to immune attacks by NETs [176]. β-glucan exposure and enhanced Dectin-1 recognition is dependent on fungal-pathogen crosstalk, as this response is dependent on neutrophil NET-mediated damage and fungal signaling cascades based on the MAP kinase Hog1. Cell wall structure in response to a neutrophil attack was found to affect more than one component, as Hog1 signaling leads to chitin deposition via the chitin synthase Chs3 and posterior cell wall remodeling via Sur7 and Phr1. Accordingly, with the enhanced immune recognition by Dectin-1 after a NET-mediated attack, macrophage cytokine response was also increased [176].

Much like *C. albicans*, *C. glabrata* resorts to cell wall remodeling in order to avoid the host's immune system, although the underlying mechanisms are mostly unknown. As indicated by increased TNFα secretion and increased efficacy of pathogen killing by macrophages, deletion mutants with disturbed cell wall integrity and altered accessibility of PAMPs caused a stronger inflammatory response. *C. glabrata* deletion mutants lacking cell surface-associated proteases (yapsins) or mutants with defective protein glycosylation were related with a stronger inflammatory response by macrophages [13,181]. Nevertheless, mutations affecting mannan, but not those affecting glucan or chitin, were found to reduce the uptake of *C. glabrata* cells by murine macrophages, suggesting that mannose side chains or mannosylated proteins are ligands recognized by macrophages [183].

Knowledge regarding the PRRs responsible for *C. glabrata* recognition by macrophages is limited. Notwithstanding, as in *C. albicans* infections, C-type lectin receptors are thought to be involved in *C. glabrata* recognition by the host immune system. Specifically, dectin-1 and dectin-2, which recognize cell wall β-glucan, and mannan and β-glucan respectively, have been reported to be involved in the recognition of this pathogen [184,185].

Unlike *C. albicans* that put effort in escaping the immune system, it is hypothesized that inducing the recruitment of macrophages to the site of infection in vivo is part of the *C. glabrata* immune system evasion strategy [21]. *C. glabrata* infection did not substantially activate any MAPK pathway, including Erk1/2 (Extracellular signal-regulated kinases), SAPK/JNK (Stress-activated protein kinases/Jun amino-terminal kinases), and NF-κB signaling. Accordingly, it was found that upon infection of macrophages, the only cytokine significantly induced is GM-CSF, whereas the induction of other proinflammatory cytokines (TNF-α, IL-1β, IL-6, IL-8, and IFN-γ) is low [21,65,186]. GM-CSF is a potent activator of macrophages and induces differentiation of precursor cells as well as the recruitment of macrophages to sites of infection. This may explain the enhanced tissue infiltration of mononuclear cells, but not neutrophils, observed in vivo [186]. Considering the ability of *C. glabrata* to survive and replicate within macrophages, it is therefore speculated that persistence within macrophages is a possible strategy of immune evasion in this pathogen [21].

Immune evasion by pathogens also entails escaping from the complement system, another pathway of the innate immune system that facilitates phagocytosis. To evade immune response, several pathogens were shown to sequester or bind complement regulators, such as factor H (FH) [187–190]. *C. albicans* expresses the glucose transporter Hgt1 that binds FH, therefore reducing complement regulatory activity and limiting phagocytosis and killing by neutrophils [191]. SAP proteases produced by *C. albicans* not only cause tissue damage [192], but also contribute to immune evasion, as Sap2 is able to cleave antimicrobial peptides and complement proteins [193,194]. *C. albicans* Sap2 also cleaves FH and its receptors (CR3 and CR4) on macrophages, thus limiting their activation [195]. Additionally, *C. albicans* was also seen to bind yet another complement regulator (vitronectin) to modulate the hosts innate response [196]. The *C. albicans* pH-regulated protein Pra1 was recently implicated in the modulation of immune response [197,198]. It was seen to cleave the complement component C3, blocking the complement effector function and interfering with killing by neutrophils [197]. Pra1 is also implicated in adaptive immune response, as it binds to mouse CD4$^+$ T cells and reduces cytokine (IFNγ and TNF) secretion and antigen stimulation [198].

Ultimately, many yeast cells are engulfed by macrophages, hence survival and replication or subsequent escape remain important features. Upon phagocytosis, the phagosome carrying the ingested microorganism, fusion with a lysosome is one central antimicrobial mechanism of macrophages [164]. In the phagolysosome, fungal pathogens have to survive the harsh environment, typically characterized by carbon source limitation, production of reactive oxygen and nitrogen species, and acidification of the phagosomal compartment [199–201]. Mature phagolysosomes are normally strongly acidified, inducing antimicrobial effector mechanisms such as the activity of hydrolytic enzymes. Nonetheless, both *C. albicans* and *C. glabrata* are able to actively limit phagosome maturation in macrophages to prevent acidification and limit hydrolytic attack [21,202]. Environmental alkalinization by amino acid use as carbon sources, which results in ammonia extrusion, has been acknowledged as a strategy of these pathogens to actively raise phagosome pH [199,203,204]. Moreover, extracellular pH-raising triggers the yeast–hyphal switch in *C. albicans* [199,204]. Regarding carbon source availability, macrophages actively deprive pathogens of glucose, and therefore alternative carbon sources must support fungal growth. Accordingly, genes coding for enzymes of glyoxylate cycle, gluconeogenesis, and β-oxidation of fatty acids were found to be upregulated in both *C. glabrata* and *C. albicans* cells ingested by macrophages [13,18,205]. Very recently, the glyoxylate cycle gene *ICL1* was demonstrated to be crucial for the survival of *C. glabrata* in response to macrophage engulfment [206]. Disruption of *ICL1* rendered

C. glabrata cells unable to utilize acetate, ethanol, or oleic acid and conferred a severe attenuation of virulence in the mouse model of invasive candidiasis [206]. Further, genes of the methylcitrate cycle, which is important for the degradation of fatty acid chains and which allows the use of lipids as alternative carbon sources, were also found to be upregulated in *C. glabrata* [13]. Downregulation of protein synthesis as well as upregulation of amino acid biosynthetic pathways and amino acid and ammonium transport genes are features of the nitrogen deprivation faced by fungal cells inside the macrophages [13].

Although being mainly known by macrophage escape, strategies to counteract oxidative and nitrosative stress have been described in *C. albicans*. For instance, flavodoxin-like proteins are part of the antioxidant response of this species by reducing ubiquinone, which acts as a membrane antioxidant [207]. Expression of four flavodoxin genes (*PST1, PST2, PST3, YCP4*) is required for *C. albicans* virulence and resist neutrophil attack [207]. Other than oxidative burst modulation, *C. albicans* was also reported to modulate the nitrosative stress exerted by macrophages [208]. Nitric oxide production is dependent on the enzyme nitric oxide synthase, which utilizes arginine as a substrate. By increasing chitin exposure, *C. albicans* induces the host arginase-1 enzyme, which competes with the nitric oxide synthase enzyme and prevents the conversion of L-arginine to nitric oxide [208]. Interestingly, the use of amino acids as carbon source is much more prominent in *C. albicans* than other fungi [199] and amino acid metabolism has been associated with more than one mechanism of phagocyte escape/survival. Arginine was also found to play a role in hyphal development upon phagocytosis by macrophages, thus contributing to escaping the phagosome [209]. Additionally, in poor glucose conditions, *C. albicans* excretes amino acid-derived ammonia that increases external pH and interferes with the acidification of the phagosome [204,210], as mentioned before. Moreover, the transcription factor Stp2 involved in amino acid acquisition, is required to prevent phagosome acidification [204,209] based on the SPS amino acid sensing system [211]. Regarding the contribution of nitrosative stress to macrophage defense against *C. glabrata*, it is known that this pathogen induces only low NO production by murine macrophages [13].

C. glabrata shows increased tolerance to oxidative stress when compared to other yeasts, including *Saccharomyces cerevisiae* and *C. albicans* [212], mostly associated with the activity of the catalase Cta1, the superoxide dismutases Sod1/Sod2 and the glutathione and thioredoxin pathways [212–214]. Despite this, it is speculated that ROS play a minor role in killing *C. glabrata* cells, since experimental inhibition of ROS production in macrophages did not result in increased fungal survival [215]. In contrast to *C. albicans*, mobilization of intracellular resources via autophagy is an important virulence factor that supports the viability of *C. glabrata* in the phagosomal compartment of innate immune cells [23]. Phagocytosis induces peroxisome production in *C. glabrata* cells, which are then degraded via pexophagy, a specialized form of autophagy [23].

Moreover, besides carbon and nitrogen, trace elements such as iron are important for yeast growth. Macrophage-engulfed *C. albicans* upregulate a set of genes involved in iron homeostasis, for example, the ferric reductase genes *FRE3* and *FRE7*, as well as uptake systems for other trace metals, such as copper (*CTR1*) and zinc (*ZRT2*) [18]. *C. glabrata* is also able to sense and respond to iron limitation, although it has not been shown to use host iron-binding proteins as iron sources and is unable to use heme or hemoglobin. Instead, *C. glabrata* uses the siderophore-iron transporter Sit1, which is essential for utilization of ferrichrome as an iron source under iron-deficient conditions, and for iron-dependent survival in macrophages [216,217]. This iron acquisition system improves the fitness of *C. glabrata* when it is subsequently exposed to macrophages [216]. Recently, PI3K-kinase (encoded by *VPS34*) signaling was revealed to play a central role in *C. glabrata* iron metabolism and host colonization. However, the strategies by which *C. glabrata* gains iron within macrophages remain unknown [218].

While *C. glabrata* is most known for a persistence strategy and survival inside macrophages, due to its high-intrinsic stress tolerance, *C. albicans* is best known for active escape via hyphal growth and phagocyte piercing. In 2014, the model of macrophage piercing due to polarized growth of hyphae was challenged by two studies [17,219]. The study by Wellington et al. (2014) [219] showed that *C. albicans* escape is not exclusively due to disruption by hyphae. The pyroptosis pathway, a proinflammatory programmed cell-death process that is dependent on caspase-1, leads to interleukin production and macrophage lysis [220] and occurs concurrently with hyphae-mediated damage. Pyroptosis has been described to occur in response to intracellular bacteria, and thus is hypothesized to achieve the goal of destroying macrophages themselves in order to eradicate phagocyted pathogens [221]. *C. albicans* yeast-to-hyphae transition induces macrophage pyroptosis, therefore indicating that phagocyte damage caused by this pathogen is a more complex mechanism than originally postulated [17,219]. Accordingly, the study by Uwamahoro et al. (2014) [17] added more evidence of this escape mechanism. The triggering of pyroptosis is necessary for full macrophage damage upon hyphal formation and is activated in early phagocytosis, followed by a more robust hyphal formation that is indeed the main mechanism of macrophage killing in a later phase [17].

Candida thrives on multiple carbon sources to survive inside macrophages, but these depend on glucose for viability. Recently, Tucey et al. (2018) [222] demonstrated that *C. albicans* exploits this limitation by depleting glucose, and triggering rapid macrophage death, in vitro. Additionally, they showed that *C. albicans* infection promotes the disruption of host glucose homeostasis in vivo and verified that glucose supplementation improves host outcomes under systemic fungal infection [222]. Thus, depriving host immune cells for glucose seems to be one mechanism of *C. albicans* to induce phagocyte cell death and actively escape from those. Moreover, it was recently discovered that candidalysin is both a central trigger for Nlrp3 inflammasome-dependent caspase-1 activation via potassium efflux, and a key driver of inflammasome-independent cytolysis of macrophages and dendritic cells upon infection with *C. albicans* [223]. This study suggests that candidalysin-induced cell damage is a third mechanism by which *C. albicans* induces phagocyte cell death in addition to damage caused by pyroptosis and the growth of glucose-consuming hyphae [223]. *C. albicans*-induced activation of the Nlrp3 inflammasome, leading to secretion of IL-1β cytokine, is a crucial myeloid cell immune response needed for antifungal host defense [224]. Very recently, Rogiers and co-workers (2019) [225] identified candidalysin as the fungal trigger for Nlrp3 inflammasome-mediated maturation and secretion of IL-1β from primary macrophages. Therefore, the expression of candidalysin is speculated to be one of the molecular mechanisms by which hyphal transformation equips *C. albicans* with its proinflammatory capacity to prompt the release of bioactive IL-1β from macrophages [225].

A more controversial evasion mechanism, based on quorum-sensing stimulation of immune recognition, has been reported in both *C. albicans* and *C. glabrata*. In the case of *C. albicans*, white cells specifically (not opaque cells) secrete E,E-Farnesol, a stimulator of macrophage chemokinesis [226]. Farnesol secretion was found to increase macrophage migration and tissue infiltration [226]. This strategy has been associated with immune evasion by the concealing of the pathogens inside immune cells themselves, an environment where *Candida* species can survive.

Conversely to *C. albicans*, *C. glabrata* is incapable of hyphal differentiation, failing at this virulence trait. Although never observed for clinical isolates, interestingly, work by Brunke et al. (2014) [227] has shown that *C. glabrata* cells co-incubated during six months with macrophages, were able to produce pseudohyphae structures and evolve into a hypervirulent phenotype characterized by higher macrophage damage and faster escape.

Overall, the two more prevalent pathogenic yeasts, *C. albicans* and *C. glabrata*, follow two main different strategies to achieve the same ultimate goal: survive host immune response. On one hand, *C. albicans* hyphal forms actively pierce the membrane of macrophages as a mechanism of killing and escape. On the other hand, *C. glabrata*, that lacks morphological plasticity, survives and replicates within macrophages due to its remarkable ability to surpass its harsh environment, ultimately leading to macrophage lysis after several days due to fungal cells overload [21] (Figure 3).

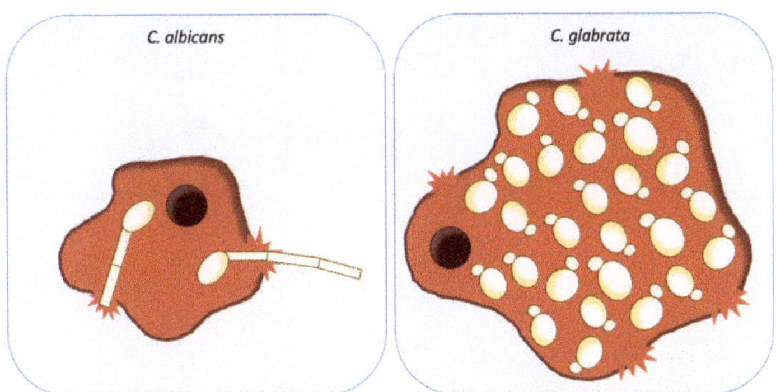

Figure 3. Schematic representation of *C. albicans* and *C. glabrata* host immune system evasion. *C. albicans* actively escape from host immune system cells through hyphal growth and phagocyte piercing. *C. glabrata* is most known for a persistence strategy, surviving and thriving inside macrophages, ultimately leading to immune cells lysis due to fungal load.

5. Conclusion and Perspectives

The pathogenic yeasts *C. albicans* and *C. glabrata* are the two most prevalent *Candida* species isolated from candidiasis patients worldwide, yet they are phylogenetically, genetically, and phenotypically very different. Indeed, each species displays divergent virulence traits, indicating differential adaptation to the human host.

A greater number of virulence mechanisms has been described in *C. albicans*, most of which are not known to occur in *C. glabrata*. Yeast-to-hyphae dimorphism is one of the most striking divergent features. Hyphal morphology is associated with several key *C. albicans* traits (tissue invasion, biofilm formation, immune evasion), but is absent in *C. glabrata*. The result is that *C. glabrata* must have acquired other molecular mechanisms to reach the same goals. While tissue invasion in *C. albicans* occurs by proteolytic enzyme secretion and hyphal penetration of host tissues, *C. glabrata*-induced tissue damage is quite negligent in comparison and is thought to occur via endocytosis induction by host cells.

A similar observation can be made regarding biofilm formation, where *C. albicans* expresses hyphal-specific adhesins and regulators required for adhesion, while *C. glabrata* biofilms are much less "bulky". Morphological dimorphism also supports noticeable differences in immune evasion, as hyphal development and phagosome piercing is the main phagocyte escape mechanism employed by *C. albicans*, while such a strategy is absent in *C. glabrata*, which rather survives (and thrives) in the phagosome.

Dissimilar traits between *C. albicans* and *C. glabrata* have been identified, and such attributes have been extensively studied in *C. albicans*. However, the molecular mechanisms specific to *C. glabrata* that allow such a disparate species to cause human candidiasis demand further scrutiny. Indeed, the up rise of *C. glabrata* as a key fungal pathogen can only be prevented with specific therapeutic options that match its specific virulence traits.

Author Contributions: M.G., P.P., M.C., D.P., R.V. and M.C.T. contributed to the literature review and manuscript drafting, each author in their specific field of expertise. M.G. wrote the manuscript and prepared the illustrative figures. M.C.T. conceived the manuscript structure, supervised its writing, and reviewed all contents thoroughly.

Funding: Work conducted in this field was supported by "Fundação para a Ciência e a Tecnologia" (FCT) (Contracts PTDC/BBB-BIO/4004/2014 and PTDC/BII-BIO/28216/2017, PhD grants to MG, PP, and MC). Funding received by iBB-Institute for Bioengineering and Biosciences from FCT-Portuguese Foundation for Science and Technology (UID/BIO/04565/2013) and from Programa Operacional Regional de Lisboa 2020 (Project N. 007317) is acknowledged.

Conflicts of Interest: The authors declare no conflict of interest.

References

1. Brown, G.D.; Denning, D.W.; Gow, N.A.R.; Levitz, S.M.; Netea, M.G.; White, T.C. Hidden Killers: Human Fungal Infections. *Sci. Transl. Med.* **2012**, *4*, 165rv13. [CrossRef] [PubMed]
2. Bassetti, M.; Peghin, M.; Timsit, J.-F. The current treatment landscape: Candidiasis. *J. Antimicrob. Chemother.* **2016**, *71*, ii13–ii22. [CrossRef]
3. Olson, M.L.; Jayaraman, A.; Kao, K.C. Relative Abundances of *Candida albicans* and *Candida glabrata* in In Vitro Coculture Biofilms Impact Biofilm Structure and Formation. *Appl. Environ. Microbiol.* **2018**, *84*, e02769-17. [CrossRef]
4. Cho, I.; Blaser, M.J. The human microbiome: At the interface of health and disease. *Nat. Rev. Genet.* **2012**, *13*, 260–270. [CrossRef] [PubMed]
5. Underhill, D.M.; Iliev, I.D. The mycobiota: Interactions between commensal fungi and the host immune system. *Nat. Rev. Immunol.* **2014**, *14*, 405–416. [CrossRef]
6. Wisplinghoff, H.; Bischoff, T.; Tallent, S.M.; Seifert, H.; Wenzel, R.P.; Edmond, M.B. Nosocomial Bloodstream Infections in US Hospitals: Analysis of 24,179 Cases from a Prospective Nationwide Surveillance Study. *Clin. Infect. Dis.* **2004**, *39*, 309–317. [CrossRef]
7. Jones, T.; Federspiel, N.A.; Chibana, H.; Dungan, J.; Kalman, S.; Magee, B.B.; Newport, G.; Thorstenson, Y.R.; Agabian, N.; Magee, P.T.; et al. The diploid genome sequence of *Candida albicans*. *Proc. Natl. Acad. Sci. USA* **2004**, *101*, 7329–7334. [CrossRef] [PubMed]
8. Hoyer, L.L. The ALS gene family of *Candida albicans*. *Trends Microbiol.* **2001**, *9*, 176–180. [CrossRef]
9. Ghannoum, M.A. Potential role of phospholipases in virulence and fungal pathogenesis. *Clin. Microbiol. Rev.* **2000**, *13*, 122–143. [CrossRef] [PubMed]
10. Calderone, R.A.; Fonzi, W.A. Virulence factors of *Candida albicans*. *Trends Microbiol.* **2001**, *9*, 327–335. [CrossRef]
11. Li, L.; Kashleva, H.; Dongari-Bagtzoglou, A. Cytotoxic and cytokine-inducing properties of *Candida glabrata* in single and mixed oral infection models. *Microb. Pathog.* **2007**, *42*, 138–147. [CrossRef] [PubMed]
12. de Groot, P.W.J.; Kraneveld, E.A.; Yin, Q.Y.; Dekker, H.L.; Groß, U.; Crielaard, W.; de Koster, C.G.; Bader, O.; Klis, F.M.; Weig, M. The Cell Wall of the Human Pathogen *Candida glabrata*: Differential Incorporation of Novel Adhesin-Like Wall Proteins. *Eukaryot. Cell* **2008**, *7*, 1951–1964. [CrossRef]
13. Kaur, R.; Ma, B.; Cormack, B.P. A family of glycosylphosphatidylinositol-linked aspartyl proteases is required for virulence of *Candida glabrata*. *Proc. Natl. Acad. Sci. USA* **2007**, *104*, 7628–7633. [CrossRef]
14. Kaur, R.; Domergue, R.; Zupancic, M.L.; Cormack, B.P. A yeast by any other name: *Candida glabrata* and its interaction with the host. *Curr. Opin. Microbiol.* **2005**, *8*, 378–384. [CrossRef] [PubMed]
15. Brunke, S.; Hube, B. Two unlike cousins: *Candida albicans* and *C. glabrata* infection strategies. *Cell. Microbiol.* **2013**, *15*, 701–708. [CrossRef] [PubMed]
16. Kasper, L.; Seider, K.; Hube, B. Intracellular survival of *Candida glabrata* in macrophages: Immune evasion and persistence. *FEMS Yeast Res.* **2015**, *15*, fov042. [CrossRef] [PubMed]
17. Uwamahoro, N.; Verma-Gaur, J.; Shen, H.H.; Qu, Y.; Lewis, R.; Lu, J.; Bambery, K.; Masters, S.L.; Vince, J.E.; Naderer, T.; et al. The pathogen *Candida albicans* hijacks pyroptosis for escape from macrophages. *MBio* **2014**, *5*, e00003-14. [CrossRef]
18. Lorenz, M.C.; Bender, J.A.; Fink, G.R. Transcriptional response of *Candida albicans* upon internalization by macrophages. *Eukaryot. Cell* **2004**, *3*, 1076–1087. [CrossRef]
19. Mayer, F.L.; Wilson, D.; Hube, B. *Candida albicans* pathogenicity mechanisms. *Virulence* **2013**, *4*, 119–128. [CrossRef]
20. Cavalheiro, M.; Teixeira, M.C. Candida Biofilms: Threats, Challenges, and Promising Strategies. *Front. Med.* **2018**, *5*, 28. [CrossRef]
21. Seider, K.; Brunke, S.; Schild, L.; Jablonowski, N.; Wilson, D.; Majer, O.; Barz, D.; Haas, A.; Kuchler, K.; Schaller, M.; et al. The facultative intracellular pathogen *Candida glabrata* subverts macrophage cytokine production and phagolysosome maturation. *J. Immunol.* **2011**, *187*, 3072–3086. [CrossRef]
22. Dementhon, K.; El-Kirat-Chatel, S.; Noël, T. Development of an in vitro model for the Multi-Parametric quantification of the cellular interactions between Candida yeasts and phagocytes. *PLoS ONE* **2012**, *7*, e32621. [CrossRef] [PubMed]

23. Roetzer, A.; Gratz, N.; Kovarik, P.; Schäller, C. Autophagy supports *Candida glabrata* survival during phagocytosis. *Cell. Microbiol.* **2010**, *12*, 199–216. [CrossRef] [PubMed]

24. Yan, L.; Yang, C.; Tang, J. Disruption of the intestinal mucosal barrier in *Candida albicans* infections. *Microbiol. Res.* **2013**, *168*, 389–395. [CrossRef]

25. Allert, S.; Förster, T.M.; Svensson, C.-M.; Richardson, J.P.; Pawlik, T.; Hebecker, B.; Rudolphi, S.; Juraschitz, M.; Schaller, M.; Blagojevic, M.; et al. *Candida albicans*-Induced Epithelial Damage Mediates Translocation through Intestinal Barriers. *MBio* **2018**, *9*, e00915-18. [CrossRef]

26. Cassone, A. Vulvovaginal *Candida albicans* infections: Pathogenesis, immunity and vaccine prospects. *BJOG Int. J. Obstet. Gynaecol.* **2015**, *122*, 785–794. [CrossRef]

27. Bokor-Bratic, M.; Cankovic, M.; Dragnic, N. Unstimulated whole salivary flow rate and anxiolytics intake are independently associated with oral *Candida* infection in patients with oral lichen planus. *Eur. J. Oral Sci.* **2013**, *121*, 427–433. [CrossRef]

28. Hibino, K.; Samaranayake, L.P.; Hägg, U.; Wong, R.W.K.; Lee, W. The role of salivary factors in persistent oral carriage of Candida in humans. *Arch. Oral Biol.* **2009**, *54*, 678–683. [CrossRef]

29. Goyer, M.; Loiselet, A.; Bon, F.; L'Ollivier, C.; Laue, M.; Holland, G.; Bonnin, A.; Dalle, F. Intestinal Cell Tight Junctions Limit Invasion of *Candida albicans* through Active Penetration and Endocytosis in the Early Stages of the Interaction of the Fungus with the Intestinal Barrier. *PLoS ONE* **2016**, *11*, e0149159. [CrossRef]

30. Dalle, F.; Wächtler, B.; L'Ollivier, C.; Holland, G.; Bannert, N.; Wilson, D.; Labruäre, C.; Bonnin, A.; Hube, B. Cellular interactions of *Candida albicans* with human oral epithelial cells and enterocytes. *Cell. Microbiol.* **2010**, *12*, 248–271. [CrossRef]

31. Phan, Q.T.; Myers, C.L.; Fu, Y.; Sheppard, D.C.; Yeaman, M.R.; Welch, W.H.; Ibrahim, A.S.; Edwards, J.E.; Filler, S.G. Als3 Is a *Candida albicans* Invasin That Binds to Cadherins and Induces Endocytosis by Host Cells. *PLoS Biol.* **2007**, *5*, e64. [CrossRef]

32. Sun, J.N.; Solis, N.V.; Phan, Q.T.; Bajwa, J.S.; Kashleva, H.; Thompson, A.; Liu, Y.; Dongari-Bagtzoglou, A.; Edgerton, M.; Filler, S.G. Host Cell Invasion and Virulence Mediated by *Candida albicans* Ssa1. *PLoS Pathog.* **2010**, *6*, e1001181. [CrossRef]

33. Zakikhany, K.; Naglik, J.R.; Schmidt-Westhausen, A.; Holland, G.; Schaller, M.; Hube, B. In vivo transcript profiling of *Candida albicans* identifies a gene essential for interepithelial dissemination. *Cell. Microbiol.* **2007**, *9*, 2938–2954. [CrossRef] [PubMed]

34. Park, H.; Myers, C.L.; Sheppard, D.C.; Phan, Q.T.; Sanchez, A.A.; Edwards, J.E.; Filler, S.G. Role of the fungal Ras-protein kinase A pathway in governing epithelial cell interactions during oropharyngeal candidiasis. *Cell. Microbiol.* **2004**, *7*, 499–510. [CrossRef] [PubMed]

35. Li, Z.; Nielsen, K. Morphology Changes in Human Fungal Pathogens upon Interaction with the Host. *J. Fungi* **2017**, *3*, 66. [CrossRef] [PubMed]

36. Jacobsen, I.D.; Wilson, D.; Wächtler, B.; Brunke, S.; Naglik, J.R.; Hube, B. *Candida albicans* dimorphism as a therapeutic target. *Expert Rev. Anti-Infect. Ther.* **2012**, *10*, 85–93. [CrossRef]

37. Feng, Q.; Summers, E.; Guo, B.; Fink, G. Ras signaling is required for serum-induced hyphal differentiation in *Candida albicans*. *J. Bacteriol.* **1999**, *181*, 6339–6346.

38. Maidan, M.M.; De Rop, L.; Serneels, J.; Exler, S.; Rupp, S.; Tournu, H.; Thevelein, J.M.; Van Dijck, P. The G Protein-coupled Receptor Gpr1 and the Gα Protein Gpa2 Act through the cAMP-Protein Kinase A Pathway to Induce Morphogenesis in *Candida albicans*. *Mol. Biol. Cell* **2005**, *16*, 1971–1986. [CrossRef]

39. Biswas, K.; Morschhäuser, J. The Mep2p ammonium permease controls nitrogen starvation-induced filamentous growth in *Candida albicans*. *Mol. Microbiol.* **2005**, *56*, 649–669. [CrossRef]

40. Stoldt, V.R.; Sonneborn, A.; Leuker, C.E.; Ernst, J.F. Efg1p, an essential regulator of morphogenesis of the human pathogen *Candida albicans*, is a member of a conserved class of bHLH proteins regulating morphogenetic processes in fungi. *EMBO J.* **1997**, *16*, 1982–1991. [CrossRef]

41. Zeidler, U.; Lettner, T.; Lassnig, C.; Müller, M.; Lajko, R.; Hintner, H.; Breitenbach, M.; Bito, A. *UME6* is a crucial downstream target of other transcriptional regulators of true hyphal development in *Candida albicans*. *FEMS Yeast Res.* **2009**, *9*, 126–142. [CrossRef]

42. Murad, A.M.A.; Leng, P.; Straffon, M.; Wishart, J.; Macaskill, S.; MacCallum, D.; Schnell, N.; Talibi, D.; Marechal, D.; Tekaia, F.; et al. NRG1 represses yeast-hypha morphogenesis and hypha-specific gene expression in *Candida albicans*. *EMBO J.* **2001**, *20*, 4742–4752. [CrossRef]

43. Braun, B.R.; Kadosh, D.; Johnson, A.D. NRG1, a repressor of filamentous growth in C.albicans, is down-regulated during filament induction. *EMBO J.* **2001**, *20*, 4753–4761. [CrossRef] [PubMed]

44. Watts, H.J.; Veacutery, A.-A.; Perera, T.H.S.; Davies, J.M.; Gow, N.A.R. Thigmotropism and stretch-activated channels in the pathogenic fungus *Candida albicans*. *Microbiology* **1998**, *144*, 689–695. [CrossRef] [PubMed]

45. Tscherner, M.; Giessen, T.W.; Markey, L.; Kumamoto, C.A.; Silver, P.A. A Synthetic System That Senses *Candida albicans* and Inhibits Virulence Factors. *ACS Synth. Biol.* **2019**, *8*, 434–444. [CrossRef]

46. Chen, H.; Fujita, M.; Feng, Q.; Clardy, J.; Fink, G.R. Tyrosol is a quorum-sensing molecule in *Candida albicans*. *Proc. Natl. Acad. Sci. USA* **2004**, *101*, 5048–5052. [CrossRef] [PubMed]

47. Naglik, J.R.; Rodgers, C.A.; Shirlaw, P.J.; Dobbie, J.L.; Fernandes-Naglik, L.L.; Greenspan, D.; Agabian, N.; Challacombe, S.J. Differential Expression of *Candida albicans* Secreted Aspartyl Proteinase and Phospholipase B Genes in Humans Correlates with Active Oral and Vaginal Infections. *J. Infect. Dis.* **2003**, *188*, 469–479. [CrossRef] [PubMed]

48. Hube, B.; Naglik, J. *Candida albicans* proteinases: Resolving the mystery of a gene family. *Microbiology* **2001**, *147*, 1997–2005. [CrossRef] [PubMed]

49. Borg-von Zepelin, M.; Beggah, S.; Boggian, K.; Sanglard, D.; Monod, M. The expression of the secreted aspartyl proteinases Sap4 to Sap6 from *Candida albicans* in murine macrophages. *Mol. Microbiol.* **1998**, *28*, 543–554. [CrossRef]

50. Bizzarri, M.; Fuso, A.; Dinicola, S.; Cucina, A.; Bevilacqua, A. Pharmacodynamics and pharmacokinetics of inositol(s) in health and disease. *Expert Opin. Drug Metab. Toxicol.* **2016**, *12*, 1181–1196. [CrossRef]

51. Tsang, P.W.-K.; Fong, W.-P.; Samaranayake, L.P. *Candida albicans* orf19.3727 encodes phytase activity and is essential for human tissue damage. *PLoS ONE* **2017**, *12*, e0189219. [CrossRef] [PubMed]

52. Lionakis, M.S.; Lim, J.K.; Lee, C.-C.R.; Murphy, P.M. Organ-Specific Innate Immune Responses in a Mouse Model of Invasive Candidiasis. *J. Innate Immunity* **2011**, *3*, 180–199. [CrossRef]

53. Saville, S.P.; Lazzell, A.L.; Monteagudo, C.; Lopez-Ribot, J.L. Engineered control of cell morphology in vivo reveals distinct roles for yeast and filamentous forms of *Candida albicans* during infection. *Eukaryot. Cell* **2003**, *2*, 1053–1060. [CrossRef] [PubMed]

54. Csank, C.; Haynes, K. *Candida glabrata* displays pseudohyphal growth. *FEMS Microbiol. Lett.* **2000**, *189*, 115–120. [CrossRef]

55. Sasani, E.; Khodavaisy, S.; Agha Kuchak Afshari, S.; Darabian, S.; Aala, F.; Rezaie, S. Pseudohyphae formation in *Candida glabrata* due to CO_2 exposure. *Curr. Med. Mycol.* **2016**, *2*, 49–52. [CrossRef] [PubMed]

56. Perlroth, J.; Choi, B.; Spellberg, B. Nosocomial fungal infections: Epidemiology, diagnosis, and treatment. *Med. Mycol.* **2007**, *45*, 321–346. [CrossRef] [PubMed]

57. Atanasova, R.; Angoulvant, A.; Tefit, M.; Gay, F.; Guitard, J.; Mazier, D.; Fairhead, C.; Hennequin, C. A mouse model for *Candida glabrata* hematogenous disseminated infection starting from the gut: Evaluation of strains with different adhesion properties. *PLoS ONE* **2013**, *8*, e69664. [CrossRef]

58. Jacobsen, I.D.; Große, K.; Berndt, A.; Hube, B. Pathogenesis of *Candida albicans* infections in the alternative chorio-allantoic membrane chicken embryo model resembles systemic murine infections. *PLoS ONE* **2011**, *6*, e19741. [CrossRef]

59. Silva, S.; Henriques, M.C.; Hayes, A.; Oliveira, R.; Azeredo, J.; Williams, D.W. *Candida glabrata* and *Candida albicans* co-infection of an in vitro oral epithelium. *J. Oral Pathol. Med.* **2011**, *40*, 421–427. [CrossRef]

60. Coco, B.J.; Bagg, J.; Cross, L.J.; Jose, A.; Cross, J.; Ramage, G. Mixed *Candida albicans* and *Candida glabrata* populations associated with the pathogenesis of denture stomatitis. *Oral Microbiol. Immunol.* **2008**, *23*, 377–383. [CrossRef]

61. Okada, K.; Nakazawa, S.; Yokoyama, A.; Kashiwazaki, H.; Kobayashi, K.; Yamazaki, Y. A Clinical Study of *Candida albicans* and *Candida glabrata* Co-infection of Oral Candidiasis. *Ronen Shika Igaku* **2016**, *31*, 346–353.

62. Tsay, S.; Williams, S.R.; Benedict, K.; Beldavs, Z.; Farley, M.; Harrison, L.; Schaffner, W.; Dumyati, G.; Blackstock, A.; Guh, A.; et al. A Tale of Two Healthcare-associated Infections: Clostridium difficile Coinfection Among Patients With Candidemia. *Clin. Infect. Dis.* **2018**, *64*, 676–679. [CrossRef]

63. Tati, S.; Davidow, P.; McCall, A.; Hwang-Wong, E.; Rojas, I.G.; Cormack, B.; Edgerton, M. *Candida glabrata* Binding to *Candida albicans* Hyphae Enables Its Development in Oropharyngeal Candidiasis. *PLoS Pathog.* **2016**, *12*, e1005522. [CrossRef]

64. Alves, C.T.; Wei, X.Q.; Silva, S.; Azeredo, J.; Henriques, M.; Williams, D.W. *Candida albicans* promotes invasion and colonisation of *Candida glabrata* in a reconstituted human vaginal epithelium. *J. Infect.* **2014**, *69*, 396–407. [CrossRef] [PubMed]

65. Li, L.; Dongari-Bagtzoglou, A. Epithelial GM-CSF induction by *Candida glabrata*. *J. Dent. Res.* **2009**, *88*, 746–751. [CrossRef] [PubMed]

66. Fidel, P.L.; Vazquez, J.A.; Sobel, J.D. *Candida glabrata*: Review of epidemiology, pathogenesis, and clinical disease with comparison to *C. albicans*. *Clin. Microbiol. Rev.* **1999**, *12*, 80–96. [CrossRef]

67. Fatahinia, M.; Halvaeezadeh, M.; Rezaei-Matehkolaei, A. Comparison of enzymatic activities in different Candida species isolated from women with vulvovaginitis. *J. Mycol. Med.* **2017**, *27*, 188–194. [CrossRef]

68. Bassyouni, R.H.; Wegdan, A.A.; Abdelmoneim, A.; Said, W.; Aboelnaga, F. Phospholipase and aspartyl proteinase activities of candida species causing vulvovaginal candidiasis in patients with type 2 diabetes mellitus. *J. Microbiol. Biotechnol.* **2015**, *25*, 1734–1741. [CrossRef]

69. De Riceto, É.B.M.; de Menezes, R.P.; Penatti, M.P.A.; dos Pedroso, R.S. Enzymatic and hemolytic activity in different Candida species. *Rev. Iberoam. Micol.* **2015**, *32*, 79–82. [CrossRef] [PubMed]

70. Atalay, M.A.; Koc, A.N.; Demir, G.; Sav, H. Investigation of possible virulence factors in Candida strains isolated from blood cultures. *Niger. J. Clin. Pract.* **2015**, *18*, 52–55. [PubMed]

71. Pandey, N.; Gupta, M.K.; Tilak, R. Extracellular hydrolytic enzyme activities of the different Candida spp. isolated from the blood of the Intensive Care Unit-admitted patients. *J. Lab. Phys.* **2018**, *10*, 392–396.

72. Canela, H.M.S.; Cardoso, B.; Vitali, L.H.; Coelho, H.C.; Martinez, R.; da Ferreira, M.E.S. Prevalence, virulence factors and antifungal susceptibility of Candida spp. isolated from bloodstream infections in a tertiary care hospital in Brazil. *Mycoses* **2018**, *61*, 11–21. [CrossRef]

73. Rossoni, R.D.; Barbosa, J.O.; Vilela, S.F.G.; Dos Santos, J.D.; Jorge, A.O.C.; Junqueira, J.C. Correlation of phospholipase and proteinase production of Candida with in vivo pathogenicity in *Galleria mellonella*. *Braz. J. Oral Sci.* **2013**, *12*, 199–204. [CrossRef]

74. Naglik, J.R.; Challacombe, S.J.; Hube, B. *Candida albicans* Secreted Aspartyl Proteinases in Virulence and Pathogenesis. *Microbiol. Mol. Biol. Rev.* **2003**, *67*, 400–428. [CrossRef]

75. Parra-Ortega, B.; Cruz-Torres, H.; Villa-Tanaca, L.; Hernández-Rodríguez, C. Phylogeny and evolution of the aspartyl protease family from clinically relevant Candida species. *Mem. Inst. Oswaldo Cruz* **2009**, *104*, 505–512. [CrossRef] [PubMed]

76. Rodrigues, C.F.; Silva, S.; Henriques, M. *Candida glabrata*: A review of its features and resistance. *Eur. J. Clin. Microbiol. Infect. Dis.* **2014**, *33*, 673–688. [CrossRef] [PubMed]

77. Sharma, Y.; Chumber, S.; Kaur, M. Studying the prevalence, species distribution, and detection of in vitro production of phospholipase from *Candida isolated* from cases of invasive Candidiasis. *J. Glob. Infect. Dis.* **2017**, *9*, 8. [CrossRef] [PubMed]

78. Rodrigues, C.; Rodrigues, M.; Silva, S.; Henriques, M. *Candida glabrata* Biofilms: How Far Have We Come? *J. Fungi* **2017**, *3*, 11. [CrossRef]

79. Kumar, K.; Askari, F.; Sahu, M.; Kaur, R. *Candida glabrata*: A Lot More Than Meets the Eye. *Microorganisms* **2019**, *7*, 39. [CrossRef]

80. Hasan, F.; Xess, I.; Wang, X.; Jain, N.; Fries, B.C. Biofilm formation in clinical *Candida* isolates and its association with virulence. *Microbes Infect.* **2009**, *11*, 753–761. [CrossRef]

81. Finkel, J.S.; Mitchell, A.P. Genetic control of *Candida albicans* biofilm development. *Nat. Rev. Microbiol.* **2011**, *9*, 109–118. [CrossRef] [PubMed]

82. Verstrepen, K.J.; Klis, F.M. Flocculation, adhesion and biofilm formation in yeasts. *Mol. Microbiol.* **2006**, *60*, 5–15. [CrossRef] [PubMed]

83. Richardson, J.P.; Ho, J.; Naglik, J.R. Candida-Epithelial Interactions. *J. Fungi* **2018**, *4*, 22. [CrossRef] [PubMed]

84. Douglas, L.J. Adhesion of Candida species to epithelial surfaces. *Crit. Rev. Microbiol.* **1987**, *15*, 27–43. [CrossRef]

85. Höfs, S.; Mogavero, S.; Hube, B. Interaction of *Candida albicans* with host cells: Virulence factors, host defense, escape strategies, and the microbiota. *J. Microbiol.* **2016**, *54*, 149–169. [CrossRef]

86. de Groot, P.W.J.; Bader, O.; de Boer, A.D.; Weig, M.; Chauhan, N. Adhesins in human fungal pathogens: Glue with plenty of stick. *Eukaryot. Cell* **2013**, *12*, 470–481. [CrossRef]

87. López-Fuentes, E.; Gutiérrez-Escobedo, G.; Timmermans, B.; Van Dijck, P.; De Las Peñas, A.; Castaño, I. *Candida glabrata*'s Genome Plasticity Confers a Unique Pattern of Expressed Cell Wall Proteins. *J. Fungi* **2018**, *4*, 67. [CrossRef] [PubMed]

88. Luo, G.; Samaranayake, L.P. *Candida glabrata*, an emerging fungal pathogen, exhibits superior relative cell surface hydrophobicity and adhesion to denture acrylic surfaces compared with *Candida albicans*. *APMIS* **2002**, *110*, 601–610. [CrossRef] [PubMed]

89. Blanco, M.T.; Sacristán, B.; Lucio, L.; Blanco, J.; Pérez-Giraldo, C.; Gómez-García, A.C. Cell surface hydrophobicity as an indicator of other virulence factors in *Candida albicans*. *Rev. Iberoam. Micol.* **2010**, *27*, 195–199. [CrossRef] [PubMed]

90. Hazen, K.C.; Plotkin, B.J.; Klimas, D.M. Influence of growth conditions on cell surface hydrophobicity of *Candida albicans* and *Candida glabrata*. *Infect. Immunity* **1986**, *54*, 269–271.

91. Hoyer, L.L.; Green, C.B.; Oh, S.-H.; Zhao, X. Discovering the secrets of the *Candida albicans* agglutinin-like sequence (ALS) gene family—A sticky pursuit. *Med. Mycol.* **2008**, *46*, 1–15. [CrossRef]

92. Cormack, B.; Zordan, R. Adhesins in Opportunistic Fungal Pathogens. In *Candida and Candidiasis*, 2nd ed.; American Society of Microbiology: Washington, DC, USA, 2012; pp. 243–259.

93. Wächtler, B.; Wilson, D.; Haedicke, K.; Dalle, F.; Hube, B. From attachment to damage: Defined genes of *Candida albicans* mediate adhesion, invasion and damage during interaction with oral epithelial cells. *PLoS ONE* **2011**, *6*, e17046. [CrossRef]

94. Nobile, C.J.; Schneider, H.A.; Nett, J.E.; Sheppard, D.C.; Filler, S.G.; Andes, D.R.; Mitchell, A.P. Complementary adhesin function in *C. albicans* biofilm formation. *Curr. Biol.* **2008**, *18*, 1017–1024. [CrossRef]

95. Younes, S.; Bahnan, W.; Dimassi, H.I.; Khalaf, R.A. The *Candida albicans* Hwp2 is necessary for proper adhesion, biofilm formation and oxidative stress tolerance. *Microbiol. Res.* **2011**, *166*, 430–436. [CrossRef] [PubMed]

96. Li, F.; Svarovsky, M.J.; Karlsson, A.J.; Wagner, J.P.; Marchillo, K.; Oshel, P.; Andes, D.; Palecek, S.P. Eap1p, an adhesin that mediates *Candida albicans* biofilm formation in vitro and in vivo. *Eukaryot. Cell* **2007**, *6*, 931–939. [CrossRef] [PubMed]

97. Li, F.; Palecek, S.P. EAP1, a *Candida albicans* gene involved in binding human epithelial cells. *Eukaryot. Cell* **2003**, *2*, 1266–1273. [CrossRef] [PubMed]

98. Semlali, A.; Killer, K.; Alanazi, H.; Chmielewski, W.; Rouabhia, M. Cigarette smoke condensate increases *C. albicans* adhesion, growth, biofilm formation, and EAP1, HWP1 and SAP2 gene expression. *BMC Microbiol.* **2014**, *14*, 61. [CrossRef] [PubMed]

99. Cormack, B.P.; Ghori, N.; Falkow, S. An adhesin of the yeast pathogen *Candida glabrata* mediating adherence to human epithelial cells. *Science* **1999**, *285*, 578–582. [CrossRef]

100. Halliwell, S.C.; Smith, M.C.A.; Muston, P.; Holland, S.L.; Avery, S.V. Heterogeneous expression of the virulence-related adhesin epa1 between individual cells and strains of the pathogen *Candida glabrata*. *Eukaryot. Cell* **2012**, *11*, 141–150. [CrossRef]

101. De Las Peñas, A.; Pan, S.J.; Castaño, I.; Alder, J.; Cregg, R.; Cormack, B.P. Virulence-related surface glycoproteins in the yeast pathogen *Candida glabrata* are encoded in subtelomeric clusters and subject to RAP1- and SIR-dependent transcriptional silencing. *Genes Dev.* **2003**, *17*, 2245–2258. [CrossRef]

102. Domergue, R.; Castaño, I.; De Las Peñas, A.; Zupancic, M.; Lockatell, V.; Hebel, J.R.; Johnson, D.; Cormack, B.P. Nicotinic acid limitation regulates silencing of Candida adhesins during UTI. *Science* **2005**, *308*, 866–870. [CrossRef] [PubMed]

103. Riera, M.; Mogensen, E.; d'Enfert, C.; Janbon, G. New regulators of biofilm development in *Candida glabrata*. *Res. Microbiol.* **2012**, *163*, 297–307. [CrossRef] [PubMed]

104. Iraqui, I.; Garcia-Sanchez, S.; Aubert, S.; Dromer, F.; Ghigo, J.M.; D'Enfert, C.; Janbon, G. The Yak1p kinase controls expression of adhesins and biofilm formation in *Candida glabrata* in a Sir4p-dependent pathway. *Mol. Microbiol.* **2005**, *55*, 1259–1271. [CrossRef]

105. Kraneveld, E.A.; de Soet, J.J.; Deng, D.M.; Dekker, H.L.; de Koster, C.G.; Klis, F.M.; Crielaard, W.; de Groot, P.W.J. Identification and Differential Gene Expression of Adhesin-Like Wall Proteins in *Candida glabrata* Biofilms. *Mycopathologia* **2011**, *172*, 415–427. [CrossRef]

106. Gómez-Molero, E.; de Boer, A.D.; Dekker, H.L.; Moreno-Martínez, A.; Kraneveld, E.A.; Chauhan, N.; Weig, M.; de Soet, J.J.; de Koster, C.G.; et al. Proteomic analysis of hyperadhesive *Candida glabrata* clinical isolates reveals a core wall proteome and differential incorporation of adhesins. *FEMS Yeast Res.* **2015**, *15*, fov098. [CrossRef]

107. Leiva-Peláez, O.; Gutiérrez-Escobedo, G.; López-Fuentes, E.; Cruz-Mora, J.; De Las Peñas, A.; Castaño, I. Molecular characterization of the silencing complex SIR in *Candida glabrata* hyperadherent clinical isolates. *Fungal Genet. Biol.* **2018**, *118*, 21–31. [CrossRef]

108. Ene, I.V.; Adya, A.K.; Wehmeier, S.; Brand, A.C.; MacCallum, D.M.; Gow, N.A.R.; Brown, A.J.P. Host carbon sources modulate cell wall architecture, drug resistance and virulence in a fungal pathogen. *Cell. Microbiol.* **2012**, *14*, 1319–1335. [CrossRef]

109. Van Ende, M.; Wijnants, S.; Van Dijck, P. Sugar Sensing and Signaling in *Candida albicans* and *Candida glabrata*. *Front. Microbiol.* **2019**, *10*, 99. [CrossRef]

110. Vale-Silva, L.; Beaudoing, E.; Tran, V.D.T.; Sanglard, D. Comparative Genomics of Two Sequential *Candida glabrata* Clinical Isolates. *G3 Genes Genomes Genet.* **2017**, *7*, 2413–2426. [CrossRef] [PubMed]

111. Ni, Q.; Wang, C.; Tian, Y.; Dong, D.; Jiang, C.; Mao, E.; Peng, Y. CgPDR1 gain-of-function mutations lead to azole-resistance and increased adhesion in clinical *Candida glabrata* strains. *Mycoses* **2018**, *61*, 430–440. [CrossRef] [PubMed]

112. Vale-Silva, L.A.; Moeckli, B.; Torelli, R.; Posteraro, B.; Sanguinetti, M.; Sanglard, D. Upregulation of the Adhesin Gene EPA1 Mediated by PDR1 in *Candida glabrata* Leads to Enhanced Host Colonization. *mSphere* **2016**, *1*, e00065-15. [CrossRef]

113. Filler, S.G. Candida-host cell receptor-ligand interactions. *Curr. Opin. Microbiol.* **2006**, *9*, 333–339. [CrossRef]

114. Salazar, S.B.; Wang, C.; Münsterkötter, M.; Okamoto, M.; Takahashi-Nakaguchi, A.; Chibana, H.; Lopes, M.M.; Güldener, U.; Butler, G.; Mira, N.P. Comparative genomic and transcriptomic analyses unveil novel features of azole resistance and adaptation to the human host in *Candida glabrata*. *FEMS Yeast Res.* **2018**, *18*, fox079. [CrossRef]

115. Vale-Silva, L.; Ischer, F.; Leibundgut-Landmann, S.; Sanglard, D. Gain-of-function mutations in PDR1, a regulator of antifungal drug resistance in *Candida glabrata*, control adherence to host cells. *Infect. Immunity* **2013**, *81*, 1709–1720. [CrossRef] [PubMed]

116. Nakamura-Vasconcelos, S.S.; Fiorini, A.; Zanni, P.D.; Bonfim-Mendonça, P.; Godoy, J.R.; Almeida-Apolonio, A.A.; Consolaro, M.E.L.; Svidzinski, T.I.E. Emergence of *Candida glabrata* in vulvovaginal candidiasis should be attributed to selective pressure or virulence ability? *Arch. Gynecol. Obstet.* **2017**, *296*, 519–526. [CrossRef]

117. Fox, E.P.; Bui, C.K.; Nett, J.E.; Hartooni, N.; Mui, M.C.; Andes, D.R.; Nobile, C.J.; Johnson, A.D. An expanded regulatory network temporally controls *Candida albicans* biofilm formation. *Mol. Microbiol.* **2015**, *96*, 1226–1239. [CrossRef]

118. Nobile, C.J.; Fox, E.P.; Nett, J.E.; Sorrells, T.R.; Mitrovich, Q.M.; Hernday, A.D.; Tuch, B.B.; Andes, D.R.; Johnson, A.D. A Recently Evolved Transcriptional Network Controls Biofilm Development in *Candida albicans*. *Cell* **2012**, *148*, 126–138. [CrossRef]

119. Nobile, C.J.; Andes, D.R.; Nett, J.E.; Smith, F.J.; Yue, F.; Phan, Q.; Edwards, J.E.; Filler, S.G.; Mitchell, A.P. Critical Role of Bcr1-Dependent Adhesins in C. albicans Biofilm Formation In Vitro and In Vivo. *PLoS Pathog.* **2006**, *2*, 0636–0649. [CrossRef] [PubMed]

120. Nobile, C.J.; Nett, J.E.; Andes, D.R.; Mitchell, A.P. Function of *Candida albicans* adhesin Hwp1 in biofilm formation. *Eukaryot. Cell* **2006**, *5*, 1604–1610. [CrossRef]

121. Finkel, J.S.; Xu, W.; Huang, D.; Hill, E.M.; Desai, J.V.; Woolford, C.A.; Nett, J.E.; Taff, H.; Norice, C.T.; Andes, D.R.; et al. Portrait of *Candida albicans* adherence regulators. *PLoS Pathog.* **2012**, *8*, 1–14. [CrossRef]

122. Silva-Dias, A.; Miranda, I.M.; Branco, J.; Monteiro-Soares, M.; Pina-Vaz, C.; Rodrigues, A.G. Adhesion, biofilm formation, cell surface hydrophobicity, and antifungal planktonic susceptibility: Relationship among Candida spp. *Front. Microbiol.* **2015**, *6*, 205. [CrossRef]

123. Bendel, C.M. Colonization and epithelial adhesion in the pathogenesis of neonatal candidiasis. *Semin. Perinatol.* **2003**, *27*, 357–364. [CrossRef]

124. Lo, H.J.; Köhler, J.R.; DiDomenico, B.; Loebenberg, D.; Cacciapuoti, A.; Fink, G.R. Nonfilamentous C. albicans mutants are avirulent. *Cell* **1997**, *90*, 939–949. [CrossRef]

125. Nobile, C.J.; Mitchell, A.P. Genetics and genomics of *Candida albicans* biofilm formation. *Cell. Microbiol.* **2006**, *8*, 1382–1391. [CrossRef]

126. Douglas, L.J. *Candida* biofilms and their role in infection. *Trends Microbiol.* **2003**, *11*, 30–36. [CrossRef]

127. Ramage, G.; Martínez, J.P.; López-Ribot, J.L. *Candida* biofilms on implanted biomaterials: A clinically significant problem. *FEMS Yeast Res.* **2006**, *6*, 979–986. [CrossRef]

128. Rodríguez-cerdeira, C.; Gregorio, M.C.; Molares-vila, A.; López-barcenas, A.; Fabbrocini, G.; Bardhi, B.; Sinani, A.; Sánchez-blanco, E.; Arenas-guzmán, R.; Hernandez-castro, R. Biofilms and vulvovaginal candidiasis. *Colloids Surf. B Biointerfaces* **2019**, *174*, 110–125. [CrossRef] [PubMed]

129. Chandra, J.; Mukherjee, P.K. *Candida* Biofilms: Development, Architecture, and Resistance. *Microbiol. Spectr.* **2015**, *3*. [CrossRef] [PubMed]

130. Kucharíková, S.; Neirinck, B.; Sharma, N.; Vleugels, J.; Lagrou, K.; Van Dijck, P. In vivo *Candida glabrata* biofilm development on foreign bodies in a rat subcutaneous model. *J. Antimicrob. Chemother.* **2015**, *70*, 846–856. [CrossRef]

131. Silva, S.; Henriques, M.; Martins, A.; Oliveira, R.; Williams, D.; Azeredo, J. Biofilms of non-*Candida albicans Candida* species: Quantification, structure and matrix composition. *Med. Mycol.* **2009**, *47*, 681–689. [CrossRef] [PubMed]

132. Seneviratne, C.J.; Silva, W.J.; Jin, L.J.; Samaranayake, Y.H.; Samaranayake, L.P. Architectural analysis, viability assessment and growth kinetics of *Candida albicans* and *Candida glabrata* biofilms. *Arch. Oral Biol.* **2009**, *54*, 1052–1060. [CrossRef]

133. Ramage, G.; VandeWalle, K.; Wickes, B.L.; López-Ribot, J.L. Characteristics of biofilm formation by *Candida albicans*. *Rev. Iberoam. Micol.* **2001**, *18*, 163–170.

134. Hawser, S.P.; Douglas, L.J. Biofilm formation by *Candida* species on the surface of catheter materials *in vitro*. *Infect. Immunity* **1994**, *62*, 915–921.

135. Lewis, R.E.; Lo, H.; Raad, I.I.; Kontoyiannis, D.P. Lack of Catheter Infection by the *efg1/efg1 cph1/cph1* Double-Null Mutant, a *Candida albicans* Strain That Is Defective in Filamentous Growth. *Antimicrob. Agents Chemother.* **2002**, *46*, 1153–1155. [CrossRef] [PubMed]

136. Ramage, G.; Vandewalle, K.; López-Ribot, J.L.; Wickes, B.L. The filamentation pathway controlled by the Efg1 regulator protein is required for normal biofilm formation and development in *Candida albicans*. *FEMS Microbiol. Lett.* **2002**, *214*, 95–100. [CrossRef] [PubMed]

137. Thein, Z.M.; Samaranayake, Y.H.; Samaranayake, L.P. In vitro biofilm formation of *Candida albicans* and *non-albicans Candida* species under dynamic and anaerobic conditions. *Arch. Oral Biol.* **2007**, *52*, 761–767. [CrossRef]

138. Fonseca, E.; Silva, S.; Rodrigues, C.F.; Alves, C.T.; Azeredo, J.; Henriques, M. Effects of fluconazole on *Candida glabrata* biofilms and its relationship with ABC transporter gene expression. *Biofouling* **2014**, *30*, 447–457. [CrossRef] [PubMed]

139. Zarnowski, R.; Westler, W.M.; de Lacmbouh, G.A.; Marita, J.M.; Bothe, J.R.; Bernhardt, J.; Lounes-Hadj Sahraoui, A.; Fontaine, J.; Sanchez, H.; Hatfield, R.D.; et al. Novel entries in a fungal biofilm matrix encyclopedia. *MBio* **2014**, *5*, e01333-14. [CrossRef] [PubMed]

140. Al-Fattani, M.A.; Douglas, L.J. Biofilm matrix of *Candida albicans* and *Candida tropicalis*: Chemical composition and role in drug resistance. *J. Med. Microbiol.* **2006**, *55*, 999–1008. [CrossRef]

141. Susewind, S.; Lang, R.; Hahnel, S. Biofilm formation and *Candida albicans* morphology on the surface of denture base materials. *Mycoses* **2015**, *58*, 719–727. [CrossRef]

142. Chandra, J.; Patel, J.P.; Li, J.; Zhou, G.; Mukherjee, P.K.; Mccormick, T.S.; Anderson, J.M.; Ghannoum, M. A Modification of Surface Properties of Biomaterials Influences the Ability of *Candida albicans* to Form Biofilms. *Appl. Environ. Microbiol.* **2005**, *71*, 8795–8801. [CrossRef]

143. Estivill, D.; Arias, A.; Torres-Lana, A.; Carrillo-Muñoz, A.J.; Arévalo, M.P. Biofilm formation by five species of *Candida* on three clinical materials. *J. Microbiol. Methods* **2011**, *86*, 238–242. [CrossRef] [PubMed]

144. Mutluay, M.M.; Oğuz, S.; Ørstavik, D.; Fløystrand, F.; Doğan, A.; Söderling, E.; Närhi, T.; Olsen, I. Experiments on in vivo biofilm formation and in vitro adhesion of *Candida* species on polysiloxane liners. *Gerodontology* **2010**, *27*, 283–291. [CrossRef] [PubMed]

145. Jin, Y.; Samaranayake, L.P.; Samaranayake, Y.; Yip, H.K. Biofilm formation of *Candida albicans* is variably affected by saliva and dietary sugars. *Arch. Oral Biol.* **2004**, *49*, 789–798. [CrossRef] [PubMed]

146. Choi, H.W.; Shin, J.H.; Jung, S.I.; Park, K.H.; Cho, D.; Kee, S.J.; Shin, M.G.; Suh, S.P.; Ryang, D.W. Species-specific differences in the susceptibilities of biofilms formed by *Candida* bloodstream isolates to echinocandin antifungals. *Antimicrob. Agents Chemother.* **2007**, *51*, 1520–1523. [CrossRef]

147. Valentín, A.; Cantó, E.; Pemán, J.; Martínez, J.P. Voriconazole inhibits biofilm formation in different species of the genus *Candida*. *J. Antimicrob. Chemother.* **2012**, *67*, 2418–2423. [CrossRef] [PubMed]

148. Kucharíková, S.; Tournu, H.; Lagrou, K.; van Dijck, P.; Bujdáková, H. Detailed comparison of *Candida albicans* and *Candida glabrata* biofilms under different conditions and their susceptibility to caspofungin and anidulafungin. *J. Med. Microbiol.* **2011**, *60*, 1261–1269. [CrossRef] [PubMed]

149. Taff, H.T.; Mitchell, K.F.; Edward, J.A.; Andes, D.R. Mechanisms of *Candida* biofilm drug resistance. *Future Microbiol.* **2013**, *8*, 1325–1337. [CrossRef]

150. Ramage, G.; Bachmann, S.; Patterson, T.F.; Wickes, B.L.; López-ribot, J.L. Investigation of multidrug efflux pumps in relation to fluconazole resistance in *Candida albicans* biofilms. *J. Antimicrob. Chemother.* **2002**, *49*, 973–980. [CrossRef]

151. Vediyappan, G.; Rossignol, T.; D'Enfert, C. Interaction of *Candida albicans* biofilms with antifungals: Transcriptional response and binding of antifungals to beta-glucans. *Antimicrob. Agents Chemother.* **2010**, *54*, 2096–2111. [CrossRef]

152. Rodrigues, C.F.; Silva, S.; Azeredo, J.; Henriques, M. *Candida glabrata's* recurrent infections: Biofilm formation during Amphotericin B treatment. *Lett. Appl. Microbiol.* **2016**, *63*, 77–81. [CrossRef]

153. Johnson, C.J.; Cabezas-Olcoz, J.; Kernien, J.F.; Wang, S.X.; Beebe, D.J.; Huttenlocher, A.; Ansari, H.; Nett, J.E. The Extracellular Matrix of *Candida albicans* Biofilms Impairs Formation of Neutrophil Extracellular Traps. *PLoS Pathog.* **2016**, *12*, 1–23. [CrossRef]

154. Xie, Z.; Thompson, A.; Sobue, T.; Kashleva, H.; Xu, H.; Vasilakos, J.; Dongari-Bagtzoglou, A. *Candida albicans* biofilms do not trigger reactive oxygen species and evade neutrophil killing. *J. Infect. Dis.* **2012**, *206*, 1936–1945. [CrossRef]

155. Lohse, M.B.; Gulati, M.; Johnson, A.D.; Nobile, C.J. Development and regulation of single-and multi-species *Candida albicans* biofilms. *Nat. Rev. Microbiol.* **2018**, *16*, 19–31. [CrossRef]

156. Medzhitov, R. Recognition of microorganisms and activation of the immune response. *Nature* **2007**, *449*, 819–826. [CrossRef]

157. Brown, G.D.; Gordon, S. Immune recognition. A new receptor for beta-glucans. *Nature* **2001**, *413*, 36–37. [CrossRef]

158. Li, M.; Chen, Q.; Tang, R.; Shen, Y.; Liu, W. Da The expression of β-defensin-2, 3 and LL-37 induced by *Candida albicans* phospholipomannan in human keratinocytes. *J. Dermatol. Sci.* **2011**, *61*, 72–75. [CrossRef]

159. Tomalka, J.; Azodi, E.; Narra, H.P.; Patel, K.; O'Neill, S.; Cardwell, C.; Hall, B.A.; Wilson, J.M.; Hise, A.G. β-Defensin 1 Plays a Role in Acute Mucosal Defense against *Candida albicans*. *J. Immunol.* **2015**, *194*, 1788–1795. [CrossRef]

160. Järvå, M.; Phan, T.K.; Lay, F.T.; Caria, S.; Kvansakul, M.; Hulett, M.D. Human β-defensin 2 kills *Candida albicans* through phosphatidylinositol 4,5-bisphosphate-mediated membrane permeabilization. *Sci. Adv.* **2018**, *4*, eaat0979. [CrossRef]

161. Hardison, S.E.; Brown, G.D. C-type lectin receptors orchestrate antifungal immunity. *Nat. Immunol.* **2012**, *13*, 817–822. [CrossRef]

162. Netea, M.G.; Maródi, L. Innate immune mechanisms for recognition and uptake of Candida species. *Trends Immunol.* **2010**, *31*, 346–353. [CrossRef]

163. Cheng, S.-C.; Joosten, L.A.B.; Kullberg, B.-J.; Netea, M.G. Interplay between *Candida albicans* and the mammalian innate host defense. *Infect. Immunity* **2012**, *80*, 1304–1313. [CrossRef]

164. Gilbert, A.S.; Wheeler, R.T.; May, R.C. Fungal Pathogens: Survival and Replication within Macrophages. *Cold Spring Harb. Perspect. Med.* **2014**, *5*, a019661. [CrossRef]

165. Free, S.J. *Fungal Cell Wall Organization and Biosynthesis*; Academic Press: Cambridge, MA, USA, 2013; pp. 33–82.

166. Graus, M.S.; Wester, M.J.; Lowman, D.W.; Williams, D.L.; Kruppa, M.D.; Martinez, C.M.; Young, J.M.; Pappas, H.C.; Lidke, K.A.; Neumann, A.K. Mannan Molecular Substructures Control Nanoscale Glucan Exposure in Candida. *Cell Rep.* **2018**, *24*, 2432–2442. [CrossRef]

167. Ballou, E.R.; Avelar, G.M.; Childers, D.S.; Mackie, J.; Bain, J.M.; Wagener, J.; Kastora, S.L.; Panea, M.D.; Hardison, S.E.; Walker, L.A.; et al. Lactate signalling regulates fungal β-glucan masking and immune evasion. *Nat. Microbiol.* **2016**, *2*, 16238. [CrossRef]

168. Flint, H.J.; Scott, K.P.; Louis, P.; Duncan, S.H. The role of the gut microbiota in nutrition and health. *Nat. Rev. Gastroenterol. Hepatol.* **2012**, *9*, 577–589. [CrossRef]

169. Lopes, J.P.; Stylianou, M.; Backman, E.; Holmberg, S.; Jass, J.; Claesson, R.; Urban, C.F. Evasion of Immune Surveillance in Low Oxygen Environments Enhances *Candida albicans* Virulence. *MBio* **2018**, *9*, e02120-18. [CrossRef]

170. Pradhan, A.; Avelar, G.M.; Bain, J.M.; Childers, D.S.; Larcombe, D.E.; Netea, M.G.; Shekhova, E.; Munro, C.A.; Brown, G.D.; Erwig, L.P.; et al. Hypoxia Promotes Immune Evasion by Triggering β-Glucan Masking on the *Candida albicans* Cell Surface via Mitochondrial and cAMP-Protein Kinase A Signaling. *MBio* **2018**, *9*, e01318-18. [CrossRef]

171. Ene, I.V.; Cheng, S.-C.; Netea, M.G.; Brown, A.J.P. Growth of *Candida albicans* cells on the physiologically relevant carbon source lactate affects their recognition and phagocytosis by immune cells. *Infect. Immunity* **2013**, *81*, 238–248. [CrossRef]

172. Klis, F.M.; de Groot, P.; Hellingwerf, K. Molecular organization of the cell wall of *Candida albicans*. *Med. Mycol.* **2001**, *39* (Suppl. 1), 1–8. [CrossRef]

173. Netea, M.G.; Brown, G.D.; Kullberg, B.J.; Gow, N.A.R. An integrated model of the recognition of *Candida albicans* by the innate immune system. *Nat. Rev. Microbiol.* **2008**, *6*, 67–78. [CrossRef] [PubMed]

174. Ruiz-Herrera, J.; Victoria Elorza, M.; ValentÃn, E.; Sentandreu, R. Molecular organization of the cell wall of *Candida albicans* and its relation to pathogenicity. *FEMS Yeast Res.* **2006**, *6*, 14–29. [CrossRef] [PubMed]

175. Wheeler, R.T.; Kombe, D.; Agarwala, S.D.; Fink, G.R. Dynamic, morphotype-specific *Candida albicans* beta-glucan exposure during infection and drug treatment. *PLoS Pathog.* **2008**, *4*, e1000227. [CrossRef] [PubMed]

176. Hopke, A.; Nicke, N.; Hidu, E.E.; Degani, G.; Popolo, L.; Wheeler, R.T. Neutrophil Attack Triggers Extracellular Trap-Dependent Candida Cell Wall Remodeling and Altered Immune Recognition. *PLoS Pathog.* **2016**, *12*, e1005644. [CrossRef] [PubMed]

177. Sherrington, S.L.; Sorsby, E.; Mahtey, N.; Kumwenda, P.; Lenardon, M.D.; Brown, I.; Ballou, E.R.; MacCallum, D.M.; Hall, R.A. Adaptation of *Candida albicans* to environmental pH induces cell wall remodelling and enhances innate immune recognition. *PLoS Pathog.* **2017**, *13*, e1006403. [CrossRef] [PubMed]

178. Wheeler, R.T.; Fink, G.R. A drug-sensitive genetic network masks fungi from the immune system. *PLoS Pathog.* **2006**, *2*, e35. [CrossRef] [PubMed]

179. Hall, R.A.; Bates, S.; Lenardon, M.D.; MacCallum, D.M.; Wagener, J.; Lowman, D.W.; Kruppa, M.D.; Williams, D.L.; Odds, F.C.; Brown, A.J.P.; et al. The Mnn2 Mannosyltransferase Family Modulates Mannoprotein Fibril Length, Immune Recognition and Virulence of *Candida albicans*. *PLoS Pathog.* **2013**, *9*, e1003276. [CrossRef]

180. Rouabhia, M.; Schaller, M.; Corbucci, C.; Vecchiarelli, A.; Prill, S.K.-H.; Giasson, L.; Ernst, J.F. Virulence of the fungal pathogen *Candida albicans* requires the five isoforms of protein mannosyltransferases. *Infect. Immunity* **2005**, *73*, 4571–4580. [CrossRef] [PubMed]

181. West, L.; Lowman, D.W.; Mora-Montes, H.M.; Grubb, S.; Murdoch, C.; Thornhill, M.H.; Gow, N.A.R.; Williams, D.; Haynes, K. Differential virulence of *Candida glabrata* glycosylation mutants. *J. Biol. Chem.* **2013**, *288*, 22006–22018. [CrossRef]

182. Zhang, S.Q.; Zou, Z.; Shen, H.; Shen, S.S.; Miao, Q.; Huang, X.; Liu, W.; Li, L.P.; Chen, S.M.; Yan, L.; et al. Mnn10 Maintains Pathogenicity in *Candida albicans* by Extending α-1,6-Mannose Backbone to Evade Host Dectin-1 Mediated Antifungal Immunity. *PLoS Pathog.* **2016**, *12*, e1005617. [CrossRef]

183. Keppler-Ross, S.; Douglas, L.; Konopka, J.B.; Dean, N. Recognition of Yeast by Murine Macrophages Requires Mannan but Not Glucan. *Eukaryot. Cell* **2010**, *9*, 1776–1787. [CrossRef] [PubMed]

184. Chen, S.M.; Shen, H.; Zhang, T.; Huang, X.; Liu, X.Q.; Guo, S.Y.; Zhao, J.J.; Wang, C.F.; Yan, L.; Xu, G.T.; et al. Dectin-1 plays an important role in host defense against systemic *Candida glabrata* infection. *Virulence* **2017**, *8*, 1643–1656. [CrossRef] [PubMed]

185. Ifrim, D.C.; Bain, J.M.; Reid, D.M.; Oosting, M.; Verschueren, I.; Gow, N.A.R.; van Krieken, J.H.; Brown, G.D.; Kullberg, B.-J.; Joosten, L.A.B.; et al. Role of Dectin-2 for Host Defense against Systemic Infection with *Candida glabrata*. *Infect. Immunity* **2014**, *82*, 1064–1073. [CrossRef]

186. Jacobsen, I.D.; Brunke, S.; Seider, K.; Schwarzmüller, T.; Firon, A.; d'Enfért, C.; Kuchler, K.; Hube, B. *Candida glabrata* persistence in mice does not depend on host immunosuppression and is unaffected by fungal amino acid auxotrophy. *Infect. Immunity* **2010**, *78*, 1066–1077. [CrossRef]

187. Lambris, J.D.; Ricklin, D.; Geisbrecht, B.V. Complement evasion by human pathogens. *Nat. Rev. Microbiol.* **2008**, *6*, 132–142. [CrossRef] [PubMed]

188. Poltermann, S.; Kunert, A.; von der Heide, M.; Eck, R.; Hartmann, A.; Zipfel, P.F. Gpm1p is a factor H-, FHL-1-, and plasminogen-binding surface protein of *Candida albicans*. *J. Biol. Chem.* **2007**, *282*, 37537–37544. [CrossRef]

189. Vogl, G.; Lesiak, I.; Jensen, D.B.; Perkhofer, S.; Eck, R.; Speth, C.; Lass-Flörl, C.; Zipfel, P.F.; Blom, A.M.; Dierich, M.P.; et al. Immune evasion by acquisition of complement inhibitors: The mould Aspergillus binds both factor H and C4b binding protein. *Mol. Immunol.* **2008**, *45*, 1485–1493. [CrossRef]

190. Luo, S.; Poltermann, S.; Kunert, A.; Rupp, S.; Zipfel, P.F. Immune evasion of the human pathogenic yeast *Candida albicans*: Pra1 is a Factor H, FHL-1 and plasminogen binding surface protein. *Mol. Immunol.* **2009**, *47*, 541–550. [CrossRef]

191. Kenno, S.; Speth, C.; Rambach, G.; Binder, U.; Chatterjee, S.; Caramalho, R.; Haas, H.; Lass-Flörl, C.; Shaughnessy, J.; Ram, S.; et al. *Candida albicans* Factor H Binding Molecule Hgt1p—A Low Glucose-Induced Transmembrane Protein Is Trafficked to the Cell Wall and Impairs Phagocytosis and Killing by Human Neutrophils. *Front. Microbiol.* **2018**, *9*, 3319. [CrossRef]

192. Staib, F. Serum-proteins as nitrogen source for yeastlike fungi. *Sabouraudia* **1965**, *4*, 187–193. [CrossRef] [PubMed]

193. Gropp, K.; Schild, L.; Schindler, S.; Hube, B.; Zipfel, P.F.; Skerka, C. The yeast *Candida albicans* evades human complement attack by secretion of aspartic proteases. *Mol. Immunol.* **2009**, *47*, 465–475. [CrossRef] [PubMed]

194. Meiller, T.F.; Hube, B.; Schild, L.; Shirtliff, M.E.; Scheper, M.A.; Winkler, R.; Ton, A.; Jabra-Rizk, M.A. A novel immune evasion strategy of *Candida albicans*: Proteolytic cleavage of a salivary antimicrobial peptide. *PLoS ONE* **2009**, *4*, e5039. [CrossRef]

195. Svoboda, E.; Schneider, A.E.; Sándor, N.; Lermann, U.; Staib, P.; Kremlitzka, M.; Bajtay, Z.; Barz, D.; Erdei, A.; Józsi, M. Secreted aspartic protease 2 of *Candida albicans* inactivates factor H and the macrophage factor H-receptors CR3 (CD11b/CD18) and CR4 (CD11c/CD18). *Immunol. Lett.* **2015**, *168*, 13–21. [CrossRef] [PubMed]

196. Hallström, T.; Singh, B.; Kraiczy, P.; Hammerschmidt, S.; Skerka, C.; Zipfel, P.F.; Riesbeck, K. Conserved Patterns of Microbial Immune Escape: Pathogenic Microbes of Diverse Origin Target the Human Terminal Complement Inhibitor Vitronectin via a Single Common Motif. *PLoS ONE* **2016**, *11*, e0147709. [CrossRef]

197. Luo, S.; Dasari, P.; Reiher, N.; Hartmann, A.; Jacksch, S.; Wende, E.; Barz, D.; Niemiec, M.J.; Jacobsen, I.; Beyersdorf, N.; et al. The secreted *Candida albicans* protein Pra1 disrupts host defense by broadly targeting and blocking complement C3 and C3 activation fragments. *Mol. Immunol.* **2018**, *93*, 266–277. [CrossRef] [PubMed]

198. Bergfeld, A.; Dasari, P.; Werner, S.; Hughes, T.R.; Song, W.-C.; Hortschansky, P.; Brakhage, A.A.; Hünig, T.; Zipfel, P.F.; Beyersdorf, N. Direct Binding of the pH-Regulated Protein 1 (Pra1) from *Candida albicans* Inhibits Cytokine Secretion by Mouse CD4+ T Cells. *Front. Microbiol.* **2017**, *8*, 844. [CrossRef]

199. Vylkova, S.; Carman, A.J.; Danhof, H.A.; Collette, J.R.; Zhou, H.; Lorenz, M.C. The fungal pathogen *Candida albicans* autoinduces hyphal morphogenesis by raising extracellular pH. *MBio* **2011**, *2*, e00055-11. [CrossRef]

200. Brown, A.J.P.; Brown, G.D.; Netea, M.G.; Gow, N.A.R. Metabolism impacts upon Candida immunogenicity and pathogenicity at multiple levels. *Trends Microbiol.* **2014**, *22*, 614–622. [CrossRef]

201. Lorenz, M.C.; Fink, G.R. The glyoxylate cycle is required for fungal virulence. *Nature* **2001**, *412*, 83. [CrossRef]

202. Fernández-Arenas, E.; Bleck, C.K.E.; Nombela, C.; Gil, C.; Griffiths, G.; Diez-Orejas, R. *Candida albicans* actively modulates intracellular membrane trafficking in mouse macrophage phagosomes. *Cell. Microbiol.* **2009**, *11*, 560–589. [CrossRef]

203. Kasper, L.; Seider, K.; Gerwien, F.; Allert, S.; Brunke, S.; Schwarzmüller, T.; Ames, L.; Zubiria-Barrera, C.; Mansour, M.K.; Becken, U.; et al. Identification of *Candida glabrata* genes involved in pH modulation and modification of the phagosomal environment in macrophages. *PLoS ONE* **2014**, *9*, e96015. [CrossRef]

204. Vylkova, S.; Lorenz, M.C. Modulation of phagosomal pH by *Candida albicans* promotes hyphal morphogenesis and requires Stp2p, a regulator of amino acid transport. *PLoS Pathog.* **2014**, *10*, e1003995. [CrossRef]
205. Rai, M.N.; Balusu, S.; Gorityala, N.; Dandu, L.; Kaur, R. Functional Genomic Analysis of *Candida glabrata*-Macrophage Interaction: Role of Chromatin Remodeling in Virulence. *PLoS Pathog.* **2012**, *8*, e1002863. [CrossRef] [PubMed]
206. Chew, S.Y.; Ho, K.L.; Cheah, Y.K.; Ng, T.S.; Sandai, D.; Brown, A.J.P.; Than, L.T.L. Glyoxylate cycle gene ICL1 is essential for the metabolic flexibility and virulence of *Candida glabrata*. *Sci. Rep.* **2019**, *9*, 2843. [CrossRef] [PubMed]
207. Li, L.; Naseem, S.; Sharma, S.; Konopka, J.B. Flavodoxin-Like Proteins Protect *Candida albicans* from Oxidative Stress and Promote Virulence. *PLoS Pathog.* **2015**, *11*, e1005147. [CrossRef] [PubMed]
208. Wagener, J.; MacCallum, D.M.; Brown, G.D.; Gow, N.A.R. *Candida albicans* Chitin Increases Arginase-1 Activity in Human Macrophages, with an Impact on Macrophage Antimicrobial Functions. *MBio* **2017**, *8*, e01820-16. [CrossRef]
209. Ghosh, S.; Navarathna, D.H.M.L.P.; Roberts, D.D.; Cooper, J.T.; Atkin, A.L.; Petro, T.M.; Nickerson, K.W. Arginine-induced germ tube formation in *Candida albicans* is essential for escape from murine macrophage line RAW 264.7. *Infect. Immunity* **2009**, *77*, 1596–1605. [CrossRef]
210. Danhof, H.A.; Lorenz, M.C. The *Candida albicans* ATO Gene Family Promotes Neutralization of the Macrophage Phagolysosome. *Infect. Immunity* **2015**, *83*, 4416–4426. [CrossRef]
211. Miramón, P.; Lorenz, M.C. The SPS amino acid sensor mediates nutrient acquisition and immune evasion in *Candida albicans*. *Cell. Microbiol.* **2016**, *18*, 1611–1624. [CrossRef]
212. Cuéllar-Cruz, M.; Briones-Martin-del-Campo, M.; Cañas-Villamar, I.; Montalvo-Arredondo, J.; Riego-Ruiz, L.; Castaño, I.; De Las Peñas, A. High Resistance to Oxidative Stress in the Fungal Pathogen *Candida glabrata* Is Mediated by a Single Catalase, Cta1p, and Is Controlled by the Transcription Factors Yap1p, Skn7p, Msn2p, and Msn4p. *Eukaryot. Cell* **2008**, *7*, 814–825. [CrossRef] [PubMed]
213. Fukuda, Y.; Tsai, H.-F.; Myers, T.G.; Bennett, J.E. Transcriptional Profiling of *Candida glabrata* during Phagocytosis by Neutrophils and in the Infected Mouse Spleen. *Infect. Immunity* **2013**, *81*, 1325–1333. [CrossRef]
214. Mahl, C.D.; Behling, C.S.; Hackenhaar, F.S.; de Carvalho e Silva, M.N.; Putti, J.; Salomon, T.B.; Alves, S.H.; Fuentefria, A.; Benfato, M.S. Induction of ROS generation by fluconazole in *Candida glabrata*: Activation of antioxidant enzymes and oxidative DNA damage. *Diagn. Microbiol. Infect. Dis.* **2015**, *82*, 203–208. [CrossRef]
215. Seider, K.; Gerwien, F.; Kasper, L.; Allert, S.; Brunke, S.; Jablonowski, N.; Schwarzmüller, T.; Barz, D.; Rupp, S.; Kuchler, K.; et al. Immune Evasion, Stress Resistance, and Efficient Nutrient Acquisition Are Crucial for Intracellular Survival of *Candida glabrata* within Macrophages. *Eukaryot. Cell* **2014**, *13*, 170–183. [CrossRef]
216. Nevitt, T.; Thiele, D.J. Host Iron Withholding Demands Siderophore Utilization for *Candida glabrata* to Survive Macrophage Killing. *PLoS Pathog.* **2011**, *7*, e1001322. [CrossRef]
217. Srivastava, V.K.; Suneetha, K.J.; Kaur, R. A systematic analysis reveals an essential role for high-affinity iron uptake system, haemolysin and CFEM domain-containing protein in iron homoeostasis and virulence in *Candida glabrata*. *Biochem. J.* **2014**, *463*, 103–114. [CrossRef]
218. Sharma, V.; Purushotham, R.; Kaur, R. The Phosphoinositide 3-Kinase Regulates Retrograde Trafficking of the Iron Permease CgFtr1 and Iron Homeostasis in *Candida glabrata*. *J. Biol. Chem.* **2016**, *291*, 24715–24734. [CrossRef]
219. Wellington, M.; Koselny, K.; Sutterwala, F.S.; Krysan, D.J. *Candida albicans* triggers NLRP3-mediated pyroptosis in macrophages. *Eukaryot. Cell* **2014**, *13*, 329–340. [CrossRef]
220. Miao, E.A.; Rajan, J.V.; Aderem, A. Caspase-1-induced pyroptotic cell death. *Immunol. Rev.* **2011**, *243*, 206–214. [CrossRef] [PubMed]
221. Chen, K.W.; Schroder, K. Antimicrobial functions of inflammasomes. *Curr. Opin. Microbiol.* **2013**, *16*, 311–318. [CrossRef] [PubMed]
222. Tucey, T.M.; Verma, J.; Harrison, P.F.; Snelgrove, S.L.; Lo, T.L.; Scherer, A.K.; Barugahare, A.A.; Powell, D.R.; Wheeler, R.T.; Hickey, M.J.; et al. Glucose Homeostasis Is Important for Immune Cell Viability during Candida Challenge and Host Survival of Systemic Fungal Infection. *Cell Metab.* **2018**, *27*, 988–1006.e7. [CrossRef] [PubMed]

223. Kasper, L.; König, A.; Koenig, P.-A.; Gresnigt, M.S.; Westman, J.; Drummond, R.A.; Lionakis, M.S.; Groß, O.; Ruland, J.; Naglik, J.R.; et al. The fungal peptide toxin Candidalysin activates the NLRP3 inflammasome and causes cytolysis in mononuclear phagocytes. *Nat. Commun.* **2018**, *9*, 4260. [CrossRef]
224. Hise, A.G.; Tomalka, J.; Ganesan, S.; Patel, K.; Hall, B.A.; Brown, G.D.; Fitzgerald, K.A. An Essential Role for the NLRP3 Inflammasome in Host Defense against the Human Fungal Pathogen *Candida albicans*. *Cell Host Microbe* **2009**, *5*, 487–497. [CrossRef]
225. Rogiers, O.; Frising, U.C.; Kucharíková, S.; Jabra-Rizk, M.A.; van Loo, G.; Van Dijck, P.; Wullaert, A. Candidalysin Crucially Contributes to Nlrp3 Inflammasome Activation by *Candida albicans* Hyphae. *MBio* **2019**, *10*, e02221-18. [CrossRef]
226. Hargarten, J.C.; Moore, T.C.; Petro, T.M.; Nickerson, K.W.; Atkin, A.L. *Candida albicans* Quorum Sensing Molecules Stimulate Mouse Macrophage Migration. *Infect. Immunity* **2015**, *83*, 3857–3864. [CrossRef]
227. Brunke, S.; Seider, K.; Fischer, D.; Jacobsen, I.D.; Kasper, L.; Jablonowski, N.; Wartenberg, A.; Bader, O.; Enache-Angoulvant, A.; Schaller, M.; et al. One Small Step for a Yeast—Microevolution within Macrophages Renders *Candida glabrata* Hypervirulent Due to a Single Point Mutation. *PLoS Pathog.* **2014**, *10*, e1004478. [CrossRef]

© 2019 by the authors. Licensee MDPI, Basel, Switzerland. This article is an open access article distributed under the terms and conditions of the Creative Commons Attribution (CC BY) license (http://creativecommons.org/licenses/by/4.0/).

International Journal of
Molecular Sciences

Review

Small Noncoding Regulatory RNAs from *Pseudomonas aeruginosa* and *Burkholderia cepacia* Complex

Tiago Pita, Joana R. Feliciano and Jorge H. Leitão *

iBB-Institute for Bioengineering and Biosciences, Departmento de Bioengenharia, Instituto Superior Técnico, Universidade de Lisboa, Av. Rovisco Pais, Torre Sul, Piso 6, 1049-001 Lisbon, Portugal; tiagopita@tecnico.ulisboa.pt (T.P.); joana.feliciano@tecnico.ulisboa.pt (J.R.F.)
* Correspondence: jorgeleitao@tecnico.ulisboa.pt; Tel.: +351-218417688

Received: 31 October 2018; Accepted: 23 November 2018; Published: 27 November 2018

Abstract: Cystic fibrosis (CF) is the most life-limiting autosomal recessive disorder in Caucasians. CF is characterized by abnormal viscous secretions that impair the function of several tissues, with chronic bacterial airway infections representing the major cause of early decease of these patients. *Pseudomonas aeruginosa* and bacteria from the *Burkholderia cepacia* complex (Bcc) are the leading pathogens of CF patients' airways. A wide array of virulence factors is responsible for the success of infections caused by these bacteria, which have tightly regulated responses to the host environment. Small noncoding RNAs (sRNAs) are major regulatory molecules in these bacteria. Several approaches have been developed to study *P. aeruginosa* sRNAs, many of which were characterized as being involved in the virulence. On the other hand, the knowledge on Bcc sRNAs remains far behind. The purpose of this review is to update the knowledge on characterized sRNAs involved in *P. aeruginosa* virulence, as well as to compile data so far achieved on sRNAs from the Bcc and their possible roles on bacteria virulence.

Keywords: cystic fibrosis; *Pseudomonas aeruginosa*; *Burkholderia cepacia* complex; small noncoding regulatory RNAs; pathogenicity

---

1. Introduction: What are Bacterial sRNAs?

Regulation of gene expression in bacteria is achieved by a diversified system of multiple regulators, from proteins to RNA molecules. Riboswitches, T Boxes, and small noncoding regulatory RNAs (sRNAs) are among the RNA molecules or sequences that regulate the expression of genes. sRNAs were first identified in *Escherichia coli* in the 1960s as part of a large group of transcriptional regulators [1,2]. Since then, sRNAs have been shown as key regulators in bacteria [3], being capable of driving the fastest responses in the bacterial cell [4]. Ranging from 50 to 400 nucleotides long in size, sRNAs have been found as encoded widespread in the bacterial genomes, mainly in intergenic regions. For example, in enterobacterial genomes, sRNAs are estimated to represent approximately 200 to 300 genes, corresponding to ~5% of the number of protein-encoding genes [4,5].

sRNAs can be multipurpose regulators, targeting a wide range of molecular structures such as DNA/chromatin, proteins, other RNA molecules, and metabolites [6]. These regulatory sRNAs are therefore involved in the regulation of diverse cellular processes, including DNA assembly, plasmid replication, phage development, transcription, translation, peptidoglycan synthesis, and bacterial virulence [1,6]. In addition, sRNAs are also known for playing a role in cellular metabolism, like carbon and amino acid metabolism, iron homeostasis, quorum sensing (QS), and biofilm formation; stress responses including acid, osmotic, and oxidative stress responses; adaptation to growth conditions like high temperature and changes in molecular oxygen availability; as well as in mechanisms related to bacterial pathogenesis [7].

Although the vast majority of the so far characterized sRNAs do not encode proteins, a few can be partially translated originating small peptides, collectively known as dual-function small regulatory RNAs. These sRNAs can exert their regulation by acting as an RNA molecule or as a peptide. The peptide can play a similar regulatory role or a distinct one, capable of acting in a distinct metabolic pathway [8].

The best described sRNAs act by antisense base pairing with a specific target mRNA. These sRNAs can be cis- or trans-encoded, depending on their genome location in relation to the mRNA target [6]. Cis-encoded antisense RNAs share a full complementarity with their target mRNAs, often leading to the formation of a near perfect sRNA-mRNA duplex. Distinctly, trans-encoded antisense sRNAs only establish a partial complementarity with their target mRNAs, forming a partial duplex. Due to their lower specificity, trans-encoded sRNAs usually have multiple targets, being part of a more complex regulatory network in the cell. In both cases, the interaction between the antisense sRNA and target mRNA can affect translation negatively or positively [9].

The base pairing of the sRNAs can occur at different regions of the target mRNA, such as the $3'$ untranslated region ($3'$-UTR), the coding region, or, like most of the cases, at the $5'$ untranslated region ($5'$-UTR) where the ribosome accommodation usually occurs. The result of the interactions between the sRNA and the target mRNA can also differ, often leading to the suppression of gene translation by inhibition of ribosome interactions and/or by the induction of degradation of the target mRNA. However, activation of gene translation has also been reported. In this case, the interaction of the sRNA with the translation initiation region of the mRNA leads to a restructure of the mRNA conformation, exposing the previously occluded ribosome biding site (RBS) [6,10]. sRNAs can also act as stabilizers of target RNAs, avoiding their degradation. In addition, sRNAs can actually lead to the cleavage of mRNAs, resulting in one or two stable coding mRNAs [9].

Sometimes a sRNA can also become a target, when another RNA molecule (usually a mRNA) binds to the sRNA, trapping it. Known cases involve a sRNA that is constitutively expressed, and it is the expression of the trap-mRNA that will impair the regulatory function of the sRNA on its target [11,12]. Nevertheless, it has also been observed that some sRNAs bind to other sRNA, like the SroC of *Salmonella* that can sequester a given sRNA, impairing its activity [13].

The interaction between sRNAs and proteins usually leads to protein sequestration [9]. One of the best characterized target protein is CrsA, which has the ability to directly bind to the $5'$ regions of mRNAs, repressing their translation [14]. Some sRNAs act like a sponge, sequestering this protein. This interaction activates the regulation of multiple pathways, related to carbon starvation and glycogen biosynthesis, affecting the physiology and virulence of several pathogenic bacteria [14]. Moreover, sRNAs can also bind to proteins in a more complex way, producing more complex outcomes, such as the modulation of enzymatic activity by inhibiting, activating or modifying the protein activity. The best known example is *E. coli* 6S sRNA, which binds to the housekeeping form of RNA polymerase (σ^{70}-RNAP), modifying its activity [3]. Nevertheless, the best characterized proteins that interact with sRNAs are the Hfq-like RNA chaperones, which play a major role in the regulatory mechanism of sRNAs base pair interactions, by stabilizing the RNA molecules and mediating the interaction between sRNAs and their targets.

Proteins of the Hfq family are conserved among bacterial genomes, being present in approximately one half of the sequenced organisms, a fact that certainly reflects its importance for the bacterial cell [15,16]. Hfq is a member of the (L)Sm protein superfamily and is present among the three domains of life. In bacteria, Hfq assembles as a homohexamer with a donut-like shape containing three distinct RNA binding sites on its surface, the proximal face, the distal face, and the convex circumferential rim of the ring. The proximal face of the ring recognizes uridine-rich sequences, a typical feature of $3'$ terminal tails of sRNAs. The distal face preferentially binds adenosine-rich sequences or repeated ARN motifs (composed of an adenosine, a purine, and any nucleotide), a characteristic usually found in mRNAs, allowing the recognition of the sRNA-target regions on the 5′ UTR of mRNAs [17,18]. The rim, which is rich in arginine residues, seems to promote RNA annealing, probably by guiding and facilitating the base pairing between the complementary strands of the sRNA and the target mRNA [19]. The mechanisms underlying the Hfq-mediated sRNA–mRNA interaction are quite diverse. Hfq increases the chance of the RNA molecules to meet by reducing their motility and flexibility; the protein can induce changes in RNA secondary structure by exposing the complementary regions of the RNA molecules, allowing the formation of a stable RNA double strand by base paring; the chaperone increases the local concentration and proximity of the pairs sRNA/mRNA by the specificity of its binding sites [20]. Moreover, Hfq is capable of accommodating multiple RNA molecules and interacting with proteins involved in RNA metabolism and translation, greatly improving the performance of sRNAs as regulators [17]. Hfq is now recognized as a major component of post-transcription regulatory mechanisms in bacterial cells. The output of Hfq-mediated sRNA-mRNA interactions is quite similar to those without mediation, occurring either an activation or a repression of gene expression, with or without mediated degradation by RNAse E [18,21]. Deletion of the gene encoding Hfq in several pathogens like *Salmonella* Typhimurium and *Pseudomonas aeruginosa* has resulted in a strong attenuation of virulence [22]. This evidences the major role played by Hfq-dependent sRNAs in virulence.

A recently characterized RNA chaperone, ProQ, is a RNA-binding protein of the ProQ/FinO domain superfamily, conserved and abundant among α-, β-, and γ-proteobacteria [23]. A study in *Salmonella* Thyphimurium revealed that the sRNAs associated to this chaperone possess extensive secondary structures, distinct from those that associate to Hfq. In addition, only 2% of the identified ProQ-interacting sRNAs also bind to Hfq [24]. The only full description of the mechanism of interaction between ProQ and a sRNA was done by Smirnov et al. [23], who described the ability of ProQ to stabilize the trans-encoded sRNA RaiZ, as well as the ability of the chaperone to mediate the interaction of RaiZ with its mRNA target *hupA*.

2. Cystic Fibrosis Lung Infections

Cystic Fibrosis (CF) is the most common and life-limiting autosomal recessive disorder among Caucasians, affecting ~70,000 individuals worldwide [25,26]. The CF disease is characterized by abnormal viscous secretions in several tissues with epithelia, like airways, pancreas, small intestine, liver, reproductive tract and sweat glands. The disease is due to mutations in the gene encoding the Cystic Fibrosis transmembrane conductance regulator (CFTR). Despite the potential damaging effect of *cftr* gene mutations on the affected organs, nowadays the great concern is centered on chronic bacterial airway infections that are responsible for 80 to 90% of mortality [26–28]. The expression of CFTR is much more pronounced in the airway's tissues, where a dysfunctional CFTR causes deficient cAMP-dependent chloride and bicarbonate secretion into airway secretions. Mucins are also secreted to the epithelial surface and then the pH decreases, a factor that impairs the host defense mechanisms. An inefficient inflammatory response, along with a defective mucus clearance, are the main causes of microbial infections in the airways of CF patients [29]. A wide array of bacteria can colonize the airways of CF patients, like *Staphylococcus aureus*, *Haemophilus influenzae*, *Stenotrophomonas maltophilia*, and *Mycobacteria* sp other than *M. tuberculosis*. Fungi like *Aspergillus fumigatus* have also been found in patients' sputum. All these microbes can contribute to the decline of lung functions [30]. However, the major threats to CF patient's survival are *P. aeruginosa*, bacteria of the *Burkholderia cepacia* complex (Bcc), and *Achromobacter xylosoxidans*. These bacteria can persist in the airways and provoke chronic infections that are the main cause of morbidity and early mortality of CF patients. *P. aeruginosa* is the leading cause of CF patients infections, being responsible for chronic infections in approximately 80% of CF adults [30,31]. Progresses recently achieved in the development of aggressive early eradication of *P. aeruginosa* led to an enhancement on prognosis [32]. Bacteria of the Bcc are the second major microbial threat to CF patients. Despite their lower incidence (only 3%–4% of CF patients are infected by Bcc), these bacteria are especially feared due to the easy transmission among the patients, the extensive antibiotic resistance, and the risk of cepacia syndrome, a fulminating pneumonia that leads to patient death in a short period of time [33,34].

This review focuses on the current knowledge of sRNAs from the CF major pathogens *P. aeruginosa* and Bcc. Since sRNAs are major regulatory molecules involved in bacteria virulence, we review in this work some sRNAs involved in virulence of *P. aeruginosa*, as well as the present scarce knowledge on Bcc sRNAs and their possible roles on bacteria virulence [35].

3. sRNAs of *Pseudomonas aeruginosa*, the Major CF Pathogen

P. aeruginosa is a Gram-negative species recovered from a wide range of environmental niches. The species is an opportunistic pathogen to animals and humans and a serious threat to public health [36]. *P. aeruginosa* is responsible for severe infections not only among CF patients, but also in other immunocompromised patients and patients with severe burn injuries [37]. It is still to be fully elucidated how an environmental species has become a major cause of nosocomial infections worldwide. However, it is known that the overuse and misuse of antibiotics have led to the selection of multiple antibiotic resistant strains, against which very few therapeutic options are available [32,36]. In a review, Potron et al. [38] showed that some markers of acquired resistance to antibiotics were found in strains isolated all over the world. The ability of this bacterium to produce a wide range of opportunistic infections and to originate strains resistant to various antibiotics can be attributed to its large genome (typically with more than 6 Mbp), containing a particularly high proportion of regulatory genes, as well as a large number of genes involved in the catabolism, transport, and efflux of organic compounds [32]. Although *P. aeruginosa* infections can be eradicated during acute lung infections in most of the cases, when a chronic infection is established, its eradication is hardly achieved [39]. The most striking examples of infections by *P. aeruginosa* are chronic lung infections in CF patients, which occur in over 60% of adult patients [40]. When invading the CF lung, *P. aeruginosa* has to surpass a heterogeneous, hostile, and stressful environment, being exposed to osmotic stress due to the viscous mucus, oxidative and nitrosative stresses due to host defenses, sublethal concentrations of antibiotics, and the presence of other microorganisms [32].

Acute infections by *P. aeruginosa* are among the best characterized; it is known that the expression of their virulence genes is controlled by extremely complex, interweaving regulatory circuits and multiple signaling systems. To achieve a successful acute infection, *P. aeruginosa* produces an impressive array of virulence factors, such as the Type 3 Secretion Systems (T3SS) to secrete various exotoxins; the quorum sensing (QS) systems to control numerous important secreted components including pyocyanin, elastase, cyanide, and rhamnolipids, exhibiting a motile phenotype [32]. On the other hand, *P. aeruginosa* virulence factors that are important in chronic infections are still to be fully elucidated. It is known that in chronic infections, *P. aeruginosa* strains became less inflammatory and less cytotoxic, and also lose structures responsible for adherence and motility, like the flagellum and pili. Other characteristics include the appearance of mucoidy, due to alginate production, the emergence of highly antibiotic resistant variants, changes in lipopolysaccharide structure including altered lipid A and loss of O-antigen, and alterations in quorum sensing [39]. The GAC system network was pointed as controlling the reversible transition from acute to chronic infections, incorporating the two-component regulatory system GacA/GacS, two other sensor kinases RetS/LadS, the small regulatory protein RsmA, and sRNAs RsmZ and RsmY [32]

To successfully infect their hosts, *P. aeruginosa* must achieve fast and precise regulation of its metabolism. sRNAs are one of the best means to achieve such a fast and tight regulation, and this is most probably why these bacteria possess a so vast array of those posttranscriptional regulators. A total of 573 sRNAs are known to be expressed by the strain PAO1 and 233 sRNAs by the strain PA14 [41]. PAO1 is a moderately virulent strain isolated from a wound, widely used as a reference strain in comparison to the highly virulent isolate PA14, a representative of the most common clonal group found worldwide [41]. 126 sRNAs are common to both strains. However, an in silico study evidenced that the strains share more sRNAs in their genomes than the ones detected [42]. This means that, although some sRNAs are encoded in the genome of both strains their expression could be strain-specific or environmental-dependent [42]. For the PAO1 strains, 149 sRNAs were already annotated and 117 were experimentally validated, the remaining predicted by similarity searching. In the case of the PA14 strains, 136 sRNAs are annotated and 66 were experimentally validated (data obtained from the sRNAs database BSRD [43]).

Although the number of characterized sRNAs is remarkably reduced compared to the total of predicted sRNAs in *P. aeruginosa*, it is still a huge number when compared to other organisms. A boost in the characterization of sRNAs from *P. aeruginosa* strains PAO1 and PA14 took place in recent years. At least one half of the characterized sRNAs were found to be dependent or mediated by the Hfq chaperone (Table 1). Among the sRNAs characterized are the RsmY and RsmZ, both were involved in the switch from the motile to the sessile modes of living, modulating the expression of T3SS and T6SS [44]. However, only the role of RsmZ in cell motility and biofilm formation was proved. Like RsmY and RsmZ, RsmW is also involved in the RsmA regulation, being implied in the regulation of biofilm formation [45,46]. Moreover, ErsA and SrbA are two sRNAs that also play a role in the regulation in biofilm formation. ErsA is part of the response to envelope stress and exerts its regulatory role by modulating the activity of the checkpoint bifunctional enzyme AlgC with phosphomannomutase/phosphoglucomutase activities and expression of the AmrZ regulon [47]. SrbA has also a role in the pathogenesis of *P. aeruginosa* PA14, although the mechanisms underlying such regulation are still to be elucidated [48]. RgsA and NrsZ are both important in swarming motility. Both are regulated by two-component regulatory systems (TCS). NrsZ is regulated by the TCS NtrB/C and perform its role through its involvement in the modulation of *rhlA* expression, and thus controlling the production of rhamnolipid surfactants [49]. RgsA is regulated by the GacS/GacA TCS, having as targets *fis* and *acpP*, leading to the regulation of pyocyanin synthesis [37].

Las, RhL and PQS are the best known *P. aeruginosa* quorum sensing systems. The Pseudomonas quinolone signal (PQS) plays a central role in the quorum sensing network in *P. aeruginosa*, linking the three systems. sRNAs PrrF1/2, ReaL, and PhrS are major players in PQS modulation, all of them playing a role in the expression of virulence genes. PhrS is involved in the regulation of expression of PqsR, a transcriptional regulator of genes involved in PQS and pyocyanin synthesis [50]. ReaL acts by positively regulating *pqsC* expression, in a response modulated by the Las system [51]. PrrF1/2 works as an iron sensor, leading to an interruption of expression of genes encoding iron-binding proteins in low iron conditions. The transcription activator *antR* is also repressed, avoiding the degradation of anthranilate, a precursor of PQS synthesis. [52,53]. The sRNA PrrH that was not linked to *P. aeruginosa* virulence also plays an important role in iron availability, by heme homeostasis [54]. In this context, the CrcZ sRNA also seems to be involved in iron regulation. Like RsmY/Z/W, CrcZ also sequesters proteins, in this case the Hfq chaperone. Although it was initially characterized as a regulator of carbon catabolite repression, recent evidence points out CrcZ as a PrrF regulator, interfering with the Hfq-mediated repression of *antR* [53,55]. Two other sRNAs were also characterized recently, PaiI plays a pivotal role in anaerobic growth conditions and in denitrification, and PesA seems to play a role in *P. aeruginosa* pathogenesis, being involved in Pyocin S3 modulation and resistance to UV radiation [56,57]. Table 1 summarizes the *P. aeruginosa* sRNAs already characterized according to the strain from which they were studied. A full description of the best characterized sRNAs playing a role in *P. aeruginosa* virulence is described below.

Table 1. Genome location and regulatory interactions of the sRNAs already characterized in *P. aeruginosa*.

Strain	sRNA	Annotation	Length	Genomic Location	Strand	Category [1]	RBP	Function/ Pathway	Regulators Direct	Regulators Indirect	Targets	Source
PAO1	RsmZ	spae4058.1	119	4057542-4057660	Rv	trans	ND	Cell motility Biofilm formation T3SS-T6SS switch	GacS/GacA PNPase	HptB; LadS; RetS; AlgR; BfiSR	RsmA	[44,46,58–60]
PAO1	RsmY	spae587.1	124	586867-586990	Fw	trans	Hfq binding	T3SS-T6SS switch	GacS/GacA PNPase	HptB; LadS; RetS; AlgR;	RsmA	[44]
PA14	RsmY	PA14_06875	124	596840-596963	Fw	trans	Hfq binding	Cell motility	GacS/GacA	HptB; LadS; RetS; AlgR;	RsmA	[59,61]
PAO1	PhrS	spae3706.1	213	3705309-3705521	Rv	trans	Hfq binding	PQS regulation Virulence gene regulation	ANR		*pqsR*	[50,62]
PAO1	PrrF1	spae5284.1	152	5283960-5284111	Fw	trans	Hfq binding	Iron acquisition and storage PQS regulation Virulence gene regulation	Fur		*antR; sodB; PA4880; acnB; m-acnB; sdhD*	[52,63,64]
PAO1	PrrH	spae5284.2	325	5283995-5284319	Fw	trans	ND	Heme homeostasis	Fur		*acnB; m-acnB; sdhD; nirL*	[52,54]
PAO1	PrrF2	spae5285.1	149	5284172-5284320	Fw	trans	Hfq binding	Iron acquisition and storage PQS regulation Virulence gene regulation	Fur		*antR; sodB; PA4880; acnB; m-acnB; sdhD*	[52,63,64]
PAO1	CrcZ	spae5309.3	407	5308587-5308993	Fw	trans	* Hfq binding	Carbon catabolite repression	CbrA/B		Hfq; Crc	[55,65,66]
PAO1	RgsA	spae3319.1	197	3318663-3318859	Fw	trans	Hfq binding	Swarming Motility Virulence	RpoS	GacS/A	*fis; acpP; rpoS*	[37,67]
PAO1	NrsZ	PA5125.1	226	5775397-5775623	Fw	trans	ND	Swarming Motility		NtrB/C	*rhlA*	[49]
PAO1	ErsA	spae6184.2	201	6183500-6183700	Rv	cis	Hfq binding	Envelope stress response Biofilm formation		σ^{22}	*algC; oprD;* AmrZ regulon	[47,68]
PA14	ErsA	spau6457.2	201	6456400-6456600	Rv	cis	ND	Envelope stress response		σ^{22}	*algC*	[68]
PA14	PaiI	PA14_13970.	126	1198928-1199053	Rv	trans	Hfq binding	Anaerobic Growth Denitrification		NarXL		[56]
PA14	SrbA	PA14_30065	239	2604298-2604536	Rv	trans	ND	Biofilm formation Virulence				[48]
PA14	ReaL	spau1600.1	201	1599900-1600100	Rv	trans	ND	PQS synthesis Regulation of virulence	LasR-3OC$_{12}$I RpoS		*pqsC rpoS*	[51,69]
PA14	PesA	spau5289.2	401	5288100-5288500	Fw	cis	ND	Pyocin S3 modulation Resistance to UV radiation Virulence				[57]

Abbreviations: ANR - Anaerobic transcriptional regulator ANR; RBP–RNA Binding Protein; Rv–reverse; Fw–forward; ND–Non-defined; * Hfq binding and sequestering. [1] chromosome location related to the mRNA target.

3.1. RsmY and RsmZ

Chronic infection seems to be the preferred mode of infection by pathogens like *P. aeruginosa*, a premise for its survival and persistence as a population. Since the cellular mechanisms involved in acute or chronic infection are different, the cell is equipped with a complex regulatory network to switch its state. In the case of *P. aeruginosa*, the Csr/Rsm system plays a major role in this switch [52]. This system controls a large variety of physiological processes, like carbon metabolism, virulence, motility, quorum sensing, siderophore production, and stress responses [45]. The system comprises a major translational regulator, the protein RsmA, which negatively regulates mRNA targets by binding to sites containing critical GGA motifs present in the 5′-UTR of mRNAs targets [45] This binding represses the translation of regulons necessary for establishing chronic infections, as the T6SS, quorum sensing systems, exopolysaccharide production, biofilm formation, and iron homeostasis [70,71]. On the other hand, RsmA exerts a positive and indirect regulation of the mechanisms linked to acute infection, like the expression of genes associated with surface motility, T3SS, type IV pili, as well as systems that operate through the cAMP/virulence factor regulator (Vfr) route [72]. RsmA appears to control T3SS gene expression by increasing *exsA* translation through an undetermined mechanism. ExsA is the master transcription factor of all T3SS genes [73]. So, RsmA regulates the switching from planktonic (acute infection) to biofilm (chronic infection) phenotype. RsmA activity is negatively regulated by the two-component regulatory system GacA/GacS that induces the expression of sRNAs antagonist of RsmA, including the RsmY and RsmZ in *P. aeruginosa* (Figure 1) [74]. These two last sRNAs have a secondary structure with several unpaired GGA motifs that sequester RsmA proteins, preventing biding to their targets, and thus enhancing the expression of specific RsmA regulons [45].

Although GacA is a major regulator of *rsmY* and *rsmZ* expression, these sRNAs can also be regulated by alternative regulators, and despite the redundancy, they can have different regulators. It was observed in *P. aeruginosa* that *rsmY* transcription steadly increases during cell growth, whereas *rsmZ* is induced harshly during the late exponential phase of growth [45]. Using swarming motility as a model, Jean-Pierre et al. [61] showed that these sRNAs are differentially regulated depending on the selected growth conditions (planktonic versus surface–grown cells), and that *rsmZ* regulation does not implicate the response regulator GacA in swarming cells. Furthermore, these authors observed that RsmY/Z expression influences swarming motility via the protein HptB, which acts as a negative regulator of these sRNAs and that they do not strictly converge to RsmA [61]. Before a biofilm can be formed, RsmZ is eliminated by the action of ribonuclease G, activated by the TCS BfiSR [45]. Other mechanisms of regulation of RsmY and RsmZ have been discovered recently. AlgR, a TCS response regulator important for *P. aeruginosa* pathogenesis in both acute and chronic infections, seems to affect the antagonizing action of RsmY and RsmZ on RsmA [75]. SuhB regulates the motile–sessile switch in *P. aeruginosa* through the Gac/Rsm pathway and c-di-GMP signaling, regulating multiple virulence factors like the T3SS, swimming motility, T6SS, and biofilm formation [44]. Li et al. [44] reported that this regulation is mediated by the GacA-RsmY/Z-RsmA. MgtE, a magnesium transporter, was demonstrated to play a role in the inhibition of the T3SS transcriptional activator ExsA. The negative regulation is mediated by expression of RsmY and RsmZ, most likely though the TCS GacA/GacS [73]. The polynucleotide phosphorylase (PNPase), which is involved in the regulation of multiple virulence factors like T3SS, T6SS and pili biosynthesis, plays an important role in RsmY/Z stability. This stabilizing ability is due to the PNPase KH-S1 domains that bind to sRNAs stabilizing them, providing a way of pathogenicity regulation by PNPase [76]. Recently, Miller et al. [45] characterized a new sRNA in *P. aeruginosa*, the RsmW. Unlike RsmY and RsmZ, this sRNA is not transcriptionally activated by GacA. RsmW was shown to be upregulated in nutrient-limiting conditions, biofilms, and at higher temperatures [45].

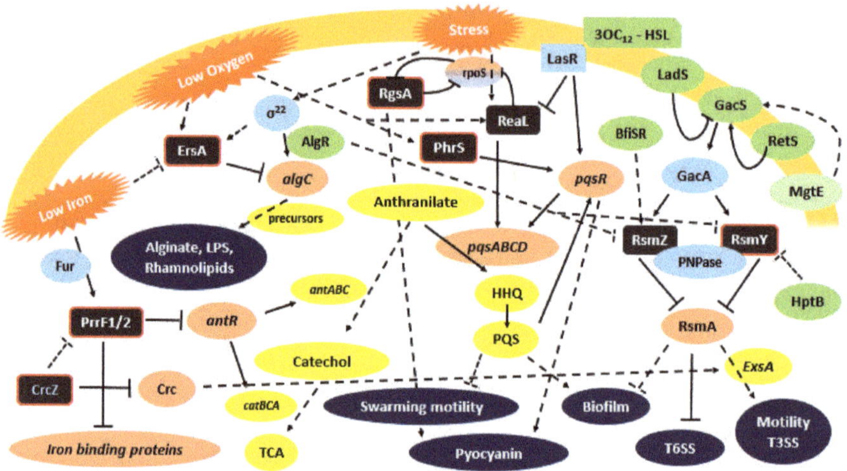

Figure 1. *P. aeruginosa* functionally characterized sRNAs and their involvement on virulence regulatory pathways. Grey box—characterized sRNAs (with red border- Hfq biding); Green box—indirect regulator, Light Blue box—direct regulator; Coral box—direct target; Yellow box—indirect target; Dark Blue—virulence factors; ↓: positive control; ⊥: negative control; dashed lines: indirect control.

3.2. ReaL

ReaL is a highly conserved *P. aeruginosa* sRNA recently characterized by Carloni et al. [51]. ReaL was shown to be a relevant element for *P. aeruginosa* pathogenesis. In the infection model *Galleria mellonella*, deletion of the ReaL encoding gene impaired *P. aeruginosa* PA14 virulence, while the sRNA overexpression resulted in a hypervirulent phenotype. ReaL is involved in the quorum sensing regulatory network architecture, interlinking the Las and PQS QS systems. Both Las and PQS are part of a major network played at least by two additional systems, the integrated quorum sensing system (IQS) and the Rhl QS system [77]. This entire network is closely related with the expression of virulence factors in *P. aeruginosa*. The Las system is composed by the transcriptional regulator LasR, its cognate autoinducer molecule N-3-oxo-dodecanoyl-homoserine lactone (3-oxo-C12-HSL), and the 3-oxo-C12-HSL synthase LasI [51]. PQS, positively regulated by the Las QS system, is an essential mediator of the shaping of the population structure of *P. aeruginosa* and of its responses and survival in hostile environmental conditions [78,79]. PQS synthesis proceeds through the condensation of an anthranilate molecule with a fatty acid to produce the 2-heptyl-4-quinolone (HHQ), which is then hydroxylated by the PqsH enzyme, whose expression is positively regulated by the LasR3-oxo-C12HSL to produce PQS [77,80]. It is also known that ReaL plays a role in the interconnection between both systems. Actually, ReaL is negatively regulated by LasR, and because it is a positive post-transcriptional regulator of the *pqsC* gene, a decrease of PQS will occur. ReaL seems to contribute to the fine co-modulation of HHQ/PQS synthesis [51]. The PQS system plays a vital role in biofilm formation and production of virulence factors such as pyocyanin, elastase, PA-IL lectin, and rhamnolipids [77]. Therefore, ReaL expression triggers the production of some virulence factors, like biofilm formation and pyocyanin synthesis, by a Las-independent way. In addition, ReaL plays a negative role in swarming motility. The upregulation of ReaL expression is observed in the conditions found when colonizing the human lung, like a temperature of 37 °C and low oxygen concentration [51].

3.3. ErsA

ErsA is the first sRNA from *P. aeruginosa* that appears to be embedded in the envelope stress response, a critical transduction pathway that impacts pathogenesis in *P. aeruginosa*. ErsA expression depends of the envelope stress responsive sigma factor σ^{22}, the major player in the envelope stress-signaling pathway like σ^E in *S.* Typhimurium and other enterobacteria. ErsA directly exerts a negative post-transcriptional regulation on the virulence associated *algC* gene encoding the bifunctional enzyme AlgC [68]. AlgC plays a central role in exopolysaccharide biosynthesis, generating the sugar precursors mannose 1-phosphate and glucose 1-phosphate, necessary for the synthesis of polysaccharides like alginate, Pel, Psl, LPS, and rhamnolipids, key components of the biofilm matrix [81]. A fine-tuned regulation exerted by ErsA on AlgC is thought to influence the dynamics of exopolysaccharide biosynthesis, underlying the development of biofilm matrix. This regulation is crucial when the bacterial cell is under envelope stress. The repressive role of ErsA on *algC* expression occurs mainly at the translational level by base pairing with the RBS in a Hfq dependent way [68]. Furthermore, Zhang et al. [82] found that ErsA represses the expression of the OprD porin, the major channel for the uptake of basic amino acids, peptides, and the carbapenem antibiotic, evidencing a role played by this sRNA in antibiotic resistance.

3.4. PrrF

PrrF sRNAs are encoded in *P. aeruginosa* genome by two highly homologous genes, *prrf1* and *prrf2*. These sRNAs are encoded in tandem sequences, separated by 95 nt, a characteristic exclusive of this species. The entire *prrF* region also encodes another sRNA, PrrH, transcribed from the 5′ end of *prrf1* to the 3′ end of *prrf2* [83]. Infection experiments in a murine model using a deletion mutant on the *prrF* locus showed that this genetic locus plays a pivotal role in *P. aeruginosa* virulence in acute murine lung infection. The encoded sRNAs are also important for iron and heme homeostasis [64]. PrrH seems to play a role in heme homeostasis and on the expression of virulence factors. However, PrrH is not vital for acute murine lung infection, as shown by Reinhart et al. [54]. Iron acquisition is essential for *P. aeruginosa* virulence and biofilm formation, highlighting the role of regulatory systems on the maintenance of iron homeostasis, with this sRNA playing a critical role in bacteria survival in the host [64]. PrrF is responsible for the iron-sparing response, an effect observed under conditions of iron limitation. PrrF exerts its regulatory activity by repressing mRNAs encoding iron-containing proteins like the iron cofactored superoxide dismutase SodB, a putative bacterioferritin, a heme-cofactored katalase and succinate dehydrogenase. Iron "sparing" is essential when the intracellular iron concentration is low [54]. Although iron is an essential metallonutrient, when it accumulates it becomes toxic, due to its ability to catalyze the formation of reactive oxygen species via Fenton reactions. The regulation of such a homeostasis is mediated by the ferric uptake repressor protein Fur [84]. Fur has the ability to bind to iron in iron-enriched environments, becoming an active repressor of PrrFs, leading to a more extensive use of iron by the proteins otherwise downregulated by PrrFs [64]. PrrF sRNAs also repress the expression of AntR, a transcription activator of genes involved in the conversion of anthranilate to catechol. By this way, the anthranilate is channeled into the biosynthetic pathway of the PQS, a system that leads to the expression of several virulence factors as already mentioned above [64,85]. Furthermore, Sonnleitner et al. [53] added another sRNA to the already extensive armory of *P. aeruginosa*, the CrcZ sRNA, which impairs the binding of PrrFs to *antR*. Since the riboregulation played by PrrFs is mediated by Hfq, the CrcZ functions like a sponge, binding to Hfq with a higher affinity compared to *antR*, inhibiting its binding. This mechanism leads to the de-repression of *antR*.

3.5. PesA

PesA, characterized by Ferrara et al. [57], is encoded in the pathogenicity island PAPI-1, being expressed by the highly virulent strain PA14, as well as by several clinical isolates obtained from CF patients. The *P. aeruginosa* strain most commonly used for research, PAO1, does not encode PesA. The sRNA is expressed under temperature and oxygenation conditions resembling those found in the CF lung, namely 37 °C and low oxygen. PesA was shown to be important for *P. aeruginosa* pathogenicity in CF bronchial cells. PesA is involved in the post-transcriptional regulation of genes involved in S-type pyocin production, like pyocin S3. Pyocin S3, like other soluble pyocins, has a killing activity that causes cell death by DNA cleavage. Pyocins are produced by more than 90% of the *P. aeruginosa* strains, helping *P. aeruginosa* in niche establishment, protecting the bacteria from competition [57,86]. By deleting *pesA*, an increased susceptibility to UV irradiation and to the fluoroquinolone antibiotic ciprofloxacin was observed, correlated with a decreased level of pyocin expression. A similar effect was observed when deleting the gene encoding pyocin S3. Although contradictory, PesA seems to play an important role in the response to DNA damage by promoting the synthesis of pyocin S3 [57]. The mechanism underlying this response is not yet clear.

4. The *Burkholderia cepacia* Complex and the Emerging Knowledge on Its sRNAs

When compared to *P. aeruginosa*, respiratory infections caused by Bcc species have a low prevalence on CF patients. However, bacteria of this complex are still one of the most feared pathogens and represent a higher risk for those patients.

Burkholderia cepacia complex (Bcc) is a group of opportunistic human pathogens presently composed of 23 closely-related species [87]. Apart from being a major threat for CF patients, Bcc bacteria can also cause lethal infections among chronic granulomatous disease (CGD) patients, immunocompromised patients (e.g., HIV infected), cancer patients, and other chronic patients [88]. *B. cenocepacia* and *B. multivorans* remain the predominant Bcc species worldwide that are responsible for infections in CF patients [89,90]. Bcc species possess genomes larger than those of *P. aeruginosa* strains, usually with a length of more than 7 Mbp. This large genome is thought to confer the bacterium an enormous ability to overcome antimicrobial therapies, as well as the host immune response, being in constant evolution inside the host lung [91].

The defective mucus clearance of the CF lung is the perfect niche for a persistent infection by *P. aeruginosa*, as well as by Bcc bacteria that tends to occur after *P. aeruginosa* colonization, usually superseding it [92,93]. Moreover, Bcc infections are easily propagated among CF patients. Bcc bacteria possess an extraordinary resistance to antibiotics [94,95]. About twenty percent of CF patients infected with Bcc bacteria develop the cepacia syndrome, a necrotizing pneumonia, often accompanied by septicemia, leading to a rapid, irreversible and deadly decline of the respiratory function [96,97].

Bcc virulence resembles several traits of *P. aeruginosa* virulence, due both to the fragility of the host and to the bacterium extraordinary evolutionary development. Once the colonization of the respiratory tract is established and the bacteria are attached to epithelial cells, they have to overcome a strong immunological response [98]. The successful invasion of host internal system is achieved by mechanisms of penetration like paracytosis and invasion as a biofilm [99,100]. Among the virulence factors expressed by Bcc species the extracellular lipase, metalloproteases and serine proteases, and some structures of the bacterial surface such as flagella, pili, and the lipopolysaccharide (LPS) can be highlighted [96,100]. Exopolysaccharide (EPS) production, namely Cepacian, also represents a major virulence factor, important for biofilm formation and to protect Bcc from the host defense machinery, as well as from antimicrobial therapy [101–104].

Like in *P. aeruginosa*, several noncoding RNAs have been associated with regulatory mechanisms involved in Bcc virulence, as is the case of the modulation of the adaptation to the host changing environment, the adherence to, and invasion of host cells, as well as the replication. However, the Bcc noncoding virulome (sRNAs playing a role in virulence) is still poorly understood compared with *P. aeruginosa*.

4.1. Discovering Bcc Noncoding Transcriptome

A pioneer systematic search for sRNAs in Bcc was first described by Coenye et al. in 2007 [105], who predicted an array of putative noncoding RNAs (ncRNAs) from *B. cenocepacia* J2315, using *R. solanacearum* GMI1000 genome as a reference. Two-hundred-and-thirteen ncRNA genes, ranging in size from 53 to 1243 nt, were predicted and described. The expression of only four of these sRNAs was confirmed by microarray experiments using total RNA from cells grown on a 10% (*w/v*) CF sputum [105]. The expression of one sRNA has confirmed by real-time PCR; none of these putative sRNAs were functionally characterized.

In order to understand the mechanisms underlying the early stage of Bcc infections, Drevinek et al. [106] performed a microarray-based transcriptomic analysis of *B. cenocepacia* grown from sputum recovered from CF patients. The strain was found out to be indistinguishable from *B. cenocepacia* J2315, a strain involved in multiple fatalities. The genes found to be upregulated were mainly involved in antibiotic resistance, the antioxidant response to reactive species, iron metabolism, and virulence factors such as flagella and metalloproteases. A thorough analysis of transcripts originating from intergenic regions was carried out, leading to the identification of 88 upregulated and 126 downregulated sequences. However, no further characterization of these regions was performed to identify putative sRNAs.

In 2009, Yoder-Himes et al. [107] used Illumina RNA-seq techniques to understand the mechanisms of transcriptional regulation associated to *B. cenocepacia* environmental adaptation. Two strains, one isolated from soil (*B. cenocepacia* HI2424) and another from a CF patient (*B. cenocepacia* AU 1054), were grown under conditions mimicking each environment, and their transcriptomes were analyzed. In both conditions a higher number of genes was transcribed in strain HI2424 than in strain AU 1054. When grown under CF-like conditions, the strains exhibited differences in the expression of genes mainly from chromosome 1, while growth under soil conditions led to differences in the expression of genes mainly encoded in chromosomes 2 and 3. Regarding sRNAs, 13 were identified (ncRNA1 to ncRNA13), but only one seems to be induced under CF-mimicking conditions. Those sRNAs have been predicted to be highly structured. The expression of 4 sRNAs was confirmed by northern blot, allowing the authors to assume that the sequences were indeed expressed in vivo. The majority of these molecules are conserved among the *B. cenocepacia* strains and the Bcc (Table S1).

Effective response to reactive oxygen species by bacteria is an important feature regarding the adoption of disinfection measures, as well as during exposure to oxidative burst by the host defenses. To evaluate the response to reactive oxygen species (ROS), Peeters et al. [108] used microarray transcriptomic analysis and qPCR to analyze whole cell responses of *B. cenocepacia* J2315 sessile cells exposed to H_2O_2 and NaOCl. The profile of genes up- and downregulated was found to be similar to that of planktonic *P. aeruginosa* PAO1 cells exposed to similar conditions. Many genes involved in prevention (counteracting or repairing of the damage from oxidative stress) were found to be upregulated, as well as some genes involved in the synthesis and assembly of flagella. Moreover, several transcripts from intergenic regions were found, 39 and 56 were upregulated upon exposure to H_2O_2 and NaOCl, respectively, whereas 54 and 68 were found to be downregulated. To avoid false positives, the authors selected as putative noncoding RNAs the regions whose expression pattern is different from the flanking genes. Eleven and 20 intergenic regions were selected as putative noncoding RNAs upregulated with H_2O_2 and NaOCl, respectively. Transcripts from intergenic regions IG1_2935724 and IG1_3008003 were upregulated in both conditions and their expression was confirmed by qPCR. The ncRNA4 and ncRNA6, found by Yoder-Himes et al. [107], are located in those intergenic regions, thus confirming their expression. A match was found between the IG1_2935724 and the secondary structure of 6S RNA consensus structure. Seven out of the 11 sRNAs upregulated upon exposure to H_2O_2 were also found significantly altered in *B. cenocepacia* J2315 cells of CF sputum [106].

Coenye et al. [109] used microarrays to study transcriptomic differences between planktonic and sessile cells of *B. cenocapacia* J2315 exposed to chlorhexidine. The authors found that sessile cells are more resistant to chlorhexidine (0.015%) than planktonic cells. Furthermore, sessile cells highly expressed genes encoding efflux systems related to drug resistance, as well as membrane-associated proteins and regulators. The results suggest that sessile cells engaged a global expression program to evade the biofilm since an adhesin was downregulated and genes encoding chemotaxis and motility-related proteins were upregulated. This study also allowed the identification of several intergenic regions upregulated in the presence of chlorhexidine. After discarding the regions with an expression level similar to the closest coding sequences (CDS) and the ones whose probes overlapped the coding sequences, 19 intergenic regions were suggested as putative sRNAs. As presented in Supplementary Table S1, these regions are largely conserved among *B. cenocepacia* strains, less conserved among the Bcc members, and only one of them is conserved in the *B. pseudomallei* group. The majority of the predicted sRNAs seems to have a stable secondary structure, however no significant match was found in the Rfam database [109].

Previous work from our research group led the experimental identification of 24 sRNAs from *B. cenocepacia*, based on co-purification of total RNA and the chaperone Hfq [110]. Expression of these sRNAs was confirmed by northern blot and characterized in silico. The only consistent results were related to Bc7, for which the *hemB* was predicted as mRNA target, and the *Salmonella* Paratyphi RybB as homolog. In addition a cis-encoded sRNA, named h2cR, was also previously reported by our research group as being involved in the regulation of the *hfq2* gene that encodes an Hfq-like chaperone in Bcc [111].

The genome of *B. cenocepacia* KC-01, a strain isolated from the coastal saline soil, was recently sequenced by Ghosh et al. [112]. Several potential sRNAs were identified by in silico analysis of the genome, and the expression of seven putative sRNAs was confirmed (Bc_KC_sr1-7). Bc_KC_sr1 and Bc_KC_sr2 were upregulated in response to iron depletion by 2,2′-bipyridyl. Bc_KC_sr3 and Bc_KC_sr4 were induced under the presence of 60 µM H_2O_2 in the culture medium. Alterations on the temperature and incubation time also induced the expression of Bc_KC_sr2, 3, and 4. Searches within the RFAM and BSRD databases led to the identification of candidate738 of *B. pseudomallei* D286, tmRNA and 6S RNA as homologs of Bc_Kc_sr4, 5, and 6, respectively. Interestingly, this group of sRNAs is extensively conserved among members of the Bcc and *B. pseudomallei* groups (Supplementary Table S1). Several targets were predicted for these sRNAs, like Fe-S cluster and siderophore biosynthesis, ROS homeostasis, porins, transcription, and translation regulators.

4.2. A Deeper Approach: Bcc sRNAs Expressed under Biofilm Formation Conditions

Based on the importance of biofilm formation to antibiotic resistance, Sass et al. [113] analyzed the transcriptome of *B. cenocepacia* J2315 grown in biofilms by differential RNA-sequencing (dRNA-seq). dRNA-seq differs from RNA-seq on a selective sequencing of primary transcripts. The techniques, used for the first time for a Bcc organism, allow mapping of the transcription start sites (TSS) based on the difference of primary and processed transcripts ends. Primary transcripts carry a 5′ triphosphate end, while processed transcripts like tRNA and rRNA, carry a 5′ monophosphate. Using the 5′ P-dependent terminator exonuclease (TEX) to degrade the 5′ monophosphate, it is possible to distinguish between primary and processed transcripts [114]. As a complementary approach, Sass et al. also carried out a global RNA-seq (gRNA-seq) to obtain a large coverage of the whole transcript length and the 3′ end of transcripts, usually lost by dRNA-seq of longer transcripts. The additional use of 5′ RACE allowed the authors to end up with 2089 genes annotated as expressed under biofilm conditions and to identify alternative start codons for some genes and novel protein sequences. A total nine sRNAs with homologs present in the Rfam database and other seven match putative sRNAs that were experimentally identified in previous studies. The sRNAs with homologs in the Rfam database were the 6S RNA; two phage-related regulatory RNAs located on genomic island BcenGI9; two conserved regulatory motifs; the SAH riboswitch located upstream of BCAL0145; an adenosylhomocysteinase; the *sucA* RNA motif located upstream of *sucA* (BCAL1515), an enzyme of the citric acid cycle; and four sRNAs from the family named "toxic small RNAs", whose expression in *B. cenocepacia* have already been confirmed by northern blot but with unknown functions. From the transcript sequences containing a TSS in intergenic regions that did not match with a gene and without a hit in the Rfam database, the authors selected only those obeying to the following criteria: strong transcription initiation with a coverage >300 reads in dRNA-Seq data, a defined 3′ end in dRNA-Seq data or a transcript appearing short (<500 nt), and truncated or missing in gRNA-Seq data. Upon selection, the sequences were compared with the experimentally confirmed sRNAs already published by other authors. Six putative sRNAs matched those found by Yoder-Himes et al. [107], five of them upregulated on soil conditions and one on cystic fibrosis sputum. Another transcript matched one sRNA previously identified by using co-purification with Hfq [110]. The expression of these sRNAs in other studies, together with their overexpression under conditions of biofilm formation, suggests a role for these sRNAs on biofilm regulation.

Sass et al. [115] refined their analysis on putative sRNAs found in association with biofilm formation. From a total of 148 TSS found in intergenic regions, 41 transcripts were classified as rho-independent (RIT), and 82 transcripts as derived from 5′ UTR of the downstream gene. Fifteen sRNAs were selected based on the following criteria, the number of starts at TSS ($\geq$250), the z-score (<-1), and the conservation on Bcc ($\geq$13 strains). 3′RACE and Northern blot analysis were used to confirm the length of sRNAs. The expression of 14 sRNAs was quantified by qPCR in different experimental conditions, all compared with planktonic cultures. Twelve sRNAs were upregulated in biofilms. Some sRNAs were slightly upregulated by osmotic or pH stress. Nutrient starvation also induced the expression of 12 sRNAs, as well as the expression of Hfq chaperone. Five of these sRNAs were also more expressed under glucose-rich medium.

Further work performed by these authors led to the identification of ncS63 as highly expressed under conditions of low-iron. This sRNA is located upstream of BCAL2297, which encodes for the hemin-uptake protein HemP, whose expression is under the regulatory control of the Fur repressor. The predicted targets of ncS63 are related to iron homeostasis like *sdhA*, a target of RyhB from *E. coli*. RyhB is a sRNA involved in iron homeostasis regulated by the Fur repressor, and a possible functional analogue of ncS63. Recently, RyhB was hypothesized to play an important role as a mediator of bacterial resistance to multiple antibiotics and stresses [116]. Further investigation is needed to check if ncS63 is a RyhB functional homolog. The overall prediction of sRNAs targets using CopraRNA and RNApredator led to similar clues about the mechanisms that can be under regulation of these sRNAs, like transcriptional regulators, carbon compound transport and metabolism, and cell envelope components such as outer membrane proteins and porins. The transcriptional regulator BCAL1948 has been predicted to be a target of nine sRNAs, exhibiting particularly extensive complementarity to ncS11. sRNAs containing double-hairpin were predicted to target genes involved on the metabolism and transport of amino acids, carbohydrates, and aromatic compounds. The ncS16 has several hits of genes related to outer membrane and cell envelope components, and genes for transport of inorganic compounds, whereas the best hits for ncS63 were genes involved in energy production and detoxication of reactive oxygen species. Regarding the putative sRNAs derived from 5′UTRs, most likely they have a cis-regulatory function, since some 3′ ends of adjacent genes are homologs of genes that possess cis-regulatory structures in other species. A possible trans-regulatory function remains to be investigated.

4.3. ncS35, a Functionally Characterized B. cenocepacia sRNA

While in other organisms sRNAs are being characterized for decades, their characterization in Bcc is starting. ncS35 was identified by Sass et al. [115] in the course of *B. cenocepacia* J2315 biofilm transcriptomic analysis, being the first trans-encoded characterized sRNA from *B. cenocepacia*. The expression of ncS35 was found to be upregulated in cells grown as biofilm and in minimal medium compared to planktonic cells grown in rich medium. Increased aggregation, higher cell metabolic activity, higher growth rate, and increased susceptibility to tobramycin were described for deletion mutant cells. Moreover, an upregulation of the phenylacetic acid and tryptophan degradation pathways was observed in the mutant cells, and the first gene of the tryptophan degradation pathway was predicted to be a putative target of ncS35. This sRNA seems to lead to an attenuation of the metabolic and growth rates, which can be a way of cells to protect themselves against stress conditions. A slow growth rate is observed on *P. aeruginosa* biofilms in CF patient's sputum, and it is also known that bacteria on slow growth rate, even in planktonic cultures, have an increased resistance to antibiotics, including tobramycin [117,118]. Slow growth rates and metabolic activity are characteristics of persister cells and probably the cause of drug resistance. Since the effect of antibiotics is mainly due to the inhibitory action of some metabolic pathways, slow-growing cells are somehow protected from antibiotics and are more likely to develop drug resistance [119]. The occurrence of persister cells in Bcc infections is known, and it has also been observed in *B. cenocepacia* J2315 after treatment with tobramycin [120]. A better characterization is required to fully understand the roles of ncS35 on *B. cenocepacia* pathogenicity, antibiotic resistance, and possibly on the formation of persister cells.

4.4. Compiling the Bcc Predicted sRNAs

Although a few sRNAs from Bcc organisms are functionally characterized, a large number of putative sRNAs have been described by several research groups. This information was gathered, and we present on Supplementary Table S1 all predicted and confirmed sRNAs, as well as the intergenic regions, which can include candidate noncoding RNAs described so far for *B. cenocepacia*. Predicted sRNAs for which data is somehow confusing and uncertain, or the methods applied were not accurate enough, were not considered. *B. cenocepacia* was chosen because it is the one of the Bcc species with the noncoding transcriptome better characterized. A total of 167 putative sRNAs were included on Table S1, seven from *B. cenocepacia* KC-01, 13 from *B. cenocepacia* AU 1054, and 147 from *B. cenocepacia* J2315, the majority of them identified by Sass et al. [115]. Putative sRNAs are unevenly distributed throughout the three chromosomes: 67.5% are located on chromosome 1, 28.1% on chromosome 2, and 4.4% on chromosome 3. The length of the replicons on the J2315 strain is 3,870,082, 3,217,062, and 875,977 bp, respectively [121]. The chromosomes 1 and 2 are approximately equally sized; however, chromosome 1 accommodates twice the sRNAs compared with the second chromosome. This suggests that chromosome 1 sRNAs are more relevant for core functions of bacterial metabolism (e.g., cell division). On the other hand, since the regulation of the other two replicons encode genes more related to accessory functions, the regulatory noncoding RNAs encoded in those replicons are probably expressed in response to specific conditions that may have not been assessed yet [121]. Wong et al. [122] used a transposon mutant pool and identified 383 possible essential genes on J2315 strain, 90% of which was present on chromosome 1. Such a strategy can not only enable the discovery of crucial coding genes expressed on different conditions, but might also unveil the Bcc noncoding transcriptome, allowing the discovery of important genes related to some features like virulence and antibiotic resistance.

Regarding the sequence conservation, 93% of the putative sRNAs listed in Table S1 are conserved or semiconserved among the species, 70% among the Bcc bacteria and 24% also in the *B. pseudomallei* group. Although the *B. pseudomallei* group is phylogenetically distant and those bacteria display a distinct pathogenicity and epidemiology, a high level of sequence homology is found in almost 25% of the putative sRNAs identified so far [123]. It is also interesting to access the conservation of the putative sRNAs in each chromosome, since a higher conservation can be observed on chromosome 1, much more extensive than chromosome 2 and 3. A huge difference is observed on the conservation of sRNAs on the Pseudomallei group, with a conservation (and semiconservation) of ~25% on sRNAs from chromosome 1, whereas sRNAs from chromosome 2 and 3 exhibited less than 4% conservation. This is also evidence that Bcc comprises more closely related species than the Pseudomallei group, mainly regarding the accessory functions.

Several sRNAs were identified by more than one experimental approach and given different names. For instance, ncRNA4, ncS17, and Bc_KC_sr6 are the same sRNA, which was found on the intergenic region IG1_2935724. This sRNA shares homology with the 6S RNA, a regulator of RNA polymerase (RNAP) that is widespread among members of the Bacteria domain. This sRNA was found as expressed in distinct situations like soil conditions, biofilm formation and response to ROS [124]. Moreover, an upregulation of the 6S RNA was found in response to oxidative stress, further implying it with a broader role than just on core metabolism, such as on the protection of the cell in the host environment [108]. The possible role of 6S RNA in pathogenesis or in host survival has been described for several bacteria [124]. The homologs ncRNA6 and ncRI12 are both located on the intergenic region IG1_3008003 and are probably the same sRNA. This sRNA was overexpressed during biofilm formation and upregulated under soil conditions, exposure to chlorhexidine, and ROS, suggesting a role on protection against stress conditions. ncS06, homolog of ncRNA7, was found to be expressed under soil conditions and biofilm formation. Prediction of ncS06 mRNA targets revealed several peptidase-encoding genes [115]. ncS11 is a homolog of ncRNA13 and was found to be upregulated on soil conditions, as well as on biofilms, rich medium and starvation conditions [115]. ncS11 has an extensive complementarity to BCAL1948, whose expression was found to be downregulated under biofilm formation. BCAL1948 is a protein of the LysR-type transcriptional regulator (LTTR) family, whose global transcriptional regulators can be found in diverse organisms. The regulated genes are hypothesized to be involved in a wide range of processes like cell division, metabolism, nitrogen fixation, oxidative stress responses, motility, quorum sensing, attachment, secretion, virulence, and toxin production [125].

5. Conclusions

There is a clear contrast between the knowledge currently available about sRNAs characterization, regulatory pathways and interactome in *P. aeruginosa* and Bcc bacteria. In *P. aeruginosa*, fourteen sRNAs have already been functional characterized. Some *P. aeruginosa* sRNAs are involved in the pathogenicity and virulence of the organism, like biofilm formation, iron metabolism, quorum sensing regulation, response to stress, and host cell invasion. On the other hand, in Bcc bacteria the first biological function of a sRNA is now being described. This sRNA is ncS35, a growth regulator that seems to do not interfere with the virulence of Bcc bacteria. Despite the scarce characterization of sRNAs in Bcc, over the last 10 years at least 167 putative sRNAs were identified in *B. cenocepacia* as being expressed on CF sputum, biofilm formation, response to oxidative stress, among others specific conditions.

In the peculiar environment of the CF lungs, it is interesting to note that both *P. aeruginosa* and Bcc bacteria can cause infections presenting a high degree of intrinsic and acquired resistance to antibiotics. Both opportunistic pathogens have two of the largest genomes known among prokaryotes, which were predicted to encode a huge number of sRNAs. The functional characterization of these sRNAs is only beginning in those CF pathogens. However, the already established or predicted roles of some sRNAs in modulating cellular processes linked to pathogenesis, suggests their yet unexplored importance for the physiology and pathology of these pathogens, as well as their likely influence in the outcome of *P. aeruginosa* and Bcc infections. The assessment of the functional roles of sRNAs in the host–pathogen interaction is expected to provide additional fundamental knowledge for the development of next-generation antibiotics inspired by sRNAs and their targets.

Supplementary Materials: Supplementary materials can be found at http://www.mdpi.com/1422-0067/19/12/3759/s1.

Author Contributions: Conceptualization, J.H.L.; Writing—Original draft preparation, T.P. and J.R.F.; Writing—Review and editing, J.H.L., T.P. and J.R.F.

Funding: This work was funded by Fundação para a Ciência e Tecnologia (Portugal) through Projects PTDC/BIA-MIC/1615/2014 and UID/BIO/04565/2013.

Acknowledgments: Tiago Pita. acknowledges a doctoral grant (PD/BD/135137/2017) and Joana R. Feliciano. acknowledges a postdoctoral grant (BL109/17 IST-ID) from FCT.

Conflicts of Interest: The authors declare no conflicts of interest.

References

1.   Mikulik, K. Structure and Functional Properties of Prokaryotic Small Noncoding RNAs. *Folia Microbiol.* **2003**, *48*, 443–468. [CrossRef]
2.   Hindley, J. Fractionation of ^{32}P-labelled ribonucleic acids on polyacrylamide gels and their characterization by fingerprinting. *J. Mol. Biol.* **1967**, *30*, 125–136. [CrossRef]
3.   Storz, G.; Vogel, J.; Wassarman, K.M. Regulation by Small RNAs in Bacteria: Expanding Frontiers. *Mol. Cell* **2011**, *43*, 880–891. [CrossRef] [PubMed]

4. Nitzan, M.; Rehani, R.; Margalit, H. Integration of Bacterial Small RNAs in Regulatory Networks. *Annu. Rev. Biophys.* **2017**, *46*, 131–148. [CrossRef] [PubMed]

5. Papenfort, K.; Pfeiffer, V.; Mika, F.; Lucchini, S.; Hinton, J.C.D.; Vogel, J. σ^E-dependent small RNAs of *Salmonella* respond to membrane stress by accelerating global omp mRNA decay. *Mol. Microbiol.* **2006**, *62*, 1674–1688. [CrossRef] [PubMed]

6. Klein, G.; Raina, S. Small regulatory bacterial RNAs regulating the envelope stress response. *Biochem. Soc. Trans.* **2017**, *45*, 417–425. [CrossRef] [PubMed]

7. Michaux, C.; Verneuil, N.; Hartke, A.; Giard, J.-C. Physiological roles of small RNA molecules. *Microbiology* **2014**, *160*, 1007–1019. [CrossRef] [PubMed]

8. Gimpel, M.; Brantl, S. Dual-function small regulatory RNAs in bacteria. *Mol. Microbiol.* **2017**, *103*, 387–397. [CrossRef] [PubMed]

9. Brantl, S. Chapter 4 Small Regulatory RNAs (sRNAs): Key Players in Prokaryotic Metabolism, Stress Response, and Virulence. In *Regulatory RNAs*, 1st ed.; Mallick, B., Ed.; Springer: Berlin/Heidelberg, Germany, 2012; pp. 73–108.

10. Park, H.; Yoon, Y.; Shinae, S.; Ji Young, L.; Younghoon, L. Effects of different target sites on antisense RNA mediated regulation of gene expression. *BMB Rep.* **2014**, *47*, 619–624. [CrossRef] [PubMed]

11. Overgaard, M.; Johansen, J.; Møller-Jensen, J.; Valentin-Hansen, P. Switching off small RNA regulation with trap-mRNA. *Mol. Microbiol.* **2009**, *73*, 790–800. [CrossRef] [PubMed]

12. Figueroa-Bossi, N.; Valentini, M.; Malleret, L.; Bossi, L. Caught at its own game: Regulatory small RNA inactivated by an inducible transcript mimicking its target. *Genes Dev.* **2009**, *23*, 2004–2015. [CrossRef] [PubMed]

13. Azam, M.S.; Vanderpool, C.K. Talk among yourselves: RNA sponges mediate cross talk between functionally related messenger RNAs. *EMBO J.* **2015**, *34*, 1436–1438. [CrossRef] [PubMed]

14. Holmqvist, E.; Vogel, J. A small RNA serving both the Hfq and CsrA regulons. *Genes Dev.* **2013**, *27*, 1073–1078. [CrossRef] [PubMed]

15. Malabirade, A.; Morgado-Brajones, J.; Trépout, S.; Wien, F.; Marquez, I.; Seguin, J.; Marco, S.; Velez, M.; Arluison, V. Membrane association of the bacterial riboregulator Hfq and functional perspectives. *Sci. Rep.* **2017**, *7*, 1–12. [CrossRef] [PubMed]

16. Brennan, R.G.; Link, T.M. Hfq structure, function and ligand binding. *Curr. Opin. Microbiol.* **2007**, *10*, 125–133. [CrossRef] [PubMed]

17. Schulz, E.C.; Seiler, M.; Zuliani, C.; Voigt, F.; Rybin, V.; Pogenberg, V.; Mücke, N.; Wilmanns, M.; Gibson, T.J.; Barabas, O. Intermolecular base stacking mediates RNA-RNA interaction in a crystal structure of the RNA chaperone Hfq. *Sci. Rep.* **2017**, *7*, 1–15. [CrossRef] [PubMed]

18. Feliciano, J.R.; Grilo, A.M.; Guerreiro, S.I.; Sousa, S.A.; Leitão, J.H. Hfq: A multifaceted RNA chaperone involved in virulence. *Future Microbiol.* **2016**, *11*, 137–151. [CrossRef] [PubMed]

19. Wroblewska, Z.; Olejniczak, M. Contributions of the Hfq protein to translation regulation by small noncoding RNAs binding to the mRNA coding sequence. *Acta Biochim. Pol.* **2016**, *63*, 701–707. [CrossRef] [PubMed]

20. Updegrove, T.B.; Zhang, A.; Storz, G. Hfq: The flexible RNA matchmaker. *Curr. Opin. Microbiol.* **2016**, *30*, 133–138. [CrossRef] [PubMed]

21. Lalaouna, D.; Simoneau-Roy, M.; Lafontaine, D.; Massé, E. Regulatory RNAs and target mRNA decay in prokaryotes. *Biochim. Biophys. Acta Gene Regul. Mech.* **2013**, *1829*, 742–747. [CrossRef] [PubMed]

22. Khoo, J.-S.; Chai, S.-F.; Mohamed, R.; Nathan, S.; Firdaus-Raih, M. Computational discovery and RT-PCR validation of novel *Burkholderia* conserved and *Burkholderia pseudomallei* unique sRNAs. *BMC Genom.* **2012**, *13* (Suppl. 7), S13. [CrossRef]

23. Smirnov, A.; Wang, C.; Drewry, L.L.; Vogel, J. Molecular mechanism of mRNA repression in *trans* by a ProQ-dependent small RNA. *EMBO J.* **2017**, *36*, 1029–1045. [CrossRef] [PubMed]

24. Smirnov, A.; Förstner, K.U.; Holmqvist, E.; Otto, A.; Günster, R.; Becher, D.; Reinhardt, R.; Vogel, J. Grad-seq guides the discovery of ProQ as a major small RNA-binding protein. *Proc. Natl. Acad. Sci. USA* **2016**, *113*, 11591–11596. [CrossRef] [PubMed]

25. Langton Hewer, S.C.; Smyth, A.R. Antibiotic strategies for eradicating *Pseudomonas aeruginosa* in people with cystic fibrosis. *Cochrane Database Syst. Rev.* **2014**, *11*, CD004197. [CrossRef]

26. Cutting, G.R. Cystic fibrosis genetics: From molecular understanding to clinical application. *Nat. Rev. Genet.* **2015**, *16*, 45–56. [CrossRef] [PubMed]

27. Stoltz, D.A.; Meyerholz, D.K.; Welsh, M.J. Origins of Cystic Fibrosis Lung Disease. *N. Engl. J. Med.* **2015**, *372*, 351–362. [CrossRef] [PubMed]

28. King, J.; Brunel, S.F.; Warris, A. *Aspergillus* infections in cystic fibrosis. *J. Infect.* **2016**, *72*, S50–S55. [CrossRef] [PubMed]

29. Cantin, A.M.; Hartl, D.; Konstan, M.W.; Chmiel, J.F. Inflammation in cystic fibrosis lung disease: Pathogenesis and therapy. *J. Cyst. Fibros.* **2015**, *14*, 419–430. [CrossRef] [PubMed]

30. Ciofu, O.; Tolker-Nielsen, T.; Jensen, P.Ø.; Wang, H.; Høiby, N. Antimicrobial resistance, respiratory tract infections and role of biofilms in lung infections in cystic fibrosis patients. *Adv. Drug Deliv. Rev.* **2015**, *85*, 7–23. [CrossRef] [PubMed]

31. Filkins, L.M.; O'Toole, G.A. Cystic Fibrosis Lung Infections: Polymicrobial, Complex, and Hard to Treat. *PLoS Pathog.* **2015**, *11*, 1–8. [CrossRef] [PubMed]

32. Winstanley, C.; O'Brien, S.; Brockhurst, M.A. *Pseudomonas aeruginosa* Evolutionary Adaptation and Diversification in Cystic Fibrosis Chronic Lung Infections. *Trends Microbiol.* **2016**, *24*, 327–337. [CrossRef] [PubMed]

33. Lewis, E.R.G.; Torres, A.G. The art of persistence-the secrets to *Burkholderia* chronic infections. *Pathog. Dis.* **2016**, *74*, 1–11. [CrossRef] [PubMed]

34. Govan, J.R.; Brown, A.R.; Jones, A.M. Evolving epidemiology of *Pseudomonas aeruginosa* and the *Burkholderia cepacia* complex in cystic fibrosis lung infection. *Future Microbiol.* **2007**, *2*, 153–164. [CrossRef] [PubMed]

35. Fröhlich, K.S.; Haneke, K.; Papenfort, K.; Vogel, J. The target spectrum of sdsr small RNA in *Salmonella*. *Nucleic Acids Res.* **2016**, *44*, 10406–10422. [CrossRef] [PubMed]

36. De Bentzmann, S.; Plésiat, P. The *Pseudomonas aeruginosa* opportunistic pathogen and human infections. *Environ. Microbiol.* **2011**, *13*, 1655–1665. [CrossRef] [PubMed]

37. Lu, P.; Wang, Y.; Zhang, Y.; Hu, Y.; Thompson, K.M.; Chen, S. RpoS-dependent sRNA RgsA regulates Fis and AcpP in *Pseudomonas aeruginosa*. *Mol. Microbiol.* **2016**, *102*, 244–259. [CrossRef] [PubMed]

38. Potron, A.; Poirel, L.; Nordmann, P. Emerging broad-spectrum resistance in *Pseudomonas aeruginosa* and *Acinetobacter baumannii*: Mechanisms and epidemiology. *Int. J. Antimicrob. Agents* **2015**, *45*, 568–585. [CrossRef] [PubMed]

39. Gellatly, S.L.; Hancock, R.E.W. *Pseudomonas aeruginosa*: New insights into pathogenesis and host defenses. *Pathog. Dis.* **2013**, *67*, 159–173. [CrossRef] [PubMed]

40. Lund-Palau, H.; Turnbull, A.R.; Bush, A.; Bardin, E.; Cameron, L.; Soren, O.; Wierre-Gore, N.; Alton, E.W.; Bundy, J.G.; Connett, G.; et al. *Pseudomonas aeruginosa* infection in cystic fibrosis: Pathophysiological mechanisms and therapeutic approaches. *Expert Rev. Respir. Med.* **2016**, *10*, 685–697. [CrossRef] [PubMed]

41. Mikkelsen, H.; McMullan, R.; Filloux, A. The *Pseudomonas aeruginosa* reference strain PA14 displays increased virulence due to a mutation in *ladS*. *PLoS ONE* **2011**, *6*. [CrossRef] [PubMed]

42. Gómez-Lozano, M.; Marvig, R.L.; Molina-Santiago, C.; Tribelli, P.M.; Ramos, J.L.; Molin, S. Diversity of small RNAs expressed in *Pseudomonas* species. *Environ. Microbiol. Rep.* **2015**, *7*, 227–236. [CrossRef] [PubMed]

43. Li, L.; Huang, D.; Cheung, M.K.; Nong, W.; Huang, Q.; Kwan, H.S. BSRD: A repository for bacterial small regulatory RNA. *Nucleic Acids Res.* **2013**, *41*, 233–238. [CrossRef] [PubMed]

44. Li, K.; Yang, G.; Debru, A.B.; Li, P.; Zong, L.; Li, P.; Xu, T.; Wu, W.; Jin, S.; Bao, Q.; et al. SuhB regulates the motile-sessile switch in *Pseudomonas aeruginosa* through the Gac/Rsm pathway and c-di-GMP signaling. *Front. Microbiol.* **2017**, *8*, 1–11. [CrossRef] [PubMed]

45. Miller, C.L.; Romero, M.; Karna, S.L.R.; Chen, T.; Heeb, S.; Leung, K.P. RsmW, *Pseudomonas aeruginosa* small non-coding RsmA-binding RNA upregulated in biofilm versus planktonic growth conditions. *BMC Microbiol.* **2016**, *16*, 155. [CrossRef] [PubMed]

46. Wang, C.; Ye, F.; Kumar, V.; Gao, Y.G.; Zhang, L.H. BswR controls bacterial motility and biofilm formation in *Pseudomonas aeruginosa* through modulation of the small RNA rsmZ. *Nucleic Acids Res.* **2014**, *42*, 4563–4576. [CrossRef] [PubMed]

47. Falcone, M.; Ferrara, S.; Rossi, E.; Johansen, H.K.; Molin, S.; Bertoni, G. The small RNA ErsA of *Pseudomonas aeruginosa* contributes to biofilm development and motility through post-transcriptional modulation of AmrZ. *Front. Microbiol.* **2018**, *9*, 1–12. [CrossRef] [PubMed]

48. Taylor, P.K.; Van Kessel, A.T.M.; Colavita, A.; Hancock, R.E.W.; Mah, T.F. A novel small RNA is important for biofilm formation and pathogenicity in *Pseudomonas aeruginosa*. *PLoS ONE* **2017**, *12*, e0182582. [CrossRef] [PubMed]

49. Wenner, N.; Maes, A.; Cotado-Sampayo, M.; Lapouge, K. NrsZ: A novel, processed, nitrogen-dependent, small non-coding RNA that regulates *Pseudomonas aeruginosa* PAO1 virulence. *Environ. Microbiol.* **2014**, *16*, 1053–1068. [CrossRef] [PubMed]

50. Sonnleitner, E.; Gonzalez, N.; Sorger-Domenigg, T.; Heeb, S.; Richter, A.S.; Backofen, R.; Williams, P.; Hüttenhofer, A.; Haas, D.; Bläsi, U. The small RNA PhrS stimulates synthesis of the *Pseudomonas aeruginosa* quinolone signal. *Mol. Microbiol.* **2011**, *80*, 868–885. [CrossRef] [PubMed]

51. Carloni, S.; Macchi, R.; Sattin, S.; Ferrara, S.; Bertoni, G. The small RNA ReaL: A novel regulatory element embedded in the *Pseudomonas aeruginosa* quorum sensing networks. *Environ. Microbiol.* **2017**, *19*, 4220–4237. [CrossRef] [PubMed]

52. Sonnleitner, E.; Romeo, A.; Bläsi, U. Small regulatory RNAs in *Pseudomonas aeruginosa*. *RNA Biol.* **2012**, *9*, 364–371. [CrossRef] [PubMed]

53. Sonnleitner, E.; Prindl, K.; Bläsi, U. The *Pseudomonas aeruginosa* CrcZ RNA interferes with Hfq-mediated riboregulation. *PLoS ONE* **2017**, *12*, e0180887. [CrossRef] [PubMed]

54. Reinhart, A.A.; Nguyen, A.T.; Brewer, L.K.; Bevere, J.; Jones, J.W.; Kane, M.A.; Damron, F.H.; Barbier, M.; Oglesby-Sherrouse, A.G. The *Pseudomonas aeruginosa* PrrF Small RNAs Regulate Iron Homeostasis during Acute Murine Lung Infection. *Infect Immun.* **2017**, *85*, 1–15. [CrossRef] [PubMed]

55. Sonnleitner, E.; Bläsi, U. Regulation of Hfq by the RNA CrcZ in *Pseudomonas aeruginosa* Carbon Catabolite Repression. *PLoS Genet.* **2014**, *10*. [CrossRef] [PubMed]

56. Tata, M.; Amman, F.; Pawar, V.; Wolfinger, M.T.; Weiss, S.; Häussler, S.; Bläsi, U. The anaerobically induced sRNA PaiI affects denitrification in *Pseudomonas aeruginosa* PA14. *Front. Microbiol.* **2017**, *8*, 2312. [CrossRef] [PubMed]

57. Ferrara, S.; Falcone, M.; Macchi, R.; Bragonzi, A.; Girelli, D.; Cariani, L.; Cigana, C.; Bertoni, G. The PAPI-1 pathogenicity island-encoded small RNA PesA influences *Pseudomonas aeruginosa* virulence and modulates pyocin S3 production. *PLoS ONE* **2017**, *12*, e0180386. [CrossRef] [PubMed]

58. Janssen, K.H.; Diaz, M.R.; Golden, M.; Graham, J.W.; Sanders, W.; Wolfgang, M.C.; Yahr, T.L. Functional analyses of the RsmY and RsmZ small noncoding regulatory RNAs in *Pseudomonas aeruginosa*. *J. Bacteriol.* **2018**, *200*, 1–15. [CrossRef] [PubMed]

59. Sorger-Domenigg, T.; Sonnleitner, E.; Kaberdin, V.R.; Bläsi, U. Distinct and overlapping binding sites of *Pseudomonas aeruginosa* Hfq and RsmA proteins on the non-coding RNA RsmY. *Biochem. Biophys. Res. Commun.* **2007**, *352*, 769–773. [CrossRef] [PubMed]

60. Petrova, O.E.; Sauer, K. The Novel Two-Component Regulatory System BfiSR Regulates Biofilm Development by Controlling the Small RNA rsmZ through CafA. *J. Bacteriol.* **2010**, *192*, 5275–5288. [CrossRef] [PubMed]

61. Jean-Pierre, F.; Tremblay, J.; Déziel, E. Broth versus surface-grown cells: Differential regulation of RsmY/Z small RNAs in *Pseudomonas aeruginosa* by the Gac/HptB System. *Front. Microbiol.* **2017**, *7*, 2168. [CrossRef] [PubMed]

62. Sonnleitner, E.; Sorger-Domenigg, T.; Madej, M.J.; Findeiss, S.; Hackermüller, J.; Hüttenhofer, A.; Stadler, P.F.; Bläsi, U.; Moll, I. Detection of small RNAs in *Pseudomonas aeruginosa* by RNomics and structure-based bioinformatic tools. *Microbiology* **2008**, *154*, 3175–3187. [CrossRef] [PubMed]

63. Wilderman, P.J.; Sowa, N.A.; FitzGerald, D.J.; FitzGerald, P.C.; Gottesman, S.; Ochsner, U.A.; Vasil, M.L. Identification of tandem duplicate regulatory small RNAs in *Pseudomonas aeruginosa* involved in iron homeostasis. *Proc. Natl. Acad. Sci. USA* **2004**, *101*, 9792–9797. [CrossRef] [PubMed]

64. Reinhart, A.A.; Powell, D.A.; Nguyen, A.T.; O'Neill, M.; Djapgne, L.; Wilks, A.; Ernst, R.K.; Oglesby-Sherrouse, A.G. The *prrF*-encoded small regulatory RNAs are required for iron homeostasis and virulence of *Pseudomonas aeruginosa*. *Infect. Immun.* **2015**, *83*, 863–875. [CrossRef] [PubMed]

65. Sonnleitner, E.; Abdou, L.; Haas, D. Small RNA as global regulator of carbon catabolite repression in *Pseudomonas aeruginosa*. *Proc. Natl. Acad. Sci. USA* **2009**, *106*, 21866–21871. [CrossRef] [PubMed]

66. Dong, Y.H.; Zhang, X.F.; Zhang, L.H. The global regulator Crc plays a multifaceted role in modulation of type III secretion system in *Pseudomonas aeruginosa*. *Microbiologyopen* **2013**, *2*, 161–172. [CrossRef] [PubMed]

67. Lu, P.; Wang, Y.; Hu, Y.; Chen, S. RgsA, an RpoS-dependent sRNA, negatively regulates *rpoS* expression in *Pseudomonas aeruginosa*. *Microbiology* **2018**, *164*, 716–724. [CrossRef] [PubMed]

68. Ferrara, S.; Carloni, S.; Fulco, R.; Falcone, M.; Macchi, R.; Bertoni, G. Post-transcriptional regulation of the virulence-associated enzyme AlgC by the σ(22)-dependent small RNA ErsA of *Pseudomonas aeruginosa*. *Environ. Microbiol.* **2015**, *17*, 199–214. [CrossRef] [PubMed]

69. Thi Bach Nguyen, H.; Romero, A.D.; Amman, F.; Sorger-Domenigg, T.; Tata, M.; Sonnleitner, E.; Bläsi, U. Negative Control of RpoS Synthesis by the sRNA ReaL in *Pseudomonas aeruginosa*. *Front. Microbiol.* **2018**, *9*, 2488. [CrossRef] [PubMed]

70. Gallique, M.; Bouteiller, M.; Merieau, A. The type VI secretion system: A dynamic system for bacterial communication? *Front. Microbiol.* **2017**, *8*, 1454. [CrossRef] [PubMed]

71. Huertas-Rosales, Ó.; Ramos-González, M.I.; Espinosa-Urgel, M. Self-regulation and interplay between Rsm family proteins modulates the lifestyles of *Pseudomonas* putida. *Appl. Environ. Microbiol.* **2016**, *82*, 5673–5686. [CrossRef] [PubMed]

72. Jakobsen, T.H.; Warming, A.N.; Vejborg, R.M.; Moscoso, J.A.; Stegger, M.; Lorenzen, F.; Rybtke, M.; Andersen, J.B.; Petersen, R.; Andersen, P.S.; et al. A broad range quorum sensing inhibitor working through sRNA inhibition. *Sci. Rep.* **2017**, *7*, 9857. [CrossRef] [PubMed]

73. Chakravarty, S.; Melton, C.N.; Bailin, A.; Yahr, T.L.; Anderson, G.G. *Pseudomonas aeruginosa* magnesium transporter MgtE inhibits type III secretion system gene expression by stimulating rsmYZ transcription. *J. Bacteriol.* **2017**, *199*. [CrossRef] [PubMed]

74. Grenga, L.; Little, R.H.; Malone, J.G. Quick change: Post-transcriptional regulation in *Pseudomonas*. *FEMS Microbiol. Lett.* **2017**, *364*, 1–8. [CrossRef] [PubMed]

75. Stacey, S.D.; Williams, D.A.; Pritchett, C.L. The *Pseudomonas aeruginosa* two-component regulator AlgR directly activates rsmA expression in a phosphorylation-independent manner. *J. Bacteriol.* **2017**, *199*. [CrossRef] [PubMed]

76. Chen, R.; Weng, Y.; Zhu, F.; Jin, Y.; Liu, C.; Pan, X.; Xia, B.; Cheng, Z.; Jin, S.; Wu, W. Polynucleotide phosphorylase regulates multiple virulence factors and the stabilities of small RNAs RsmY/Z in *Pseudomonas aeruginosa*. *Front. Microbiol.* **2016**, *7*, 247. [CrossRef] [PubMed]

77. Lee, J.; Zhang, L. The hierarchy quorum sensing network in *Pseudomonas aeruginosa*. *Protein Cell* **2014**, *6*, 26–41. [CrossRef] [PubMed]

78. Tipton, K.A.; Coleman, J.P.; Pesci, E.C. Post-transcriptional regulation of gene PA5507 controls *Pseudomonas* quinolone signal concentration in *P. aeruginosa*. *Mol. Microbiol.* **2015**, *96*, 670–683. [CrossRef] [PubMed]

79. Häussler, S.; Becker, T. The *pseudomonas* quinolone signal (PQS) balances life and death in *Pseudomonas aeruginosa* populations. *PLoS Pathog.* **2008**, *4*. [CrossRef] [PubMed]

80. Florez, C.; Raab, J.E.; Cooke, A.C.; Schertzer, J.W. Membrane Distribution of the *Pseudomonas* Quinolone Signal Modulates Outer Membrane Vesicle Production in *Pseudomonas aeruginosa*. *mBio* **2017**, *8*, e01034-17. [CrossRef] [PubMed]

81. Mann, E.E.; Wozniak, D.J. Pseudomonas matrix biofilm composition and niche biology. *FEMS Microbiol. Rev.* **2012**, *36*, 893–916. [CrossRef] [PubMed]

82. Zhang, Y.; Han, K.; Chandler, C.E.; Tjaden, B.; Ernst, R.K.; Lory, S. Probing the sRNA regulatory landscape of *Pseudomonas aeruginosa*: Post-transcriptional control of determinants of pathogenicity and antibiotic susceptibility. *Mol. Microbiol.* **2017**, *106*, 919–937. [CrossRef] [PubMed]

83. Oglesby-Sherrouse, A.G.; Murphy, E.R. Iron-responsive bacterial small RNAs: Variations on a theme. *Metallomics* **2013**, *5*, 276–286. [CrossRef] [PubMed]

84. Osborne, J.; Djapgne, L.; Tran, B.Q.; Goo, Y.A.; Oglesby-Sherrouse, A.G. A method for in vivo identification of bacterial small RNA-binding proteins. *Microbiologyopen* **2014**, *3*, 950–960. [CrossRef] [PubMed]

85. Oglesby, A.G.; Farrow, J.M.; Lee, J.H.; Tomaras, A.P.; Greenberg, E.P.; Pesci, E.C.; Vasil, M.L. The influence of iron on *Pseudomonas aeruginosa* physiology: A regulatory link between iron and quorum sensing. *J. Biol. Chem.* **2008**, *283*, 15558–15567. [CrossRef] [PubMed]

86. Duport, C.; Baysse, C.; Michel-Briand, Y. Molecular characterization of pyocin S3, a novel S-type pyocin from *Pseudomonas aeruginosa*. *J. Biol. Chem.* **1995**, *270*, 8920–8927. [CrossRef] [PubMed]

87. Martina, P.; Leguizamon, M.; Prieto, C.I.; Sousa, S.A.; Montanaro, P.; Draghi, W.O.; Stämmler, M.; Bettiol, M.; de Carvalho, C.C.C.R.; Palau, J.; et al. *Burkholderia puraquae* sp. nov., a novel species of the *Burkholderia cepacia* complex isolated from hospital settings and agricultural soils. *Int. J. Syst. Evol. Microbiol.* **2018**, *68*, 14–20. [CrossRef] [PubMed]

88. Sousa, S.A.; Ramos, C.G.; Leitão, J.H. *Burkholderia cepacia* complex: Emerging multihost pathogens equipped with a wide range of virulence factors and determinants. *Int. J. Microbiol.* **2011**, *2011*, 9. [CrossRef] [PubMed]

89. Leitão, J.H.; Sousa, S.A.; Ferreira, A.S.; Ramos, C.G.; Silva, I.N.; Moreira, L.M. Pathogenicity, virulence factors, and strategies to fight against *Burkholderia cepacia* complex pathogens and related species. *Appl. Microbiol. Biotechnol.* **2010**, *87*, 31–40. [CrossRef] [PubMed]

90. LiPuma, J.J. The changing microbial epidemiology in cystic fibrosis. *Clin. Microbiol. Rev.* **2010**, *23*, 299–323. [CrossRef] [PubMed]

91. Sousa, S.A.; Feliciano, J.R.; Pita, T.; Guerreiro, S.I.; Leitão, J.H. *Burkholderia cepacia* complex regulation of virulence gene expression: A review. *Genes (Basel)* **2017**, *8*, 43. [CrossRef] [PubMed]

92. Schwab, U.; Abdullah, L.H.; Perlmutt, O.S.; Albert, D.; William Davis, C.; Arnold, R.R.; Yankaskas, J.R.; Gilligan, P.; Neubauer, H.; Randell, S.H.; et al. Localization of *Burkholderia cepacia* complex bacteria in cystic fibrosis lungs and interactions with. *Pseudomonas aeruginosa in hypoxic mucus. Infect. Immun.* **2014**, *82*, 4729–4745. [CrossRef] [PubMed]

93. Boucher, R. An overview of the pathogenesis of cystic fibrosis lung disease. *Adv. Drug Deliv. Rev.* **2002**, *54*, 1359–1371. [CrossRef]

94. Drevinek, P.; Mahenthiralingam, E. *Burkholderia cenocepacia* in cystic fibrosis: Epidemiology and molecular mechanisms of virulence. *Clin. Microbiol. Infect.* **2010**, *16*, 821–830. [CrossRef] [PubMed]

95. Leitão, J.H.; Sousa, S.A.; Cunha, M.V.; Salgado, M.J.; Melo-Cristino, J.; Barreto, M.C.; Sá-Correia, I. Variation of the antimicrobial susceptibility profiles of *Burkholderia cepacia* complex clonal isolates obtained from chronically infected cystic fibrosis patients: A five-year survey in the major Portuguese treatment center. *Eur. J. Clin. Microbiol. Infect. Dis.* **2008**, *27*, 1101–1111. [CrossRef] [PubMed]

96. Leitão, J.H.; Feliciano, J.R.; Sousa, S.A.; Pita, T.; Guerreiro, S.I. *Burkholderia cepacia* Complex Infections Among Cystic Fibrosis Patients: Perspectives and Challenges. *Prog. Underst. Cyst. Fibros.* **2017**. [CrossRef]

97. Golshahi, L.; Lynch, K.H.; Dennis, J.J.; Finlay, W.H. In vitro lung delivery of bacteriophages KS4-M and φKZ using dry powder inhalers for treatment of *Burkholderia cepacia* complex and *Pseudomonas aeruginosa* infections in cystic fibrosis. *J. Appl. Microbiol.* **2011**, *110*, 106–117. [CrossRef] [PubMed]

98. Boucher, R.C. Evidence for airway surface dehydration as the initiating event in CF airway disease. *J. Intern. Med.* **2007**, *261*, 5–16. [CrossRef] [PubMed]

99. Schwab, U.; Leigh, M.; Ribeiro, C.; Yankaskas, J.; Burns, K.; Gilligan, P.; Sokol, P.; Boucher, R. Patterns of epithelial cell invasion by different species of the *Burkholderia cepacia* complex in well-differentiated human airway epithelia. *Infect. Immun.* **2002**, *70*, 4547–4555. [CrossRef] [PubMed]

100. McClean, S.; Callaghan, M. *Burkholderia cepacia* complex: Epithelial cell-pathogen confrontations and potential for therapeutic intervention. *J. Med. Microbiol.* **2009**, *58*, 1–12. [CrossRef] [PubMed]

101. Sousa, S.A.; Ulrich, M.; Bragonzi, A.; Burke, M.; Worlitzsch, D.; Leitão, J.H.; Meisner, C.; Eberl, L.; Sá-Correia, I.; Döring, G. Virulence of *Burkholderia cepacia* complex strains in gp91$^{phox-/-}$ mice. *Cell. Microbiol.* **2007**, *9*, 2817–2825. [CrossRef] [PubMed]

102. Caraher, E.; Reynolds, G.; Murphy, P.; McClean, S.; Callaghan, M. Comparison of antibiotic susceptibility of *Burkholderia cepacia* complex organisms when grown planktonically or as biofilm in vitro. *Eur. J. Clin. Microbiol. Infect. Dis.* **2006**, *26*, 213–216. [CrossRef] [PubMed]

103. Cunha, M.V.; Sousa, S.A.; Leitão, J.H.; Moreira, L.M.; Videira, P.A.; Sá-Correia, I. Studies on the involvement of the exopolysaccharide produced by cystic fibrosis-associated isolates of the *Burkholderia cepacia* complex in biofilm formation and in persistence of respiratory infections. *J. Clin. Microbiol.* **2004**, *42*, 3052–3058. [CrossRef] [PubMed]

104. Richau, J.A.; Leitão, J.H.; Correia, M.; Lito, L.; Salgado, M.J.; Barreto, C.; Cescutti, P.; Sá-Correia, I. Molecular typing and exopolysaccharide biosynthesis of *Burkholderia cepacia* isolates from a Portuguese cystic fibrosis center. *J. Clin. Microbiol.* **2000**, *38*, 1651–1655. [CrossRef] [PubMed]

105. Coenye, T.; Drevinek, P.; Mahenthiralingam, E.; Shah, S.A.; Gill, R.T.; Vandamme, P.; Ussery, D.W. Identification of putative noncoding RNA genes in the *Burkholderia cenocepacia* J2315 genome. *FEMS Microbiol. Lett.* **2007**, *276*, 83–92. [CrossRef] [PubMed]

106. Drevinek, P.; Holden, M.T.G.; Ge, Z.; Jones, A.M.; Ketchell, I.; Gill, R.T.; Mahenthiralingam, E. Gene expression changes linked to antimicrobial resistance, oxidative stress, iron depletion and retained motility are observed when *Burkholderia cenocepacia* grows in cystic fibrosis sputum. *BMC Infect. Dis.* **2008**, *8*, 121. [CrossRef] [PubMed]

107. Yoder-Himes, D.R.; Chain, P.S.G.; Zhu, Y.; Wurtzel, O.; Rubin, E.M.; Tiedje, J.M.; Sorek, R. Mapping the *Burkholderia cenocepacia* niche response via high-throughput sequencing. *Proc. Natl. Acad. Sci. USA* **2009**, *106*, 3976–3981. [CrossRef] [PubMed]

108. Peeters, E.; Sass, A.; Mahenthiralingam, E.; Nelis, H.; Coenye, T. Transcriptional response of *Burkholderia cenocepacia* J2315 sessile cells to treatments with high doses of hydrogen peroxide and sodium hypochlorite. *BMC Genom.* **2010**, *11*, 90. [CrossRef] [PubMed]

109. Coenye, T.; Van Acker, H.; Peeters, E.; Sass, A.; Buroni, S.; Riccardi, G.; Mahenthiralingam, E. Molecular mechanisms of chlorhexidine tolerance in *Burkholderia cenocepacia* biofilms. *Antimicrob. Agents Chemother.* **2011**, *55*, 1912–1919. [CrossRef] [PubMed]

110. Ramos, C.G.; Grilo, A.M.; da Costa, P.J.P.; Leitão, J.H. Experimental identification of small non-coding regulatory RNAs in the opportunistic human pathogen *Burkholderia cenocepacia* J2315. *Genomics* **2013**, *101*, 139–148. [CrossRef] [PubMed]

111. Ramos, C.G.; da Costa, P.J.P.; Döring, G.; Leitão, J.H. The Novel Cis-Encoded Small RNA h2cR Is a Negative Regulator of *hfq2* in *Burkholderia cenocepacia*. *PLoS ONE* **2012**, *7*, e47896. [CrossRef] [PubMed]

112. Ghosh, S.; Dureja, C.; Khatri, I.; Subramanian, S.; Raychaudhuri, S.; Ghosh, S. Identification of novel small RNAs in *Burkholderia cenocepacia* KC-01 expressed under iron limitation and oxidative stress conditions. *Microbiology* **2017**. [CrossRef] [PubMed]

113. Sass, A.M.; Van Acker, H.; Förstner, K.U.; Van Nieuwerburgh, F.; Deforce, D.; Vogel, J.; Coenye, T. Genome-wide transcription start site profiling in biofilm-grown *Burkholderia cenocepacia* J2315. *BMC Genom.* **2015**, *16*, 775. [CrossRef] [PubMed]

114. Sharma, C.M.; Vogel, J. Differential RNA-seq: The approach behind and the biological insight gained. *Curr. Opin. Microbiol.* **2014**, *19*, 97–105. [CrossRef] [PubMed]

115. Sass, A.; Kiekens, S.; Coenye, T. Identification of small RNAs abundant in *Burkholderia cenocepacia* biofilms reveal putative regulators with a potential role in carbon and iron metabolism. *Sci. Rep.* **2017**, *7*, 15665. [CrossRef] [PubMed]

116. Zhang, S.; Liu, S.; Wu, N.; Yuan, Y.; Zhang, W.; Zhang, Y. Small Non-coding RNA RyhB mediates persistence to multiple antibiotics and stresses in uropathogenic *Escherichia coli* by reducing cellular metabolism. *Front. Microbiol.* **2018**, *9*, 136. [CrossRef] [PubMed]

117. Høiby, N.; Bjarnsholt, T.; Givskov, M.; Molin, S.; Ciofu, O. Antibiotic resistance of bacterial biofilms. *Int. J. Antimicrob. Agents* **2010**, *35*, 322–332. [CrossRef] [PubMed]

118. Walters, I.I.I.M.C.; Roe, F.; Bugnicourt, A.; Franklin, M.J.; Stewart, P.S. Contributions of antibiotic penetration, oxygen limitation. *Society* **2003**, *47*, 317–323. [CrossRef]

119. Maisonneuve, E.; Gerdes, K. Molecular mechanisms underlying bacterial persisters. *Cell* **2014**, *157*, 539–548. [CrossRef] [PubMed]

120. Scoffone, V.C.; Chiarelli, L.R.; Trespidi, G.; Mentasti, M.; Riccardi, G.; Buroni, S. *Burkholderia cenocepacia* infections in cystic fibrosis patients: Drug resistance and therapeutic approaches. *Front. Microbiol.* **2017**, *8*, 1592. [CrossRef] [PubMed]

121. Holden, M.T.G.; Seth-Smith, H.M.B.; Crossman, L.C.; Sebaihia, M.; Bentley, S.D.; Cerdeño-Tárraga, A.M.; Thomson, N.R.; Bason, N.; Quail, M.A.; Sharp, S.; et al. The genome of *Burkholderia cenocepacia* J2315, an epidemic pathogen of cystic fibrosis patients. *J. Bacteriol.* **2009**, *91*, 261–277. [CrossRef] [PubMed]

122. Wong, Y.C.; El Ghany, M.A.; Naeem, R.; Lee, K.W.; Tan, Y.C.; Pain, A.; Nathan, S. Candidate essential genes in *Burkholderia cenocepacia* J2315 identified by genome-wide TraDIS. *Front. Microbiol.* **2016**, *7*, 1288. [CrossRef] [PubMed]

123. Ho, C.-C.; Lau, C.C.Y.; Martelli, P.; Chan, S.-Y.; Tse, C.W.S.; Wu, A.K.L.; Yuen, K.Y.; Lau, S.K.; Woo, P.C. Novel Pan-Genomic Analysis Approach in Target Selection for Multiplex PCR Identification and Detection of *Burkholderia pseudomallei*, *Burkholderia thailandensis*, and *Burkholderia cepacia* Complex Species: A Proof-of-Concept Study. *J. Clin. Microbiol.* **2011**, *49*, 814–821. [CrossRef] [PubMed]

124. Wassarman, K.M. 6S RNA, A Global Regulator of Transcription. *Microbiol. Spectr.* **2018**, *6*. [CrossRef]

125. Maddocks, S.E.; Oyston, P.C.F. Structure and function of the LysR-type transcriptional regulator (LTTR) family proteins. *Microbiology* **2008**, *154*, 3609–3623. [CrossRef] [PubMed]

© 2018 by the authors. Licensee MDPI, Basel, Switzerland. This article is an open access article distributed under the terms and conditions of the Creative Commons Attribution (CC BY) license (http://creativecommons.org/licenses/by/4.0/).

International Journal of
Molecular Sciences

Review

Catching a SPY: Using the SpyCatcher-SpyTag and Related Systems for Labeling and Localizing Bacterial Proteins

Daniel Hatlem, Thomas Trunk, Dirk Linke and Jack C. Leo *

Bacterial Cell Surface Group, Section for Evolution and Genetics, Department of Biosciences, University of Oslo, 0316 Oslo, Norway; daniel.hatlem@ibv.uio.no (D.H.); thomas.trunk@ibv.uio.no (T.T.); dirk.linke@ibv.uio.no (D.L.)
* Correspondence: j.c.leo@ibv.uio.no; Tel.: +47-228-59027

Received: 2 April 2019; Accepted: 26 April 2019; Published: 30 April 2019

Abstract: The SpyCatcher-SpyTag system was developed seven years ago as a method for protein ligation. It is based on a modified domain from a *Streptococcus pyogenes* surface protein (SpyCatcher), which recognizes a cognate 13-amino-acid peptide (SpyTag). Upon recognition, the two form a covalent isopeptide bond between the side chains of a lysine in SpyCatcher and an aspartate in SpyTag. This technology has been used, among other applications, to create covalently stabilized multi-protein complexes, for modular vaccine production, and to label proteins (e.g., for microscopy). The SpyTag system is versatile as the tag is a short, unfolded peptide that can be genetically fused to exposed positions in target proteins; similarly, SpyCatcher can be fused to reporter proteins such as GFP, and to epitope or purification tags. Additionally, an orthogonal system called SnoopTag-SnoopCatcher has been developed from an *S. pneumoniae* pilin that can be combined with SpyCatcher-SpyTag to produce protein fusions with multiple components. Furthermore, tripartite applications have been produced from both systems allowing the fusion of two peptides by a separate, catalytically active protein unit, SpyLigase or SnoopLigase. Here, we review the current state of the SpyCatcher-SpyTag and related technologies, with a particular emphasis on their use in vaccine development and in determining outer membrane protein localization and topology of surface proteins in bacteria.

Keywords: autotransporter; covalent labeling; bacterial surface protein; SpyCatcher; topology mapping; virulence factor

1. Introduction

The peptide bond is essential to biology. This secondary amide bond is produced by the condensation of the carboxyl group of one amino acid residue with the α-amino group of the following residue and results in the polymerization of amino acids to produce polypeptides. This central reaction is catalyzed by the ribosome and the sequence of the polypeptide chain is defined by the genetic code. In proteins, the peptide bond is sometimes referred to as the eupeptide bond. In addition, peptide bonds can be found in non-ribosomal peptides, which are secondary metabolites such as antibiotics produced by non-ribosomal peptide synthases [1]. A minority of proteins contain a second class of peptide bonds, referred to as isopeptide bonds. This is an amide bond similar to the peptide bond, but it is produced by the condensation of an amine group with a carboxyl group or a primary amide group located in amino acid side chains, rather than the main chain-building amino or carboxyl groups (Figure 1A).

In eukaryotes, the most prevalent example of isopeptide bond formation is the ubiquitination of proteins targeted for degradation by the 26S proteasome [2]. The isopeptide is formed by the C-terminal carboxyl group of ubiquitin and the side-chain amine of a lysine residue in the target protein. This reaction is catalyzed by E3 enzymes, or ubiquitin-protein ligases, which also impart specificity to the system [3]. Similarly, members of the small ubiquitin-like modifiers (SUMO) family, which regulate a number of cellular processes, are ligated to their target proteins via an isopeptide bond [4].

In bacteria, isopeptides are mostly known from surface proteins of Gram-positive bacteria. Two classes of proteins have notable isopeptide bonds: the microbial surface components recognizing adhesive matrix molecules (MSCRAMMs) and pili [5]. In pili, pilin subunits are covalently bound to each other via an intermolecular isopeptide bond between the carboxyl group of a C-terminal threonine and the ε-amine of a lysine in the next subunit, a reaction catalyzed by a pilin-specific sortase [6]. In MSCRAMMs and some pilins, isopeptides are formed spontaneously between residues within the same domain. These domains belong to the immunoglobulin fold superfamily. Isopeptide bond formation increases the stability of the proteins or pili and may contribute to stronger binding to ligands by MSCRAMMs, since in some cases the isopeptide only forms upon ligand binding to lock the protein in a bound conformation [7].

The SpyCatcher-SpyTag system was developed by the Howarth laboratory based on the internal isopeptide bond of the CnaB2 domain of FbaB, a fibronectin-binding MSCRAMM and virulence factor of *Streptococcus pyogenes* [8,9]. An internal isopeptide bond forms spontaneously in this domain between the ε-amine of lysine K31 and the side chain carboxyl of aspartic acid D117. The reaction is catalyzed by the spatially adjacent glutamate E77 (Figure 1B). The resulting isopeptide bond confers high stability to the CnaB2 domain [10]. The CnaB2 domain can be stably split into two components: a larger, incomplete immunoglobulin-like domain (termed SpyCatcher) of 138 residues (15 kDa) and a shorter peptide (SpyTag) of 13 residues (see Table 1 for peptide sequences). SpyCatcher contains the reactive lysine and catalytic glutamate, whereas SpyTag includes the reactive aspartate. The two components can still recognize each other with high affinity (0.2 μM) and the isopeptide can form between SpyCatcher and SpyTag to form a covalently bound complex (Figure 1C). Under experimental conditions relevant to life science research (room temperature, dilute protein concentrations), the reaction rates allow the bonds to form at high efficiency within minutes [9]. SpyTag in particular is equivalent in size to a number of epitope tags and can be genetically fused to a number of proteins and is able to react with SpyCatcher when inserted at the N- or C-termini of target proteins, as well as internal sites [9,11]. SpyCatcher itself can also be produced as a fusion protein, allowing the formation of covalently bonded protein partners that might otherwise be difficult to produce as protein fusions [12–14].

A similar system has been developed based on another Gram-positive surface protein, the pilus adhesin RrgA of *S. pneumoniae* [15,16]. The D4 domain of this protein is stabilized by an isopeptide forming between a lysine (K742) and an asparagine (N854), catalyzed by the spatially adjacent E803 [15] (Figure 1D). This domain was split into a scaffold protein called SnoopCatcher and a 12-residue peptide termed SnoopTag, which can spontaneously form a covalent isopeptide bond upon mixing [16] (Figure 1E). In contrast to SpyCatcher-SpyTag, the reactive lysine is present in SnoopTag and the asparagine in SnoopCatcher. This system is orthogonal to SpyCatcher-SpyTag; that is, SnoopCatcher does not react with SpyTag and SpyCatcher does not react with SnoopTag. This allows the use of both systems simultaneously to produce "polyproteams," programmed modular polyproteins, for use in biotechnological applications [16].

The Howarth lab further modified these technologies by making tripartite systems, where the isopeptide-forming lysine and aspartate/asparagine are located on separate peptides, and the catalytic glutamate is present on a larger scaffold protein [17]. This was first attempted with the SpyCatcher system to produce SpyLigase. Here, a second peptide containing the reactive lysine (KTag) was separated from SpyCatcher, which itself was modified to produce the stable SpyLigase protein containing the catalytic glutamate [18]. When SpyLigase was mixed with two proteins containing one of the reactive peptides each, SpyLigase was able to catalyze the fusion of the two tags. However, although SpyLigase could mediate the fusion of KTag and SpyTag located at both N- and C-terminal and even internal positions, the reactions had a ~50% efficiency at best and were dependent on specific buffer conditions and low temperature [16]. In contrast, the recently developed SnoopLigase system appears more robust and efficient [19]. SnoopLigase was engineered similarly to SpyLigase, and catalyzes the isopeptide formation between the lysine of a modified SnoopTag (SnoopTagJr) and the asparagine in a second peptide termed DogTag (Figure 1F). This system can have efficiencies over 95% that are less sensitive to temperature and reaction conditions than SpyLigase. Furthermore, as SnoopTagJr and DogTag have relatively high affinity for SnoopLigase, immobilizing SnoopLigase allows washing away unconjugated reactants followed by elution of essentially pure fusion products [19]. The various Catcher-Tag systems and their development are summarized in Table 1.

Below, we review the various applications of the SpyCatcher/SnoopCatcher systems in biotechnology. Our emphasis will be on using this methodology to label surface-exposed proteins, especially for mapping the topology of outer membrane-embedded virulence factors of Gram-negative bacteria.

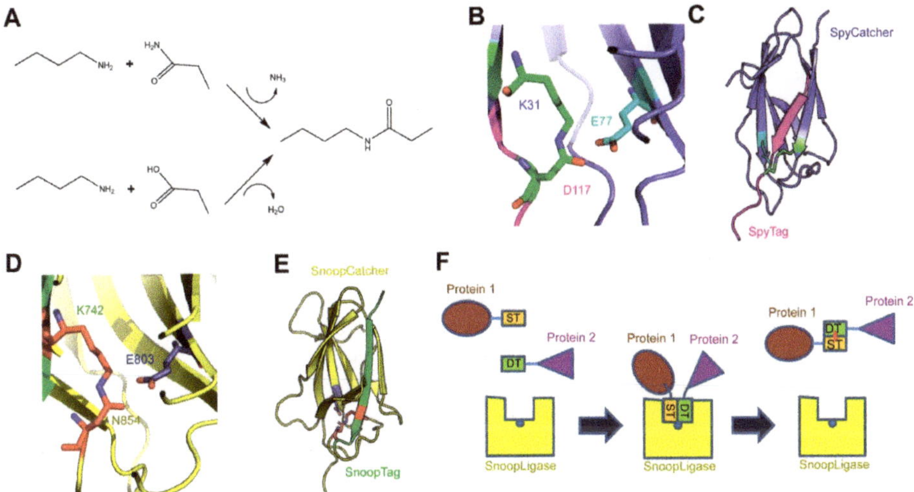

Figure 1. The SpyCatcher-SpyTag, SnoopCatcher-SnoopTag, and SnoopLigase systems. (**A**) Formation of an isopeptide bond. The primary amine of a lysine side chain condenses with either the side-chain amide of an asparagine (upper reaction) or the side-chain carboxyl of an aspartate (lower reaction) to produce the isopeptide bond, releasing either ammonia or water, respectively; (**B**) the isopeptide bond between SpyCatcher-SpyTag. The isopeptide formed by the reactive lysine (K31) in SpyCatcher and aspartate (D117) in SpyTag is shown in green, and the catalytic glutamate (E77) in cyan; (**C**) crystal structure of SpyCatcher-SpyTag. SpyCatcher is in blue and SpyTag is in magenta. The structures shown in panels b and c are based on the Protein Data Bank (PDB) entry 4MLI [20]; (**D**) the isopeptide bond between SnoopCatcher-SnoopTag, formed by lysine (K742) in SnoopTag and the asparagine (N854) in SnoopCatcher, is shown in red, and the catalytic glutamate (E803) is in blue; (**E**) crystal structure of SnoopCatcher-SnoopTag. SnoopCatcher is in yellow and SnoopTag in green. The structures shown in panels d and e are based on the RrgA D4 domain structure (PDB ID: 2WW8) [15]; (**F**) schematic of the function of SnoopLigase. A protein (in brown) containing SnoopTagJr (ST, in orange) and another protein (purple) with DogTag (DT, green) are mixed in the presence of SnoopLigase (yellow). The tags bind to SnoopLigase, which contains the catalytic glutamate (blue dot) and catalyzes the formation of an isopeptide between SnoopTagJr and DogTag (red dot). The ligated proteins can then be eluted from SnoopLigase.

Table 1. Summary of Catcher-Tag technologies and their development.

Catcher	Tag	Tag Sequence	Description	Publication Year	Reference
SpyCatcher	SpyTag	AHIVMVDAYKPTK [1]	Original Catcher-Tag technology.	2012	[9]
SpyCatcher ΔN1ΔC1	SpyTag	AHIVMVDAYKPTK [1]	Minimal SpyCatcher construct that still binds efficiently to SpyTag.	2014	[20]
SpyLigase [2]	SpyTag KTag	AHIVMVDAYKPTK [1] ATHIKFSKRD	Rationally engineered system for ligating two peptides.	2014	[18]
SnoopCatcher	SnoopTag	KLGDIEFIKVNK [1]	Orthogonal technology to SpyCatcher.	2016	[16]
SpyCatcher002	SpyTag002	VPTIVMVDAYKRYK [1]	Improved SpyCatcher-SpyTag system with faster reaction rate.	2017	[21]
SnoopLigase [2]	SnoopTagJr DogTag	KLGSIEFIKVNK [1] DIPATYEFTDGKHYITNEPIPPK	Rationally engineered system for ligating two peptides.	2018	[19]
SpyDock	SpyTag002	VPTIVMVDAYKRYK [1]	Protein affinity purification system (Spy&Go) based on SpyCatcher.	2019	[22]

[1] The reactive residues are shown in bold. [2] SpyLigase and SnoopLigase have two target peptides.

2. Applications of the SpyCatcher-SpyTag System

Being a quick and reliable coupling tool for irreversible peptide-protein ligation, the SpyTag-SpyCatcher system is ideal for a wide range of applications, ranging from increasing protein stability to antigen delivery during vaccination [14,23]. Before focusing on the use of the SpyTag-SpyCatcher system in the investigation of bacterial virulence factors, we give a short overview of those applications. For more detailed information, a list of publications and patents using the SpyTag-SpyCatcher and related technologies is accessible at the SpyInfo web page (available online: https://www.bioch.ox.ac.uk/howarth/info.htm) and the corresponding sequences and expression routes are listed in the SpyBank database [24].

A major application of the SpyTag-SpyCatcher system is the formation of so-called SpyRings in order to increase the intrinsic resilience of proteins to denaturation. SpyRings are generated by circularization of a single protein, which is accomplished by fusing an N-terminal SpyTag with a C-terminal SpyCatcher, or vice versa (Figure 2A). Enzyme circularization increases resistance to hyperthermal denaturation and aggregation, as well as alkali tolerance of individual enzymes without a loss in enzymatic activity [25–29]. Enzymes with a short distance between termini (<15 Å), and an active site which is not in direct proximity to one of the termini, are considered ideal candidates for protein circularization [30]. However, efficient SpyRing circularization and increased thermal stability have been shown for proteins where the termini are even farther apart [28].

Another frequently used application for the SpyCatcher-SpyTag system is the decoration of protein hydrogels (Figure 2B). Hydrogels have found application, for example as an artificial extracellular matrix material in medicine [31], tissue engineering [32,33], and cell culturing [34]. Hydrogels are polymeric materials engineered to resemble the extracellular environment of specific tissues with defined functional and structural properties. They have been used for decades as molecule delivery devices and as carriers for cells in tissue engineering due to their ability to mimic aspects of the native cellular environment [35]. However, hydrogels fail to fully imitate the complexity of biological systems. Fusing SpyCatcher to the polymeric material used in hydrogel synthesis allows the decoration with SpyTagged proteins post-hydrogelation to mimic specific microenvironments [36–39]. Thus, the SpyTag-SpyCatcher system is used as a quick and simple molecular tool for the simultaneous incorporation and presentation of different target molecules into and on hydrogels, thereby avoiding an otherwise laborious engineering process.

The SpyTag-SpyCatcher system has also been used for the modular assembly of proteins onto nanoparticles [40] (Figure 2C) and bacterial outer membrane vesicles [41] (Figure 2D; see also Section 3.3). The same system can be used for the in vivo encapsulation of enzymes fused to phage capsid proteins in order to create a protein nanocompartment [42] (Figure 2E) and for the decoration of virus-like particles (VLPs) for antigen delivery to the immune system [43,44] (Figure 2F; see also Section 3.3), while successfully preserving the structure and the function of the assembled proteins.

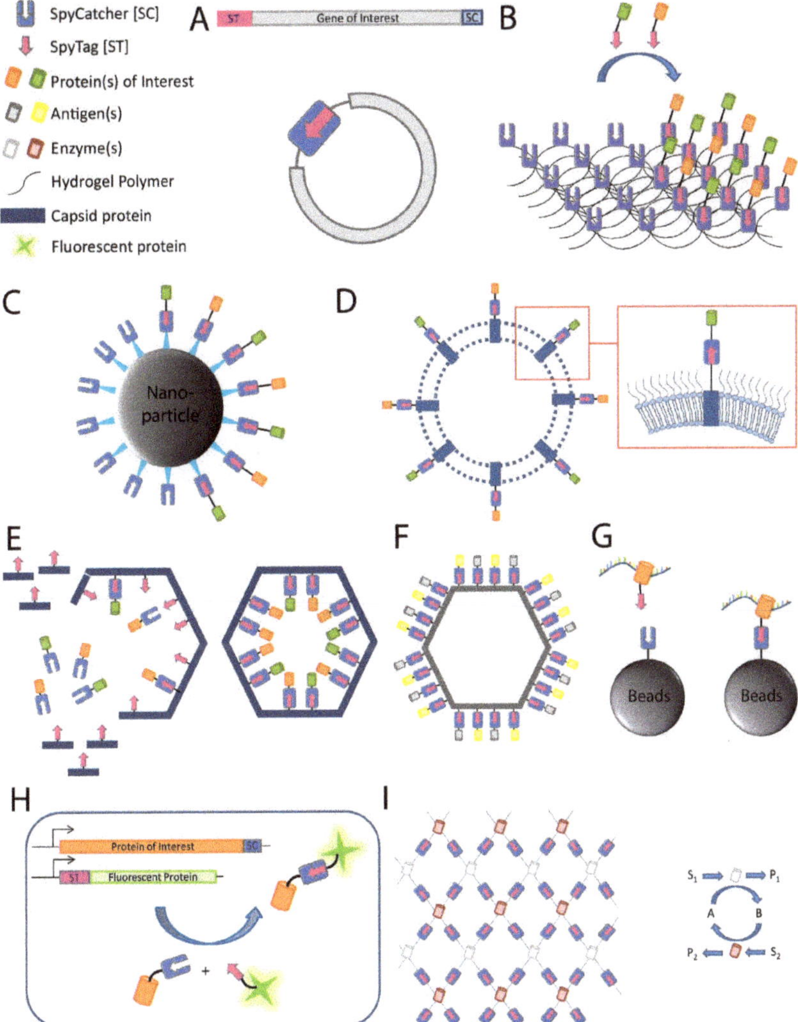

Figure 2. Applications of the SpyCatcher-SpyTag system. (**A**) SpyRing: SpyCatcher and SpyTag are fused to the terminal ends of the protein of interest resulting in protein cyclization and thereby conferring an increased resilience to denaturation; (**B**) post-hydrogelation decoration of protein hydrogels: SpyCatcher is fused to the polymeric material used in hydrogel synthesis, which allows the posthydrogelation decoration with proteins of interest fused to SpyTag; (**C–F**): (**C**) bioconjugation of target proteins to nanoparticles, (**D**) outer membrane vesicles, (**E**) phage capsid proteins to create a proteinaceous nanocompartment, and (**F**) virus-like particles; (**G**) SpyCLIP: SpyTagged RNA-binding protein interacts with RNA and is covalently attached to beads with fused SpyCatcher for use in pull-down assays; (**H**) fluorescent protein labeling for use, for example, in microscopy (ST represents SpyTag sequence); (**I**) artificial multi-enzyme nanodevices for increased efficiency, stability, and reusability. Schematic of mesh-like nanodevice is shown on the left and an enzymatic reaction scheme on the right. S represents substrate; P represents product; A and B represent cofactors.

Recently, the SpyTag-SpyCatcher system was used for the characterization of protein–RNA interactions as an alternative for ultraviolet (UV) crosslinking and immunoprecipitation (CLIP) [45]. The CLIP method relies on the limited specificity of antibody–antigen interactions which cannot withstand harsh washing conditions resulting in insufficient purity of ribonucleoprotein complexes after immunoprecipitation. Therefore, additional gel purification steps are necessary to further purify these complexes, which results in loss of protein. The disadvantages of CLIP are the labor-intensive gel purification steps as well as a high number of false-positive signals due to non-specific interactions. In comparison, the SpyTag-SpyCatcher technology allows the method to be performed with beads, skips the gel purification steps altogether, and withstands harsh washing steps, thus reducing non-specific interactions. The improved method using the Spy technology was termed SpyCLIP and requires SpyCatcher being fused to beads for immunoprecipitation and fusion proteins of the protein of interest, in this case the RNA-binding protein, to SpyTag [45] (Figure 2G). UV-crosslinking and affinity purification (uvCLAP) [46] as well as gel-omitted ligation-dependent CLIP (GoldCLIP) [47], using the covalent HaloTag-HaloLink purification system for immunoprecipitation [48], are both improvements to the traditional CLIP protocol, omitting the labor-intensive gel purification steps. The advantage of the SpyCatcher system over those methods is the ability of SpyTag-SpyCatcher to withstand even harsher washing conditions than the biotin-streptavidin coupling used during uvCLAP, and the small size of SpyTag compared to the 33 kDa protein tag used in GoldCLIP.

The system has also found increasing usage in conventional and super-resolution microscopy [49–51] by providing an easy and reliable way to label a SpyTagged protein with SpyCatcher coupled to a fluorescent dye (Figure 2H). Additionally, the limited accessibility of large antibodies, classically used for detection of epitope-tagged molecules, can be avoided by using the relatively small SpyCatcher protein during the detection of SpyTagged targets.

The Spy technology can also be used for the targeting of chemically synthesized voltage-sensitive dyes to specific cells for optical measurement of voltage dynamics in living cells. To this end, SpyTag is linked via a polyethylene linker to voltage-sensitive dyes while expressing SpyCatcher on the target cells. These SpyTag–dye conjugates, termed Voltage Spy, display improved targeting, good voltage sensitivity, and fast-response kinetics [52].

The system can also be used for the spatial and functional coordination of enzyme functions, creating a microenvironment normally associated with biological compartmentalization and conferring some of the same benefits (e.g., a local increase in enzyme and metabolite concentrations). Linkage of SpyTag with SpyCatcher is utilized for enzyme organization, allowing construction of artificial multi-enzyme nanodevices with a controlled spatial arrangement for increased efficiency of enzyme cascades while maintaining or even increasing stability and reusability of biocatalysts [53] (Figure 2I).

3. Using SpyCatcher-SpyTag to Investigate Bacterial Virulence Factors

The SpyCatcher-SpyTag system has recently been used for investigating bacterial virulence factors. So far, these have been limited to studying surface proteins of Gram-negative bacteria, especially autotransporter proteins. Autotransporters, also called type V secretion systems, are a widespread family of secreted proteins from Gram-negative bacteria, and many of these mediate virulence-related functions [54]. These proteins, which are divided into several subclasses (type Va through type Ve), have an outer membrane-embedded β-barrel domain and an extracellular region or passenger, which harbors the specific activity of each protein. Although SpyCatcher-SpyTag has been used mostly for studying autotransporters, the methods described below would be applicable to other surface-exposed proteins, both in Gram-negative and Gram-positive bacteria.

The cell surface of Gram-negative bacteria is represented by the outer membrane (OM), which is a complex asymmetric lipid bilayer. The extracellular face of the OM is mainly composed of lipopolysaccharides and outer membrane proteins that include several common virulence factors responsible for adhesion, mobility, and secretion, such as autotransporter adhesins, flagella, and type I-VIII secretion systems, among others [55,56]. Studying the expression, secretion, migration, and interactions of OM proteins often requires labeling by reporter proteins or other fluorophores. Labeling OM proteins, however, is challenging. Many fluorescent reporter proteins fail to mature in the periplasm [57]; thus, genetic fusions with fluorescent proteins are limited to a handful of options [58–61]. An alternative approach is non-covalent affinity-based labeling by using antibody–reporter protein fusions, affinity tags, or other high-affinity interactions, for example, colicins ColE9 and ColIa for labeling of the vitamin B12 transporter BtuB [62]. Alternatively, labeling methods using small organic molecules include amine-reactive fluorescence labeling [63,64], cysteine-reactive labeling [65], or site-specific labeling with unnatural amino acids, and tag-specific labeling [66,67]. Each of these techniques, however, has shortcomings. Non-covalent labeling requires high (nM) affinities for the label to remain associated with its target for a significant amount of time, which often makes such labeling unsuitable for time-resolved imaging on the minute scale. Small-molecule labeling often requires harsh treatment of the cells with reactive molecules, and site-specific orthogonal labeling requires co-expression of additional OM proteins that can disturb the system. In this section, we present a few recent examples showing how the SpyCatcher system can overcome many of these challenges by covalently labeling OM-bound virulence factors with high specificity, both with the purpose of studying the virulence factors and of exploiting their properties in vaccine development.

3.1. Using SpyCatcher to Investigate Membrane Protein Topology and Secretion

Trimeric autotransporter adhesins (TAAs), or type Vc secretion systems, constitute a group of surface-displayed virulence factors in Gram-negative bacteria that are responsible for adhesion to organic and inorganic surfaces. The prototypical TAA is *Yersinia* adhesin A (YadA), which consists of a C-terminal transmembrane β-barrel domain, where three protomers form a 12-stranded β-barrel, and an N-terminal trimeric passenger that is translocated through the β-barrel to the bacterial surface [68]. Once the passenger is translocated, it adopts a lollipop-like structure, forming a fibrous trimeric coiled-coil stalk that ends in an N-terminal globular head domain, responsible for adhesion and autoaggregation (Figure 3A). Autotransporters belonging to the type Va and Ve secretion systems initiate their translocation process via a hairpin intermediate, meaning that the part of the passenger most proximal to the β-barrel is initially inserted into the pore and then exported in a hairpin-like loop from the β-barrel until the entire passenger is outside the cell [69,70].

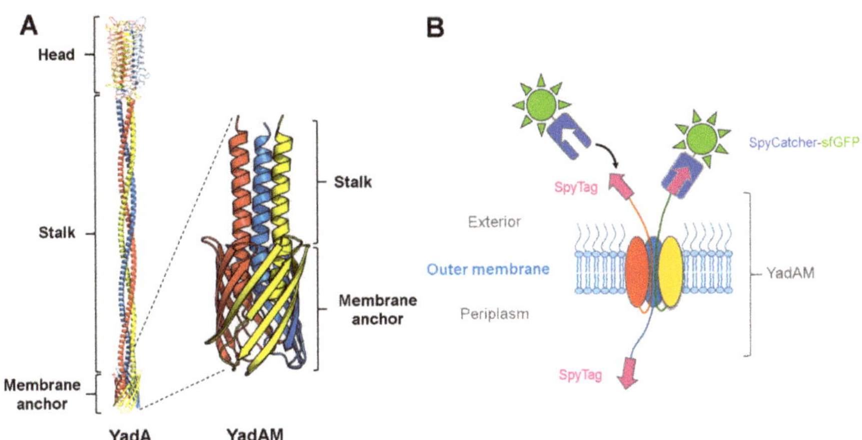

Figure 3. Using SpyCatcher to determine the surface topology of the *Yersinia* adhesin YadA. (**A**) Model of full-length YadA [71] and solid-state NMR structure of YadAM (PDB ID: 2LME); (**B**) schematic of the trimeric SpyTag-YadAM$_{A354P}$ construct with two secreted chains. The secreted chains fused to SpyTag are located on the bacterial surface and are able to bind to the SpyCatcher-sfGFP, while the chain located in the periplasm is inaccessible to the SpyCatcher fusion.

In a recent study, we investigated whether the hairpin model also holds true for TAAs using YadA as a model system, and aided by the SpyCatcher-SpyTag technology [13]. Earlier studies had identified a flexible region at the start of the coiled coil embedded within the lumen of the β-barrel, termed the ASSA region, due to the sequence consisting of alanines and serines. The ASSA region was hypothesized to be important for the formation of the hairpin [72]. To test this hypothesis, a truncated YadA construct was used, containing only the β-barrel and a small part of the passenger, termed YadAM (for YadA membrane anchor; see Figure 3A). Single proline substitutions were introduced to the ASSA region to probe its importance for translocation. Prolines are known secondary structure disruptors due to their rigid backbone, and we expected the substitutions to disrupt the flexible ASSA region sufficiently to stall the autotransporter process and trap the N-termini in the periplasm. To test this, a representative mutant, $YadAM_{A354P}$, was chosen for further studies, and a SpyTag was inserted in the wild-type YadAM and $YadAM_{A354P}$ constructs at the N-terminus of the mature proteins, following the signal peptide required for secretion. For easy detection, we fused purified SpyCatcher fused to sfGFP (superfolder green fluorescent protein). *Escherichia coli* cells expressing the different constructs were incubated with SpyCatcher-sfGFP (that is too large to penetrate the OM), allowing the SpyCatcher domain to bind to the exported SpyTags presented on the bacterial surface, but not to any stalled intermediates where the SpyTags would still be located in the periplasm. Bacterial cells producing the wild-type construct displayed higher fluorescence compared with cells producing the A354P mutant, indicating that the mutation was interfering with secretion of the passenger domain [13]. The effect of the A354P mutation on the surface topology of YadA was further investigated by isolating OM fractions of the two constructs after SpyCatcher-sfGFP treatment and analyzing them by semi-native SDS-PAGE (i.e., without heating the samples beforehand) (Figure 4A). GFP is stable in SDS at ambient temperatures [73], allowing instant visualization of GFP-containing bands by in-gel fluorescence under blue light. The gel showed a ladder-like pattern of SpyTag-YadAM complexes bound to 1–3 SpyCatcher-GFP molecules, corresponding to the number of strands exported to the bacterial surface (Figures 3B and 4A). Comparison of the band intensities indicates that the majority of YadAM exports all three passengers, whereas $YadAM_{A354P}$ mainly exports 1–2 SpyTags. These experiments demonstrated that the proline substitution is a partially stalled intermediate where one or two passenger strands are exported, but not a fully blocked autotransport intermediate.

As an alternative to using reporter molecules such as GFP to study the surface topology of OM proteins, the binding of SpyCatcher alone can be utilized for assessing surface exposure. Because SpyCatcher forms a covalent bond with SpyTag, the added molecular weight of SpyCatcher causes a change in electrophoretic mobility, giving a shift of 16 kDa that can be observed either by Coomassie staining or Western blotting. As an example, we studied the surface exposure of the full-length, SpyTagged YadA construct. YadA was treated with SpyCatcher alone (i.e., with no fusion partner) and compared with YadA treated with an inactive SpyCatcher, where the catalytic glutamate was changed to glutamine ($SpyCatcher_{EQ}$), by SDS-PAGE. In this example, unlabeled YadA remains trimeric even when heated to 95 °C and is seen as a single band at 200 kDa [74]. By contrast, the sample treated with active SpyCatcher forms a ladder-like pattern corresponding to YadA alone and YadA plus 1–3 bound SpyCatchers (Figure 4B). The experiments with truncated and full-length YadA demonstrate how SpyCatcher can readily bind to SpyTags located in close proximity to the bacterial surface, as shown for the YadAM constructs, as well as for the more exposed SpyTag protruding far from the surface located on the full-length protein. Future work will determine whether this technology is also suitable for detecting OM proteins without protruding domains, for example, where SpyTag is inserted into the loops of transmembrane β-barrel domains.

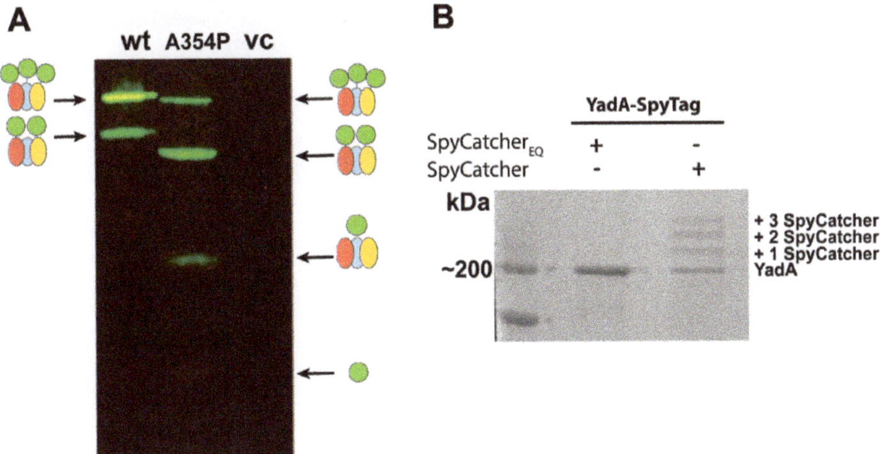

Figure 4. SpyCatcher assays with YadA. (**A**) Semi-native SDS-PAGE of OM preparations of wild-type SpyTag-YadAM and SpyTag-YadAM$_{A354P}$ after treatment with SpyCatcher-sfGFP, visualized under blue light. The ladder-like pattern indicates 1–3 bound SpyCatcher-sfGFP molecules. The gel shown is based on results from [13]. The wild-type YadAM binds mainly three SpyCatcher-sfGFP molecules, demonstrating full surface exposure of the N-terminal SpyTags, whereas the A354P mutant mainly binds to 1–2 SpyCatcher-sfGFP molecules. Note that YadAM$_{A354P}$ migrates anomalously compared with the wild-type. A schematic of the species visible in the gel is given on the side: SpyCatcher-sfGFP in green and YadAM monomers in red, blue, and yellow. vc represents vector control; (**B**) SDS-PAGE of OM preparations of full-length, trimeric SpyTag-YadA (~200 kDa) treated with the inactive SpyCatcher$_{EQ}$ and normal SpyCatcher, stained with Coomassie blue. The SpyCatcher-treated SpyTag-YadA shows a similar ladder-like pattern as in panel a, corresponding to the change in electrophoretic mobility as 1–3 SpyCatchers are bound.

3.2. Using SpyCatcher to Investigate Membrane Dynamics

The OM of Gram-negative bacteria consists of a mixture of OM proteins, lipopolysaccharides, and phospholipids. Both OM proteins and lipopolysaccharides are inserted into the OM in localized patches around the cell center where they diffuse to form a uniform distribution across the bacterial surface [75]. When the bacterium grows in preparation for cell division, the outer membrane proteins migrate towards the poles as a result of new membrane material and peptidoglycan being incorporated near the cell center in a process that is still not well understood [62].

In a recent paper, Keeble et al. used the SpyCatcher system to study the OM dynamics in *E. coli*. To this end, they used a novel phage display method to develop an improved SpyCatcher-SpyTag pair, termed SpyTag002 and SpyCatcher002 [21]. The rate constant of the improved 002 pair was increased by an order of magnitude compared to the original SpyCatcher–SpyTag interaction (2.0×10^4 M^{-1} s^{-1} vs. 1.7×10^3 M^{-1} s^{-1}). The efficiency of the newly developed pair was demonstrated by labeling the peptidoglycan-binding autotransporter intimin in order to study the membrane dynamics during cell growth and division. Intimin, a virulence factor of enterohemorrhagic *E. coli*, is a surface-exposed protein belonging to the type Ve secretion systems, also called inverse autotransporters [76]. Intimin consists of three major parts: an N-terminal periplasmic domain that binds to peptidoglycan under acidic conditions [77], a transmembrane β-barrel anchor [78], and the C-terminal passenger that is translocated to the bacterial surface through the β-barrel [79]. Keeble et al. prepared an intimin–SpyCatcher002 construct with SpyCatcher002 fused to the C-terminus of a truncated intimin variant (Figure 5). Upon overexpression in *E. coli* cells, the SpyCatcher002 domain is secreted together with the truncated passenger and presented on the bacterial surface, where it can be labeled using SpyTag002 fused to the fluorescent reporter protein mClover3 [80], in a similar fashion to that described for YadAM in Section 3.1. By imaging the cells using wide-field fluorescence microscopy after labeling, the movement of the mClover-tagged intimins toward the poles could be tracked to study the dynamics of the OM. Intimin was initially uniformly spread across the bacterial surface, but migrated towards the poles during cell division [21], suggesting that the polar movement results from incorporation of new peptidoglycan and OM material in preparation for cell division. This notion was further tested by repeating the experiment in the presence of cefalexin, a β-lactam antibiotic that prevents bacterial division by inhibiting septum formation and peptidoglycan synthesis. This time, the cells became elongated in preparation for division but were unable to divide. After 45 min, the fluorescent signal was localized in patches around the bacterium, thus demonstrating how inhibited peptidoglycan synthesis prevents the polar migration of OM proteins. Keeble et al. demonstrated here how an intimin–SpyCatcher fusion can be used to study the effect of peptidoglycan synthesis during division upon outer membrane dynamics [21]. This method could be used for tracking the migration of other OM proteins in the membrane during different stages of the cell cycle.

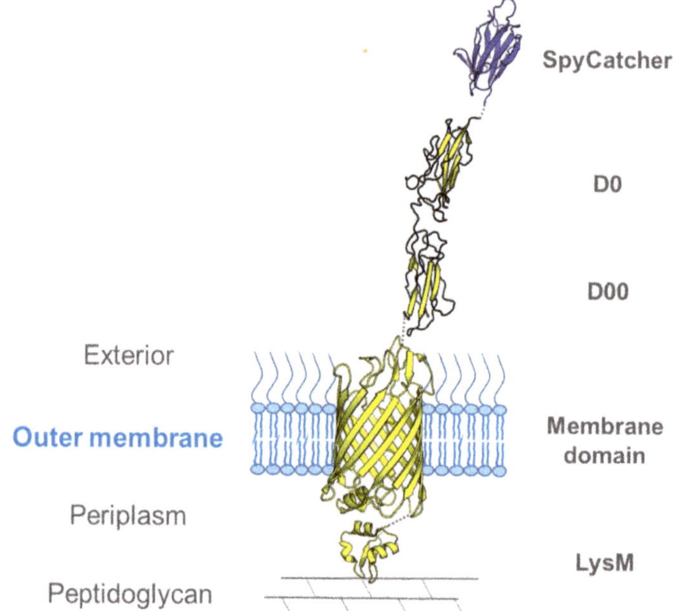

Figure 5. Intimin–SpyCatcher002 fusion for studying membrane dynamics. Structural model of the intimin construct used in [21]. The model is based on the crystal structures of the intimin transmembrane domain (PDB ID: 4E1S) and SpyCatcher (PDB ID: 4MLI), the solution structure of the peptidoglycan-binding LysM domain (PDB ID: 2MPW), and the homology model structures of the extracellular D00 and D0 domains [81].

3.3. Exploiting Virulence Factors and Virus-Like Particles for Vaccine Development using SpyCatcher-SpyTag

Outer membrane vesicles (OMVs) are ubiquitously produced by Gram-negative bacteria and are often responsible for delivering virulence factors to the host cells during infection [82]. Derived from the outer membrane, OMVs contain many naturally occurring immunogenic components such as lipopolysaccharides, lipoproteins, outer membrane proteins, peptidoglycan, and other periplasmic components with intrinsic adjuvant properties [83]. As opposed to live and attenuated bacteria, OMVs are non-replicating and therefore pose no risk of infection after vaccination of immunocompromised individuals. Consequently, OMVs are promising candidates for modern vaccine development, both as adjuvants and as delivery vehicles for antigens. Vaccines utilizing OMVs as adjuvants have already been on the market for more than two decades; however, the development of heterologous antigen-presenting OMV-based vaccines using recombinant technology is still in its infancy [83,84]. Several strategies have been employed to present antigens on OMV surfaces, most notably as heterologous fusions with autotransporters [85–87]. The use of autotransporters for heterologous antigen presentation on OMVs has recently been reviewed in detail elsewhere [88], so we will only provide a short overview here.

The Luirink group has pioneered the use of recombinant OMVs for vaccine development. As a scaffold, they earlier utilized a classical (type Va) autotransporter, the *E. coli* hemoglobin protease (Hbp). Hbp consists of two major parts: a C-terminal transmembrane β-barrel that is inserted into the outer membrane, and an N-terminal passenger that is translocated through the β-barrel to the bacterial surface where it is released to the environment through autoproteolysis (Figure 6) [89,90]. The Luirink group developed an Hbp variant suitable for heterologous antigen export and display by introducing a point mutation that prevents autoproteolysis into a truncated Hbp passenger domain (HbpD(Δd1); from now referred to only as Hbp) [86]. By combining the new Hbp variant with a hypervesiculating *Salmonella* Typhimurium Δ*tolRA* strain, they developed a method to isolate large amounts of Hbp-presenting OMVs [87]. To utilize Hbp as an antigen-presenting tool, several loops in the passenger were identified that allowed the insertion of heterologous protein antigens without impeding the autotransport process. However, even though the autotransporter accepts substantial changes to its passenger, there are still considerable limitations to the size and folding properties of passenger fusions [91].

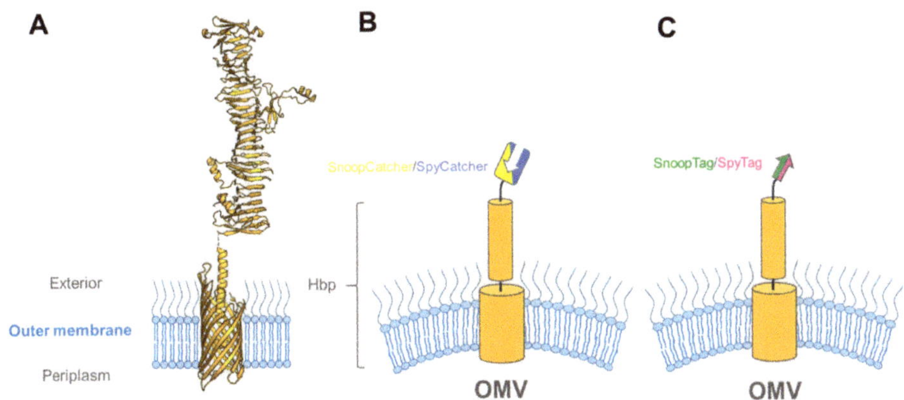

Figure 6. Hbp constructs used for vaccine display on outer membrane vesicles. (**A**) Structural model of full-length HbpD(Δd1) based on the crystal structures of the Hbp transmembrane domain (PDB ID: 3AEH) and passenger (PDB ID: 1WXR); (**B**) and (**C**) schematic drawings of Hbp fusions used in [92] to determine binding efficiencies between the SpyTag-SpyCatcher and SnoopTag-SnoopCatcher pairs on Hbp-fusion-presenting OMVs.

In a recent paper, this problem was circumvented by using the SpyCatcher-SpyTag technology to decorate the OMVs after expression of Hbp [92]. In the process, the investigators comprehensively tested the applicability of the SpyCatcher and SnoopCatcher systems for modular vaccine building, and we will only cover the highlights here. To fully test the range of the new technology, the authors prepared OMVs presenting Hbp N-terminal fusions with SpyTag, SpyCatcher, SnoopTag, and SnoopCatcher (Figure 6). By treating the different constructs with their respective binding partners, and comparing the changes in electrophoretic mobility upon ligation by SDS-PAGE, they showed that all four Hbp-variants exported their passenger to the surface and were amenable to ligation. As a proof-of-concept for vaccine development, they proceeded to treat Hbp-SpyTag-presenting OMVs with a relatively large (24 kDa) domain of the antigenic *S. pneumoniae* surface protein, PspAα, fused to both SpyCatcher and SnoopTag (SpyCatcher–PspAα–SnoopTag) (Figure 7A). Earlier attempts to fuse PspAα with the Hbp passenger required the antigen to be divided into two smaller domains, which still resulted in significantly reduced export and lower antigen presentation on the OMV surface [93]. The SpyCatcher–PspAα—SnoopTag, on the other hand, was ligated to the Hbp-presenting OMVs with similar efficiency as the earlier controls, demonstrating that the SpyCatcher system is a robust tool suitable for coating OMVs with recombinant antigens, completely circumventing the passenger export problem.

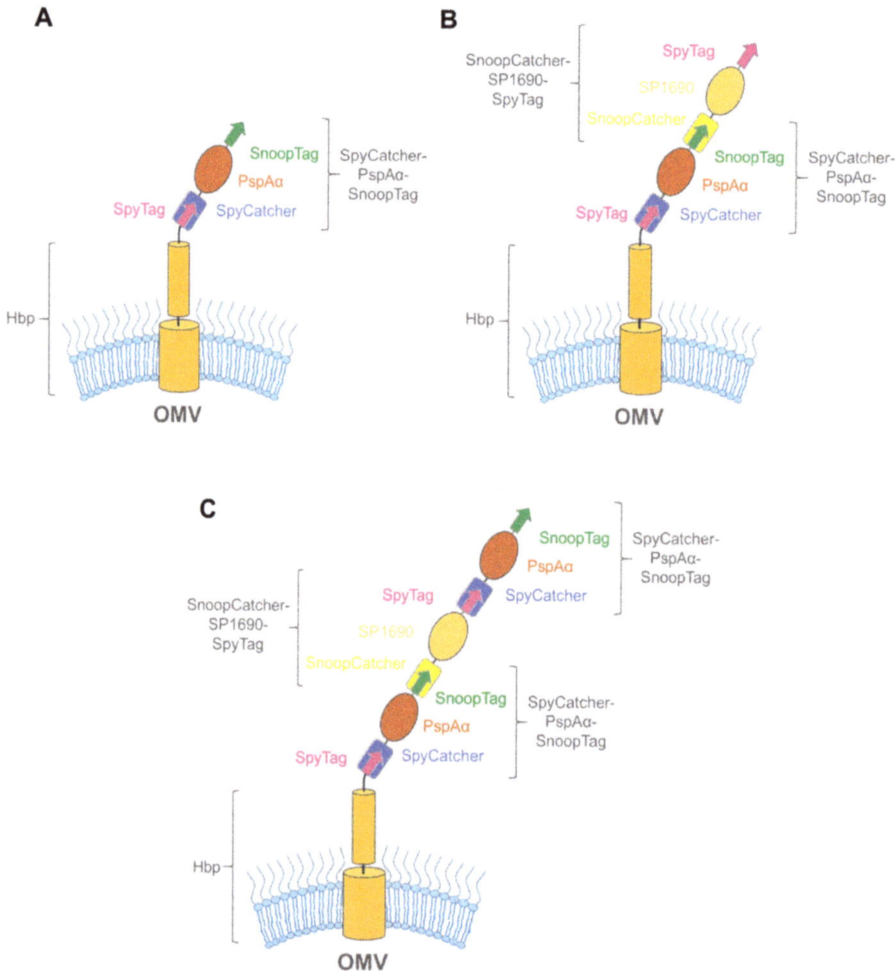

Figure 7. Hbp-antigen complexes made through Spy/SnoopCatcher fusions. (**A**) Schematic of Hbp bound to SpyCatcher–PspAα–SnoopTag; (**B**) schematic of Hbp bipartite binding adduct SpyCatcher–PspAα–SnoopTag and SnoopCatcher–SP1690–SpyTag; (**C**) schematic of Hbp tripartite binding adduct with two SpyCatcher–PspAα–SnoopTags flanking SnoopCatcher–SP1690–SpyTag.

The next step was to check whether multiple antigens could be coupled to the OMV surface in an iterative fashion by using the alternating combinations of the SpyCatcher and SnoopCatcher systems flanking the antigen proteins. Combinations of antigens to form multivalent OMV-based vaccines can significantly improve their efficiency, either by eliciting a stronger immune response to a particular pathogen, or by providing a broader response by combining antigens from multiple pathogens [94,95]. For this approach, they treated the OMVs with PspAα fusion and a SnoopCatcher–SP1690–SpyTag fusion with wash steps in between, successfully forming bipartite (Figure 7B) and tripartite (Figure 7C) coupling adducts. This work demonstrated how various recombinant antigens can be coupled to the OMV surface using the SpyCatcher system.

A similar application for the SpyCatcher system in vaccine development has been used for VLPs. VLPs are virus-derived proteins that self-organize into noninfectious virus-like structures that are capable of eliciting strong immune reactions [96]. Much effort has been put into designing VLP–antigen fusions; however, this approach has proven to be labor-intensive and time-consuming, since the fusions often misalign, or are unable to form stable capsids [97]. Brune et al. from the Howarth group circumvented this problem by using the SpyCatcher technology in an analogous fashion to the OMV-based vaccines [44]. They were able to create stable SpyCatcher-presenting VLPs by using a fusion between the bacteriophage AP205-derived capsid and SpyCatcher. They further demonstrated how the VLPs could be decorated using SpyTag-fused malarial antigens (Figure 2F) in a plug-and-play fashion, and how this product elicited an antibody response after a single injection in mice. In similar work, Thrane et al. fused either SpyTag or SpyCatcher to the C- and/or N-termini of the AP205 capsid protein, and subsequently treated the resulting VLPs with 11 different antigen–SpyCatcher/SpyTag fusions [43]. In both works, the authors confirmed the immunogenic properties of the novel vaccines by immunizing mice, thus confirming the functional activity of the novel vaccines. The results from the OMV and VLP experiments demonstrated the versatility of the SpyCatcher system in a novel approach for designing vaccines. By presenting one of the binding SpyCatcher/SpyTag partners on the particle surface, linking antigens in a plug-and-play fashion circumvents the need for designing complicated fusion constructs, and could soon pave the road for modular vaccines that can easily be tailored for any need. For further information, Brune and Howarth recently published a more comprehensive review on the design of VLP-based vaccines, including the use of the SpyCatcher platform for this purpose [14].

4. Conclusions and Future Perspectives

The SpyCatcher system and its close relatives have been successfully used for protein engineering purposes, including the cyclic polymerization of enzymes, or the decoration of hydrogels. In our hands, the system has been extremely useful for showing cell surface localization, and to elucidate the topology of membrane proteins. Other authors have used the technology for decorating vesicles and virus particles with antigens. In all cases, the benefits of the system are the same: irreversible, covalent coupling and the fact that the two partners of the system are relatively small entities that do not interfere strongly with the native systems that they label.

It is somewhat ironic that the system, derived from a bacterial surface protein, has been widely used for studying and modifying other bacterial or viral surface proteins. We strongly believe that this system will have many more applications in basic science, and that this is only limited by the fact that many researchers are not aware of its existence. This prompted us to write this review.

Some of the approaches related to vaccines will need more testing before they can potentially be used in patients. For example, it remains unclear whether SpyCatcher (or its relatives) elicits a strong immune response by itself, and whether that would limit its usefulness in vaccines. The work on VLPs suggests that this could be problematic for some applications, where the SpyCatcher-VLPs did elicit an immune response of their own that was then later masked after decorating the particles with the target antigen [44].

Our own experiments were until now limited to surface-localized proteins in bacteria, but it is conceivable that the system could also be used to label proteins in other cellular localization (e.g., after membrane permeabilization) as an alternative to antibody-based approaches. We also believe that the system can be used in versatile ways to modify and functionalize artificial or biological surfaces of almost any kind (e.g., to develop ELISA-like assays that require covalent immobilization), and possibly to functionalize surfaces used in lab-on-a-chip approaches. Another conceivable use of the system includes in situ labeling of proteins for added density to identify individual proteins (e.g., in cryo-electron micrographs or tomograms). As this review shows, the SpyCatcher-SpyTag system has proven to be an extremely versatile tool for both basic research and applied science, and it will undoubtedly spur further innovations in the future.

Author Contributions: Conceptualization, J.C.L. and D.L.; writing—original draft preparation, D.H., T.T., D.L., and J.C.L.; writing—review and editing, D.L. and J.C.L.

Funding: This research was funded by the Research Council of Norway, grant numbers 249793 (to J.C.L.) and 240483 (to D.L.).

Acknowledgments: The authors wish to thank Nandini Chauhan for providing the gel image shown in Figure 4B. They also acknowledge the members of the Leo and Linke groups for fruitful scientific discussions.

Conflicts of Interest: The authors declare no conflict of interest.

Abbreviations

CLIP	UV crosslinking and immunoprecipitation
GoldCLIP	Gel-omitted ligation-dependent CLIP
Hbp	Hemoglobin protease
MSCRAMMs	Microbial surface components recognizing adhesive matrix molecules
OM	Outer membrane
OMV	Outer membrane vesicle
PDB	Protein Data Bank
SDS-PAGE	Sodium dodecylsulfate polyacrylamide gel electrophoresis
sfGFP	Superfolder green fluorescent protein
SUMO	Small ubiquitin-like modifier
TAA	Trimeric autotransporter adhesin
UV	Ultraviolet
uvCLAP	UV-crosslinking and affinity purification
VLP	Virus-like particle
YadA	*Yersinia* adhesin A
YadAM	YadA membrane anchor

References

1. Koglin, A.; Walsh, C.T. Structural insights into nonribosomal peptide enzymatic assembly lines. *Nat. Prod. Rep.* **2009**, *26*, 987–1000. [CrossRef] [PubMed]

2. Collins, G.A.; Goldberg, A.L. The logic of the 26S proteasome. *Cell* **2017**, *169*, 792–806. [CrossRef]

3. Zheng, N.; Shabek, N. Ubiquitin ligases: Structure, function, and regulation. *Annu. Rev. Biochem.* **2017**, *86*, 129–157. [CrossRef] [PubMed]

4. Wilson, V.G. Introduction to Sumoylation. *Adv. Exp. Med. Biol.* **2017**, *963*, 1–12. [PubMed]

5. Baker, E.N.; Squire, C.J.; Young, P.G. Self-generated covalent cross-links in the cell-surface adhesins of Gram-positive bacteria. *Biochem. Soc. Trans.* **2015**, *43*, 787–794. [CrossRef]

6. Hendrickx, A.P.; Budzik, J.M.; Oh, S.-Y.; Schneewind, O. Architects at the bacterial surface—Sortases and the assembly of pili with isopeptide bonds. *Nat. Rev. Microbiol.* **2011**, *9*, 166–176. [CrossRef]

7. Sridharan, U.; Ponnuraj, K. Isopeptide bond in collagen- and fibrinogen-binding MSCRAMMs. *Biophys. Rev.* **2016**, *8*, 75–83. [CrossRef] [PubMed]

8. Terao, Y.; Kawabata, S.; Nakata, M.; Nakagawa, I.; Hamada, S. Molecular characterization of a novel fibronectin-binding protein of *Streptococcus pyogenes* strains isolated from toxic shock-like syndrome patients. *J. Biol. Chem.* **2002**, *277*, 47428–47435. [CrossRef]

9. Zakeri, B.; Fierer, J.O.; Celik, E.; Chittock, E.C.; Schwarz-Linek, U.; Moy, V.T.; Howarth, M. Peptide tag forming a rapid covalent bond to a protein, through engineering a bacterial adhesin. *Proc. Natl. Acad. Sci. USA* **2012**, *109*, E690–E697. [CrossRef]

10. Hagan, R.M.; Björnsson, R.; McMahon, S.A.; Schomburg, B.; Braithwaite, V.; Bühl, M.; Naismith, J.H.; Schwarz-Linek, U. NMR spectroscopic and theoretical analysis of a spontaneously formed Lys-Asp isopeptide bond. *Angew. Chem. Int. Ed. Engl.* **2010**, *49*, 8421–8425. [CrossRef]

11. Zhang, W.-B.; Sun, F.; Tirrell, D.A.; Arnold, F.H. Controlling macromolecular topology with genetically encoded SpyTag-SpyCatcher chemistry. *J. Am. Chem. Soc.* **2013**, *135*, 13988–13997. [CrossRef]

12. Bedbrook, C.N.; Kato, M.; Kumar, S.R.; Lakshmanan, A.; Nath, R.D.; Sun, F.; Sternberg, P.W.; Arnold, F.H.; Gradinaru, V. Genetically encoded Spy peptide fusion system to detect plasma membrane-localized proteins in vivo. *Chem. Biol.* **2015**, *22*, 1108–1121. [CrossRef]

13. Chauhan, N.; Hatlem, D.; Orwick-Rydmark, M.; Schneider, K.; Floetenmeyer, M.; van Rossum, B.; Leo, J.C.; Linke, D. Insights into the autotransport process of a trimeric autotransporter, *Yersinia* Adhesin A (YadA). *Mol. Microbiol.* **2019**, *111*, 844–862. [CrossRef]

14. Brune, K.D.; Howarth, M. New routes and opportunities for modular construction of particulate vaccines: Stick, click, and glue. *Front. Immunol.* **2018**, *9*, 1432. [CrossRef]

15. Izoré, T.; Contreras-Martel, C.; El Mortaji, L.; Manzano, C.; Terrasse, R.; Vernet, T.; Di Guilmi, A.M.; Dessen, A. Structural basis of host cell recognition by the pilus adhesin from *Streptococcus pneumoniae*. *Structure* **2010**, *18*, 106–115. [CrossRef]

16. Veggiani, G.; Nakamura, T.; Brenner, M.D.; Gayet, R.V.; Yan, J.; Robinson, C.V.; Howarth, M. Programmable polyproteams built using twin peptide superglues. *Proc. Natl. Acad. Sci. USA* **2016**, *113*, 1202–1207. [CrossRef]

17. Veggiani, G.; Zakeri, B.; Howarth, M. Superglue from bacteria: Unbreakable bridges for protein nanotechnology. *Trends Biotechnol.* **2014**, *32*, 506–512. [CrossRef]

18. Fierer, J.O.; Veggiani, G.; Howarth, M. SpyLigase peptide-peptide ligation polymerizes affibodies to enhance magnetic cancer cell capture. *Proc. Natl. Acad. Sci. USA* **2014**, *111*, E1176–E1181. [CrossRef]

19. Buldun, C.M.; Jean, J.X.; Bedford, M.R.; Howarth, M. SnoopLigase catalyzes peptide-peptide locking and enables solid-phase conjugate isolation. *J. Am. Chem. Soc.* **2018**, *140*, 3008–3018. [CrossRef]

20. Li, L.; Fierer, J.O.; Rapoport, T.A.; Howarth, M. Structural analysis and optimization of the covalent association between SpyCatcher and a peptide Tag. *J. Mol. Biol.* **2014**, *426*, 309–317. [CrossRef]

21. Keeble, A.H.; Banerjee, A.; Ferla, M.P.; Reddington, S.C.; Anuar, I.N.; Howarth, M. Evolving Accelerated Amidation by SpyTag/SpyCatcher to Analyze Membrane Dynamics. *Angew. Chem. Int. Ed. Engl.* **2017**, *56*, 16521–16525. [CrossRef]

22. Anuar, I.N.; Banerjee, A.; Keeble, A.H.; Carella, A.; Nikov, G.I.; Howarth, M. Spy&Go purification of SpyTag-proteins using pseudo-SpyCatcher to access an oligomerization toolbox. *Nat. Commun.* **2019**, *10*, 1743.

23. Reddington, S.C.; Howarth, M. Secrets of a covalent interaction for biomaterials and biotechnology: SpyTag and SpyCatcher. *Curr. Opin. Chem. Biol.* **2015**, *29*, 94–99. [CrossRef]

24. Keeble, A.H.; Howarth, M. Insider information on successful covalent protein coupling with help from SpyBank. *Meth. Enzym.* **2019**, *617*, 443–461.

25. Si, M.; Xu, Q.; Jiang, L.; Huang, H. SpyTag/SpyCatcher cyclization enhances the thermostability of firefly luciferase. *PLoS ONE* **2016**, *11*, e0162318. [CrossRef]

26. Schoene, C.; Fierer, J.O.; Bennett, S.P.; Howarth, M. SpyTag/SpyCatcher cyclization confers resilience to boiling on a mesophilic enzyme. *Angew. Chem. Int. Ed. Engl.* **2014**, *53*, 6101–6104. [CrossRef]

27. Sun, X.-B.; Cao, J.-W.; Wang, J.-K.; Lin, H.-Z.; Gao, D.-Y.; Qian, G.-Y.; Park, Y.-D.; Chen, Z.-F.; Wang, Q. SpyTag/SpyCatcher molecular cyclization confers protein stability and resilience to aggregation. *New Biotechnol.* **2018**, *49*, 28–36. [CrossRef]

28. Schoene, C.; Bennett, S.P.; Howarth, M. SpyRing interrogation: Analyzing how enzyme resilience can be achieved with phytase and distinct cyclization chemistries. *Sci. Rep.* **2016**, *6*, 21151. [CrossRef]

29. Wang, J.; Wang, Y.; Wang, X.; Zhang, D.; Wu, S.; Zhang, G. Enhanced thermal stability of lichenase from *Bacillus subtilis* 168 by SpyTag/SpyCatcher-mediated spontaneous cyclization. *Biotechnol. Biofuels* **2016**, *9*, 79. [CrossRef]

30. Schoene, C.; Bennett, S.P.; Howarth, M. SpyRings declassified: A blueprint for using isopeptide-mediated cyclization to enhance enzyme thermal resilience. *Methods Enzym.* **2016**, *580*, 149–167.

31. Langer, R.; Tirrell, D.A. Designing materials for biology and medicine. *Nature* **2004**, *428*, 487–492. [CrossRef]

32. Guilak, F.; Butler, D.L.; Goldstein, S.A.; Baaijens, F.P.T. Biomechanics and mechanobiology in functional tissue engineering. *J. Biomech.* **2014**, *47*, 1933–1940. [CrossRef]

33. Lee, K.Y.; Mooney, D.J. Hydrogels for tissue engineering. *Chem. Rev.* **2001**, *101*, 1869–1879. [CrossRef]

34. Caliari, S.R.; Burdick, J.A. A practical guide to hydrogels for cell culture. *Nat. Methods* **2016**, *13*, 405–414. [CrossRef]

35. Burdick, J.A.; Murphy, W.L. Moving from static to dynamic complexity in hydrogel design. *Nat. Commun.* **2012**, *3*, 1269. [CrossRef]

36. Gao, X.; Fang, J.; Fu, L.; Li, H. Engineering protein hydrogels using SpyCatcher-SpyTag chemistry. *Biomacromolecules* **2016**, *17*, 2812–2819. [CrossRef]

37. Wang, R.; Yang, Z.; Luo, J.; Hsing, I.-M.; Sun, F. B_{12}-dependent photoresponsive protein hydrogels for controlled stem cell/protein release. *Proc. Natl. Acad. Sci. USA* **2017**, *114*, 5912–5917. [CrossRef]

38. Gao, X.; Lyu, S.; Li, H. Decorating a blank slate protein hydrogel: A general and robust approach for functionalizing protein hydrogels. *Biomacromolecules* **2017**, *18*, 3726–3732. [CrossRef]

39. Liu, X.; Yang, X.; Yang, Z.; Luo, J.; Tian, X.; Liu, K. Versatile engineered protein hydrogels enabling decoupled mechanical and biochemical tuning for cell adhesion and neurite growth. *ACS Appl. Nano Mater.* **2018**, *1*, 1579–1585. [CrossRef]

40. Ma, W.; Saccardo, A.; Roccatano, D.; Aboagye-Mensah, D.; Alkaseem, M.; Jewkes, M.; Di Nezza, F.; Baron, M.; Soloviev, M.; Ferrari, E. Modular assembly of proteins on nanoparticles. *Nat. Commun.* **2018**, *9*, 1489. [CrossRef]

41. Alves, N.J.; Turner, K.B.; Daniele, M.A.; Oh, E.; Medintz, I.L.; Walper, S.A. Bacterial nanobioreactors—Directing enzyme packaging into bacterial outer membrane vesicles. *ACS Appl. Mater. Interfaces* **2015**, *7*, 24963–24972. [CrossRef]

42. Giessen, T.W.; Silver, P.A. A catalytic nanoreactor based on in vivo encapsulation of multiple enzymes in an engineered protein nanocompartment. *ChemBioChem* **2016**, *17*, 1931–1935. [CrossRef]

43. Thrane, S.; Janitzek, C.M.; Matondo, S.; Resende, M.; Gustavsson, T.; de Jongh, W.A.; Clemmensen, S.; Roeffen, W.; van de Vegte-Bolmer, M.; van Gemert, G.J.; et al. Bacterial superglue enables easy development of efficient virus-like particle based vaccines. *J. Nanobiotechnol.* **2016**, *14*, 30. [CrossRef]

44. Brune, K.D.; Leneghan, D.B.; Brian, I.J.; Ishizuka, A.S.; Bachmann, M.F.; Draper, S.J.; Biswas, S.; Howarth, M. Plug-and-Display: Decoration of Virus-Like Particles via isopeptide bonds for modular immunization. *Sci. Rep.* **2016**, *6*, 19234. [CrossRef]

45. Zhao, Y.; Zhang, Y.; Teng, Y.; Liu, K.; Liu, Y.; Li, W.; Wu, L. SpyCLIP: An easy-to-use and high-throughput compatible CLIP platform for the characterization of protein-RNA interactions with high accuracy. *Nucleic Acids Res.* **2019**, *6*, 3. [CrossRef]

46. Maticzka, D.; Ilik, I.A.; Aktas, T.; Backofen, R.; Akhtar, A. uvCLAP is a fast and non-radioactive method to identify in vivo targets of RNA-binding proteins. *Nat. Commun.* **2018**, *9*, 1142. [CrossRef]

47. Gu, J.; Wang, M.; Yang, Y.; Qiu, D.; Zhang, Y.; Ma, J.; Zhou, Y.; Hannon, G.J.; Yu, Y. GoldCLIP: Gel-omitted Ligation-dependent CLIP. *Genom. Proteom. Bioinform.* **2018**, *16*, 136–143. [CrossRef]

48. England, C.G.; Luo, H.; Cai, W. HaloTag technology: A versatile platform for biomedical applications. *Bioconjugate Chem.* **2015**, *26*, 975–986. [CrossRef]

49. Moon, H.; Bae, Y.; Kim, H.; Kang, S. Plug-and-playable fluorescent cell imaging modular toolkits using the bacterial superglue, SpyTag/SpyCatcher. *Chem. Commun.* **2016**, *52*, 14051–14054. [CrossRef]

50. Hinrichsen, M.; Lenz, M.; Edwards, J.M.; Miller, O.K.; Mochrie, S.G.; Swain, P.S.; Schwarz-Linek, U.; Regan, L. A new method for post-translationally labeling proteins in live cells for fluorescence imaging and tracking. *Protein Eng. Des. Sel.* **2017**, *30*, 771–780. [CrossRef]

51. Pessino, V.; Citron, Y.R.; Feng, S.; Huang, B. Covalent protein labeling by SpyTag-SpyCatcher in fixed cells for super-resolution microscopy. *ChemBioChem* **2017**, *18*, 1492–1495. [CrossRef]

52. Grenier, V.; Daws, B.R.; Liu, P.; Miller, E.W. Spying on neuronal membrane potential with genetically targetable voltage indicators. *J. Am. Chem. Soc.* **2019**, *141*, 1349–1358. [CrossRef]

53. Yin, L.; Guo, X.; Liu, L.; Zhang, Y.; Feng, Y. Self-assembled multimeric-enzyme nanoreactor for robust and efficient biocatalysis. *ACS Biomater. Sci. Eng.* **2018**, *4*, 2095–2099. [CrossRef]

54. Fan, E.; Chauhan, N.; Udatha, D.B.; Leo, J.C.; Linke, D. Type V Secretion Systems in Bacteria. *Microbiol. Spectr.* **2016**, *4*. [CrossRef]

55. Costa, T.R.; Felisberto-Rodrigues, C.; Meir, A.; Prevost, M.S.; Redzej, A.; Trokter, M.; Waksman, G. Secretion systems in Gram-negative bacteria: Structural and mechanistic insights. *Nat. Rev. Microbiol.* **2015**, *13*, 343–359. [CrossRef]

56. Chagnot, C.; Zorgani, M.A.; Astruc, T.; Desvaux, M. Proteinaceous determinants of surface colonization in bacteria: Bacterial adhesion and biofilm formation from a protein secretion perspective. *Front. Microbiol.* **2013**, *4*, 303. [CrossRef]

57. Feilmeier, B.J.; Iseminger, G.; Schroeder, D.; Webber, H.; Phillips, G.J. Green fluorescent protein functions as a reporter for protein localization in *Escherichia coli*. *J. Bacteriol.* **2000**, *182*, 4068–4076. [CrossRef]

58. Meiresonne, N.Y.; Consoli, E.; Mertens, L.M.Y.; Chertkova, A.O.; Goedhart, J.; den Blaauwen, T. Superfolder mTurquoise2ox optimized for the bacterial periplasm allows high efficiency in vivo FRET of cell division antibiotic targets. *Mol. Microbiol.* **2019**, *111*, 1025–1038. [CrossRef]

59. Dinh, T.; Bernhardt, T.G. Using superfolder green fluorescent protein for periplasmic protein localization studies. *J. Bacteriol.* **2011**, *193*, 4984–4987. [CrossRef]

60. El Khatib, M.; Martins, A.; Bourgeois, D.; Colletier, J.-P.; Adam, V. Rational design of ultrastable and reversibly photoswitchable fluorescent proteins for super-resolution imaging of the bacterial periplasm. *Sci. Rep.* **2016**, *6*, 18459. [CrossRef]

61. Meiresonne, N.Y.; van der Ploeg, R.; Hink, M.A.; den Blaauwen, T. Activity-Related Conformational Changes in d,d-Carboxypeptidases Revealed by In Vivo Periplasmic Förster Resonance Energy Transfer Assay in *Escherichia coli*. *MBio* **2017**, *8*, e01089-17. [CrossRef]

62. Rassam, P.; Copeland, N.A.; Birkholz, O.; Tóth, C.; Chavent, M.; Duncan, A.L.; Cross, S.J.; Housden, N.G.; Kaminska, R.; Seger, U.; et al. Supramolecular assemblies underpin turnover of outer membrane proteins in bacteria. *Nature* **2015**, *523*, 333–336. [CrossRef]

63. Toseland, C.P. Fluorescent labeling and modification of proteins. *J. Chem. Biol.* **2013**, *6*, 85–95. [CrossRef]

64. Morales, A.R.; Schafer-Hales, K.J.; Marcus, A.I.; Belfield, K.D. Amine-reactive fluorene probes: Synthesis, optical characterization, bioconjugation, and two-photon fluorescence imaging. *Bioconjugate Chem.* **2008**, *19*, 2559–2567. [CrossRef]

65. Jose, J.; von Schwichow, S. "Cystope tagging" for labeling and detection of recombinant protein expression. *Anal. Biochem.* **2004**, *331*, 267–274. [CrossRef]

66. Zhou, Z.; Cironi, P.; Lin, A.J.; Xu, Y.; Hrvatin, S.; Golan, D.E.; Silver, P.A.; Walsh, C.T.; Yin, J. Genetically encoded short peptide tags for orthogonal protein labeling by Sfp and AcpS phosphopantetheinyl transferases. *ACS Chem. Biol.* **2007**, *2*, 337–346. [CrossRef]

67. Zhang, G.; Zheng, S.; Liu, H.; Chen, P.R. Illuminating biological processes through site-specific protein labeling. *Chem. Soc. Rev.* **2015**, *44*, 3405–3417. [CrossRef]

68. Mühlenkamp, M.; Oberhettinger, P.; Leo, J.C.; Linke, D.; Schütz, M.S. *Yersinia* adhesin A (YadA)—Beauty & beast. *Int. J. Med. Microbiol.* **2015**, *305*, 252–258.

69. Oberhettinger, P.; Leo, J.C.; Linke, D.; Autenrieth, I.B.; Schütz, M. The inverse autotransporter intimin exports its passenger domain via a hairpin intermediate. *J. Biol. Chem.* **2015**, *290*, 1837–1849. [CrossRef]

70. Junker, M.; Besingi, R.N.; Clark, P.L. Vectorial transport and folding of an autotransporter virulence protein during outer membrane secretion. *Mol. Microbiol.* **2009**, *71*, 1323–1332. [CrossRef]

71. Koretke, K.K.; Szczesny, P.; Gruber, M.; Lupas, A.N. Model structure of the prototypical non-fimbrial adhesin YadA of *Yersinia enterocolitica*. *J. Struct. Biol.* **2006**, *155*, 154–161. [CrossRef]

72. Shahid, S.A.; Bardiaux, B.; Franks, W.T.; Krabben, L.; Habeck, M.; van Rossum, B.-J.; Linke, D. Membrane-protein structure determination by solid-state NMR spectroscopy of microcrystals. *Nat. Methods* **2012**, *9*, 1212–1217. [CrossRef]

73. Saeed, I.A.; Ashraf, S.S. Denaturation studies reveal significant differences between GFP and blue fluorescent protein. *Int. J. Biol. Macromol.* **2009**, *45*, 236–241. [CrossRef]

74. Grosskinsky, U.; Schütz, M.; Fritz, M.; Schmid, Y.; Lamparter, M.C.; Szczesny, P.; Lupas, A.N.; Autenrieth, I.B.; Linke, D. A conserved glycine residue of trimeric autotransporter domains plays a key role in *Yersinia* adhesin A autotransport. *J. Bacteriol.* **2007**, *189*, 9011–9019. [CrossRef]

75. Ursell, T.S.; Trepagnier, E.H.; Huang, K.C.; Theriot, J.A. Analysis of surface protein expression reveals the growth pattern of the gram-negative outer membrane. *PLoS Comput. Biol.* **2012**, *8*, e1002680. [CrossRef]

76. Leo, J.C.; Oberhettinger, P.; Schütz, M.; Linke, D. The inverse autotransporter family: Intimin, invasin and related proteins. *Int. J. Med. Microbiol.* **2015**, *305*, 276–282. [CrossRef] [PubMed]

77. Leo, J.C.; Oberhettinger, P.; Chaubey, M.; Schütz, M.; Kühner, D.; Bertsche, U.; Schwarz, H.; Götz, F.; Autenrieth, I.B.; Coles, M.; et al. The Intimin periplasmic domain mediates dimerisation and binding to peptidoglycan. *Mol. Microbiol.* **2015**, *95*, 80–100. [CrossRef] [PubMed]

78. Fairman, J.W.; Dautin, N.; Wojtowicz, D.; Liu, W.; Noinaj, N.; Barnard, T.J.; Udho, E.; Przytycka, T.M.; Cherezov, V.; Buchanan, S.K. Crystal structures of the outer membrane domain of intimin and invasin from enterohemorrhagic *E. coli* and enteropathogenic *Y. pseudotuberculosis*. *Structure* **2012**, *20*, 1233–1243. [CrossRef] [PubMed]

79. Oberhettinger, P.; Schütz, M.; Leo, J.C.; Heinz, N.; Berger, J.; Autenrieth, I.B.; Linke, D. Intimin and invasin export their C-terminus to the bacterial cell surface using an inverse mechanism compared to classical autotransport. *PLoS ONE* **2012**, *7*, e47069. [CrossRef]

80. Bajar, B.T.; Wang, E.S.; Lam, A.J.; Kim, B.B.; Jacobs, C.L.; Howe, E.S.; Davidson, M.W.; Lin, M.Z.; Chu, J. Improving brightness and photostability of green and red fluorescent proteins for live cell imaging and FRET reporting. *Sci. Rep.* **2016**, *6*, 20889. [CrossRef] [PubMed]

81. Leo, J.C.; Oberhettinger, P.; Yoshimoto, S.; Udatha, D.B.; Morth, P.J.; Schütz, M.; Hori, K.; Linke, D. Secretion of the intimin passenger domain is driven by protein folding. *J. Biol. Chem.* **2016**, *291*, 20096–20112. [CrossRef]

82. Tan, K.; Li, R.; Huang, X.; Liu, Q. Outer Membrane Vesicles: Current Status and Future Direction of These Novel Vaccine Adjuvants. *Front. Microbiol.* **2018**, *9*, 783. [CrossRef]

83. Gnopo, Y.M.; Watkins, H.C.; Stevenson, T.C.; DeLisa, M.P.; Putnam, D. Designer outer membrane vesicles as immunomodulatory systems—Reprogramming bacteria for vaccine delivery. *Adv. Drug Deliv. Rev.* **2017**, *114*, 132–142. [CrossRef]

84. Acevedo, R.; Fernández, S.; Zayas, C.; Acosta, A.; Sarmiento, M.E.; Ferro, V.A.; Rosenqvist, E.; Campa, C.; Cardoso, D.; Garcia, L.; et al. Bacterial outer membrane vesicles and vaccine applications. *Front. Immunol.* **2014**, *5*, 121. [CrossRef]

85. Jose, J.; Meyer, T. The autodisplay story, from discovery to biotechnical and biomedical applications. *Microbiol. Mol. Biol. Rev.* **2007**, *71*, 600–619. [CrossRef]

86. Jong, W.S.; Soprova, Z.; de Punder, K.; ten Hagen-Jongman, C.M.; Wagner, S.; Wickström, D.; de Gier, J.-W.; Andersen, P.; van der Wel, N.N.; Luirink, J. A structurally informed autotransporter platform for efficient heterologous protein secretion and display. *Microb. Cell Fact.* **2012**, *11*, 85. [CrossRef]

87. Daleke-Schermerhorn, M.H.; Felix, T.; Soprova, Z.; Ten Hagen-Jongman, C.M.; Vikström, D.; Majlessi, L.; Beskers, J.; Follmann, F.; de Punder, K.; van der Wel, N.N.; et al. Decoration of outer membrane vesicles with multiple antigens by using an autotransporter approach. *Appl. Environ. Microbiol.* **2014**, *80*, 5854–5865. [CrossRef]

88. Van Ulsen, P.; Zinner, K.M.; Jong, W.S.P.; Luirink, J. On display: Autotransporter secretion and application. *FEMS Microbiol. Lett.* **2018**, *365*, 1093. [CrossRef]

89. Tajima, N.; Kawai, F.; Park, S.-Y.; Tame, J.R.H. A novel intein-like autoproteolytic mechanism in autotransporter proteins. *J. Mol. Biol.* **2010**, *402*, 645–656. [CrossRef]

90. Otto, B.R.; Sijbrandi, R.; Luirink, J.; Oudega, B.; Heddle, J.G.; Mizutani, K.; Park, S.-Y.; Tame, J.R.H. Crystal structure of hemoglobin protease, a heme binding autotransporter protein from pathogenic *Escherichia coli*. *J. Biol. Chem.* **2005**, *280*, 17339–17345. [CrossRef]

91. Jong, W.S.P.; Ten Hagen-Jongman, C.M.; den Blaauwen, T.; Slotboom, D.J.; Tame, J.R.H.; Wickström, D.; de Gier, J.-W.; Otto, B.R.; Luirink, J. Limited tolerance towards folded elements during secretion of the autotransporter Hbp. *Mol. Microbiol.* **2007**, *63*, 1524–1536. [CrossRef]

92. Houben, D.; de Jonge, M.I.; Jong, W.S.; Luirink, J. Display of recombinant proteins on bacterial outer membrane vesicles by using protein ligation. *Appl. Environ. Microbiol.* **2018**, *84*, e02567-17.

93. Kuipers, K.; Daleke-Schermerhorn, M.H.; Jong, W.S.P.; Ten Hagen-Jongman, C.M.; van Opzeeland, F.; Simonetti, E.; Luirink, J.; de Jonge, M.I. *Salmonella* outer membrane vesicles displaying high densities of pneumococcal antigen at the surface offer protection against colonization. *Vaccine* **2015**, *33*, 2022–2029. [CrossRef]

94. Leitner, D.R.; Lichtenegger, S.; Temel, P.; Zingl, F.G.; Ratzberger, D.; Roier, S.; Schild-Prüfert, K.; Feichter, S.; Reidl, J.; Schild, S. A combined vaccine approach against Vibrio cholerae and ETEC based on outer membrane vesicles. *Front. Microbiol.* **2015**, *6*, 823. [CrossRef]

95. Mi, K.; Ou, X.; Guo, L.; Ye, J.; Wu, J.; Yi, S.; Niu, X.; Sun, X.; Li, H.; Sun, M. Comparative analysis of the immunogenicity of monovalent and multivalent rotavirus immunogens. *PLoS ONE* **2017**, *12*, e0172156. [CrossRef]
96. Chackerian, B. Virus-like particles: Flexible platforms for vaccine development. *Expert Rev. Vaccines* **2007**, *6*, 381–390. [CrossRef]
97. Smith, M.T.; Hawes, A.K.; Bundy, B.C. Reengineering viruses and virus-like particles through chemical functionalization strategies. *Curr. Opin. Biotechnol.* **2013**, *24*, 620–626. [CrossRef]

© 2019 by the authors. Licensee MDPI, Basel, Switzerland. This article is an open access article distributed under the terms and conditions of the Creative Commons Attribution (CC BY) license (http://creativecommons.org/licenses/by/4.0/).

MDPI
St. Alban-Anlage 66
4052 Basel
Switzerland
Tel. +41 61 683 77 34
Fax +41 61 302 89 18
www.mdpi.com

International Journal of Molecular Sciences Editorial Office
E-mail: ijms@mdpi.com
www.mdpi.com/journal/ijms

www.ingramcontent.com/pod-product-compliance
Lightning Source LLC
LaVergne TN
LVHW071610170726
843515LV00008B/2365